Precast Prestressed Concrete for Building Structures

This guide to precast prestressed concrete (PSC) introduces and applies principles for the design of PSC slabs, thermal slabs, beam and block flooring and main beams, including (where appropriate) cantilevers, and composite and continuous construction.

This book provides numerous worked examples for a wide range of PSC elements and covers the innovative use of PSC on several projects in the UK over the past 10 years, drawing on the authors' first-hand experience in the design and manufacture of special products. The contents are in line with the latest revisions of the Eurocodes and European Product Standards.

Precast Prestressed Concrete for Building Structures is ideal for consulting structural engineers, clients, PSC manufacturers, and advanced undergraduate and graduate students, both as a guide and a textbook.

Kim S. Elliott is a consultant for the precast industry worldwide, and a former Senior Lecturer at Nottingham University, UK, and a structural design engineer and a site agent at Trent Concrete Structures, UK. He was the Chairman of the European research project COST C1 on *Semi-Rigid Connection in Precast Structures* and is a member of *fib* Commission 6 on Prefabrication. He is the author of *Precast Concrete Structures*, now in its second edition.

Mark Magill is a chartered engineer and the Engineering and Technical Director for Creagh Concrete Products Ltd in Northern Ireland.

Precast Prestressed Concrete for Building Structures

Kim S. Elliott and Mark Magill

CRC Press
Taylor & Francis Group
Boca Raton London New York

CRC Press is an imprint of the
Taylor & Francis Group, an **informa** business

Cover image: Creagh Concrete Products Ltd.

First edition published 2024
by CRC Press
2385 NW Executive Center Drive, Suite 320, Boca Raton FL 33431

and by CRC Press
4 Park Square, Milton Park, Abingdon, Oxon, OX14 4RN

CRC Press is an imprint of Taylor & Francis Group, LLC

ISBN: 978-1-032-33391-5 (hbk)
ISBN: 978-1-032-33392-2 (pbk)
ISBN: 978-1-003-31945-0 (ebk)

DOI: 10.1201/9781003319450

Typeset in Sabon
by codeMantra

Dedication from Kim Elliott

Since the publication of my last book *Precast Concrete Structures* in 2018 we have lost two people who helped to influence and shape my career in the analysis and design of precast/prestressed concrete structures. Mr Arnold van Acker was Chairman of FIP Commission on Prefabrication when he asked me to join the Commission in 1992. He was Technical Director at Partek & Partek Ergon in Belgium and received the Gold Medal for his contributions to FIP/FIB. He taught me to see the prefabrication of concrete structures in a holistic way, as a concept and entity, which should not be treated any differently from other forms of concrete construction. Dr Howard Taylor was Technical Director at Dowmac (later Costain-Dowmac and Tarmac Precast). During his term as President of The Institution of Structural Engineers he presented me with the Institution's Henry Adams Award for the best pair of research papers published in The Structural Engineer in 1992, on the subject of horizontal diaphragm action in precast concrete floor units without a structural topping. Having spent many years fending off consulting engineers who insisted on adding a structural topping, he summarised my papers and said, "You see, that's all there is to it". I dedicate this book to the memory of these truly exceptional gentlemen.

Contents

Foreword by Seamus McKeague

I congratulate the co-authors of this book, Kim S Elliott (KSE) and Mark Magill (MM) for their dedication in capturing the essence of precast prestressed concrete and presenting it in a comprehensive and accessible manner. I first came to know KSE 30 years ago when he was Lecturer of Civil Engineering at The University of Nottingham. We met at a British Precast event when he was explaining and promoting the use of precast concrete for high-rise buildings. We at Creagh Concrete Ltd. (Creagh) were in the early stage of creating and growing our business to become a specialist precast manufacturer of prestressed hollow core floors and structural precast elements. We engaged KSE to become an advisor and mentor to our young team of engineer and designers. MM was a young enthusiastic (Creagh) engineer, and he benefited from the help and knowledge provided by KSE who is a highly respected international precast concrete expert with great practical experience and an inspiring academic career. It is very pleasing that by their friendship and partnership developed over the years, and they are now combining their efforts in compiling this knowledge, which will undoubtedly inspire and educate readers, spurring further innovation and pushing the boundaries of what can be achieved.

This book is a comprehensive exploration of precast and prestressed concrete. It is a testament to its transformative power and immense value in the realms of construction. It delves deep into its innovative applications and advanced techniques that have elevated it to the forefront of the construction industry.

One of the key strengths of precast and prestressed concrete lies in its ability to embrace both tradition and innovation. With each passing year, new advancements in materials, design techniques and production processes pushed the boundaries of what is possible. This book serves as a guide to these cutting-edge technologies and methodologies, highlighting the incredible strides made in the field and demonstrating how they have revolutionised the concrete industry. It showcases the myriad of benefits utilising precast concrete such as enhanced quality control, reduced construction time, cost-effectiveness and sustainability; advantages that sets it apart from conventional construction methods.

Within these pages, you will find a wealth of knowledge, insights and practical guidance from experts who have dedicated their lives to the advancement and application of precast and prestressed concrete. Their experience and passion are evident in the depth of information shared, making this book an invaluable resource for engineers, architects, students and anyone fascinated by the transformative potential of precast.

Finally, I invite you, the reader, to immerse yourself in the world of precast and prestressed concrete. A world where vision becomes reality, where tradition meets innovation and where the future of construction is shaped. May this book serve as a guiding light, empowering you to contribute to the creation of a built environment that is both awe-inspiring and sustainable.

Seamus McKeague
COO Creagh Concrete Ltd. | President of IPHA - International Prestressed Hollowcore Association| President of IAE (Irish Academy of Engineers)

Preface

When Kim Elliott used to lecture to students and run workshops to professional structural engineers on the 'Design of precast concrete structures' he'd break the sessions with a number of 15 minute videos, short case studies mainly. At the beginning of one such case study filmed in 1990 by Costain-Dowmac of 'The Shires' retail development in Leicester, UK, narrator Hugh Laurie quips with alacrity "I know what you're thinking, precast concrete, ha!, it's only good for railway sleepers and bridge beams ...". The same is true of prestressed concrete, where many structural engineers have expressed surprise that the beams used in beam and block flooring for example are prestressed, and why prestressed hollow core floor units do not have transverse reinforcement or shear links. In fact the majority of text books on prestressed concrete focus on the generic design of prestressed elements, without any particular use in mind, or bridge-beam type beams. This is probably because the authors do not come from a precast/prefabrication background, preferring to leave the design of typical prestressed concrete elements found in multi-storey buildings to in-house design teams, at the producer's control.

This book aims to redress this imbalance towards the use of prestressed concrete in buildings such as offices, retail/shopping, houses/apartments, hotels, and parking garages. Its main objective is to fill the gap where prestressed concrete is considered from not only the actual design, i.e. calculating bending and shear capacities, but also from the manufacturer's production angle, quality control, testing and on-site erection methods, all of which play a major role in the final products. At the end of said video the narrator concludes "Well who said that precast concrete was only good for bridge beams and railway sleepers. It just goes to show that the right blend of design, manufacture and erection skills can achieve some outstanding results!" The authors of this book hope that the reader will agree, and think along the same lines.

However, most structural design engineers and civil engineering contractors still think of prestressed concrete as 'just another technique of squeezing more out of the concrete', which of course it is; refer to Freyssinet, or more importantly for long-line prestressing refer to Hoyer[1]. Unlike reinforced concrete, where the concrete is in compression in simply supported spans only in the top of the beam (and close to the bearing supports), prestressed concrete is in compression almost everywhere, see Figure 1, as well as making considerable contributions to the shear capacity and reducing deflections by preventing the beam from loosing stiffness by cracking. Prestress is to concrete what lead crystal is to a wine glass.

Highly tuned and working efficiently in this way, some designers do not realize that there are a number of prestressed concrete elements that do not comply with the basic rules afforded to reinforced concrete. It was surprising to learn that during an expert investigation involving prestressed hollow core floor units (hcu), Figure 2*, the consultant could not understand why hcu did not have (i) shear stirrups for the ultimate limit state, (ii) links surrounding the main

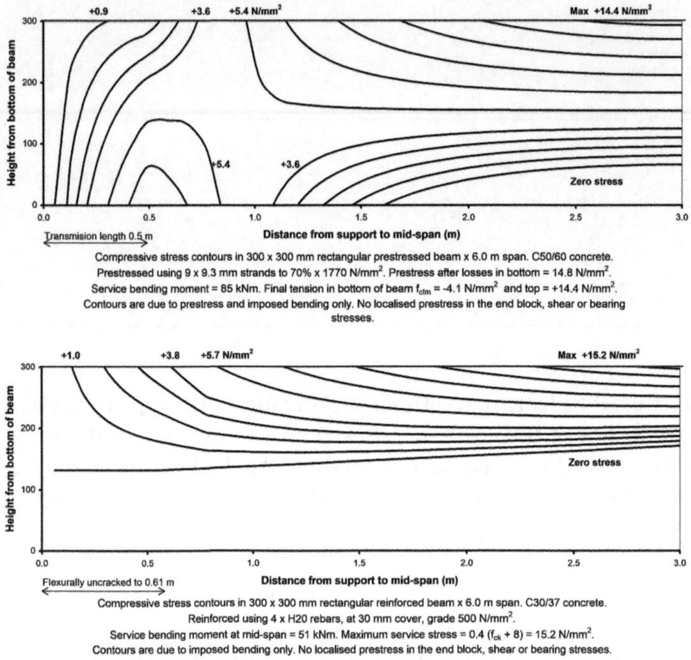

Figure 1 Compressive stress contours in one-half of a simply supported prestressed concrete (top) and reinforced concrete beam subjected to uniform loading.

Figure 2 Prestressed concrete hollow core floor unit (*not the unit under investigation).

tendons, (iii) transverse bars (across the width of the hcu), (iv) transverse ties between individual units, (vi) a proper bend or hook where the tendons were cut and exposed at the ends of the hcu, and (v) the hcu's were able to distribute point or line loads transversely between units. Hollow core slabs and wall panels, as we know them today, were born c.1950 in the factory, not in the design office. Recently the floor area of prestressed hcu supplied in Europe passed one billion sq.m. (equivalent to 125,000 football pitches)!

But being 'just another technique of squeezing more out of the concrete' explains why prestressed concrete is not taught widely at undergraduate level, certainly in the UK. You can hear course directors: "It's too specialized, in any case in-house production designers do that". This is not the case in Scandinavia or the Benelux, or many Schools in North America, where prestressed concrete is not thought of as *just another technique*. The result of this is evident in Figure 3 from The Netherlands.

The global trend for de-carbonisation has reignited interest in the benefits of prestressed concrete elements. In particular, prestressed hollow core slabs are being specified at early design stage due to the minimisation of the use of materials within the product itself and the overall reduction in loading applied to the superstructure and substructure. Prestressed design can offer a competitive advantage due to production efficiency and reduced costs but specifiers and customers are now looking to source products and use forms of construction that also provide a sustainable solution. Offsite manufacturing can contribute to social sustainability by ensuring for example that prestressed factories are safe, controlled environments to work in. Precast concrete manufacturing uses cement, which receives considerable attention regarding the amount of energy consumed to produce it. To address this, manufacturers are looking at cement replacement products and carbon capture technology to strengthen their credentials in the environmental aspect of sustainability. The building shown in Figure 4 makes use of the benefits of prestressed hcu to provide a sustainable design solution in the fully precast structure.

Figure 3 55 m long × 106 tonne prestressed concrete beam at VBI, Netherlands.

Figure 4 Fully precast and prestressed car park structure. Newcastle, UK.

This book aims to fill many of these shortcomings. It is based on the Eurocodes (EC2 mainly but with passing references to EC0 and EC1 for basic principles and loads/factors) and where relevant the EC Product Standards, e.g. pull-in of tendons and splitting stresses in hcu in EN 1168, sequences for deflections in beam and block flooring. Finding the relevant clauses in Eurocode EC2 is a tedious exercise; e.g. (i) the service loading and permissible service stress in prestressed concrete in tension for internal exposure XC1 (so clearly given in the former British code BS 8110) requires clauses 7.3.2.(4), 7.3.2(2) and Table 3.1, and (ii) 26 clauses and/or equations are needed in Chapters 3, 5 and 10 and Appendix B to determine losses of prestress. This book pieces everything together, both strategically throughout the chapters and logically in the text and via the design examples.

The author's are aware that the structural Eurocodes are being revised. Kim Elliott was commissioned by British Precast to identify all clauses in pr EN 1992-1-1:2018 version D3 relating to precast and prestressed concrete, preparing model calculations and worked examples for key elements, and so is familiar with many of the proposed changes. However, it is unlikely that the revised code will be made available to BSI before 2024, in preparation for the UK National Annex, and that coexistence of the present and new codes may last until 2028. For this reason it was decided to continue with the present code BS EN 1992:2004+A1:2014.

This book has been made possible because the authors have had different backgrounds in their professional careers – Kim Elliott, formerly of Trent Concrete Ltd. and Nottingham University, is known throughout the world for his text books and workshops and contributions to bulletins attributed to FIB Commission 6 on Prefabrication, and Mark Magill, in the front line of the design of millions of pounds worth of precast and prestressed concrete buildings at Creagh Concrete Ltd. This is Elliott's sixth book, e.g.[2,3], in which the work draws on R&D, analysis and the loop that binds the manufacture-design-construction of precast elements in building frames. Magill leads a design team responsible for the design of precast and prestressed elements and entire structures for the residential, commercial, industrial and agricultural sectors. This includes buildings such as the 18 storey fully precast concrete student accommodation scheme shown in Figure 5.

Figure 5 Fully precast concrete 18 storey student accommodation building in Leeds, UK.

REFERENCES

1. Web site. https://civilengineeringbible.com/article.php?i=100. What is the difference between Hoyer system and Freyssinet system of Prestressing?
2. Elliott, K. S. and Jolly, C. K. 2013. Multi-Storey Precast Concrete Framed Structures, 2nd Edition, John Wiley, London, UK, 752 pages.
3. Elliott, K. S. 2017. Precast Concrete Structures, 2nd Edition, CRC Press / Taylor Francis, Florida, USA, 694 pages.

Acknowledgement

Kim Elliott is grateful to the contributions and assistance made by the following individuals and organisations; to the members of the fib Commission 6 on Prefabrication in particular the late Arnold Van Acker (Belgium), to John Cotton and Karl Timmins (Bison Precast Ltd., UK), Peter Hackett (Coltman Precast Ltd., UK), Guy Schofield (Litecast Ltd., UK), John Thornberry (Moulded Foams Ltd., UK), Richard Morton (Taranto Ltd., UK), Simon Hughes (xx, Australia), Wim Jansze (Consolis, Belgium and IPHA), Karen Ludgren (University of Chalmers, Sweden), Aimee Wragg (Taylor & Francis publishers) and to my co-author Mark Magill (Creagh Concrete Ltd., UK). Permission to reproduce extracts from Eurocode EC2 (BS EN 1992-1-1) is granted by the British Standards Institute. Permission to use photographic images is granted by Bison Precast Ltd., Litecast Ltd., VTT (Espoo, Finland), Echo nv/sa (Belgium), Simon Hughes (Australia), BIBM / IPHA (International) and FIB (International).

Mark Magill would like to thank his family members Fiona, Liam, Eva and Odhrán Magill for their patience while this book was being put together. I would also like to thank the following individuals who contributed to the contents of the book: Adrián Oliva Munoz, Diarmuid Dallat, Mark Gilliland and Moussa Chidiac (Creagh Concrete Products Ltd.). There have been many people who have given me assistance and support over the years, but in particular I want to thank Dimitris Sagias, Durlav Khanal, Martin Breen, Ramón Escriva Plá, Sean Rocks, Sean Toal, the McKeague family (Creagh Concrete Products Ltd.) and last but not least Kim Elliott.

Notation

ABBREVIATIONS AND GENERAL NOMENCLATURE

A	Prefix for square structural steel mesh (bars at 200 mm spacing) followed by area of bars per m width
A1	Euroclass fire rating for non-combustible material
A-frame	Support frame used when transporting precast elements, typically walls
Acrow	Adjustable steel prop
BIM	Building Information Modelling
BSI	British Standards Institution
CARES	Certification Authority for Reinforcing Steels
CE marked	Conforming with European health, safety, and environmental protection standards
CEM	Types of cements
CofG	Centre of gravity
EHF	Equivalent horizontal forces
EPDM	Ethylene Propylene Diene Monomer
EPS	Expanded Polystyrene insulation
Exp.	Expression=Equation from a Code of Practice
Eq.	Equation in this book
Fig.	Figure from a Code of Practice
Figure	Figures in this book
FEM	Fixed end moments
fi	Fire situation (subscript)
GA	General Arrangement drawings
i	*Insitu* infill concrete, mortar or grout (subscript)
IFC	Industry Foundation Class
hcu	Hollow core unit
M&E	Mechanical and electrical services
MEP	Mechanical electrical plumbing
MEWPS	Mobile elevating work platforms
PSF	Partial safety factor
RH, RHs	Relative humidity; and in service (%)
RSA	Rolled steel angle
S2, S3, S4	Slump classes
sd	Standard deviation
SCC	Self compacting concrete
T	*Insitu* concrete structural topping (subscript)

Toast rack	Support frame used when transporting precast elements
TWC	Temporary Works Co-ordinator
TWD	Temporary Works Designer
TWDC	Temporary Works Design Checker
TWS	Temporary Works Supervisor

NOTATION

a	Axis distance to tendon or bar; shear span; lever arm from centroid of topping to centroid of composite section
a'	Distance to point load in cantilever
a_b	Edge distance or half of centre-to-centre distance between bars
a_{Ed}	Distance to zero ultimate moment in the main span to a cantilever
a_h	Horizontal distance from wall
a_i	Distance to a tendon in row *i*
a_{mean} a_m	Mean axis distance of a group of bars or tendons
a_{sd} $a_{sd,m}$	Side axis distance to bars or groups of bars or tendons
a_v	Distance to the primary shear force from support
a_2	Distance to zero moment in a continuous span due to stage 2 loads
b	Height to prop position on panel, base panel height.
b b_1 b_2	Width, and in stage 1 and 2 loading
b_{ave}	Average width of a hcu
b_c	Total width of cores (in hollow core slab)
b_e	Width of precast floor unit
b_{ef}	Effective width of *insitu* topping at ultimate
b_{eff}	Effective width of *insitu* topping in service, effective width of webs in punching shear
b_f	Width of flanges in beams or slabs
b_{fi}	Width of beams or slabs in a fire situation
b_i	Interface contact length between beams and slabs or infill
b_{min}	Minimum width of section in fire
b_p	Width of wall panel
b_s	Net width of the reduced section at sloping or notched ends of slab
b_t	Mean width of section in the tension zone
b_u b_w	Width of upstand in beam
b_w $b_w(y)$	Width of web(s); width at height *y* from bottom of section
$b_{w,o}$ b_y	Width of the outermost web; total width at centroidal axis
b'	Width of soffit in a cantilever
c	Cohesion factor in interface shear stress
c_{alt}	Altitude factor
c_d	Edge distance to rebars or gap between rebars
c_{dir}	Directional factor
$c_e(z)$	Exposure factor
c_f	Force coefficient
c_o	Site orography factor
c_{pa}	Net pressure coefficient Zone A
c_{pb}	Net pressure coefficient Zone B
c_{prob}	Probability factor
c_{season}	Season factor

d	Effective depth to tendons or bars from compression face
$d_1\ d_2\ d_3$	Effective depth to tendons in rows 1,2, and 3
$d'\ d_2'$	Effective depth to compression bars; effective depth to top cantilever and continuity bars; and in stage 2 ditto
d''	Effective depth to bottom continuity bars
d_c	Depth of filled core from top of section
d_{ef}	Equivalent rectangular depth of an I section for spalling stresses
d_{eff}	Effective height of bottom flange of beam
d_g	Maximum size of coarse aggregate
$d_n\ d_{n2}\ d_{n2,3}$	Depth to centroid of concrete in compression; in composite section stage 2; in composite section stage 3
d_s	Depth at the shear plane at sloping or notched ends of slab
d_T	Effective depth to tendons in tension (ultimate limit state)
$d_1\ d_2$	Effective depth in stage 1 and 2 loading
$e\ e_{nom}$	Eccentricity; nominal eccentricity (torsion calculation)
$e_{max}\ e_{min}$	Imbalanced load eccentricity (torsion calculation)
e_o	Eccentricity to tendons (spalling calculation)
f_{adh}	Adhesion to mould
f_b	General term for final stress in bottom fibre of prestressed section
$f_{b1}\ f_{b2}$	Bottom fibre stress due to stage 1 and 2 bending moments
$f_{bd}\ f_{b,req}$	Design and required concrete bond strength for rebars
f_{bpd}	Design concrete bond strength for tendons
$f_{b,max}' f_{t,max}'$	Maximum combined prestress and flexural stress in bottom and top fibres of a cantilever or continuous unit near to supports
f_c	Chain angle factor;
$f_{cd}\ f_{cdi}$	Design strength of concrete; and design strength of *insitu* concrete
$f_{cd,fi}(20)$	Design strength of concrete in a fire situation
f_{ck}	Characteristic cylinder strength of concrete determined at 28 days
$f_{ck,cube}\ f_{cu}$	Characteristic cube strength of concrete ditto (f_{cu} is an outdated term but still used in many situations)
f_{cki}	f_{ck} of *insitu* or infill concrete
f_{ckT}	f_{ck} of *insitu* topping
$f_{ck}(t_i)\ f_{ck}(t_i)$	Value of f_{ck} at time t and at installation of elements t_i
$f_{cm}\ f_{cm,cube}$	Mean value of f_{ck} and $f_{ck,cube}$
$f_{cm}(t)$	Value of f_{cm} after time t
f_{ct}	Cracking tensile strength
$f_{ctd}\ f_{ctdi}\ f_{ctdT}$	Design strength of concrete, *insitu* concrete infill and topping in tension
$f_{ctd}(t)$	Value of f_{ctd} at transfer or at time t
$f_{ct,eff}$	Value of f_{ctm} in crack width calculations
$f_{ctk,0.05}$	The (lowest) 5% fractile value of f_{ctm}
f_{cmo}	Reference mean strength of concrete for shrinkage
$f_{ctm}\ f_{ctmi}$	Mean strength of concrete; and *insitu* in tension
$f_{ctm,fl}$	Flexural strength of concrete in tension
$f_{c1}\ f_{c2}$	First and second specimen compressive cylinder strength prior to release of tendons
f_d	Dynamic factor
$f_{min,p}$	Required value of f_{c1}
f_n	Natural frequency

f_p $f_{p,max}$	Final stress in tendons (ultimate limit state); maximum ultimate stress in tendons
$f_{LOP} f'_p$	Stress in tendons at ε_{LOP}; at intercept in σ–ε diagram
$f_p f'_p$	Final stress in tendons at ultimate; and in a cantilever
$f_{p0,1k}$	0.1% proof strength of prestressing tendon
$f_{p1} f_{p2} f_{p3}$	First, second and third trial values for f_p in iteration methods
f_{bpt}	Bond stress at transfer
f_{pd}	Design strength of prestressing tendon$=f_{pk}/\gamma_m$
f_{pk} f_{pki}	Characteristic strength of prestressing tendon; strength of tendon in row *i*
f_{pu} $f_{p0,1}$	Measured values of f_{pk} and $f_{p0,1k}$ in a tensile tests
f_t	General term for final stress in top fibre of prestressed section
f_{t1} f_{t2}	Top fibre stress due to stage 1 and 2 bending moments
f'_{t2}	Top of topping stress due to stage 2 bending moment
$f_u f_{ub} f_{uw}$	Ultimate tensile strength of hot rolled structural steel, bolts and welds
f_w	Wind pressure
f_{yd}	Design strength of steel rebar$=f_{yk}/\gamma_m$
f_{yk} f_{yki}	Characteristic yield strength of steel rebar; strength of bar in row *i*
f_{ywk}	Value of f_{yk} of stirrups
f_{ybk} f_{yw}	Yield strength of bolts, dowels and weld
g_k	Pressure of permanent (dead) load per unit area
h	Depth of section; storey height of column, design height of retaining wall
h_2	Design height
h_{agg}	Maximum nominal size of coarse aggregate
h_b	Distance from ground to base of wall
h_{ft} h_{fb} h_f	Depth of top and bottom flanges; the smaller of h_{ft} or h_{fb}
h_o	Notional depth or thickness of element
h_p	Wall panel height
h_s	Depth of the reduced section at sloping or notched ends of slab; slab depth in shrinkage effects calculation; depth of upstand in beams
k	Depth factor for shear strength; 'kernel' of a section $k=Z_{b,c}/A_{c,c}$; torsional constant for rectangular areas; rotational stiffness of supports in continuous spans
k_c k	Factors for minimum area of reinforcement
k_n	Size coefficient for shrinkage
$k_c(\theta)$ $k_s(\theta)$ $k_p(\theta)$	Reduced compressive strength of factors for concrete, rebars and tendons in fire
$k_p(\theta_{cr})$	Stress ratio in prestressing tendons in fire
k_T	Limit for prestress ratio after initial losses
l	Length of edge line loads in hcu
l_b	Bearing length (parallel with the span of the element) (termed *a* in the code)
l_{disp}	Length over which prestress gradually disperses to become a linear distribution
l_{bd} $l_{b,rqd}$	Design anchorage (bond) length; structural anchorage length required
l_{core}	Length of the top opened cores from the supports

l_e	Effective span of wall panels
l_{eff}	Effective span of horizontal elements
$l_n l'_n$	Clear distance between the faces of the supports; and to cantilever
l_o	Effective span in effective width calculations
l_p	Length of wall panel;
l_{pt}	Transmission length of tendon
l_{pt1} l_{pt2}	Lower bound value of l_{pt}; upper bound design value of l_{pt} at ultimate limit state
l'_{pt2}	Increase in l_{pt2} for tendons with slippage
l_{bpd}	Ultimate anchorage length for tendons
$l_v l'_v$	Effective shear span; and in a cantilever
l_x	Distance to shear plane from end of prestressed element
m	The gradient of the straight line in the Magnel diagram
m	Modular ratio (long-term)
m_x s_x	mean and standard deviation for a number of samples for quality control
m^- m^+	Dimensionless terms in Magnel diagram
n	Number of lifting anchors;
n_p	Number of props;
p	Dimensionless abscissa $P_{pi}/A\ \sigma_{max}$ in Magnel diagram
p	Probability of exceedance
p_a	Average pressure (in wall panels)
p_{a2}	Average horizontal pressure
p_m	Maximum horizontal pressure
p_1	Pressure at top of base panel, Horizontal pressure at bottom of panel
p_2	Horizontal pressure at top of panel
p_q	Shear strengths of bolts and dowels
p_{ybd} p_{ywd}	Design tensile strengths of bolts and dowels, and welds
$p_{yk}(20°C)$	Strength of prestressing tendons at ambient temperature
q_b	Basic velocity pressure
q_{Ed}	Variable (live) ultimate load per unit length
q_k	Variable (live) load per unit area
q_p, $q_p(z)$	Peak velocity pressure
r	Radius of beam curvature
r_c	Radius of gyration
r_{inf} r_{sup}	Partial safety factor for variations in prestressing
s	Cement factor; distance between bars, tendons, stirrups or links
s_l	Variable surcharge load
$s_{r,max}$	Crack spacing
s_v s_h	Vertically and horizontal spacing between tendons
t	Time; thickness; depth of topping; width of the support; minimum of h_f or b_w in hcu edge loads
t_e	Equivalent depth or thickness in fire of hollow core slabs
t_{ef} $t_{ef,i}$	Effective thickness; t_{ef} of a wall i (torsion calculation)
t_o	Time when load is first applied
t_s	Time to removal from mould (or detensioning for prestress)
t_{0T}	Equivalent age at transfer for creep
u u_1	Perimeter of element; u for sub-section 1
u_k	Perimeter of area A_k (torsion calculation)

v	Shear stress
$v\ v_1$	Strength reduction factor for concrete cracked in shear
v_b	Basic wind velocity
$v_{b,0}$	Fundamental value of the basic wind velocity
$v_{b,map}$	Fundamental basic wind velocity
$v_{Ed}\ v_{Ed1}$	Ultimate shear stress, and due to stage 1 loads
v_{Edi}	Design interface shear stress
v_{min}	Minimum ultimate shear resistance
$v_{Rdi}\ v_{Rdi,calc}$	Horizontal interface shear resistance; and calculated value
v_{Rdj}	Critical shear capacity
$v_{Rdj1}\ v_{Rdj2}$	Shear capacity of flange, an shear capacity of joint
v_{ui}	Experimental horizontal interface shear stress
$w\ w'$	Uniformly distributed load per unit length, and in a cantilever
w	Width of footprint of load in punching shear
w_c	Temporary construction traffic
w_d	Peak line load to distribute
$w_{Ed,ls}$	Ultimate load on bottom panel
w_{eq}	Equivalent line load, bottom panel
w_{eq2}	Equivalent line load, second panel from bottom
$w_G\ w_Q\ w_{G,sup}$	Service dead and live loads in torsion calculations; and (superior) dead load in a cantilever
w_0	Uniformly distributed service load due to self weight of precast unit
w_{0c}	Ditto due to self weight of filled cores in hcu
w_1	Ditto due to self weight of slab after grouting
w_2	Ditto due to dead loads
w_3	Ditto due to imposed live loads
w_{Ed}	Ultimate uniformly distributed load per unit length
$w'_{Ed}\ w'_{Ed1}\ w'_{Ed2}$	w_{Ed} in a cantilever, and stage 1 and 2 in composite cantilever
$w_{Ed,max}\ w_{Ed,min}$	Maximum and minimum values of w_{Ed} for patch loading
w_{max}	Limiting crack width
$w_s\ w_{s1}\ w_{s2}$	Service loads; and stage 1 and imposed stage 2 in composite design
$w_{s,ls}$	Service load on bottom panel
$w_{s,qp}$	Service loads due to quasi-permanent loads
$w_T\ w_{T1}\ w_{T2}$	Self weight of topping; ditto at first and second floors in dual propping
w_v	Width of voids in hcu
x	Distance from centre of support; depth to the centroid of areas in a section, length of sloping end of slab
$x_u\ x_c$	Depth to centroidal axis, flexurally uncracked and cracked sections
$x_d\ x_{Ed,d}$	Distance from the simple support where $M_s = M_{s,d}$ in the main span of a cantilever beam; ditto distance for ultimate
x_{pc1}	Depth to centroidal axis in partially cracked beam section
$\overline{x}$	Distance to the centroid of the areas
x_{fi}	Depth to neutral axis in a fire situation
y	Dimensionless ordinate $P_{pi}\ z_{cp}/Z_t\ \sigma_{max}$ in Magnel diagram; depth of sloping end of slab
$y_b\ y_t$	Distances to centroid of section from bottom and top
$y_{b,c}$	Distance to centroid of composite section from bottom
$y_{b,co}\ y_{t,co}$	Compound values of y_b and y_t

$y_{b,c,co}$ $y_{t,c,co}$	Compound composite values of y_b and y_t
$y_{b,cr}$	Height to centroid from bottom of flexurally cracked section
y_i	Height to centroid of tendons in row *i*
y_s	Height to centroid of all tendons
y_{sT}	Height to centroid of tendons in tension zone
z	Reference height above ground for wind pressure
z z_1 z_2 z'	Lever arm, and in stage 1 and 2 loading; in a cantilever
z_{cp} $z_{cp,max}$	Eccentricity of prestressing force; maximum possible value of z_{cp}
z_{fi} $z(\theta)$	Lever arm in a fire situation; ditto at temperature θ
z_i	Side length of the wall *i* between the intersection of the walls (torsion calculation)
A_{alt}	Site altitude
A_c	Cross sectional area
$A_{c,co}$	Compound value of A_c
$A_{c,c}$	Cross sectional area of composite precast and *insitu* topping
$A_{c,c,co}$	Compound value of A_{cc}
$A_{c,eff}$ A_{ct}	Concrete area in tension zone
A_{ci}	Cross sectional area of *insitu* concrete
A_{cT}	Area of *Insitu* concrete topping
$A_c(y)$	Concrete section area above height *y* from bottom of section
A_{gt}	Percentage total elongation of a rebar at maximum force
A_j	Area of interface joint
A_k	Area enclosed by the centre lines of the connecting walls (torsion calculation)
A_p A_{p1} A_{p2}	Area of prestressing tendons, and in stage 1 and 2 loading
A'_p	Equivalent area subtracted from each tendon A_{p1}/N; A_p in a cantilever
$A_{p,min}$	Minimum area of prestressing tendons
A_{pT} A_{pT2}	Area of prestressing tendons in tension (ultimate limit state); and in composite section stage 2
$A_{pT,net}$	Net area of A_{pT} averaged over multiple units
A_s A_{s1} A_{s2}	Area of rebars in tension and interface shear, and in stage 1 and 2 loading
A'_s A'_{s2}	Area of rebars in compression; top bars in cantilever or continuous spans
A_{si}	Area of interface shear dowels, loops or stirrups; area of a tendon in row *i*
A_{sl}	Area of longitudinal reinforcement due to torsion
$A_{s,min}$ $A'_{s,min}$	Minimum area of reinforcement in tension or compression required
$A_{s,max}$	Maximum area of reinforcement allowed
$A_{s,req}$ $A_{s,prov}$	Area of rebar 'required' and 'provided'
A_{sw}	Area of shear or torsion stirrups
$A'_s(x)$	A'_s required at distance *x* from the support (continuous spans, fire)
C	A constant in the Magnel diagram
C C'	Cover to bars and tendons, and to top bars in cantilever or continuous spans
C_{dev}	Allowance for deviation of cover for durability
C_{min}	Minimum cover for durability
C_{nom}	Nominal cover for durability

$C_{pt}(y)$	Factor according to position of tendons ($V_{Rd,c}(y)$ calculation)
$C_{Rd,c}$	Shear stress constant
E_{cm} E_c E_d	Concrete Young's modulus - secant value; tangent value; dynamic value
$E_{cm}(t)$	E_{cm} at time t
E_{cmi} E_{cmT}	E_{cm} of *insitu* or infill concrete, and structural topping
$E_{cm,long}$ $E_{c,eff}$	Long term value of E_{cm}
E_d	Ultimate action
$E_{d,fi}$ $E_{Ed,fi}$	Fire action (e.g. load); ultimate fire action
$E_{Rd,t,fi}$	Fire resistance
E_p	Young's modulus for prestressing strand or wire
E'_p	Young's modulus after ε_{LOP}
E_s	Young's modulus for steel or rebars
F	End reaction
F_B F_C	End reactions due to distributed moments in continuous spans
F_c F_{c1}	Ultimate force in concrete in compression; first trial of F_c in M_{Rd} iteration
F_{cap}	Lifting anchor capacity;
$F_{c,T}$	Horizontal interface shear force in topping due to stage 2 loads
F_{Ed}	Ultimate end reaction; ultimate point load for edge loads in hcu
$F_{Ed,sup}$	Ultimate support reaction at both sides of the support
F_{Eq}	End reaction in main span for equilibrium (overturning) of a cantilever
F_h	Tension load on prop connection
F_k	Service point load for edge loads in hcu
F_{la}	Load on each lifting anchor;
F_p	Ultimate tension force in tendons
F_{pk}	Strand breaking load
$F_p(\theta)$	Tension force in tendons in a fire situation
F_{Rd}	Point load capacity anywhere in a hcu floor area
F_s F_{s2}	Ultimate force in bars and tendons; in composite section stage 2
F_{sh}	Restoring force due to restraint of free shrinkage
F_{soil}	Soil force
$F_{soil,b}$ $F_{soil,t}$	Soil force at bottom and top of wall
F_{sur} $F_{sur,t\&b}$	Surcharge force, and ar top & bottom of wall
G	Shear modulus for concrete
G_{25}	25% of unit self weight
G_{fs}	Factored self weight load;
G_k $G_{k.j}$	Characteristic permanent (dead) load
$G_{k,stage2}$	Value of G_k for imposed loads, e.g. stage 2 in composite design
$G_{k,sup}$ $G_{k.inf}$	Upper superior value and lower inferior value of G_k
G_{unit}	Unit self weight
I I_c I_{xx}	Second moment of area of section, concrete and about axis x-x
$I_{c,co}$ $I_{c,c,co}$	Compound and composite compound values of I_c
$I'_{c,co}$ $I''_{c,co}$	Value of $I_{c,co}$ in a cantilever with filled cores top rebars, and long-term ditto
I_{beam} I_{column}	Second moment of area for beam and column in cantilever beam flexibility calculation
I_u I_{cr} I_{ef}	Uncracked, cracked and effective values of I
I_{zz}	Second moment of area in transverse direction per unit length.

J	Torsional constant for solid sections
K	Bending moment factor $M_{Ed}/f_{ck}\ bd^2$; curve fitting function for creep coefficients; flexural beam stiffness; St. Venant's torsional constant
K_a	Active pressure coefficient
$K_1\ K_2\ K_{1w}$	Value of K for stage 1 and 2; K for stage 1 including webs
$K'K_2'$	Limiting value of K; value of K in a cantilever
$K_1\ K_2\ K_{eff}$	Uncracked, cracked and effective flexural rigidity
L	Strand extension required
$L\ L_{qp}$	Effective span; and due to quasi-permanent loads
L'	Cantilever effective span
L_b	Bearing length
L_c	Effective span in continuous design between the centres of supports
L_{ef}'	Apparent effective span of the cantilever (from span/deflection check)
L_{shore}	Distance to shoreline
L_t	Transmission length in slippage calculations
L_{0c}	Length of filled cores from end of the span
M	Bending moment
M_{aw}	Moment applied to wall;
$M_B\ M_C$	Continuity moments at supports in continuous spans
$M_{B1}\ M_{B2}\ M_{B3}$	Values of M_B for self weight, dead and quasi-permanent live loads (ditto for C)
$M_c\ M_{c,c}$	Experimental cracking moment of resistance in tests; and for composite tests
$M_{cr}\ M_{cr,c}$	Cracking moment of resistance; and for composite section
$M_{cr,calc}\ M_{cr,c,calc}$	Calculated cracking moment of resistance in experimental tests; and for composite tests
$M_{cr,d}'$	Value of M_{cr} after debonding tendons in a cantilever
$M_{Ed}M_{Ed}'$	Ultimate design moment; and in a cantilever
$M_{Ed1}M_{Ed2}M_{Ed1}'M_{Ed2}'$	Value of M_{Ed} for stage 1 and 2 loading (composite design); and in a cantilever
$M_{Ed\ max}'$	Maximum ultimate design moment in a cantilever
$M_{Ed,fi}\ M_{0Ed,fi}$	Design moment in fire situation
M_{Eq}'	Maximum cantilever moment for equilibrium (overturning)
M_k	Applied bending moment at serviceability
$M_G\ M_Q$	Applied bending moment due to dead and live loads at serviceability
M_P	Internal couple due to prestress $= P\ z_{cp}$
M_{prop}	Service moment at position of props
$M_{Rd}\ M_{Rd,c}\ M_{Rd,c2}$	Ultimate moment of resistance; for composite section; for composite section in stage 2
$M_{Rd,fi}$	Ultimate moment of resistance in fire situation
$M_{Rd2}\ M_{Rd3}$	M_{Rd} for stages 2 and 3 (slab with a topping)
M_sM_s'	Service moment; and in a cantilever
$M_{s,freq}\ M_{s2,freq}$	Service moment due to frequent combination of loads; stage 2 value of $M_{s,freq}$
$M_{s,min}\ M_{s,max}$	Minimum and maximum service moments at critical sections along the span of the element
M_{soil}	Moment due to soil
M_{sur}	Moment due to surcharge

$M_{s,qp}$ $M_{s2,qp}$	Service moment due to quasi-permanent combination of loads; stage 2 value of $M_{s,qp}$
$M_{s,qt}$	Value of M_s at quarter span
M_{sR} M_{sR2} $M_{sR,c}$	Serviceability moment of resistance, and in stage 2; and for composite section
$M_{sR,b}$ $M_{sR,t}$	Serviceability moment of resistance at bottom and top fibres
$M_{sR,x}$	Value of M_{sR} at distance x from support
$M_{sR,net}$	Net average value of M_{sR} across multiple units
M'_{sR}	Serviceability moment of resistance in a cantilever
$M_{sR,d}$ $M_{Rd,d}$	Values of M_{sR} and $M_{Rd,d}$ after debonding tendons
$M_{s,test}$	imposed bending moment at the elastic limit in the tests
M_{sw} M_{self}	Service moment due to self weight
$M_{sw,qt}$	Value of M_{sw} at ¼ span
M_{s1} M_{s2} M_{s3}	Value of M_s for stages 1, 2 and 3 loading (composite design);
$M_{s1,max}$ $M_{s2,max}$	Maximum values of M_{s1} and M_{s2}
M'_{s0} M'_{sT}	Service moment due to self weight and topping in a cantilever
$M'_{s,lpt}$	Service moment at the end of the transmission zone in a continuous span
M_{test}	Applied test bending moment, based on BS 8110-2, Section 9
$M_{s,x}$	Service moment at distance x from support
M_{s0}	Service moment due to self weight
$M_{s0,x}$	Value of M_{s0} at distance x from support
M_u M_{uc}	Maximum failure bending moment in experimental test; and in Composite tests
M_1 M_2	Out of balance moments in continuous spans
N	Number of tendons
N_i	Number of tendons in a row i
N_T	Number of tendons in tension (ultimate limit state)
P P'	Prestressing force (action); point load in cantilever
P_{Ed}	Ultimate point load
P_{max} P_{min}	Imbalanced ultimate reactions (torsion calculation)
$P_{max,0}$ $P_{min,0}$	P_{max} and P_{min} at construction stage
P_{pi}	Initial prestressing force
P_r	Prestressing force at release
P_{pm0}	Prestressing force after initial loses
P_{pmi}	Prestressing force at installation
P_{po}	Final prestressing force after all loses (at transfer)
$P_t(l_x)$	Prestressing force at distance l_x from end of element
Q	Dimensionless notation Z_b/Z_t in Magnel diagram
Q_k	Characteristic variable (live) load
$Q_{k.1}$	Leading live load
$Q_{k.i}$	Accompanying live loads
R R' R_1 R_2	Prop reaction; ditto in a cantilever; ditto at first and second floors in dual propping
R_e	Yield strength of reinforcing bars
R_m	Tensile strength of reinforcing bars
R_{tr}	Ratio of prestressing force after initial losses (at transfer)
R_w	Weighted sound reduction index
R_{wk}	Ratio of prestressing force after final losses (in service)

S_c $S_{c,c}$	First moment of area of section about centroidal axis; and in composite section
S_y	First moment of area above y from bottom or top of section
T	Mean curing temperature
T_c	Tension capacity M16 solid rod socket;
T_{Ed} T_{Edi}	Ultimate design torque; T_{Ed} in a sub-section or wall i
$T_{Ed,0}$	T_{Ed} at the construction stage
T_{py}	Return period
T_{Rd}	Predicted torsion moment of resistance (Holcotors tests)
$T_{Rd,c}$ $T_{Rd,max}$ $T_{Rd,c,i}$	Cracking and maximum ultimate torsional moments of resistance; $T_{Rd,c}$ in sub-sections i
$T_{Rd,c,top}$	$T_{Rd,c}$ in top flange
T_G T_Q	Service torque due to dead and live loads
T_u	Maximum torsion moment in experimental tests
V	Shear force in general
V_{base}	Ultimate shear at base of wall
V_c	Shear capacity M16 solid rod socket;
V_{cw}	Ultimate shear resistance (in American code ACI-318)
V_{Ed} V_{Ed1} V_{Ed2}	Ultimate design shear force; in composite design stages 1 and 2
$V_{Ed,fl}$ $V_{Ed,boot}$	V_{Ed} in a flange and in boot of a beam (torsion calculation)
$V_{Ed,i}$	Design ultimate shear in the wall i (torsion calculation)
$V_{Ed,out}$	Shear force in outer webs (torsion calculation)
$V_{Ed,p}$	Reaction from the point load in punching shear
$V_{Ed,t}$	Shear force in all webs (torsion calculation)
V_{Rd}	Ultimate shear resistance
$V_{Rd,c,fi}$	V_{Rd} of in a fire situation
$V_{Rd,c}$	V_{Rd} due to flexurally uncracked concrete section
$V_{Rd,calc}$	Calculated shear capacity in exterimental tests
$V_{Rd,cr}$	V_{Rd} due to flexurally cracked section
$V'_{Rd,c}$ $V'_{Rd,cr}$	$V_{Rd,c}$ and $V_{Rd,cr}$ in a cantilever
$V_{Rd,c,d}$ $V_{Rd,cr,d}$	Values of $V_{Rd,cr}$ and $V_{Rd,cr}$ after debonding tendons
$V_{Rd,c}(y)$	$V_{Rd,c}$ at distance l_x from the end of the unit and at height y
$V_{Rd,c,c}$ $V_{Rd,c,rc}$ $V_{Rd,c,c2}$	Values of $V_{Rd,c}$ and $V_{Rd,cr}$ for composite section; in composite section stage 2
$V_{Rd,cr,c,calc}$	Calculated flexurally cracked composite shear capacity
$V_{Rd,c+t}$	Ultimate shear capacity section plus filled core(s)
$V_{Rd,max}$	$V_{Rd,fl,max}$ Maximum ultimate shear resistance; in a flange (torsion calculation)
$V_{Rd,n}$	Net shear capacity $V_{Rd,c}$ - $V_{Ed,t}$
$V_{Rd,s}$	V_{Rd} due to steel stirrups
$V_{Rd,c,splay}$ $V_{Rd,c,solid,splay}$	Value of $V_{Rd,c}$ at splayed ends of slabs; and solid spalyed ends
$V_{Rd,t}$	Ultimate shear capacity of filled core(s)
V_{top}	Ultimate shear at top of wall
V_u V_{uc}	Maximum failure shear reaction measured in experimental tests; and in composite tests
$V_{u,cold}$ $V_{Rd,cold}$	Average test shear capacity of hcu at ambient temperature; and calculated capacity (fire tests)
$V_{u,fi}$	Maximum shear reaction measured in fire tests
W	Elastic section modulus at serviceability
W_{Ed}	Ultimate load from floor loads

W_{lb} W_{lt} W_l	Section modulus at bottom and top fibre in transverse direction per unit length; minimum of W_{lb} and W_{lt}
W_p	Load on prop due to wind;
W_t	Torsional section modulus
$W_{t,top}$ $W_{t,bottom}$ $W_{t,web}$	Torsional section modulus of top and bottom flanges and web
W_w	Wind load
X X_1 X_2	Depth to neutral axis; in stage 1 and 2 composite design
X_1 X_2 X_3	First, second and third trial values for X in iteration methods
X'	Height to neutral axis for negative moment at continuous supports
$X(\theta)$	Depth to neutral axis in a fire situation
Y_c	Height from bottom to position to calculate $V_{Rd,c}(y)$
Z_b Z_t	Section modulus at bottom and top fibre
$Z_{b,c}$ $Z_{t,c}$	Composite section modulus at bottom and top fibre
$Z_{b,co}$ $Z_{t,co}$	Compound values of Z_b and Z_t
$Z_{b,c,co}$ $Z_{t,c,co}$	Compound composite values of Z_b and Z_t
$Z_{b,co,net}$	Net average value of $Z_{b,co}$ across multiple units
$Z'_{b,co,ef}$	Effective value for $Z_{b,co}$ in a partially flexurally cracked section at the support of a continuous span
$Z'_{tc,co}$	Compound values of Z_t at top of the topping
$Z_{b,c3,co}$ $Z_{t,c3,co}$	Compound section moduli for stage 3 if topping is present
Z_z	Section modulus at centroid of tendons
α	Modular ratio (short-term) E_s/E_{cm}, E_{cmi}/E_{cm}; ratio Z_t/Z_b; ratio $I'_{c,co} / I_{c,co}$; ratio $I_{c,co,1}/I_{c,c,co2}$
α	Cement class factor
α α'	Angles of inclined interface shear loops
α_{cc}	Concrete strength factor
α_{cw}	Axial stress parameter for shear stress
α_{ds1} α_{ds2}	Coefficients of cement for shrinkage strains
α_e	Elastic modular ratio E_p/E_{cm}; relative eccentricity (spalling calculation)
α_l	Ratio l_x/l_{pt2}
α_p	Angle of prop to horizontal
α_2	Transmission length factor for tendons; modular ratio (long-term) in stage 2 design
α_2 to α_5	Bond length factors
α_I α_{II}	Deformation parameters in serviceability design
α_1 α_2	Anchorage length parameters
α_1 α_2 α_3	Coefficients of concrete for creep
β	Proportion of the total imposed load resisted by each beam in lateral load tests; factor for stage 1 moment of resistance including webs; proportion of self weight needed to cause zero deflection at a prop; beam-to-column flexibility ratio
β (f_{cm})	Strength factor for creep
β_{ds} (t,t_s)	Age factor for shrinkage
β_{RH}	Relative humidity factor for shrinkage
β (t_i)	Age at installation loading factor for creep
β (t_0)	Age at release loading factor for creep
χ	Stress reduction factor
δ δ'	Deflections, movement after application of finishes and camber; deflections in a cantilever

δ_c	Deflection at first cracking
δ_{cc}	Deflection due to crack closure in top of beam in flexural tests
δ_{eq}	Equivalent deflection for comparisons with and without creep
δ_{free}	Free elastic shortening over half the length of the unit
δ_{test}	Measured mid-span deflection in flexural tests
$\delta_{s,mean}$ δ_{mean}	Mean measured slippage of tendons in a unit at transfer; mean of all tendons ditto
δ_u δ_y	Deformation, deflection at ultimate and yielding
ε	Strain
ε_{cu} ε_{cu3}	Ultimate crushing strain in concrete
ε_{ca}	Autogenous shrinkage strain
ε_{cd}	Drying shrinkage strain
$\varepsilon_{cd,o}$	Basic drying shrinkage strain
ε_{cm}	Concrete strain in crack width calculation
ε_{cs}	Shrinkage strain
ε_{cx}	Concrete strain at distance x after transfer
$\varepsilon_{\mathrm{LOP}}$	Strain in tendons at limit of proportionality (elastic limit)
ε_p	Final strain in tendons
ε_{px}	Steel strain at distance x after transfer in slippage calculation
ε_{po}	Prestrain in tendons after losses
ε_{p1} ε_{p2} ε_{p3}	First, second and third trial values for ε_p in iteration methods
ε_{sh}	Relative shrinkage between precast and *insitu* concretes
$\varepsilon_{sc\mathrm{T}}$	Nominal unrestrained, or 'free', shrinkage of topping
ε_{sm}	Steel bar or tendon strain in crack width calculation
ε_{ud}	Limiting strain for rebars
ε_{uk}	Elongation of rebars at the breaking load
ε_s	Elastic strains in reinforcing bars
ϕ	Internal angle of shearing resistance
$\phi'_{\mathrm{r.k}}$	Effective angle of shearing resistance
γ	Density of material
γ_{G} $\gamma_{\mathrm{G},j}$	Partial safety factors for dead load
$\gamma_{\mathrm{G},sup}$ $\gamma_{\mathrm{G},inf}$	Superior and inferior values of γ_{G}
$\gamma_{\mathrm{G}kj,inf}$ $\gamma_{\mathrm{G}kj,sup}$ γ_{G}	for inferior dead load in a back-span, and (superior) dead load in a cantilever
γ_{inf} γ_{sup}	Partial safety factor for variations in prestressing
γ_{Q}	Partial safety factors for live load
γ_m	Material partial safety factor (general)
γ_c	Material partial safety factor for concrete
γ_{P}	Partial safety factor for prestressing action
$\gamma_{p,fav}$	Favourable value of γ_{P} for effect of prestressing
γ_s	Material partial safety factor for steel bars
γ_{sr}	Saturated density of retained material
η	Rectangular concrete stress block factor; initial factor for prestressing; normalised ratio M_{Rd}/M_{sR}
η	Percentage of strand tension capacity
η_1 η_2	Casting condition and bar diameter parameters for bond
η_{fi}	Fire load ratio $E_{d,fi}/E_d$
φ	Creep coefficient (general) for calculating loss of prestress
φ_{RH}	Creep coefficient for relative humidity ditto
$\varphi(\mathrm{t},\mathrm{t}_0)$	Creep coefficient from time t_0 to time t ditto

$\varphi(\infty,t_0)$	Long-term creep coefficient from time t_0 to time ∞
λ	Rectangular concrete stress block factor
μ	Coefficient of friction; ratio of initial prestress
ν	Poisson's ratio
θ θ'	Temperature; angle of rotation; slab or beam end rotation; end rotation due to cantilever; angle of twist
θ_{cr}	Critical fire temperature for rebars and prestressing tendons
ρ	Reinforcement ratio A_s/A_c, air density
ρ_{cc}	Solid rod socket component check
ρ_{con}	Concrete density
ρ_{eff}	Value of ρ in crack width calculation$=A_p/A_{c,eff}$
ρ_1	Reinforcement ratio for tendons extending beyond shear plane
ρ_o	Concrete strength factor
ρ_{1000}	Tendon relation loss at 1000 hours
σ	Stress; prestress
σ_b σ_t	Final prestress in concrete at bottom and top fibres
$\sigma_b(t)$ $\sigma_t(t)$	Prestress in concrete at transfer at bottom and top fibres
$\sigma_{b,net}$	Net average stress in bottom fibre across multiple precast units
$\sigma_{b,x}$ $\sigma_{t,x}$	Value of σ_b and σ_t at distance x from support
σ_c, σ_{c0}	Concrete stress at centre of tendons; bending stress σ_c for service moment due to self weight
σ_c	Limiting compressive stress in prestressed concrete
σ_{cp}	Prestress at centroidal axis after all losses (include $\gamma_{p,fav}$)
σ_{cpx}	Value of σ_{cp} at x from end of element
$\sigma_{cp(y)}$	Value of σ_{cp} at height y from bottom of section
σ_{ct}	Limiting tensile stress in prestressed concrete; Axial concrete tensile stress due to prestress and flexural stress
σ_{Ed}	Ultimate bearing stress
σ_{max}	Maximum compressive stress σ_c allowed
σ_n	Normal stress
σ_{pi}	Initial prestress in tendons
σ_{pd}	Ultimate stress in tendons (used in bond design)
$\sigma_{p,fi}$	Stress in tendons in a fire situation
σ_{pm0}	Prestress in tendons after initial losses (at transfer)
σ_{po}	Final prestress in tendons after all losses (in service). Note EC2 uses $\sigma_{p\infty}$
σ_r	Prestress at release
$\sigma_{Rd,max}$	limiting compressive strength or strut strength of concrete
σ_s	Stress in rebars; stress variation in tendons after decompression in crack width calculations
σ_{sh}	Free shrinkage stress due to ε_{sh}
$\sigma_{sh,b}$ $\sigma_{sh,t}$	Restoring stress in bottom and top of precast slab due to shrinkage
σ_{sp}	Spalling (or splitting) stress in webs of prestressed units
$\sigma_{s,0}$	Bending stress in bottom fibre due to self weight
$\sigma_{s0,x}$	Value of $\sigma_{s,0}$ at distance x from support
$\sigma_{t,x}$	Tensile stress due to $M_{s,x}$ at distance x from support
τ	Shear stress; elastic shear stress distribution (shape) function; total vertical shear stress $\tau_t+\nu$
$\tau_{cp}(y)$	Concrete shear stress at height y from bottom of section

$\tau_t\ \tau_{ti}\ \tau_{up}$	Torsional shear stress; τ_t in each part i of the thin walled section; torsional shear stress in an upstand
$\tau_{T,top} \leq \omega$	Reinforcement ratio
$\psi_i\ \psi_{28}$	Creep coefficient for deflections at t or 28 days
ψ_0	Characteristic combination load factor
$\psi_{0.i}$	Live load factor
ψ_1	Frequent combination load factor
$\psi_2\ \psi_{2t}$	Quasi-permanent combination load factor; ψ_2 at time t
$\psi_{2.i}$	Quasi-permanent live load factor
ψ	Creep coefficients for deflection
ψ_{eq}	Equivalent creep coefficient for comparisons with and without creep
ψ_1	Creep coefficient for deflection from transfer to installation
$\psi_{28}\ \psi_{28t}$	Creep coefficient for deflection from installation to long-term; ψ_{28} at time t
$\psi_\infty\ \psi_{\infty,t}$	Creep coefficient for deflection at long-term; ψ_∞ at time t
ψ'	Coefficient of development for creep coefficients
ξ	Loss ratio after initial losses; reduction factor for dead loads
ξ_1	Factor for tendons in calculation of $A_{p,min}$
ς	Characteristic combination dead load factor, ratio of solid material (including infilled joints) to the whole of a hcu (fire design)
X	Concrete aging coefficient
Δa	Reduced additional axis distance to tendons in fire
$\Delta a_2\ \Delta a_3$	Deviations of bearing lengths
ΔC_{dev}	Dimensional deviation for cover to bars
$\Delta L_o\ \Delta L_{o,limit}$	Slippage, or pull-in, of tendons; limiting code value
ΔM_{Ed}	Reduction in the design support moment at the cusp of the support
$\Delta\sigma_p$	Stress variation in tendons from the state of zero strain of the concrete
$\Delta\sigma_{pr}$	Prestress loss due to initial relaxation of tendons
$\Delta\sigma_{el}$	Prestress loss due to elastic shortening
$\Delta\sigma_{p,c}$	Initial prestress losses $\Delta\sigma_{pr}+\Delta\sigma_{el}$
$\Delta\sigma_{pci}$	Prestress loss due to creep before installation
$\Delta\sigma_{pc}$	Prestress loss due to creep after installation
$\Delta\sigma_{p,s}$	Prestress loss due to shrinkage
$\Delta\sigma_{pr}$	Stress due to final relaxation of tendons (note same notation as for initial relax loss)
$\Delta\sigma_{p,r}$	Prestress loss due to final relaxation of tendons
Δ_s	Horizontal interface deformation, or slippage, between precast unit and structural topping
Δx	Incremental length (interface shear stress calculation)
Δ_1	Deflection at first floor due to prop load R_2 in dual propping
Φ	Reinforcing tendon or bar diameter
Ψ	Exponential creep growth factor

Chapter 1

Commercial use of precast and prestressed concrete in buildings

1.1 COMMON PRECAST PRODUCT TYPES

1.1.1 Benefits of precast concrete construction

All of the typical structural components that make up a building, such as piles, pile caps, ground beams, walls, lift shafts, slabs, stairs, balconies, columns and beams, can be manufactured as precast concrete units. This range of products allows for the entire structure of a building to be constructed using precast units, and the architectural intent for the façade can be accommodated using precast cladding and sandwich panels. Figure 1.1 shows a classical precast skeletal frame for use in a multi-storey car park. Figure 1.2 shows the extraordinary potential for architectural precast concrete.

Figure 1.1 Car park constructed with a precast concrete beam and column frame and long-span prestressed hollow core slabs, Blackpool, UK.

DOI: 10.1201/9781003319450-1

Figure 1.2 Precast single skin cladding with brick slip finish fixed to steel framed carpark with precast stability cores, Manchester, UK.

One of the major benefits of constructing a fully precast building, both structurally and architecturally, is that there are minimal interfaces between other forms of construction to be designed and coordinated.

A fully precast superstructure rising off an *insitu* concrete substructure is a very common method of construction, detailed coordination between the design teams and the sub-contractors occurs at the level where the *insitu* and precast meet. Once the first level of precast is installed, the precast-to-precast construction sequence is simplified and designed to be repetitive, which yields fast build times and rapid progress on site. The building can be made weathertight as it is erected, with windows fitted offsite into the precast cladding and/or sandwich panels prior to delivery.

Precast balcony units can be installed with the balustrade fixed in position, as shown in Figure 1.3. The balustrade contractor carries out this work at a set-down area on site rather than in the precast yard due to the practical limitations of transporting balconies with balustrades already fixed onto them. Prefixing of the balustrade to the balcony units at ground level on site prior to erection improves health and safety by reducing the exposure to work at height for the sub-contractor and generally speeds up the overall process as well.

A common feature in fully precast structures is the use of bathroom pods; their delivery is scheduled in line with the precast delivery and erection programme. The pods are lifted into their approximate position at each level prior to installing the next level of precast flooring. Final positioning and leveling are carried out by the sub-contractor responsible for this package of works. The position of the pods is added to the precast sub-contractors' temporary works drawings so they can be avoided when setting out the push-pull propping used for the precast walls.

Figure 1.3 Precast balcony with balustrade pre-installed.

For multi-storey buildings, the waterproofing of the floors in the form of tanking can be carried out by the main contractor at selected levels. This waterproofing strategy allows follow-on fit-out trades to commence very early in the overall build programme, usually working a minimum of three levels below the precast erection team.

The benefits of a fully precast building include

- improved Health & Safety on site
- excellent fire resistance
- extremely durable with a long design life
- excellent acoustic performance due to the density of the floors and walls
- greater security
- robust structure
- inherent thermal mass
- reduced heating and maintenance costs
- reduced site coordination as many trades are no longer required
- reduced preliminary costs due to reduced programme times

1.1.2 Precast columns and beams

Precast columns and beams are classified as 'linear structural elements' according to the product standard BS EN 13225 (BS EN 13225, 2013). They are used to form the structural frame for many types of building – worldwide they are most commonly used in car parks, industrial buildings, educational buildings, offices and retail developments. Columns and beams work well in conjunction with other precast concrete products such as prestressed hollow core units (hcu) and wall panels. This gives an option to use a fully precast solution for the building. Table 1.1 presents a sample of some recent precast frame projects in the UK.

Precast transfer structures can also be formed using a column and beam arrangement. These are typically required when apartment or hotel structures are built over car parks or open plan office/commercial spaces. The term 'transfer' literally means to transfer from one structural layout to another.

Table 1.1 Sample of recent precast frame projects in the UK

Precast frames and miscellaneous	*Number of storeys and floor area*	*Description*	*Build programme*
Blackpool MSCP	6 storeys, footprint 3790 m^2	Fully precast multi-storey car park frame, precast lift shafts and stair cores, prestressed hcu slabs with *insitu* structural toppings - 1120 spaces	19 weeks
Newcastle MSCP	9 storeys, footprint 2370 m^2	Fully precast multi-storey car park frame, precast lift shafts and stair cores, prestressed hcu slabs with *insitu* structural toppings -900 spaces	34 weeks
MMU Birley Fields	5 storeys, footprint 5544 m^2	Precast frame consisting of columns, beams, staircore and lift shaft shear walls, stair flights and landings and plate flooring. Feature cantilever stair and Spanish Steps area. Steel plate girder transfer beam at 3rd floor and sports hall steel roof trusses included in build programme	20 weeks
Leeds Arena	12,000 seat arena with an upper and lower tier	Precast units supported on a steel frame - 1044 terrace units, 522 step units, 67 vomitory wall panels, 508 acoustic envelope wall panels, 63 stair flights, 10484 m^2 of hcu slabs	28 weeks

Concrete strength of C50/60, i.e. characteristic compressive cylinder strength $f_{ck}=50$ N/mm^2 and cube strength $f_{ck,cube}=60$ N/mm^2 according to BS EN 1992-1-1, Table 3.1 (BS EN 1992-1-1, 2004) is readily available when manufacturing precast columns and beams. This gives the designer the potential to reduce column and beam section sizes and reinforcement quantities when compared with *insitu* concrete structures. Improved control of cover in the factory environment allows the available effective depth in section sizes to be maximised in design and also helps to ensure that the requirements for durability are met. A very important feature of precast concrete frames is the inherent fire resistance and long design life – these characteristics are particularly useful in car parks: for example, the precast columns, beams, floors, walls and stairs used to build a fully precast car park can all be readily designed to achieve a design life in excess of 50 years in an XC3/4 exposure class (see Table 5.4), and they can also be designed to provide 120 minutes fire resistance (see Table 5.11) with no additional protection works necessary.

Multi-storey precast columns can be cast to improve the speed of erection on site, two or three storeys are the most common sizes but a greater number of storeys could be cast if this was beneficial to the project. Alternatively, single-storey columns are used, often with multispan precast beams. Columns with cast-in steel base plates give a detail that is practical on site and allows for quick erection without the need for temporary propping. The base plates are connected to the foundations using holding down bolts, the column can be plumbed and aligned by adjusting the holding down bolts. Columns can be cast with corbels or proprietary hidden connections to support incoming beams; these support methods work equally well with precast beams and steel beams. These details are extensively dealt with by the FIB (FIB, 2008, 2014), Elliott & Jolly (Elliott and Jolly, 2013) and Elliott (Elliott, 2017).

Precast columns also work very well in hybrid construction schemes, with the precast columns often being used in conjunction with an *insitu* concrete reinforced flat slab or an *insitu* post-tensioned slab.

Precast beams are usually cast in the form of an inverted T or an L profile, the downstand that is formed in these profiles supports the floor slabs or other secondary beams. If the downstand on precast beams causes an issue with head height or complicates the routing of M&E, one alternative detail that can be used to remove a downstand completely, or at least reduce its depth, is the addition of RSA's bolted to the sides of rectangular profiled beams. This detail usually requires the beam to be made wider, and deflection limits tend to govern the feasibility of its use.

A very well-proven and practical solution for car park structures that gives fast coverage of large floor areas is the use of long-span hcu with a structural topping. Typically 15–16 m slab spans onto an 8 m grid of precast columns and beams. This is a very efficient structure and provides large open spaces for the car park layout. A useful feature on precast car parks is that the beams around the perimeter can be designed and cast with an upstand wall on them to act as the barrier for accidental impact loads. The precast stair core walls can also be designed for accidental impact loading. This additional function of the precast elements can be utilised to remove the need for steel or other types of vehicle impact barriers on the project.

1.1.3 Crosswalls

Crosswall construction is a term used to describe a form of construction where the load-bearing walls of the structure are arranged perpendicular to the building axes, the crosswalls transfer the gravity and lateral loads to foundation or transfer slab level. Their design is according to the product standard BS EN 14992 (BS EN 14992, 2007). All of the internal partitions can be used as crosswalls; they are designed to support the gravity loads from the floors and roof and act as shear walls to provide horizontal stability against wind and EHF loads. EHFs are 'notional lateral loads' applied to the structure to account for global imperfections such as lack of verticality or plumbness. They must be applied in all Eurocode load combinations. Figure 1.4 is an illustration of a crosswall frame comprising crosswalls and prestressed solid or hcu flooring. Crosswall construction can be used with precast external cladding and precast sandwich panels or lightweight cladding alternatives to provide a variety of elevation styles and finishes. Table 1.2 presents a sample of some recent precast crosswall projects in the UK.

When used with precast sandwich panels, the structural load-bearing leaf of the sandwich panels can also be included as shear walls in the lateral stability system. Further efficiency in design can be achieved when using precast crosswall and sandwich panels with the utilisation of vertical shear connection between walls. Connecting perpendicular walls, for example where crosswalls and sandwich panels meet on the façade line, will create a T-shaped structural element. These T-shaped elements, in combination with C-shaped or box-shaped lift shafts and stair cores, can be modelled using analysis software to create a very efficient, economical and robust structure. The vertical shear transfer through the joints between each of the individual elements that form these composite shapes must be checked and verified to ensure the structure behaves as modelled.

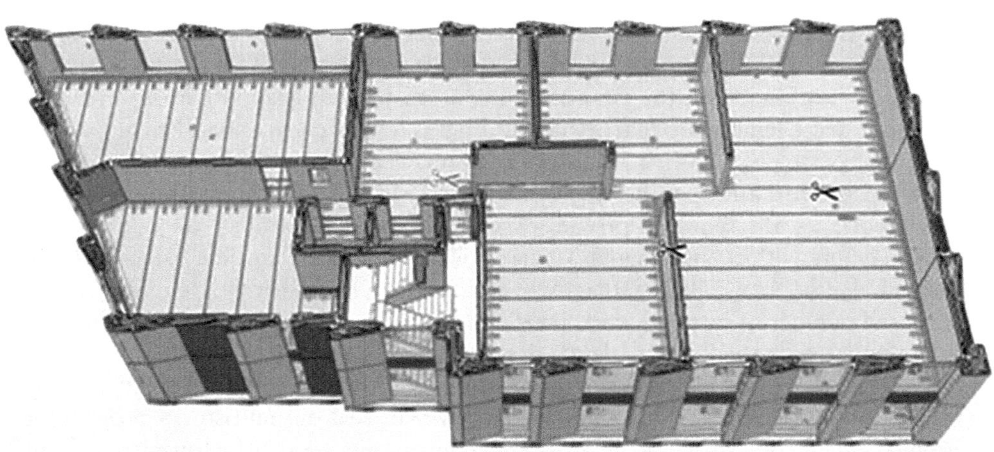

Figure 1.4 Fully precast structure comprising crosswalls, prestressed hcu and sandwich panels.

Table 1.2 Sample of recent crosswall projects in the UK

Crosswall – residential, student accommodation, hotels, prisons	*Number of storeys and height*	*Number of apartments*	*Build programme*
Guildford Crescent, Cardiff	32 storeys, 94 m	265 apartments	46 weeks
Arena Central, Birmingham	19 and 24 storeys, 67 m	297 apartments	Both apartment blocks erected concurrently, 50 weeks
Gulson Rd, Coventry	16 storeys, 53 m	110 apartments	26 weeks
GH20, Glasgow	12 and 15 storeys, 49 m	320 apartments over commercial space and underground car park. 10,000 m² fully precast underground car park, including two levels of precast retaining walls around perimeter and precast transfer structure at ground level	Underground car park, 25 weeks Both apartment blocks erected concurrently, 32 weeks
Leeds Oasis	21 storeys, 62 m	212 student accommodation studios	25 weeks

Crosswall is an inherently stable, efficient and economical form of construction. Utilising most, if not all, of the load-bearing walls available within the stability system maximises the potential to distribute lateral loads from wind and EHF. This avoids, as much as possible, highly concentrated loading being applied to particular walls within the structure. Distributing the loading in this way assists in minimising wall thickness required and can also make for a more efficient and economical transfer slab or foundation design. Precast prestressed solid or hcu floor slabs can be used to provide a rigid diaphragm without the need for a structural topping. Peripheral and internal ties are contained within the depth of the hcu slab; the ties are designed and detailed to ensure the transfer of all lateral loads at each level into the shear wall stability system.

Using the total frame solution construction speed is greater than alternative systems. Units are manufactured off site with a high standard of finish, walls are mist coated and delivered to site suitable for direct application of paint. M&E services can be incorporated into the production process; back boxes, conduits, ducting, builders' works openings, etc.

Crosswall frames are the ideal solution for the residential market, student accommodation and hotels as well as commercial offices, schools and prisons. This form of construction has been used worldwide for the design of fully precast buildings up to 30 storeys.

A common arrangement for apartment buildings is to use internal load-bearing crosswalls at a maximum spacing of 10 m. This allows 250 mm deep prestressed hcu to support the typical loading specified and span between supporting walls. The minimised number of internal load-bearing walls required provides increased set-out flexibility for apartments and the room layouts within apartments. Using solid or ordinary reinforced slabs will require greater slab depths than 250 mm to achieve a 10 m span; alternatively, there will need to be an increased number of internal walls provided to reduce the span to a dimension that can be accommodated by the 250 mm depth slab.

Crosswalls can be readily manufactured using C50/60 concrete to minimise reinforcement and/or wall thickness. This gives a benefit when used on multistorey projects, as the additional cement cost can be offset by the savings in reinforcement. However, it should be noted that wall thickness is also influenced by the fire resistance and acoustic performance required, it is often the case that the load-bearing capacity is only critical at the bottom levels of medium to high rise multistorey buildings. In fact, in low-rise structures where reinforcement content in walls is often at a minimum and load-bearing capacity is not critical the use of higher strength concrete is not economical or a benefit.

Precast crosswalls provide an excellent barrier to sound transmission, see example R_w calculation below for a t=200 mm thickness element

Average density (conservative); $\rho_{con}=2350\,kg/m^3$;
Mass of wall; $w=\rho_{con}\,t=470\ kg/m^2$
Weighted sound reduction index; $R_w=37.5\ \log(w/1\ kg/m^2)-42=58.2\,dB$.

Table 5.4 of BS EN 1992-1-2 (BS EN 1992-1-2, 2004) sets out the thickness and axis distance required to satisfy the fire resistance for load-bearing walls *exposed on one side* and *exposed on two sides*. Care needs to be taken when determining the exposure each crosswall needs to be designed for. For example, the fire strategy for an apartment building states a requirement of REI 120 for load-bearing walls. A crosswall containing a door opening forms the dividing wall between an apartment and the corridor. If the apartment door does not provide 120 minutes of fire resistance, then this crosswall may be exposed to fire on two sides during a fire, i.e. the fire inside the apartment breaks through the door and into the corridor, exposing the wall to fire on both sides simultaneously. The utilisation factor of crosswalls can be checked to see if the wall thickness and cover provided satisfy the fire resistance requirements. Interpolation can be applied to the tabulated data in Table 5.4 of BS EN 1992-1-2. If tabulated data does not provide adequate fire resistance for the proposed wall details, further detailed design can be carried out in accordance with the simplified or general calculation methods given in BS EN 1992-1-2 to determine if the wall details will satisfy the fire resistance specified in the fire strategy.

1.1.4 Precast stair cores and lift shafts

Precast stair cores and lift shafts are a useful form of construction that can be readily used in conjunction with all other forms of construction and as part of a fully precast structure. Their design is according to the product standard for walls BS EN 14992 (BS EN 14992, 2007). It is common to have precast cores and lift shafts where the main building structure is constructed using timber frame, steel frame, *insitu* concrete frame or masonry. The precast core and lift structures can be designed to provide stability to the main building structure or to be stand-alone and self-supporting only. Figure 1.5 shows such a stability core with cast-in break-out bars for connection to the floor slab and cast-in steel plates for connection to the steel frame.

Precast cores and lift shafts are typically built at full height in one continuous visit. This helps to minimise costs as the precast contractor is on site for as short a period as possible, and it also provides immediate access and egress to the main building structure as each of the floor levels is constructed.

When the design intent is for the precast cores and/or lift shaft to provide stability to the main building structure, there are a number of typical connection details and ways to robustly tie the structures together. Steel frames are generally connected to the precast structure using cast-in plates within the precast walls. The cast-in plates are detailed and fabricated to allow a bolted connection to be made between the steel beam and the wall. *Insitu* concrete frames are generally connected using breakout reinforcement bars; these bars are cast into the precast walls in preformed cassettes. The cassettes are available from several manufacturers and can be purchased in standard forms or with bespoke reinforcement arrangements to suit particular project details. The bars within the cassettes are bent out on site to tie into the reinforcement cages of the *insitu* concrete. Breakout bar connection details also work with metal deck flooring and all of the precast floor slab types when combined with an *insitu* structural topping.

Figure 1.5 Stability core with cast-in break-out bars for connection to the floor slabs and cast-in steel plates for connection to the steel frame, Warrington, UK.

1.1.5 Precast concrete cladding

Precast concrete single skin cladding, with or without integral insulation, can be designed and manufactured as the façade finish for *insitu* structures, steel-framed structures and of course, precast structures. Cladding panels are designed according to BS 8297 (BS 8297, 2017). Precast cladding can be manufactured with a very wide range of finishes, e.g. brick and/or stone, coloured concrete, etched, polished, patterned or a combination of different finishes, with examples as shown in Figures 1.2 and 1.6. The cladding can be designed to be stacked from foundation level or supported on the building structure at each level. The typical support arrangement for cladding supported at each level consists of steel brackets or corbels projecting from the cladding at its base; these supports carry the vertical loading and provide horizontal restraint, with additional fixings positioned at the head of the cladding to provide horizontal restraint only.

The window and door systems for the project can be installed and sealed in the cladding panels off-site at the precast yard. The completed assembly is then transported and craned into position on site. Typically, a double mastic seal is applied to the vertical and horizontal joints between precast units on site to provide a fully weathertight façade.

1.1.6 Precast concrete sandwich panels

A popular and extremely versatile element of precast construction is the use of the sandwich panel system. Their design is according to the product standard for walls BS EN 14992

Figure 1.6 Precast brick-faced cladding on a steel-framed office building, Coventry, UK.

(BS EN 14992, 2007). These panels consist of an inner load-bearing structural leaf, an insulation layer and an outer non-load-bearing façade leaf. No external scaffolding is required. The sandwich panel is manufactured using a proprietary system that connects the outer leaf and insulation to the inner load-bearing leaf. These connections are usually manufactured from a fibre composite material or stainless steel. The sandwich panel façade leaf can also have the same finish as for cladding panels.

Windows and doors can be installed and sealed in the panels off-site at the precast yard. The completed assembly is then transported and craned into position on site. Figure 1.7. Typically, a double mastic seal is applied to the vertical and horizontal joints between precast units on site to provide a fully weathertight façade.

Balcony support assemblies, usually in the form of cast-in plates or stubs, can be cast into the panels during production to allow the connection of precast or metal balconies on site. The structural leaf of the panels can be designed to support the loading from the balcony units. Alternatively, the balcony connection can be taken into the floor slab although this generally requires additional site works and *insitu* concrete stitching. Depending on the details of the support and the connection method used, the precast balconies may need to be erected in sequence along with the panels. Typically metal balconies are fixed to projecting stubs at the outer face of the non-load-bearing façade leaf and can be erected in any sequence independent of the panels.

The structural connections between sandwich panels are formed using proprietary loop boxes in the vertical joints; the panels are connected to each other across the horizontal joint using vertical tie bars in grouted sleeves or bolted connections; and horizontal ties are provided between the panels and the precast hcu or solid floor slabs. All joints are filled with high-strength grout or structural mortar to provide a robust construction that will satisfy the requirements of fire compartmentation and acoustic performance.

Figure 1.7 Sandwich panel loaded on to a trailer in a precast yard ready for delivery to site with windows and aluminium cladding installed.

A very high standard of finish is achieved by power floating the internal face of the structural leaf. The internal face is given a light mist coating of paint to highlight any blemishes, and cosmetic finishing of the surface is carried out in the factory prior to being placed in the stockyard. The sandwich panels are delivered to the site suitable for the direct application of paint to the internal face.

The most common insulation used in precast sandwich panels for buildings greater than 18 m in height is Class A1 mineral wool. The cavity is fully filled as part of the production process. Sandwich panels can be readily designed to achieve the typical 120-minute fire resistance required (REI 120) for structures in the Building Regulations and Table 5.4 of BS EN 1992-1-2.

Generally, minimum wall thicknesses for sandwich panels are 150 mm structural inner leaf and 80mm non-structural outer leaf. These values satisfy the minimum dimensional requirements shown in Diagram 9.2 of ADB Volume 2 (Building Regulations 2010) for the exclusion of a cavity barrier. Should a cavity barrier be a specific requirement, the mineral wool insulation used in the sandwich panel system complies with the requirements of 9.14(c) of ADB Volume 2 as the insulation is placed under compression due to the production process, i.e. the concrete forming the structural leaf is poured onto and compresses the insulation during production.

Provision for mechanical & electrical (M&E) services can be cast into sandwich panels in the form of back boxes, conduits and ducts, and builders' work openings can also be formed in both leaves of the panel. Figure 1.8. It is also possible to provide a cast-in conduit and access detail specifically for the installation and connection of lightning protection tapes. It is very important to have early input at the design stage from the Architect and

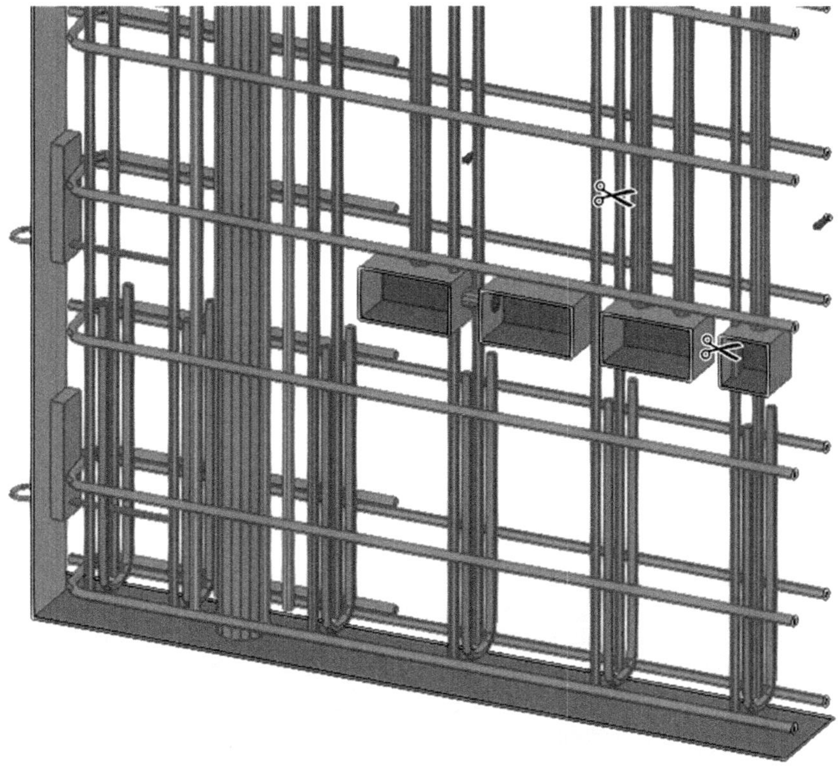

Figure 1.8 Backboxes and conduit for electrical services within a wall panel.

M&E Consultant to allow coordination of the position of any M&E with reinforcement details and other cast-in items such as lifting anchors, grout sleeves, propping sockets, etc. A common problem that occurs is clashing between the vertical conduits and the horizontal reinforcement. This can be resolved by using back box extension pieces or conduit elbows to move the position of the vertical conduit further into the panel.

1.1.7 Precast retaining walls

The product standard BS EN 15258 (BS EN 15258, 2008) for precast retaining wall elements focuses mainly on vertical cantilever walls retaining soil and granular materials. There are numerous other methods of utilising precast walls to retain various solid and liquid materials. Design guidance that is applicable to generic retaining wall arrangements can also be found in the standards BS EN 13369 (BS EN 13369, 2018) 'common rules' and BS EN 14992 'wall elements'.

Precast vertical cantilever walls can be designed and manufactured to be prestressed and reinforced. A common construction method is the use of precast walls with an *insitu* concrete base. The precast wall is cast with holes at the required centres along its width for reinforcement bars to be fitted through on site. These bars are tied into the main reinforcement cage for the base and provide a connection between the precast and *insitu* sections. The main design checks for this form of construction are overturning, sliding and ground-bearing pressure. The base and the wall stem are both checked for bending and shear capacity. The vertical joints between the individual units are formed with a tongue and groove detail so that the wall panels interlock with each other when installed and concreted into the base. Once constructed, the face of the wall and base on the retaining side can be tanked to provide a watertight structure. Figure 1.9 shows an example of a prestressed wall for retaining soil.

Another use for vertical precast cantilever walls with an *insitu* base that is cost-effective and rapid on site is the construction of boundary walls. The height of the wall can be manufactured to suit the project requirements, and all of the finishes described in the section above can be provided on the wall faces. The walls may have a retaining function, and this can be accounted for in the design, but generally the main loading is from wind. Prison walls and other tall security-sensitive boundary walls are particularly well suited to precast vertical cantilever construction. On these projects, the minimum wall height is generally 5.2 m, which is readily manufactured in modular sections. This size of wall panel can be transported and installed with standard haulage and site craneage.

Rather than cast an integral *insitu* concrete base with the wall panels, fully precast retaining walls can be manufactured as an L shape. These L-wall units can be freestanding for low retained loads and are useful for temporary storage and as a moveable storage option, for higher loading the L-walls can be manufactured with pockets in the base leg of the L so that they can be bolted down to a foundation or slab. The L-walls can also be manufactured with projecting reinforcement from the base leg of the L, this reinforcement is normally tied into the main cage of an *insitu* slab that forms the floor of the storage structure. L-walls are available in various lengths and in heights up to 6 m.

An alternative to the L-wall profile is the precast rocket wall. These are designed to be bolted down to an existing foundation or slab and have the benefit of being suitable for use as a dividing wall that can be loaded on either side.

Figure 1.9 Prestressed concrete vertical cantilever wall panels and *insitu* base constructed as a soil retaining wall.

Figure 1.10 400 mm thick prestressed solid retaining walls, spanning 6 m, this thickness was required at the base of a 16 m high storage facility for wood pellets.

Horizontal spanning precast walls can be used to retain material when supported by precast, steel or *insitu* concrete frames. Storage of products as diverse as grain, wood pellets, vegetables, aggregates, and silage are just some examples of the use of this form of construction. Prestressed solid and hcu wall panels are particularly suited to support the high pressures generated by retained materials in storage buildings. Figure 1.10 shows a large wood pellet store constructed with a steel frame and horizontally spanning prestressed solid wall panels. The walls spanned 6 m, were 400 mm thick and supported the pressure from 16 m of heaped material. The wall panels are designed with a tongue and groove or shear key joint between adjacent panels; this joint is designed to transfer load laterally between panels. This capacity to transfer load allowed the most heavily loaded bottom panels to share load with the panels above and give a more economic design for the wall thicknesses and the amount of prestressing strand required. Another feature of the design for some storage buildings is impact loading. The use of very heavy-duty shovel machines in the storage facility requires the precast to be designed for relatively high impact loading, which generally removes the possibility of using prestressed hcu wall panels at the levels where impact could occur.

1.2 HORIZONTAL ELEMENTS – SLABS, TERRACES AND STAIRS

1.2.1 Solid and hollow core units for floor and roof slabs

Hollow core units (hcu) are now one of the primary solutions selected by engineers, contractors and increasingly by developers themselves to meet the demands of fast-track site programmes.

The units can be reinforced or prestressed, solid or hcu; but by far the most common slab type is prestressed hcu, due to efficiencies in the manufacturing process as well as load/span performance as shown in Sections 2.2.3–2.2.6. The product standard BS EN 1168 (BS EN 1168, 2005) also applies to solid slabs that are manufactured using the same methods as hcu slabs. The structural efficiency of hcu precast slabs facilitates a very cost-effective floor that is suitable for use in all types of buildings. The design flexibility and load-carrying capacity provided by the range of slab depths available see this precast element being used as a versatile flooring solution in residential, commercial and industrial projects.

Hcu slabs are manufactured within a factory environment to a very strict quality process, which gives the control required over prestressing and concrete mix. Figure 1.11 shows a typical production plant containing 13 no. steel beds up to 227 m long. The longest bed extends beyond the internal envelope of the factory by 80 m, the typical internal beds are 120–150 m long.

The steel bed gives a smooth-formed finish to the soffit that can be left exposed in car parks, for example. The soffit of the hcu slab is also suitable to receive a painted finish. Typical 28 days concrete strength is f_{ck}=40 to 50 N/mm^2 and cube strength $f_{ck,cube}$=50–60 N/mm^2. The strength of the concrete at detensioning at about 20 hours is $f_{ck}(t)$=25–30 N/mm^2, obtained by crushing cubes cured adjacent to the production beds.

Automated plotters can be used to mark up the slabs with cut lines, open cores, notches, holes, etc. These details are preformed in the factory prior to dispatch to site. Hcu can also be manufactured with integral heating and cooling pipes cast in; this requires the bottom flange of the section profile to be thickened so that it can accommodate the pipework.

Common prestressed solid slab depths range from 100 to 150 mm, hcu slab depths typically range from 150 to 500 mm in 50 mm increments. Figure 1.12 shows three such units, pretensioned using helical strands shown in Figure 1.13. The most common standard

Figure 1.11 Prestressed hollow core factory showing long line production beds.

Figure 1.12 Prestressed precast hollow core floor units of nominally 1200 mm width and 200, 300 and 400 mm depth produced by the Echo slipformer.

Figure 1.13 Close up of cross section of 12.5 mm diameter seven-wire helical strand.

slab width is 1200 mm although some manufacturers produce slabs up to 2400 mm wide. Narrow width units can be cut from full width units with some restrictions to the dimensions of the widths available due to the hcu geometry.

Some adjustment of the machines used in the manufacturing process is possible; this allows bespoke depths to be manufactured. For example, a 260 mm depth unit can be formed using a 250 mm hcu profile by making an adjustment to the machine and increasing the depth of the top flange by 10 mm. The majority of prestressed hcu are manufactured on long line beds of over 100 m, this means that a full bed of each hcu depth needs to be cast, and this can have an influence on costs for bespoke units.

Using the values given in Table 5.8 of BS EN 1992-1-2 and noting the clause regarding axis distance in clause 5.2 (5), a fire resistance of REI 90 can be achieved for units 100 mm thick or greater; it is possible for slabs of 150 mm thick or greater to achieve REI 120 with no requirement to provide any additional fire protection to the unit. If the units do not satisfy the criteria given in Table 5.8 of BS EN 1992-1-2 a more in-depth approach can be undertaken, which involves the calculation of the moment and shear capacity in the fire situation and can result in an improved fire resistance period being demonstrated. See Section 5.5 for further details.

Prestressed hcu can be designed and manufactured for very long spans. See Sections 2.2.2–2.2.6. This gives excellent flexibility to create structures with large open spaces that can be divided up using lightweight partitions in numerous configurations. The prestressing of the concrete enhances deflection performance and the hcu section minimises the self-weight of the slab. These are two very important factors in providing a slab solution that can be used in spans of 17 m and greater depending on the specific loading requirements of the project. Hcu can also be designed as composite slabs with the application of an *insitu* concrete structural topping once installed on site. See Section 12.3. This can further enhance the load-carrying capacity and increase the spans that can be achieved.

The minimised self-weight of hcu (typically 60% of that of a solid slab) reduces the loading that needs to be supported by beams, columns, walls, etc., and taken down to the substructure of the building. This reduction in overall loading can provide cost savings on the other structural elements and parts of the building. The efficient section profile of hcu reduces material usage and contributes to sustainability and reduction in the carbon footprint of the building.

Hcu can be used to provide diaphragm action and transfer horizontal forces into the stability system. The diaphragm can be designed using the tie reinforcement provided as part of the robustness details, and a structural topping is not required. The peripheral, internal and horizontal ties are all provided within the depth of the hcu, as shown in Figure 1.14. These ties are anchored with *insitu* concrete into open cores and pockets formed in the slabs during production. The quantity of tie reinforcement provided is checked to ensure that it satisfies the requirements of the diaphragm and robustness in accordance with BS EN 1992-1-1, whichever forces are greater will govern the reinforcement details provided.

Precast hcu can be erected very quickly on site; this gives a significant reduction to site durations and the construction programme can be shortened. There are a number of common methods used to install these units; they can be fitted using lifting anchors that are cast into the slabs during the production process, fitted using clamps or fitted using chains as shown in Figure 1.15. Lifting anchor pins are most often used as they provide a very quick method to install slabs directly to their final position. The use of clamps requires manual adjustment of the slabs into their final positions, while the use of chains can cause aesthetic damage to the slab edges and soffit and also requires the manual adjustment of the slabs into their final positions.

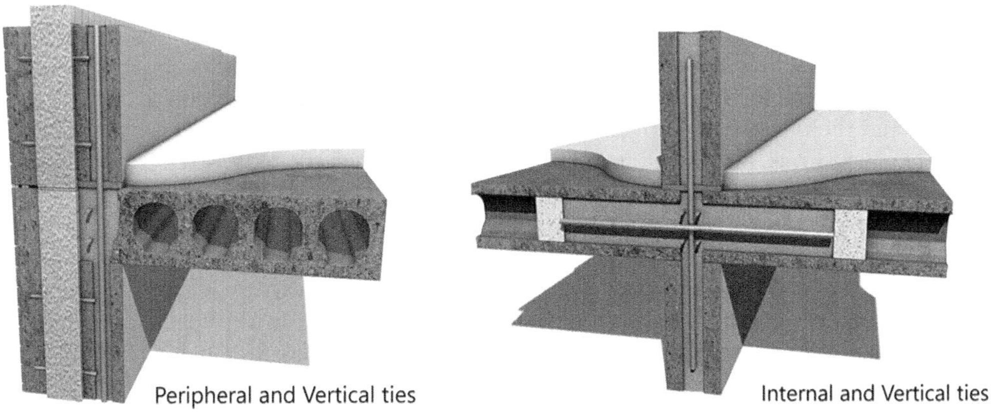

Figure 1.14 Horizontal and vertical tie details on a fully precast structure. (a) Peripheral and vertical ties. (b) Internal and vertical ties.

Figure 1.15 Lifting hollow core units on site with chains and cast-in lifting pins.

On the majority of projects, hcu do not require any temporary propping works on site; the exception to this is the occasional requirement for a composite slab to be touch propped at mid-span (or two props at ⅓ span) prior to pouring the structural topping. Propping is not common though, unless there is a limitation on the depth of the composite slab, as the need for temporary propping is usually designed out by increasing the depth of the hcu, the uplift in cost for a thicker slab is usually much less than the cost of the props and the time lost installing and removing the temporary works. Once installed onto the supporting structure, the slabs immediately create a safe working platform, the tie steel required for the design is fixed, and the *insitu* concreting works to the longitudinal joints between units and at the ends are carried out. When the *insitu* concrete has gained adequate strength access to the floor can be given to follow on trades, normally a 72-hour period is programmed, but this can be reduced by using specific *insitu* concrete mixes and taking cubes to confirm the concrete strength has reached the value required. Design for temporary loading from MEWPS, scissor lifts, etc. can be carried out if this is a requirement of the construction process. The number of square metres fitted in a day will vary depending on the specific details of the project and any site constraints, but an average of 40 lifts per day for 8 m long 200 mm depth hcu would be typical. This rate of fitting gives an installation area of approximately 400 m^2 per day.

The apartment building shown in Figure 1.16 is a fully precast structure using 250 mm deep hcu, supported on precast crosswalls and sandwich panels. The hcu spanned up to 10 m which allows (i) the number of internal load-bearing walls to be minimised, (ii) flexibility in the apartment configuration and room layouts and (iii) rapid construction of each level.

1.2.2 Plate flooring

Prestressed and standard reinforced plate flooring consists of a thin precast section, typically 50–100 mm thick, with longitudinal steel girder cast into it at regular spacing across

Figure 1.16 Fully precast apartment buildings under construction utilising prestressed hollow core slabs, Glasgow, UK.

the width of the unit. Their design is according to the product standard BS EN 13747 (BS EN 13747, 2005). The girders' primary function is to deal with handling stresses in the thin precast section, the loading applied at installation and the following construction process. In virtually all projects, the girder is used when lifting the plate flooring, and chains are hooked on at designated points along the span. For long-span units, a lifting frame is used to ensure the slab is not cracked during handling. Figure 1.17 shows an example of a 12 m span floor plate, with lifting girders.

Plate flooring receives an *insitu* topping on site; the permanent works design is based on the composite section and the main reinforcement cast within the precast section; top steel can be added over supports to create a continuous spanning slab. The temporary works design focuses mainly on the steel girder capacity; generally, the limiting factor for prop spacing is the buckling of the top bar in the girder. The temporary works design must consider loading from the concrete, site fixed reinforcement, any other project-specific 'dead' loading and an agreed 'live' load allowance for construction traffic. Propping is typically provided at 2.5 m centres for standard reinforced plate flooring but longer spans are achievable depending on girder details and overall slab depth. Prestressed plate flooring typically spans up to 4 m before propping needs to be introduced. It is essential that care is taken when pouring the *insitu* concrete topping, and the method adopted ensures that concrete is not heaped anywhere on the floor. On floors with large section depths, the use of polystyrene void formers within the *instiu* topping becomes practical, and this can significantly reduce the self-weight of the floor.

Figure 1.17 Installing a 12m span precast floor plate with a lifting frame, polystyrene void formers used to reduce the self-weight of the completed composite floor, heating and cooling pipes were cast into the precast elements.

Heating and cooling pipework can be cast into the plate flooring during the manufacturing process. The pipework is pressure tested before and after casting to ensure no damage or leakage occurs. Pipework 'tails' are left projecting from the precast so the M&E contractor can complete the pipework connections on site once the floors are installed. Backboxes and conduits for light fixings, smoke alarms, thermostats, etc. can also be cast into the precast units during manufacture.

Due to the cost of this floor type compared to other forms of construction, plate flooring is typically used in buildings that require a very high-quality finish to the slab soffit. It tends to be specified for universities, public buildings such as libraries, and office developments where the design intent is to have the concrete surfaces exposed to view, as shown in Figure 1.18.

1.2.3 Precast ribbed floor slabs

Although they are sometimes thought of and referred to as beams in many countries, one of the most commonly used ribbed floor slab forms is prestressed double-tee (as shown in Figure 1.19) and triple-tee 'beams'. See also Section 2.2.7. These slabs are an efficient way of building car park decks. Their design is according to the product standard BS EN 13747 (BS EN 13224, 2011). Long-span units up to 16 m are commonly used. The double-tee and triple-tee units are typically 2.4 and 3.6 m wide, respectively; this allows for coverage of large floor areas in very short programme durations. One major advantage is that these floor units do not require a finish screed. Typical overall structural depth for prestressed composite ribbed floor units at 16 m span would be 575 mm for standard imposed car park loading in an XC3/4 exposure class environment. See Figure 2.22. It is worth noting that the

Figure 1.18 Precast plate flooring manufactured with rebated soffit profiles.

Figure 1.19 Prestressed double-tee floor units.

top level of a car park often has a layer of finishes for waterproofing and this additional dead load can require an increase in the overall structural depth of the floor unit. Double-tee and triple-tee units are most often used on steel and precast-framed car parks. The connection details between the floor slabs and frames of steel, precast or *insitu* concrete are virtually identical making for a simple build methodology that can be easily adapted to suit the form of construction used for the main structure.

A specific form of ribbed slabs, termed as minor floor elements in the product standard, is designed for use at ground floor and can be manufactured with integral insulation, as shown in Figure 1.20. The rib sections can be designed as reinforced or prestressed. This floor slab, when produced as a 375 mm deep unit of concrete and EPS insulation, can achieve U-values as low as 0.11 W/m^{2}K based on the exposed perimeter length (m) divided by the floor area (m^2) of 0.4. Typical Psi values of 0.11 W/mK can be achieved at junctions between elements such as floors and walls. The Psi value is a measure of the heat loss at a thermal bridge. Note that U-values and Psi values are dependent on the details of the wall configuration and construction materials.

The exceptional overall thermal performance and speed of installation sets this type of flooring apart from traditional beam and block and *insitu* concrete ground floor slabs. This product is particularly suited to residential and commercial ground floors; site labour is radically reduced, service penetrations are preformed during manufacture, 80–100 m^2 of flooring can be installed in 90 minutes using a lorry-mounted crane (often called a Hiab, although this is actually a brand name), building of walls can commence 3 hours after installation of the floor is completed. For reinforced concrete slabs, a 25 mm self-levelling screed can be applied as a finish, but for prestressed slabs with upward camber a much deeper finish screed will normally be required.

Figure 1.20 Ground floor insulated ribbed slab.

The thickness of the top section of the slab spanning between the ribs is generally 50 mm for 1200 mm wide floor units. This top section of the slab can be designed and manufactured with a light reinforcement mesh or macro synthetic fibres depending on the project specification. Load span tables are available from manufacturers. The use of a prestressed slab gives greater spans and load-carrying capacity, but the additional depth of finishes required to deal with camber needs to be considered.

1.2.4 Terracing – sports stadiums and entertainment venues

Precast terracing is an excellent solution for constructing sports stadiums and entertainment venues such as arenas and cinemas. Figure 1.21. Sometimes known as 'bleachers', the terraces are typically manufactured as single L-shaped units spanning between bearing points on a supporting structure. An alternative, but less common approach, that reduces the number of pieces and therefore installation time on site is to manufacture the units as multiple L-shaped units with two or three terraces cast as one unit. The supporting structure for the precast terrace units can be constructed in precast, *insitu* concrete, steel or in smaller projects masonry can be used. The terrace units are typically designed as two separate reinforced elements, a beam section and a horizontal slab section. The natural frequency of the units is checked to ensure the calculated value is in accordance with the requirements of the specification. This check relates to the vibration or perceived motion that can be felt by people sitting or standing on the terracing. Note that the minimum natural frequency required will occasionally govern the design of the terrace units beam section size and reinforcement content. A typical arrangement of supporting structure will have terrace unit spans of 7.5 m; terrace beam depth of 475 mm × 150 mm thick,

Figure 1.21 Precast concrete terrace units spanning up to 10.3 m installed onto a steel frame.

slab section depth is usually a minimum of 100 mm (note for external stadiums the slab thickness varies as it is generally cast with a fall of 1:40), the slab span is normally in the region of 800 mm. The typical geometry described in the preceding sentence will provide a terrace unit with a minimum natural frequency in excess of 6 Hz for a typical loading arrangement.

The presence of step blocks also needs to be considered in the design; these step blocks will be supported by the terrace units and may be positioned anywhere along the span depending on the seating and access arrangement.

Allowance for thermal movement is particularly important in outdoor terracing; this can be accommodated by movement joints in the supporting structure at the necessary spacing and a sliding connection between the terrace unit and its support. Waterproofing of the joints between units is normally in the form of a mastic sealant with a secondary protection layer beyond this seal to the details provided in the project specification.

1.2.5 Precast concrete stairs

Precast stairs and landings are some of the most common precast products. Their popularity is driven by the fact that on site *insitu* works are complicated and skill intensive. Precast stairs are used in all types of buildings and forms of construction; masonry, steel frames, *insitu* concrete and precast frames. Figure 1.22 shows a precast stair flight being lifted by the crane for installation into position on a steel-framed structure. Their design is according to product standard BS EN 14843 (BS EN 14843, 2007). Straight stair flights are the most common shape, winders and curved flights are also possible, and many very impressive bespoke stair flight shapes can be manufactured. Depending on the support options

Figure 1.22 Stair flight being lifted for installation.

available and the design intent, separate landing slabs or stair flights cast with top and/or bottom landings are provided.

The stair core shown in Figure 1.23 and noted in Table 1.3 was manufactured as individual precast walls, landings, stair flights and capping slabs. The walls that formed each box were bolted together in the precast yard with a landing or half landing also fixed within the box. Each assembled box was delivered to the site on wide loads and craned into position, followed by the stair flights and finally the capping slabs. This allowed a very rapid build programme on site.

In some buildings where a staircase is designed to be an architectural feature, the precast stairs can be manufactured, for example, using coloured concrete, textured finishes, upstands and many other details.

Typical waist thicknesses are 150, 175 and 200 mm. From a structural perspective the thickness of the waist is governed by the span, the loading, the fire performance and the durability requirements.

Risers can be cast square or at a raked angle, nosings can be cast into the stair or recesses can be formed during casting for post-fixed nosings. Sockets and proprietary systems can be cast into the stairs for connection of the balustrade. Commonly a 175 mm waist is used to allow space between the main top and bottom reinforcement for side fixing of balustrades using concrete screws or other post fixings.

A proprietary temporary edge protection system can be clamped to the stairs and landings at ground level on site prior to installation. It is also possible to fix the permanent balustrade to the stair at ground level on site, this balustrade needs protection during construction but

Figure 1.23 Nissan Factory stair core under construction, Sunderland, UK.

Table 1.3 Examples of recent projects using precast concrete cores for stability

Precast cores	*Number of storeys and height*	*Stability or stand alone*	*Build programme*
Working MSCP	11 storeys, 44 m	Stability to steel frame	2 main cores erected concurrently, 12 weeks
Paddington MSCP	14 storeys, 45 m	Stand alone	2 cores erected concurrently, 15 weeks
Salford MSCP	10 storeys, 31 m	Stability to steel frame	2 cores erected concurrently, 10 weeks
Nissan Factory	27 m	Stand alone	Sent to site as boxes weighing up to 24 tonnes, 2 weeks

has the health and safety benefit of reducing work at height, and gives time and cost saving due to the removal of the need for temporary edge protection installation and removal.

A typical standard detail for supporting precast stairs is the use of cast-in sockets and rolled steel angle(s) (RSA) bolted onto the flights. The RSAs are detailed with a hole in the horizontal leg so that they can also be used to fix the stair to its support and satisfy the robustness requirements. Where separate precast landing slabs are used, there are two very common standard details for support in precast and *insitu* concrete structures. A RSA is bolted to the precast or *insitu* concrete wall and the landing slab bears onto it. Notches in the precast landing slab and shear studs welded to the RSA are designed to satisfy the robustness tie requirement.

An alternative to RSA supports is provided by proprietary hidden connections; these are cast within the precast landing slabs and a corresponding pocket is cast in the precast or *insitu* concrete wall. Robustness can be satisfied by the use of a connection pin between the cast-in unit and the supporting wall or an arrangement of the connections.

REFERENCES

BS 8297. 2017. Design, manufacture and installation of architectural precast concrete cladding, BSI, London, UK.

BS EN 1168. 2005. Precast concrete products - Prestressed concrete hollow core units, +A3:2011, BSI, London, UK.

BS EN 13224. 2011. Precast concrete products - Ribbed floor elements, BSI, London, UK.

BS EN 13225. 2013. Precast concrete products - Linear structural elements, BSI, London, UK.

BS EN 13369. 2018. Common rules for precast concrete products, BSI, London, UK.

BS EN 13747. 2005. Precast concrete products - Floor plates for floor systems, +A2:2010, BSI, London, UK.

BS EN 14843, 2007. Precast concrete products - Staircases, BSI, London, UK.

BS EN 14992, 2007. Precast concrete products - Wall elements, BSI, London, UK.

BS EN 15258, 2008. Precast concrete products - Retaining wall elements, BSI, London, UK.

BS EN 1992-1-1. 2004. Eurocode 2: Design of concrete structures - Part 1-1: General rules and rules for buildings, +A1:2014, BSI, London, UK.

BS EN 1992-1-2. 2004. Eurocode 2: Design of concrete structures, Part 1-2: General rules - Structural fire design, BSI, London, UK.

Building Regulations. 2010. Approved Document B: Fire Safety - Volume 2- Diagram 9.2 HM Government 2020.

Elliott, K. S., 2017. *Precast Concrete Structures*, 2nd ed., CRC, Taylor & Francis, London, UK, 694 p.

Elliott, K. S. and Jolly, C. K. 2013. *Multi-Storey Precast Concrete Framed Structures*, 2nd ed., John Wiley, London, UK, 750 p.

FIB. 2008. *Bulletin 43, Structural Connections for Precast Concrete Buildings, Guide to Good Practice*, Fédération Internationale du Béton, Lausanne, Switzerland, 370 p.

FIB. 2014. *Bulletin 74, Planning and Design Handbook on Precast Building Structures*, Fédération Internationale du Béton, Lausanne, Switzerland, 299 p.

Chapter 2

Why precast, why prestressed?

2.1 USE OF PRESTRESSED CONCRETE IN BUILDINGS

Precast prestressed concrete (psc) is used extensively in buildings and structural frames, particularly floor slabs on load-bearing masonry for housing and apartments, and in multi-storey precast concrete, steelwork and timber frames for offices, retail, hotels, car parking, and other residential buildings. In about 50% of projects, psc floor units are designed and constructed compositely with a cast *insitu* reinforced concrete topping (known as a 'structural topping' to distinguish it from a non-structural 'finishing screed'). The UK market for psc floors in 2022 is around £250 million, or 2,240,000 sq.m. area per year. At present, this represents about 40% of the UK's suspended flooring market (excluding cast *insitu* ground floors). These floors are manufactured by around 12 companies in the UK. One of the most common types is the hollow core unit (hcu) for floors, roofs and even walls. Figure 1.12 shows some typical prestressed hcu. The units are produced by approximately 125 companies in Europe. The International Prestressed Hollowcore Association (IPHA) [https://hollowcore.org] was formed in 1969 to bring the hcu industry together and drive it forward with R&D and innovation. It has 57 member companies worldwide that produce the majority of hcu. Figure 2.1 shows a 16 m long × 400 mm depth unit destined for a multi-storey car park (commonly known as parking structures). These units are pretensioned longitudinally using tendons such as seven-wire helical strand, shown in Figures 1.13 and 2.2, or individual plain or indented circular wire, shown in Figure 2.3, placed in the lower quarter of the depth. Note the (approximately) 40 mm upward camber due to eccentric prestress in Figure 2.1. These units may be designed compositely using a structural topping. Figure 2.4 shows the preparation for such a topping. Elsewhere the units may receive a non-composite finishing screed, such as in offices and housing, or a raised access floor lattice with floor tiles in offices or retail buildings.

Prestressed concrete beams are designed in two major types:

- small beams of inverted-tee shape, 150, 180 to 225 mm deep, used in beam and block flooring for housing (residential, congregation, garages) as shown in Figure 2.5.
- main beams, typically 300 to 600 mm wide × 300 to 1000 mm deep, used to support floor slabs or terraces over medium to large spans, of rectangular, inverted-tee or I shape, as shown in Figures 2.6 and 2.7.

So-called 'house beams' form suspended floors together with infill blocks, either of solid concrete (density ≈ 1950 kg/m^3), aerated concrete (density ≈ 1350 kg/m^3) or expanded polystyrene foam (EPS, density ≈ 20 kg/m^3) for thermally insulated ground floors (Figure 2.5c). The beams are prestressed simply to overcome the problems of large deflections for such shallow depths

DOI: 10.1201/9781003319450-2

Figure 2.1 Lifting 16m long × 400mm depth prestressed concrete hollow core units using lifting pins with safety chains in the stockyard.

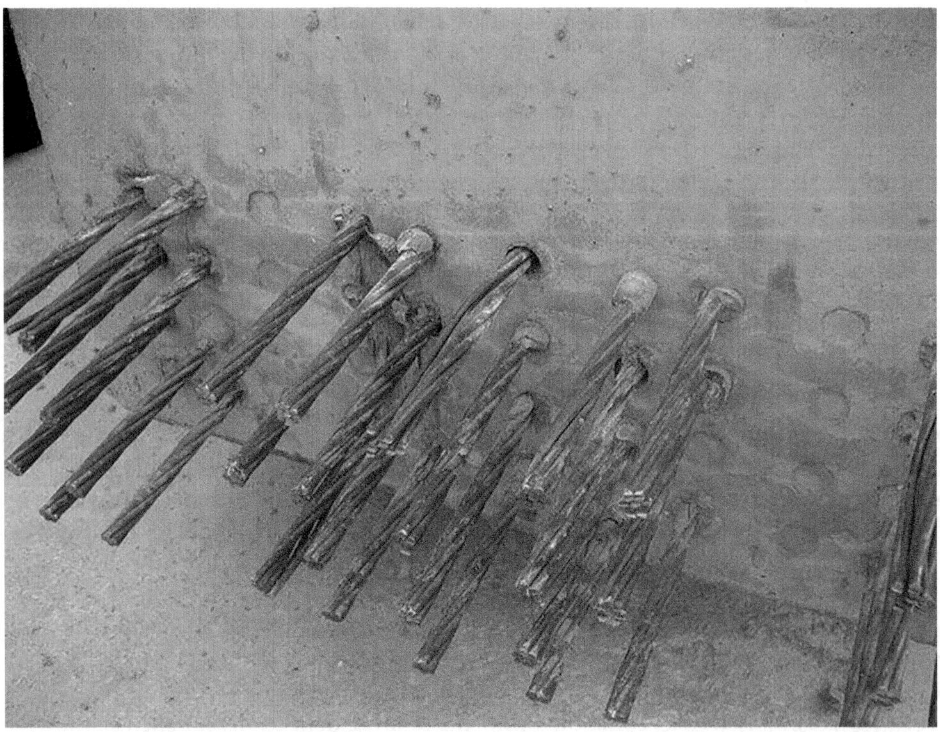

Figure 2.2 Examples of pretensioning helical strand.

Figure 2.3 Circular indented wire pretensioning tendons used in prestressed floor beams and slabs.

Figure 2.4 Preparation and completion of a power-floated structural topping onto prestressed hollow core floor units.

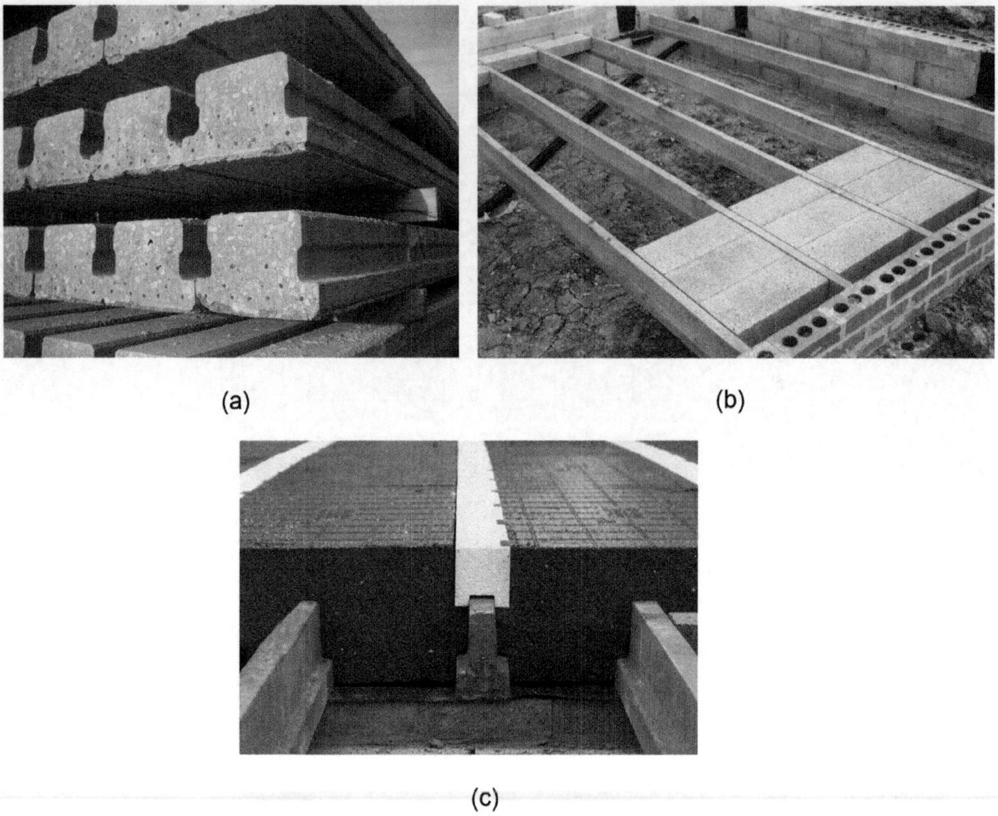

Figure 2.5 Beam and block flooring, using (a) wide beams, (b) narrow beams with concrete infill blocks and (c) narrow beams with extruded polystyrene blocks and rail (white polystyrene) for insulated ground floors ((b) & (c) c. Litecast Ltd., UK).

having a span/depth ratio up to about 40:1. Today the UK market is around 9,500,000 linear m, yielding 4,800,000 sq.m. floor area, approximately 40% of the housing market.

Figure 2.7 shows why inverted-tee beams are the most popular form of main beam -

i. the floor slab is recessed into the top of the beam, reducing the overall depth of construction compared to a rectangular beam where the depth of slab adds to the depth of the beam
ii. the beam is highly pretensioned to increase clear span
iii. the floor slab, with or without a structural topping, may act compositely with the beam, further increasing capacity or reducing beam depth.

Note that the sides of the upstand have been scabbled to remove surface laitance in order to enhance the interface shear resistance between the beam and floor units.

Bate and Bennett (Bate and Bennett, 1976) commence their book on the design of psc as:

> Prestressing is a technique whereby the performance of a structure is improved by the introduction of a permanent stress (prestress) so as to cancel out some of the stress produce by the dead and imposed loading. Concrete, with its inherent weakness in tension, is a particularly suitable material for prestressing since a compressive prestress can be used to reduce the tensile stress to an acceptably low value or, if desired, to eliminate it altogether.

Figure 2.6 Prestressed rectangular beams (at Creagh Concrete Ltd, UK).

Figure 2.7 Prestressed concrete inverted-tee beam (at Hollowcore Pty, Melbourne, Australia). Note the roughened surface at the sides of the upstand of the beam.

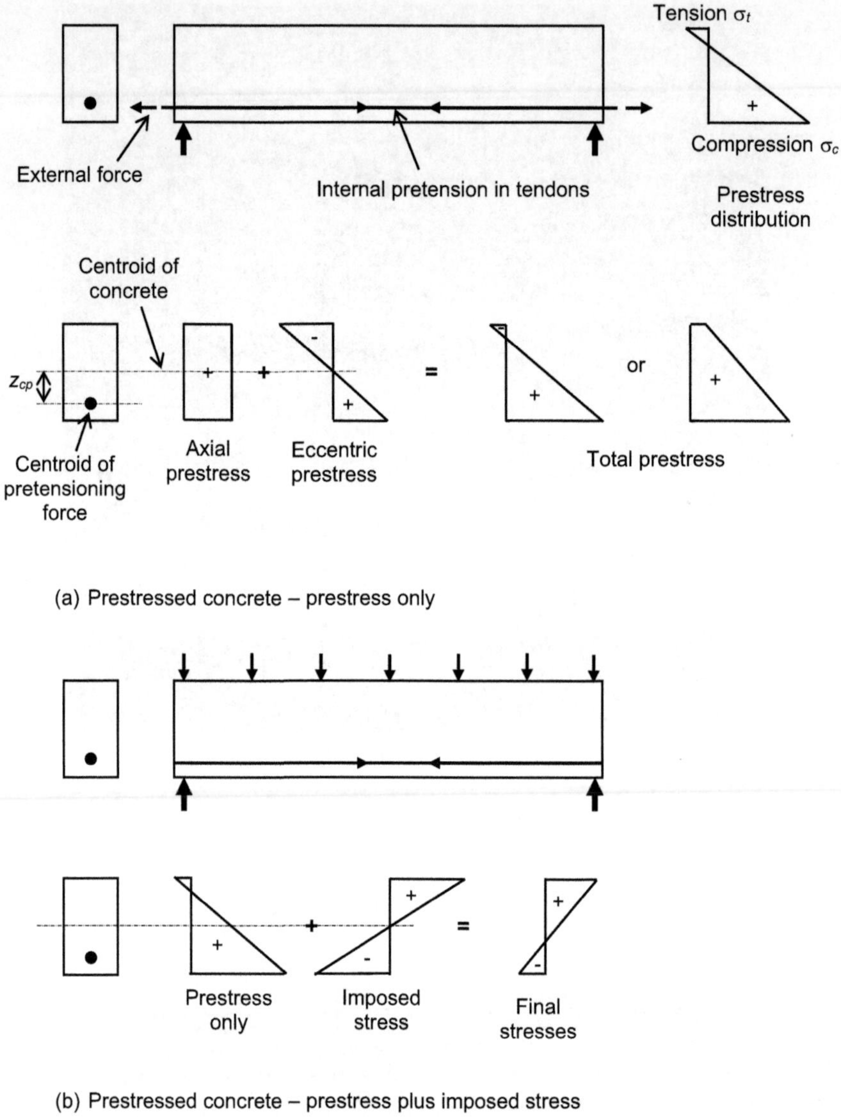

Figure 2.8 Principles of prestressing concrete with eccentric pretensioning force. (a) Prestressed concrete–prestress only. (b) Prestressed concrete – prestress plus imposed stress.

Figure 2.8 shows the key features of prestressing – to pre-compress the concrete in the tension face, in this case the soffit (but is equally applicable to the top face in a cantilever) in order to partially or totally nullify internal tension due to imposed loads with the aim of controlling or eliminating cracking. The final stresses in the top and bottom surfaces (often called top and bottom 'fibres'), which are due to the addition of prestress and bending loads, are limited by values given in national Codes of Practice – in this book, the values are based on the Eurocode EC2, namely BS EN 1992, Part 1-1 (BS EN 1992-1-1, 2004). Design procedures for stress analysis are given in Chapters 6 and 7.

Bate and Bennett add interesting notes of the history of the development of a Code of Practice for psc:

> Prestressed concrete became a practical possibility in the 1930's when the phenomena of creep and shrinkage began to be understood, and it was realized that the early failures with mild steel tendons could be avoided by the use of high tensile steel. After the war, development in Europe, and a little later in America, was rapid and the first attempt in the UK to codify a design procedure was made in 1951 when the Institution of Structural Engineers published its First Report on Prestressed Concrete (Institution of Structural Engineers, 1951). The Report formed the basis of the British Standard Code of Practice CP 115 (CP 115, 1959). ...supervised by the unified Code CP 110 (CP 110, 1972) ... The main difference in concept between design to CP 115 and that to CP 110 is the introduction of the limit state approach in the latter Code. The changes in procedure between the two Codes were however small and produced none of the difficulties experienced with changes in procedure required for reinforced concrete in adopting limit state design. The reason for this was that by its very nature, prestressed concrete require separate consideration of the behaviour under normal conditions of loading of service from its performance under exceptional conditions such as over loading ... and hence CP 115 anticipated the need to consider serviceability and ultimate limit states, which were later to become an inherent part of limit state design philosophy.

Statically reinforced concrete, known simply as 'reinforced concrete' (rc), contains steel rebars and shear stirrups to resist internal tensile forces, and together with the compressive resistance of the concrete, it provides flexural (bending) and shear strength for beams and slabs. However, in practical designs with acceptable load-span capabilities, static rebars cannot prevent cracking or a loss of flexural stiffness, leading to increased deflections. Prestressed concrete satisfies both strength and stiffness by pre-compressing the concrete in the otherwise tensile zone using high-strength steel tendons, such as individual wires or helical strands composed of seven wires (three wire strands are available but not common), as shown in Figures 1.13 and 2.2, that are first stretched or elongated by about 0.6% of their original length (about 1360 mm for the 227 m long pretensioning bed shown in Figure 1.11). After the concrete is cast and hardened, typically after 18 hours, the tendons are either cut or otherwise released to transfer their pretensioning force to the concrete by bond. This is known as the 'transfer' stage (sometimes known as the 'detensioning' stage). In this way, the reinforcement is *dynamic* rather than *static*.

The compressive strength of the concrete must be greater than that used in rc (which is typically f_{ck}=30–35 N/mm^2 cylinder strength) for two reasons:

1. if the transfer strength $f_{ck}(t)$ is around 25–30 N/mm^2 after 18 hours (t means *time* not transfer) then the 28 days characteristic strength will inevitably be f_{ck}=40–55 N/mm^2
2. in order to create a fairly well balanced section at the ultimate limit state, with the depth to the neutral axis X being about 0.5 times the effective depth to the tendons in the tension zone, the strength must also be around f_{ck}=45–50 N/mm^2. As a simple example (the details of which will be given in later chapters), a 300×600 mm deep psc beam is pretensioned using 12 no. strands (area A_p=93 mm^2 of characteristic strength f_{pk}=1770 N/mm^2). Aiming for $X=0.5\times500=250$ mm gives $f_{ck}\geq12\times93\times0.87\times1770/(0.8\times0.567\times300\times250)=50.6$ N/mm^2.

If the flexural tension due to external loads at the serviceability limit state does not exceed the pre-compression plus an allowance for the flexural tensile strength of the concrete, known in Eurocode EC2, Part 1-1 (BS EN 1992-1-1, 2004) as f_{ctm}, the section remains uncracked, retaining its full cross-sectional stiffness to resist a much greater imposed load than a comparable reinforced concrete section.

The complete flexural, shear and deflection analyses of psc slabs and beams are given in Chapters 7–9, 10 and 11, respectively.

Prestressing therefore reduces the net cross section of members, and/or gives longer spans for the same imposed loads compared with rc. Figure 2.9 shows the differences in the simply supported span-to-depth requirements for psc hcu and rc ribbed floor units (with 3 ribs×150 mm wide and a top flange of 60 mm depth), both 1200 mm wide and of similar self-weight, subjected to an imposed dead load of 2 kN/m² and live loads of 1.5, 5.0 and 10.0 kN/m². The hcu design satisfies service and ultimate moments of resistance and 'active' deflection limit of span/500 after construction (relevant to 1.5 kN/m² load only). The rc ribbed slab satisfies the ultimate moment of resistance and span/depth ratio according to BS EN 1992-1-1, Exp. 7.16a or 7.16b, simultaneously. Taking, for example, a clear span of 10.0 m with the live load=5 kN/m² the required depth of hcu is 238 mm, which would be rounded to the nearest commercially available unit of 250 mm depth (f_{ck}=40 N/mm², self-weight=3.27 kN/m² with 11 no. 12.5 mm helical strands of 1023 mm² area at 40 mm cover), but the required depth of the rc unit is 435 mm (f_{ck}=30 N/mm², self-weight=5.03 kN/m² with 9 no. H16 rebars of 1809 mm² area at 30 mm cover). In both cases, the exposure is 'internal', class XC1 to BS 8500-1 (BS 8500-1, 2023), and floor type is for 'offices' using an ultimate load combination factor ψ_0=0.7 (as explained in Chapters 3 and 5 and in Tables 3.1 and 5.4). To make a more accurate comparison, because the actual limiting span of the 250 mm deep hcu is 10.23 m, the depth of the rc ribbed units at this same span would be 448 mm, with self-weight of 5.16 kN/m². The reduction in the depth of the slab would be 45%, or almost 200 mm, and a reduction in self-weight of 37%.

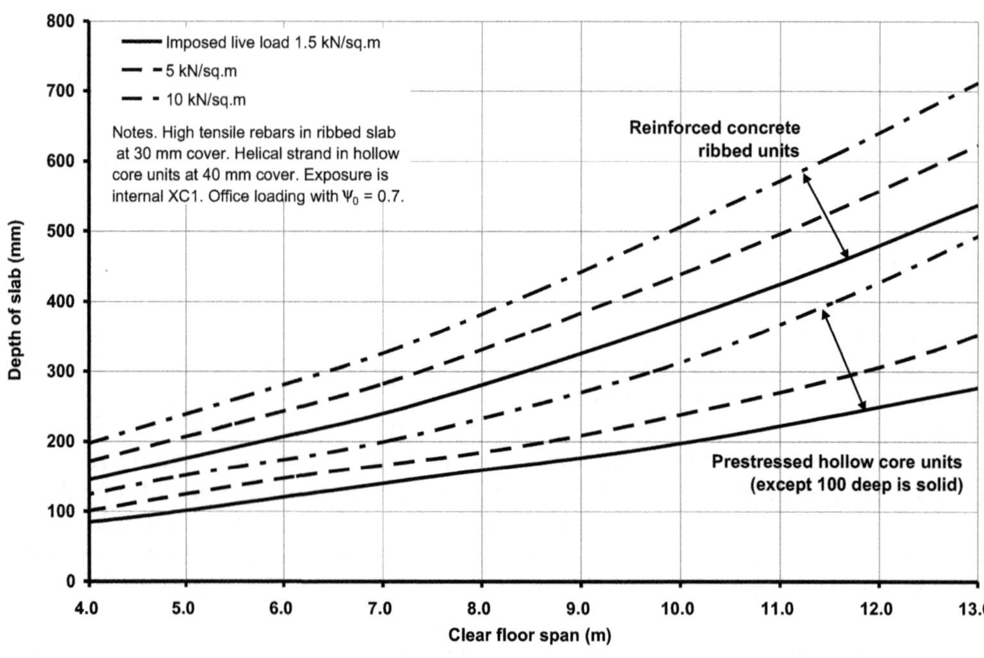

Figure 2.9 Comparison of span vs. depth for prestressed concrete hollow core floor units and reinforced concrete ribbed floor units.

Figure 2.10 shows the same differences, this time for the mass of strands in the hcu and rebars in the rc ribbed unit (including mesh in the top flange) expressed in kg/m^2. The sudden increases in mass of rebars are due to the diameter of the main bars jumping to the next available size of rebar, e.g. from 16 to 20 mm diameter, and where the graph levels are due to the same size of rebars being required over a range of span, e.g. from 10.0 to 11.5 m. Nevertheless, the greater amount of rebars than strands, typically three times, is obviously a consequence of the much greater design stress in strands at the ultimate limit state ($f_{pd} \approx 1370$ to 1400 N/mm^2) than of rebar ($f_{yd} = 435$ N/mm^2), together with the action of prestress. Further, there are additional U bars at the supports and A98 mesh (5 mm diameter bars at 200 mm spacing gives 98 mm^2/m) in the top flange of the rc ribbed units, not present in hcu. The economic implications of the data in Figures 2.9 and 2.10 are as follows:

The budget cost for hcu and ribbed slabs is built up from direct material costs, production costs, transportation and installation costs. The ground floor of a residential building could be constructed with prestressed hcu or reinforced concrete ribbed floor slabs: a typical example being a clear span of 5.9 m, domestic imposed loading of 1.5 kN/m^2, 1 kN/m^2 for lightweight partitions and 1.3 kN/m^2 for finishes, services, etc. A prestressed hcu is compared to thermal ribbed units in terms of concrete and steel material usage in Table 2.1. The typical cost, at time of writing, of prestressing strand is £1030 per tonne and reinforcement bars £750 per tonne. The concrete mixes for both slab types would be similar in cost except for the addition of polypropylene fibres in the top flanges of thermal ribbed units. The size of the project, location and site logistics has an impact on the overall costs of these products when supplied and fitted. The limiting design criteria for the hcu is moment capacity which is at approximately 90% for both service and ultimate, the limiting design criteria for the ribbed slab is deflection, which is at 93% of capacity,

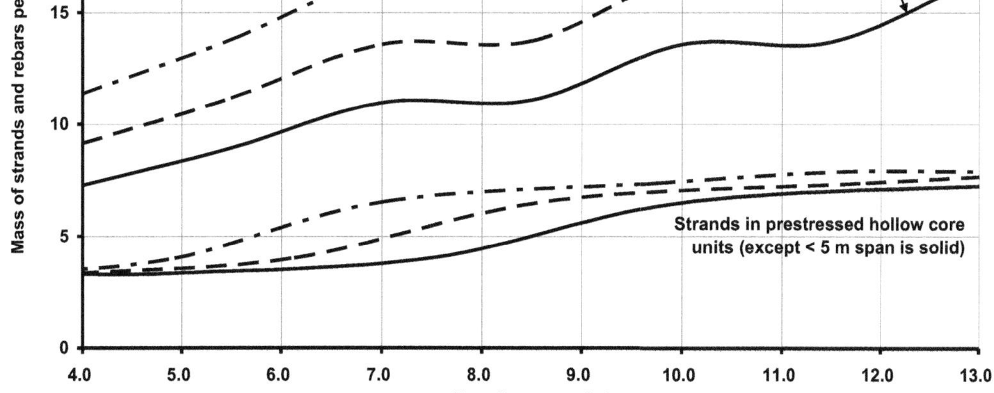

Figure 2.10 Comparison of span vs. mass of strands and rebars (including mesh in top flange) for prestressed concrete hollow core floor units and reinforced concrete ribbed floor units.

Table 2.1 Economic comparison of prestressed hcu and reinforced concrete thermal ground floor units

Slab type	*Slab length (mm)*	*Slab width (mm)*	*Slab depth (mm)*	*Concrete weight (kg/m²)*	*Steel weight (kg)*	*Strand & reinforcement costs (£/tonne)*
Prestressed hollow core units	6100	1200	150	257	14.9	1030
Reinforced thermal ribbed units	6100	1200	300	233	46.5	750

Data supplied by Creagh Concrete Ltd., UK and is based on the same load carrying capacity and span for the hcu and thermal ribbed units.

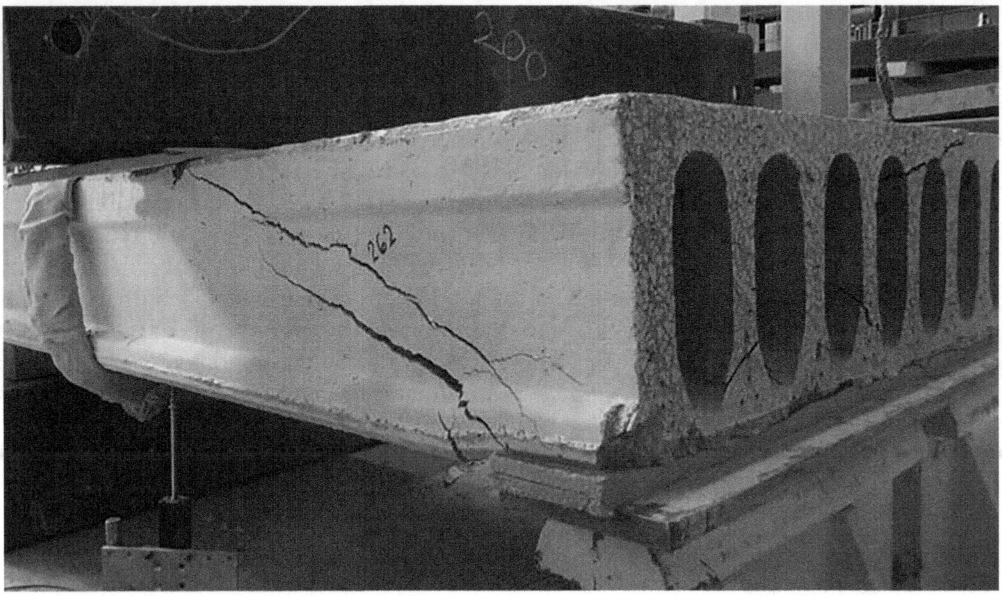

Figure 2.11 Shear failure in prestressed hollow core floor unit without shear reinforcement, by diagonal tensile splitting in the 35° shear plane.

Prestressed members may also contain static rebars, in the form of shear stirrups, to cater for ultimate shear forces that are greater than the shear capacity of the prestressed section. Prestressed beams may also contain confinement and strut-and-tie rebars at the supports, particularly where recessed half-joints are used, but these rebars are not influenced by prestress. Floor slabs such as hcu (Figure 1.12) and house beams (Figure 2.5) do not contain shear reinforcement for production reasons. Their ultimate shear capacity $V_{Rd,c}$ is limited by the ultimate design tensile strength of concrete f_{ctd} acting in the diagonal shear plane, as shown in Figure 2.11. The mode of shear failure is often sudden and brittle; therefore, design capacity equations contain several hefty factors of safety. In most cases, e.g. uniformly distributed load, the shear capacity of these units is rarely critical in favour of the limiting flexural tension f_{ctm}, so there is no incentive to disrupt production by adding shear stirrups.

Longitudinal statical rebars may also be added to supplement pretensioned tendons in so-called 'partially prestressed members'. The total ultimate moment capacity is the addition, considering the strain compatibility, of the strength of the tendons plus the statical rebars. This has a particular advantage where, in the case of a totally prestressed member without additional rebars, the transfer stresses would be much greater than allowed, even causing considerable cracking close to the ends of the member where the imposed flexural stresses are small. As an alternative to reducing the amount of prestress, at the detriment

Figure 2.12 Thermal ribbed ground floor units (these were manufactured as reinforced not prestressed concrete, although the cross section is the same).

of flexural strength, statical reinforcement is added in the regions of maximum imposed bending moments. The design of partially prestressed members is extensively dealt with by Gilbert et al. (Gilbert et al., 2017) and is therefore not covered in this book.

Prestressed concrete 'thermal' slabs, shown in Figure 2.12, are variations on standard psc units, but the principles and design methods are the same. These members may have wide and shallow flanges, which have to be carefully considered when working alongside highly prestressed webs, usually by adding lateral rebars where the two sections meet, but the requirements are usually catered for by adding steel mesh, such as A142 (6 mm diameter bars at 200 mm spacing gives 142 mm^2/m), at the interface.

This book concerns precast prestressed concrete units, pretensioned in a factory (although units are manufactured under a weather shield outdoor facility in many countries and nearly all tropical countries) to form predominantly floor slabs (solid, voided, ribbed, thermally insulated units), main beams, floor beams and terraces/bleachers, but lesser production of stair flights, landings, columns and wall panels. This book does not deal with post-tensioned prestressed concrete, a subject covered in other text books on prestressed concrete design (e.g. Gilbert et al., 2017).

2.2 PRODUCT DEVELOPMENT AND LOAD V SPAN TABLES & GRAPHS

2.2.1 Introduction

This section introduces the major precast prestressed concrete floor slab and beam elements used in buildings. The main objective is to achieve longer spans or shallower depths, for the same loading arrangement, compared to comparable rc designs, and to utilise the innovations made over the past 70 years to produce elements in the controlled environment of a factory that cannot be achieved on site. The extruded or slip-formed hcu (Figures 1.12 and 2.1) is the perfect example. In terms of load carrying capacity and span/depth ratio, these one-way spanning units can often rival post-tensioned flat slabs, particularly if they are supported on the shallow bottom flanges of inverted-tee beams or steel box section beams, and are designed continuously over internal beams and/or columns (see Chapter 13). For example, using office loading (2 kN/m^2 dead plus 5 kN/m^2 live load) floor bay areas of 10×6 m can be achieved within a 300 mm structural depth (200 mm for 10 m span hcu plus 100 mm for the bottom flange of a psc beam) giving a bay area/depth ratio of 200 m^2/m.

The psc slabs and beams discussed below, together with a range of load v span tables and graphs, include

- Solid plank floor units
- Dry-cast, slip-formed or extruded hcu
- Wet-cast voided units
- Double-tee units
- Thermal ground floor units
- Beam and block flooring
- Main beams – chiefly inverted-tee beams

2.2.2 Composite slabs and beams

The structural capacity of psc units may be increased by adding a layer of structurally reinforced concrete to the top of the unit. Providing that the topping concrete is fully anchored and bonded to the precast unit the two concretes – precast and cast *insitu*, may be designed as monolithic. The section properties of the precast unit plus the topping are used to determine the structural performance of the composite floor. A composite floor may be made using any of the floor units mentioned above. The minimum thickness of the topping is given in some texts as 40 mm, but in the author's design and site experience, this should not be less than 50–75 mm depending on the span of the floor. There is no limit to the maximum thickness, although 100 mm is a practical limit. Longer term problems due to the movement of the bottom of psc units relative to their support have been reported where the thickness of topping is around 150 mm. Toppings are reinforced using steel mesh reinforcement such as A98 or A142.

Composite floor design is carried out in two stages, before and after the *insitu* topping hardens and becomes structural. The precast unit carries its own weight plus the self-weight of the wet *insitu* concrete (plus a construction traffic allowance of 1.5 kN/m^2 typically). The composite floor carries imposed dead and live loads. Propping the floor unit prior to casting the topping increases 'performance' in one of two ways: either increased span or reduced depth. In the final analysis, the stresses and forces resulting from the two cases (minus the construction traffic allowance, which is temporary) are additive. Chapter 12 gives the full analysis.

Composite beams are likewise designed to either increase the span of the beam or, more commonly, reduce the overall construction depth, where the ends of hcu may be recessed into the top of inverted-tee beams and designed compositely using small quantities of cast *insitu* infill or a structural topping. Composite beam design is also carried out in two (or even three) stages, before and after the *insitu* infill and/or the topping become structural. The precast unit carries its own weight plus the self-weight of the floor units and wet *insitu* concrete (plus a construction traffic allowance). The composite beam carries the imposed loads. Propping the beam prior to placement of the floor slab can increase capacity considerably for shallow beams with deeper floor units.

2.2.3 Non-composite solid and hollow core floor units

2.2.3.1 Product development

The hollow cores in the earliest hcu's were circular rubber tubes filled with small stones, around which wet-cast concrete was poured and vibrated into a 4 feet wide mould, probably 6, 8 and 10 inches deep. When the concrete had hardened sufficiently to resist its own weight in bending, the unit was tilted upwards allowing the stones to fall out! In other developments, circular pneumatic tubes were pressurised and later deflated to form circular cores. Although circular cores certainly reduced the self-weight of the unit, giving a void ratio (void-to-total cross-sectional area) of around 50%, to improve on this, the webs and top and bottom flanges were too thin. This led to the oval or 'tunnel' (inverted 'pear' narrow at the bottom)-shaped voids when the present generation of hcu was developed in the 1950s, when the dual techniques of long line prestressing and concrete production through machines using semi-dry concrete (you can squeeze the concrete into a ball in your hand) were developed by companies such as *Spiroll* in the USA and *Roth* in Europe. Chapter 4 describes the manufacturing methods.

The following references give wider information on the development and design on the range of hcu currently available: The Association of Manufacturers of Prestressed Hollow Core Floors (ASSAP, 2002), FIB Bulletins (FIB, 2000), FIP recommendations (FIP, 1988), Elliott (Elliott & Jolly 2013, Elliott 2017), Prestressed Concrete Institute (PCI, 1991) and Walraven (Walraven & Mercx, 1983).

Deeper hcus, such as the recently developed 750 and 1000 mm deep units, have void ratios nearer to 70%. These units may require steel mesh to stabilise the slender edge webs. They have a limited market in buildings, more suited to civil engineering works such as cut and cover tunnels. The most common depths range from 150 to 400 mm. Most units are 1200 mm wide; Figure 1.12 shows three very efficient profiles comprising 6 or 11 cores. The 'tunnel' shape 6 core unit has the benefit of 32×32 mm triangular fillets giving additional cover to individuals or groups of strands. Hcu are reinforced only by longitudinal strands or wires, and, in contravention of standard rc or psc Code requirements, do not contain transverse reinforcement, shear or torsion stirrups, U bars in the sides or ends, or end hooks or bends for shear-tension anchorage failures. Instead they are designed and type-tested according to national product standards, such as BS EN 1168 (BS EN 1168, 2005).

Table 2.2 lists a range of hcu from 100 to 500 mm depth based on units containing a range of larger cores, e.g. 4 to 6 no. per 1200 mm width, or smaller (narrower) oval-shaped cores, e.g. 9 to 11 per 1200 mm width, together with the same service M_{sR} and ultimate M_{Rd} moments of resistance and ultimate shear capacity $V_{Rd,c}$ (without shear links).

The height of voids should not exceed h –50 mm, where h is the overall depth of the unit. The diameter of circular voids is usually h –75 mm. The minimum flange thickness depends on the overall depth of the unit h, given by 1.6 $\sqrt{h}$, e.g. 25.3 mm for h=250 mm would be considered too thin to manufacture. Because of cover requirements, it is usually necessary for the bottom flange to be at least 30 mm thick. The minimum width of a web should not be less than 30 mm. There is other guidance on the shape of the top flange above cores in FIP (FIP, 1988). The edge profile, typically 40–60 mm in height×10–15 mm depth, provides an important shear key, which enables the hcu to transfer shear forces, and effectively bending moments, between adjacent units to form a pseudo two-way spanning slab, i.e. the hcu do not physically span in two directions but lateral load spreading gives the same structural effect. See BS EN 1168, Annex C and Elliott (Elliott, 2017) for a wider discussion and analysis on this.

Table 2.2 Structural properties of prestressed hollow core floor units

Depth (mm)	*Number and breadth (mm) of cores*	*Self-weight (kN/m²)*	*Service moment of resistance M_{sR} (kNm)*	*Ultimate moment of resistance M_{Rd} (kNm)*	*Ultimate shear capacity $V_{Rd,c}$ (kN)*
150	11×60	2.69	67.5	100.1	120.5
200	6×140	3.33	123.7	179.0	101.1
	11×60	3.33	120.8	175.7	158.5
250	6×140	3.71	183.9	258.1	131.1
	11×60	4.03	180.9	256.8	200.1
300	6×135	4.22	249.0	342.3	174.0
	11×60	4.75	246.8	341.7	238.2
350	6×135	4.62	326.9	448.0	212.0
400	6×135	5.03	407.9	551.2	247.0
450	6×135	5.43	507.6	677.8	283.0
500	6×135	5.84	598.7	792.7	317.1

Cover or average cover to pretensioning strands=30 mm (9.3 mm dia.) and 35 mm (12.5 mm dia.). Final prestress at bottom (excluding self-weight) σ_b= 12.5 to 13.0 N/mm² depending on the availability of positions of tendons. Self-weight includes infilled joints between units. Exposure=XC1, f_{ck}=45 N/mm², f_{ctm}=3.80 N/mm².

2.2.3.2 *Load v span tables*

Prestressed solid units are typically in the range from 75 to 150 mm in depth, although depths of less than 100 mm are rarely used alone. Hcu's range from 150 to 500 mm in depth, although the maximum depth for some manufacturers is 250 or 400 mm. The width of units is commonly 1200 mm, with 600 and 2400 mm wide units available for special situations. And 1500 mm width is (uncommonly) used in some countries in Eastern Europe.

The derived data in the load v span graph shown in Figure 2.13 are based on the following. Units are pretensioned using 9.3 and/or 12.5 mm diameter 7-wire helical strand comprising of 7 no. 3 and 4 mm diameter clustered wires, respectively. Cover to the soffit is 35 mm. In units of high prestress top wires are added to control tensile cracking at transfer and upward camber prior to installation – top wires comprise 2 or 4 no. 5 or 7 mm diameter wires at 30 mm top cover. The characteristic tensile strength of strands and 5 mm wires is f_{pk}=1770 N/mm², and 1670 N/mm² for 7 mm wires, pretensioned to 70% f_{pk}. Concrete strength is f_{ck}=45 N/mm² at 28 days and $f_{ck}(t)$=30 N/mm² at transfer. For hcu, the maximum prestress at transfer in the bottom of the unit is limited to $\sigma_b(t) \leq 15$ N/mm², about 0.5 $f_{ck}(t)$, but $\sigma_b(t) \leq 11$ N/mm² for the solid units, while at the same time the stress in the top is limited $\sigma_t(t) \geq -2.5$ N/mm² at transfer. No debonding or deflecting of tendons is allowed, so these are the limiting criteria. Although the clear span is mostly limited by the permissible tension in the soffit f_{ctm} at the serviceability limit state for exposure class XC1, for longer spans the criteria may be limiting active deflection after construction of effective span/500, using a quasi-permanent factor ψ_2=0.3.

As a guide for consultants and architects, 'load v span' tables are published by hcu manufacturers in terms of the limiting clear span against the imposed live load, typically for 1.5 (domestic), 2.5 (car parks), 3.0 (congregation), 5.0 (offices), 7.5 and 10.0 (general storage of 3 to 4 m in height) kN/m². The data already include an allowance for finishes of 1.5 kN/m², although they should really include a further 0.5 kN/m² for services and ceilings, etc. Note that, in these tables, the quasi-permanent live load factor for domestic and offices is ψ_2=0.3, for storage it is ψ_2=0.8, and otherwise it is ψ_2=0.6. The ultimate combination live load factor for storage ψ_0=1.0, otherwise is ψ_0=0.7. Some manufacturers limit the clear span to the criteria that the natural frequency of the floor unit is at least $f_n \geq 4$ Hz, although this may be unconservative in special cases where $f_n \geq 7$ Hz. $f_n < 4$ Hz may be critical for live loads up to 2.5 kN/m². Exposure is internal class XC1 (although for car parking it should be external class XC3 (See Table 5.4)).

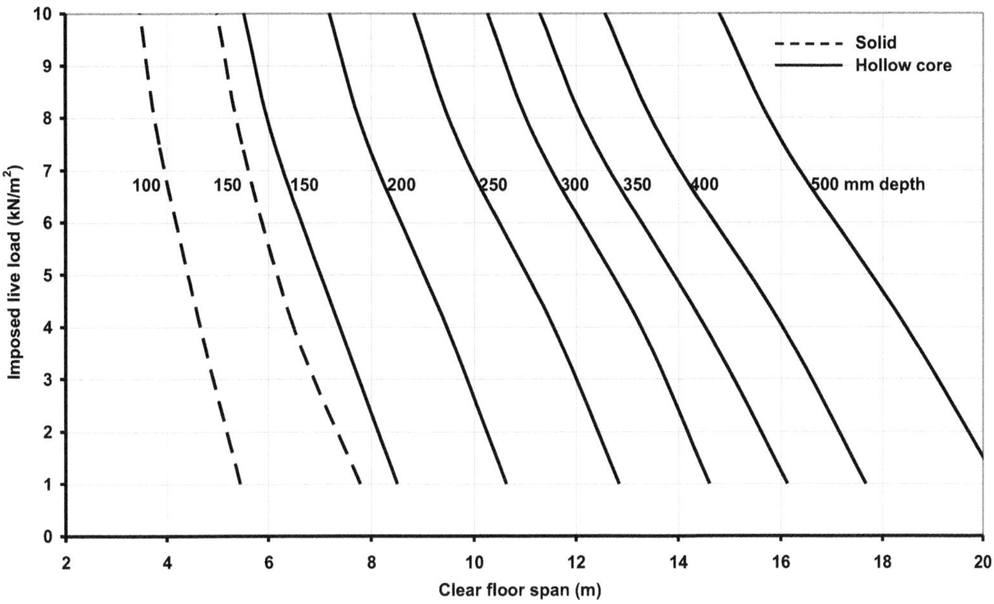

Figure 2.13 Imposed uniformly distributed live load vs. clear floor span for non-composite prestressed solid and hollow core floor units. The data include a 65 mm deep screed of self-weight 1.5 kN/m^2.

Table 2.3 gives an example of this table for the same units used to generate the load v span graph in Figure 2.13, but note that, as given above, ψ_0 and ψ_2 may be different in the table.

2.2.4 Composite solid and hollow core floor slabs

2.2.4.1 Load v span tables

Composite slabs comprise a psc solid or hcu unit together with a cast *insitu* reinforced (usually steel mesh) structural topping. The derived data in the load v span graph shown in Figure 2.14 are based on the same solid and hcu's as above, together with a 75 mm (at the crown of the span) deep reinforced structural topping. The average self-weight of the topping depends on an allowance for the upward camber at installation, typically span/500, and is therefore between 1.85 and 2.0 kN/m^2. The *insitu* concrete topping strength is f_{ckT}=25 N/mm^2. Table 2.4 gives an example for the same composite slabs used to generate Figure 2.14.

The precast unit is unpropped during casting of the topping. If the units are propped, using 2 no. props at one-third span, the increase in clear span would be between 1.09 and 1.14 for the 100 mm deep solid unit, between 1.02 and 1.06 for the 200 mm deep hcu, but only 1.01 for the 500 mm deep hcu, where the negative propping moment has little effect on the service moment of resistance. See Section 12.3.3 for further details about propping.

2.2.5 Wet-cast voided units

2.2.5.1 Product development

Not all hcu are manufactured using 'dry cast' long line extrusion or slip forming methods with water/cement ratios w/c≈0.4, described above, and in more detail in Chapter 4. Prestressed concrete 'wet-cast' voided units with w/c≈0.5 to 0.55 are manufactured to the

Table 2.3 Example of load v span tables published for hollow core floor units

Depth (mm)	Self-weight[a] (kN/m²)	Clear span (m) for imposed live load (kN/m²) includes 1.5 kN/m² for finishes					
		1.5 (ψ_2=0.3)	2.5 (ψ_2=0.6)	3.0 (ψ_2=0.6)	5.0 (ψ_2=0.3)	7.5 (ψ_2=0.8)	10.0 (ψ_2=0.8)
150	2.69	8.30 *(7.78)*	7.25	7.02	6.95	4.68	4.24
200	3.33	10.41 *(9.28)*	9.22 *(9.19)*	8.96	8.94	6.22	5.64
250	3.71	12.59 *(10.60)*	11.22 *(10.55)*	10.92 *(10.53)*	10.92 *(10.44)*	7.64	6.93
300	4.22	14.35 *(11.74)*	12.90 *(11.69)*	12.57 *(11.66)*	12.59 *(11.57)*	9.04	8.21

The limiting span is either for the serviceability limit state of stress in the soffit of the unit or active deflection after construction of span/500. Italic figures in brackets are limiting spans due to natural frequency f_n=4 Hz. ψ_2 values are for the *type of floor loading* defined in the text.

[a] Self-weight of finishes floor including grouted joints.

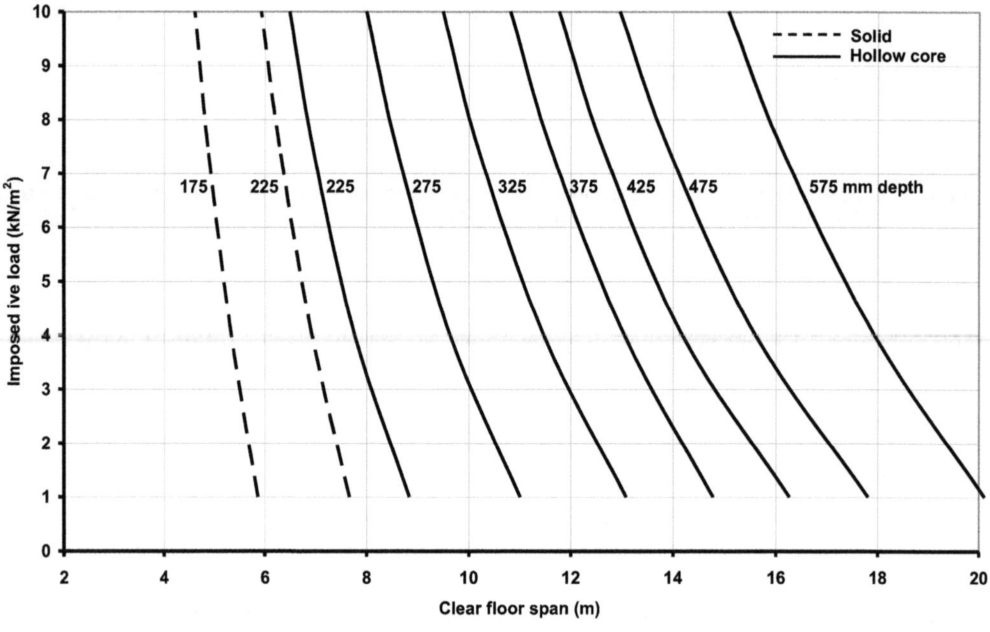

Figure 2.14 Imposed uniformly distributed live load vs. clear floor span for composite prestressed solid and hollow core floor slabs. The data include 75 mm deep structural topping of self-weight 1.85–2.0 kN/m² depending on precamber at installation, and a 65 mm deep screed of self-weight 1.5 kN/m². The precast unit is unpropped during casting of the topping.

typical profiles shown in cross section in Figure 2.15, and in practice in Figure 2.16, where concrete is manually placed into moulds around void formers, either psc or reinforced. In this book, they are referred to as 'voided' units to distinguish from hcu, but are commercially known as wet-cast hollow core slabs.

The structural 'voids' are formed using prismatic blocks of high-density polystyrene (approx 20 kg/m³) in lengths of about 2 m, with short solid sections of 300 mm length between the ends. The ends of the units are solid for a distance of about 500 mm and partially solid by reducing the number of voids to accommodate lifting devices for 1200 mm. The casting process involves three stages

Table 2.4 Example of load v span tables published for composite hollow core floor slabs

		Clear span (m) for imposed live load (kN/m²) includes 1.5 kN/m² for finishes plus 1.85 to 2.0 kN/m² topping					
Total composite depth (mm)	*Total self-weight[a] (kN/m²)*	*1.5 (ψ_2=0.3)*	*2.5 (ψ_2=0.6)*	*3.0 (ψ_2=0.6)*	*5.0 (ψ_2=0.3)*	*7.5 (ψ_2=0.8)*	*10.0 (ψ_2=0.8)*
225	4.64	8.62	8.24	8.08	7.51	6.58	6.14
275	5.24	10.72	10.11	9.91	9.29	8.05	7.52
325	5.72	12.82	11.88	11.65	11.05	9.51	8.89
375	6.25	14.57	13.48	13.22	12.57	10.86	10.17

The limiting span is either for the serviceability limit state of stress in the soffit of the composite slab or active deflection after construction of span/500. Limiting spans are not critical due to natural frequency f_n=4 Hz.

[a] Self-weight including 75 mm depth topping at the crown (mid-span) of the prestressed unit.

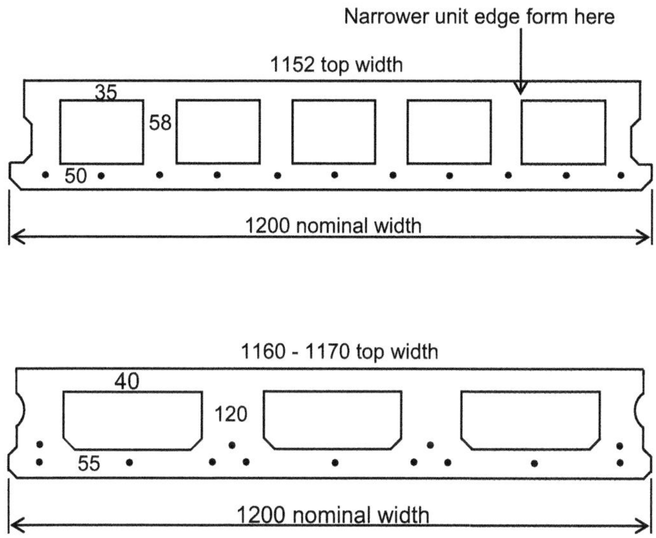

Figure 2.15 Examples of cross section of wet cast voided floor units. 9.3 mm diameter strands are most common, but 12.5 mm strands and 7 mm wire, with 5 mm top wire may also be used.

i. casting the bottom flange, typically 50 mm minimum thickness to enable adequate cover both to the soffit and infill block
ii. placing the infill blocks to form the voids, typically 150 to 300 mm wide
iii. second casting of the webs, 50 to 70 mm wide and top flange 35 to 50 mm depth.

The time delay between (i) and (iii) should not be more than about 45 minutes (depending on the consistency of the mix) as it is important to achieve coarse aggregate penetration between the two layers. The development of prestress through the flange-web interface must be assured.

The introduction of self-compacting concrete (SCC, see Chapter 4, Section 4.2.3) has made the operational task of using such delicate materials easier; however, the casting joint after stage (i) can be smooth due to the thixotropic nature of SCC, the absence of bleeding and the tendency of coarse aggregate to settle beneath the surface. Consequent horizontal cracks may be seen along the casting line at the ends of the unit, where bond stress in

Figure 2.16 Manufacture of wet-cast voided floor units using polystyrene void forms. Self-compacting concrete has been used to cast the bottom flange prior to fixing void formers. Units may be cast solid for cantilevers by debonding some bottom tendons and adding top rebars before casting.

the transmission zone and vertical splitting stresses are greatest. Voided units manufactured using SCC should therefore be tested for shear flow across the casting line at transfer, and tested in shear. However, there is a tendency for the buoyancy of the polystyrene to cause problems in maintaining the thickness of the top flange. Straps across the top of the moulds are used to support the blocks. Tying the blocks to the bottom tendons should not be practiced.

The UK market for wet-cast voided units is small, around 5% of the precast flooring market, but up to 40% of the North European market for domestic hcu is psc (or rc) wet-cast into moulds of fixed dimensions (length and width) and stockpiled for building contractors to purchase as the need arises.

Section properties and moment and shear capacities are comparable with the dry cast units and self-weights are lower in the deeper units. The big advantage of the wet-cast method is that additional reinforcement may be added to standard cages for any special requirement, e.g. point loads, large holes or notches, high shear, cantilevers, etc.

Wet-cast voided units are typically in the range from 150 to 300 mm in depth. The standard mould width in the UK is 1200 mm, but by virtue of the casting process, block-outs may be placed to create any reasonable width, although it is preferred if the moulded edge which is flat without a shear key coincides with the edge of a web. To meet sound attenuation values, a higher density unit of 150 mm depth is produced with fewer (or narrower) void formers to obtain a surface mass of 300 kg/m^2 (compared to the standard surface mass of 250 kg/m^2).

2.2.5.2 Load v span tables

The derived data in the load v span graph shown in Figure 2.17 are based on the following. Units are pretensioned using 9.3 mm diameter 7-wire helical strand. Cover to the soffit is controlled by the axis distance of 30 mm, and if required in a second row placed in the webs at 50 mm. The pretensioning is arranged so that no top wires/strands are required. The characteristic tensile strength of strands and 5 mm wires is f_{pk}=1770 N/mm^2 pretensioned to 70% f_{pk}. Concrete strength is f_{ck}=45 N/mm^2 at 28 days and $f_{ck}(t)$=30 N/mm^2 at transfer. The maximum prestress at transfer in the bottom of the unit is limited to $\sigma_b(t) \leq 13$–15 N/mm^2, up to 0.5 $f_{ck}(t)$, while at the same time the stress in the top $\sigma_t(t) \geq -2.7$ N/mm^2 at transfer. No debonding of tendons is carried out (this could be a commercial option for greater spans with heavier loads) so these are the limiting criteria. Although the clear span is mostly limited by the permissible tension f_{ctm} in the soffit at the serviceability limit state for exposure class XC1, for longer spans with imposed live loads <3 kN/m^2 the criteria may be limiting active deflection after construction of effective span/500, using a quasi-permanent factor ψ_2=0.3. For these conditions, the age of the unit at installation is taken as 28 days (reducing this to 7 days will reduce the limiting clear span by 2%). The usual maximum handling length-to-depth ratio is 55, which may be limiting for the longer spans and/or shallow depths (e.g. 150 mm).

As a guide for consultants and architects, 'load v span' tables are published by wet-cast manufacturers on the same basis as hcu, with the same values for ψ_2 and ψ_0. The data already include an allowance for finishes of 1.5 kN/m^2. Some manufacturers limit the clear span to the criteria that the natural frequency of the floor unit is at least $f_n \geq 4$ Hz. f_n<4 Hz may be critical for live loads up to 3.0 kN/m^2. Table 2.5 gives an example of this table for the same units used to generate the load v span graph in Figure 2.17.

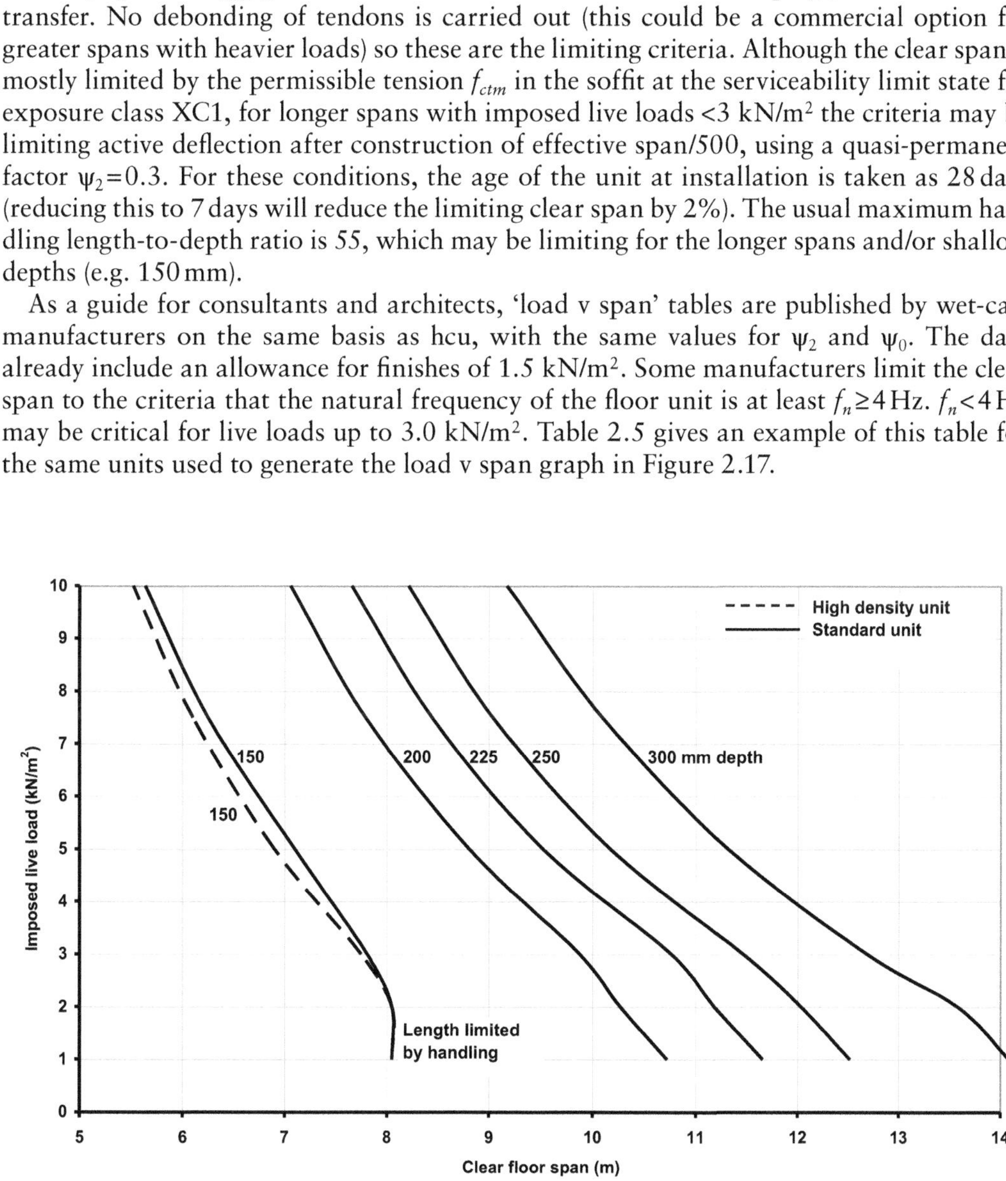

Figure 2.17 Imposed uniformly distributed live load vs clear floor span for wet-cast non-composite prestressed voided floor units. The data include a 65 mm deep screed of self-weight 1.5 kN/m^2.

Table 2.5 Example of load v span tables published for wet-cast hollow core floor units

Depth (mm)	Self-weight[a] (kN/m²)	Clear span (m) for imposed live load (kN/m²) includes 1.5 kN/m² for finishes					
		1.5 (ψ_2=0.3)	2.5 (ψ_2=0.6)	3.0 (ψ_2=0.6)	5.0 (ψ_2=0.30)	7.5 (ψ_2=0.80)	10.0 (ψ_2=0.8)
150	2.59	8.05[b] *(7.87)*	7.26	7.03	7.10	4.67	4.24
150H	3.00	8.05[b] *(7.75)*	7.23	7.02	6.90	4.72	4.28
200	3.02	10.48 *(9.38)*	9.23	8.95	8.81	6.11	5.54
225	3.23	11.41 *(10.04)*	10.09 *(9.99)*	9.80	9.52	6.80	6.17
250	3.44	12.27 *(10.66)*	10.90 *(10.61)*	10.60	10.17	7.47	6.78
300	3.87	13.83 *(11.79)*	12.38 *(11.73)*	12.06 *(11.71)*	11.32	8.78	7.97

The limiting span is either for the serviceability limit state of stress in the soffit of the unit or active deflection after construction of span/500. Italic figures in brackets are limiting spans due to natural frequency f_n=4 Hz. ψ_2 values are for the *type of floor loading* defined in the text.

[a] Self-weight of finishes floor including grouted joints. 150H=high density unit 300 kg/m².
[b] Limited by handling length/depth=55.

The data in Table 2.5 are for internal exposure class XC1 (although for car parking it should be external class XC3 (See Table 5.4)). For external exposure class XC3 the effect on the final clear span for the range of cases given would be from –11% to +5% as follows. The ratio of the clear span for classes XC3/XC1 for the 150 mm deep voided unit is from 1.00 to 1.02. The latter is for an imposed live load>5 kN/m² where the greater load, being multiplied by ψ_2=0.3, has a beneficial effect compensating for the allowable zero tension in the soffit. The ratio for the 300 mm deep voided unit is from 0.97 to 1.00. Here the latter is for live loads>5 kN/m² where the slab is critical at the ultimate moment of resistance, not affected by the limiting tensile stress nor ψ_2.

2.2.6 Composite wet-cast voided slabs

2.2.6.1 *Load v span tables*

Composite slabs comprise a psc wet-cast voided unit together with a cast *insitu* reinforced (usually steel mesh) structural topping. The derived data in the load v span graph shown in Figure 2.18 are based on the same wet-cast units as above, together with a 75 mm (at the crown of the span) deep reinforced structural topping. The average self-weight of the topping depends on an allowance for the upward camber at installation, typically span/500, and is therefore between 1.9 and 2.0 kN/m². The *insitu* concrete topping strength is f_{ckT}=25 N/mm². Table 2.6 gives an example for the same composite slabs used to generate Figure 2.18.

The data in Table 2.6 are for internal exposure class XC1. For external exposure class XC3 the effect on the final clear span for the range of cases given would be from –11% to +5% as follows. The ratio of the clear span for classes XC3/XC1 for the 225 mm deep slab is from 0.98 to 1.05. The latter is for an imposed live load of 10 kN/m² for the same reasons as for the non-composite units above. The ratio for the 375 mm deep slab is from 0.89 to 1.00. Here the latter is also for the greater live loads for the same reasons as above for the non-composite units.

The precast unit is unpropped during the casting of the topping. If the units are propped, using one prop at mid-span, the increase in clear span would be between 1.05 and 1.08 for the 150 mm deep unit, but only 1.01 to 1.06 for the 300 mm deep unit, where the negative propping moment has a smaller effect on the service moment of resistance.

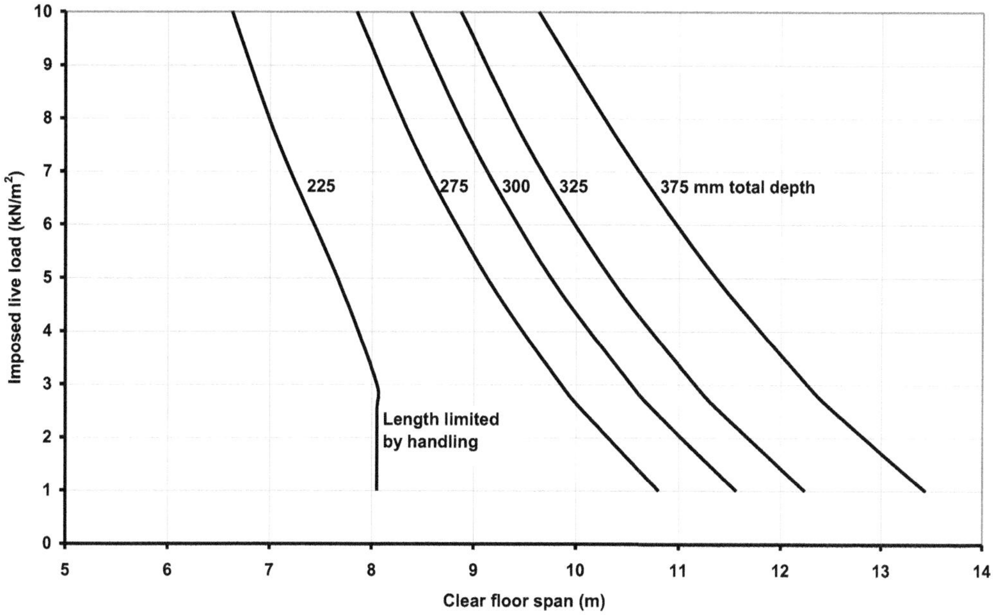

Figure 2.18 Imposed uniformly distributed live load vs clear floor span for wet-cast composite prestressed voided floor slabs. The data include 75 mm deep structural topping of self-weight 1.9–2.0 kN/m^2 depending on precamber at installation, and a 65 mm deep screed of self-weight 1.5 kN/m^2. The precast unit is unpropped during casting of the topping.

Table 2.6 Example of load v span tables published for composite wet-cast voided floor units

		Clear span (m) for imposed live load (kN/m²) includes 1.5 kN/m² for finishes plus 1.9 to 2.0 kN/m² topping					
Total depth[a] *(mm)*	*Self-weight*[a] *(kN/m²)*	*1.5 (ψ_2=0.3)*	*2.5 (ψ_2=0.6)*	*3.0 (ψ_2=0.6)*	*5.0 (ψ_2=0.3)*	*7.5 (ψ_2=0.8)*	*10.0 (ψ_2=0.8)*
225	4.56	8.05	8.05	8.05	7.66	6.29	5.71
275	5.00	10.54	10.06	9.85	9.13	7.61	6.90
300	5.21	11.27	10.76	10.53	9.75	8.25	7.49
325	5.42	11.94	11.40	11.16	10.33	8.89	8.07
375	5.82	13.11	12.52	12.26	11.36	10.13	9.19

The limiting span is either for the serviceability limit state of stress in the soffit of the unit or active deflection after construction of span/500. Limiting spans are not critical due to natural frequency f_n=4 Hz.

[a] Total depth and self-weight including 75 mm depth topping at the crown (mid-span) of the prestressed unit.

2.2.7 Composite double-tee floor slabs

2.2.7.1 Product development for double-tee units

Precast psc double-tee floor units evolved from a single-tee T section beam where the deep web, 400–1200 mm depth (and sometimes 2000 mm depth where transport permits, e.g. USA) is reinforced with helical strands. The top flange, typically 1200 mm wide (from the old 4 feet modular width) × 75 to 150 mm depth, depending on the total depth, and whether there is a structural topping to cater for transverse bending between the webs. As shown in Figures 2.19 and 2.20, the sides of the webs are tapered, typically 40–50 mm,

Figure 2.19 Prestressed 'double-tee' floor units.

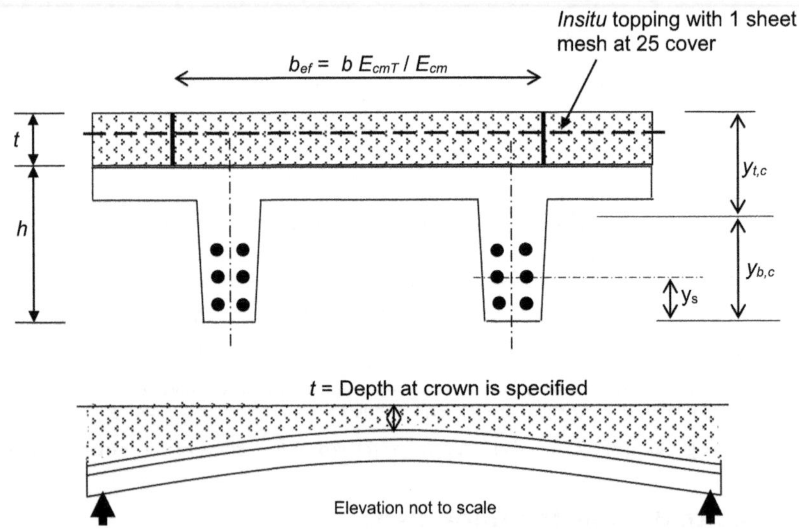

Figure 2.20 Cross section and elevation of 'double-tee' units, also showing the option for a structural topping in a 'double-tee' slab.

and the strands are placed in pairs from the bottom of the web upwards depending on project requirements for the level of prestress in the bottom fibre, cracking and deflections at the serviceability limit state and flexural strength at the ultimate limit state. The strand pattern is not governed by the ultimate shear force V_{Ed}, nor the ultimate shear capacity

$V_{Rd,c}$, as shear reinforcement in the form of vertical stirrups is provided for a distance from the supports where $V_{Ed} > V_{Rd,c}$ (see Section 10.2). Some stirrups, or more correctly termed 'links', are provided close to the support irrespective of the shear capacity, to cater for localised end-bearing effects, particularly at half-joints as shown at the ends of the units in Figure 2.19.

The axis distance to the first row of strands is 40–50 mm, depending on durability requirements and the manufacturer's preference, i.e. cover to links beneath the strands and avoidance of localised splitting cracks along the line of the strands. The vertical distance between subsequent rows is at least 3×diameters of strand, typically 40–50 mm, and the horizontal axis distance is 60–100 mm depending on the width of the web. Due to the taper of the webs the width of the soffit increases as block-outs are added within the full depth mould. To avoid stress concentrations at the top of the web a small 40–50 mm triangular or circular fillet is added.

The next stage in the development was to double the width to 2400 mm and add a second web (or even third), giving the units immediate stability on site, at 1200 mm spacing with 600 mm wide overhangs at each edge, and name the unit a 'double-tee floor slab' (or 'triple-tee'). The equivalent void ratio (i.e. 1- area of section/area of enveloping rectangle) is about 0.65–0.75 for 400 and 900 mm deep units, respectively, allowing the unit to span over longer spans and with less weight per floor area than hcu. Mechanical connectors at the edge of the outer flange (usually welded plates) enabled the top flange to be designed transversely between the webs and between units, as well as providing floor diaphragm action together with a structural topping.

Double-tee units achieved even greater spans and reduced mass compared with hcu. However, there are two major drawbacks with double-tees

i not being able to provide sufficient pretensioning force to satisfy the available section as strands can only be placed in pairs at the bottom of the webs, forcing more strands to be positioned higher in the web thereby reducing eccentricity
ii the centroidal axis is high in the section, typically $y_b = 0.7h \times$depth from the bottom, resulting in a low value of the elastic modulus at the bottom Z_b relative to the cross-sectional area, e.g. $Z_b/A_c\ h \approx 0.12$ to 0.16 compared with 0.22 to 0.25 for hcu. This offers the potential for composite action with a structural topping.

Debonded (or less commonly deflected) tendons are used in some cases to overcome transfer stress problems in long span units. To optimise the transfer prestress more accurately about $^1/_6$ of the total number of strands present at mid-span are debonded at a distance of about span/4, and half this number at about span/5.

Figure 2.21 shows a variation of the traditional double-tee unit where the keyed end section is made solid to facilitate continuous and composite action between spans, the splayed end is reinforced to resist large negative bending moments on either side of the interior main beam(s). These units may also be found as triple-tee, up to 2.4 m wide.

2.2.7.2 Load v span tables for composite double-tee slabs

Prestressed double-tee units are typically in the range from 300 to 900 mm in depth, and 2400 mm in width, comprising a 75–100 mm depth top flange and two tapered webs (or ribs) of mean width of 200–240 mm. The width of the web varies as the depth of the unit decreases, typically by 6–8 mm per 100 mm depth. The width of units is commonly 2400 mm, although if transportation and site access allow, up to 3600 mm with three webs (triple-tee) are possible.

Figure 2.21 So-called 'multi-rib' double-tee floor units with special end details for continuous spans.

The derived data in the load v span graph shown in Figure 2.22 are based on the following. Top flange depth = 75 mm, and the mean width of each web = 220 mm. Units are pretensioned using pairs of 12.5 mm diameter 7-wire helical strand in each of two webs, with axis distance from the bottom of the webs at 50, 100, 150, 200 mm, etc. using as many strands as necessary to satisfy two conditions

i using full prestress, the serviceability limit state for the final composite stress, after losses, for the limiting tension in the soffit of f_{ctm}, or active deflection after construction of span/500. Ultimate flexural strength is not critical.

ii using debonded strands, typically net transfer stress at the bottom of the webs due to compressive prestress and tensile self-weight alone $\leq 0.6\ f_{ck}(t)$, where $f_{ck}(t) = 32$ N/mm^2 based on a 28 days characteristic strength $f_{ck} = 50$ N/mm^2.

In units of high prestress top strands, say 2 no. 9.3 mm diameter, are added to control tensile cracking at transfer and upward camber prior to installation at 40 mm top cover. The characteristic tensile strength of strands is $f_{pk} = 1770$ N/mm^2 pretensioned to 70% f_{pk} (top strands are often stressed to 40% f_{pk}).

Table 2.7 gives an example for the same composite double-tee slabs used to generate Figure 2.22. Compared with composite hcu, taking for example 400 mm deep units and an imposed live load of 5.0 kN/m^2 the limiting clear span for the double-tee and hcu are 11.45 and 15.05 m, respectively, a difference of 31%. One reason for this is that the radius of gyration of the composite second moment of area/cross-sectional area $r_c = \sqrt{(I_{c,c}/A_{c,c})} = 122$ and 161 mm, respectively, also a difference of 31%. r_c is a *measure* of the flexural efficiency of a section. Further, the ratio of composite section modulus at

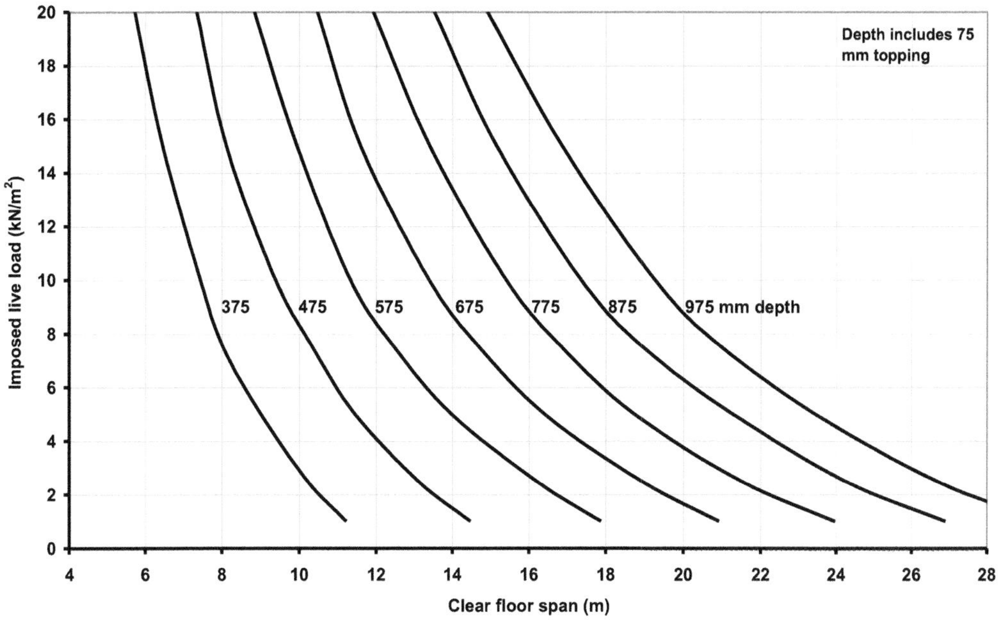

Figure 2.22 Imposed uniformly distributed live load vs. clear floor span for composite prestressed double-tee floor slabs. The data include 75 mm deep structural topping of self-weight 1.85–2.0 kN/m^2 depending on precamber at installation, and a 65 mm deep screed of self-weight 1.5 kN/m^2. The precast unit is unpropped during the casting of the topping.

Table 2.7 Example of load v span tables published for composite double-tee floor slabs

Precast double-tee depth (mm)	*Total composite depth (mm)*	*Total self-weight*[a] *(kN/m²)*	*Clear span (m) for imposed live load (kN/m²) includes 1.5 kN/m² for finishes plus 1.85 to 2.0 kN/m² topping*					
			2.5 (ψ_2=0.6)	*5.0 (ψ_2=0.3)*	*7.5 (ψ_2=0.8)*	*10.0 (ψ_2=0.8)*	*15.0 (ψ_2=0.8)*	*20.0 (ψ_2=0.8)*
300	375	4.82	10.21	9.01	8.03	7.44	6.45	5.73
400	475	5.26	13.11	11.45	10.34	9.39	8.12	7.34
500	575	5.70	16.20	13.98	12.47	11.34	9.94	8.85
600	675	6.14	18.92	16.41	14.70	13.40	11.60	10.47
700	775	6.58	21.50	18.74	16.86	15.41	13.41	11.93
800	875	7.02	24.25	21.27	18.95	17.37	15.16	13.53
900	975	7.46	26.68	23.45	20.98	19.27	16.87	14.92

The limiting span is either for the serviceability limit state of stress in the soffit of the composite slab or active deflection after construction of span/500. The span is also limited by the net stress due to prestress and self-weight in the debonded tendons at ¼ span at transfer $\leq 0.6\ f_{ck}(t)$.

[a] Self-weight including 75 mm depth topping at the crown (mid-span) of the prestressed unit.

the bottom fibre to area $k = Z_{b,c}/A_{c,c}$, often termed the 'kernel' *(the central or important part of something)* of the section that controls flexural stress, is 46 and 100 mm, respectively, more than double, due in part to the much greater height to the centroidal axis of 328 mm for the double-tee compared with 258 mm for the hcu. However, the self-weight of the composite double-tee is 5.26 kN/m^2 compared with 7.03 kN/m^2 for the composite hcu, 33% heavier.

The precast unit is unpropped during the casting of the topping. If the units are propped, using 2 no. props at one-third span, the increase in clear span would be between 1.02 and 1.09 for the 300 mm deep unit (the greater figure for the smallest live load), between 1.01 and 1.04 for the 500 mm deep unit, but only 1.01 for the 900 mm deep double tee, where the negative propping moment has little effect on the service moment of resistance.

2.2.8 Non-composite thermal ground floor units

2.2.8.1 *Product development*

Thermally insulated, wide floor slabs for domestic and residential use, and possibly certain offices with limited spans (span/depth ratio around 27 compared with 43–47 for non-composite hcu), are a recent development to compete in the housing market for ground slabs. Based on achieving a reference U value of 0.13 W/m²K by virtue of the preformed polystyrene thermal coat, the psc (or reinforced concrete) units are structurally similar to double-tee units, i.e. a shallow top flange, typically 50–60 mm deep and reinforced with either steel mesh (A142) or steel or polypropylene fibres, spanning transversely between two webs of 200–300 mm depth. Figure 2.23 shows a cross section for a typical unit. The chamfered profile at the edges (50–60 mm depth×10–12 mm recess) is grouted to form a longitudinal joint acting as a shear key. In some other variations, the profile of the soffit of the top flange is curved, with a depth of 50 mm at mid-width and 70–75 mm at the intersection with the webs.

2.2.8.2 *Load v span tables*

The derived data in the load v span graph shown in Figure 2.24 are based on the following. Nominal width 1200 and 600 mm, with actual top flange width 1146 and 546 mm,

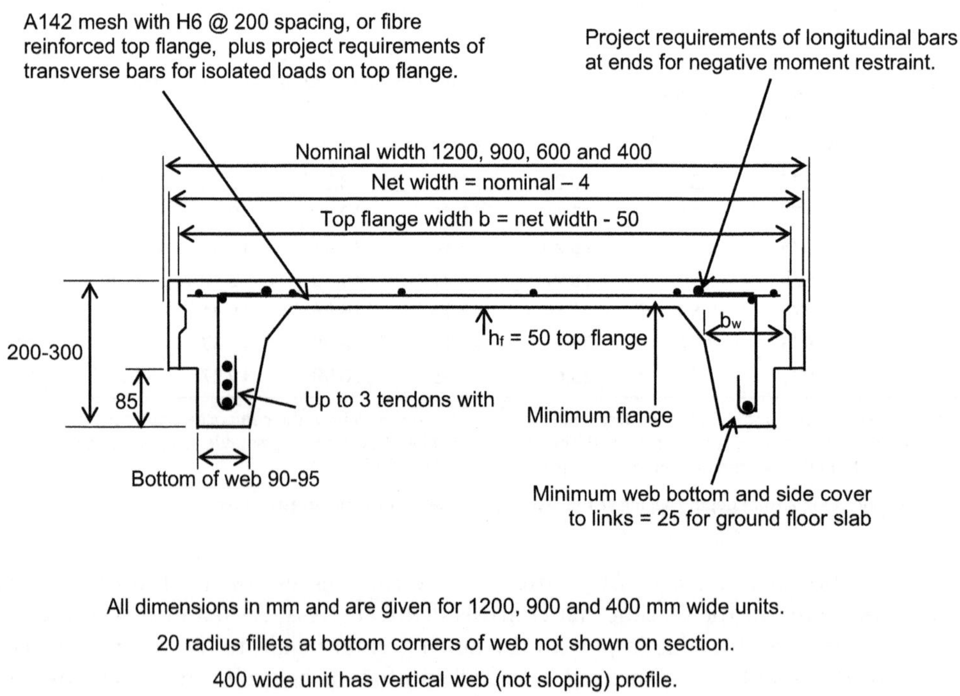

Figure 2.23 Cross section and manufacturing details for thermally insulated ground floor units.

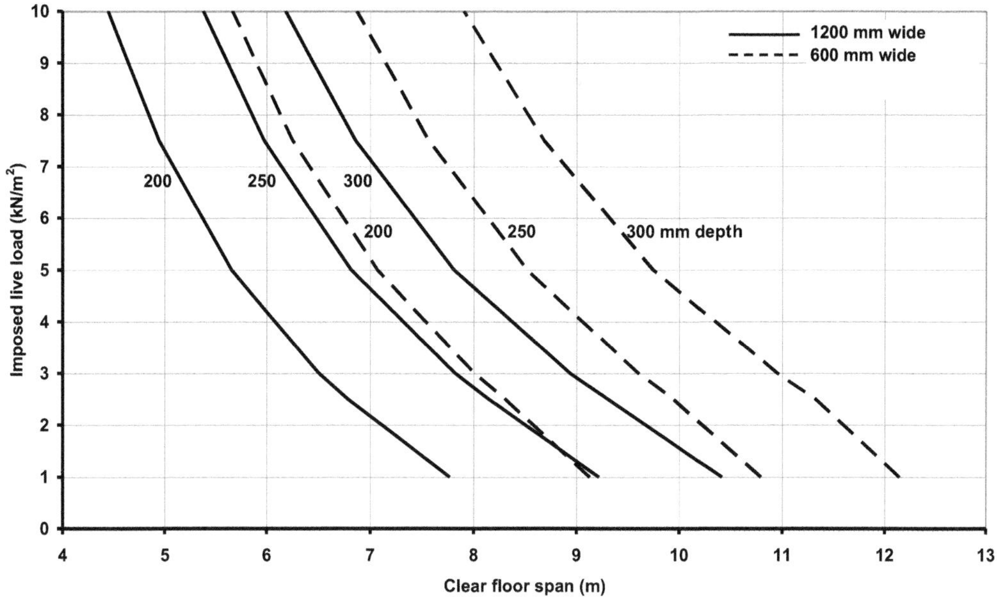

Figure 2.24 Imposed uniformly distributed live load vs clear floor span for 1200 and 600 mm wide non-composite prestressed thermally insulated floor slabs (900 and 400 mm wide units are also available). The data include a 65 mm deep screed of self-weight 1.5 kN/m^2. Exposure is for external ground slabs class XC3.

respectively. The 600 mm wide units have greater capacity as they have a greater number of prestressed webs per unit width, i.e. 3⅓ per m width. Top flange depth = 50 mm, and the mean width of each web = 113 mm (measured at the top of the slope minus the shear key). Units are pretensioned using three single 9.3 and/or 12.5 mm diameter 7-wire helical strands in each of two webs, with cover to the bottom of the webs 35, 75 and 115 mm, plus 2 no. 5 mm diameter wires at 30 mm top cover. The characteristic tensile strength of strands and 5 mm wires is $f_{pk} = 1770$ N/mm^2, pretensioned to 70% f_{pk}. Concrete strength is $f_{ck} = 45$ N/mm^2 at 28 days and $f_{ck}(t) = 30$ N/mm^2 at transfer. The maximum prestress at transfer in the bottom of the unit is limited in the same way as for hcu. No debonding or deflecting of tendons is carried out. The durability criteria is for external ground floor exposure, class XC3. For information, the differences in the spans for exposure classes XC1 and XC3 are only about 2%. The clear span is always limited by zero tension in the soffit at the serviceability limit state. Limiting tensile stress, rather than active deflection, is critical due mainly to the larger height to the centroidal axis (0.7 × depth) in these (and double-tee) units, compared to hcu (roughly 0.5 × depth), so the flexural stress in the soffit is that much greater.

As a guide for consultants and architects, 'load v span' tables are published by manufacturers in terms of the limiting clear span against the imposed live load, typically 1.5, 2.5, 3.0, 5.0, 7.5 and 10.0 kN/m^2. The data already include an allowance for finishes of 1.5 kN/m^2. The quasi-permanent live load factor for domestic rooms is set as $\psi_2 = 0.3$. The ultimate combination live load factor is $\psi_0 = 0.7$. Some manufacturers limit the clear span to the criteria that the natural frequency of the floor unit is at least $f_n \geq 4$ Hz, which may be critical for live loads up to 2.5 kN/m^2. Table 2.8 gives an example of this table for the same units used to generate the load v span graph in Figure 2.24.

Table 2.8 Example of load v span tables published for thermally insulated floor units

Nominal width (mm)	Depth of unit (mm)	Self-weight[a] (kN/m²)	*Clear span (m) for imposed live load (kN/m²) includes 1.5 kN/m² for finishes*					
			1.5	*2.5*	*3.0*	*5.0*	*7.5*	*10.0*
1200	200	2.08	7.47	6.79	6.51	5.66	4.95	4.45
	250	2.30	8.95	8.15	7.82	6.82	5.98	5.38
	300	2.47	10.19	9.31	8.94	7.81	6.86	6.19
600	200	2.96	8.98 *(8.38)*	8.31	8.01	7.07	6.26	5.67
	250	3.40	10.63 *(9.63)*	9.95 *(9.58)*	9.61	8.52	7.57	6.87
	300	3.74	11.97 *(10.74)*	11.33 *(10.69)*	10.95	9.75	8.68	7.90

The limiting span is for the serviceability limit state of zero stress in the soffit of the unit because exposure class=XC3. Italic figures in brackets are limiting spans due to natural frequency f_n=4 Hz.

[a] Includes *insitu* filled edge joint.

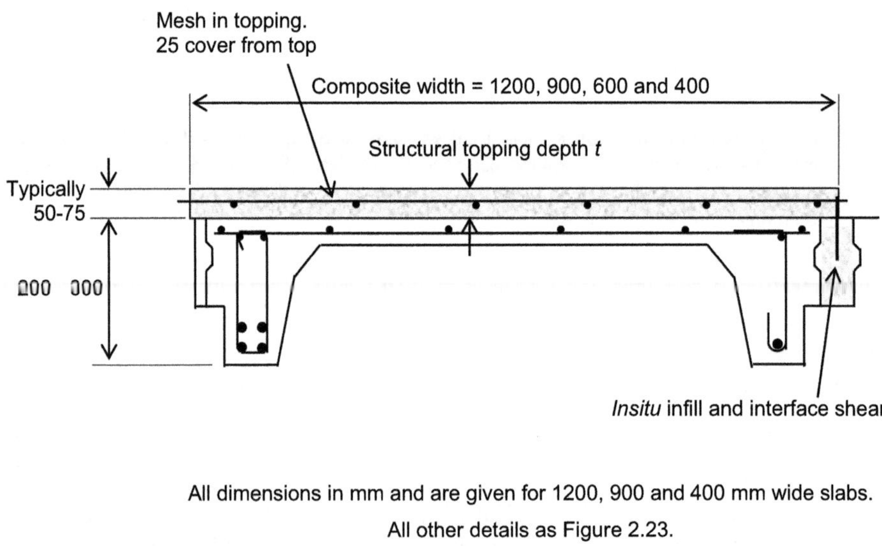

Figure 2.25 Cross section and manufacturing details for thermally insulated composite ground floor slabs.

2.2.9 Composite thermal ground floor slabs

2.2.9.1 Load v span tables

Composite thermal ground slabs comprise a psc unit together with a cast *insitu* reinforced (usually steel mesh, although steel fibres are a popular option) structural topping, as shown in Figure 2.25. The derived data in the load v span graph shown in Figure 2.26 are based on the same data as for the non-composite units above, together with a 75 mm (at the crown of the span) deep reinforced structural topping. The average self-weight of the topping depends on an allowance for the upward camber at installation, typically span/500, and is therefore between 1.85 and 2.0 kN/m². The *insitu* concrete topping strength is f_{ckT}=25 N/mm². Table 2.9 gives an example for the same composite slabs used to generate Figure 2.26. The precast unit is unpropped during the casting of the topping unless the ground floor slab is sufficiently elevated.

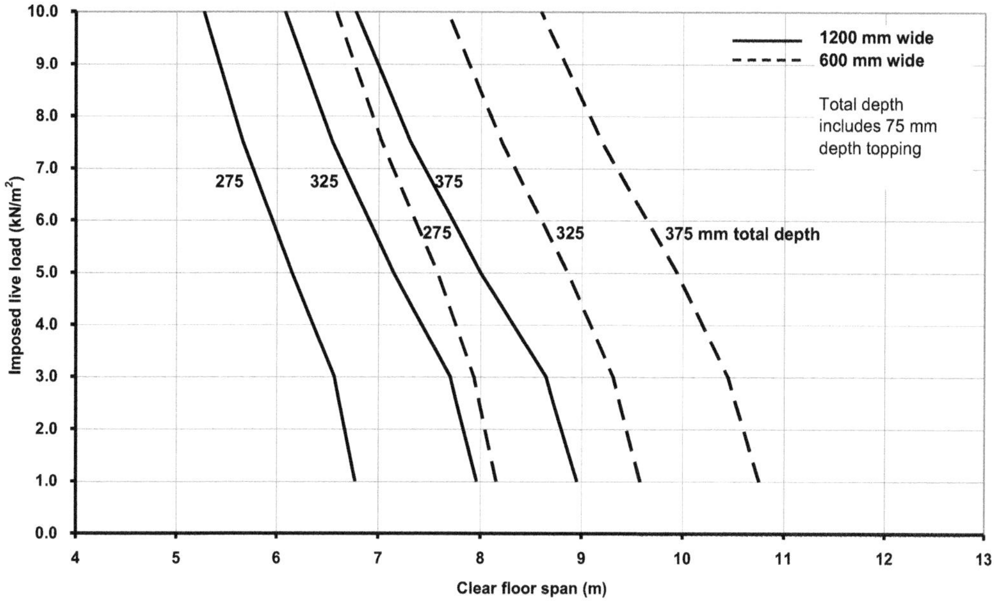

Figure 2.26 Imposed uniformly distributed live load vs. clear floor span for composite prestressed thermally insulated floor slabs. The data include 75 mm deep structural topping of self-weight 1.85–2.0 kN/m² depending on precamber at installation. The precast unit is unpropped during casting of the topping.

Table 2.9 Example of load v span tables published for thermally insulated composite floor slabs

Nominal width (mm)	Depth of unit (mm)	Total depth (mm)	Total self-weight[a] (kN/m²)	*Clear span (m) for imposed live load (kN/m²) includes 1.5 kN/m² for finishes plus 1.85 to 2.0 kN/m² topping*					
				1.5	*2.5*	*3.0*	*5.0*	*7.5*	*10.0*
1200	200	275	4.11	6.72	6.62	6.57	6.15	5.66	5.28
	250	325	4.34	7.90	7.76	7.70	7.14	6.55	6.08
	300	375	4.52	8.87	8.72	8.65	8.00	7.31	6.78
600	200	275	4.99	8.10	7.99	7.94	7.58	7.03	6.59
	250	325	5.47	9.51	9.38	9.31	8.86	8.20	7.67
	300	375	5.83	10.67	10.52	10.45	9.95	9.20	8.59

All values for domestic loading where $\psi_2=0.3$

The limiting span is for the serviceability limit state of zero stress in the soffit of the unit, because exposure class=XC3. Limiting spans are not critical due to natural frequency $f_n=4$ Hz.

[a] Self-weight including 75 mm depth topping at the crown (mid-span) of the prestressed unit.

The benefit of a composite slab, in terms of load or span capacity, is only realised for the imposed live loads greater than 3 kN/m², for example, the ratio of the limiting clear spans for composite/non-composite unit 1200 mm wide×250 mm deep is in the range of 1.13–0.86 for imposed live loads of 10.0–1.0 kN/m², respectively. For 600×250 mm unit, the ratio is in the range 1.12–0.89.

2.2.10 T beams ('house beams') for beam and block floors

2.2.10.1 Product development

The concept of using structural members spaced apart with concrete infill blocks spanning transversely between them has been used in buildings since the 1930s. Victorian rail bridges used steel RSJs 18 inches apart with concrete or clay blocks between them. The current generation of beam and block flooring, known originally as 'beam and pot', because the block was hollow and may have resembled a pot, used X-shaped beams (with semi-circular webs) of 7 and 9 inches depth. Today the structural members are prestressed (less frequently reinforced using lattice cages) inverted T beams, typically 150–225 mm in depth in order to course with brickwork, and 92–140 mm wide at the bottom, respectively. A slip-formed version of the T beam, with a good strength/mass ratio, is 155 mm depth × 115 mm wide, pretensioned using 3 no. 7 mm wires. Wider beams of 225 mm width are also used. (The beams are known as T beams to distinguish them from the main inverted tee beams.)

T beams, for beam and block flooring, are specified and designed according to the product standard BS EN 15037, Part 1: Beams (BS EN 15037-1, 2008), in particular the clear spacing between the pretensioning wires, cover to wires for durability and fire resistance (see Sections 5.2.2 and 5.5.4). Structural design is carried out according to BS EN 1992-1-1.

The beams are most commonly manufactured by moulding, in inverted moulds as shown in Figure 2.27, but may also be slip-formed by machines, as shown in Figure 2.28. In residential buildings, the beams are positioned with a direct bearing onto load-bearing blockwork, typically with 100 mm bearing length, as shown in Figure 2.5b. Bearing pads are not used. BS EN 15037-1, clause D.2.1 allows a minimum bearing length of $l_b = 60$ mm, but this cannot be justified according to BS EN 1992-1-1, clause 10.9.5.2 for isolated bearings, which gives $l_b \geq 95$ mm on masonry, $l_b \geq 85$, 80 and 75 mm on *insitu* concrete, precast, and steelwork, respectively. A non-structural screed of between 65 and 75 mm thickness is cast on to the beams and blocks. At ground floors, a damp proof membrane is laid on to the beams prior to casting the screed, as shown in Figure 2.29.

Figure 2.27 Long line wet casting of T beams by moulding.

Figure 2.28 Manufacture of slip formed T beams.

Figure 2.29 Casting a finishing screed on to beam and block ground floor (c. Litecast Ltd., UK).

Depending on the structural requirements of imposed loads and span, floors are designed using single or double beams, as shown in Figure 2.30a, or even triple or quadruple beams if necessary. Infill concrete blocks, 440 or 215 mm wide, or alternate 440/215 mm

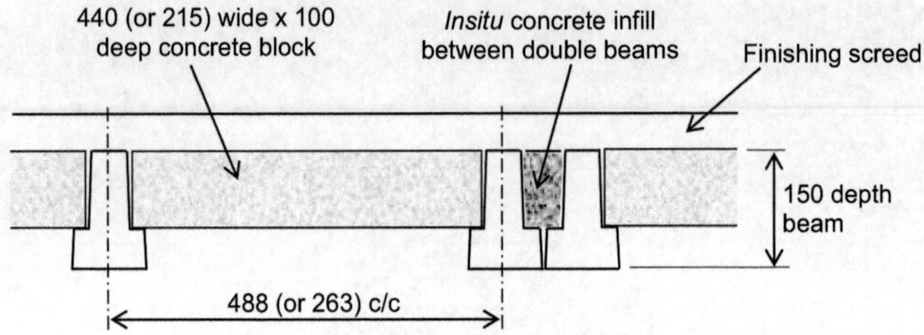

(a) Example shows non-composite floor using single and double beam arrangement. Triple and quadruple beams are also used.

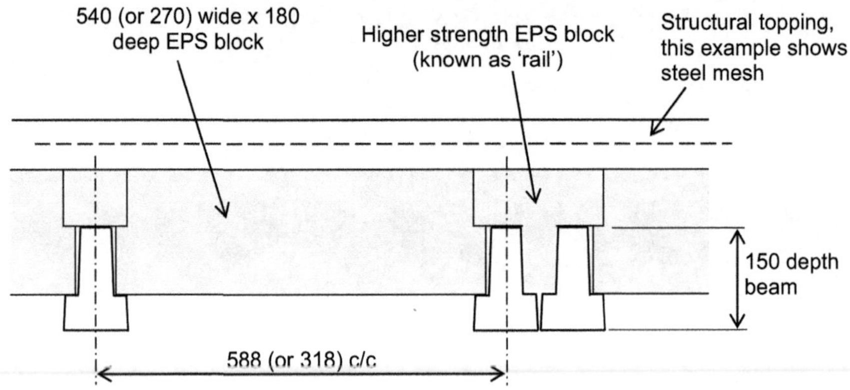

(b) Example shows non-composite thermal floor using single and double beam arrangement. In some cases the insulation blocks pass beneath the beams and are level with the tops of beams.

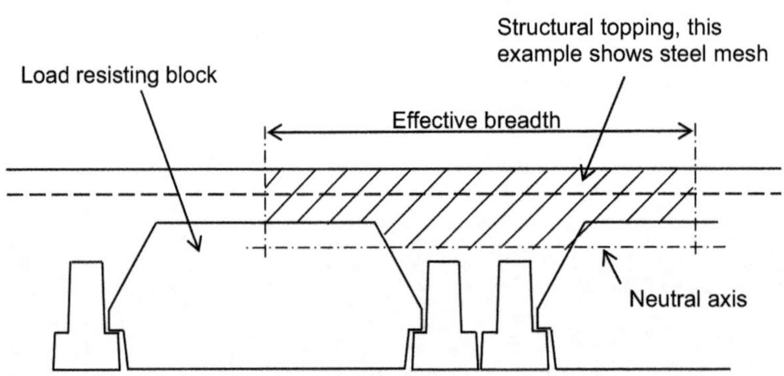

(c) Example shows principle of composite floor using single and double beam arrangement. Triple and quadruple beams are also used.

Figure 2.30 Cross sections of non-composite and composite T beams in beam and block flooring. (a) Example shows non-composite floor using single and double beam arrangement. Triple and quadruple beams are also used. (b) Example shows non-composite thermal floor using single and double beam arrangement. In some cases the insulation blocks pass beneath the beams and are level with the tops of beams. (c) Example shows principle of composite floor using single and double beam arrangement. Triple and quadruple beams are also used.

(also lightweight blocks 540 or 360 mm wide), are placed on the beam's bearing ledge and grouted. As these blocks are classed as 'resisting blocks' the floor is completed using a typically 65 mm depth (non-structural) finishing (or floating) screed, or a wooden floor. Concrete blocks are specified according to BS EN 15037, Part 2: Concrete blocks (BS EN 15037-2, 2009).

Thermally insulated ground floors are provided with 540 or 270 mm wide expanded polystyrene (EPS) blocks, as shown in Figures 2.5c, 2.30b and 2.31. The first substitution of concrete blocks by EPS occurred in Europe, particularly France and Holland in the mid-1980s, and in the UK in the early 1990s. These insulating blocks passed beneath the beams, but later, c.2013, projected above the beam with a stiffer rail to transfer loads directly to the top of the beam. See Figure 2.30b. If the EPS rail, or an EPS over-sheet, is used the long-term creep strength and deformation of the EPS is checked. EPS blocks are specified according to BS EN 15037, Part 4: Expanded polystyrene blocks (BS EN 15037-4, 2010). EPS blocks are 'non-resisting' and therefore require a structural topping to span transversely between beams, typically 70–75 mm thickness, reinforced using steel mesh or steel or polypropylene fibres.

Lateral distribution of isolated point loads and/or longitudinal wall loads (parallel to span) is carried out over several T beams in beam and block flooring, i.e. a point load acting on top of a screeded floor with solid blocks may be distributed over 3 single T beams, or up to 6 double beams (See Section 7.1.3 for the results of full scale testing). If the blocks are EPS, the numbers are 2 and 4, respectively. For longitudinal wall loads the load spread width is limited to about 700 mm and therefore the number of beams is also 2 and 4, respectively. Note that lateral distribution is not permitted beyond an edge beam where the load is positioned between the first and second beams from the edge.

The beams are defined in BS EN 15037-1 as 'self-bearing beams - which provide the final strength of the floor system independently of any other constituent part of the floor system', i.e. the beams do not require additional structural topping or reinforcements at the supports.

Figure 2.31 Site fitting of ground floor beam and block flooring using EPS blocks (c. Litecast Ltd., UK).

The beams are therefore designed as non-composite, but composite beams are possible using a structural topping which is adequately keyed into the gaps between the beams and blocks, as shown in Figure 2.30c.

2.2.10.2 Load v span tables

The derived data in the load v span graph shown in Figures 2.32 and 2.33 are based on the following. 150 and 225 mm depth T beams are pretensioned using 5 mm diameter wires (less commonly 7 mm wires) with up to 4 wires in 150 mm deep beams, and up to 8 wires in 225 mm deep beams. Cover to the soffit is 20 and 30 mm, respectively. Top wires are not added because the highest wire is at about ⅔ depth from the bottom. The characteristic tensile strength $f_{pk}=1770$ N/mm^2 (1670 N/mm^2 for 7 mm wires), pretensioned to 70%–75% f_{pk}. Concrete strength is $f_{ck}=50$–55 N/mm^2 at 28 days and $f_{ck}(t)=25$ to 32 N/mm^2 at transfer. The maximum prestress in the bottom of the unit is limited to $\sigma_b(t) \leq 18.5$ N/mm^2 at transfer, about 0.6 $f_{ck}(t)$. The stress in the top $\sigma_t(t) \geq -0.2$ N/mm^2 at transfer, which is not governing. The durability criteria are for internal exposure, class XC1. Although the clear span is mostly limited by the permissible tension f_{ctm} in the soffit at the serviceability limit state, only 150 mm beams with spans greater than 5.2 m with the live load = 1.5 kN/m^2 are limited by the active deflection after construction of effective span/350 based on non-brittle finishes, using a quasi-permanent factor $\psi_2=0.3$.

The type of blocks specified in the load v span graph in Figure 2.32 are 440 or 215 mm wide x 100 mm depth concrete blocks of density 1950 kg/m^3. These would be used in upper, internal floor spaces with a 65 mm depth finishing screed (not a structural topping).

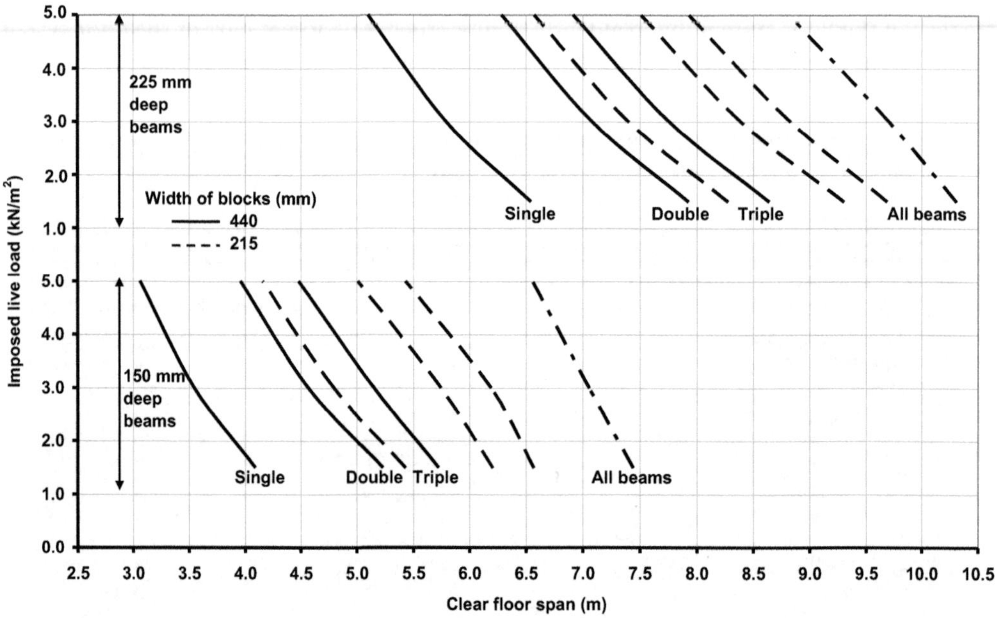

Figure 2.32 Imposed uniformly distributed live load vs clear floor span for prestressed 'house beam and block flooring'. The data include for a 65 mm deep screed of self-weight 1.495 kN/m^2. Check with manufacturers for maximum handling lengths. 'Single' etc. refers to the number of beams supporting the infill blocks, e.g. 'double' has two beams concreted on site side-by-side supporting either 440 or 215 mm width concrete blocks of 1950 kg/m^3 density. The dashed lines are in the same sequence as the solid lines. 'All beams' are concreted solid on site without blocks.

Comparing the load v span performance of single 'wide' beams shown in Figure 2.5a (215 mm width × 150 mm depth) with the data for narrow beams (92 mm width) in Figure 2.30, for single beams with 440 mm wide blocks limiting clear spans for 'wide' beams are 1.24 to 1.28 times greater for imposed live load of 1.5 and 5.0 kN/m^2, respectively. Although the self-weight of the floor is only 9% greater, the 'wide' beam is 2.9 times heavier than the narrow beam, so site handling may be an important factor. Unsurprisingly, the limiting clear spans for single 'wide' beams are almost the same as for double narrow beams, the ratios being 0.97–0.99, respectively, and the self-weight of beams and blocks is identical (2.38 kN/m^2). However, 'wide beams' use 9 no. wires compared with 8 no. for double narrow beams.

The blocks in the load v span graph in Figure 2.33 are 540 or 270 mm wide × 180 mm depth expanded polystyrene (EPS) blocks of grade 80 kPa (i.e. 0.08 N/mm^2 compressive strength) with a density of 20 kg/m^3. These would be used for thermally insulated ground floors with a 75 mm depth structural reinforced (steel mesh or fibres) topping laid on top of the EPS blocks, designed to span transversely between the beams.

As a guide for consultants and architects, 'load v span' tables are published by manufacturers in terms of the limiting clear span against an imposed live load of 1.5 and 3.0 kN/m^2 for residential and congregation use. The data for floors with concrete blocks include an allowance of 1.5 kN/m^2 for a 65 mm depth screed, and data with EPS blocks an allowance of 1.725 kN/m^2 for a 75 mm screed. The factors are $\psi_2 = 0.3$ and 0.6 for domestic and congregation use, and $\psi_0 = 0.7$. Single, double and triple beams are given here, but alternate single-double beams, quadruple beams or even solid beams (no concrete blocks) are possible. Some manufacturers limit the clear span to the criteria that the natural frequency of the floor unit is at least $f_n \geq 4$ Hz, which may be critical for live loads up to 2.5 kN/m^2. Table 2.10a and b gives an example of this table for the same units used to generate the load v span graph in Figures 2.32 and 2.33 for concrete blocks and EPS blocks, respectively.

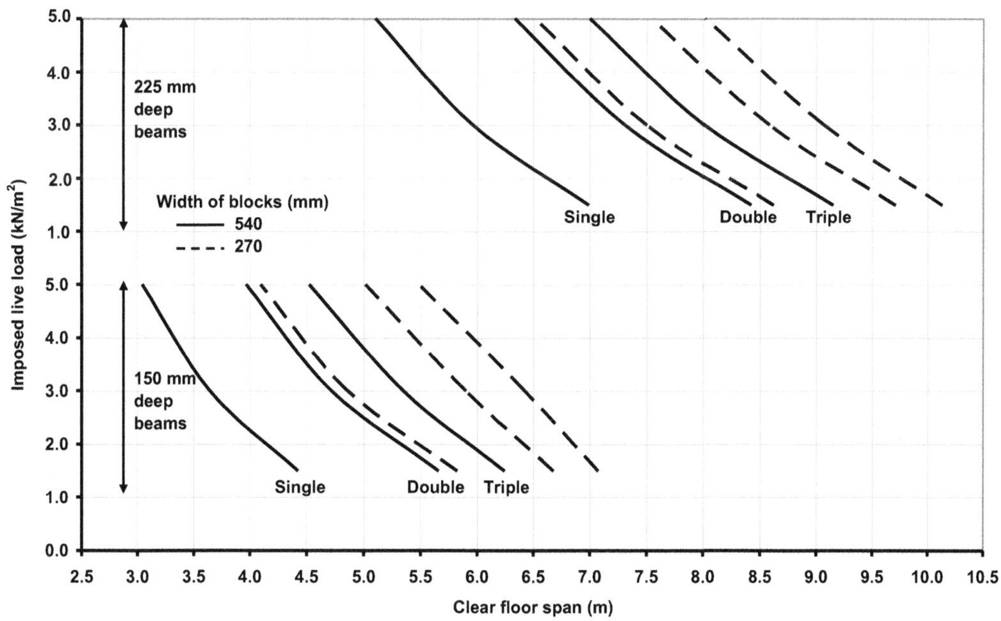

Figure 2.33 Imposed uniformly distributed live load vs clear floor span for prestressed insulated ground floor 'house beam and block flooring'. The data include for a 75 mm deep screed of self-weight 1.725 kN/m^2. Beams support either 540 or 270 mm width polystyrene (EPS) blocks of 20 kg/m^3 density. All other information is as per Figure 2.32.

Table 2.10 Example of load v span tables published for beam and block flooring for (a) concrete blocks and (b) EPS blocks

(a)

		Clear span (m) for imposed live load (kN/m²) includes 1.5 kN/m² for finishing screed					
		150 mm deep beams			*225 mm deep beams*		
Beam arrangement	*Block width (mm)*	*Self-weight (kN/m²)*	*1.5 (ψ_2=0.3)*	*3.0) (ψ_2=0.6)*	*Self-weight (kN/m²)*	*1.5 (ψ_2=0.3)*	*3.0 ψ_2=0.6)*
Single	440	2.17	4.09	3.55	2.72	6.54	5.79
	215	2.40	5.43	4.60	3.32	8.28	7.40
Double	440	2.38	5.23	4.43	3.25	7.93	7.08
	215	2.68	6.21	5.29	3.93	9.30 *(8.56)*	8.24
Triple	440	2.54	5.72	4.86	3.59	8.64 *(8.28)*	7.75
	215	2.85	6.57 *(6.39)*	5.62	4.25	9.69 *(8.74)*	8.56

Self-weight is for beams, blocks and *insitu* infill.

Concrete blocks of density 1950 kg/m³.

(b)

		Clear span (m) for imposed live load (kN/m²) includes 1.725 kN/m² for finishing screed					
		150 mm deep beams			*225 mm deep beams*		
Beam arrangement	*Block width (mm)*	*Self-weight (kN/m²)*	*1.5 (ψ_2=0.3)*	*3.0 (ψ_2=0.6)*	*Self-weight (kN/m²)*	*1.5 (ψ_2=0.3)*	*3.0 (ψ_2=0.6)*
Single	540	0.39	4.42	3.65	0.96	6.99	5.96
	270	0.70	5.83	4.67	1.67	8.62	7.50
Double	540	0.65	5.66	4.56	1.56	8.42	7.31
	270	1.06	6.68	5.42	2.39	9.71 *(8.90)*	8.38
Triple	540	0.85	6.24	5.02	1.97	9.15 *(8.69)*	7.95
	270	1.29	7.07 *(6.79)*	5.79	2.80	10.13 *(9.05)*	8.74

Self-weight is for beams and blocks only.

EPS blocks of density 20 kg/m³.

The limiting span is either for the serviceability limit state of stress in the soffit of the unit or active deflection after construction of span/350 for stud portioning or non-brittle finishes. Italic figures in brackets are limiting spans due to natural frequency f_n=4 Hz. ψ_2 values are for the *type of floor loading* defined in the text.

The reduction in the clear span by specifying 'congregation' floor use (where ψ_2=0.6) for a live load of 3 kN/m² compared with 'domestic' use (where ψ_2=0.3) occurs only for 150 mm beams where active deflection is limiting in some cases. This does not occur with 225 mm deep beams. The reduction for double beams is 3% to 8%, and for triple beams is 6% to 10% – the latter value is for narrower blocks. The values are roughly the same for both concrete and EPS blocks.

2.2.11 Non-composite main beams

2.2.11.1 Product development for main beams

Main beams are defined as load-bearing beams supporting floor slabs on one or two sides. Prestressed beams are mostly internal (spine) beams supporting slabs on two sides. Some may be used as external beams, but reinforced L-shape or deep spandrel beams supporting slabs and the external façade are preferred. It is possible for slabs to be supported perpendicular to the span of the beam on one side and run parallel to the beam on the other side, where a notional load-bearing span of 1.0 m is taken on this side. In this case, beams are subjected to bending moments, shear forces and torsional moments (see Section 10.6). Main beams are specified and designed as linear elements according to the product standard BS EN 13225 (BS EN 13225, 2013). Structural design is carried out to BS EN 1992-1-1.

There are three main types of psc main beams, shown in Figure 2.34

1. rectangular cross section – the simplest beam to design and structurally efficient, although the total construction depth is the sum of the depth of the beam and slab.
2. inverted-tee beams – where the slab is recessed into the top of the beam such that the total construction depth is the depth of the beam alone, and maybe the thickness of a structural topping on the slab. To achieve this, the width is typically 500–600 mm, although shallower and wider beams of up to 1200 mm have been used.
3. I beams, comprising a wide bottom flange (600–1000 mm width), a narrow web (200–300 mm width) and either a wide or narrow top flange (similar to 'M-type' bridge beams) depending on whether the beam is designed non-compositely or compositely, respectively. Here they resemble the top part of inverted-tee beams, such that the minimum depth of the beam is about 900 mm. For this reason, I beams are not commonly used in most commercial or parking structures, where head-room is at a premium. Flanges may have tapered inside faces; typically 30 mm slope, to assist with demoulding, or may be curved with a radius to blend with the vertical face of the web.

As an example of the structural efficiency of the three types of beams, Table 2.11 gives the service M_{sR} and ultimate M_{Rd} moments of resistance for beams of 900 mm depth. The characteristic tensile strength of strands is f_{pk}=1770 N/mm^2, pretensioned to 70% f_{pk}. Concrete strength is f_{ck}=45 N/mm^2 at 28 days and $f_{ck}(t)$=30 N/mm^2 at transfer.

Although the inverted-tee beam (self-weight 12.25 kN/m) has the greatest value of M_{sR} the number of strands to achieve this is N=42, such that M_{sR}/N=35.1 kNm/strand. The rectangular beam (self-weight 7.88 kN/m) has M_{sR} 51% lower, but M_{sR}/N=37.7 kNm/strand. The I beam (self-weight 8.53 kN/m) has M_{sR} 21% lower, but M_{sR}/N=42.1 kNm/strand. If these ratios are expressed in terms of the self-weight of the beams, the reader will see that both the rectangular and I beam outperform the inverted-tee beam by a factor of about 1.65. In spite of this, the inverted-tee beam is the easiest beam to design, reinforce and manufacture. For this reason, the inverted-tee beam is presented in the load v span graphs.

2.2.11.2 *Load v span graph*

The load v span graph shown in Figure 2.35 for non-composite inverted-tee beams is presented in terms of the total imposed *ultimate* dead and live UDL w_{Ed} (excluding the self-weight of the beam). It is presented as the ultimate load, rather than the service load, due to the large number of permutations for dead and live floor loads. However, in order to obtain the data for the most typical combination, the imposed dead load is set as g_k=30 kN/m, based on a 200 mm deep hcu (say 3 kN/m^2, plus finishes and services of 2 kN/m^2 acting over spans of 6.0 m on either side of the beam). The service live load q_k is obtained from the total ultimate using BS EN 1990 (BS EN 1990, 2002) Exp. 6.10b 1.25 g_k+1.5 q_k. See Section 3.1.3. Note that Exp. 6.10b gives the critical service load q_k as low as w_{Ed}≈45 kN/m, whence q_k is only about 5 kN/m. Should g_k differ from 30 kN/m, by say ±20%, the clear spans are still valid to ±0.7% as the live load q_k compensates for the change in g_k, as $q_k=(w_{Ed}-1.25\ g_k)/1.5$.

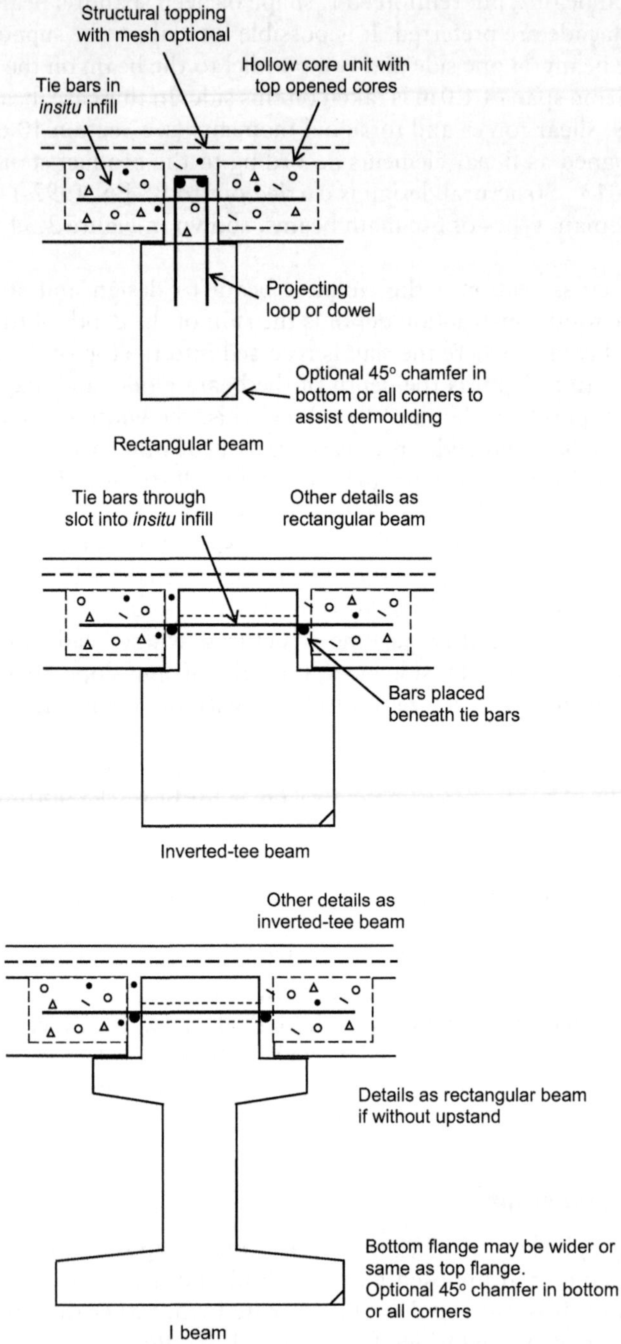

Figure 2.34 Cross sections of rectangular, inverted-tee and I beams, showing connection details to precast hcu.

It is not usual for manufacturers to publish load v span tables for beams due to the large permutations in beam sizes, slab depths, range of dead and live loads, etc.

The derived data in the load v span graph shown in Figure 2.35 are based on the following. The total width of beams is 600 mm, with a 350 mm wide×200 mm deep upstand in order to recess 200 mm deep floor units (assumed to be hcu). The net bearing length for

Table 2.11 Comparison of service and ultimate moments of resistance for rectangular, inverted-tee beam and I beam, all 900 mm depth, with identical materials

Type	*No. of strands at mid-span*	*Bottom width (mm)*	*Upstand width (mm)*	*Web width (mm)*	*Bottom flange depth (mm)*	*Service M_{sR} (kNm)*	*Ultimate M_{Rd} (kNm)*
Rectangular	26	350				979	1605
Inverted-tee	42	600	350		400[a]	1475	2320
I beam	29	600	350	200	150	1223	2148

[a] Normally called the 'boot'.

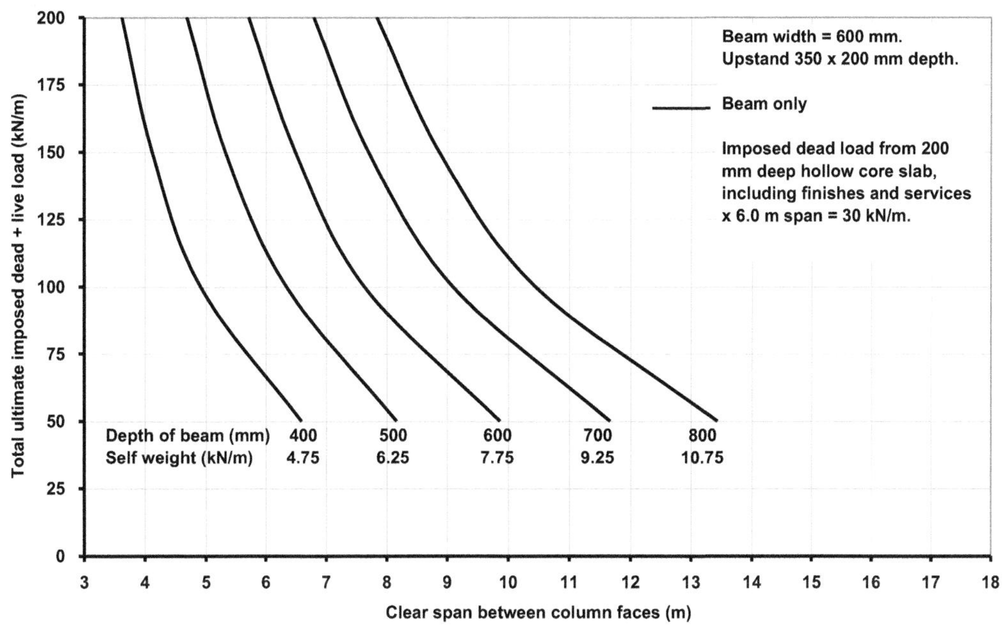

Figure 2.35 Imposed ultimate uniformly distributed dead plus live load vs. clear span (between column faces) for non-composite prestressed inverted-tee beams. Data include self-weight of beam as indicated in the graph. Note that the service dead load is $G_k = 30$ kN/m and the service live load Q_k is obtained from the ultimate load based on BS EN 1990, Exp. 6.10(b), i.e. Q_k = (Total ultimate load in graph − 1.25 G_k)/1.5.

the beam to column or wall support is taken as 140 mm. Beams are pretensioned using 12.5 mm diameter 7-wire helical strands, with axis distance from the bottom at 50, 100, 150, 200 mm, etc. using as many strands as necessary to satisfy two conditions.

i. Using full prestress, the serviceability limit state for the final stress, after losses of 0.45 f_{ck} in the top of the beam (and not f_{ctm} in the soffit, often the situation with inverted-tee sections). The active deflection after construction of span/500, or ultimate flexural strength are critical.
ii. Using debonded strands, typically net transfer stress at the bottom of the webs due to compressive prestress and tensile self-weight alone ≤ 0.6 f_{ck}(t), where $f_{ck}(t)$. = 30 N/mm².

The strands are debonded at ¼ and ¾ of the effective span. There are 8 no. strands in the bottom row (9 no. for 800 mm depth beam) of which 4 are debonded, and 4 or 6 no. in the second row of which 2 are debonded. There are 2 no. strands at the top of the boot and 2 no. at the top of the upstand, all at axis distances of 50 mm from edges. Note that, in generating the load v span data for different beam depths and ultimate loads, it is not possible for the final net prestress at the point of debonding to be exactly the same in all cases because the strands are positioned at 50 mm increments, so there is a small scatter in the results.

The characteristic tensile strength of strands is $f_{pk}=1770$ N/mm^2, pretensioned to 70% f_{pk}. Concrete strength is $f_{ck}=45$ N/mm^2 at 28 days and $f_{ck}(t)=30$ N/mm^2 at transfer. For maximum prestress at transfer in the bottom of the unit is limited to $\sigma_b(t)\leq 15$ N/mm^2, about 0.5 $f_{ck}(t)$, while at the same time the stress in the top is limited $\sigma_t(t)\geq -2.7$ N/mm^2 at transfer. The clear span (between column or wall faces) is limited by the permissible tension in the soffit f_{ctm} at the serviceability limit state for exposure class XC1. Due to the deep and large cross section of the beam the criteria of active deflection after construction of effective span/500 is not critical, using $\psi_2=0.3$ for domestic floor use. Even if $\psi_2=0.8$ for storage the active deflection is less than 0.8 times its limiting value.

2.2.12 Composite inverted-tee main beams

2.2.12.1 *Load v span graphs*

Composite beams comprise a psc inverted-tee beam together with precast floor slabs, either hcu or solid (but not double-tee or other ribbed units), with cast *insitu* infill at the ends of the floor units. For hcu the infill is placed in opened cores for a distance of 500–600 mm depending on tying requirements for tie steel rebars. Composite action may further be increased using a reinforced structural topping on the floor units; however, the data in Figures 2.36 and 2.37 are for beams with the hcu alone without topping. The *insitu* concrete topping strength is $f_{ckT}=25$ N/mm^2. It is difficult to form composite action using solid units without the use of a structural topping.

Unlike the non-composite beam which is critical in the top fibres, the serviceability limit state for the final stress of composite beams is f_{ctm} in the bottom of the beam, due to the increase in the height of the composite centroidal axis, from about 0.41 to 0.62×depth of beam.

The derived data in the load v span graphs shown in Figures 2.36 and 2.37 are based on the same non-composite beams, but designed compositely with 200 and 300 mm depth hcu, respectively, cast solid at their ends using *insitu* concrete $f_{cki}=25$ N/mm^2. Data include the self-weight of the beam as indicated in the graph. The service dead load is $G_k=30$ and 36 kN/m, comprising 18 and 24 kN/m for the self-weight of the 200 and 300 mm depth hcu, respectively, plus 12 kN/m for finishes & services. The service live load Q_k is obtained from the ultimate load based on BS EN 1990, Exp. 6.10b, i.e. Q_k=(Total ultimate load in graph $-1.25\ G_k)/1.5$.

The increases in span, due to composite action are summarised in Table 2.12 (referred to as 'C') for the 400 and 800 mm deep beams.

The precast beam is unpropped during the placement of the floor units and casting of the infill. The flexural stresses resulting from removing the props, applied finishes/services, etc. and live loads are carried by the composite section. If the beams are propped, using 2 no. props at one-third span, these are also summarised in Table 2.12 (referred to as 'CP') for the 400 and 800 mm deep beams. Note that for the reasons mentioned above (*viz.* small scatter in results due to strands positioned at 50 mm increments) the trends over the range of imposed loads may vary a little. However, it is clear that for the deeper beam, the ratio of the increases in clear span is almost independent of the imposed ultimate load, whereas for the shallower beam, composite action (and propping) is more beneficial for the greater

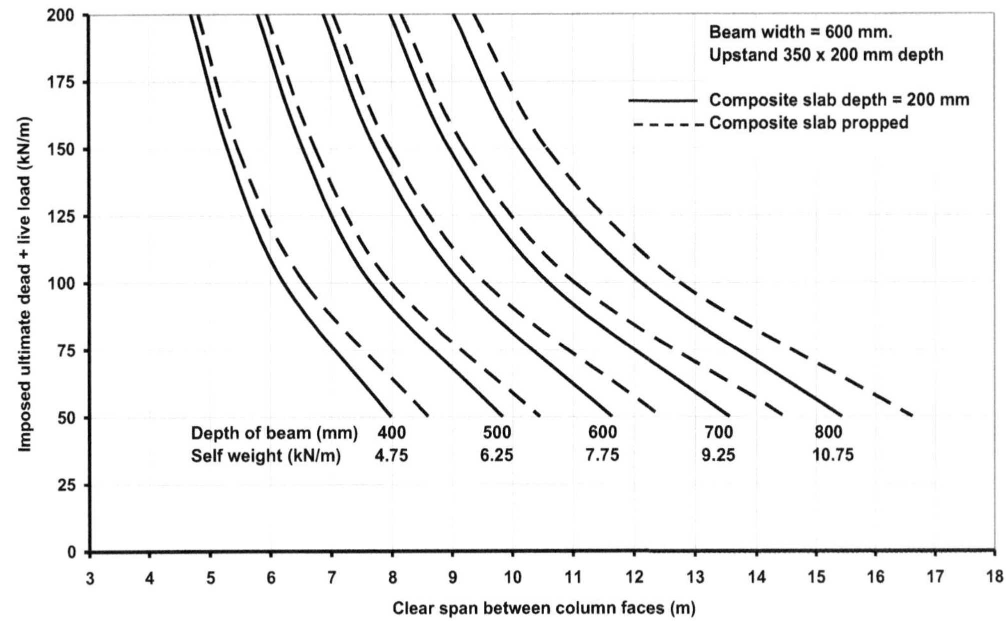

Figure 2.36 Imposed ultimate uniformly distributed dead plus live load vs clear span (between column faces) for composite prestressed inverted-tee beams. The beam is designed compositely with 200 mm deep hollow core floor slab, cast solid at its ends using *insitu* concrete $f_{cki}=25$ N/mm². Data include self-weight of beam as indicated in the graph. Note that the service dead load is $G_k=30$ kN/m, comprising 18 kN/m for the self-weight of the slab plus 12 kN/m for finishes & services, and the service live load Q_k is obtained from the ultimate load based on BS EN 1990, Exp. 6.10(b), i.e. $Q_k=$(Total ultimate load in graph – 1.25 G_k)/1.5. The precast unit is also propped during placement of the floor unit and *insitu* concrete infill using two props at ⅓ span. All other notes are as per Figure 2.35.

loads with the greater imposed live loads being carried by the composite section, a feature increased further for the 300 mm depth hcu. The maximum ratio for the 400 mm depth beam is 1.60, justifying the placement of small quantities of cast *insitu* infill at the ends of the hcu, and, if commercially viable, i.e. additional time and resources, the use of propping.

2.3 COMPARISON OF THE DESIGN OF A PRECAST PRESTRESSED FLOOR SLAB AND CAST *INSITU* FLAT SLAB

To conclude this chapter a comparison of the design of a rectangular floor bay, 6.0 m×9.6 m floor span, as part of the building layout in a multi-storey building shown in Figure 2.38 (grid 1–2,C-D) is carried out using reinforced and prestressed concrete beams and prestressed hcu slab solution and a cast *insitu* flat slab. These were designed by, and published in Worked Examples to Eurocode 2, Section 3.4, Flat slab (The Concrete Centre, 2012). The loading used for the design was self-weight, 1.0 kN/m² for finishes, 4.0 kN/m² for imposed live load, and 10 kN/m perimeter dead load. The precast hcu floors also had an additional 1.2 kN/m² for a levelling screed, included in the values for the weight of concrete in Table 2.13. The weight of steel for the flat slab solution was calculated from the information given in Figure 2.39 where some assumptions had to be made regarding exact reinforcement bar lengths for the flat slab, but these are not overly conservative and are based on typical detailing methods.

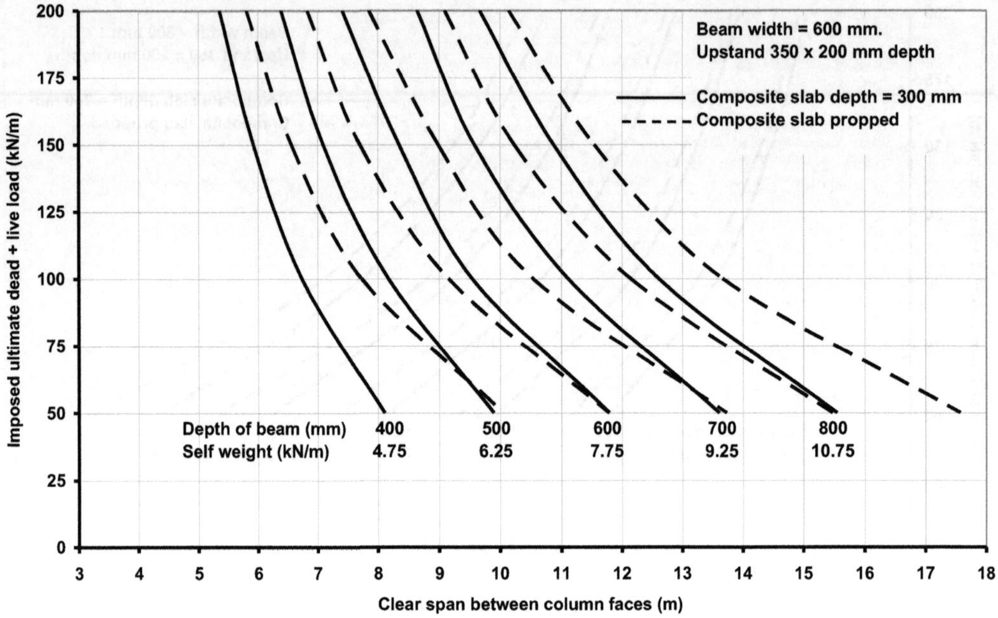

Figure 2.37 Imposed ultimate uniformly distributed dead plus live load vs. clear span (between column faces) for composite prestressed inverted-tee beams. The beam is designed compositely with 300 mm deep hollow core floor slab. All other notes as per Figure 2.36 except $G_k = 36$ kN/m, including 24 kN/m for the self-weight of the 300 mm depth slab.

Table 2.12 Ratio of clear spans for composite, and composite-propped vs non-composite inverted-tee beams designed compositely with 200 and 300 mm depth hcu

Composite or composite & propped	*Imposed dead and live UDL (kN/m)*	*Ratio of clear span for composite vs non-composite inverted-tee beam for hcu and beam depths (mm)*			
		200 depth hcu		*300 depth hcu*	
		400	*800*	*400*	*800*
C	200	1.30	1.15	1.48	1.23
	150	1.29	1.15	1.44	1.22
	100	1.27	1.16	1.37	1.21
	50	1.21	1.16	1.23	1.16
CP	200	1.33	1.20	1.60	1.30
	150	1.33	1.21	1.59	1.30
	100	1.32	1.22	1.57	1.31
	50	1.31	1.24	1.52	1.31

Beam width = 600 mm, and upstand = 350 mm wide × 200 mm deep.

C, composite; CP, composite and propped.

Using Figure 2.13, the precast solution used 250 mm depth prestressed hcu of self-weight of 3.63 kN/m^2 supported by simply supported precast reinforced and prestressed beams with a 200 mm downstand below the soffit of the hcu. Using Figure 2.35, this beam has an imposed ultimate load = 80 kN/m and clear span between columns = 5.6 m, the depth

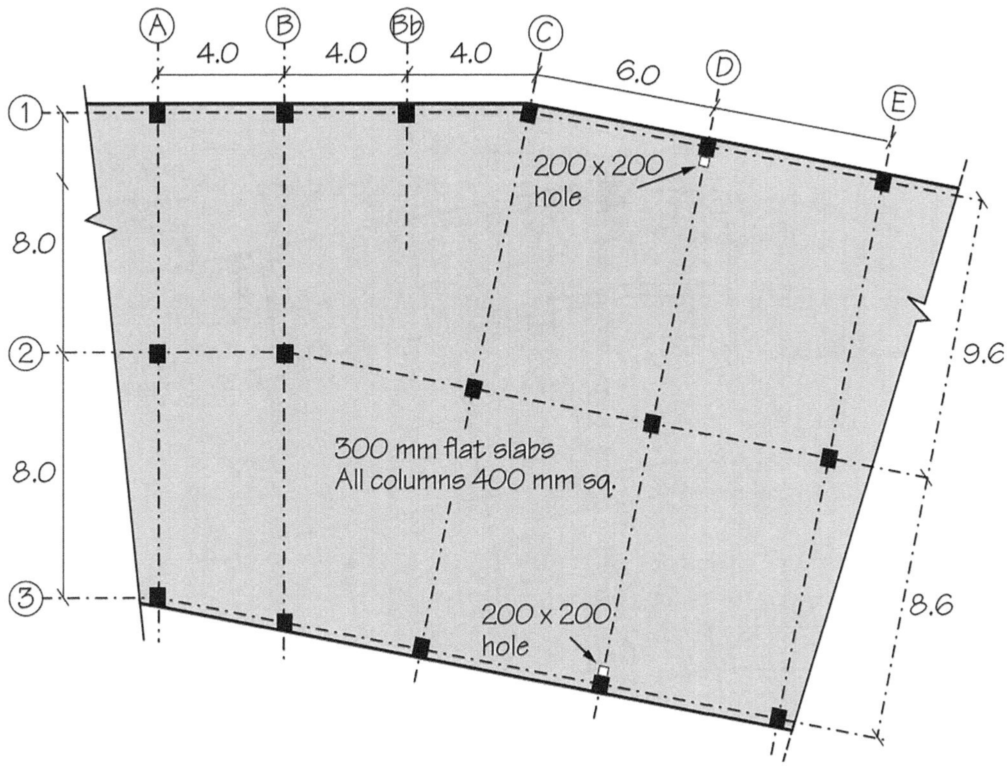

Figure 2.38 Dimensions for the floor bay within gridlines 1-2,C-D used for each design solution (courtesy The Concrete Centre, UK).

Table 2.13 Summary of reinforcement and concrete weights (kg) for slab types

Slab type	*Weight of reinforcement bars in reinforced & prestressed beams*	*Weight of strand in prestressed beams*[a]	*Weight of strand in hcu*[b]	*Total weight of reinforcement steel*	*Weight of concrete*
Insitu flat slab	-	-	-	1537	43218
Prestressed hcu with reinforced beams	1008	-	236	1244	31206
Prestressed hcu with prestressed beams	112	94	234	440	31650

[a] Weight of strand in edge beam = 49 kg and internal beam = 90 kg. Net weight in the floor bay = 49 + 90/2 = 94 kg.
[b] 7 no. 12.5 mm diameter strands with total area 651 mm² per unit.

of beam is 450 mm. Allowing 10 mm for precamber of the hcu the overall structural zone for the precast concrete solutions is 460 mm compared with 300 mm for the flat slab. Figures 2.40 and 2.41 show the geometry and rebar/strand arrangement of the reinforced and prestressed beams. The exposure category is XC1. Concrete strength for prestressed elements $f_{ck} = 45$ N/mm^2 with a limiting tensile stress in the soffit of prestressed elements $f_{ctm} = 3.80$ N/mm^2. The axis distance to the strands = 50 mm from external surfaces.

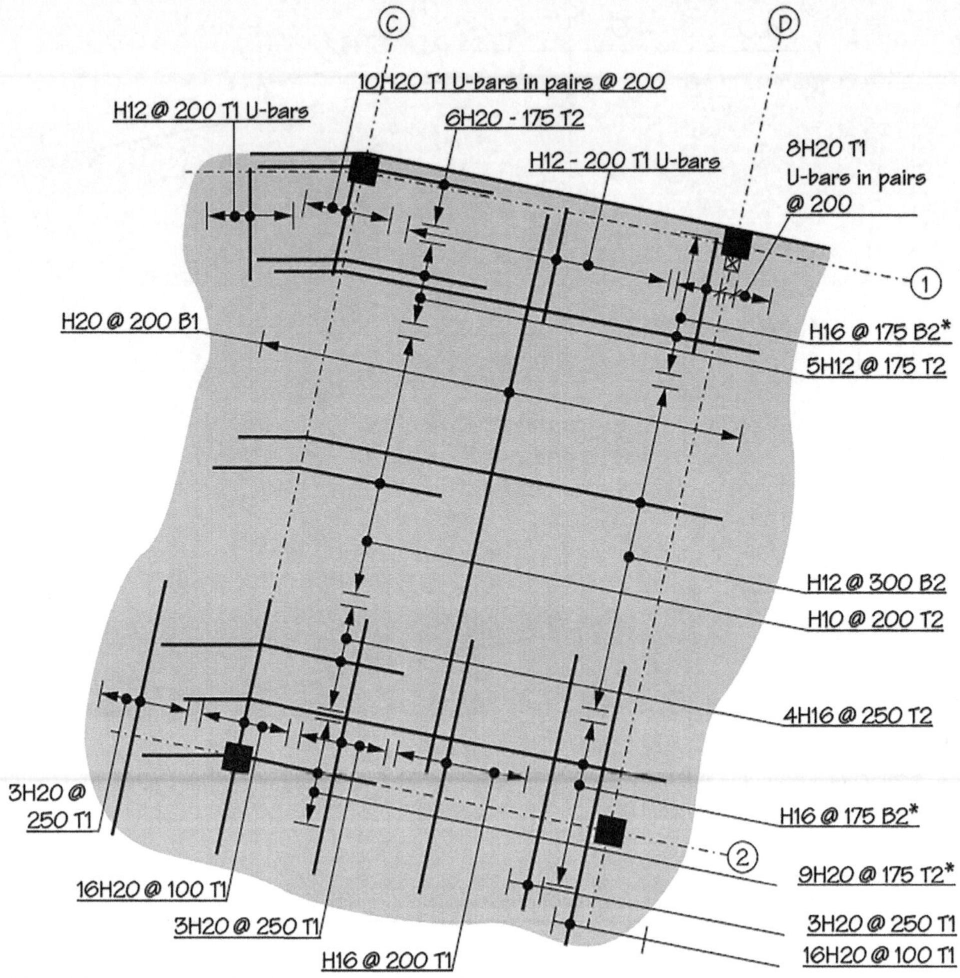

Figure 2.39 Reinforcement details for *insitu* concrete flat slab bay (courtesy The Concrete Centre, UK).

The reinforcement and concrete weights are based on the material that falls within the 6.0×9.6 m footprint on plan, i.e. half of the internal beam is considered in the totals. Concrete density taken as 2450 kg/m^3. The typical cost, at time of writing, of prestressing strand is £1030 per tonne and reinforcement bars £750 per tonne.

The concrete weight can be reduced by approx 27% when using a prestressed option instead of the *insitu* flat slab. The reduction in steel usage is greatest using the fully prestressed option as the weight of steel is reduced by approximately 68% compared to the flat slab. The design of the flat slab is slightly conservative, so the design could be value-engineered to reduce material usage in this solution. Even with some savings on the flat slab, this exercise demonstrates the potential that prestressed concrete has to deliver benefits in terms of reduced material usage, reduction in loading to superstructures and substructures and the positive contribution to the sustainability of a building.

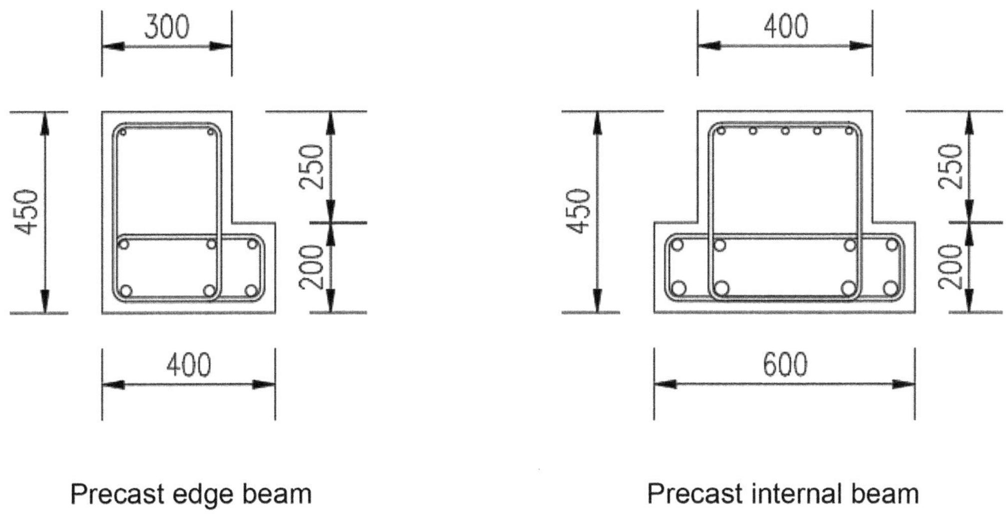

Figure 2.40 Precast reinforced concrete edge and internal beams (courtesy The Concrete Centre, UK).

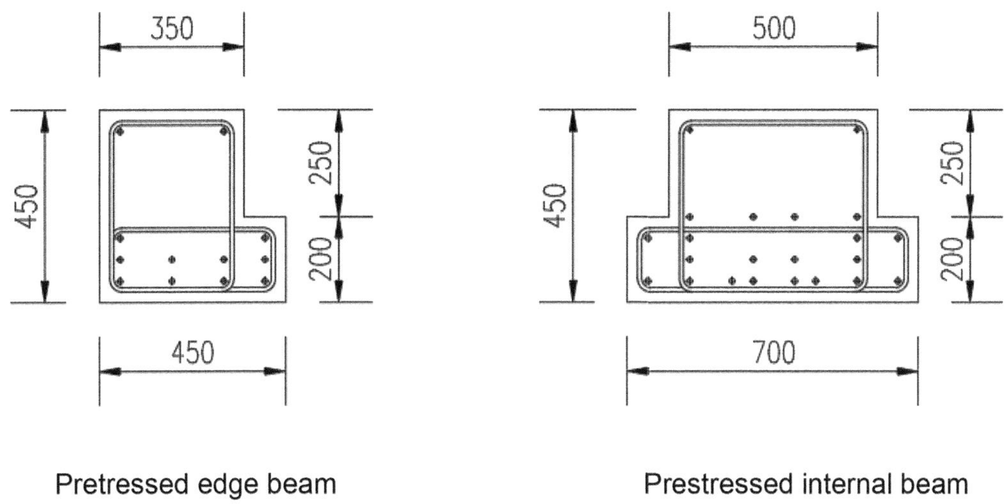

Figure 2.41 Prestressed concrete edge and internal beams (courtesy The Concrete Centre, UK).

REFERENCES

ASSAP. 2002. *The Hollow Core Floor Design and Applications, Manual 1st Edition*, Association of Manufacturers of Prestressed Hollow Core Floors (ASSAP), Verona, Italy.

Bate, S. C. C. and Bennett, E. W. 1976. *Design of Prestressed Concrete*, Surrey University Press & International Textbook Company Ltd., London, UK, 138 p.

BS 8500-1:2023. Concrete. Complementary British Standard to BS EN 206: Method of specifying and guidance for the specifier, BSI, London, UK.

BS EN 1168. 2005. Precast Concrete Products - Prestressed concrete hollow core units, +A3:2011, BSI, London, UK.

BS EN 13225. 2013. Precast concrete products – Linear structural elements, BSI, London, UK.

BS EN 15037-1. 2008. Precast Concrete Products - Beam and block flooring systems - Part 1: Beams, BSI, London, UK.

BS EN 15037-2. 2009. Precast Concrete Products - Beam and block flooring systems - Part 2: Concrete blocks, +A1:2011, BSI, London, UK.

BS EN 15037-4. 2010. Precast Concrete Products - Beam and block flooring systems - Part 4: Expanded polystyrene blocks, +A1:2013, BSI, London, UK.

BS EN 1990. 2002. Eurocode 0- Basis of structural design, +A1:2005, BSI, London, UK.

BS EN 1992-1-1. 2004. Eurocode 2: Design of concrete structures - Part 1-1: General rules and rules for buildings, +A1:2014, BSI, London, UK.

CP 110. 1972. British Standards Institute, CP 110, The Structural Use of Concrete, BSI, London, UK.

CP 115. 1959. British Standards Institute, CP 115, The Structural Use of Prestressed Concrete in Buildings, BSI, London, UK.

Elliott, K. S. 2017. *Precast Concrete Structures,* 2nd ed., CRC, Taylor & Francis, London, UK, 694 p.

Elliott, K. S. and Jolly, C. K. 2013. *Multi-Storey Precast Concrete Framed Structures*, 2nd ed., John Wiley, London, UK, 750 p.

FIB. 2000. Bulletin 6, Special design considerations for precast prestress hollow core floors, Guide to good practice, Fédération Internationale du Béton, Lausanne, Switzerland, 180 p.

FIP. 1988. *Precast Prestressed Hollow Cored Floors*, FIP Commission on Prefabrication, Thomas Telford, London, UK, 31 p.

Gilbert, R. I., Mickleborough, N. C. and Ranzi, G. 2017. *Design of Prestressed Concrete to Eurocode 2*, 2nd ed., CRC, Taylor & Francis Group, Boca Raton, FL, 665 p.

Institution of Structural Engineers. 1951. *First Report on Prestressed Concrete*, Institution of Structural Engineers, London, UK, 31 p.

PCI. 1991. *PCI Manual for the Design of Hollow Core Slabs*, Precast/Prestressed Concrete Institute, Chicago, USA, 88 p.

The Concrete Centre. 2012. *Worked Examples to Eurocode 2, Section 3.4, Flat Slab*, The Concrete Centre, Gilllingham House, London, UK.

Walraven, J. C. and Mercx, W. 1983. The bearing capacity of prestressed hollow core slabs, *Heron,* Vol. 28, No. 3, pp. 3–46.

Chapter 3

Basis for the design of prestressed concrete elements

3.1 BASIS FOR DESIGN

3.1.1 Basis for the design of precast buildings

The basis for the design of precast concrete buildings and prestressed elements is no different from that for the design of any other form of concrete construction, except perhaps for particular aspects of the design for temporary stability and the quality control exercised in the controlled environment of the factory, assuming, of course, that the factory is in a temperature- and humidity-controlled environment.

Precast concrete buildings are characterised in the vast majority of projects by simply supported spans, most often by prestressed floor slabs spanning perpendicular to prestressed beams, producing a two-dimensional grid of discrete one-way spanning elements. Figure 1.1 shows a typical example of prestressed floor slabs and beams. Beams are either simply supported at pinned joints at the faces of multi-storey columns, or on top of single-storey columns, or are designed and constructed as continuous elements at moment-resisting joints. In the former, stability is achieved using shear walls, cores or cross-bracing (depending on the height of the building), and in the latter, by frame action, where the moment of resistance of prestressed beams is usually made composite by the contribution from prestressed floor slabs with or without a cast *insitu* structural topping. This is a key factor in the preliminary design. Unlike simply supported prestressed beams, continuous and cantilever prestressed beams must be carefully analysed and designed to cater for the negative bending moment in a section that is already eccentrically prestressed (see Chapter 13). The *fib* handbook Bulletin 74 (FIB, 2014) is an important publication for the planning and preliminary design of precast concrete structures.

The principles and requirements for safety, serviceability and durability are given in Eurocode EC0 (BS EN 1990, 2002) where Section 2, Clause 2.1 establishes the basis for design – that building structures shall be designed to have structural resistance, serviceability and durability, fire resistance for the required period of time, and not be damaged by explosion, impact and human errors that are disproportionate to the original cause. The latter is commonly known as 'structural integrity'. Clause 2.1(5) is particularly relevant to precast buildings and individual prestressed elements, i.e. 'selecting a structural form and design that can survive adequately the accidental removal of an individual member, or limited part of the structure, or the occurrence of acceptable localized damage' … 'tying the structural members together'. The *fib* Bulletin 63 (FIB, 2012) devotes an entire publication of 72 pages to this topic.

In prestressed concrete, the term 'serviceability' has dual meaning, in that it also implies that due to the presence of pretensioning forces giving rise to upward camber, certain stresses may not be exceeded under the characteristic service load.

DOI: 10.1201/9781003319450-3

BS EN 1990 clauses 3.1 to 3.5 establish the principles of limit state design outlined here in Sections 3.1.2–3.1.5. Clause 4.1 establishes the variables for actions, e.g. permanent (dead loads), variable (live loads) and accidental loads, and Clause 4.2 deals with materials. Clauses 6.3 to 6.5 define the design values of actions and resistance, and the values for ultimate and serviceability limit states. Annex A.1 (for buildings) provides the substance to the above with regard to the load combinations given here in Sections 3.1.2–3.1.5) and ψ factors in Table 3.1. Finally, Informative Annexes B, C and D expand on structural reliability, reliability factors, e.g. expressions to determine ψ_0 factors for two variable actions, and for the design assisted by testing, respectively.

National Annexes, e.g. the UK National Annex for Eurocode 0 (NA to BS EN 1990, 2002) provide nationally determined parameters; those relevant to precast buildings are

i. design working life, i.e. 50 years for buildings and other common structures (Table NA.2.1.)
ii. ψ factors for the characteristic, frequent and quasi-permanent load combinations for buildings (Table NA.A1.1) are reproduced here in Table 3.1.
iii. partial safety factors γ for equilibrium and over-turning of structures or elements (Table NA.A1.2 (A))
iv. γ partial safety factors for structural actions, i.e. loads (Table NA.A1.2 (B))
v. design values for accidental combinations of actions (Table NA.A1.3). Seismic actions are not dealt with in this book.
vi. for the function of and damage to elements, the characteristic combination of loads is used for the serviceability limit state (clause NA.2.2.6, and Table A1.4), i.e. the full leading live load with the characteristic value ψ_0 for accompanying live loads.
vii. guidance on the use of Informative Annexes B, C and D on structural reliability, reliability factors, e.g. obtaining ψ_0 factors, and design assisted by testing is generally accepted (clause NA.3.1 for buildings).

3.1.2 Partial safety factors for serviceability limit states

The principles and requirements for safety given in BS EN 1990 include related aspects of structural reliability in defining the so-called 'Consequences classes' for examples of buildings and civil engineering works given in Table B.1 of BS EN 1990. For residential and office buildings, this is defined as CC2. The lower consequences class CC1 is for agricultural buildings and is

Table 3.1 Partial live load factors, NA to BS EN 1990, Table NA.A1.1

Usage	ψ_0	ψ_1	ψ_2
Domestic, residential, offices	0.7	0.5	0.3
Shopping, congregation	0.7	0.7	0.6
Storage	1.0	0.9	0.8
Traffic area ≤ 3t vehicle weight	0.7	0.7	0.6
Traffic area > 3t vehicle weight	0.7	0.5	0.3
Roof[a]	0.7	0	0
Snow at altitude > 1000 m	0.7	0.5	0.2
Snow at altitude ≤ 1000 m	0.5	0.2	0
Wind pressure	0.5	0.2	0

[a] Imposed, snow or wind loads should not be applied together simultaneously (EN 1991-1, clause 3.3.2 (1)).

not applicable to the type of buildings that are the subject of this book. The corresponding reliability class for CC2 is defined in clause 3.2 as RC2. The target reliability index for Class RC2 structural members is defined in Table C.2 as an irreversible serviceability limit state. Thus, the combination of actions for the serviceability limit state of stress is given in BS EN 1990, clause 6.5.3, Exp. 6.14 as the 'characteristic' combination.

$$\Sigma G_{k,j} + P + Q_{k.1} + \Sigma \psi_{0.i} Q_{k,i} \tag{3.1}$$

where

+ implies 'to be combined with'
Σ implies 'the combined effect of'
Suffix j is for all loads, 1 is for a leading live load, i is for all other accompanying live loads, and k is the characteristic value.
$G_{k.j}$ is for dead loads, known in BS EN 1990 as 'permanent' actions
P is the prestressing 'permanent' action
$Q_{k.1}$ is leading live load, known in BS EN 1990 as 'variable' action
$\psi_{0.i}$ is the live load factor given in the UK National Annex (NA to BS EN 1990. 2002) Table NA.A1.1, reproduced here in Table 3.1.
$Q_{k.i}$ are accompanying live loads

For imposed loads in all uses of buildings ψ_0=0.7, except that for storage ψ_0=1.0. The origin of the statistical combination of variable loads may be traced back to Turkstra's rule (Turkstra 1970). For the combination of two independent time variant loads, the maximum effect of the combination is when the leading action takes its maximum value during the reference period and the other, an accompanying action, is at an arbitrary point-in-time value. Figure 3.1.

A single value is used for G_k if its coefficient of variability is small, between 0.05 and 0.1. This is certainly the case for precast prestressed concrete where the mass of elements (*viz.* cross-sectional area) is controlled to be less than ±0.03 of the design value for slabs and less than ±0.01 for beams. Otherwise, two values of G_k are specified: an upper value $G_{k,sup}$ and lower value $G_{k,inf}$. Variations in cast *insitu* structural toppings may be up to ±0.05 of the design value due to both thickness and concrete density, even though allowances for the change in thickness due to upward camber of prestressed slabs, possibly 20 mm in a 6–8 m span, are included in the design value.

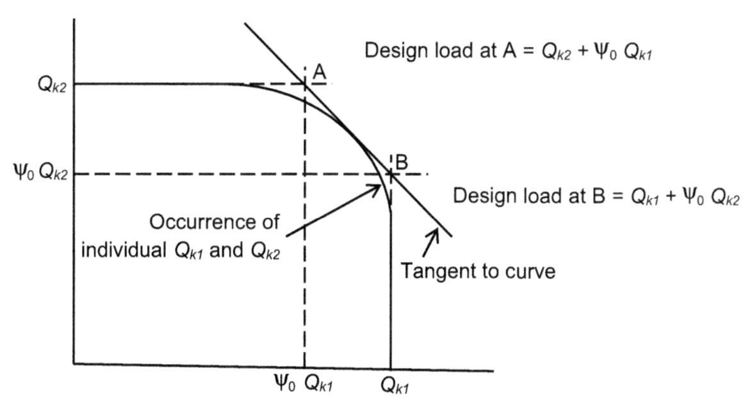

Figure 3.1 Basis of Turkstra's rule for the combination of independent imposed loads Q_1 and Q_2.

Each characteristic live load Q_k is considered in turn as the leading action when calculating service bending moments. The exception is stair landings where Q_k for the flight and landing are both leading, and this also applies to ultimate combinations. The greatest value for each permutation is taken as the maximum design value. In continuous spans, the minimum load $\Sigma G_{k,j}+P$ is considered in alternate spans to obtain the greatest bending moments, or in every third span to obtain the greatest shear force. In cantilevers, the minimum load is considered in both the cantilever and main span to obtain the greatest bending moments and shear forces, but not applied for the greatest support reaction at the start of the cantilever. The above cases are summarised in Figure 3.2.

The characteristic combination is used in the calculation of limiting tensile service stresses, after all losses of prestress, according to BS EN 1992-1-1, clause 7.3.2 (4) (BS EN 1992-1-1, 2004) as not greater than σ_{cp}. This is defined in clause 7.3.2 (2) as $\sigma_{cp}=f_{ct,eff}=f_{ctm}$, where f_{ctm} is the mean axial tensile strength defined in Table 3.1. This is the basis of the design procedure used in Section 7.2.

Some designers ignore the term $\Sigma\psi_{0.i}\ Q_{k,i}$ in Eq. 3.1 on the grounds that if $\psi_{0.}$should increase due to a change of use of the floor the design is unaffected. The characteristic service load is then simply the sum of all dead and all live loads.

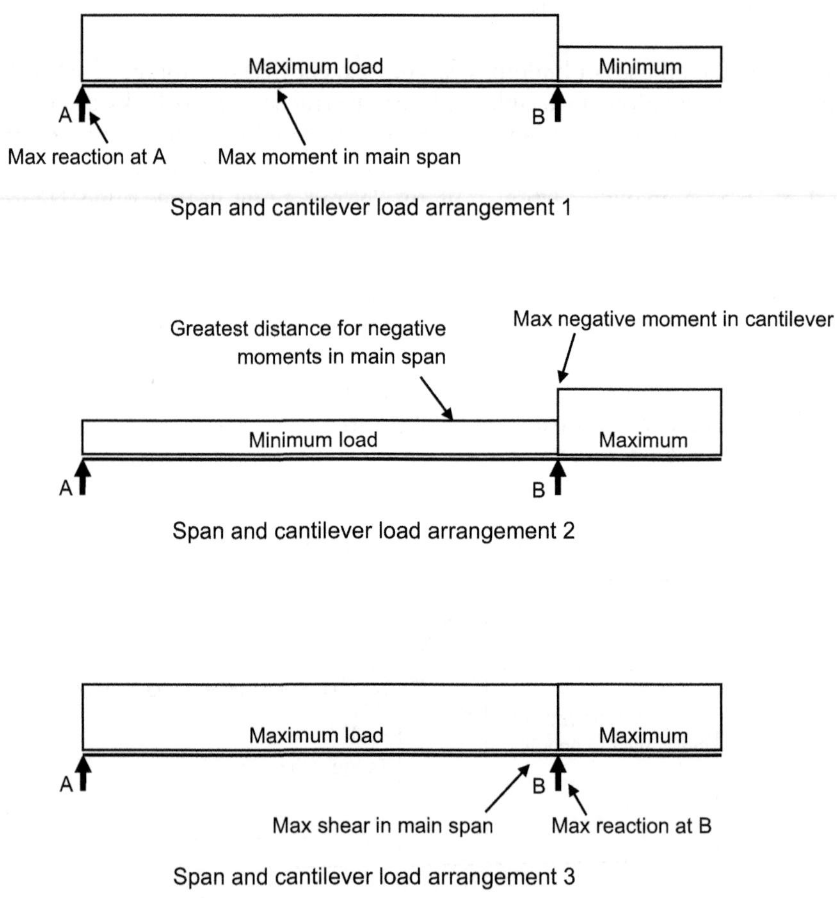

Figure 3.2 Load arrangements to obtain maximum moments, shear forces and support reactions for slabs and beams with cantilevers.

The combination of actions used for the calculation of crack width, defined in BS EN 1992-1-1, Table 7.1N and the UK National Annex to BS EN 1992-1-1, Table NA.4 (NA to BS EN 1992-1-1, 2004) is known as the 'frequent' combination. It is distinguished from the characteristic combination as a 'reversible' limit state; Eurocode BS EN 1990 explains this as 'there are no consequences of actions exceeding the specified service requirements will remain when the actions are removed'.

Thus, the combination of actions for the serviceability limit state of cracking is given in clause 6.5.3, Exp. 6.15 as the 'frequent' combination

$$\sum G_{k,j} + P + \psi_{1.1} Q_{k.1} + \sum \psi_{2.i} Q_{k,i} \tag{3.2}$$

where

$\psi_{1.1}$ is the frequent live load factor for leading actions
$\psi_{2.i}$ is the quasi-permanent live load factor for accompanying actions also given in Table NA.A1.1.

For imposed loads in all uses of buildings $\psi_1=0.5–0.7$, except for storage it is $\psi_1=0.9$. For imposed loads in all uses of buildings $\psi_2=0.3–0.6$ for all uses of buildings, except for storage it is $\psi_1=0.8$. Some designers ignore the term $\Sigma\psi_{2.i}\ Q_{k,i}$ in Eq. 3.2 for the same reasons as above.

Each live load is considered in turn as the leading action when calculating the frequent value for service bending moments used to calculate crack widths. The greatest value for each permutation is taken as the design value.

For exposure class XC1 the limiting crack width is $w_{max}=0.2$ mm, according to Table NA.4 of the NA to BS EN 1992-1-1. The limiting tensile stress is defined above as f_{ctm}. This is also summarised in Table 11.2 for the maximum crack width for prestressed concrete elements with bonded tendons.

However, for exposure classes XC2, XC3 and XC4, the limiting crack width is also $w_{max}=0.2$ mm, but the section is also checked for 'decompression', i.e. zero tensile stress is allowed, or $f_{ctm} \rightarrow 0$, under quasi-permanent loads. Thus, the combination of actions for the serviceability limit state is given in BS EN 1990, clause 6.5.3, Exp. 6.16 as the 'quasi-permanent' combination.

$$\sum G_{k,j} + P + \sum \psi_{2.i} Q_{k,i} \tag{3.3}$$

where

$\psi_{2.i}$ is the quasi-permanent live load factor for all actions

Note that the limiting tensile stress of f_{ctm} is also checked under the characteristic load, so two calculations are necessary. The characteristic combination may be critical for 'large' live loads, say 7.5 kN/m^2 or more, where $\psi_2=0.3$, although it is likely that for such live loads ψ_2 will be 0.6 and is less critical.

For exposure classes XD and XS, the limiting crack width is also $w_{max}=0.2$ mm, but the section is also checked for 'decompression' under frequent loads, defined in Eq. 3.2. The Code also specifies that for the decompression limit, all parts of tendons should lie at least 25 mm within concrete in compression, i.e., a nominal cover from the surface of ≥25 mm. This is satisfied for exposure classes greater than XC1 according to BS 8500-1 (BS 8500-1, 2023), Table A.4, summarised here in Table 3.2.

Table 3.2 **Relationship between exposure classes, concrete strength, maximum water/cement ratio, minimum cement content and nominal cover for prestressed concrete using cement CEM I according to BS 8500-1:2023 for 50 years working life.**

	Nominal cover (mm)		
Exposure class	*≥C28/35*	*≥C40/50*	*≥C45/55*
XC1	20	←	←
XC2	20	←	←
XC3/4	60	35	30

Minimum cement content ≥ 200 kg/m^3

Exposure class			*Nominal cover (mm)*			
Max w/c ratio	*0.60*	*0.55*	*0.50*	*0.45*	*0.40*	*0.35*
Min cement (kg/m^3)	*300*	*320*	*340*	*360*	*380*	*380*
XD1	40	35	←	30	←	←
XD2		45	40	←	35	←
XD3				60	50	45
Concrete grade				>C28/35	>C32/40	>C35/45
XS1		70	65	60	55	50
XS2			85	80	70	65
XS3						85

Resistance to chloride ingress is mainly dependent on the type of cement and water/cement (w/c) ratio. The recommended grade of concrete for normal weight concrete is C20/25, except where shown for XD3.

Compressive stresses due to characteristic (i.e. full) service loads and prestress are limited according to BS EN 1992-1-1, clauses 7.2 (2) and 7.2 (3) to guard against longitudinal horizontal splitting or spalling. Clause 7.2 (2) only specifies a limiting compressive stress of 0.6 f_{ck} for exposure classes XD and XS, not XC. However, it is common practice to limit compression to 0.6 f_{ck} for exposure XC, although it is rarely critical against the unfavourable situation in tension, except for some inverted-tee sections with relatively deep bottom flanges. Clause 7.2 (3) specifies a limiting compressive stress of 0.45 f_{ck} for all exposure classes under quasi-permanent loads in order to avoid non-linear creep. However, it will be shown in Section 7.2.1 that the presence of non-linear creep is avoided if all other limitations mentioned above are satisfied. Checking the compressive stress of 0.45 f_{ck} under quasi-permanent loads is rarely critical for exposure XC1, unless the imposed live load is very small and self-weight and dead loads dominate. In design, it is often, but not always, convenient to limit the compressive stress to 0.45 f_{ck} for all situations.

The above combinations of actions, limiting stresses and crack widths are summarised in Table 3.3 for prestressed concrete members with bonded tendons. Note that this book does not address unbonded tendons.

Table 3.3 Load combinations used for limiting tensile stress and crack widths

Exposure class	*Load combination*	*Limiting service stress in tension*	*Crack width (mm)*	*Limiting service stress in compression*
XC1	Characteristic			
	Frequent	f_{ctm}	0.2	0.6 f_{ck}[a]
XC2, XC3, XC4	Characteristic			
	Quasi-permanent	f_{ctm}		0.6 f_{ck}
	Frequent	*Zero*	0.2	0.45 f_{ck}
XD, XS	Characteristic	f_{ctm}		
	Frequent	*Zero*	0.2	0.6 f_{ck}[a]

[a] In design it is often, but not always, convenient to limit the compressive stress to 0.45 f_{ck} for all situations.

3.1.3 Partial safety factors for ultimate limit state

BS EN 1990, clause 6.4.1 (1)P defines the following ultimate limit states (relevant to this book) as

- STR: internal failure or excessive deformation of the structure or structural members where the strength of construction materials governs, i.e. structural strength such as flexural, shear, bearing or compressive strength
- EQU: loss of static equilibrium of the structure or any part of it, i.e. overturning, uplift

Load combinations are less complicated than the serviceability limit state. Fundamental (as distinct from accidental) combinations for STR are given in clause 6.4.3.2 as either Exp. 6.10

$$\Sigma \gamma_{G,j} G_{k,j} + \gamma_p P + \gamma_{Q,1} Q_{k.1} + \Sigma \gamma_{Q,i} \psi_{0.i} Q_{k,i} \tag{3.4}$$

or the greater of Exp 6.10a or 6.10b

$$\Sigma \gamma_{G,j} G_{k,j} + \gamma_p P + \gamma_{Q,1} \psi_{0.i} Q_{k.1} + \Sigma \gamma_{Q,i} \psi_{0.i} Q_{k,i} \tag{3.5a}$$

$$\Sigma \xi \gamma_{G,j} G_{k,j} + \gamma_p P + \gamma_{Q,1} Q_{k.1} + \Sigma \gamma_{Q,i} Q_{k,i} \tag{3.5b}$$

where ξ is a reduction factor for dead loads. The value of this in the UK NA to BS EN 1990, Table NA.A1.2(B) is $\xi=0.925$. (Note that in EN 1990 it is $\xi=0.85$).

There is no obvious reason to use Exp. 6.10 except for the same reason as mentioned for service loads – that if $\psi_{0.}$ increases due to a change in use of the floor the design is unaffected.

The partial safety factors for dead and live loads are given in Table NA.A1.2 (B) as

- $\gamma_{G,j}=1.35$ and $\gamma_{Q,1}=\gamma_{Q,i}=1.5$ for unfavourable situations, i.e. maximum load
- $\gamma_{G,j}=1.0$ and $\gamma_{Q,1}=\gamma_{Q,i}=0$ for favourable situations, i.e. minimum load

Simplified, and using $\psi_0=0.7$ for offices and residential buildings, the maximum loads are

Exp 6.10a is 1.35 G_k+1.05 Q_k (3.6a)

Exp 6.10b is 1.25 G_k+1.5 Q_k (3.6b)

and the minimum load is 1.0 G_k (3.6c)

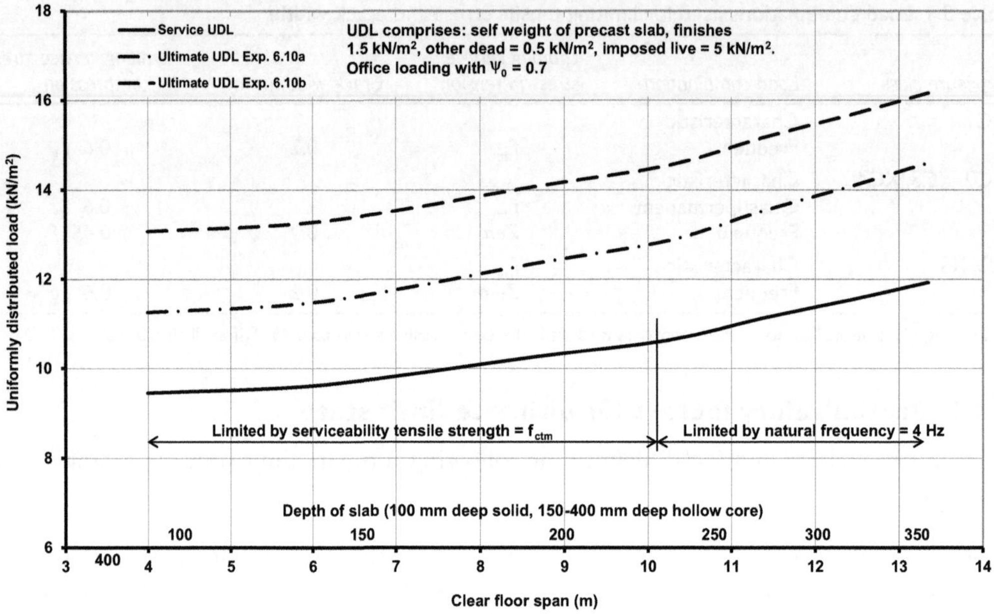

Figure 3.3 Service and ultimate floor loads as a function of the simply supported clear spans for prestressed floor slabs.

Exp. 6.10b governs if the ratio between the dead and live load is $G_k/Q_k<4.5$, a common situation. Floor loads for a typically 6 m span prestressed hollow core floor slab for office loading are (say) $G_k=4.6$ kN/m² and $Q_k=5.0$ kN/m² (including partitions), then w_{Ed} (Exp. 6.10a) = 11.46 kN/m² and w_{Ed} (Exp. 6.10b) = 13.25 kN/m². The global factor of safety being $\gamma_F=13.25/9.6=1.38$.

Figure 3.3 shows service and ultimate floor loads as a function of the simply supported clear spans of typical prestressed hollow core slabs (except 100 mm deep solid slab), where the service dead and live UDL are 2.0 and 5.0 kN/m², respectively. It shows that Exp. 6.10b governs, where w_{Ed} (Exp. 6.10b) ≈ 1.13 w_{Ed} (Exp. 6.10a). The criteria for the limiting spans at the serviceability limit state is the tensile stress f_{ctm} up to a span of 10.3 m, thereafter the span is controlled by the natural frequency of the completed floor being taken as ≤4 Hz, for office conditions.

The partial safety factor for prestressing γ_P is not applied to the ultimate limit state for flexure, but for the ultimate shear capacity a factor of $\gamma_P=0.9$ is applied to the axial prestress (see Section 10.2.2).

3.1.4 Partial safety factors for equilibrium

Fundamental combinations for EQU are given in clause 6.4.3.2, Exp. 6.10

$$\Sigma \gamma_{G,j} G_{k,j} + \gamma_{Q,1} Q_{k.1} + \Sigma \gamma_{Q,i} \psi_{0.i} Q_{k,i} \tag{3.7}$$

The partial safety factors are given in Table NA.A1.2 (A) as (called here Set 1)

- $\gamma_{G,j}=1.1$ and $\gamma_{Q,1}=\gamma_{Q,i}=1.5$ for unfavourable situations, i.e. maximum load
- $\gamma_{G,j}=0.9$ and $\gamma_{Q,1}=\gamma_{Q,i}=0$ for favourable situations, i.e. minimum load

The clause also states that if the verification of equilibrium also involves the resistance of structural members, i.e. in a cantilever where the moment of resistance at the start of the cantilever is also critical, a combined verification should be adopted as follows (called here Sets 2 and 3)

- $\gamma_{G,j}=1.35$ and $\gamma_{Q,1}=\gamma_{Q,i}=1.5$ for unfavourable situations
- $\gamma_{G,j}=1.15$ and $\gamma_{Q,1}=\gamma_{Q,i}=0$ for favourable situations

As well as $\gamma_{G,j}=1.0$ and $\gamma_{Q,1}=\gamma_{Q,i}=1.5$ for unfavourable situations

- $\gamma_{G,j}=1.0$ and $\gamma_{Q,1}=\gamma_{Q,i}=0$ for favourable situations

There is a problem here, as Set 1 may be more critical than Set 2. Consider a cantilever floor if span L' subjected to the same characteristic loads as in the previous example where the main span $L=6.0$ m. Equilibrium is checked for the unfavourable cantilever load w_2 and favourable main span w_1.

- Set 1. $w'=1.1\times4.6+1.5\times5.0=12.56$ kN/m². $w=0.9\times4.6=4.14$ kN/m². The critical span L' is given when the reaction at the remote end of the main span is zero, i.e. $-12.56\ L'^2/2/6.0+4.14\times6/2\geq0$. Then $L'\leq3.44$ m.
- Set 2. $w'=1.35\times4.6+1.5\times5.0=13.71$ kN/m². $w=1.15\times4.6=5.29$ kN/m². Then $-13.71\ L'^2/2/6.0+5.29\times6/2\geq0$. Then $L'\leq3.73$ m
- Set 3. $w'=1.0\times4.6+1.5\times5.0=12.1$ kN/m². $w=1.0\times4.6=4.6$ kN/m². Then $-12.1\ L'^2/2/6.0+4.6\times6/2\geq0$. Then $L'\leq3.70$ m

For this reason, Set 1 is more often considered to give the critical case. A point load, due to a balustrade or wall, at the end of the cantilever may change these relative values. Furthermore, the critical span of the cantilever slab is found to be 3.44 m, much greater than commonly used in practice.

If there are two cantilevers, one at each end of the main span, then the favourable load is also used in the second cantilever in turn.

3.1.5 Partial safety factors for accidental actions, including exposure to fire

Accidents are unintended actions, such as loss of bearing, impact, or explosions, and may be of short duration, and a low probability of occurrence. Accidents generally occur when the full characteristic live load is not present, and so the fundamental combination of the leading live load $Q_{k.1}$ is multiplied by the frequent factor $\psi_1<1$, and the accompanying live loads $Q_{k,i}$, each considered in turn, are multiplied by the quasi-permanent factor $\psi_2<1$.

Accidental combinations are given in Table NA.A1.3 (Exp. 6.11 is not used) as the fundamental combination

$$\Sigma G_{k,j} + A_d + \psi_{1.1}Q_{k.1} + \Sigma\psi_{2.i}Q_{k,i} \tag{3.8}$$

where A_d is the leading accidental action, such as tie forces when a support fails and slabs and beams sag in a catenary model, indirect thermal action due to fire, or the impact force.

There are no partial safety factors, effectively all $\gamma=1.00$.

The frequent factor $\psi_{1.1}$ is also given in the National Annex to BS EN 1991-1-2 (NA BS EN 1991-1-2, 2002) clause 4.3.1 (2). Note that EN 1992-1-2 recommends the factor as quasi-permanent $\psi_{2.1}$.

For prestressed concrete slabs and beams in multi-storey concrete, precast concrete or steel frames, the preferred design method to resist accidental floor loads is the use of catenary ties. Refer to Elliott (Elliott, 2017) Chapters 11 and 12.

3.2 PARTIAL FACTORS FOR MATERIALS

The partial safety factor at the ultimate limit state for concrete $\gamma_c = 1.5$ and steel rebars and prestressing tendons $\gamma_s = 1.15$ are given in BS EN 1992-1-1, Table 3.1N. The factors at service and for accidental actions are all 1.0. The UK National Annex uses the same values.

A further factor is applied to concrete in axial and/or flexural compression of $\alpha_{cc} = 0.85$, but not to other phenomena (shear, anchorage, bearing, etc.) according to the UK NA to BS EN 1992-1-1, clause 3.1.6.1(P). Therefore, the design strength of concrete in axial and/or flexural compression is

$$f_{cd} = \alpha_{cc} f_{ck} / \gamma_c = 0.567 f_{ck} \quad (3.9a)$$

otherwise

$$f_{cd} = f_{ck} / \gamma_c = 0.667 f_{ck} \quad (3.9b)$$

In passing, it is interesting to note that the use of $\alpha_{cc} = 0.85$ brings it into line with the design strength of $0.45 f_{cu}$ in BS 8110-1 (BS 8110-1 1997), e.g. for grade C40/50 concrete f_{cd} to EC2 is $0.567 \times 40 = 22.7$ N/mm^2 and to BS8110 is $0.45 \times 50 = 22.5$ N/mm^2.

The design strength of rebars and prestressing tendons is

$$f_{yd} = f_{yk} / \gamma_s = 0.87 f_{yk} \quad (3.10a)$$

$$f_{pd} = f_{pk} / \gamma_s = 0.87 f_{pk} \quad (3.10b)$$

3.3 LOAD CASES AND ARRANGEMENTS

3.3.1 Ultimate and service loads for cantilevers and continuous spans

The purpose of load arrangements is to arrange the unfavourable (maximum) and favourable (minimum) loads on different parts of a structure in order to establish critical design conditions at all sections. These include bending moments, shear forces, support reactions, and deflections (although this will be the same arrangement as for moments). In this book, this applies to cantilever and continuous beams and slabs, rather than 2-D frames.

Guidance on load arrangements is given in the UK NA to BS EN 1992-1-1, clause 5.1.3 1(P). Accepting that the prestressing action γ_p P is present in all spans, the following ultimate load arrangements should be considered, as summarised in Figure 3.4. The serviceability load arrangements are the same without the γ factors. The load arrangements commonly known as 'max-max-max', 'max-min-max', 'min-max-min', 'max-max-min', etc. are for stage 2 imposed dead and live loads after the prestressed slab has been positioned in stage 1. Stage 2 imposed dead loads include finishes, services, ceilings, etc. but do not include structural toppings. The authors have added load combination (c) as this

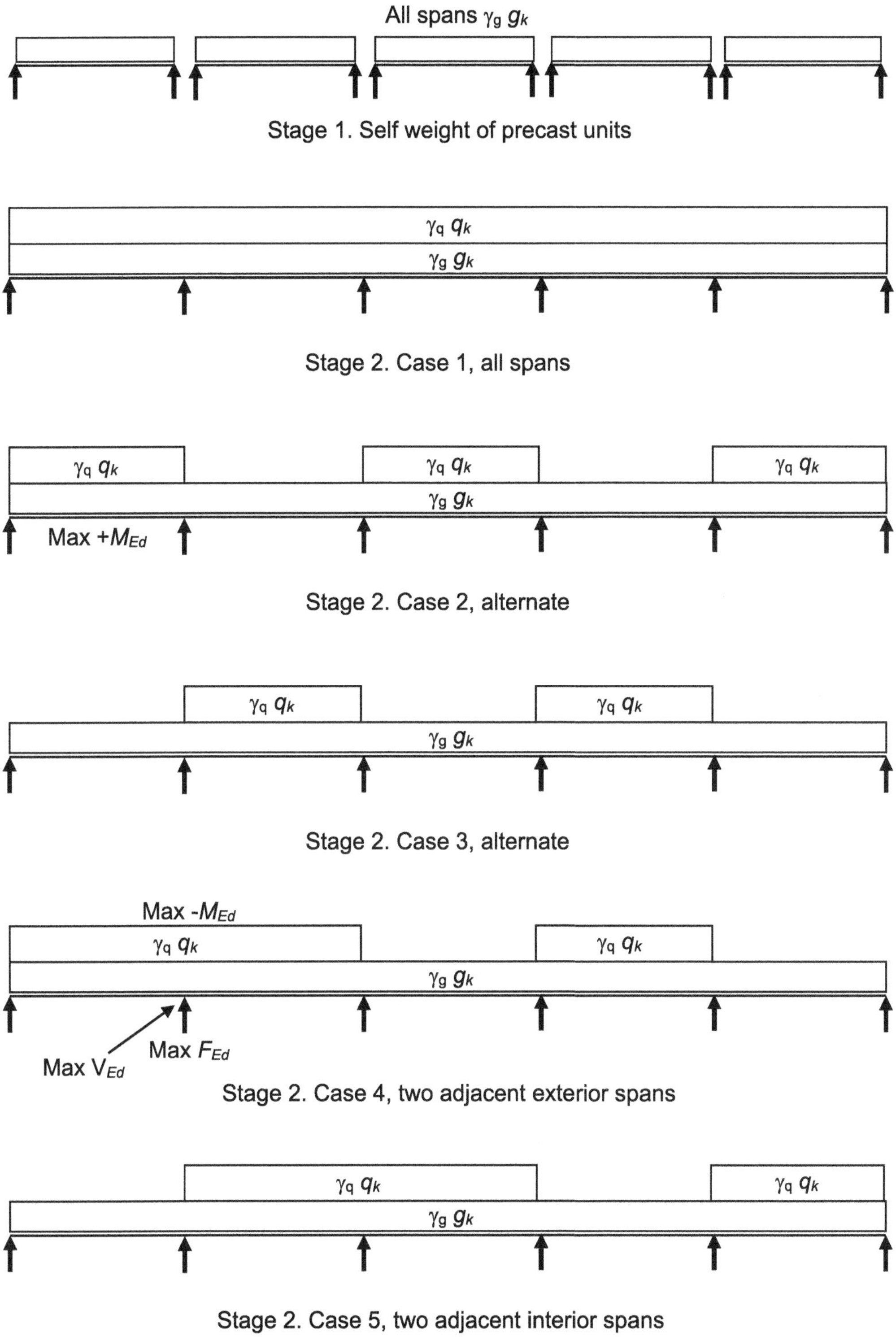

Figure 3.4 Load arrangements for self-weight (stage 1) and imposed dead and live loads (stage 2) in continuous spans are necessary to establish maximum bending moments and shear forces.

gives a slightly greater maximum hogging moment at that support between the maximum loaded spans (see the worked Example 13.7 for three-span continuous floor slab).

a. all spans carrying stage 2 design permanent and variable loads

$$\Sigma \gamma_{G,j} G_{k,j} + \gamma_{Q,1} Q_{k.1} + \Sigma \gamma_{Q,i} \psi_{0.i} Q_{k,i}$$

b. alternate spans carrying $\Sigma\gamma_{G,j}\, G_{k,j}+\gamma_{Q,1}\, Q_{k.1}+\Sigma\gamma_{Q,i}\, \psi_{0.i}\, Q_{k,I}$ with other spans carrying the dead load only $\Sigma\gamma_{G,j}\, G_{k,j}$ with the same $\gamma_{G,j}$ used for all spans.
c. any two adjacent spans carrying $\Sigma\gamma_{G,j}\, G_{k,j}+\gamma_{Q,1}\, Q_{k.1}+\Sigma\gamma_{Q,i}\, \psi_{0.i}\, Q_{k,I}$ with other spans carrying the dead load only $\Sigma\gamma_{G,j}\, G_{k,j}$. Note if the outer two spans carry the maximum, the third and fifth spans are minimum and the fourth span is maximum, and so on.

The terms are given for Exp. 6.10, but they are also applicable to the greater of Exp. 6.10a or 6.10b, e.g. if $\gamma_{G,j}=1.25$ in Exp. 6.10b is used for the maximum load it should also be used for the minimum load.

Figure 3.4 shows that for five equal spans, with the ratio of imposed live/dead=1.25, the maximum sagging moment $+M_{Ed}$ is, as expected, due to alternate spans Case 2, and the maximum hogging moment $-M_{Ed}$ is at the first interior support due to Case 4. The maximum shear force V_{Ed} and support reaction F_{Ed} are both at the first interior support. Surprisingly, Case 1 (all spans loaded) is not critical, although $-M_{Ed}$, V_{Ed} and F_{Ed} are only 8%, 1% and 4% lower, respectively. Case 5 is not critical at all, but of course if the spans differ, then Case 5 may become critical. If the ratio of imposed live/dead is increased to 2.5, the critical locations are the same.

Further simplifications are given for slabs to use the 'all spans' arrangement described in (a) if

i. in a one-way spanning slab the area of each bay, defined as the full width of the building bounded on the other sides by lines of supports, is at least 30 m^2
ii. the ratio $Q_k/G_k \leq 1.25$
iii. $Q_k \leq 5$ kN/m^2, excluding partitions

It is not common for stage 2 loads to satisfy (ii) for office loading, e.g. $Q_k=5$ kN/m^2, $G_k\approx 2$ to 3 kN/m^2, except for residential loading where $Q_k=1.5$ kN/m^2, $G_k\approx 2$ kN/m^2. If this option is used 20% moment redistribution is allowed at internal supports, excluding cantilevers.

At first sight, this UK National Annex option (it is not in BS EN 1992-1-1) appears to be at odds with common sense for two-stage precast/prestressed floor slabs, in that

- alternate span arrangement is not used to establish the maximum sagging moments, or the two adjacent spans to establish the maximum hogging moment at supports
- why it is only applicable for a floor bay area > 30 m^2, which is not a particularly large area, e.g. 8 m span × 4 no. 1.2 m wide units
- it is applicable for relatively low live loads $Q_k \leq 1.25\, G_{k,stage2}$

However, consider stage 2 imposed loads as $G_k=2$ kN/m^2 and $Q_k=2.5$ kN/m^2, using Exp. 6.10b ($\gamma_G=1.25$, $\gamma_Q=1.5$) for the ultimate loads, in five continuous spans of 6.0 m between centres of supports, the two arrangements (a) and (b) give reasonably close ratios of 0.97 and 0.94, particularly for the exterior span that has the critical bending moment, to within 3% of each other, as shown in Table 3.4a. This arrangement has a ratio $Q_k/G_k=1.25$. For residential loading, if $Q_k/G_k=0.75$ the ratios are 0.99 and 0.98. If Exp 6.10 is used ($\gamma_G=1.35$

Table 3.4a Maximum sagging moments in a continuous floor slab comprising five simply supported precast prestressed units made continuous using a cast *insitu* infill and site placed rebars at the interior supports

	Ultimate bending moment (kNm/m width)		*Service bending moment (kNm/m width)*	
Load arrangement	*Exterior span*	*Interior span*	*Exterior span*	*Interior span*
All spans	32.4[a]	27.0[a]	23.2	18.0
Alternate	33.4	28.8	24.9	21.5
Ratio of moments	0.97	0.94	0.93	0.84

Note all values include the stage I bending moment for the self-weight of the prestressed unit weighing 2.6 kN/m² with a simply supported span of 5.7 m (i.e. 6.0–0.3m). Service moment=$2.6\times5.7^2/8=10.56$ kNm/m width. Ultimate moment=$1.25\times10.56=13.2$ kNm/m width.

[a] After 20% moment redistribution at supports.

Table 3.4b Maximum hogging (negative) moments in a continuous floor slab

	Ultimate negative bending moment (kNm/m width)	
Load arrangement	*First interior support*	*Second interior support*
All spans	−18.9[a]	−14.2[a]
Two external spans with maximum load	−25.5	−10.7
Ratio of moments	0.74	1.33

[a] After 20% moment redistribution at supports. Stage I moments are zero at the supports. All other details as per Table 3.4a.

with $Q_k/G_k=1.25$) the moments are around 5% greater but ratios are almost the same as for Exp. 6.10b, i.e. 0.97 and 0.95.

For service loads, there is no moment redistribution. Using elastic moments, the results for stage 2 loads of $G_k=2.0$ kN/m² and $Q_k=2.5$ kN/m² are given in Table 3.4a. There are differences of up to 16% using load arrangements (a, without redistribution) and (b). Clearly, 'all spans' cannot be used to establish service moments; alternate spans should be used with $\gamma=1$.

Regarding ultimate negative moments at the first interior support, the two options are (a) and (c) for stage 2 loads alone (there are no stage 1 moments at supports). Following the same example as above the results are given in Table 3.4b. The critical location is at the first interior support. These results show that the 'all spans' option (a) does not establish the obviously greater support negative moment obtained from the 'two-external-span' (max-max-min-max-min) option (c). However, there is no reason why load arrangement (c) cannot also benefit from 20% redistribution, such that $M_{Ed}=0.8\times-25.6=-20.5$ kNm/m width, just 8% less than arrangement (a). Designers will often apply 20% moment redistribution at supports, whichever load arrangement is used.

Furthermore, where a beam or slab is continuous over a support that offers no resistance to rotation, BS EN 1992-1-1, clause 5.3.2.2 (4) allows a reduction in the design ultimate support moment at the cusp of the support according to Exp. 5.9 of

$$\Delta M_{Ed} = F_{Ed,sup}t / 8 \tag{3.11}$$

where

$F_{Ed,sup}$=the design ultimate support reaction
t=width of the support, such as the width of the top of a beam, typically 300–600 mm, or wall

This should not be carried out before end reactions, shear forces and the span moments are calculated, but at the end of the analysis.

Collecting data from load arrangements (b) and (c) Figure 3.5 shows the bending moment envelope in a five-span continuous slab for the maximum moments due to stage 2 loads (solid line) and the total moment (dashed line) including the stage 1 moment due to the self-weight of the slab. The left half of the diagram includes for 20% moment redistribution of stage 2 moments, together with the reduction ΔM_{Ed} based on a beam width t=400 mm. The right half of the diagram is for the elastic moments.

Figure 3.5 shows that the effect of 20% moment redistribution at the supports from load combination (c), i.e. support moments reduced by 20% with a consequential increase in span moments, hardly makes any difference in the outer spans because load combination (b) is dominant. The positive moments before and after redistribution are 33.4 and 34.9 kNm, 4% more. The corresponding moments in the interior spans are 28.8 and 31.3 kNm, 8% more. For this reason, and the explanation about moment redistribution for the positive sagging moments in psc elements given in Section 13.5.3.2, redistribution is not carried out in psc elements.

3.3.2 Service loads for deflections of spans with cantilevers and continuous spans

The purpose of load arrangements is to arrange the unfavourable (maximum) and favourable (minimum) loads on different parts of a structure in order to establish maximum deflections in the main span of slabs with cantilevers, and in continuous spans. The prestressing action P plays a major role, due to the instantaneous upward camber caused by eccentric prestress, and the long-term viscoelastic camber due to creep, where, for hollow core slabs,

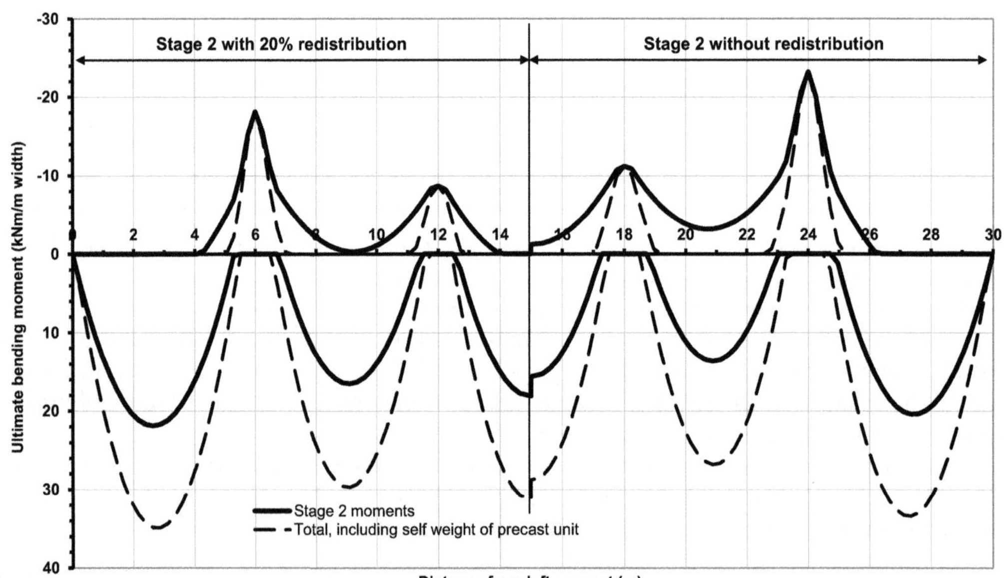

Figure 3.5 Ultimate bending moments envelope in a five-span continuous floor slab subject to load arrangements (b) and (c), with 20% moment redistribution in the left half and without redistribution in the right half. Load arrangement (b), for alternate spans carrying (1) max-min-max-min-max loads, and (2) min-max-min-max-min loads, gives maximum sagging moments. Load arrangement (c), two adjacent spans carrying maximum loads, gives maximum negative hogging moments.

the creep factor from transfer at 1 day to installation (at say 28 days) is ψ_1=0.72 and after installation is ψ_{28}=1.20 (see Section 11.2.1).

The load arrangements for both slabs with cantilevers and continuous spans, commonly known as 'max-min-max', are for stage 2 imposed dead and live loads after the prestressed slab has been positioned in stage 1.

Alternate spans carry $\Sigma G_{k,j}+P+\Sigma\psi_2\, Q_{k,j}$ with other spans carrying the dead load only $\Sigma G_{k,j}+P$.

In slabs with a cantilever(s), the prestressing effect causes upward camber

$$\delta_1 = -P(t)\ z_{cp}L^2(1+\psi)/8E_{cm}I_{c,co} \tag{3.12}$$

where

$P(t)$=prestressing force at the time (t) at which deflections are calculated
z_{cp}=eccentricity from the centroid of the concrete section to the centroid of the force
L = effective span of the main span
ψ=creep factor at time t
E_{cm}=Young's modulus of concrete at 28 days, or at time t<28 days
$I_{c,co}$=compound second moment of area of the section (including transformed area of tendons). See Section 6.2.4 for the calculation of $I_{c,co}$.

The camber has a major effect on the deflection of the cantilever itself due to the end rotation as

$$\theta = P(t)\ z_{cp}L\ (1+\psi)/2E_{cm}I_{c,co} \tag{3.13}$$

If the length of the cantilever is L' the downward deflection at the end of the cantilever due to rigid body rotation is

$$\delta = +\theta L' \tag{3.14}$$

The final deflections are due to the addition of those due to gravity loads $\Sigma G_{k,j}+\Sigma\psi_2\, Q_{k,j}$ or $\Sigma G_{k,j}$ as given above.

For a typical hollow core slab, say 150 mm deep×span of 6.0 m, subjected to G_k=2.6+2=4.6 kN/m^2 and Q_k=5 kN/m^2 (on both main and cantilever spans) the long-term deflection (with ψ_2=0.3 for office loading) of the simply supported slab alone is δ=4.5 mm. The long-term upward camber for this example is δ_1=–27.2 mm, virtually eliminating the effects of gravity loads. If the slab includes a cantilever of length L'=1.0 m, the deflection reduces to δ=3.1 mm. The limiting deflection is given in BS EN 1992-1-1, clause 7.4.1 (4) as $L/250$=24.0 mm.

Similarly, the so-called 'active' deflection (sometimes called 'movement' δ_2) after construction of finishes and partitions, is the greater of (i) ψ_{28} times the dead loads+$\Sigma(1+\psi_{28})\ \psi_2\, Q_{k,j}$+visco-elastic increase in camber after construction, or (ii) $\Sigma(1+\psi_{28})\ \psi_2\, Q_{k,j}$ alone. This is usually greater than the final long-term deflection because they do not include the instantaneous camber. In the same example δ_2=7.3 and 6.5 mm without and with the cantilever. The limiting active deflection is given in BS EN 1992-1-1, clause 7.4.1 (5) as $L/500$=12.0 mm, although other limits such as $L/350$ are used for non-brittle finishes or ceilings. This example shows that deflections in slabs containing cantilevers are not usually critical, at the expense of limiting tension at serviceability, at least up to span/depth ratios of around 40 (at which point the natural frequency f_n<4 Hz may be critical).

In continuous spans, the end moments due to the continuity of adjacent spans subjected to stage 2 dead loads considerably reduce main span deflections. For example, in the example for a five span×6.0 m continuous slab, the simply supported active deflection of one span alone (L=5.7 m) when subjected to G_k=4.6 kN/m^2 and $Q_k=\psi_2$ Q_k=0.3×5=1.5 kN/m^2 is δ_2=5.1 mm. The final stage 2 end moments due to dead and live loads at the first support are M_G=–9.1 kNm and M_Q=–3.4 kNm per 1.2 m wide floor unit, respectively (where G_k=2 kN/m^2 and ψ_2 Q_k=1.5 kN/m^2). In this case the less favourable negative deflection is due to the live load alone $\delta_2=-M\ L^2\ (1+\psi)/16\ E_{cm}\ I_{c,co}$=–1.4 mm giving a net active deflection δ_2=3.7 mm<limit of span/500=12.0 mm. This example also shows that deflections in continuous slabs are not usually critical, at the expense of limiting tension at serviceability, up to span/depth ratios of around 65, much greater than is possible to manufacture.

REFERENCES

BS 8110-1. 1997. The Standard for Structural use of concrete – Part 1: Code of practice for design and construction, BSI, London, UK.

BS 8500-1:2023. Concrete. Complementary British Standard to BS EN 206: Method of specifying and guidance for the specifier, BSI, London, UK.

BS EN 1990. 2002: Eurocode 0- Basis of structural design, +A1:2005, BSI, London, UK.

BS EN 1992-1-1. 2004. Eurocode 2: Design of concrete structures - Part 1-1: General rules and rules for buildings, +A1:2014, BSI, London, UK.

Elliott, K. S., 2017. *Precast Concrete Structures,* 2nd ed., CRC, Taylor & Francis, London, UK, 694 p.

FIB. 2012. *Bulletin 63. Design of Precast Concrete Structures against Accidental Actions, Guide to Good Practice,* Fédération Internationale du Béton, Lausanne, Switzerland, 72 p.

FIB. 2014. Bulletin 74. *Planning and Design Handbook on Precast Building Structures, Manual-Textbook,* Fédération Internationale du Béton, Lausanne, Switzerland, 313 p.

NA to BS EN 1990. 2002. UK National Annex for Eurocode 0- Basis of structural design, +A1:2005, BSI, London, UK.

NA to BS EN 1991-1-2. 2002. UK National Annex for Eurocode 1: Actions on structures - Part 1-2: General actions - Actions on structures exposed to fire, BSI, London, UK.

NA to BS EN 1992-1-1. 2004. UK National Annex for Eurocode 2: Design of concrete structures - Part 1-1: General rules and rules for buildings, +A1:2014, BSI, London, UK.

Turkstra, C. J. 1970. *Theory of Structural Design Decisions,* SM Studies Series No. 2, Solid Mechanics Division, University of Waterloo, Ontatio, Canada.

Chapter 4

Pretensioning and casting methods

4.1 PRETENSIONING AND CASTING METHODS

There are two commonly used methods to manufacture precast prestressed concrete elements (i) wet-cast into moulds for beams, wall panels and floor slabs and (ii) semi-dry slip-forming and extrusion of floor slabs, wall panels, T-beams (for beam and block flooring) and ground beams. Figure 4.1 shows a prestressed beam mould with the strand tensioned and ready for the concrete to be poured, these units are classified as wet-cast. Figure 4.2 shows steel moulds for casting prestressed T beams for beam and block flooring, 12 abreast and typically 80 m long. Figure 4.3 shows a wet-cast floor slab during the production process. The wet-cast concrete has been poured into the steel mould and the top surface is being levelled off and floated. Additional details regarding wet-cast production are given in Section 4.2.3.

Figure 4.1 Prestressed beam moulds with strand in place.

DOI: 10.1201/9781003319450-4

Figure 4.2 Steel moulds for wet casting of T beams (house beams).

Figure 4.3 Wet-cast floor slab being manufactured. (Courtesy of Coote Engineering, N.Ireland.)

The prestressing forces need to be anchored to a bulkhead capable of dealing with the very high tension loading, but the moulds themselves can be constructed from steel, timber, cardboard, and in some cases, the base of the mould is formed with concrete.

Long line production uses a semi-dry mix, and the methods are often classified as dry-cast. This method is suitable for linear products such as slabs, walls and T beams. Figure 4.4 shows the long beds that are used to cast the product in a prestressed hcu factory. Additional details on dry-cast production are given in Section 4.2.1.

In precast prestressed production, there are two widely used methods of applying tension to the steel strands before casting the concrete. These will be described in the following two sections.

4.1.1 Method 1 – individual strand tensioning

The first method for pretensioning is the use of hand-operated hydraulic stressing jacks where each strand is tensioned individually to the required load, as shown in Figure 4.5. The strands are anchored through a bulk head at each end, one end being the jacking point.

The stressing jack can be connected to a calibrated control unit with a digital display to show the load applied to the strand during pretensioning. Alternatively, the strand is lightly tensioned to engage the barrel and wedges, a line is marked on the strand close to the end at the jacking point and the strand is then extended by a calculated length to provide the desired pretension force. Figure 4.5 shows a hydraulic stressing jack in use tensioning individual strands for hollow core unit (hcu) production. This jack is connected to a calibrated unit, so the operator applies the required tension to each strand by reading off the digital display. The display is mounted out of shot behind the hydraulic jack. The tension required for each strand type used in design is specified for the production bed. The prestressed

Figure 4.4 Prestressed hollow core production facility.

Figure 4.5 Hydraulic stressing jack.

hollow core factory shown has 13 production beds of lengths up to 227 m, each bed can be set up with a bespoke strand pattern using a variety of quantities and diameters of wire and/or strand. Slabs and walls of different thicknesses can be manufactured on each bed.

Figure 4.6 shows the barrel and wedge connectors that anchor each individual strand at the bulkhead during pretensioning. These are critical safety features in prestressed production, and they are used widely in virtually every factory that manufactures prestressed concrete elements.

4.1.2 Method 2 – simultaneous strand tensioning

The second method for pretensioning uses a sliding bulkhead and hydraulic rams, where all of the strands are anchored through the bulkhead and the rams are extended to tension all strands simultaneously.

The required extension of the rams is calculated, and the subsequent movement of the sliding bulkhead applies the desired pretension force to the strands. For slabs and wall panels, typical sliding bulkhead and ram prestressing systems can operate at total pretension forces of up to 325 tonnes. It is also possible to use stressing jacks on individual stands in conjunction with a sliding bulkhead.

In both of the pretensioning methods, care must be taken not to overload the capacity of the bulkhead or casting bed when calculating ram extensions and/or using stressing jacks.

The bulkhead shown at the end of the prestressed concrete mould in Figure 4.7 has holes drilled through it to provide the capacity to receive 2 rows of 24 no. 9.3 mm diameter strand, with each strand stressed to an average of 60 kN giving a total pretension force of 2880 kN or approximately 294 tonnes. This bed, bulkhead and rams are rated to carry a total of load up to 300 tonnes. The bed is 85 m long, can cast units at a maximum width of 2400 mm and is used to produce solid prestressed slabs and wall panels.

Figure 4.6 Strand anchored through fixed bulkhead using barrel and wedges.

Figure 4.7 Sliding bulkhead at end of the casting bed.

The bulkhead is moved using hydraulic rams positioned at the outer left and right sides of the mould. The required load is applied by calculating the extension required in the strand based on the overall length of the mould. At the opposite end of the mould, there is a fixed bulkhead with the same hole arrangement through it. Figures 4.8 and 4.9 show

Figure 4.8 Hydraulic ram used to move sliding bulkhead and apply tension to strands. (Courtesy of Coote Engineering, N. Ireland.)

Figure 4.9 Strand anchored through sliding bulkhead using barrel and wedges. (Courtesy of Coote Engineering, N. Ireland.)

one of the hydraulic rams and the steel frame structure that it is mounted within on the side of the mould.

In theory, any length of mould is possible – when using rams the limit for mould length is governed by the maximum extension of the hydraulic rams and the movement of the sliding bulkhead as this will limit the amount of prestress that can be applied. Of course, factory size and production requirements also influence the length of mould required, and a number of mould manufacturers provide a modular system to allow variance in mould length to suit the changing requirements of production.

4.1.3 Pretensioning strand based on extension

The following example shows how to calculate the extension required to pretension the steel strand or wire in relation to the length of the mould or bed used for production.

The amount of prestress in the concrete is calculated for the slab, wall, beam or other unit that is being designed. The strand pattern necessary to provide the required prestress is defined based on a percentage of the strands breaking load, known as the characteristic ultimate tensile strength f_{pk} (see Section 5.1.1 and Table 5.8), and this is generally restricted to a maximum of 70%. Some manufacturers of T beams pretension 5 mm and 7 mm diameter wires to 75% f_{pk}.

For example, a 1.2 m wide slab requires 10 no. 9.3 mm diameter strands stressed to give a total prestress force of 620 kN, therefore 62kN per strand. The mould is 85 m long.

Young's modulus = stress/strain
Stress = Load in strand (kN)/strand area (mm^2)
Strain = extension of strand/mould length
Young's modulus of strand $E_p = 195000$ N/mm^2
Area of 9.3 mm diameter, 7 wire strand $A_p = 52$ mm^2
Yield strength $f_{pk} = 1860$ N/mm^2
Length of mould $L = 85000$ mm
Pretension required per strand $F_p = 62$ kN
Strand breaking load $F_{pk} = A_p\, f_{pk} = 96.7$ kN
Percentage of tension capacity $\eta = F_p/F_{pk} = 64\% < 70\%$ OK
Extension required $= [(F_p/A_p)/E_p]\, L = 520$ mm

Note that the type of stressing jack used with the strand has an influence on the amount of extension and needs to be considered. A keyed jack prevents the strand from rotating, so it behaves as a solid wire. If a jack with no key is used the strand can rotate during stressing which will result in up to 5% greater extensions being required.

4.1.4 Anchorage of strand and wire during pretensioning

In both of the methods described above, the strands are restrained at the fixed or sliding bulkhead using barrel and wedge anchors. Figure 4.10. The safe anchorage of pretensioned steel strand is a very important consideration in the manufacturing process, and subsequently, strict control procedures must be adhered to when cleaning, inspecting and using barrel and wedge anchors. These two components are used to grip and hold the strand during and after tensioning. The barrel is fitted over the strand and has a tapered internal profile to receive the wedge. The wedge is fitted over the strand and has an external taper in the same profile as the barrel, but internally it has serrations to grip the strand. As the tension force is applied to the strand, the wedge is drawn into the barrel and compressed due to

Figure 4.10 Barrels and wedges.

the tapered profile. An increase in tension will increase the grip on the strand and prevent slippage as the wedge locks tighter in relation to the applied force. The specific type of barrel and wedge used depends on the type and diameter of the strand or wire being tensioned.

The most common types of strand used in prestressed manufacture in the UK and Ireland are the seven wire profile with diameters of 9.3 mm, 12.5 and 15.7 mm, the most common types of wire used are those with diameters 5 mm and 7 mm. Some European countries favour 9.6 and 12.9 mm diameter strands. Three wire strand of 8 mm diameter is sometimes used in South America. Seven wire strands comprise a central straight wire with six outer wires twisted into a spiral (with a pitch of 200 mm) to help increase bond stress as the cement paste fills the crevices between the outer wires. Hence, the name 'helical strand'. Table 4.1 presents some commercial literature supplied by manufacturers (see also Table 5.8). Wire is available with a smooth or indented surface profile, the profile affects the bond characteristic of the wire and therefore the amount of pretension that can be applied.

4.2 MANUFACTURE OF PRESTRESSED CONCRETE ELEMENTS

The manufacture of prestressed precast units can be categorised into two separate methodologies based on the properties of the concrete mix used – the methods are described in the industry as dry-cast and wet-cast. The concrete mix type dictates the casting method that can be used. A dry mix is used for the production of long line products such as hcu and solid planks. A wet mix is used to cast for example solid slabs, walls, beams and lintels.

4.2.1 Dry-cast concrete – S1 slump class

In simple terms, the wetter the concrete, the higher the slump will be. The slump range is often associated with water content, but it is more accurately defined as a measurement of

Table 4.1 Prestressed strand and wire data

Common mechanical characteristics	
Modulus of elasticity:	≈195 GPa +–10%
Elongation	≥3.5% w L>500 m
Very low relaxation:	≤2.5% after 1000 hours stressed at 70% of breaking load

7 wire helical strands: Tensile strength 1860 MPa									
Nominal diameter		*Nominal cross-sectional area*		*Nominal weight per 1000 m*		*Breaking load F_{pk}* ≥		*Yield strength $F_{p0,1}$* ≥	
inch	*mm*	*mm²*	*inch²*	*kg*	*lbs*	*kN*	*1000 lbs*	*kN*	*1000 lbs*
1/4	6.9	29.0	0.045	226.5	499	53.9	12.12	47.4	10.66
3/8	9.3	52.0	0.081	406,1	895	96,7	21,75	85,1	19,14
3/8	9,6	55,0	0,085	429,6	946	102	22,93	89,8	20,00
4/9	11,0	70,0	0,108	546,7	1205	130	29,24	114	25,64
1/2	12,5	93,0	0,144	726,3	1601	173	38,91	152	34,19
1/2	12,9	100	0,155	781,0	1722	186	41,84	164	36,89
0,6	15,3	140	0,217	1093	2410	260	58,48	229	51,51
0,62	15,7	150	0,232	1172	2584	279	62,76	246	55,34

Single wire. Resistance 1770 MPa									
Nominal diameter		*Nominal cross-sectional area*		*Nominal weight*		*Breaking load F_{pk}* ≥		*Yield strength $F_{p0,1}$* ≥	
inch	*mm*	*mm²*	*inches²*	*kg/m*	*lbs/m*	*kN*	*1000 lbs*	*kN*	*1000 lbs*
0.157	4	12,6	0,019	0,098	0,216	22,3	5,01	19,6	4,40
0.197	5	19,6	0,030	0,153	0,337	34,7	7,81	30,5	6,86
0.236	6	28,3	0,044	0,221	0,487	50,1	11,27	44,1	9,92
0.276	7	38,5	0,060	0,301	0,663	68,1	15,32	59,9	13,47

$F_{p0.1}$ is the 0.1% proof load, see Section 5.3.1 and Figure 5.22.

F_{pk} is the characteristic ultimate tensile load.

the concretes consistence. The workability or consistency of the concrete is given by the slump class, an S1 mix is a dry mix with a slump range of 10–40 mm, further information is given in BS 8500-1:2023.

A long line bed is used where the casting machine moves along the full length of the bed and the concrete is delivered from the batching plant into a hopper on the continuously moving casting machine. Figure 4.4. The geometry of the unit is governed by the set-up and shape of the casting machine, e.g. 200 mm depth, 1200 mm wide, 11 no. hollow cores in a hcu. These are manufactured as one unit the entire length of the bed, typically 120–150 m. Individual units are then cut from the continuous unit once the concrete has cured sufficiently. There are limitations on what shapes can be cast, essentially a continuous basic rectangular profile is formed, and any details such as notches, holes, splays and narrow width units need to be formed or cut post-casting. The beds are typically constructed on the ground and sheeted with steel. Heating pipework is installed under the beds to aid rapid curing so that the precast elements can be removed in as short a time period as possible.

Curing times for hcu vary depending on the geometry of the unit, the type of machine used to produce it and the ambient conditions. Typically, 16–20 hours are sufficient to achieve adequate strength for transfer of prestress to the concrete. Note that there are manufacturers of extrusion machines that suggest workable mixes can be used with their machines that will give curing time to adequate transfer strength in under 6 hours. Prestressed factories have testing facilities and cubes are crushed prior to detensioning, concrete strength at transfer, $f_{ck}(t)$, for typical levels of pretressing is between C25/30 and C30/40. The cast units can also be covered by hessian sheeting while on the beds during curing to retain heat and assist rapid curing times. The most common bed widths are 1200 and 2400 mm. These widths fit conveniently onto the trailer of a lorry making for straightforward and efficient transportation to site. Lifting pins can be installed during the manufacturing process, and slabs can be lifted from the production bed using the cast-in lifting pins or with the use of clamps. The main advantage of this form of manufacturing is the efficiency of production with minimal waste and low labour costs.

Top wire or strand can be placed in the slabs to assist with handling stresses during the lifting and installation of slabs. These may be pretensioned to 70% f_{pk} but, as this may reduce the flexural bending strength of the unit, they are often pretensioned to 40% f_{pk}. Another feature of using a top strand is that it can be designed to act as the cantilever reinforcement in slabs. Also the units can be manufactured with the same strand pattern at the top and bottom of the section, for example wall panels that are used as retaining walls that will be subjected to loading on both sides. The use of the same strand pattern at the top and bottom of the section is not that common in dry-cast manufacture, most often units that require this detail as part of the design are manufactured using a wet-cast method.

It is possible to manufacture units with pipework cast into the bottom flange in long line dry-cast production; this pipework can be used as part of the heating and cooling system for the completed building. The height of the hollow core profiles and their position vertically on the slab section are adjusted to give a thicker bottom flange that can accommodate the pipework during the production process. A hcu has a 35 mm bottom flange thickness, but a bespoke profile was manufactured as shown in Figure 4.11 where the bottom flange was increased to 90 mm to accommodate the pipework.

Two main machine types in long line hollow core slab production are slipformers and extruders. Slipforming machines first place the concrete to form the bottom flange of the slab and then the cores 'slip' over the concrete to compact it around the bottom strand and form the lower section of the ribs. Concrete is then placed over the cores to form the upper section of the ribs and the top flange of the slab. A vibrating plate compacts the rib and top flange concrete. Extruder machines use augers or screws to force the concrete through a profile former to produce a compacted full-depth section.

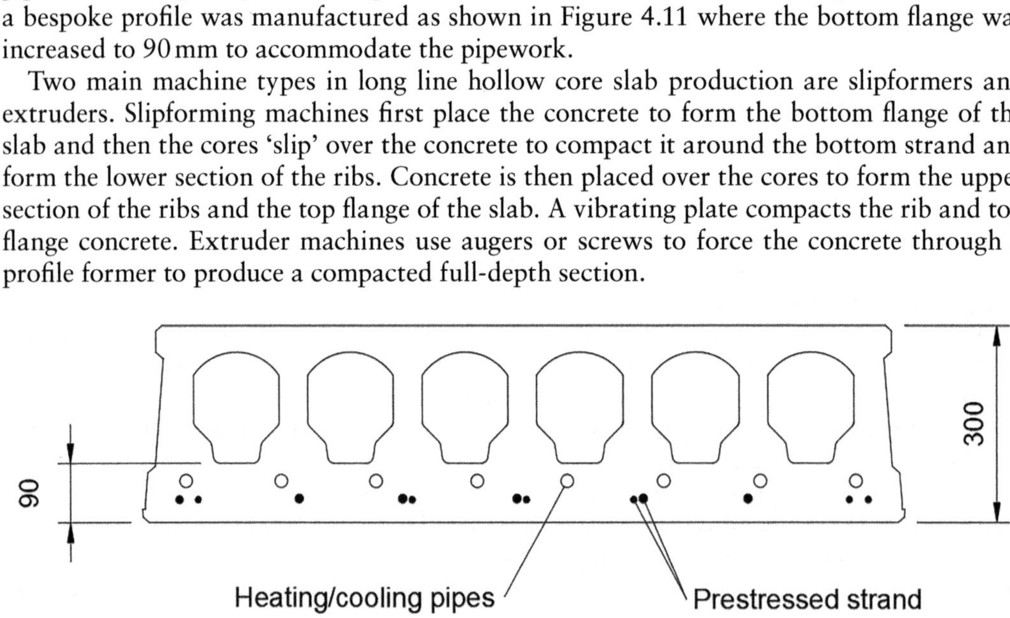

Figure 4.11 Hollow core unit cross section containing heating and cooling pipes.

The choice between slipformer and extrusion machines can be influenced by a number of factors. Both types of machines can produce slabs and walls of various profiles and thicknesses with the slipfomer machine being the least expensive to purchase. The cost of the main components used by each machine also varies, so the profile range that the manufacturer wants to produce will have an effect on set-up cost. For example, the cores used to form each profile on a slipformer machine are less expensive than the augers used in extrusion machines. Operationally, slipformer machines are quick to set up for different slab or wall profiles, therefore reducing downtime on the production line. A useful advantage of slipformer machines is the relative ease of modifying the cores to create a bespoke slab profile, extruder machines will require a bespoke set of augers if there is a requirement to create a modified profile. Maintenance and running costs for extrusion machines are higher than for slipformer machines due to the cost associated with the components. Increased operational costs for extrusion machines are caused by wear on the augers due to the coarse dry mix used to produce hcu. There are also positive points associated with extrusion machines; the finish of hcu is generally much more aesthetically pleasing than that of a slipformed element. Extruded elements are typically more dimensionally accurate with a consistent uniform compaction of the concrete throughout the full depth of the section. The development of hydraulic extrusion machines has resolved issues such as wear on the augers and slow machine change over to different slab or wall profiles, the different extrusion methods used in these machines has helped to reduce the operational costs.

Both types of machine run continuously along the bed on side rails and are fitted with a hopper to receive concrete from the batching plant. Once the concrete has reached the required strength, the strands are detensioned by cutting or relaxing the bulkhead and the prestress is 'transferred' into the concrete. This transfer of stress occurs as each individual unit is cut in turn from the continuous unit that has been cast on the long line bed.

The key equipment that makes up a prestressed dry-casting factory is

1. Long line heated steel sheeted beds
2. Batching plant and concrete delivery system
3. Rail-mounted casting machine – slipformer or extruder
4. Rail-mounted bed cleaning and oiling machine
5. Stressing jacks or rams and fixing bulkheads at each end of the bed
6. Rail-mounted concrete saws
7. Vacuum aspirator
8. Bed plotter
9. Crane for transferring precast units to trailers

The first step in the process for the manufacture of prestressed long line units is to prepare the bed. The bed is cleaned with a rotating brush to remove any residue or debris from the previous production run, and a rubber blade follows the brush sliding over the surface to remove water from the bed. Release oil is then applied using rear-mounted spray nozzles to evenly coat the bed. The same machine is used to pull the strands down the bed ready for anchoring and pretensioning. Figure 4.12 shows a typical rail-mounted machine used to prepare the casting bed.

The unit depth and section profile required are set up in the casting machine and once the strand has been pretensioned the casting machine is craned into position on the rails. Figure 4.13 shows an 11-core cassette for a slipforming machine; various shapes, sizes and numbers of cores are used to create the profiles that are available from the different manufacturers of the casting machines. In extrusion machines, the slipformers shown in Figure 4.13 are replaced by rotating augers and circular core formers connected to the ends of the augers,

Figure 4.12 Rail-mounted cleaning and oiling machine.

Figure 4.13 11-core cassette for a slipforming casting machine.

as shown in Figures 4.14 and 4.15. Adjacent augers rotate in opposite directions to increase the compaction of concrete, which has a greater density and quality of finish compared with slipformers.

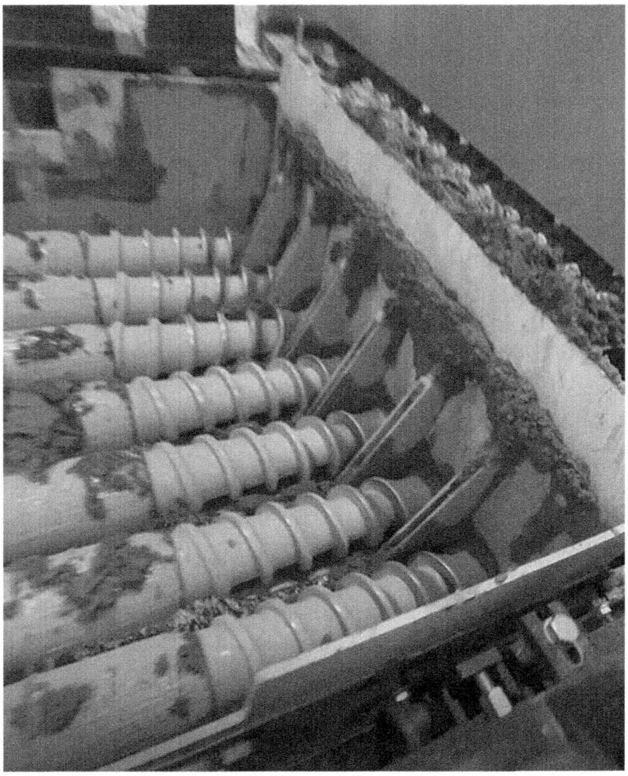

Figure 4.14 Extrusion machine augers and core formers.

Figure 4.15 Extrusion machine core formers, note that the machine is raised up and supported on a steel frame.

Solid slabs and walls can also be manufactured using long line casting machines. The concrete is delivered to the hopper on the casting machine as it moves continuously along the bed, as shown in Figure 4.16.

Once the production bed has been cast, or while it is being cast, a laser-guided rail-mounted plotter moves along the bed and marks up details on the surface of the concrete, as shown in Figure 4.17. This includes lines to indicate the cutting position for slab length, narrow width cuts, splayed ends, holes, notches, open cores, lifting pin positions, solid ends, chamfered ends and any other details that are required on the slab. Solid ends are formed by infilling the cores at the end of the slab by the specified distance required on the project. The infill concrete is typically added manually while the slabs are on the production bed. Chamfered ends are formed in the uncured concrete while the slabs are on the production bed; this is done using either an angled steel plate mounted on a machine or manually. The dimensions for the chamfer can be a bespoke arrangement if agreed upon with the precast manufacturer, but they are typically a reduction of approx. 75 mm over a length of 250 mm.

An example of the lines drawn by the plotter can be seen in Figure 4.18. The cut line for the slab length can be seen running across the width of the slab, this is also a narrow width slab so a longitudinal line is visible. The position of the cast in lifting pins is marked, although the marks are not visible as the vacuum aspirator has removed the top flange concrete from four locations and is removing it at the fifth one. The spherical head lifting pins are installed manually into the slabs by operatives following a strictly controlled process to ensure adequate anchorage within the concrete. These pins can be used to lift the slabs off the production bed, handle them in the stockyard and also fit the slabs on site. Figure 4.19 shows the finished lifting pin installation in a hcu; typically 4 no. pins are installed in a standard width unit at approximately one metre from the end of the unit.

Figure 4.16 Concrete being delivered to the hopper on a slipformer.

Figure 4.17 Plotter mounted on rails, guided by laser to mark up the slab details. Vacuum aspirator machine can be seen behind the plotter.

Figure 4.18 Detail being formed in slab using vacuum aspirator, in this case a void for locating lifting pins.

Figure 4.19 Lifting pins installed in hollow core unit.

Holes and notches are formed in the slabs or walls after casting using a vacuum aspirator to remove the concrete. Any strand that is exposed is burnt out with a gas torch, for large openings the strand is often left intact and removed on site once the unit is installed. This helps prevent cracking at the opening during handling and transportation. Typically, openings with a length and width between 150 and 300 mm can be formed in the factory during production. See Figures 7.14–7.17. Larger openings up to 400 mm in size are normally cored on site, but it is important to ensure that all holes are accounted for in the design. The slab drawings for a project must show all hole sizes and positions that will affect the structural performance of the slab. For small holes, it is normal practice to agree on a safe drilling zone within the slab that ensures no strands are cut or damaged and the effect on the slab's performance is negligible.

Top open cores (often known as slots) are a very common detail in hcus. These are used to locate horizontal tie reinforcement bars on site that form part of the robustness method in precast construction. Another purpose is to allow the placement of transverse reinforcement required in composite steel beam design. The top flange of the hcu is removed during production, the two outside cores are not usually opened as the web will be vulnerable to damage. It is possible to open every other core on a unit, for example an 11 core unit could have 5 cores opened making it particularly well suited for use in composite steel beam design. A 6 core unit could have 2 cores opened as shown in Figure 4.20. If required, additional adjacent cores could be cut on site although this is rarely required.

The chamfered end detail is useful when installing precast slabs on the bottom flange of steel beams. Figure 4.21 shows the chamfered end slab installed on site; in this case, the 200 mm depth slab bears onto the bottom flange of a structural steelwork 203×203 mm serial size Universal column. The beam can have holes drilled through its web to allow

Figure 4.20 A 500 mm deep hollow core unit with 2 no. top opened cores and lifting pins installed.

Figure 4.21 Prestressed hollow core unit with chamfered end

horizontal reinforcement bars to pass through the beam and provide the necessary robustness ties. The bars are positioned in the open cores of the slabs, which are co-ordinated with the steel beam fabrication and hole positions. This detail can avoid the requirement for a steel plate to be welded to the bottom of the beam. The slab length is detailed to ensure adequate bearing can be achieved, and the slab can be fitted into position between two beams if required.

With all holes, notches, top opened cores, chamfers, etc. formed according to the lines made by the plotter, the full length casting is cut to the lengths required by shop drawings. Figure 4.22 shows a rail-mounted transverse saw used to cut units to length on the production bed. The accuracy of the length of individual units is typically ±25 mm, but this tolerance can be reduced subject to agreement with the manufacturer. Occasionally, there are issues when saw cutting hcu to length on the production bed, in units manufactured with a high level of pretension the camber induced in the slab during the cutting process can cause the saw blade to become jammed. This can cause cracking to occur in the slab as well as being a potential health and safety risk. To prevent this problem from occurring, weight packs can be loaded onto the slabs during cutting. This reduces the camber during the cutting process and prevents the top flanges of the slab from pinching together to jam the saw.

In situations where the supporting structure of beams or walls is not on a rectangular grid, the end of the unit is cut at an angle. Known as splayed end units, as shown in Figure 4.23. The limitation in the angle of the cut is between 30° and 90° to the longitudinal axis of the unit.

4.2.2 Narrow width units in long line dry-cast production

There are limitations on the width of units that can be cut from the standard width hcu. Non-standard widths are commonly termed 'rip units' by precast manufacturers. They can be cut while the slab is on the production bed, as shown in Figure 4.24 before the concrete

Figure 4.22 Rail-mounted saw used to cut slabs to length on the production bed.

Figure 4.23 Splay cut hollow core unit.

Figure 4.24 Rail-mounted saw cutting rip units on the production bed.

has cured or afterwards in the stockyard once cured prior to delivery to site. The core profile dictates what rip unit width can be provided. Figure 4.25 shows the typical widths available for 11 core and 6 core slab profiles. The limitation for solid planks is the rip line should not be closer than 1.5 × the diameter of the pretensioned strand or wire when the centre-to-centre distance of adjacent strands is greater than 3 × the strand diameter. This limitation also applies to hcu. Clause 4.3.1.2.2 of BS EN 1168 gives further details regarding minimum cover dependent on spacing and strand diameter. If the minimum cover requirements are not satisfied, the strand can be discounted when assessing the capacity of the slab at the design stage to ensure there are no issues.

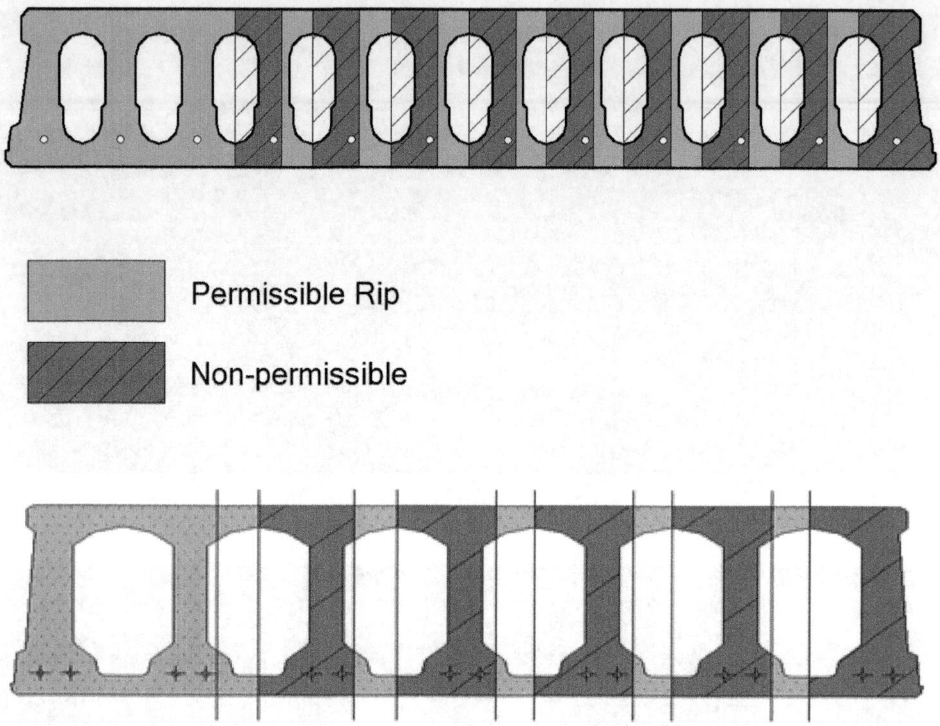

Figure 4.25 Permissible cutting positions for rip units for a standard width hollow core unit.

Rip units should be used such that waste is minimised. It is important to remember that when a rip is specified, a corresponding rip of the same length is produced on the other side of the bed, e.g. for a standard 1200 mm width unit if a 700 mm rip is detailed a 500 mm rip is left on the other side of the bed. It follows that if a number of 700 mm rips are required on a particular contract, a number of 500 mm rips are potentially wasted unless they can be utilised elsewhere or on another project.

Finally, the units are lifted from the casting bed using clamps, as shown in Figure 4.26. The sideways pressure exerted on the outer web is another reason why top opened cores are not formed in the external cores.

Summary of the key process during casting prior to concrete curing

1. Plotter marks lengths, widths, lifter points, open cores, etc.
2. Vacuum aspirator follows plotter creating holes, open cores, etc.
3. Operatives install lifting pins
4. Operatives from chamfered ends, solid ends and other details required
5. Narrow width units are cut with the rail-mounted saw

Concrete cures for the time required to reach transfer strength

1. Cubes are crushed to check the strength of concrete at transfer
2. Unit is cut to length at plotted lines once the concrete transfer strength is adequate
3. Units are lifted by pins or clamps and stacked on trailers in sequence

Figure 4.26 Lifting slab off production bed using a clamp.

4.2.3 Wet-cast concrete – S3 to S4 slump class

An S3 slump class is a wet mix with high flowability and a slump range of 100–150 mm, S4 is a wet mix with very high flowability and has a high slump range of 160–210 mm, further information is given in BS 8500-1:2023.

The more common and familiar wet-cast mix is used for casting into moulds. The moulds are constructed to retain the wet concrete and form the geometry of the unit. Greater flexibility of geometry is possible with this method although some limitations exist due to the practicalities of stressing the strand in a straight line between bulk heads. Slabs (solid and voided), walls, T beams (for beam & block flooring), main beams, inverted T-beams and lintels are common prestressed products manufactured using a wet mix. The strands are fixed and tensioned along the length of the bed or mould. Holes and notches can be formed during casting by adding formwork, or frequently polystyrene block-outs (held in place against buoyancy). Cast in lifting anchors, sockets, channels, etc. can also be cast in using this method for production. The units can be cast as one long unit similar to the dry mix method but it is more common for stop ends to be used to cast the units to the desired length. The strand is cut or burnt out between stop ends once the concrete has cured. Beds can be vibrated and/or concrete pokered to ensure satisfactory compaction and flow throughout the mould extents. Self-compacting concrete (SCC) can be used in the production of precast elements. This is a flowing concrete mix that does not require the use of vibration to ensure compaction and release of air. The slump flow rate is typically 750 mm. The mix contains superplasticisers, stabilisers and a high binder content. The costs associated with the materials used in the mix can be offset by the reduction in labour for vibration/pokering and finshing works.

The moulds can be constructed at ground level, in the same manner as the dry mix long line method, or constructed at a raised level from steel frames that are sheeted and installed

Figure 4.27 Raised steel bed with adjustable width side formers. (Courtesy of Coote Engineering, N. Ireland.)

into factories at the length required, as shown in Figure 4.27. This gives the advantage of being able to disassemble and move the casting beds to different factories or sites.

Steel is typically used to form stop ends, as shown in Figures 4.28 and 4.29. Once the concrete has reached the required strength, the strands are cut at each stop end, which transfers the pretensioning force into the concrete sections.

Box-outs for holes and notches are typically formed using timber. Steel formers are used for these details only if there is very high repetition due to the cost of this type of formwork. The use of mixes with a relatively high slump aids the concrete flow around box outs and other details with minimal vibration required.

Narrow width slabs and walls can be formed by placing formwork to suit, but it should be noted that this can be wasteful as the strand pattern will run the full length and width of the bed or mould.

Wet-cast manufacturers can also easily accommodate additional standard reinforcement bars if required, such as shear stirrups, and bars close to supports. The bars can be tied in and set in the mould with the strand prior to casting the concrete.

A feature of wet-cast production that is not practical in dry-cast production is the capability to debond (or less frequently to deflect) a number of strands at the ends of the units by sheathing them with polymer pipe for the required length, as shown in Figure 4.30. Debonding deals with the high stresses at the transfer of prestress into the concrete and prevents splitting failure into units that have a relatively concentrated strand pattern within the section. This is common practice in the design and manufacture of prestressed double-tee floor units and main beams. The length of the debonding ensures all of the strand will have full anchorage for the extent required to provide the necessary moment capacity and deflection and crack control. Any issues caused by reduction in section capacity due to debonding are normally resolved by fixing standard reinforcement in the affected zones.

Figure 4.28 The strands are pretensioned and the mould is ready for the concrete to be poured. (Courtesy of Coote Engineering, N.Ireland.)

Figure 4.29 Steel stop ends for wall panels in a wet-cast mould. (Courtesy of Coote Engineering, N.Ireland.)

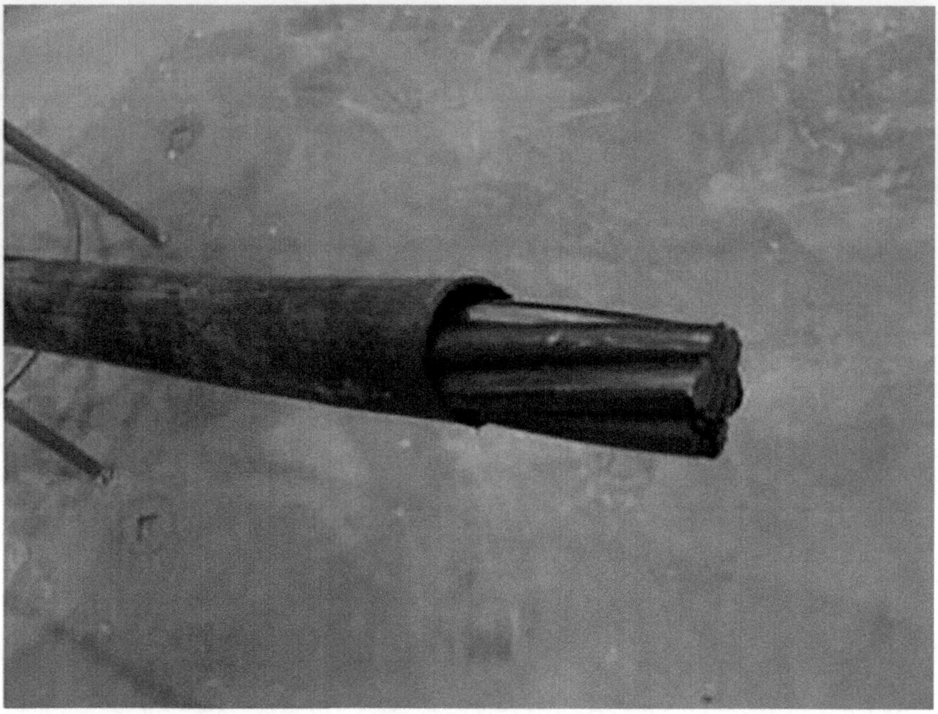

Figure 4.30 Strand sheathed with polymer pipe to debond it from concrete.

Summary of the process for wet-cast production of prestressed units

1. Stop ends set in the mould at the required positions
2. Strand pattern pulled along the bed, anchored at each bulkhead and pretensioned
3. Box outs for holes, notches, etc., fixed into the mould
4. Side shutters fixed into the mould for narrow width units
5. Any additional reinforcement is fixed in the mould
6. Any lifting pins or cast-in items required fixed in the mould
7. Concrete is poured

Concrete cures for the time required to reach transfer strength

1. Cubes are crushed to check concrete strength
2. Strand is cut at stop ends once concrete strength is adequate
3. Slabs are lifted by pins or clamps and stacked on trailers in sequence

Cantilevered units

Cantilevers are unsupported ends of units, such as the slab examples shown in Figure 4.31. The negative hogging moment over the support creates stress at the top of the concrete section. To deal with this stress some form of reinforcement is required at the top of the units.

Prestressed units can be designed and manufactured with reinforcement on the top surface using two common methods.

Figure 4.31 Cantilevered prestressed hollow core slabs.

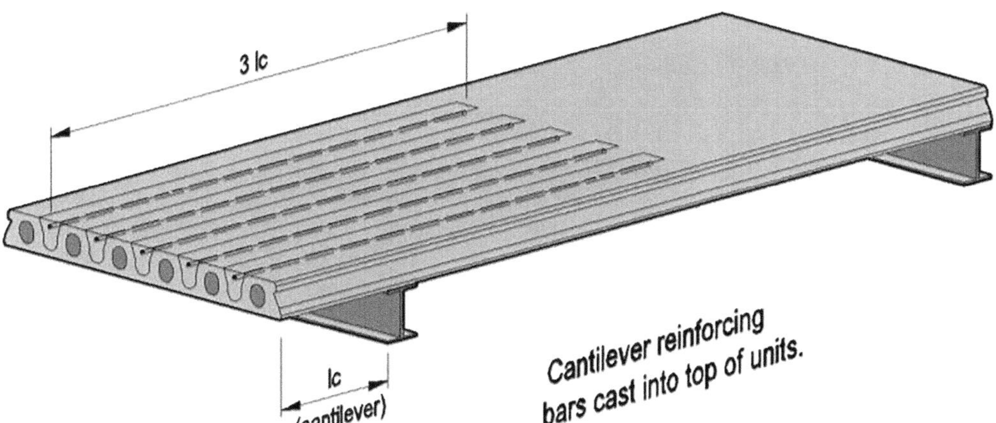

Figure 4.32 Cantilever reinforcement bars added to dry-cast hollow core units during manufacture.

The first method is specifically used for dry-cast hcu. In this method, cores are manually opened in the factory immediately following casting, and reinforcement bars are placed in the opened cores and concreted into place. See Figure 4.32 showing the arrangement. An 11-core hcu is very well suited to this method as it allows reinforcing bars to be placed at 200 mm centres in 5 no. opened cores. The reinforcing bars need to extend into the back-span for the distance required to deal with the hogging moment, usually at least 2 × the cantilever but must be checked during design. This method is used if only a small number of units are required and for relatively short cantilevers typically up to 1200 mm long for 150–200 mm deep units. A major drawback of this method is the increase in self-weight due to the infilling of the cores.

If a large number of cantilever units are required and the cantilever span is greater than about 1200 mm, the units are designed and manufactured with the top strand. This is much more economical for hcu as no manual work is required and the design is more efficient due to the self-weight of the unit being unaffected. The strand pattern in the slab is designed for the cantilever moment and the back-span sagging moment, shear and deflection. The strands required in the top and bottom flanges of the unit are tensioned to the required prestressing forces prior to casting the concrete. This method is suitable for dry-cast and wet-cast production. Figure 4.31 shows a 6-core prestressed cantilever hollow core slab, which was dry-cast using a slipforming machine. The effective cantilever span is 3.45 m, using a bespoke 285 mm depth unit. Refer to Chapter 13 for further information on cantilever slabs.

Chapter 5

Materials, durability and fire resistance

5.1 MATERIAL PROPERTIES FOR PRESTRESSED CONCRETE ELEMENTS

5.1.1 Introduction to concrete and tendons used in prestressed concrete

Prestressed concrete (psc) is high-strength concrete, inherently so because of the compressive strength required at 'transfer', the point of detensioning the tendons when the pretensioning force in transferred to the concrete by the action of bond. It is also high-performance concrete in terms of its resistance to the passage of moisture, low porosity and surface hardness in the soffit and sides of elements owing to the use of steel moulds or timber moulds with medium density overlay, such as WISA®-Form.

High strength is particularly important near to the ends of members where there is a rapid increase in stress transfer in the transmission zone. Compressive strength is usually obtained from 100 mm cubes (rather than 150×300 mm long cylinders) for convenience, where the cube strength from three to six samples $f_{ck,cube}(t)$ is obtained daily, depending on the volume of production. Note that (*t*) refers to time, less than 28 days, rather than to transfer. Typically, at an age of 20 hours, $f_{cm,cube}=35-45$ N/mm^2 for corresponding 28 days design cylinder strengths of $f_{ck}=40-55$ N/mm^2 (or $f_{ck,cube}=50-65$ N/mm^2). In some product standards, such as BS EN 15037-1 ((BS EN 15037-1, 2008) for beam and block flooring, the first specimen compressive strength prior to release f_{c1} should be 5 N/mm^2 greater than the required, called $f_{min,p}$; otherwise, a waiting period is required until the second $f_{c2} \geq f_{min,p}+5$, and so on. The margin of 5 N/mm^2 represents the anticipated standard deviation (sd) when testing several specimens, *viz.* $f_{ck}(t)=f_{min,p}$ -1.64 sd. If cubes are used $f_{ck,cube}(t)$ is converted to $f_{ck}(t)$ using the range of values given in BS EN 1992-1-1, Table 3.1 (BS EN 1992-1-1, 2004). For the range of strengths used in psc elements, the conversion of cube to cylinder strength is approximately $f_{ck}/f_{ck,cube}=0.83$ for higher strengths ($f_{ck}>45$); otherwise, it is 0.78–0.80. Although BS EN 1992-1-1, Table 3.1 states that the characteristic value f_{ck}=mean strength f_{cm} −8 N/mm^2, this is not respected for the transfer strength.

The behaviour of psc elements, throughout their full life cycle, depends on more than just the compressive strength of the concrete. It is surprising how much of the deformation characteristics of psc slabs and beams, as well as reinforced concrete columns and walls, etc., rely on the humble 'cube strength'. Only in research (and by a manufacturer when a new product is introduced, e.g. high strength or self-compacting concrete) is the true uni-axial compressive cylinder strength, tensile strength, bond strength, splitting strength, Young's modulus of elasticity, shrinkage, creep coefficient, flow, and so on, truly measured. In design, the ultimate strength of elements in bending, compression, tension, shear and torsion, as well as checks for serviceability stresses, short-term and time-dependent long-term deflections and

DOI: 10.1201/9781003319450-5

crack widths, fatigue, etc., are all required. The long-term effects are particularly important to psc where losses of prestress, due to elastic shortening, creep and shrinkage, (as well as relaxation of the steel tendons, see below) are required to predict the stresses, deflections and cracking at the serviceability limit state.

Similarly, the ultimate and service strength of steel reinforcement and pretensioned tendons (wires, strands and bars), as well as their modulus of elasticity, relaxation of pretensioning force, ductility and fatigue behaviour, together with compatibility of strains when bonded to concrete, are all required to determine the structural behaviour of psc. Relaxation of tendons is quoted at 1000 hours after tensioning at a mean temperature of 20°C as ρ_{1000}. Here the characteristic ultimate tensile strength f_{pk} is the basis for specifications and design, although BS EN 1992-1-1, clause 3.3.2 (1) states tendons should be classified according to the characteristic 0.1% proof stress in axial tension $f_{p0.1k}$ and the ratio of $f_{pk}/f_{p0.1k}$, together with the elongation strain at maximum load ε_{uk}. Values quoted in BS EN 1992-1-1 give $f_{pk}/f_{p0.1k}=1.1$, $\varepsilon_{uk}=0.0222$ and $\rho_{1000}=2.5\%$ for the vast majority of tendons used in elements in buildings. Again only in research are these strengths, modulus of elasticity, relaxation and ductility, etc. fully measured.

Table 5.1 lists typical mean concrete strengths f_{cm} (usually obtained from 2 or 3 cubes) for precast production at 20 hours (detensioning prestress or stripping moulds), 7 and 28 days, and the characteristic values for cement cubes f_{cu} according to BS EN 197 Part 1 (BS EN 197-1, 2011). The table shows that, even allowing for $f_{cu}=f_{cm}-4$ N/mm², concrete at 20 hours just about exceeds the requirements for cement strength at 2 days.

5.1.2 Strength and quality control of concrete

Using materials that have passed strict quality control procedures, rapid hardening cement type CEM I class 42.5R or 52.5R according to BS EN 197 Part 1 (BS EN 197-1, 2011) is mixed with excellent quality aggregates of known source and purity, in computer-controlled batching and mixing plant, to produce concrete of specified workability and strength. Even the introduction of small quantities of uncontaminated recycled concrete, usually from the factory's own waste production, superplasticisers and pozzolanic materials (such as pulverised fuel ash) has not reduced this standard. To quote standard deviations of less than 2 N/mm² on concrete of 28 days compressive cube strength of between $f_{ck,cube}=50-65$ N/mm² would not be an exaggeration. Figure 5.1 shows variations in the mean m_x and standard deviation s_x for the production of prestressed hollow core floor units (hcu) where the target characteristic cube strength was 60 N/mm². The author has chosen to show m_x and s_x based on samples taken from the last 6 days' production; in fact, the trends are not very different based on the last 8 or 10 days' production. The mean characteristic strength is $f_{ck}=60.1$ N/mm², and the average $s_x=3.1$ N/mm².

Table 5.1 Strength requirements for cement based on BS EN 197-1 and typical concrete strengths achieved by precast producers

Cement strength class	Early strength (N/mm²) 20 hours concrete Producer	2 days cement BS EN 197-1	7 days concrete Producer	Standard concrete strength at 28 days (N/mm²) BS EN 197-1	Producer
42.5N	15–20	*≥10.0*	40–55	*42.5–62.5*	50–70
42.5R	25–35	*≥20.0*	50–70		60–90
52.5N	20–30	*≥20.0*	50–70	*≥52.5*	60–90
52.5R	30–35	*≥30.0*	50–80		70–100

Key: BS EN 197-1 characteristic value for cement (italics); producer's mean cube strength.

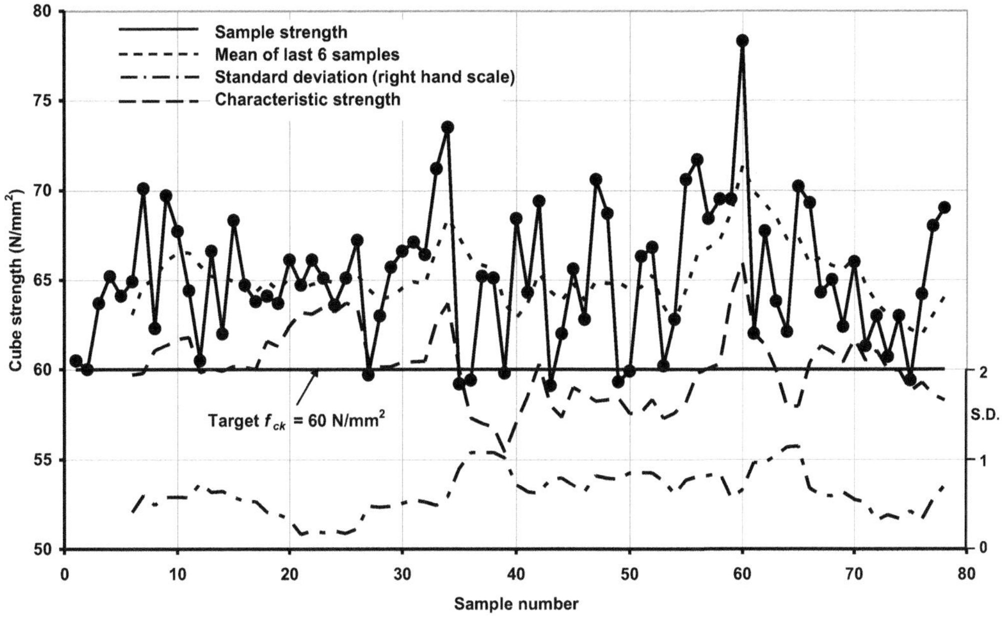

Figure 5.1 Variation in daily 28 days cube strengths, mean of 6, standard deviation and characteristic strength for prestressed hollow core floor units.

Table 5.2 Concrete strengths and modulus of elasticity used in prestressed elements

Component	f_{ck} *at 28 days (N/mm²)*	$f_{cm,cube}$ *demould (N/mm²)*	*Design* ηf_{cd} *(N/mm²)*	*Tensile strength* f_{ctm} *(N/mm²)*	E_{cm} *at 28 days (kN/mm²)*	$E_{cm}(t)$ *(kN/mm²)*
Beams, columns	40–60	35–55	22.7–32.3	3.50–4.35	35.2–39.1	30.0–32.3
Dry cast slabs	40–50	35–42	22.7–28.3	3.50–4.07	35.2–37.3	31.5–32.3
Beams in beam & block floors	50–55	42–45	28.3–30.4	4.07–4.21	37.3–38.2	32.3–33.3

E_{cm} and $E_{cm}(t)$ based on gravel aggregates. For limestone or sandstone divide by 1.1 or 1.3, respectively.

$f_{cm,cube}$ is based on $f_{ck}(t)/0.8+5$ N/mm².

Table 5.2 lists typical characteristic strength f_{ck} at 28 days, mean concrete strengths $f_{cm,cube}$ at 20 hours transfer, design strength for compression and bending ηf_{cd}, design mean tensile strength f_{ctm}, and moduli of elasticity E_{cm} at 28 days and at transfer. The factor η is defined in BS EN 1992-1-1, clause 3.1.7 (3). See Section 5.1.4 for modulus of elasticity.

Concrete is specified according to BS EN 206 (BS EN 206, 2013). The design strength for normal weight concrete (>2000 kg/m³) is given in BS EN 1992-1-1, clause 3.1.2 (1) by strength classes (commonly known as 'grade') given in BS EN 1992-1-1, Table 3.1 as the characteristic (5% failures) compressive cylinder strength f_{ck} and the cube strength $f_{ck,cube}$ in accordance with BS EN 206, e.g. grade C35/45 means $f_{ck}=35$ N/mm² and $f_{ck,cube}=45$ N/mm². The additional strength of the cube is due to the lateral confinement of the loading platens during a compression test, and the small height-to-width shape of a cube, both of which contribute to an artificially higher strength of around 20%–25%.

The strength class used in design is the cylinder strength f_{ck} up to a maximum of 90 N/mm² (105 cube strength) with a minimum 20 N/mm² for structural grades. The true stress *vs.* strain (σ_c *vs.* ε_c) relationship for concrete of normal and higher mean strength f_{cm} is shown in Figure 5.2a, based on BS EN 1992-1-1, Fig. 3.2. Note the reduced ductility after the peak stress of 'higher' strength, defined as greater C50/60, where the design concrete crushing strain reduces from $\varepsilon_{cu3}=0.0035$ to 0.0026. In prestressed design, this has a small effect on the ultimate moment of resistance, typically 0.3%, unlike in reinforced concrete where this has a much greater effect on the balanced section, i.e. the point at which compression reinforcement is required.

The constitutive stress-strain relationship is idealised as a parabola-rectangle in Figure 5.2b where the design strength $f_{cd}=f_{ck}/\gamma_m$, where γ_m is the material partial safety factor of 1.5 at the ultimate limit state and 1.2 for accidental situations based on the UK National Annex to BS EN 1992-1-1, Table 2.1N (NA to BS EN 1992-1-1, 2004). Figure 5.2c shows the idealised rectangular stress block used in design, where the distance $\bar{x}$ to the centroid of the areas beneath the curves in Figure 5.2b and c due to the design strength ηf_{cd} are established by national codes to be as close as possible. BS EN 1992-1-1, Exp. 3.19–3.22 gives

$$\eta=1 \text{ for } f_{ck}\leq 50 \text{ N/mm}^2$$

$$\eta=1-(f_{ck}-50)/200 \text{ for } f_{ck}\leq 90 \text{ N/mm}^2 \tag{5.1}$$

e.g. for $f_{ck}=90$ N/mm^2 then f_{cd} loses 20%.

In Figure 5.2c if the x-axis represents the depth to the neutral axis x, then the design rectangular stress block has a depth λx, where

$$\lambda=0.8 \text{ for } f_{ck}\leq 50 \text{ N/mm}^2$$

$$\lambda=0.8-(f_{ck}-50)/400 \text{ for } 50<f_{ck}\leq 90 \text{ N/mm}^2 \tag{5.2}$$

i.e. 0.8 reduces to 0.7 at $f_{ck}=90$ N/mm^2

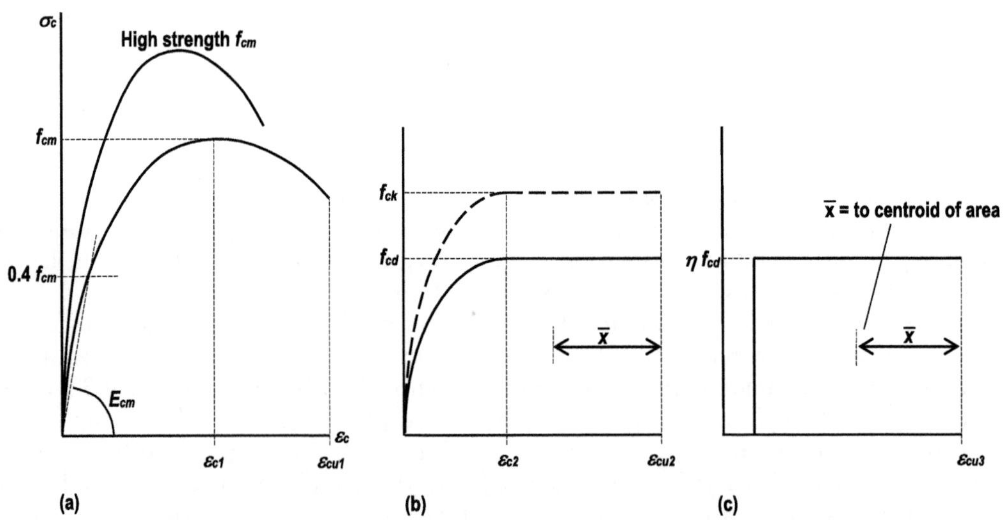

Figure 5.2 Constitutive stress vs. strain relationships for concrete in compression. (a) Actual relationships for normal and high strengths, (b) parabola-rectangle idealisation, and (c) idealised rectangular stress block used in design.

e.g. the ultimate force in a concrete beam of width b is F_c=design stress×area=η f_{ck} b λx=0.8 f_{ck} bx for f_{ck}=50 N/mm^2 reducing to 0.56 f_{ck} bx for f_{ck}=90 N/mm^2. Coupled with the reduced depth at the balanced section, the total force at the balanced section is F_c=24.68 bd and 27.47 bd for f_{ck}=50 and 90 N/mm^2, respectively, i.e. only 11% greater for 80% increase in f_{ck}!

The strength of concrete f_{ck}(t) between 3 and 28 days can be determined either by testing cubes or cylinders, or estimated using semi-empirical equation for an average temperature of 20°C and standard curing conditions, given in BS EN 1992-1-1, clause 3.1.2 (6) as follows:

$$f_{cm}\left(t\right) = f_{cm}e^{s(1 - \sqrt{28/t})} \tag{5.3}$$

where

$$f_{cm}\left(t\right) = f_{ck}\left(t\right) + 8 \text{ N/mm}^2 \tag{5.4}$$

and f_{cm}=f_{ck}+8 N/mm^2 (representing a margin of 1.64 sd (standard deviations) of strength for 5% characteristic values). Precast production quality measures achieve early strength sd≈2–3 N/mm^2, making the above conservative.

s=0.20, 0.25 and 0.38 for cement Class R, N and S

e.g. if f_{ck}=40 N/mm^2 at 28 days using CEM 52.5N (Class R), at say t=3 days $f_{cm}(t)$=48 $e^{0.2(1-\sqrt{28/3})}$=48×0.663=31.8 N/mm^2 and $f_{ck}(t)$=23.8 N/mm^2. If the margin at 3 days is taken as 3 N/mm^2, then for the same example $f_{ck}(t)$=28.8 N/mm^2, giving a better estimate of more than ⅔ at 3 days.

When precasting in a modern factory, it is vital to achieve the dual requirements of good workability and early strength. For wet-cast work, the workability of the concrete is still specified according to slump value, e.g. 25–100 mm in factory production, except in the case of concrete provided for slip forming or extrusion which has no workability requirements (zero slump). Concrete is often transported automatically from an automatic batching plant to the mould leaving no opportunity for workability tests. It is unlikely that such tests are required thanks to the experience of batching staff.

As explained in Section 4.2.3, a huge impact on precast production methods has been the use of self-compacting concrete (SCC), which, as the name suggests, will flow and compact without external vibration. SCC admixtures conform to BS EN 206 Part 9 (BS EN 206-9, 2010). After the problems of low early strength gain were resolved, this has been widespread in wet-cast voided units (Figures 2.15 and 2.16), double-tee floor units (Figure 2.19) and beams. Figure 5.3 shows SCC flowing up to 1.5 m in an outdoor steel mould on quite a cold day in the UK (around 7°C), achieving 18 hours of detensioning cube strength of $f_{ck,cube}$=25–30 N/mm^2. The workability of SCC is measured using a flow table in which the diameter of flow, as shown in Figure 5.4 is usually 600–700 mm, neglecting the small ring of cement slurry at the edges. The precast industry has exploited the advancements made in concrete technology more than in any other part of the construction industry.

For the production of standard elements such as beams, concrete is cast into clean steel or timber (sometimes concrete) moulds, accurate to±3 mm, or less, in cross section. Figure 5.5 shows an adjustable steel mould of around 100 m in length for casting rectangular and inverted-tee beams, or for casting double-tee slabs of fixed width (usually 2.4 m) and variable length and depth as shown in Figure 5.6. The use of clamped vibrators tuned to the correct oscillations for the size and weight of the filled mould ensures correct compaction to a density of around 2400 kg/m^3 (excluding reinforcement). The resulting surface finish

Figure 5.3 Self-compacting concrete, flowing and compacting without segregation, used to manufacture prestressed floor slabs.

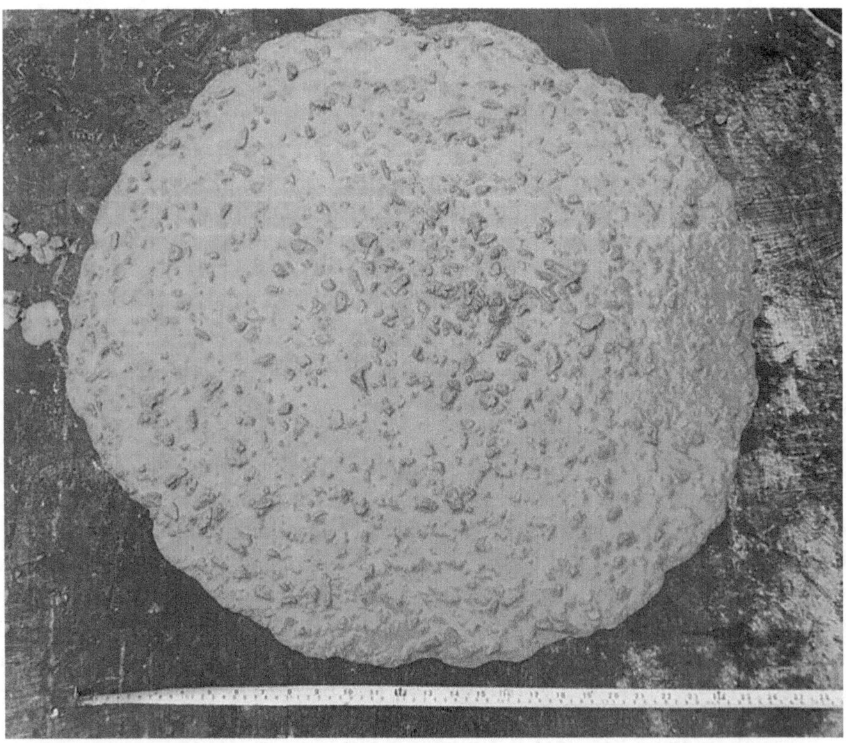

Figure 5.4 Measurement of the workability of self-compacting concrete by measuring the diameter of flow on a flow table.

Figure 5.5 Steel mould for casting inverted-tee beams.

Figure 5.6 Stripping of steel moulds and lifting of prestressed double-tee unit 20 hours after casting.

results in minimum porosity for maximum durability. Concrete strengths are made to match the optimum performance of each prestressed element, such that hcu, wet-cast voided units, double-tee floor units and beams are produced using f_{ck}=40–45 N/mm^2, while T beams for beam and block flooring use f_{ck}=50–55 N/mm^2.

5.1.3 Concrete for prestressed floor units

Precast floor units are manufactured using either 'wet cast' (semi-wet with less than 50 mm slump, or wet-cast with 75 mm slump) or 'dry cast' ('semi-dry' with zero slump and 5–10 seconds V-B time). Steel moulds are used in the production of prestressed voided units, as shown in Figure 5.7. The concrete is $f_{cm,cube}$=40 N/mm^2 or $f_{ck}(t)$=25 N/mm^2 at 20 hours and hence achieves at least f_{ck}=45 N/mm^2.

Machine extruded concrete is not, by definition, cast and compacted. It is instead forced through rotating augers to pass out through steel formers as shown in Figures 4.14 and 4.15. An alternative method is for the moving steel bed to pass beneath a fixed position extruder, as shown in Figure 5.8. Good quality hcu are also manufactured using slip-formed concrete, produced through a machine that compacts concrete by rows of small hammers. Today slip-formers are less popular than extruders, due mainly to the better compaction and surface finish of the latter, but they are generally cheaper to purchase and maintain. To enable both of these processes to happen, the concrete is 'earth-dry', having a water-cement (or water-cementitious material) ratio of around 0.35–0.4. Aggregates are carefully selected, especially the sand grading and content, and the shape and size of the coarse aggregates, typically 14 mm down. Because prestressed concrete benefits from additional strength in flexural compression, it is advantageous to use f_{ck}=40–45, although for certain geometry

Figure 5.7 Casting wet-cast voided floor units.

Figure 5.8 Double hollow core units produced by extruder at Bison Precast Ltd., UK.

and in order to balance the limiting compressive stress (0.45 f_{ck}) and tensile stress (f_{ctm}) the common grade is C45/55. For internal exposure, XC1, BS EN 1992-1-1, clause 7.3.2 (4) limits the service condition to the mean axial tensile stress f_{ctm} given in Table 3.1 as

$$f_{ctm} = 0.3 f_{ck}^{2/3} \tag{5.5a}$$

for concrete grade ≤ C50/60, otherwise

$$f_{ctm} = 2.12 \ \ln\left[1 + \left(f_{ck} + 8\right)/10\right] \quad \left(\text{natural log to base } e\right) \tag{5.5b}$$

For exposure > XC1, BS EN 1992-1-1 Table 7.1 (superceded by the NA in Table A.4) requires that the service condition is 'decompression', i.e. zero tension. (See Section 7.2.1 for full analysis.)

Each producer will modify their mix according to variations in local supplies. The grading of fine aggregate is often the most critical mix parameter, especially in the dry mixes used in slip-formed and extruded concrete. Figure 5.9 shows typical grading curves for fine aggregate used in machine cast units; note the high percentage passing the 0.5 mm sieve size for the extrusion machine.

5.1.4 Modulus of elasticity and modular ratios

The static value for the modulus of elasticity is calculated according to BS EN 1992-1-1, clause 3.1.3, and given in Table 3.1 as

$$E_{cm} = 22\left[\left(f_{ck} + 8\right)/10\right]^{0.3} \tag{5.6}$$

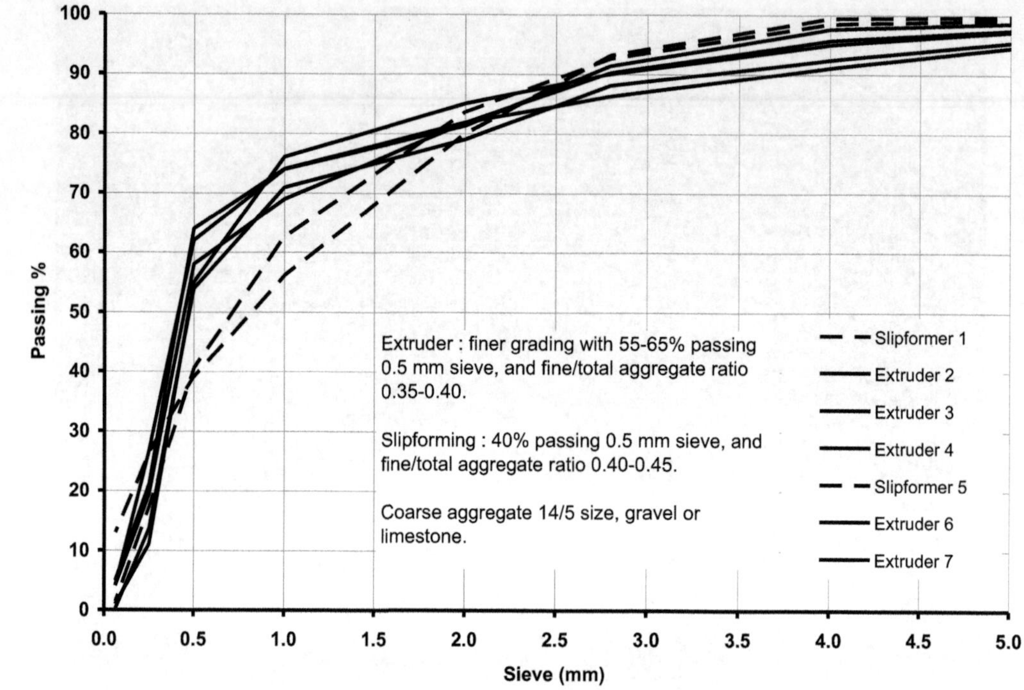

Figure 5.9 Grading curves for fine aggregate used in the extrusion and slipforming of prestressed hollow core slabs (date collected in the UK by the author 2004–2008).

for siliceous gravel aggregate. For limestone and sandstone aggregates, the value should be reduced by 10% and 30% respectively. For basalt aggregates, the value should be increased by 20%. Table 5.3 gives the relevant information for prestressed elements and the modular ratio when designed compositely with *insitu* concrete.

The tangent modulus E_c is measured at zero stress, i.e. it is the tangent of the stress v strain curve at zero stress. BS EN 1992-1-1, clause 3.1.4 (2) gives this as $E_c = 1.05\ E_{cm}$. E_c is also known as the dynamic modulus E_d because it may be determined experimentally by dynamic vibration or pulse velocity methods when the concrete is unstressed. In the author's experience of testing E_d and E_{cm} (at an applied stress of 0.33 $f_{ck,cube}$ using Trent Valley gravel), the relationship in clause 3.1.4 (2) under estimates E_d, (the relationship given by Neville (Neville, 1995) $E_d = (E_{cm} + 19)/1.25$ (N/mm^2 units) is reliable) although the expression for E_{cm} in Eq. 5.6 tends to over estimate by about 10%.

Where *insitu* concrete of strength f_{cki} is added to psc slabs and beams to form composite behaviour (see Chapter 12) the relative strengths (f_{ck}) and stiffness (E_{cm}) of the two materials must be considered using the modular ratio method. The most common combination of *insitu*-to-precast strengths is $f_{cki}/f_{ck} = 25/45$. This leads to the modular ratio $m = E_{cmi}/E_{cm} = (3.3/5.3)^{0.3} = 0.87$.

5.1.5 Shrinkage and creep

Shrinkage and creep are time-dependent properties of concrete. They depend on the ambient humidity in service (RH$_s$ %); the notional depth or thickness of element=twice the cross-sectional area/exposed perimeter, $h_o = 2A_c/u$; concrete composition (water/cement ratio (w/c),

Table 5.3 Strengths and short-term modulus of elasticity for typical concrete used in composite construction, from BS EN 1992-1-1

Type of concrete	f_{ck} *(N/mm²)*	f_{ctm} *(N/mm²)*	E_{cm} *(kN/mm²)*	*Modular ratio*
Insitu	25	N/A	31.48	N/A
	30	N/A	32.84	N/A
Prestressed	40	3.51	35.22	0.89–0.93
	45	3.80	36.28	0.87–0.90
	50	4.07	37.28	0.84–0.88
	55	4.21	38.21	0.82–0.86

$E_{cm}=22\ (f_{ck}+8)/10^{0.3}$ based on gravel aggregates.

For limestone or sandstone divide by 1.1 and 1.3, respectively

cement type, e.g. rapid hardening Class R, f_{ck}); maturity of the concrete when the elements are stored; age (t_o) when load is first applied; and the duration and magnitude of the loading.

BS EN 1992-1-1, clause 2.3.2.2 notes: *Shrinkage and creep effects should generally be taken into account for the verification of serviceability limit states. The effects should be considered at ultimate limit states only where their effects are significant, for example in the verification of ultimate limit states of stability where second-order effects are of importance. In other cases, these effects need not be considered for ultimate limit states, provided that ductility and rotation capacity of the elements are sufficient. Creep effects should be evaluated under the quasi-permanent combination of actions irrespective of the design situation considered, i.e. persistent, transient or accidental. In most cases, the effects of creep may be evaluated under permanent loads and the mean value of prestress.* In other words, creep is considered for changes in secondary moments at supports in continuous slabs and beams, but otherwise is used only in the determination of long-term deformations of deflection and crack width.

In prestressed elements, some of the creep and shrinkage deformations take place prior to installation, something that is not considered in the Code's equations, and therefore these effects are calculated at the time of installation (typically 7–28 days) and in the long term of 500,000 hours (20833 days, or 57 years).

5.1.5.1 Shrinkage strains

Shrinkage causes shortening of concrete, around 400×10^{-6} (μm) ≈ 4 mm in 10 m length, and is considered mainly in the long-term, although in composite design the short-term shrinkage of the prestressed element is determined so that the effects of the relative shrinkage of the *insitu* to the precast can be determined for shrinkage-induced curvature. First introduced in BS EN 1992-1-1 as clause 2.3.2.2, shrinkage actions are evaluated in clause 3.1.4 and Appendix B as follows. The shrinkage strain ε_{cs} at time t (20833 days for long-term) after removal from mould or detensioning at time t_s=typically 1 day is (Code clauses are listed in the left-hand column)

Exp. 3.8 Total shrinkage strain $\varepsilon_{cs}=\varepsilon_{cd}+\varepsilon_{ca}$ (5.7)

where

Exp. 3.9 Drying shrinkage strain $\varepsilon_{cd}=\beta_{ds}(t,t_s)k_n\varepsilon_{cd,o}$ (5.8)

Exp. 3.10 Age factor $\beta_{ds}(t,t_s)=(t-t_s)/[(t-t_s)+0.04h_o^{1.5}]$ (5.9)

Table 3.3. Size coefficient $k_n = 1.0-0.0015\ (h_o-100)$ for $100 \le h_o < 200$ mm

$0.85-0.001\ (h_o-200)$ for $200 \le h_o < 300$ mm

$0.75-0.0005\ (h_o-300)$ for $300 \le h_o < 500$ mm (5.10)

where h_o=notional size of the cross section (in mm)$=2A_c/u$ (5.11)

where A_c=cross-sectional area, and u=perimeter exposed to drying.

Before the installation of wide floor slabs, this is the full perimeter, but afterwards, only the soffit as the sides and top of the element are protected by another unit or the finishing screed or topping.

Exp. B.11 $\varepsilon_{cd,o}=0.85[(220+110\alpha_{ds1})\ \exp-(\alpha_{ds2}f_{cm}/f_{cmo})]\times 10^{-6}\beta_{RH}$ (5.12)

Exp. B.12 RH factor $\beta_{RH}=1.55\left[1-(RH_s/100)^3\right]$ (5.13)

Exp. 3.11 Autogenous shrinkage strain $\varepsilon_{ca}(t)=\beta_{as}\,\varepsilon_{ca}$ (5.14)

10.3.1.2 (3) In precast elements subjected to heat curing $\varepsilon_{ca}(t)$ is negligible.

Exp. 3.13 $\beta_{as}=1-e^{-0.2\sqrt{t}}=1$ for long term (t is indays) (5.15)

Exp. 3.12 $\varepsilon_{ca}=2.5\ (f_{ck}-10)\times 10^{-6}$ (5.16)

f_{ck} and f_{cm} in N/mm^2, $f_{cmo}=10$ N/mm^2

3.1.2(6) Cement classes and α_{ds} factors are

B.2 Class R=CEM 42.5 R, CEM 52.5 N and CEM 52.5 R ($\alpha_{ds1}=6$, $\alpha_{ds2}=0.11$)

Class N=CEM 32.5 R, CEM 42.5 N ($\alpha_{ds1}=4$, $\alpha_{ds2}=0.12$)

Class S=CEM 32.5 N ($\alpha_{ds1}=3$, $\alpha_{ds2}=0.13$)

Where shrinkage actions are required at the ultimate limit state, a partial factor $\gamma_{SH}=1.0$ is used.

Example 5.1

Calculate the long-term shrinkage strain ε_{cs} for a prestressed concrete wet-cast voided unit of 1200×200 mm cross section shown in Figure 5.10 manufactured using grade C35/45 concrete using cement CEM I class 52.5R. The unit is demoulded at 1 day and is used inside a building with RH_s=50%. Cross-sectional area=141210 mm^2. Repeat using CEM I class 42.5N and comment on the answer.

Solution

Eq. 5.4 $f_{cm}=45+8=53$ N/mm^2

Eq. 5.13 $\beta_{RH}=1.55\ [1-(50/100)^3]=1.356$

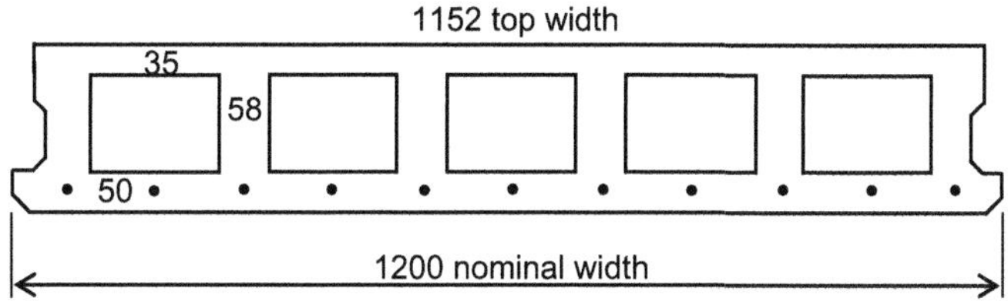

Figure 5.10 Cross section of wet cast voided floor unit in worked examples for shrinkage and creep coefficients.

cl. B2	Cement class R where $\alpha_{ds1}=6$ and $\alpha_{ds2}=0.11$
Eq. 5.12	$\varepsilon_{cd,o}=0.85\times[(220+110\times6)\ e^{-(0.11\times5.3)}]\times10^{-6}\times1.356=566\times10^{-6}$
Eq. 5.9	$h_o=2\times141210/1200=235\,\text{mm}$
Eq. 5.10	$k_n=0.815$
Eq. 5.9	$\beta_{ds}(t,t_s)=(20833-1)/[(20833-1)+0.04\times235^{1.5}]=0.993$
Eq. 5.8	$\varepsilon_{cd}=0.993\times0.815\times566\times10^{-6}=458\times10^{-6}$
Eq. 5.14	$\varepsilon_{ca}=0$
Eq. 5.7	$\varepsilon_{cs}=458+0=\mathbf{458\times10^{-6}}$

This is an average value for psc hcu where $\varepsilon_{cs}=430\times10^{-6}$ for 400 mm depth units and $\varepsilon_{cs}=500\times10^{-6}$ for 150 mm depth. However, if the cement is changed to class 42.5N of class N ($\alpha_{ds1}=4$, $\alpha_{ds2}=0.12$) $\varepsilon_{cd,o}=0.85\times[(220+110\times4)\ e^{-(0.12\times5.3)}]\times10^{-6}\times1.356=403\times10^{-6}$ a reduction of 40% compared with Class R. Note that $\varepsilon_{cd,o}$ values are tabulated in BS EN 1992-1-1, Table 3.2 for Class N cement, giving 398×10^{-6} for this example.

5.1.5.2 Creep strains

Concrete is a visco-elastic material, i.e. undergoes an increase in strain (and) when it is subjected to constant stress. It is mainly due to the deformability of the cement paste, which increases rapidly over the first few months after loading, and slows after 2 years to around 10% of that at 3 months. In addition to the environmental and mix design parameters that affect creep strains, e.g. lower water-to-cement ratio and higher aggregate proportions reduce creep, a major factor is the age when the concrete is loaded, and loading history during unloading and reloading. For this reason, creep deflections due to imposed transient live loads are modified by the quasi-permanent factor as explained in Chapter 3, Section 3.1.1. The development of creep strains is shown in Figure 5.11, and the influence of age at loading is shown in Figure 5.12. This is important when calculating creep deflections in prestressed slabs and beams where imposed dead and live loads are applied at different ages, and particularly so in composite construction where there are different concrete strengths in the precast and *insitu* concretes.

Creep causes changes in the modulus of elasticity from the immediate tangent value (as above) E_c to $E_c/(1+\varphi(t,t_0))$ where $\varphi(t,t_0)$=creep coefficient from time t_0 to t. Similarly, the long-term E_{cm} is $E_{cm}/(1+\varphi(t,t_0))$. This means that $\varphi(t,t_0)$ may be calculated for any time interval as

Exp. B.2 Creep coefficient $\varphi(t,t_0)=\varphi_{RH}\cdot\beta(f_{cm})\cdot\beta(t_0)\cdot\beta_c(t,t_0)$ (5.17)

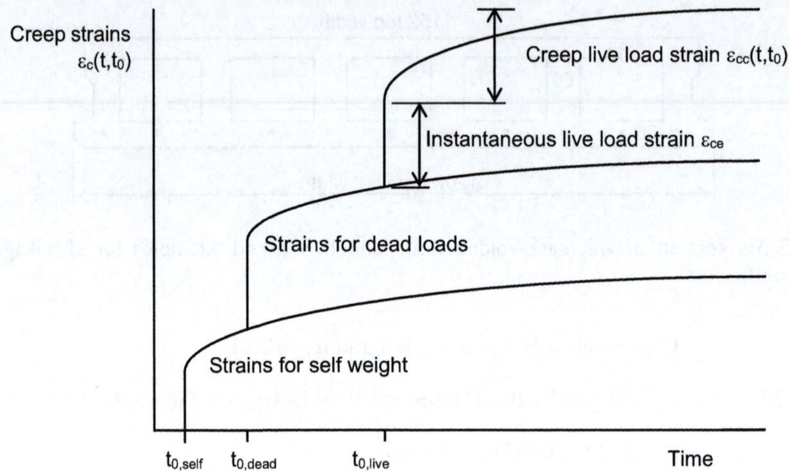

Figure 5.11 Instantaneous and creep strains for self-weight, and imposed dead and live loads.

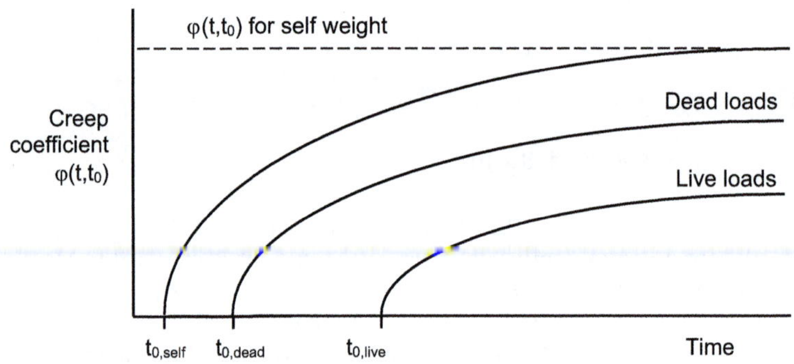

Figure 5.12 Effect of age at loading on creep coefficients.

where

Exp. B.3 $$\varphi_{RH} = \left\{1+\left[\left(1- RH_s/100\right)\alpha_1 /0.1h_o^{1/3}\right]\right\}\alpha_2 \quad (5.18)$$

Exp. B.8c $$\alpha_1 = \min\left\{1;\ \left(35/f_{cm}\right)^{0.7}\right\} \quad (5.19a)$$

$$\alpha_2 = \min\left\{1;\ \left(35/f_{cm}\right)^{0.2}\right\} \quad (5.19b)$$

Exp. B.4 Strength factor $\beta(f_{cm}) = 16.8\ /\sqrt{f_{cm}}$ (5.20)

Exp. B.5 Age at installation loading factor $\beta(t_i) = 1/\left(0.1+t_i^{0.2}\right)$ (5.21a)

Exp. B.7 $$\beta_c(t,t_0) = \left\{(t-t_0)/\left[(t-t_0)+\beta_H\right]\right\}^{0.3} \quad (5.22)$$

Exp. B.8b RH factor β_H (days) $= 1.5\ [1+(0.012\ RH_s)^{18}]\ h_o + 250\ \alpha_3$ (5.23)

Exp. B.8c $$\alpha_3 = \min\left\{1;\ \left(35/f_{cm}\right)^{0.5}\right\} \quad (5.19c)$$

High-temperature steam or otherwise curing of precast elements gives an equivalent age of maturity age when it is demoulded at time t_0, as follows:

Exp. B.5 Age at release loading factor $\beta(t_0) = 1/\left(0.1 + t_0^{0.2}\right)$ (5.21b)

Exp. B9 Here may use $t_0 = t_{0T}\left(\frac{9}{2+t_T^{1.2}} + 1\right)^{\nu} \geq 0.5$ days (5.24)

where α=1, 0, ×1 for Class R, N, and S cement, respectively

Exp. B10 Equivalent age at transfer $t_{0T} = \Sigma e^{-[4000/(273+T)-13.65]}.\Delta t_i$ (5.25)

where T=mean curing temperature °C (typically 50°C) during curing time Δt_i (typically 20 hours) in days.

Example 5.2

Calculate the equivalent age at release t_0 for concrete matured for 20 hours under a mean temperature of 50°C using cement class R.

Solution

Eq. 5.25 $t_{0T}=e^{-[4000/(273+50)-13.65]}(20/24)=3.547\times0.833=2.96$ days

α=1 for Class R

Eq. 5.24 $t_0=2.96\times\{[9/(2+2.96^{1.2})]+1\}^1=2.96\times2.582=$ **7.65 days**

Example 5.3

Calculate the long-term creep coefficient $\varphi(t,t_0)$ for the wet-cast voided section in Example 5.1 and curing maturity in Example 5.2. Repeat at t=28, 56 and 90 days and comment on the progression of creep in the short term.

Solution

Eq. 5.19 $\alpha_1=\min\{1;(35/53)^{0.7}\}=0.748;\ \alpha_2=0.920;\ \alpha_3=0.813$

Eq. 5.18 $\varphi_{RH}=\{1+[(1-50/100)\ 0.748/0.1\times235^{1/3}]\}\times0.920=1.478$

Eq. 5.20 $\beta(f_{cm})=16.8/\sqrt{53}=2.308$

Eq. 5.21 $\beta(t_0)=1/(0.1+7.65^{0.2})=0.624$

Eq. 5.23 $\beta_H=1.5\ [1+(0.012\times50)^{18}]\times235+250\times0.813=556$ days

Eq. 5.22 $\beta_c(t,t_0)=\{(20833-7.65)/[(20833-7.65)+556]\}^{0.3}=0.992$

Eq. 5.17 $\varphi(t,t_0)=1.478\times2.308\times0.624\times0.992=2.129\times0.992=$ **2.11**

Eq. 5.22 If t=28 days, $\beta_c(t,t_0)=\{(28-7.65)/[(28-7.65)+556]\}^{0.3}=0.367$

Eq. 5.17 $\varphi(t,t_0)=2.129\times0.367=$ **0.78**

Eq. 5.17 If t=56 days, $\beta_c(t,t_0)=0.468$, $\varphi(t,t_0)=$ **1.00**

Eq. 5.17 If t=90 days, $\beta_c(t,t_0)=0.541$, $\varphi(t,t_0)=$ **1.15**

The short-term creep is known to have a logarithmic relationship with time, and the above results suggest that $\varphi(t,t_0)$ is a function of $K \log_e t^x$ (by curve fitting the above the standard deviation of the differences is found to be minimum when $x=1.3$, from which the $K=0.163$ is then found) as follows:

At $t=28$ days, $0.163\times(\log_e 28)^{1.3}=0.78$ (cf. 0.78)
At $t=56$ days, $0.163\times(\log_e 56)^{1.3}=1.00$ (cf. 1.00)
At $t=90$ days, $0.163\times(\log_e 90)^{1.3}=1.15$ (cf. 1.15), and so on.

BS EN 1992-1-1, Fig. 3.1 presents a graphical route to $\varphi(t,t_0)$ as reproduced in Figure 5.13. The additional (dashed) line gives the solution to the long-term value above as $\varphi(t,t_0)\approx 2.2$, in reasonable agreement with their equations in Appendix B.

5.1.5.3 Creep coefficients and creep deflections, and shrinkage strains in hollow core floor units

Creep coefficients, instantaneous and creep deflections and shrinkage strains have been calculated for the 1200×265 mm depth prestressed hcu shown in the cross section in Figure 5.14. Relative humidity in the factory and stockyard is $RH_s=70\%$ for the period up to installation at 28 days when dead loads are applied. Thereafter, the floor slab is in an enclosed environment where $RH_s=50\%$ for the long-term at 500,000 hours (or 20833 days). The age of the hcu is 84 days when imposed live loads are first applied. Use $f_{ck}=40$ N/mm²

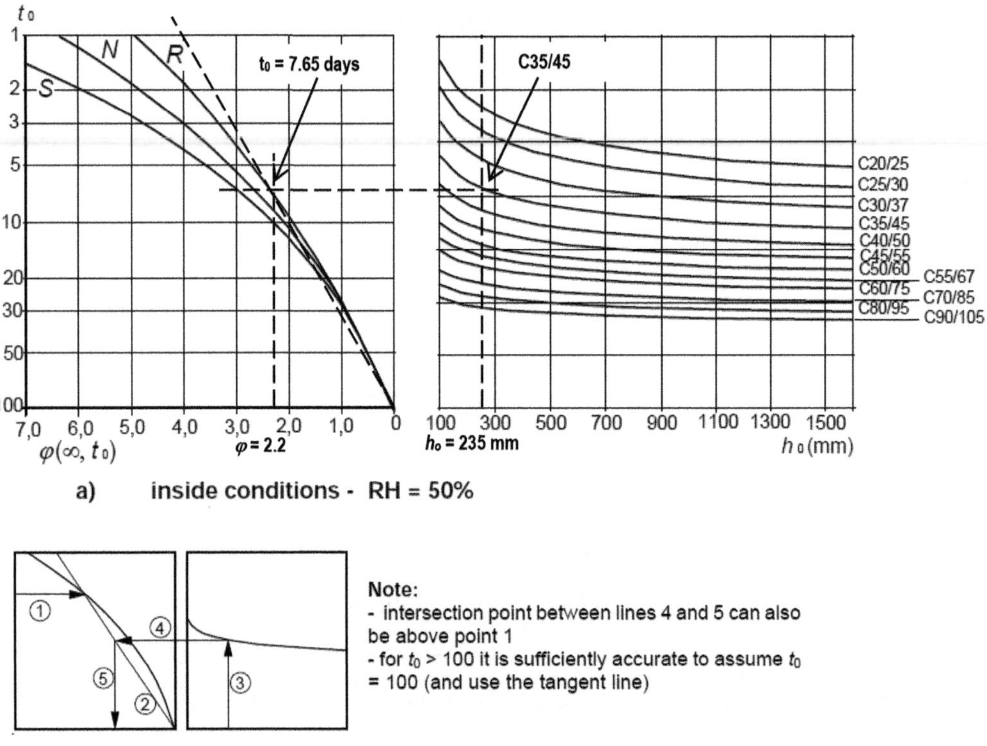

Figure 5.13 Method for determining the long-term creep coefficient $\varphi(\infty,t_0)$ for concrete under normal environmental conditions. $t_0=7.65$ days and C35/45 annotated by the author. Based on BS EN 1992-1-1, Figure 3.1. (Permission to reproduce extracts from British Standards is granted by BSI. British Standards can be obtained in PDF or hard copy formats from BSI Knowledge: https://knowledge.bsigroup.com or by contacting BSI Customer Services for hardcopies only: Tel: +44 (0)20 8996 9001, Email: cservices@bsigroup.com.)

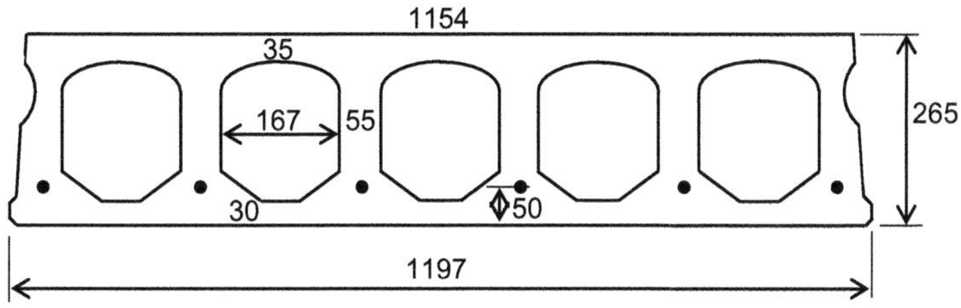

Figure 5.14 Cross section of hcu used to calculate creep and shrinkage strains, creep deflections, and fire moment and shear calculations.

and $f_{ck}(t)=28$ N/mm^2 at transfer, gravel aggregates, $E_{cm}=35220$ N/mm^2 and $E_{cm}(t)=32310$ N/mm^2 at transfer. The age at transfer is 20 hours. Self-weight of hcu including infilled joints=3.93 kN/m^2. The floor has an effective span of 8.0 m and is subjected to imposed dead and live UDL of 3.0 and 5.0 kN/m^2, respectively. Use office loading with $\psi_2=0.3$. Use $\beta(t_0)=1/(0.1+7.65^{0.2})=0.624$ as in Example 5.3.

Calculated creep coefficients $\varphi(t,t_0)$ are shown in Figure 5.15a for the long-term to 20833 days, and in Figure 5.15b up to 1 year. For the self-weight of the hcu $\varphi(t,t_0)$ is first calculated using $f_{ck}(t)$, $RH_s=70\%$ and the notional depth $h_o=126$ mm (for full perimeter exposed, hollow cores are ignored), but at 28 days the parameters change to f_{ck}, $RH_s=50\%$ and $h_o=321$ mm (for the soffit perimeter only as the hcu is grouted to adjacent units and finishing screed is applied). Using the former $\varphi(t,t_0)=0.91$, but the latter produces lower creep deflections and so the 'effective' creep coefficient is modified to allow for this to $\varphi(t,t_0)=0.75$. The long-term effective value of $\varphi(t,t_0)=1.96$. The creep coefficient for dead loads, applied at 28 days, is $\varphi(t,t_0)=1.74$, and for imposed live load, applied at 84 days, is $\varphi(t,t_0)=1.41$. (These values may be compared with the effective creep factors for deflections of prestressed elements in Section 11.2.1).

The quasi-permanent compressive stress in the top of the slab at mid-span is 0.16 f_{cm}, and the prestress at transfer is 0.22 $f_{cm}(t)$. Although these are about half the value where E_{cm} is measured (i.e. at 0.4 f_{cm}, Figure 5.2a) deflections are calculated using E_{cm} and not E_c – the difference would only be a factor of about 1.025. The following deflections may also be compared with those in Section 11.2.1.

Calculated instantaneous and creep deflections are shown in Figure 5.16a for the long-term to 20833 days, and the same data in Figure 5.16b up to 1 year. The data also include the instantaneous and creep camber due to prestress (using 6 no. 12.5 mm strands at an axis height of 50 mm). The creep coefficient for camber at installation is $\varphi(t,t_0)=0.91$ using $E_{cm}(t)$ at 20 hours increasing gradually to E_{cm} at 28 days. In the same period, the pretensioning force reduces by 2% (note that relaxation and elastic shortening have already taken place immediately after transfer). The final creep coefficient for camber is $\varphi(t,t_0)=2.23$, an additional coefficient of 1.32, but note that the pretensioning force reduces by a further 15% in the long term due to creep, shrinkage and relaxation of strands. When camber and self-weight deflections are added together, the conflicting parameters cancel out and the long-term deflection of the slab remains virtually the same, as shown in Figure 5.16a. Surprisingly, because camber is a function of span^2 and self-weight deflection a function of span^4 the result is the same for 6 and 10 m spans.

Deflections for imposed live loads are subjected to the quasi-permanent factor ψ_2, in this case $\psi_2=0.3$. The maximum long-term deflection of 8.3 mm is well below the limit of span/250=32 mm. The active deflection after installation of finishes of 7.6 mm is below the limit of span/500=16 mm (for brittle partitions or finishes).

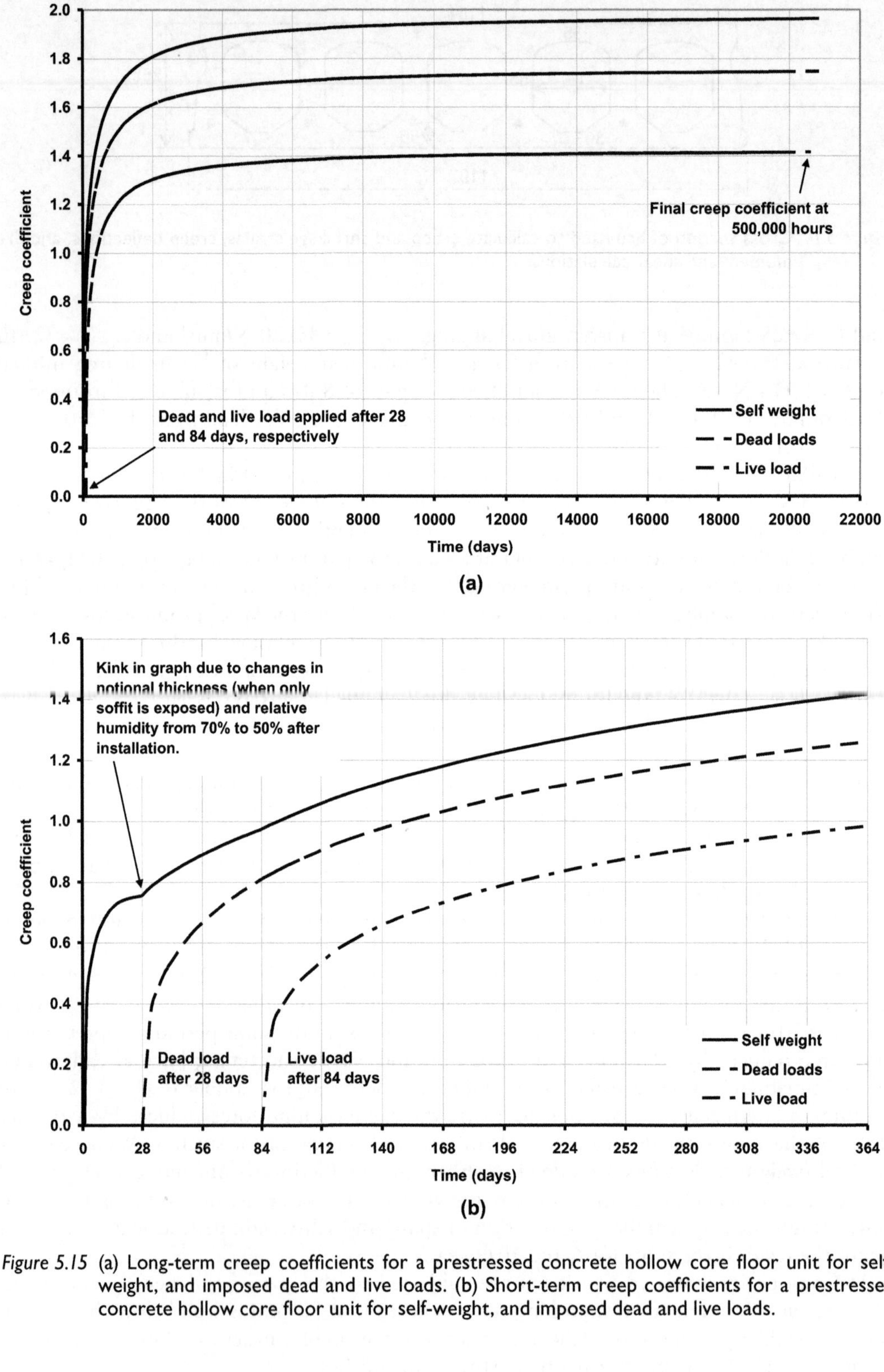

Figure 5.15 (a) Long-term creep coefficients for a prestressed concrete hollow core floor unit for self-weight, and imposed dead and live loads. (b) Short-term creep coefficients for a prestressed concrete hollow core floor unit for self-weight, and imposed dead and live loads.

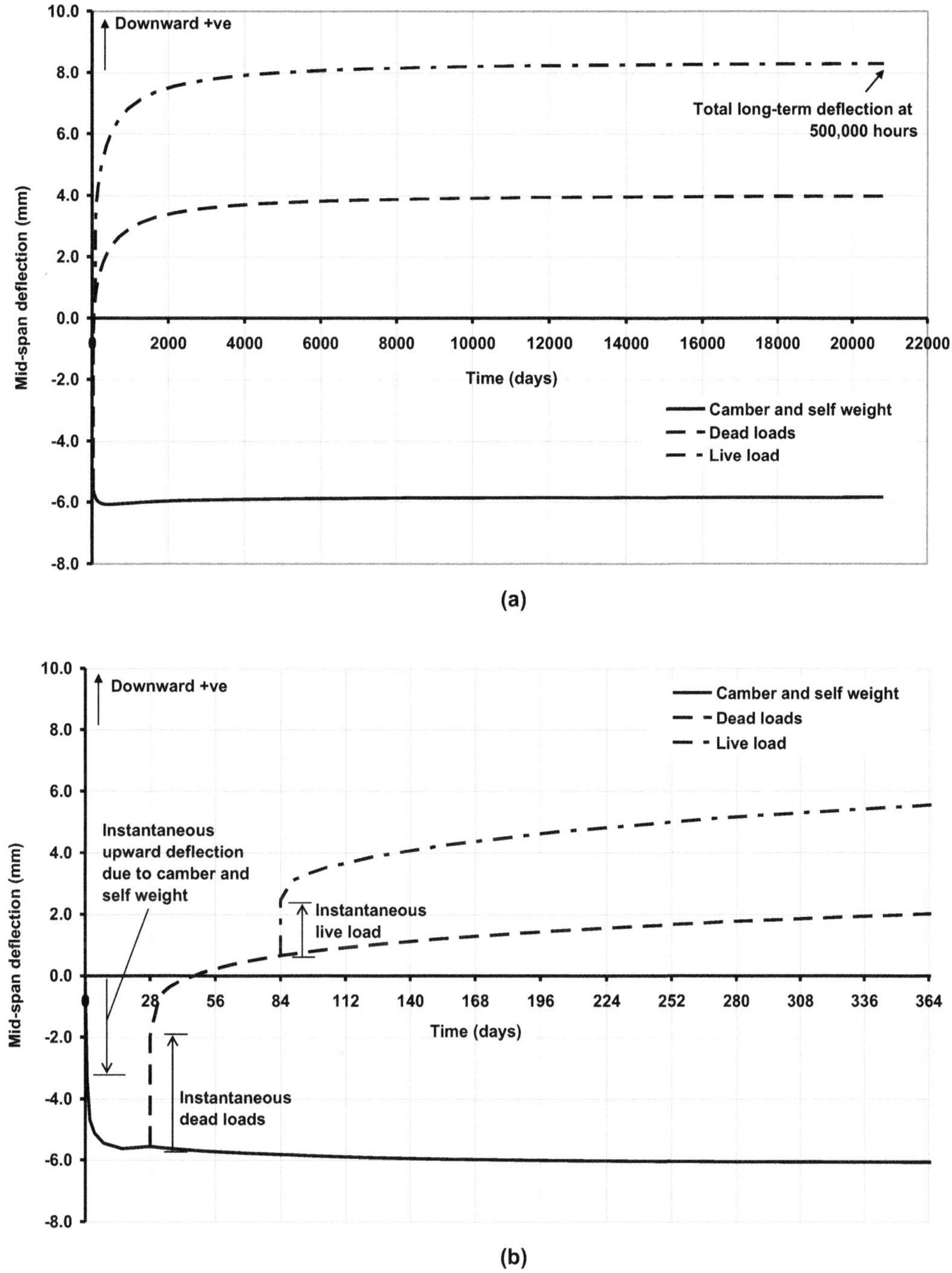

Figure 5.16 (a) Long-term instantaneous and creep deflections for a prestressed concrete hollow core floor unit for camber & self-weight, and imposed dead and live loads. Long-term camber also includes a reduction in prestress due to shrinkage and relaxation of strands. (b) Short-term instantaneous and creep deflections for a prestressed concrete hollow core floor unit for camber & self-weight, and imposed dead and live loads.

Calculated shrinkage strains are shown in Figure 5.17 for up to 3 years; beyond this age, the shrinkage curve is almost linear attaining a long-term value of $\varepsilon_{cs}=0.00044$. Note that the rate of strain during time in the stockyard prior to installation at 28 days is roughly 5.2×10^{-6} per day, but within 2 years the rate is 0.3×10^{-6} per day. The loss of prestress due to shrinkage $\Delta_{p,s}$ has also been calculated, but not presented as the increase of $\Delta_{p,s}$ is almost an exact image of ε_{cs}. It is interesting to note that the ratio $\Delta_{p,s}/\varepsilon_{cs}$ is constant (within 3 percentage points) ≈ 185000 N/mm² per unit strain, close to the modulus of elasticity of strands ($E_p=195000$ N/mm²).

5.2 MIX DESIGN

5.2.1 Exposure, cover to tendons and concrete strength

The producer specifies the mix design according to BS EN 206 and the quantities of cement, aggregates (coarse and fine), water and admixtures & additives. The producer will specify the early strength gain expected. The quantity of mixing water, including the use of superplasticisers and self-compacting admixtures, is the minimum required to ensure the correct workability and strength intended for purpose. If a new family of concrete mixes is specified, trial mixes will be carried out to the satisfaction of the producer to demonstrate early handling and 28-day strengths.

Unlike previous British codes, environmental exposure is directly linked to concrete strength and vice versa, in addition to the usual mix design parameters of water/cement ratio (w/c) and cement type and content, and nominal cover (C_{nom}) to reinforcement. Exposure classes are for reinforcement corrosion induced by

- carbonation (XC1 to XC4)
- chlorides (XD1 to XD3)

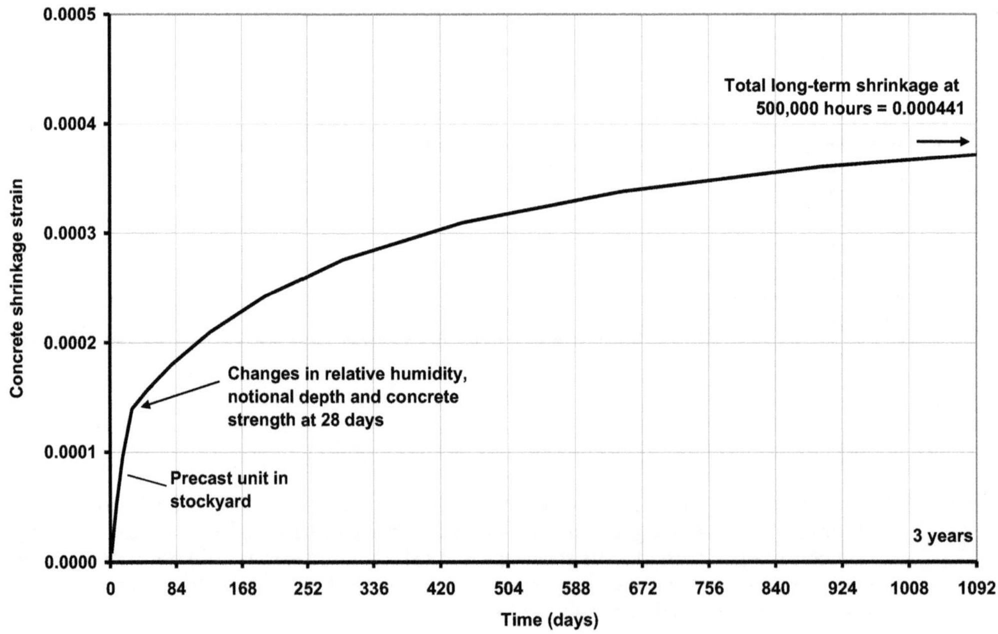

Figure 5.17 Shrinkage strains calculated for a prestressed concrete hollow core floor unit.

- chlorides from sea water (XS1 to XS3)
- freeze/thaw and chemical attack (XF1 to XF4, and XA1 to XA3)

The relationship between f_{ck} and C_{nom} is given in

- BS 8500-1 (BS 8500-1, 2023) Tables A.4 (50 years life) and A.5 (100 years)
- BS EN 1992-1-1 with the UK National Annex to BS EN 1992-1-1, Tables NA2 and NA3.

Note that in Table 5.4 only the grade of concrete is specified for exposure classes XC1-4, and although it is noted that the minimum cement content CEM I is 200 kg/m^3. Conversely in Table 5.5 the grade of concrete is not specified for exposure classes XD and XS further to a note stating that the resistance to chloride ingress is mainly dependent on the type of cement and w/c ratio. The recommended grade of concrete for normal weight concrete is C20/25, except where shown for XD3.

Referring to BS EN 1992-1-1, clauses 4.4.1.1 (1) and 4.4.1.3 (3), nominal cover $C_{nom} = C_{min} + C_{dev}$. Where cover is regulated by quality control in the factory C_{dev} may be taken as 5 mm. \$C_{nom} = C_{min} + 5$ mm. Tables 5.4 and 5.5 give C_{nom} according to exposure class for grade of concrete, or maximum w/c ratio and minimum cement content, respectively. Table 5.6 combines the information in these tables showing the requirements for w/c ratio and cement content for grades of concrete from C20/25 to C45/55, which is also based on data by the UK Concrete Centre (Concrete Centre, 2024) for exposure up to XD3.

5.2.2 Exposure, cover and strength for T beams in beam and block flooring

The recommendations given in the product standard BS EN 15037-1 (BS EN 15037-1, 2008) are slightly different to the above. BS EN 15037-1, clause 4.3.7 states 'Alternative conditions described in Annex A of BS EN 13369 (BS EN 13369, 2004) may apply', i.e. cover to the lower face may comply with the values given for slabs in Table A.2 of the common rules for precast concrete BS EN 13369.

For internal exposure XC1, Table A.1 refers to Environmental Condition B. For external moderate exposure XC3, Table A.1 refers to Condition C.

Table 5.4 Relationship between exposure conditions and classes, concrete strength and nominal cover for prestressed concrete using cement CEM I according to BS 8500-1:2023, Table A.4a for 50 years working life for corrosion induced by carbonation.

		Nominal cover (mm) for minimum grade of concrete		
Exposure condition	*Exposure class*	*C28/35*	*C40/50*	*≥C45/55*
Inside enclosed buildings, low humidity. Permanently wet.	XC1	20	20	20
Long term contact with water, not containing chlorides.	XC2	20	20	20
Inside building, moderate and high humidity.	XC3/4	60	35	30
External, sheltered from rain (ground floors). Cyclic wet and dry.				

Table 5.5 Relationship between exposure conditions and classes, maximum water/cement ratio and minimum cement content and nominal cover for prestressed concrete using cement CEM I with 20 mm aggregate size according to BS 8500-1:2023, Table A.4b and A.4c for 50 years working life for corrosion induced by chlorides

Exposure Condition	*Exposure class*	*Nominal cover (mm)*			
	Max w/c ratio	*0.50*	*0.45*	*0.40*	*0.35*
	Min cement (kg/m³)	*340*	*360*	*380*	*380*
Moderate humidity. ≤10 m from carriageway. Slight chloride conditions. Airborne chlorides (open car parks).	XD1	35	30	30	30
Wet, rarely dry. Immersed in water containing chlorides.	XD2	40	40	35	35
Cyclic wet and dry. ≤10 m from carriageway. Bridges parapet edge beams. Reinforced pavements and car park slabs.	XD3 Concrete grade		60 >C28/35	50 >C32/40	45 >C35/45
Airborne salt. Not direct contact to sea water. Surfaces in coastal areas.	XS1	65	60	55	50
Permanently submerged. Marine permanently wet. Below mid-tide level.	XS2	85	80	70	65
Spray containing chlorides. Marine tidal, splash and spray, and upper tidal zones. Exposed soffits above sea water.	XS3				85

Resistance to chloride ingress is mainly dependent upon the type of cement and water/cement (w/c) ratio. The recommended grade of concrete for normal weight concrete is C20/25, except where shown for XD3. Adjustment to cement content for different size of aggregate is given in BS 8500-1, Table A.7.

Table 5.6 Mix design requirements for XC and XD exposure classes using CEM I according to BS 8500-1:2023

Concrete grade	*Maximum w/c ratio : minimum cement content (kg/m³) for exposure classes*				
	XC1/2	*XC3/4*	*XD1*	*XD2*	*XD3*
C20/25	N/A : 200				
C25/30				0.55 : 320	
C28/35			0.6 : 300	0.55 : 320	0.45 : 360
C32/40			0.6 : 300	0.55 : 320	0.4 : 360
C35/45			0.6 : 300	0.5 : 340	0.35 : 380
C40/50		0.45 : 340	0.55 : 320	0.4 : 360	0.45 : 360 0.35 : 380
C45/55			0.45 : 360		

N/A = maximum w/c ratio not applicable to XC1/2 classes

Table A.2 For condition B, exposure XC1, concrete > C30/37 and design life = 50 years C_{min} = 15 mm. ∴C_{nom} = 20 mm
Table A.2 For condition C, exposure XC2/3, concrete > C35/45 and design life = 50 years C_{min} = 20 mm. ∴C_{nom} = 25 mm

However, BS EN 13369, clause A.2 states that for the alternative conditions to be applied (i) concrete class ≥ C40/50 and (ii) characteristic value of water absorption is less than 6.0%, and the concrete cover given in Table A.2 may be reduced by 5 mm. Under private arrangements, using an independent testing house in the UK, characteristic values of water absorption from two sets of six samples were found to be 4.05% and 2.52%, both well below 6%, together with mean compressive cube strengths of 77 and 61 N/mm^2, respectively. Cover may therefore be reduced by 5 mm such that C_{nom} = 25 − 5 = 20 mm for exposure XC3. The reduction of C_{nom} to 15 mm for exposure XC1 is not considered necessary as 20 mm is the smallest cover used in practice. Values of C_{nom} for other exposure conditions up to XD1 are given in Table 5.7.

5.2.3 Mortar

Mortar, a sand-cement semi-dry mix, is used during the construction of prestressed concrete elements for wet bedding of wide slab units and but infrequently for double-tee slabs and beams, as these are usually erected onto elastomeric (or felt) pads due to high concentrated bearing forces. Mortar is also used for completing the infill in the longitudinal joints between floor units (hcu, voided and double-tee), although it may be termed infill 'concrete' as it contains small sizes of gravel (typically 6 mm) rather than sand (up to 4 mm). In both cases of bearings and infill, the mortar is therefore structural. Mortar is also brushed into the joints between T beams and concrete blocks (not EPS blocks) in beam and block flooring in order to stabilise the elements. In this context, it is a non-structural mortar.

Mix design is according to BS EN 206 as quantities of cement, fine aggregates, water and admixtures & additives. The sand is typically graded at 0/4 mm, but may include 6 mm pea gravel for joints wider than 50 mm. BS EN 13139 (BS EN 13139, 2002) lists the preferred sizes of aggregates for mortar as: (min/max) 0/2 mm, 0/4 mm, 2/4 mm and 2/8 mm. Sieve sizes and other information are according to PD 6682 Part 3 (PD 6682-3, 2003). The strength of mortar in compression is generally not less than the parent material at the joint,

Table 5.7 Relationship between exposure classes and nominal cover for prestressed T beams according to BS EN 15037-1 and BS EN 13369

Exposure condition	*Condition to BS EN 13369*	*Class designation*	*Nominal cover (mm) for grade C55/65 for 50 years working life*
Inside building, low humidity. Permanently wet.	B	XC1	20
Long-term contact with water.	C	XC2	20
Inside building, moderate to high humidity. External, sheltered from rain (ground floors).	C	XC3	20
Cyclical contact with water.	D	XC4	25
Moderate humidity. Airborne chlorides (open car parks).	E	XD1	30

but in other cases such as longitudinal joints between slabs may be C25/30. The workability of mortar is classed as 'very low' with no slump.

Dry pack mortar may be used where it is not possible to pour; it has $w/c \approx 0.25-0.30$ and is therefore mouldable in the hand. It may be placed by hand and trowel and hammered into position, which removes free air. It may be either 'structural' or 'non-structural' as above, but it is not possible to make control cubes.

5.2.4 Aggregates

EN 206 clause 5.1.3 states that normal weight coarse aggregates from natural land sources conform to BS EN 12620 (BS EN 12620, 2002). Sea dredge aggregates are not used. Moisture content is measured according to BS EN 1097 Part 6 (BS EN 1097-6, 2000). Samples should not be taken from the base of the stock pile. BS EN 206 clause 5.5.2 states that normal weight aggregates should have an oven-dry density between 2000 and 2600 kg/m^3. BS EN 206 clause 5.2.7 states that aggregate chloride content is ≤0.2% by mass of cement. Chloride content is certified by the aggregate supplier, both in bulk or in bags.

Annex E of BS EN 13139 prescribes minimum test frequencies for the various properties and the appropriate test methods. Aggregates are checked for reactive silica content, which may cause alkali-silica reactions with cement, so they remain below a threshold value that has to be carefully experimentally assessed by the aggregate producer. If this is not satisfied low alkali cement with less than 0.75% alkalis may be used such that the total alkali content of the concrete is less than 3 kg/m^3.

5.2.5 Substitute materials

Recycled concrete aggregate (RCA) may be used as a substitute for natural aggregates up to a maximum replacement of 20% by weight of coarse aggregate not passing a 4 mm sieve. Waste hcu, Figure 5.18, produce a well-graded coarse RCCA but a very fine RCFA, too fine to be reintroduced in machine cast units. RCCA is stockpiled as shown in Figure 5.19; note that the angle of repose shows that this is as well graded and shaped as natural aggregate. RCFA can be used in wet-casting if not passing a 1 mm sieve. Recycled fines as dust is not used. The water absorption of all RCA is determined according to BS EN 1097 Part 6 (BS EN 1097-6, 2000) in the surface saturated dry condition. The chloride content of all RCA is determined according to BS EN 1744 Part 1 (BS EN 1744-1, 2009), clause 7, and is not greater than 0.2% CaCl content by weight. Recycled crushed brick and mortar, or any other crushed material of unknown parentage, is not used as replacement aggregate.

5.2.6 Admixtures

Admixtures comply with BS EN 934 Parts 1 to 3 (BS EN 934-1, 2008, BS EN 934-2, 2009, BS EN 934-3, 2009). Dosages are permitted not to be within the range given in proprietary data sheets providing that the performance of the concrete is verified by the manufacturer or, ready-mix supplier. Calcium chloride is not used as an admixture.

EN 206 clause 5.1.5 states that air entrainment agents and superplasticisers, excluding SCC, may be used to comply with BS EN 934-2, clause 4.2 as a water-reducing agent where total air content < 6%. Superplasticisers should comply with the testing standard BS EN 480 Part 1 (BS EN 480-1, 2006).

Figure 5.18 Waste or rejected hollow core slabs prepared for crushing towards recycled concrete aggregate (courtesy Bison Precast Ltd., UK).

Figure 5.19 Stockpiled recycled concrete aggregate, the angle of repose shows the RCA is as well graded and shaped as natural aggregate (courtesy Bison Precast Ltd., UK).

5.2.7 Water

Mixing water conforms to BS EN 1008, clause 3.1 (BS EN 1008, 2002) and should be potable from land sources. Sea water is not permitted. Water reclaimed from cement slurry must be filtered and the cement content is calculated and known from the specific gravity of the filtered slurry. Crushed ice may be used to replace 50% of the mixing water, or according to the manufacturer's specification. The maximum particle size for crushed ice is not more than about 20 mm. No particles of crushed ice are visible in the final mix.

5.3 STEEL REINFORCEMENT FOR PRESTRESSED CONCRETE

5.3.1 Pretensioning tendons

Two main types of steel tendons are used for pretensioning – (i) plain or indented (or crimped) wire, and (ii) 7-wire helical strand. Although 3-wire helical strand is used in some countries prestressing bars, such as 'Macalloy', are not common in precast factories. Figure 2.2 shows an array of 12.5 mm diameter 7-wire strand in a prestressed beam, and Figure 5.20 shows the same cross section in a hcu. Note the excellent compaction of concrete around the strand, particularly in the crevices between the outer wires. This is vital for the good transfer of stress to the concrete. Prestressing tendons conform to BS EN 10138 Parts 1 to 3 (BS EN 10138-1, 2015, BS EN 10138-2, 2015, BS EN 10138-3, 2015) (Parts 1 and 3 are under 'technical review'). Standard 7-wire helical strand of 9.3, 10.9, 12.5 and 15.7 mm diameter; standard 3-wire helical strand of 6.5 mm diameter, and super strand (Dyform) of 9.6, 12.9 and 15.2 mm diameter, may be used. Section properties are listed in Table 5.8.

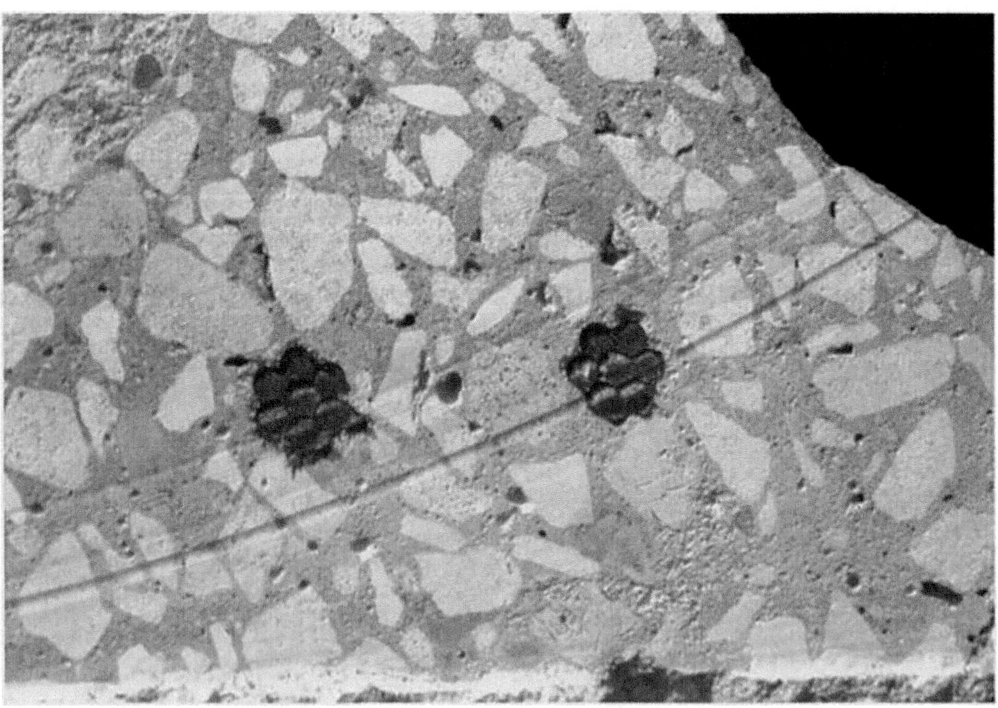

Figure 5.20 Cross section of 7-wire helical strand in a hollow core unit.

Table 5.8 Types and specification of pretensioning tendons

Type	*Nominal diameter (mm)*	*Cross-sectional area (mm²)*	*Characteristic load (kN)*	*Characteristic strength f_{pk} (N/mm²)*	*Modulus of elasticity (kN/mm²)*
Wire	5.0	19.6	34.7	1770	205
	7.0	38.5	64.2	1670	
Standard strand	9.3	52	92	1770	195
	10.9	71	126	1770	
	12.5	93	165	1770	
	15.2	138	244	1770	
Super strand	12.9	100	186	1860	195
	15.7	150	265	1770	
Drawn strand	12.7	112	209	1860	195
	15.2	165	300	1820	

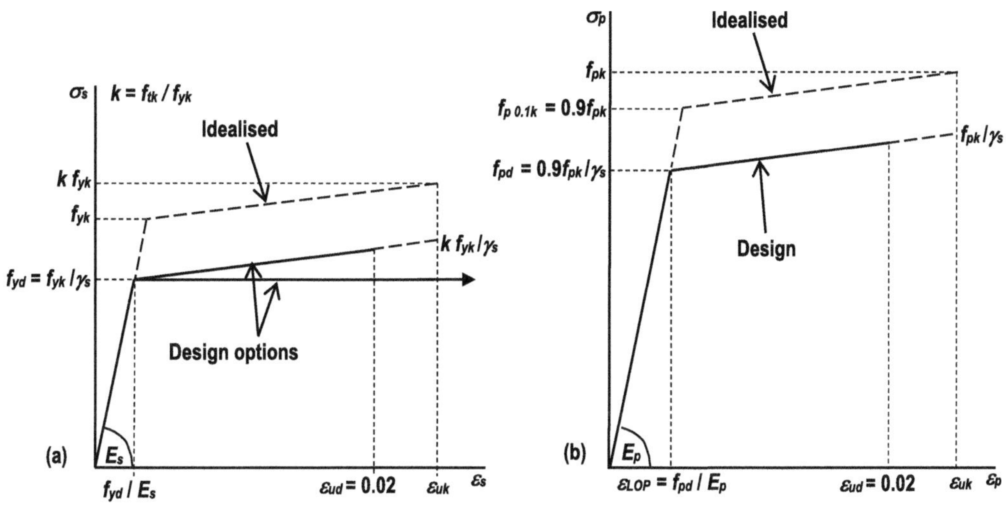

Figure 5.21 Idealised and design stress vs. strain relationships. (a) Reinforcing bars (subscript *y*), and (b) prestressing tendons (subscript *p*) (based on BS EN 1992-1-1, Figures 3.8 and 3.10).

The 5% characteristic tensile strength is taken as f_{pk}=1770 N/mm², except for 7 mm wire where f_{pk}=1670 N/mm². Super-stress strand is also available with f_{pk}=1770 to 1860 N/mm², with reported strengths of up to 2100 N/mm², and relaxation at 1000 hours ρ_{1000}=2.5%. The modulus of elasticity E_p is taken as 195 kN/mm² for helical strand, and 205 kN/mm² for indented wire. The elongation at the breaking load (obtained from BS 5896, 1980*) is taken as 3.5% or $\varepsilon_{uk}=35\times10^{-3}>0.0222$ used in BS EN 1992-1-1 as shown in the idealised stress v strain relationship in Figure 5.21b (Figure 5.21a shows the same for rebars for comparison). For hcu, the maximum diameter of strand is 16 mm and wire 11 mm according to BS EN 1168 (BS EN 1168, 2005), clause 4.1.1.1. The diameter of wires used in T beams for beam and block is 5 mm and less commonly 7 mm. (*Although this standard is being superceded by BS EN 10138 the material properties of strand supplied by the manufacturers still conform to Tables 4–6 of this standard.)

The choice of tendon is often a matter of the arrangement of tendons and the correct distribution of pretensioning force in a section. Large tendons should not be placed in thin wall sections; to avoid localised splitting and bond failure, the edge cover to tendons is at least twice the diameter. For this reason, helical strand is preferred in larger units or where the level of prestress is high.

Helical strand comprising of 3 or 7 hot rolled wires, and individual hot rolled plain or Belgian indented wire, is used according to BS EN 1992-1-1, clause 3.3.2 (4) as low relaxation *Class 2*. There are immediate and long-term losses of force called 'relaxation' due to the stress-relieving heat treatment process. *Class 2* tendons have 2.5% low relaxation, meaning that the residual stress after 1000 hours of relaxation is 97.5% of the original, although manufacturers may quote characteristic relaxation as low as 1.6% at 1000 hours. *Class 1* relaxation is greater but is rarely used today. When calculating relaxation losses, BS EN 1992-1-1 splits the losses into immediate and long term as discussed in Chapter 7, Section 7.2.4. Plain and indented wire is of 5 and 7 mm diameter, and ribbed wire of 9 mm diameter may be used. Table 5.8 lists the range of strength f_{pk}, modulus of elasticity E_p and area A_p of tendons used in prestressed elements.

The actual and idealised stress vs. strain relationships for tendons are given in BS EN 1992-1-1, Figs. 3.9 and 3.10. Figure 5.22 shows real tensile force vs. elongation for 7-helical strand of 12.5 mm diameter carried out by the manufacturer using grade 1860 strand. The measured cross-sectional area is 92.04 mm^2 (1.04% less than 93 mm^2 nominal). The mean ultimate tensile strength f_{pu} in 9 such tests is 1970 N/mm^2, with a standard deviation sd=8.5 N/mm^2, giving a characteristic f_{pk}=1956 N/mm^2, 5.2% greater than nominal. The mean tensile strength at 1% elongation $f_{p0.1}$ is 1856 N/mm^2, with sd=9.5 N/mm^2 and $f_{p0.1k}$=1840 N/mm^2. The mean ratio in the tests $f_{p0.1k}/f_{pk}$=0.94 is in reasonable agreement, yet conservative, with the idealised value of 0.9 given in BS EN 1992-1-1. Interestingly, the characteristic elastic modulus in the tests E_p=206 kN/mm^2 is 6% greater than the nominal 195 N/mm^2 for strand. Paradoxically the greater E_p increases elastic and creep losses of prestress, although this is only by 0.8% in the worked Example 7.1 in Chapter 7, giving a reduction of 0.4% in the service moment of resistance.

Referring to Figure 5.21b, the design value is f_{pk}/γ_s, where γ_s=1.15 for ultimate and 1.00 for accidental limit states from NA to BS EN 1992-1-1, Table 2.1N. Adequate ductility in tension may be assumed if $f_{pk}/f_{p0,1k}$≥k=1.1 from clause 3.3.6 (7) and the UK National Annex, then

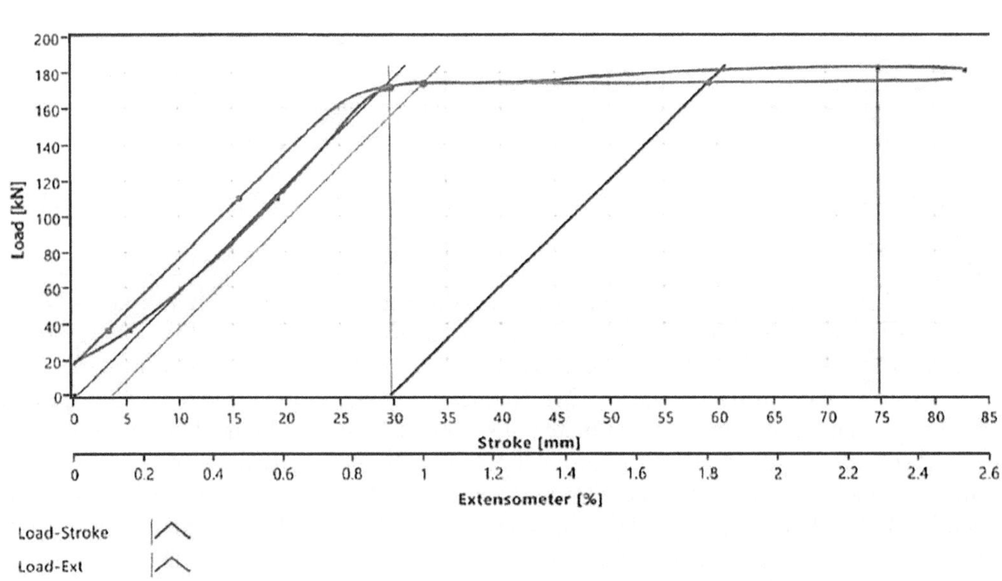

Figure 5.22 Real tensile load vs. elongation curves for 12.5 mm diameter 7-wire pretensioning strand. Axial load of 171 kN is equivalent to the design ultimate strength of 1860 N/mm^2.

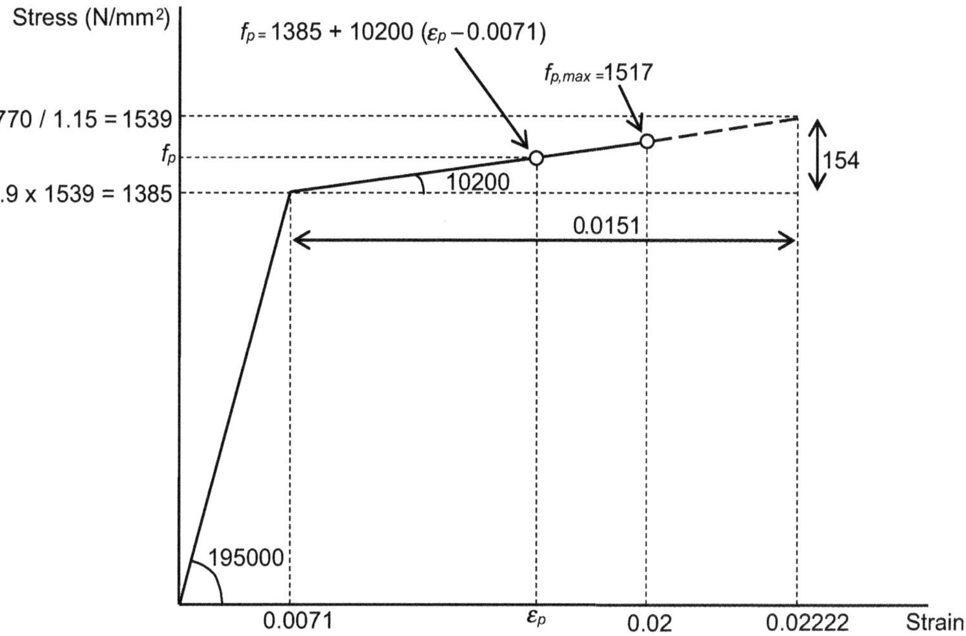

Figure 5.23 Design stress vs. strain diagram for standard helical strand using fpk = 1770 N/mm2. See Chapter 9, Figure 9.8 for the explanation of fp and εp.

$f_{p0,1k}=0.9\,f_{pk}$. The design strength at the limit of proportionality (LOP) is therefore $f_{pd}=0.9\,f_{pk}/\gamma_s$. The inclined branch is used for prestressing tendons as the additional strength between f_{pd} and f_{pk}/γ_s is worthwhile, even if the strain is limited to 0.02. Figure 5.23 shows the design values for strand for example $f_{pk}=1770$ N/mm^2. Then $f_{pd}=0.9\,f_{pk}/1.15=1385$ N/mm^2 and $f_{pk}/\gamma_s=1539$ N/mm^2, an increase of stress of 154 N/mm^2 for an increase in a strain ε_{uk} - $\varepsilon_{LOP}=(0.0222–1385/195000)=0.0151$. This gives $f_{p,max}=1517$ N/mm^2, an additional 132 N/mm^2 or 10% extra. The design may also be based on the actual stress vs. strain relationship, if this is known, with stress above the elastic limit reduced analogously with BS EN 1992-1-1, Fig. 3.10.

The example shown in Figure 5.23 shows the final ultimate limit state design point at f_p about half-way between f_{pd} and $f_{p,max}$ – a typical situation for elements with rows of tendons at the bottom of the section, such as hcu, voided units, thermal slabs, where the final strain $\varepsilon_p\approx0.014$. However in T beams for beam and block flooring, due to the narrow upstand and wires at a relatively higher position (than say hcu) the final strain $\varepsilon_p\approx0.007$ is close to the LOP, i.e. the wires remain almost elastic at the ultimate limit state.

5.3.2 Rebars

Prestressed concrete elements may contain high tensile hot rolled ribbed bar (HT rebar), for example as shear stirrups, top steel for lifting purposes, as confinement or strut-and-tie reinforcement or links at supports, and steel mesh in top flanges of double-tee or ribbed thermal units. They may also contain main rebars in partially prestressed elements, but these are not considered in this book, as they are dealt with extensively by Gilbert et al. (2017).

Rebars designated 'H' are used in 95% of cases, even in shear links where mild steel would be suitable. The small cost difference compared to the additional strength, *viz.* 500 vs. 250 N/mm^2, and the need for consistency of habit when assembling cages, makes it more economical. High tensile steel bars conform to BS EN 10080 (BS EN 10080, 2005) as B500A, B500B and B500C (some designers specify the bars as 'B' rather than 'H'). The technical specification according to the UK National Annex gives values of

- R_e=yield strength=500 N/mm^2 (note that strengths of 600 N/mm^2 are available in some countries)
- R_m/R_e=ratio of tensile strength/yield strength (also known as $k=f_{tk}/f_{yk}$)=1.05, 1.08 and 1.15 to 1.35 for classes A, B and C in BS EN 1992-1-1, Annex C.
- A_{gt}=percentage total elongation at maximum force=2.5% ($\varepsilon_{uk}=25\times10^{-3}$), 5.0% and 7.5%, respectively.

The standard also gives fatigue strength, bend performance, weldability, bond strength, shear and weld strength for use in welded fabric, tolerances and dimensions. B500 6 mm diameter reinforcing steel does not comply with BS EN 1992-1-1 with respect to ductility. Reinforcement shape codes are according to BS 8666 (BS 8666, 2005) and details according to ISO 3766 (BS EN ISO 3766, 2003).

The actual and idealised stress vs. strain relationships for reinforcing bars are given in BS EN 1992-1-1, Figs. 3.7 and 3.8. Typically $k=f_{tk}/f_{yk}=1.08$. The design value for bars f_{yk}/γ_s, where $\gamma_s=1.15$ and 1.00 as before. Then at ultimate $f_{yd}=0.87\ f_{yk}=435$ N/mm^2 for H500 rebar. The bi-linear idealised and design curves are shown here in Figure 5.21a. There are two design options known as the 'inclined' or 'horizontal' branches. The horizontal branch (having a constant value f_{yd}=435 N/mm^2 for H500 rebar) is more convenient for the design of reinforced concrete sections and has no limitation of strain in under-reinforced sections. The drawback with the inclined branch is that even if the strain in the bars at ultimate is say 0.01 the design stress f_{yd}=449 N/mm^2, i.e. only 14 N/mm^2 greater than 435 N/mm^2. Second, the strain is limited to $\varepsilon_{ud}=0.02$.

Mild steel is now less frequently used for projecting loops etc. even though it is easier to hand bend on site. Bar diameters commonly used are 8 and 10 mm for column links, 10 and 12 mm for beam stirrups and other distribution or anti-crack bars, and 16, 20, 25, 32 and 40 mm for main flexural bars. Plain round mild steel bar of yield strength 250 N/mm^2 does not conform to BS EN 10080 and BS EN 1992-1-1, clause 3.2.2 (3).

5.3.3 Mesh

Welded fabric, or 'mesh', is used in the top flanges of double-tee and thermal units. It is also used to reinforce structural toppings to floor slabs on site. The popular mesh sizes are A98, A142 or A193 (5, 6 or 7 mm bars at 200 mm centres in both directions). A rectangular (so called 'long') mesh C283 (6 mm at 100 mm centres × 5 mm at 400 mm centres) is often used in one way spanning units such as the flanges of double tee slabs. For rebar and mesh, the characteristic strength f_{yk}=500 N/mm^2. The design strength $f_{yd}=f_{yk}/\gamma_s$=500/1.15=438 N/mm^2. Young's modulus is taken as 200 kN/mm^2. There is no European standard for welded mesh; it still conforms to BS 4483 (BS 4483, 2005).

5.4 STRUCTURAL STEEL, WELDING, INSERTS AND BOLTS

5.4.1 Structural steelwork

Structural steelwork sections are used in end connections of prestressed beams, particularly in highly stressed half joints where the strut-and-tie resistance of the concrete and rebars is structurally or practically inadequate. Steel plates, rolled angles, channels and small rectangular and square hollow sections (RHS, SHS) may be used and anchored into the beams by welded rebars, loops, etc. Solid steel billets may be required to complete end connections, but as these are required in columns and walls they are beyond the scope of this book. See Elliott and Jolly (Elliott and Jolly, 2013) and Elliott (Elliott, 2017) for further information.

Structural steelwork conforms to BS EN 10025 Part 1 (BS EN 10025-1, 2004) and is designed according to BS EN 1993 Part 1-1 (BS EN 1993-1-1, 2005). The grade of hot-rolled structural steel is S275 (f_y/f_u=yield/ultimate=275/430 N/mm^2), S355 (355/510) or S450 (440/550) according to BS EN 10025 Part 2 (BS EN 10025-2, 2004). (These used to be known as grade 43, 51 and 55, respectively.) The grade of rectangular hollow sections is S275H (275/430) or S355H (355/510) according to BS EN 10210-1 (BS EN 10210-1, 1994). Its specification includes product analysis, carbon equivalent, yield strength, tensile strength, stress v strain data, and elongation. Steel strength is according to Table 5.9.

5.4.2 Welding

Rolled steel sections and flat steel plates are welded to form steel connectors in many highly stressed support situations where direct contact between concrete surfaces is to be avoided. Permitted procedures for lap or butt welding of rebars and spot welding of cruciform rebars to produce mesh is given in BS EN 1992-1-1, clause 3.2.5. This limiting ratio of diameters of 0.57 is not contravened for rectangular mesh.

Welding conforms to BS EN 1011, Part 1 (BS EN 1011-1, 2009). Welding equipment conforms to BS EN 60974, Part 1 (BS EN 60974-1, 2005) and may use proprietary plant such as Murex, Thermadyne, Rand, Migatronic, Cebora, etc. This BS was published in January 2006, but is currently being revised.

Welding electrodes conform to BS EN 1011 Parts 2 and 3 (BS EN 1011-2, 2001, BS EN 1011-3, 2000). Rod sizes should vary between 2.5 and 6 mm. Welded electrodes are mostly grade E275. When used to join grade S275 steel (previously known as grade 43) this gives the yield strength of the weld as f_{yw}=275 N/mm^2 and ultimate tensile strength f_{uw}=430 N/mm^2. When used in combination with grade S355 steel grade E355 electrodes gives f_{yw} and f_{uw}=355 and 510 N/mm^2. The values are divided by γ_{M2}=1.25, so that the design strengths are p_{ywd}=220 and 284 N/mm^2, respectively. Intermittent fillet welds are rarely used as weld lengths tend to be short. Weld strength is according to Table 5.10.

Welding of ferrous reinforcing bars uses either metal inert gas (MIG) welding or metal active gas (MAG) welding. The welding electrode rod for high tensile bar is a low hydrogen type to cover a wide range of applications for welding carbon manganese and low alloy steels.

Table 5.9 Yield and ultimate tensile strength of steel sections

Plate grade		*Plate yield strength f_y (N/mm^2)*	*Plate ultimate tensile strength f_u (N/mm^2)*
BS5950[a]	*BS EN 1993-1*		
	S235	235	360
43	S275	275	430
50	S355	355	510
	S420	420	520
	S450	450	550

Design same as yield strength $p_{yd}=f_y/\gamma_{M0}$. γ_{M0}= 1.0.

[a] Included for comparison of BS notation.

Table 5.10 Yield and ultimate tensile strength of welds

	BS EN 1993-1-8		
Weld grade	*Design strength p_{ywd} (N/mm²)*	*Weld yield strength f_{yw} (N/mm²)*	*Weld ultimate tensile strength f_{uw} (N/mm²)*
S235	188	235	360
S275	220	275	430
S355	284	355	510

Design $p_{ywd}=f_{yw}/\gamma_{M2}$. $\gamma_{M2}=1.25$

Welding of reinforcing steel bars to structural steel in load-bearing joints is in accordance with BS EN ISO 17660 Part 1 (BS EN ISO 17660-1, 2006). The processes used are shielded metal arc, gas metal arc. The welding electrode rods for B500 rebar is low hydrogen type electrodes to cover a wide range of applications for welding carbon manganese and low alloy steels. Spot welding of non-load-bearing bars is according to BS EN ISO 17660 Part 2 (BS EN ISO 17660-2, 2006).

Butt and fillet welds are fully fused fillet welds to make lap joints, corner joints, and T joints, having a convex profile, and may be either continuous or intermittent. Rebar T-joints, i.e. bars of equal diameter crossing at right angles, are a flared V groove weld. Butt welds may be used along a single edge in a single plane. The throat thickness of the weld, run-out lengths, under cutting, cracks, and inaccuracies of position are inspected for compliance, recorded and if necessary rejected. When thicker materials are welded, the edges are bevelled to form a single or double V-groove. No backup weld or backing strip is used. Preheating is not required, and post-weld heat treatment is normally not required unless specified on the production drawings. Site welding is carried out using an intermediate reinforcing bar provided in V-grooved butting joints. The diameter of the bar is equal to 0.7×the depth of V groove.

5.4.3 Cast-in fixings and lifting devices

Cast-in fixings, such as threaded sockets, lifting loops and sockets, spherical head anchors, rotating anchors, flat steel and tail anchors, etc. are used according to manufacturers' instructions, including edge distances, depth of embedment, anchorage reinforcement, concrete strengths, etc.

Lifting systems are designed according to their intended use and the mode of lifting, number of lifters and angle of lifting chains, concrete strength, element dimensions, type of lifting equipment, speed of lifting, and ground conditions. Spherical head lifting anchor and plate anchors are available in steel minimum grade S355 or stainless steel 304 or hot dip spun galvanised. Anchors are manufactured, batch-checked and tested by an independent testing house. Lifting loops are manufactured using galvanised steel wire minimum grade S500, and supplied with anchorages. All galvanised work is according to BS EN ISO 1461 (BS EN ISO 1461, 2022). Polypropylene loops may be used where steel is not appropriate. Care is taken not to pass these loops over sharp edges. Anchors have a minimum factor of safety of 3. The factor of safety against concrete failure is taken as 2.5. Bespoke lifting devices cast into concrete may be designed and manufactured by a manufacturer or frame erector. Load tests can carried out according to BS 5080 Part 1 (BS 5080-1, 1993) and evaluated according to BS EN 1990 (BS EN 1990, 2002).

5.4.4 Bolting

Bolting in prestressed concrete elements is uncommon. Making a fixing to a cast-in socket is the usual situation, but rarely to floor units. Metric bolts, set screws and nuts conform to BS 3692 (BS 3692, 2001). These are current documents with no BS EN. The minimum diameter for a structural bolt or threaded rod is 12 mm. Clearance holes should not be more than 2 mm greater than the nominal bolt or rod diameter. Holes should not be enlarged.

Strength grades for carbon steel are preferably limited to grades 4.6 and 8.8. The code, e.g. 4.6: 4 means ultimate strength f_{ub}=400 N/mm^2; 6 means yield strength f_{ybk}=0.6×400=240 N/mm^2. The design strength is $p_{ybd}=f_{ybk}/\gamma_{M2}$ where γ_{M2} =1.25. Grade strength of nuts is 6 and 8, respectively. Black bolts are grade 4:6 (p_{ybk}=240 N/mm^2 in tension and p_q=0.6×240=144 N/mm^2 in shear) and grade 8:8 (p_{ybk}=480 N/mm^2 and p_q=0.6×480=288 N/mm^2).

High strength friction grip bolts (HSFG) are used in special circumstances where the integrity and safety (both temporary and permanent) of connections made with ordinary bolts in clearance holes cannot be guaranteed. HSFG bolts conform to the UK National Annex to BS EN 1993 Part 1–8 (NA to BS EN 1993-1-8, 2005) using grades 8.8, 10.9 and 12.9. Preloading of HSFG bolts conforms to BS EN 14399 Part 1 (BS EN 14399-1, 2005). HSFG bolts are tightened using power tools showing specified torque or be tightened by hand using spanners according to the load indicating washer.

5.5 FIRE RESISTANCE OF PRESTRESSED ELEMENTS

5.5.1 Fire resistance and axis distance to rebars and tendons

This section deals with the relationship of fire resistance with the cover, or more precisely the mean axis distance, to the centroid of tendons contributing to the bending strength of prestressed slabs and beams. In the Eurocodes, fire is classed as an 'accidental situation of fire exposure', where the assessment of fire and its consequential conformity is according to BS EN 1992, Part 1–2 (BS EN 1992-1-2, 2004) and the UK National Annex (NA to BS EN 1992-1-2, 2004) 'General rules - Structural fire design'. The fire resistance of certain precast products is covered in product standards such as BS EN 13369 for the 'Common Rules', BS EN 1168 for hollow core and voided slabs and BS EN 15037-1 for T beams for beam and block flooring. Double-tee units and thermal units are classed as ribbed slabs and, apart from some thermal contours in the tapered ribs given in BS EN 13224 (BS EN 13224, 2004), are covered in BS EN 1992-1-2 as 'beams' (for the ribs) and 'slabs' for the top flanges. It is important to note that product standards are intended to be used for the product itself (beams, floor units) and not for the design of the completed building or floor slab.

The fire resistance of structural elements has three components

1. R: fire resistance class for the load-bearing criterion – the capacity of an element to resist loads in the fire limit state, such as bending moments $M_{Ed,fi} \leq M_{Rd,fi}$ where the fire load is taken as dead load plus the frequent fraction of the live load, $G_k+\psi_1 Q_k$ (see Chapter 3, Section 3.1.2). The fire load ratio is given in BS EN 1992-1-2, Exp. 2.4 as $\eta_{fi}=E_{d,fi}/E_d$. Then $M_{Ed,fi}=\eta_{fi}\, M_{Ed}$. The resistance $M_{Rd,fi}$ is based on the reduced strength of steel and concrete at elevated temperature.
2. E: fire resistance class for the integrity criterion – otherwise known as the separating criterion based on the wall thickness or slab depth of fire-resisting compartments, including plastering, screeds, finishes or other non-combustible layers, e.g. a 150 mm

deep concrete slab with 50 mm insulating layer plus a 75 mm sand/cement screed would have an equivalent fire thickness of 225 mm.

3. I: fire resistance class for the insulation criterion – this is also a separating criterion where the average temperature rise over the whole of the non-exposed surface is limited to 140°C, and the maximum temperature rise at any point of that surface does not exceed 180°C.

Fire resistance is measured by the ability of a structure or elements to satisfy the above for 30, 60, 90, 120, 180 or 240 minutes in standard fire exposure. Therefore if an element achieves $M_{Ed,fi}=M_{Rd,fi}$ after say 100 minutes of fire exposure its fire class is R 90. If the element also has sufficient thickness/depth to satisfy that requirement after say 75 minutes fire exposure its fire class is E 60, and is able to satisfy a separating criterion for 120 minutes its fire class is I 120. The element's combined class is then REI 60.

The design for fire resistance can be satisfied either by

i. member analysis in BS EN 1992-1-2, Sections 2 to 4, in which the criterion for the fire action $E_{Ed,fi} \leq$ fire resistance $E_{Rd,fi}$ where E represents shear, bending, bearing, buckling, etc.
ii. indirectly by the use of tables in BS EN 1992-1-2, Section 5 giving the requirements for member thickness or depth h_{eff}, width of beam b_{min}, width of webs b_w and centroidal cover distance (called 'axis' distance a) to steel bars/tendons, satisfying fire class R and E
iii. specific precast floors are covered in EC Product Standards

For (i) BS EN 1992-1-2, Section 4.2 member capacity is based on the reduced compressive strength of concrete factor $k_c(\theta)$ (Fig. 4.1), for rebars $k_s(\theta)$ (Fig. 4.2), and for prestressing tendons $k_p(\theta)$ (Fig. 4.3) at elevated temperature. BS EN 1992-1-2 Annexes B.1 and B.2 provide a systematic approach to spalling breadths as well as the calculation model for reinforced concrete $M_{Rd,fi}$ in Fig. B.2. Note that the strength of concrete in a fire situation $f_{cd,fi}(20)$ is in fact the strength of concrete at normal temperature $f_{ck}/\gamma_{c,fi}$ because $\gamma_{c,fi}=1.0$, so $k_c(\theta)$ is not used. Values for λ and η are the same as in BS EN 1992-1-1, clause 3.1.7 even though the stress-strain profile at elevated temperature given in BS EN 1992-1-2, Fig. 3.1 is quite different to BS EN 1992-1-1, Fig. 3.3. The net width of concrete in compression subjected to fire b_{fi}, i.e. sides of beams but not the top of slabs, is given in Fig. B.1 at the position of the 500°C isotherm. The calculation for $M_{Rd,fi}$ according to Fig. B.2 can be written for a singly reinforced section as

$$M_{Rd,fi} = A_{s,prov} k_s(\theta) f_{yk} z_{fi} \tag{5.26a}$$

where $A_{s,prov}$ means the area of rebar 'provided'

$$z_{fi} = d - 0.4x_{fi} \leq 0.95d \tag{5.27}$$

$$x_{fi} = A_s k_s(\theta) f_{yk} / \left(0.8 b_{fi} f_{ck}\right) \tag{5.28}$$

Then

$$M_{Rd,fi} = A_{s,prov} k_s(\theta) f_{yk} \left[d - \left(0.5 A_{s,prov} k_s(\theta) f_{yk} / b_{fi} f_{ck}\right)\right] \tag{5.26b}$$

But BS EN 1992-1-2, clause 4.2.1 (3) states that Annex E also provides a simplified calculation for the design of beams and slabs, subjected to predominantly UDL and linear analysis, i.e. the traditional methods used in BS 1992-1-1. Exp. E.3 is

$$M_{Rd,fi} = \gamma_s k_s(\theta)\left(A_{s,prov}/A_{s,req}\right)M_{Ed} \qquad (5.29)$$

where $A_{s,req}$ means the area of bar 'required' at ultimate

Note that $A_{s,prov}$, $A_{s,req}$ and M_{Ed} are for ambient temperature design.

A good approximation is that $A_{s,prov}/A_{s,req} \approx M_{Rd}/M_{Ed}$. In fact for the hcu analysed in Section 5.5.4, at almost full capacity $M_{Rd}/M_{Ed}=1.088$ and $A_{s,prov}/A_{s,req}=1.095$, a difference of 0.6%. If the hcu is say 30% under designed where $M_{Rd}/M_{Ed}=1.42$ then $A_{s,prov}/A_{s,req}=1.45$, a difference of 2.1%.

Substituting $M_{Ed,fi}=\eta_{fi}\, M_{Ed} \le M_{Rd,fi}$ Eq. 5.29 may be simplified even further as

$$M_{Ed}\ /\ M_{Rd} \le \gamma_s k_s(\theta)\ \ /\ \eta_{fi} \qquad (5.30)$$

A worked example is given in Section 5.5.4.

As BS EN 1992-1-2, Section 4.2 refers also to prestressing tendons it may be assumed that Exp. E.3 is also applicable to prestressed concrete as follows

Now

$$A_{s,req} = M_{Ed}/\left(z\ f_{yk}\ /\ \gamma_s\right) \qquad (5.31)$$

then

$$z = M_{Ed}\gamma_s/A_{s,req}f_{yk} \qquad (5.32)$$

If it is assumed, as suggested in BS EN 1168 for hcu, that the lever arm $z_{fi}=z$, which for most wide slabs is $z=0.95d$, then

$$M_{Rd,fi} = A_{s,prov}k_s(\theta)f_{yk}z \qquad \text{(as 5.26a above)}$$

Substituting $M_{Rd,fi} = A_{s,prov}k_s(\theta)f_{yk}M_{Ed}\gamma_s\ /\ A_{s,req}f_{yk}$

or

$$M_{Rd,fi} = \gamma_s k_s(\theta)\ \left(A_{s,prov}\ /\ A_{s,req}\right)M_{Ed} \qquad \text{(as 5.29 above)}$$

is the same as Exp. E.3.

For prestressed concrete Eqs. 5.26 and 5.29 are not exactly the same because there is a limiting value to the maximum stress in tendons which is lower than p_{yk}/γ_s, e.g. for $p_{yk}=1770$ N/mm^2 it is 1516 N/mm^2 < 1770/1.15 = 1539 N/mm^2. See Figure 5.23. This means that $\gamma_p=1770/1516 \approx 1.17$. However, the simplified design methods given by either BS EN 1992-1-2, Fig. B.2 or Exp. E.3 may be used.

For (ii), tabulated data for h_{eff}, b_{min}, b_w and a are given in BS EN 1992-1-2 Section 5 for beams and slabs as shown in Figure 5.24. Axis distance a=cover plus link diameter, neglecting the 2Φ radius bend at the corners of the links which often causes the main bars to rise up the section. For (iii), the fire resistance of prestressed hcu is obtained from BS EN 1168, Annex G, and for T beams and the completed beam and block floor is obtained indirectly from tables in BS EN 15037-1.

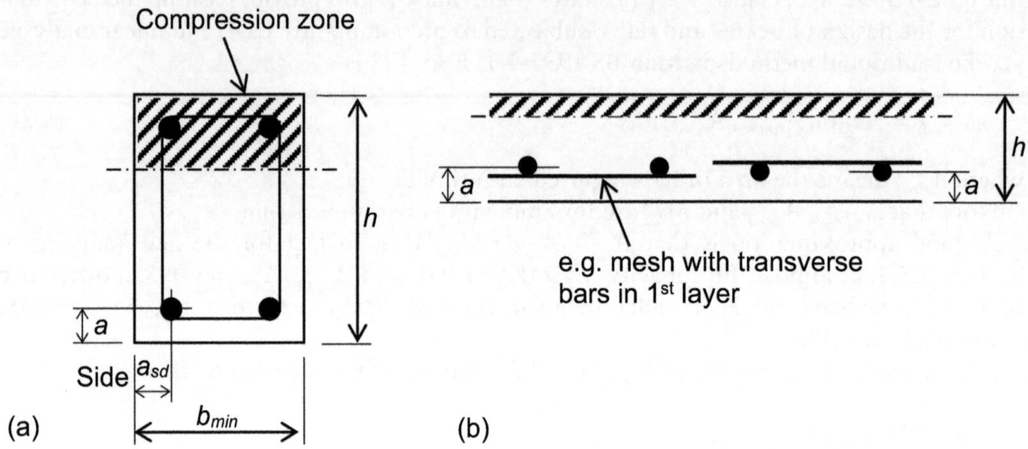

Figure 5.24 Definition of breadth b_{min}, depth *h* and axis distance *a* in fire situation (a) beams, (b) slabs (based on BS EN 1992-1-2, Section 5)

5.5.2 Axis distances to pretensioned tendons

The axis distance to tendons *a* is according to BS EN 1992-1-2, Table 5.8 for one-way spanning slabs, e.g. for R 90 a=30 mm, meaning that cover to larger strands may be 30-12.5/2=23.75 mm, which is rather close to the soffit to avoid splitting cracks beneath the strands at transfer. Therefore a=40 mm is more commonly used.

The axis distance *a* is according to BS EN 1992-1-2, Table 5.5 for simply supported beams, together with a minimum width b_{min}, e.g. for R 90 and a width of 400 mm a=35 mm, the smallest allowable.

Note that *a* is the mean axis distance to centroid of tendons according to area (not prestressing force) in the tension zone, i.e. at depths below $h-x_{fi}$, where x_{fi} is depth to neutral axis for fire moment $M_{Ed,fi}=\eta_{fi}\ M_{Ed}$ (See Section 5.5.1). Then $x_{fi}=(d-z_{fi})/0.4$, where $z_{fi}=d$ min{0.95; [0.5+√(0.25−K_{fi}/1.134)]}, where $K_{fi}=M_{Ed,fi}/f_{ck}\ b\ d^2$. BS EN 1992-1-2, clause 5.2 (14) states that *a* is a nominal value – allowance for tolerance need not be added.

For prestressed elements Tables 5.5, 5.6 and 5.8 state that *a* should be increased according to clause 5.2 (5) as follows: *For prestressing tendons the critical temperature for bars is assumed to be 400°C and for strands and wires to be 350°C. This assumption corresponds approximately to* $E_{d,fi}=0.7\ E_d$, $f_{p0,1k}/f_{pk}=0.9$ *and* $\gamma_s=1.15$ *(stress level* σ_s, $f_i/f_{p0,1k}=0.55$*). If no special check according to* clause 5.2 (7) *is made in prestressed tensile members, beams and slabs the required axis distance a should be increased by*

- *10 mm for prestressing bars, corresponding to* $\theta_{cr}=400°C$
- *15 mm for prestressing wires and strands, corresponding to* $\theta_{cr}=350°C$

BS EN 1992-1-2, Fig. 5.1, shows the relationship between the reduction factor $k_p(\theta_{cr})$ and critical temperature θ_{cr} for prestressing tendons. (Do not confuse this with $k_p(\theta)$ from Fig. 4.3.) Clause 5.2 (6 to 8) and Exps. 5.2 and 5.3 together with clause 2.4.2 allows the axis distance *a* to be increased by Δa (instead of 15 mm for wires and strands) if θ_{cr} is greater than 350°C and less than 500°C as

$$\Delta a = 0.1\left(500-\theta_{cr}\right)\ \text{mm} \tag{5.33}$$

e.g. if θ_{cr}=500°C, Δa=0 mm therefore a is not increased. But if θ_{cr}=400°C, Δa=10 mm a should be increased by only 10 mm.

The critical temperature in the tendons θ_{cr} is obtained from BS EN 1992-1-2, Fig. 5.1, or the following curve fitting equations. For prestressing tendons, curve 3 may be expressed as follows:

$$\begin{aligned}
\theta_{cr} &= 655.5 - 555.5k_p(\theta_{cr}) \quad \text{if } k_p(\theta_{cr}) = 0.55 \text{ to } 1.0 \\
\theta_{cr} &= 594.4 - 444.4k_p(\theta_{cr}) \quad \text{if } k_p(\theta_{cr}) = 0.1 \text{ to } 0.55 \\
\theta_{cr} &= 1200 - 6500k_p(\theta_{cr}) \quad \text{if } k_p(\theta_{cr}) < 0.1
\end{aligned} \tag{5.34}$$

where

$$k_p(\theta_{cr}) = \sigma_{p,fi}/p_{yk}(20°C) \tag{5.35}$$

and

$$\sigma_{p,fi} = (E_{d,fi}/E_d)(p_{yk}(20°C)/1.15)\ (A_{p,req}/A_{p,prov}) \tag{5.36}$$

The fire load ratio $\eta_{fi}=E_{d,fi}/E_d$ is given in BS EN 1992-1-2, clause 2.4.2, is the least of Code's Expressions 2.5a or 2.5b for the following load combinations.

Exp. 2.5a $$\eta_{fi} = (G_k + \psi_1 Q_k)/(\gamma_G G_k + \gamma_Q \psi_0 Q_k) \tag{5.37a}$$

Exp. 2.5a $$\eta_{fi} = (G_k + \psi_1 Q_k)/(\zeta\gamma_G G_k + \gamma_Q Q_k) \tag{5.37b}$$

where γ_G=1.35, $\zeta\gamma_G$=1.25 and γ_Q=1.5. Refer to Table 3.1 for values of ψ_0 and ψ_1.

BS EN 1992-1-2, clause 2.4.2 (4) states that as only thermal deformations across the cross section need be considered, and axial and in-plane expansions are ignored, η_{fi} may be taken as $M_{Ed,fi}/M_{Ed}$.

For example, if the fire load ratio in a beam is $E_{d,fi}/E_d$=0.60, and the area ratio $A_{p,req}/A_{p,prov}$=1.00, $k_p(\theta_{cr})$=0.6×1/1.15=0.522, θ_{cr}=594.4−444.4×0.522=362°C and Δa=13.7 mm. Therefore for R 90 fire rating axis distance needs to be a≥35+13.7=48.7 mm. For a greatly under-designed beam (greater area of tendons have been provided than required), the area ratio might be say 0.70, then $k_p(\theta_{cr})$=0.365, θ_{cr}=432°C, Δa=6.8 mm, and a≥41.8 mm.

Note that clause 5.2 (7) states that the Δa method is not applicable to simply supported floor slabs, Table 5.8 is excluded. However it is known that $M_{Rd,fi}$ calculated using the fire-moment method given in BS EN 1992-1-2, clause 4.2 and Figure B.2 gives similar (or better) results. Therefore the Δa method may be used to achieve the same result.

Note that BS EN 1992-1-2, clause E.2 (2) states that the simplified calculations given by Eq. 5.26 or 5.29 provide an extension to the use of Tables 5.5–5.11. The calculations determine bending resistance for situations where the axis distance a to bottom reinforcement is less than required from the tables, and Exp. 5.2 (5) for Δa.

5.5.3 Floor slabs

The fire resistance of solid, ribbed, one-way spanning slabs, etc. is given in BS EN 1992-1-2, Section 5.7. Common Rules for Precast Elements BS EN 13369 clause 4.4.4.2 states tabulated data can be found in EN 1992-1-2, but when applicable complementary rules may be given in product standards.

Floor slabs are required to meet the criterion REI in terms of h_s given in BS EN 1992-1-2, Fig. 5.7 and Table 5.8 to ensure an adequate separating function (Criterion E and I). If a load-bearing function (Criterion R) is required alone the necessary slab thickness h_1, assumed for design to BS EN 1992-1-1, may be taken. Values for h_s and axis distance a are according to Table 5.11 for simply supported one-way spanning slabs and for the top flanges of double-tee units (based on BS EN 1992-1-2, Table 5.8).

5.5.4 Solid plank and hollow core units

The fire resistance of prestressed concrete hcu is a function of depth h and mean axis distance a. Depth h corresponds with the minimum solid floor thickness given in Table 5.8 of BS EN 1992-1-2, and has been calculated using a conversion equation for hcu as $t_e = h\sqrt{\zeta}$ where h=actual depth and ζ=ratio of solid material (including filled joints) to the whole. This explains the differences of depth required for solid units and hcu in Table 5.11, e.g. a certain 150 mm depth hcu has a fire rating of R 60 compared with a solid slab of R 180. This is because the voids occupy 40% of the gross cross-sectional area and therefore $\zeta=0.6$. A solid slab of equivalent depth $\sqrt{0.6}\times150=116$ mm would have a rating of R 90. Note that the data in BS EN 1168, Table G.1 are for siliceous gravel aggregates: granite, sandstone, greywacke perform no worse. If calcareous or dolomitic limestone is used the dimensions h and a may be increased by 10%. For example, a 150 mm deep hcu made of limestone aggregate with $h=1.1\times150=165$ mm has a rating of R 90. The data given in BS EN 1992-1-2 and BS EN 1168 are therefore still appropriate to isolated simply supported units with no inherent longitudinal or transverse boundary conditions representative of floor panels tied into internal or peripheral beams.

Over the past 20 years, there has been a huge effort to establish the behaviour of prestressed hcu, both as isolated units and in floor panels, exposed to fire, culminating in a European research project led by BIBM (Bureau International du Beton Manufacture) and IPHA (International Prestressed Hollow Core Association), known as 'Holcofire'. The project leader Wim Jansze (of Consolis and IPHA) presented and coordinated the report (BIBM/IPHA, 2014), which included

i. a state-of-the-art review, database and analysis of 162 laboratory fire tests carried out between 1966 and 2010
ii. the results from fire tests to validate the shear and anchorage capacity of hcu (i) under restrained conditions (i.e. with longitudinal and/or transversal steel tie bars), and (ii) according to BS EN 1168, Annex G

Table 5.11 The total thickness and axis distance for solid plank and hollow core units and composite slabs required for fire resistance

	Total thickness for solid, hollow core units and composite slab (mm)		
Fire resistance (mins)	*Solid unit to BS EN 1992-1-2*	*Hollow core to BS EN 1168*	*Axis distance a to EC2 or EN1168 (mm)*
30	80	130	10
60	80	130	20
90	100	160	30
120	120	200	40
180	150	250	55

For prestressed slabs increase a according to BS EN 1992-1-2, clause 5.2(5 to 8).

iii. a review of the failure of hcu due to cars on fire at the parking garage at Lloydstraat in Rotterdam in 2007 (Efectis Nederland 2007, TNO Bouw en Onderground 2008)
iv. the shear capacity of hcu on flexible supports, such as shallow steel beams, but the conclusions were similar to hcu on rigid supports
v. a simulation model for buckling and horizontal web cracking

Figure 5.25 shows a photograph of the soffit of a prestressed hcu subjected to a standard ISO 834 fire (ISO 834-1 1999), with a soffit temperature of 1050°C after 122 minutes, which was later tested to shear failure as part of the Holcofire project. The original colour photograph shows a slight discolouration (pinking) of parts of the soffit and a tiny patch of exposed aggregate.

The most significant findings of the state-of-the-art review are summarised in Figure 5.26, showing the relationship between the calculated shear capacity $V_{Rd,c,fi}$ (y axis) according to BS EN 1168, Annex G.1.3 and the failure shear capacity in 42 fire tests. Data lying above the 1:1 line are considered to be unsafe, of which there are 20 in number. However, 12 of these had been carried out using 'double-web' hcu, i.e. units split along the line of two cores. Although the side webs were insulated from fire there was no two-dimensional system effect that is present in full 1.2 m wide units, or full floor panels arresting deformation by horizontal blocking.

Tests from 31 no. single slabs (i.e. 1200 mm wide hcu) and 102 no. tests on floor systems (I.e. hcu bearing onto supporting beams, similar to those shown in Figure 5.27) all lie below

Figure 5.25 The soffit of a prestressed hcu and supporting beam after 122 minutes of a standard ISO fire (courtesy BIBM/IPHA).

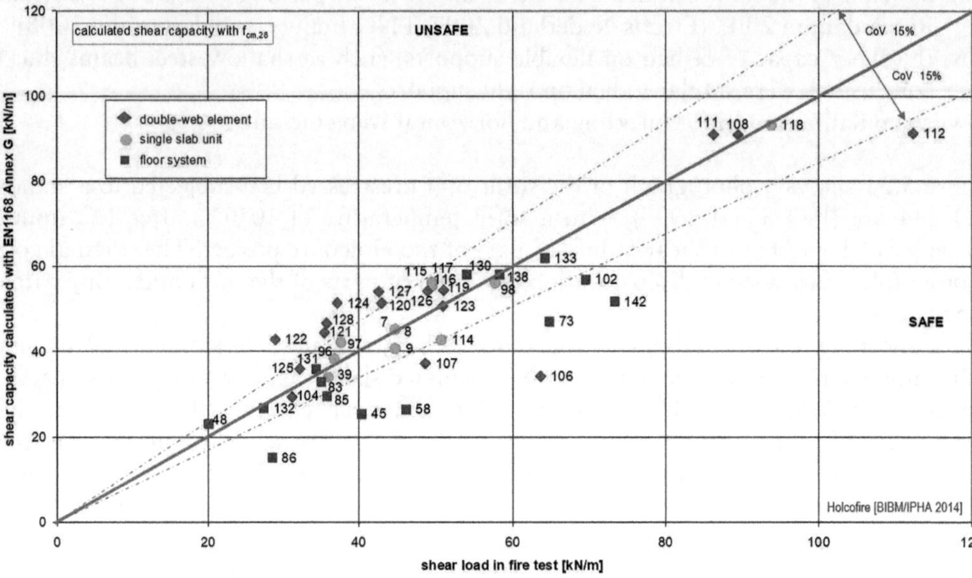

Figure 5.26 Relationship between calculated shear capacity according to BS EN 1168, Annex G and shear load in fire tests reported by BIBM/IPHA (Fig. 2.12 in BIBM/IPHA, 2014) (courtesy BIBM/IPHA).

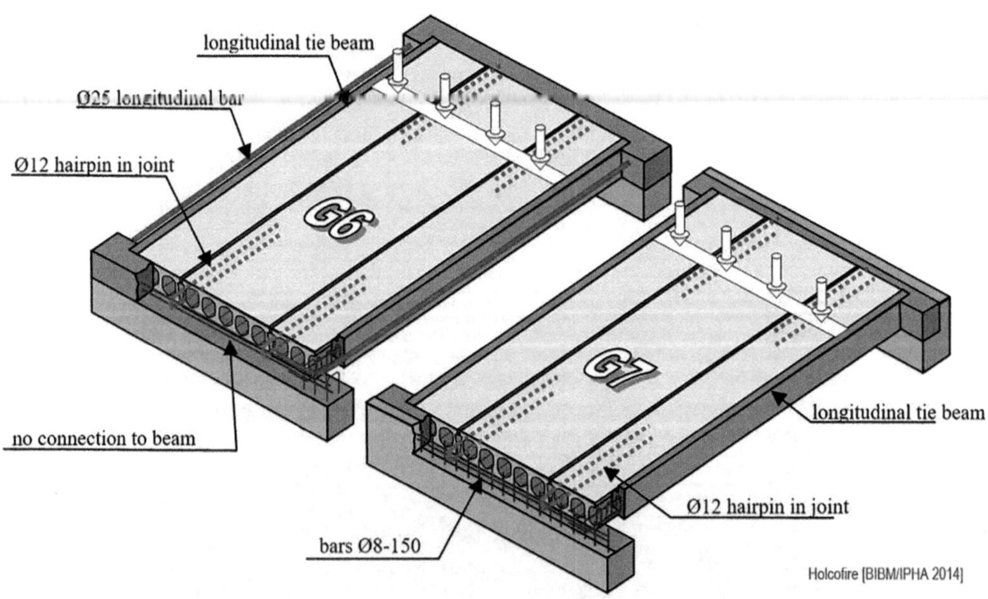

Figure 5.27 Arrangement of hollow core floor slabs in two (out of 6) fire tests under shear load with transverse tie beams reported by BIBM (Fig. 3.4 in BIBM/IPHA, 2014) (courtesy BIBM/IPHA).

the 1:1 line in the safe zone. It is not surprising to find that 'double-web' hcu have so-called 'unsafe' shear failures:- a significant observation made in all hollow core fire tests is the presence of vertical splitting cracks in the webs in a region of high shear force. These are due to a thru-thickness non-linear temperature gradient causing differential expansion, resulting in compression in the top and bottom flanges and compatibility tension in the webs, reducing the principal tensile capacity of the concrete in the webs (Van Acker, 2010). Double-web units would therefore be susceptible to such splitting cracks as deformation would not be arrested by horizontal blocking in these test specimens.

Figure 5.27 shows two of the seven floor system fire test arrangements as part of the 'Holcofire' project. Five tests have longitudinal steel tie bars, representing either gable edge tie beams or the presence of adjacent units, and transverse support beams. Two tests have 50 mm structural topping reinforced using A142 mesh. The floors were 3.9 m in length, 2.4 m width and h=265 mm depth. Shear loads were activated in the furnace for at least 120 minutes at a distance of 2.5h from the support. Test G6 is not structurally connected to the support beam; however, the hcu are tied at their ends using 2 no. 12 mm diameter hairpin bars cast *insitu* into the longitudinal joint. G7 therefore represents a typical floor system. One test had hairpin bars cast into 4 no. top opened cores.

Prior to the fire tests 10 no. hcu were tested under ambient conditions (referred to as 'cold' in BS EN 1168, Annex G) returning an average test shear capacity of $V_{u,cold}$=218.3 kN/m width. The design value according to BS EN 1992-1-1, Exp. 6.4 is $V_{Rd,c,cold}$=97.3 kN/m width. Thus the ratio $V_{u,cold}/V_{Rd,c,cold}$=2.24 is not unexpected for hcu of this depth. The maximum shear reaction measured in the fire tests ranged from $V_{u,fi}$=82.5–134.7 kN/m width, such that $V_{u,fi}/V_{Rd,c,cold}$=0.78–1.38 with an average of 1.09. As the design data for the ratio $V_{Rd,c,fi}/V_{Rd,c,cold}$ given in Annex C, Table G.2 for this slab at 120 minutes is 0.55 is a characteristic value, then the test ratio should also be characteristic – but this is not possible due to the test parameters all being different. Note that the ratios in Table G.2 include for an area of longitudinal tie steel of 188 mm²/m (based on 2 no H12 bars per unit), whereas the tie steel in the tests was less than this. However, it is unlikely that the characteristic value associated with the lowest test ratio of 0.78 would not be less than 0.55. Therefore it is concluded that Table G.2 is conservative, at least for the depth of hcu and fire period in these tests, and using the formula will result in a more optimal design. Finally, the calculated shear capacity in fire, according to Annex G.1.3 at 120 minutes, varied from $V_{Rd,c,fi}$=40.7–63.7 kN/m width, such that $V_{u,fi}/V_{Rd,c,fi}$=1.58–2.66 with an average of 2.09. The results show considerable reserve of shear strength compared with $V_{Rd,c,fi}$. Table 5.12 summarises the data in Annex G, Table G.2. To conclude, the Holcofire project confirmed the shear capacity according to BS EN 1168.

The UK BRE carried out two fire tests in 2001 on floor assemblies comprising 5 no.×6.0 m span×200 mm depth hcu (self-weight 2.67 kN/m²) with a 50 mm structural topping in Test 1 and 500 mm long hooked end rebars in the longitudinal joints Test 2 (Lennon, 2003). Axis distance to 9.3 mm diameter strands was based on 40 mm cover for 2 hours fire resistance according to BS 8110, Part 2 (BS 8110, 1985), then a=44.6 mm. Subjecting the slabs to at least 60 minutes exposure, under an imposed UDL of 3.66 kN/m², there was no loss of bearing capacity, minimal spalling to the soffit, and the results of temperature profiles were in agreement with BS EN 1992-1-2, Annex A for compartment temperature. Using load factors with ψ_1=1 and ultimate design based on Eurocodes

Table 5.12 Ratio of the ultimate shear capacity of prestressed hollow core units exposed to fire and ambient (cold) conditions, from BS EN 1168, Table G.2

Fire resistance (mins)	*Ratio of $V_{Rd,c,fi}/V_{Rd,c,cold}$ for depth of hcu (mm)*				
	160	*200*	*240*	*280*	*320*
60	0.70	0.65	0.60	0.60	0.60
90	0.65	0.60	0.60	0.60	0.55
120	0.60	0.60	0.55	0.55	0.50
180	0.45	0.50	0.50	0.50	0.45

for R 120 minutes the author has calculated: $M_{Ed,fi}$=28.5 kNm/m<$M_{Rd,fi}$=34.5 kNm/m; $V_{Ed,fi}$=18.1 kN/m width<$V_{Rd,c,fi}$=0.6×62.3=37.4 kNm/m; confirming that the slabs should not have failed in bending or shear.

Example 5.4

Calculate the shear capacity $V_{Rd,c,fi}$ and moment of resistance $M_{Rd,fi}$ in a fire situation for the 1200×265 mm depth prestressed hcu shown in cross section in Figure 5.14 exposed to 120 minutes fire. The hcu is similar to that in the 'Holcofire' tests and is pretensioned using 6 no. 12.5 mm diameter strands (area A_p=93 mm²) stressed to 70%×1770 N/mm²=1239 N/mm² at 50 mm axis distance from the soffit. Use f_{ck}=40 N/mm², siliceous gravel aggregates, 70 mm bearing length, height to centroid y_b=126 mm. Assume that the final loss of prestress is 20%. E_p for strands is 195000 N/mm². The ultimate moment of resistance M_{Rd}=168.7 kNm. Self-weight of hcu including infilled joints=3.93 kN/m². The floor has an effective span of 8.0 m and is subjected to imposed dead and live UDL of 3.0 and 5.0 kN/m², respectively. Use office loading with ψ_1=0.5.

Note that the axis distance of 50 mm suggests that Δa=10 mm, i.e. from Table 5.8 a=40 mm, therefore Δa=50−40=10 mm. However, the results from the fire tests and the fire moment calculation show that this is adequate.

Solution

Service UDL w_s=(3.93+3.0)+5.0=11.93 kN/m²
Ultimate UDL (Exp. 6.10b of BS EN 1990) w_{Ed}=1.25×6.93+1.5×5.0=16.15 kN/m²
Ultimate moment M_{Ed}=1.2×16.15×8.0²/8=155.0 kNm per unit.
Shear span=8.0–2×(0.07/2+0.126)=7.678 m
Ultimate shear force V_{Ed}=1.2×16.15×7.678/2=74.4 kN.

Shear capacity check

BS EN 1992-1-1, Exp. 6.4 gives $V_{Rd,c,cold}$=121.3 kN. Table 5.12 gives $V_{Rd,c,fi}$=0.55 $V_{Rd,c,cold}$=66.7 kN. The ultimate shear force in a fire situation $V_{Ed,fi}=\eta_{fi}\,V_{Ed}$, where the reduction factor η_{fi} is according to BS EN 1992-1-2, Exp. 2.5b as

$$\eta_{fi} = G_k + \psi_1 Q_k / (\xi \gamma_G G_k + \gamma_Q Q_k) \tag{5.37b}$$

i.e. η_{fi}=frequent load/ultimate load

$$\eta_{fi} = \left[(3.93+3.0)+(0.5\times 5.0)\right]/16.15 = 9.43/16.15 = 0.584$$

$V_{Ed,fi}$=0.584×74.4=43.4 kN<66.7 kN. OK

Ultimate moment of resistance check

The first check is whether the bottom zone of the hcu may be classed as a solid section as defined in BS EN 1168, Fig. G.1. If the tendons are located in a zone of depth $a_{50\%}$, i.e. the height at which the sum of the widths of the cores is 50% of the total width of the slab (50%×1200=600 mm, or 120 m per core) the temperature around the strands may be assumed to be equal to that of a solid section. The profile of the hcu used in the 'Holcofire' tests gives this height as $a_{50\%}\approx$70 mm>50 mm axis distance to strands is satisfactory.

Method 1

The moment of resistance in a fire situation $M_{Ed,fi}$ is calculated using BS EN 1992-1-2, Section 4.2 and Annex B. The ultimate moment of resistance $M_{Rd,fi}$ for a prestressed section is approached in the same way as a reinforced section, i.e. the final pre-strain after losses in the strands is not additive to the compatibility strain. The ultimate design moment M_{Ed}=155.0 kNm per unit. Then $M_{Ed,fi}$=0.584×155.0=90.5 kNm.

BS EN 1992-1-2, Fig. A.2 gives the temperature at the level of the tendons at a=50 mm for R120 minutes as θ_s=385°C. The temperature of the concrete at the top of the slab is taken as θ_c=160°C, based on BS EN 1168, Annex G.1.2.

BS EN 1992-1-2, Fig. 4.1 gives the concrete strength reduction factor

$$k_c(\theta) = 0.97 \text{ for } \theta_c = 160°\text{C}$$

$$\text{Then} \quad f_{ck}(\theta) = 0.97 \times 40 = 38.8 \text{ N/mm}^2$$

(Note that the result of the calculation is the same if $k_c(\theta)$ is ignored.)

BS EN 1992-1-2, Fig. 4.3, curve 1a gives the Class A tendon strength reduction factor

$$\theta_s = 385°\text{C}$$

$$k_p(\theta) = 0.524$$

$$\text{Then} \quad f_p(\theta) = 0.524 \times 1770 = 926.7 \text{ N/mm}^2$$

$$\text{Tension force} \quad F_p(\theta) = 926.7 \times 558 = 517100 \text{ N}$$

$$\text{Depth to neutral axis} \quad X(\theta) = 517100 / (0.8 \times 1200 \times 38.8) = 13.9 \text{ mm}$$

Effective depth d=265−50=215 mm

Lever arm $z(\theta)=d-0.4\,X(\theta)$=min {215−0.4×13.9; 0.95×215}=204.3 mm

$M_{Rd,fi}=F_p(\theta)\,z(\theta)$=517100×204.3/10^6=105.6 kNm>90.5 kNm. OK.

Note that in order to achieve $M_{Rd,fi}=M_{Ed,fi}$=90.5 kNm, the axis distance may be reduced to a=44.8 mm, i.e. Δa=4.8 mm. For this situation θ_s=431°C, $k_p(\theta)$=0.438, $f_p(\theta)$=775.2 N/mm², $X(\theta)$=11.6 mm, d=220.2 mm, $z(\theta)$=209.2 mm, $M_{Rd,fi}=F_p(\theta)\,z(\theta)$=432547×209.2/10^6=90.5 kNm.

Method 2

Using BS EN 1992-1-2, Annex E, Exp. E.3

$$\text{Eq. 5.29} \quad M_{Rd,fi}=\gamma_p\, k_p(\theta)\,(A_{p,prov}/A_{p,req})\, M_{Ed}$$

The area of strands $A_{p,req}$ to satisfy $M_{Ed}=M_{Rd}$=155.0 kNm can be achieved in one of 2 ways (i) by trial and error inputting $A_{p,req}$ (but as shown in Chapter 9, Section 9.4.1 it is too complex to back-calculate $A_{p,req}$ from M_{Rd}) or (ii) approximation using traditional bending design and assuming a value for the limiting strength of strands.

By (i) it is found that $A_{p,req}$=511 mm²

By (ii)

$$K=155.0\times10^6/(40\times1200\times215^2)=0.070$$

$$z=215 \min\{0.95;\ (0.5+\sqrt{(0.25-0.070/1.134)})\}=200.8 \text{ mm}$$

$$x=(215-200.8)/0.4=35.5 \text{ mm (level with the top flange}=35 \text{ mm)}$$

Compatibility strain in strands $\varepsilon_{pu}=0.0035\times(215-35.5)/35.5=0.017697$

Prestrain $\varepsilon_{po}=0.8\times1239/195000=0.0050831$

Total $\varepsilon_p=0.222780$, but$>e_{ud}=0.02$

$f_p=1517$ N/mm^2

Then $A_{p,req}=155.0\times10^6/200.8\times1517=509$ mm^2. Remarkably close to 511 mm^2. Then 588 / 509 = 1.15 < 1.3.

Eq. 5.29 $M_{Rd,fi}=1.15\times0.524\times(588/509)\times155.0=102.4$ kNm>90.5 kNm (compared with 105.6 kNm in Method 1)

Then 588 / 509 = 1.15 < 1.3.

The condition may also be checked using Eq. 5.30, $M_{Ed}/M_{Rd}\leq\gamma_s\, k_p(\theta)/\eta_f$

$$M_{Ed}/M_{Rd}=155.0/168.7=0.918$$

$$\gamma_s k_p(\theta)/\eta_f=1.15\times0.524/0.584=1.032. \text{ OK}$$

It is interesting to note that in the 'Holcofire' tests without a structural topping the maximum bending moment (beneath the applied load) was 102 kNm per unit, in good agreement with the predicted $M_{Rd,fi}=105.6$ or 102.4 kNm above. No bending failure was reported in the tests.

Finally, for $\theta=160°$C and siliceous gravel aggregates, the strain in the concrete at the top of the hcu is given in BE EN 1992-1-2, Table 3.1 as $\varepsilon_{c1,\theta}=0.0049$.

The compatibility strain in the strands is $\varepsilon_{pu,\theta}=\varepsilon_{c1,\theta}\,(d-X(\theta))/X(\theta)$. But note that to achieve a lever arm $z(\theta)=0.95d$ the depth to the neutral axis is

$$X(\theta)=d(1-0.95)/0.4=0.125d.$$

Therefore

$$X(\theta)=0.125\times215=26.9 \text{ mm} \left(\text{not } 13.9 \text{ mm from above}\right).$$

Then

$$\varepsilon_{pu,\theta}=0.0049(215-26.9)/26.9=0.0343.$$

For $\theta=388°$C the limiting strain in the strands is given in BE EN 1992-1-2, Table 3.1 as $\varepsilon_{pu,\theta}=0.1094>0.0343$.

5.5.5 T beams in beam and block flooring

5.5.5.1 General considerations

The fire resistance of beam and block flooring, REI, is a function of the behaviour of both prestressed concrete beams and the infill blocks, and in certain cases whether a reinforced screed is present. Solid concrete blocks, known as semi-resisting (SR), contribute to the structural system, give fire protection to the narrow upstand of beams, and therefore the width of beams in BS EN 1992-1-2, Table 5.5 is based on the width at the soffit, or for beams with tapered sides at the top of the flange, e.g. single 150 mm beams with a soffit width of 92 mm or 88 mm at the top of the flange; double beams, grouted together, having a soffit width of 184 mm or 180 mm at the top of the flange, and so on. Figure 5.28 shows the equivalent profile of the exposed soffit of a double beam and fire-resistant blocks.

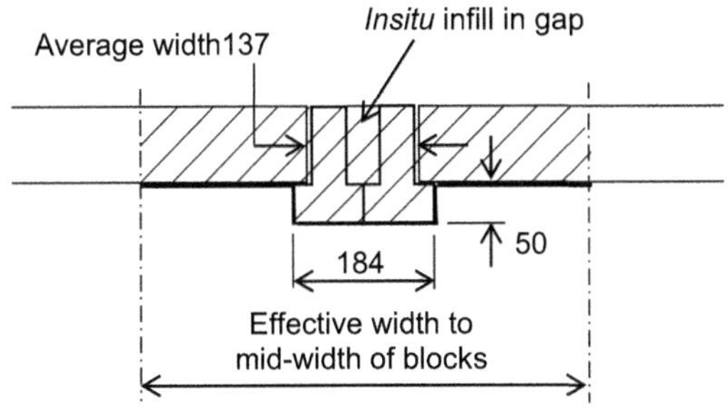

Figure 5.28 Equivalent profile of the exposed soffit for beam and block flooring up to 60 minutes fire resistance.

Polystyrene (EPS) blocks do not offer fire protection to the narrow upstand of beams and therefore the mean width of the upstand, e.g. 45 mm for 150 mm deep beams, is limiting with zero fire resistance. However, there are special recommendations in BS EN 15037-1, Annex K, for both of these situations. Clause K4.1 refers to the calculation in BS EN 1992-1-2 for bending resistance (shear resistance is not necessary). Clause K.5 and Table K.1 provide tabulated data based on tests of floor systems. The integrity criteria I require a screed reinforced with steel mesh.

In BS EN 1992-1-2 T beams are classed as beams and critical nominal axis distance a and minimum width b_{min} are given in Table 5.5. Clause 5.2 (5) states that 15 mm should be added to a and clause 5.2 (7) states that 15 mm may be replaced using Δa in the Code's Exp. 5.3 (see Section 5.5.2, Eqs. 5.33 to 5.35). As before, a=mean axis distance to centroid of tendons according to area (not prestressing force) in the tension zone. The beams are pretensioned with more than one row of wires (typically 3 or 4) so the additional side axis distance a_{sd} is not applicable.

Web thickness class is given in the NA to BS EN 1992-1-2 as WA in clause 5.6.1 (1). If fire-resistant blocks are specified the web thickness is replaced by the width of bottom flange. The required a and b_{min} may be reduced by 10% for calcareous aggregates, according to clause 5.1 (2). For multiple layers of wires, a_{min} ≥ half the average a according to clause 5.2 (17), e.g. if cover to 5 mm diameter wires = 20 mm, a_{min} = 22.5 mm and therefore average $a \leq 45$ mm (see Example 5.5).

The examples below are based on single or alternate single/double beams of 150 mm depth. For double, triple or quad beams, multiply the widths by 2, 3 and 4, respectively. Multiple beams should be touching and grouted between the webs

- For 30 minutes with b_{min} = 88 mm at top of the bottom flange, $a \geq 25+15^* = 40$ mm or 40/1.1 ≥ 36 mm for limestone, with $a_{min} \geq 20$ or 18 mm. Web width (upstand) ≥ 80 mm (or 73 mm using limestone) unless protected by fire-resistant solid blocks where bottom flange width is used.
- For 60 minutes beams with b_{min} = 88 mm do not comply (<100 mm required). Double beams with grout between the beams $b_{min} = 2 \times 88 + 4 = 180$ mm > 100 mm, do comply.

Tables 5.13a and 5.13b give a summary of the fire resistance for combinations of beams, blocks and allowable spans for siliceous gravel and limestone aggregates, respectively. Note there are situations where the maximum span is governed by deflection, a function of E_{cm}, and as this is reduced by 10% for limestone the maximum spans may be reduced in Table 5.13b.

Table 5.13a Fire resistance and clear spans of simply supported beam and block flooring using gravel aggregates

Depth of beam (mm)	*Beam arrangement*	*Fire resistance (minutes)*	*Maximum clear span (m) for imposed live load. Data includes 1.5 kN/m² for screed*	
			1.5 kN/m²	*3.0 kN/m²*
150	Single	R30[a]	4.08	3.55
	Double	R30[a]	5.20	4.56
		R60	4.34	3.96
	Triple	R30[a]	5.72	5.15
		R60	5.44	4.96
225	Single	R60	6.53	5.80
	Double	R60	7.90	7.08
		R90	7.71	7.08
	Triple	R90	8.58	7.74
		R120	6.42	5.98

[a] R60 is achieved using 40 mm minimum thickness screed reinforced using steel mesh or fibres.

Table 5.13b Fire resistance and clear spans of simply supported beam and block flooring using limestone aggregates

Depth of beam (mm)	*Beam arrangement*	*Fire resistance (minutes)*	*Maximum clear span (m) for imposed live load. Data includes 1.5 kN/m² for screed*	
			1.5 kN/m²	*3.0 kN/m²*
150	Single	R30[a]	4.08	3.55
	Double	R30[a]	5.20	4.56
		R60	5.16	4.56
	Triple	R60	5.84	5.15
		R90	4.74	4.34
225	Single	R60	6.53	5.80
	Double	R90	7.90	7.08
		R120	6.06	5.62
	Triple	R90	8.58	7.74
		R120	7.61	7.08

[a] R60 is achieved using 40 mm minimum thickness screed reinforced using steel mesh or fibres.

Beams with polystyrene blocks are governed by the average web width b_w (in Table 5.5) = 45 mm and 83 mm for 150 and 225 mm deep beams, respectively. 150 mm deep single beams do not comply as $b_w < 80$ mm. 225 mm deep single beams comply with R 30 minutes. Some 150 mm deep double beams are grouted in the 47 mm wide gap between the beams where $b_w = 45 + 92 = 137$ mm and comply with R 60 minutes, but still have to be checked for axis distance.

5.5.5.2 Fire resistance based on BS EN 15037-1

For beam and semi-resisting block floors with a cast *insitu* structural topping, the fire-resistance in BS EN 15037-1, clause K.5 and Table K.1, and clause B2.3 is satisfied if (i) beam centres are less than 700 mm, (ii) topping is 40 mm minimum thickness, (iii) reinforced with

(a) steel mesh or (b) reinforcing fibres (steel or polypropylene) if the imposed load ≤ 2.5 kN/m² and clear span ≤ 6.0 m. Toppings of less than 50 mm are not used in practice. The mechanical resistance depends on block type and the protection to the narrow web (upstand). Fire resistance REI are as follows:

Block types	*Resistance (mins)*
Polystyrene (over crawl spaces only)	30
Hollow concrete or clay	30
Solid concrete or clay	60

The data for non-solid blocks conflicts with BS EN 1992-1-2 for beams having less than 80 mm width of web, which have zero fire rating, irrespective of the positions of wires. This applies to 150 deep beams used singly.

Example 5.5

Determine the fire resistance for single and double 150 mm deep beams with 440 mm fire-resistant concrete infill blocks of 1950 kg/m³ density. The cross section of the beam and the positions of wires are shown in Figure 5.29. Due to 4 mm taper of the bottom flange b_{min} at the top of the flange is 88 mm. At the ultimate limit state the area of three tendons in the tension zone $A_p = 58.9$ mm² at a centroidal height of $(2 \times 22.5 + 40)/3 = 28.3$ mm from the soffit, stressed to 75% of $p_{yk} = 1770$ N/mm². Use $f_{ck} = 55$ N/mm² and siliceous gravel aggregates. The floor is subjected to dead UDL $g_k = 1.50$ kN/m² and live UDL $q_k = 1.50$ kN/m² over a simply supported span of 4.0 m. Use residential loading with $\psi_1 = 0.5$ and $\psi_0 = 0.7$. Average beam spacing is 488 and 290 mm, and the self-weight of beam and blocks is 2.173 and 2.385 kN/m² for single and double beams, respectively.

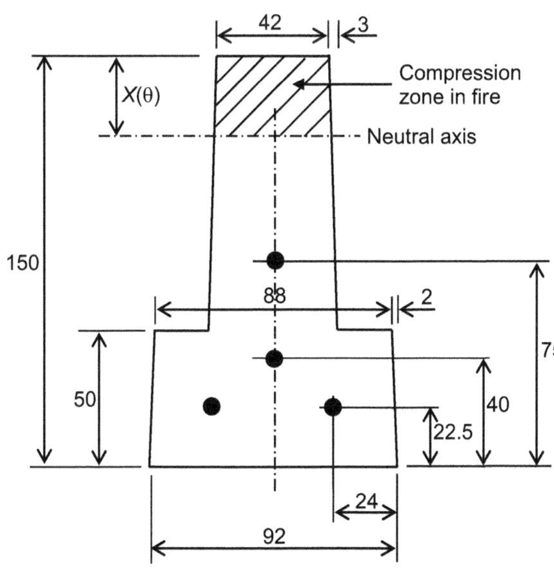

Figure 5.29 Cross section of 150 mm depth beams and nominal positions of wires in Example 5.5.

Solution

Single 150 mm beams. Double beams in brackets.

Dead UDL = 2.173 + 1.50 = 3.673 kN/m². (3.885)

Frequent live load = 0.5 × 1.50 = 0.75 kN/m²

Fire UDL $w_{d,fi}$ = 3.673 + 0.75 = 4.423 kN/m². (4.635)

Ultimate UDL w_{Ed} = 1.25 × 3.673 + 1.5 × 1.5 = 6.841 kN/m². (7.101)

Fire load ratio η_{fi} = 4.423/6.841 = 0.647. (0.653)

$M_{Ed,cold}$ = 0.488 × 6.841 × 4.0²/8 = 6.67 kNm per beam. (4.12)

d = 150 − 28.3 = 121.7 mm

K_{cold} = 6.67 × 10⁶/(55 × 42 × 121.7²) = 0.195. (0.120)

z = 93.8 mm and x = 70.8 mm. (106.5; 38.5)

Total strain ε_p = 0.006964 from which f_p = 1387 N/mm². (0.011474; 1432)

$A_{p,req}$ = 6.67 × 10⁶/(93.8 × 1387) = 51.3 mm². (27.0)

$A_{p,req}/A_{p,prov}$ = 51.3/58.9 = 0.871. (0.458)

$k_p(\theta_{cr})$ = 0.647 × 0.871/1.15 = 0.49. (0.26)

θ_{cr} = 594.4 − 444.4 × 0.49 = 377°C. (478°C)

Δa = 0.1 (500-377) = 12.3 mm. (2.1)

For $M_{Ed,fi}$ = 0.647 × 6.67 = 4.32 kNm, K_{fi} = 0.126 and x_{fi} = 40.7 mm from the top of the beam or 109.3 mm from the soffit, allowing four wires to contribute to the fire axis distance.

Average axis distance to four wires beneath the neutral axis in a fire situation a_{mean} = (2 × 22.5 + 40 + 75)/4 = 40 mm.

Axis height a_{min} = 22.5 > a_{mean}/2 = 40/2 = 20 mm. OK

Therefore, net axis distance $a = a_{mean}$ − Δa = 40 − 12.3 = 27.7 mm (37.9)

Single beam

Check whether a = 27.7 mm is adequate according to BS EN 1992-1-2, Table 5.5. The effective width of beam is at the top of the bottom flange b_{min} = 88 mm because the upstand is protected by the blocks.

For R 30, b_{min} ≥ 80 mm, a ≥ 25 mm. OK

For R 60, b_{min} ≥ 120 mm, a ≥ 40 mm. Fails on both criteria.

Final resistance = R 30.

To achieve R 60 according to BS EN 15037-1, Annex K, 40 mm minimum thickness reinforced screed with steel mesh or fibres is required.

Checking the single beam to determine $M_{Rd,fi}$ according to BS EN 1992-1-2, Annex B. Temperatures are obtained from BS EN 1992-1-2, Fig. A.3 for 150 × 80 mm beam for R 30 fire resistance.

Position of wire	*Row 1 corner*	*Row 1 corner*	*Row 2*	*Row 3*	*Totals*
Axis from soffit a (mm)	22.5	22.5	40.0	75.0	
Axis from side a_{sd} (mm)	23.0	23.0	44.0	44.0	
Temperature θ	540	540	380	300	
$k_p(\theta)$	0.236	0.236	0.540	0.700	
Strength $\sigma_p(\theta)$ (N/mm^2)	418	418	956	1239	
Force $F_p(\theta)$ (N)	8202	8202	18767	24328	**59499**
Effective depth (mm)	127.5	127.5	110.0	75.0	
Lever arm $z(\theta)$ (mm)	114.6	114.6	97.1	62.1	
$M_{Rd,fi}$ per wire (kNm)	0.940	0.940	1.823	1.511	**5.214**

Depth to neutral axis $X(\theta)=59499/(55\times0.8\times42)=32.2$ mm
Lever arm $z(\theta)=d-0.4\ X(\theta)$
$M_{Rd,fi}=F_p(\theta)\ z(\theta)=5.214$ kNm>4.32 kNm by a margin of 1.20

If the temperatures in rows 2 and 3 were taken as 400°C, rather than extrapolating below 400°C, then $M_{Rd,fi}=4.74$ kNm still>4.32 kNm.

Double beam

Check whether $a=37.9$ mm is adequate, $b_{min}=(2\times92)-4=180$ mm (the joint between beams is grouted)
For R 30, $b_{min}\geq160$ mm, a≥15 mm. OK
For R 60, $b_{min}\geq160$ mm, a≥35 mm. OK
For R 90, $b_{min}\geq150$ mm, a≥55 mm. Fails on axis distance.
Final resistance=R 60.

It is not possible to calculate $M_{Rd,fi}$ for R 60 as accurately as R 30 above as there are no temperature profiles for beams with a 50 mm deep bottom flange. However, if the concrete blocks offer a lower, but nonetheless active fire resistance, Fig. A.4 may be used for a 300×160 mm deep beam. The R 60 contours are used for the first and second rows of wires at $a=22.5$ and 40 mm, and the third row at $a=75$ mm is estimated. Note that for double beams the side axis distance for one of the wires in row 1 is $a_{sd}=90-23=67$ mm.

Position of wire	*Row 1 corner*	*Row 1*	*Row 2*	*Row 3*	*Totals*
Axis from soffit a (mm)	22.5	22.5	40.0	75.0	
Axis from side a_{sd} (mm)	23.0	67.0	44.0	44.0	
Temperature θ	710	560	480	450	
$k_p(\theta)$	0.085	0.204	0.340	0.400	
Strength $\sigma_p(\theta)$ (N/mm^2)	150	361	602	708	
Force $F_p(\theta)$ (N)	2954	7090	11816	13902	**35762**
Effective depth (mm)	127.5	127.5	110.0	75.0	
Lever arm $z(\theta)$ (mm)	114.6	114.6	97.1	62.1	
$M_{Rd,fi}$ per wire (kNm)	0.339	0.813	1.148	0.864	**3.162**

Depth to neutral axis $X(\theta)=35762/(55\times0.8\times42)=19.4$ mm
$M_{Rd,fi}=3.162$ kNm$>M_{Ed,fi}=2.69$ kNm by a margin of 1.18

It appears that the calculation for Δa gives a reasonable estimate of the required axis distance.

5.5.6 Main beams

The fire resistance of reinforced and prestressed beams are required to meet the structural criterion R (only) in terms of minimum width of beam b_{min}, width of web b_w and mean axis distance a_{mean} according to BS EN 1992-1-2, Section 5.6. For non-rectangular beams, such as tapered or I section beams, beam width is defined in BS EN 1992-1-2, Fig. 5.4. To avoid spalling of shallow bottom flanges the effective height $d_{eff} \geq d_1 + 0.5\, d_2 \geq b_{min}$ where b_{min} is according to Table 5.14. There are some other modifications necessary to a for narrow webs and for shallow bottom flanges given in BS EN 1992-1-2, Exp. 5.10. The calculation for multiple bars $a_{mean} = a_m = \Sigma\, A_{si}\, a_i / \Sigma\, A_{si}$ (1–4) and at the side $a_{sd,m} = \Sigma\, A_{si}\, a_i / \Sigma\, A_{si}$ (5–8) is as illustrated in Figure 5.30. The rebar in the middle of the bottom row may be included in $a_{sd,m}$ although technically its influence is less than the outer bars and it is shared with $a_{sd,m}$ from the other side. When reinforcement consists of rebars and tendons with different characteristic strength A_{si} should be replaced by $A_{si}\, f_{yki}$ (or $A_{si}\, f_{pki}$).

There is a major division of data between simply supported and continuous beams. Options for the least values of b_{min} in combination with a_{mean} and are summarised in Table 5.14 for simply supported beams (BS EN 1992-1-2, Table 5.5). The axis distance a to prestressing tendons given in Table 5.14 should be increased according to BS EN 1992-1-2, clause 5.2 (5).

Table 5.14 Fire resistance of simply supported reinforced or prestressed beams adapted from BS EN 1992-1-2, Table 5.5

Fire resistance (mins)	*Minimum dimensions (mm)*				
	Combinations of beam width b_{min} and axis distance a				*Web width b_w* [a]
R60	120/40	160/35	200/30	300/25	100
R90	150/55	200/45	300/40	400/35	110
R120	200/65	240/60	300/55	500/50	130
R180	240/80	300/70	400/65	600/60	150
R240	280/90	350/80	500/75	700/70	170
	For prestressed beams increase *a* according to 5.2 (5 to 8)				

b_{min} may be increased, and axis distance *a* reduced by 10% using limestone aggregates.

[a] Class WA according to NA to BS EN 1992-1-2.

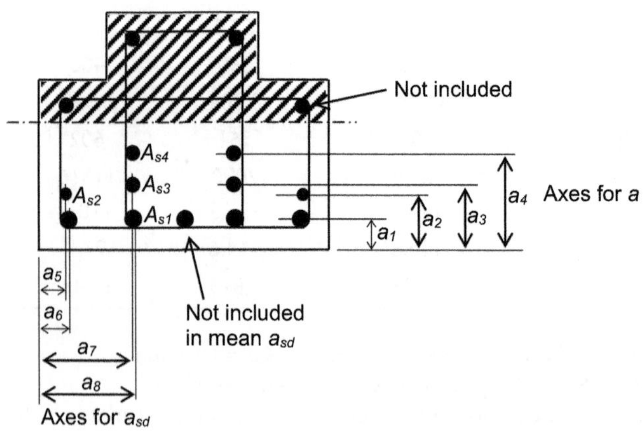

Figure 5.30 Calculation of mean axis distance for multiple bars (based on BS EN 1992-1-2, Section 5).

The 'special' check according to clause 5.2 (7) allows the increase in a, known as Δa, to be reduced according to the ratio of the fire load $E_{d,fi}$ to the ultimate load E_d and the ratio of the area of reinforcement required/provided, e.g. if $M_{Ed,fi}=80$ kNm and $M_{Ed}=125$ kNm, and if the designed area of strands $A_{p,req}=280\,mm^2$ and 6 no. 9.3 mm strands $A_{p,prov}=6\times52=312\,mm^2$ are provided, then the stress used to evaluate the critical temperature of reinforcement $\sigma_p(\theta)=(1770/1.15)\times(80/125)\times(280/312)=884$ N/mm^2. Then $k_p(\theta)=0.50$, $\theta_{cr}=372°C$ and $\Delta a=12.8$ mm, a reduction of 2.2 mm from the maximum.

The axis distance to the side of beam for the corner bars (or tendon or wire) of beams with only one layer of reinforcement is $a_{sd}=a+10$ mm. No increase is required for values of b_{min} greater than those in the third column of options (from the left) in Table 5.14, e.g. for R 60 if $b_{min}=200$ mm then $a_{sd}=30+10=40$ mm, but if $b_{min}=250$ mm then $a_{sd}=30$ mm.

Example 5.6

Determine the mean axis distance a required for a 300 mm wide prestressed concrete beam having R 120 minutes fire resistance. The beam is subjected to dead UDL $G_k=40$ kN/m and live UDL $Q_k=30$ kN/m. The area of tendons required is known to be $A_{p,req}=1000\,mm^2$. Use 12.5 mm diameter strands of 94 mm^2 per strand with $p_{yk}=1770$ N/mm^2. Use office loading with $\psi_1=0.5$ and $\psi_0=0.7$.

Solution

$A_{p,req}=1000\,mm^2$. Number of strands required $>1000/94=10.6$
Use 11 no. $A_{p,prov}=11\times94=1034\,mm^2$
$A_{p,required}/A_{p,\,provided}=1000/1034=0.967$
$w_{Ed,fi}=40+0.5\times30=55.0$ kN/m
$w_{Ed}=\max\{1.35\times40+0.7\times1.5\times30;\ 1.25\times40+1.5\times30\}=\{85.5;\ 95.0\}=95.0$ kN/m
$\eta_{fi}=55.0/95.0=0.579$
$\sigma_p(\theta)=0.579\times(1770/1.15)\times0.967=862$ N/mm^2.
$k_p(\theta_{cr})=862/1770=0.487$ (note this can be reached by $0.579\times0.967/1.15=0.487$)
Then if $k_p(\theta_{cr})=0.1$ to 0.55, $\theta_{cr}=594.4-444.4\,k_p(\theta_{cr})=594.4-444.4\times0.487=378°C$
$\Delta a=0.1\,(500-378)=12.2$ mm

From Table 5.14 for R 120, use $b_{min}=300$ mm, a $=55+12.2=67.2$ mm

The first row of strands (5 no.) could be placed at $a=40$ mm, second and third rows (3 each) at 80 and 120 mm giving $a=(5\times40+3\times80+3\times120)/11=72$ mm >67.2 mm.

Check $a_{min}=40$ mm $>a/2=72/2=36$ mm. OK.

REFERENCES

BIBM/IPHA. 2014. *Structural Behaviour of Prestressed Concrete Hollow Core Floors Exposed to Fire*, Bureau International du Beton Manufacture and International Prestressed Hollow Core Association, Brussels, Belgium, 226 p.

BS 3692. 2001. ISO metric precision hexagon bolts, screws and nuts. Specification, BSI, London, UK.

BS 4483. 2005. Steel fabric for the reinforcement of concrete, BSI, London, UK.

BS 5080-1. 1993. Structural fixings in concrete and masonry - Part 1: Method of test for tensile loading, BSI, London, UK.

BS 5896. 1980. High tensile steel wire and strand for the prestressing of concrete, BSI, London, UK.

BS 8110-2. 1985. Structural use of concrete - Part 2: Code of practice for special circumstances, BSI, London, UK.

BS 8500-1:2023. Concrete. Complementary British Standard to BS EN 206: Method of specifying and guidance for the specifier, BSI, London, UK.

BS 8666. 2005. Scheduling, dimensioning, bending and cutting of steel reinforcement for concrete, BSI, London, UK.

BS EN 10025-1. 2004. Hot rolled products of structural steels. General technical delivery conditions, BSI, London, UK.
BS EN 10025-2. 2004. Hot rolled products of structural steels, Part 2: Technical delivery conditions for non-alloy structural steels, BSI, London, UK.
BS EN 1008. 2002. Mixing water for concrete, BSI, London, UK.
BS EN 10080. 2005. Steel for the reinforcement of concrete, BSI, London, UK.
BS EN 1011-1. 2009. Welding. Recommendations for welding of metallic materials. General guidance for arc welding, BSI, London, UK.
BS EN 1011-2. 2001. Arc welding of ferritic steels, BSI, London, UK.
BS EN 1011-3. 2000. Arc welding of stainless steels, BSI, London, UK.
BS EN 10138-1. 2015. Prestressing steel - Part 1: General requirements, BSI, London, UK, under technical review.
BS EN 10138-2. 2015. Prestressing steel - Part 2: Stress relieved cold drawn wire, BSI, London, UK, under review (accepted).
BS EN 10138-3. 2015. Prestressing steel - Part 3: strand, BSI, London, UK, under technical review.
BS EN 10210-1. 1994. Hot finished structural hollow sections of non-alloy and fine grain structural steels - Part 1: Technical delivery requirements, BSI, London, UK.
BS EN 1097-6. 2000. Tests for mechanical and physical properties of aggregates, BSI, London, UK.
BS EN 1168. 2005. Precast concrete products - Hollow core slabs, +A3:2011, BSI, London, UK.
BS EN 12620. 2002. Aggregates for concrete, BSI, London, UK.
BS EN 13139. 2002. Aggregates for mortar, BSI, London, UK.
BS EN 13224. 2004. Precast concrete products - Ribbed floor elements, BSI, London, UK.
BS EN 13369. 2004. Common rules for precast concrete products, +A1:2018, BSI, London, UK.
BS EN 14399-1. 2005. High-strength structural bolting assemblies for preloading. General requirements, BSI, London, UK.
BS EN 15037-1. 2008, Precast Concrete Products - Beam and Block Floor Systems, Part 1 Beams, BSI, London, UK.
BS EN 1744-1. 2009. Tests for chemical properties of aggregates - Chemical analysis, BSI, London, UK.
BS EN 197-1. 2011. Cement - composition, specifications and conformity criteria for common cements, BSI, London, UK.
BS EN 1990. 2002. Eurocode 0, Basis of Structural Design, +A1:2005, BSI, London, UK.
BS EN 1992-1-1. 2004. Eurocode 2: Design of concrete structures. Part 1-1: General rules and rules for buildings, +A1:2014, BSI, London, UK.
BS EN 1992-1-2. 2004. Eurocode 2: Design of concrete structures, Part 1-2: General rules - Structural fire design, +A1:2019, BSI, London, UK.
BS EN 1993-1-1. 2005. Eurocode 3, Design of Steel Structures - Part 1-1: General Rules and Rules for Buildings, +A1:2014, BSI, London, UK.
BS EN 206. 2013. Concrete - Specification, performance, production and conformity, +A2:2021, BSI, London, UK.
BS EN 206-9. 2010. Concrete - Additional rules for self-compacting concrete, BSI, London, UK.
BS EN 480-1. 2006. Admixtures for concrete, mortar and grout. Test methods. Reference concrete and reference mortar for testing, +A1:2011, BSI, London, UK.
BS EN 60974-1. 2005. Arc welding equipment - welding power, BSI, London, UK.
BS EN 934-1. 2008. Admixtures for concrete, mortar and grout. Common requirements, BSI, London, UK.
BS EN 934-2. 2009. Admixtures for concrete, mortar and grout. Concrete admixtures. Definitions, requirements, conformity, marking and labelling, +A1:2012, BSI, London, UK.
BS EN 934-3. 2009. Admixtures for concrete, mortar and grout. Admixtures for masonry mortar. Definitions, requirements, conformity, marking and labelling, +A1:2012, BSI, London, UK.
BS EN ISO 1461. 2022. Hot dip galvanized coatings on fabricated iron and steel articles - Specifications and test methods, BSI, London, UK.
BS EN ISO 17660-1. 2006. Welding of reinforcing steel - Part 1: Load-bearing welded joints, BSI, London, UK.
BS EN ISO 17660-2. 2006. Welding of reinforcing steel - Part 2: Non-load-bearing welded joints, BSI, London, UK.

BS EN ISO 3766. 2003. Construction drawings. Simplified representation of concrete reinforcement, BSI, London, UK.

Concrete Centre. 2024. How to design concrete structures using Eurocode 2, The Concrete Centre, London, UK.

Efectis, Nederland. 2007. Onderzoek brand parkeergagarge (investigate parking garage fire) Lloydstraat Rotterdam, in opdracht voor Veiligheidsregio (Commisioned for security region) Rotterdam-Rijnmond, Report 2007-Efectis-R0894, 51 p.

Elliott, K. S. 2017. *Precast Concrete Structures*, 2nd ed., CRC, Taylor & Francis, London, UK, 694 p.

Elliott, K. S. and Jolly, C. K. 2013. *Multi-Storey Precast Concrete Framed Structures*, 2nd ed., John Wiley, London, UK, 750 p.

Gilbert, R. I., Mickleborough, N. C. and Ranzi, G. 2017. *Design of Prestressed Concrete to Eurocode 2*, 2nd ed., CRC, Taylor & Francis Group, Boca Raton, FL, 665p.

ISO 834-1. 1999. *Fire-Resistance Tests - Elements of Building Construction - Part 1: General Requirements*, International Organization for Standardization, Geneva, Switzerland.

Lennon, T. 2003. Precast concrete hollow core slabs in fire, The Structural Engineer, Vol. 81, pp. 30–35.

NA to BS EN 1992-1-1. 2004. UK National Annex to Eurocode 2: Design of concrete structures. Part 1-1: General rules and rules for buildings, BSI, London, UK.

NA to BS EN 1992-1-2. 2004. UK National Annex to Eurocode 2: Design of concrete structures, Part 1-2: General rules - Structural fire design, BSI, London, UK.

NA to BS EN 1993-1-8. 2005. UK National Annex to Eurocode 3. Design of steel structures. Design of joints, BSI, London, UK.

Neville, A. 1995. *Properties of Concrete*, 4th ed., Longman, London, UK.

PD 6682-3. 2003. Aggregates for mortar- Guidance on the use of BS EN 13139, BSI, London, UK.

TNO Bouw en Onderground. 2008. Onderzoek naar het constructieve gedrag tijens brand van een kanaalplaat zoals toegepast aan de (Research into the constructive behavior during fire of a hollow-core slab as applied to) Lloydstraat te Rotterdam, in opdracht voor Veiligheidsregio (Commisioned for security region) Rotterdam-Rijnmond, Report TNO-2007-D-R1236/C, 12 p.

Van Acker, A. 2010. Fire safety of prestressed hollowcore floors, Concrete Plant International, Vol. 1, pp. 182–196.

Chapter 6

Prestressing and detensioning stresses

6.1 CHOICE OF INITIAL PRESTRESS

Prestressing aims to induce a flexural compressive stress in those parts of precast concrete elements where tension would otherwise exist due to external loads and, in some instances, internal effects such as thermal restraint and/or shrinkage. If a slab or beam is simply supported and subjected to gravity loads, then compressive stress is induced into the bottom of the section, and vice versa for a cantilever that is at the top of the section. Figure 6.1 shows the basic principles. Prestressing to induce axial compressive stress is also possible, but less common, for example, in lintels containing a single strand or a couple of wires, by positioning the tendons at the centroid of the concrete section, theoretically of course, but as close as is practically possible. Figure 6.2 shows the basic principles. Prestressing induces compressive stress in order to

i. reduce or nullify tensile stress, or the flexural tensile stress f_t if the section is subjected to bending, or combined bending and axial tension.
ii. increase the flexural stiffness of the section by preventing cracking when f_t exceeds the mean value of the flexural tensile strength of concrete, defined in BS EN 1992, Part 1-1 (BS EN 1992-1-1, 2004), clause 3.1.6 as $f_{ctm,fl}$. However, this value is seldom used in design in favour of the mean axial tensile strength defined in clauses 3.1.2 (9) and Table 3.1 as f_{ctm}. Preventing cracking allows deflections to be calculated using the flexurally uncracked cross section, which is the major advantage of prestressed concrete over reinforced concrete.
iii. reduce downward deflections by inducing an upward deflection, known as camber, to nullify deflections due to gravity loads. In perfect conditions, the final deflection is about span/1000, although this changes over the lifetime of an element because of creep and other changes in the elastic modulus of concrete. Inducing compressive stress on the top of a cantilever has the same effect, just reversed.
iv. increase the shear capacity of a section that is flexurally uncracked due to (ii) to a point where shear reinforcement, in the form of designed shear stirrups or less commonly bent-up rebars, *may* not be required. The term *may* means that links may be provided even though they are not required in design, e.g. at the ends of beams and at bearings where localised tensile and/or splitting forces are present.
v. increase the torsional capacity of a section likewise, although the savings in closed torsional stirrups may be small.
vi. improve the handling and lifting of long slender sections, such as tall columns or thin panels, by preventing cracking as in (ii) using axial prestress.

 DOI: 10.1201/9781003319450-6

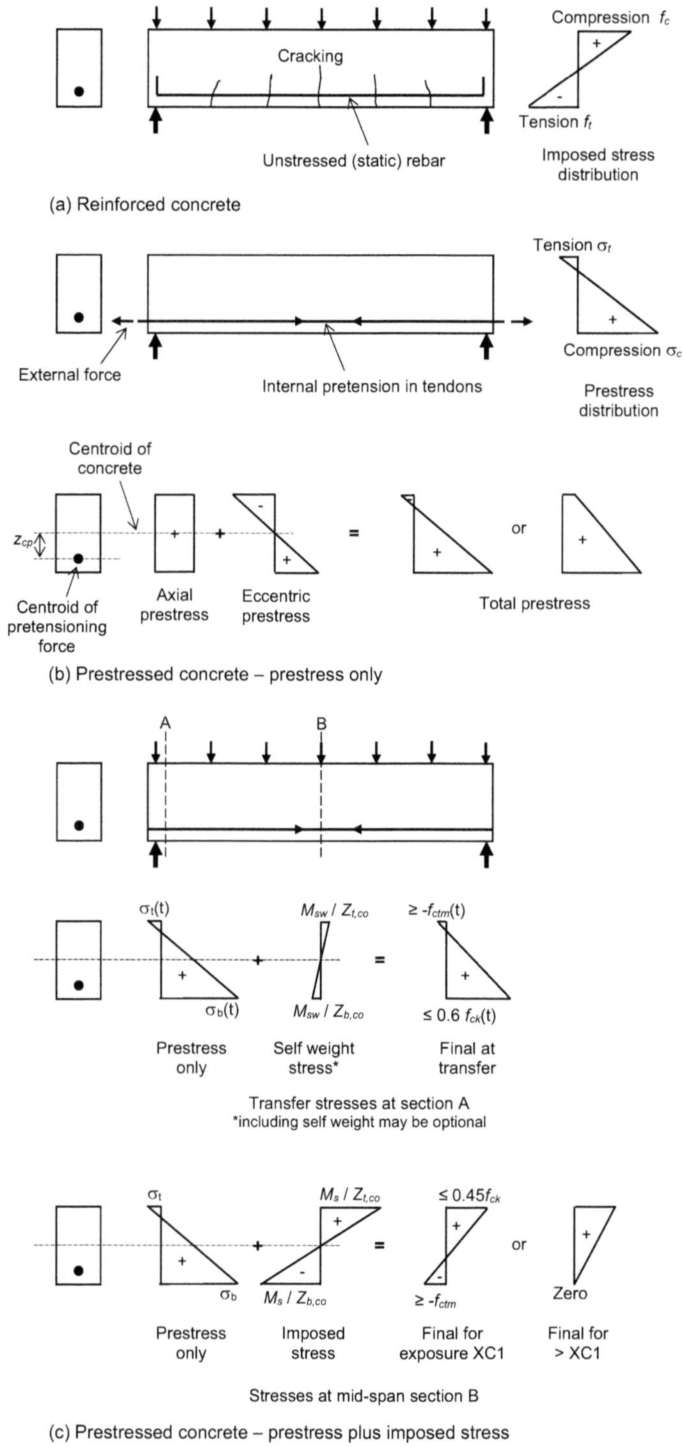

Figure 6.1 Principles of prestressing concrete with eccentric pretensioning force. (a) Reinforced concrete. (b) Prestressed concrete – prestress only. (c) Prestressed concrete – prestress plus imposed stress.

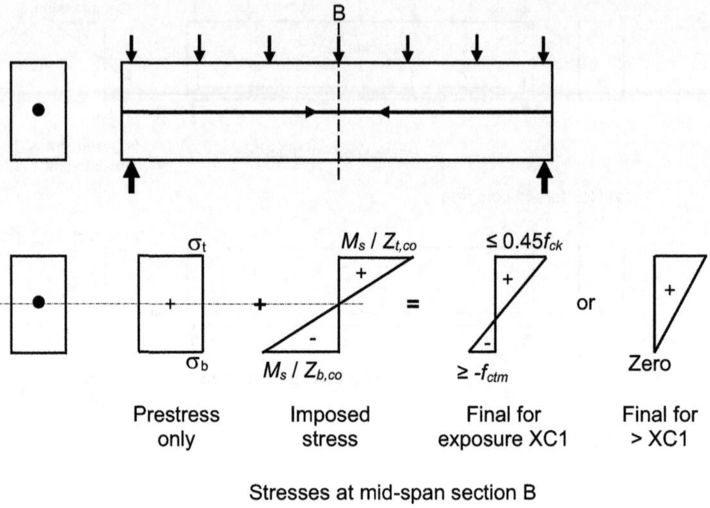

Figure 6.2 Principles of prestressing concrete with axial pretensioning force.

In tests, f_{ctm} is obtained by splitting a cylinder across its diagonal (so-called Brazilian test). The tensile stress is (almost) uniform over the diameter of the cylinder and is able to find points of weakness anywhere in its path. However, $f_{ctm,fl}$ is obtained by subjecting a prismatic section (usually 500 mm long × 150 mm square) to four-point bending (two point loads and two supports) under hydraulic load until the prism cracks suddenly at a peak load. The concrete is only fully in tension at the bottom fibre, reducing rapidly to zero at the centroid of the prism and thereafter it is in compression. The resulting strength is greater than f_{ctm} because there are fewer points of weakness at the bottom of the prism and in the mid-span region where failure occurs. For this reason, the ratio of $f_{ctm,fl}/f_{ctm}$ is typically 1.4–1.5, reducing as the prism is deeper and the tensile stress gradient shallower.

The aim of prestressing is also to satisfy the dual requirements of

a. limiting concrete compressive and tensile stresses when the pretensioning forces in the tendons are first released into the concrete, typically at the age of 12–24 hours when the concrete has a compressive strength $f_{ck}(t)$. See Chapter 4 for the techniques used. Known as 'detensioning' or 'transfer', BS EN 1992-1-1, clause 8.10 refers to 'transfer of prestress' and sets out guidance for its success, e.g. clear distances between tendons, anchorage of tendons, avoidance of tensile stress causing cracking or splitting, as shown in Figures 6.3 and 6.4. Transfer stresses are calculated at the ends of the elements where bending stresses due to self-weight are zero. However, because the build-up of prestress from zero at the end of the element to a maximum at a point at the end of the transmission zone, when the element lifts off the casting bed opposing stresses due to self-weight reduce the transfer stresses a little and may, if the designer wishes (particularly in longer elements) be used. The 'point' in question is at the 'basic' transmission length l_{pt} (see also Section 10.2.2 for the derivation). See Example 6.1. l_{pt} is typically 400–650 mm depending on the diameter of the tendons.
b. limiting the final (in service) compressive and tensile stresses in the section due to the addition, or combined effects of prestress and imposed bending stresses, as well as satisfying the ultimate limit state of bending and shear (and maybe torsion) due to ultimate loads.

Figure 6.3 Flexural tension transfer cracks in the top of hollow core floor unit (possibly weakened by lifting hooks).

Figure 6.4 Splitting in the transmission zone in prestressed concrete hollow core floor units (courtesy Bison Precast Ltd., UK).

Point (b) provides an internal couple with a magnitude $M_P = P\, z_{cp}$ at least equal to, and often greater than, the externally imposed serviceability bending moment M_s at a particular cross section of the element, as shown in Figure 6.5. Here P is the force in the tendons at any time after casting. Of course, the M_P is much greater than that which is required closer to the supports (in a simply supported span) and has to be checked when the concrete is at an early age, e.g. at transfer. The buildup of M_P from the end of the slab is known to be roughly parabolic, although BS EN 1992-1-1, Fig. 8.17 shows it as linear. Defining and measuring the transmission length and build-up of stress is difficult and variable, as reflected in the widely different values given in international codes. It has been the subject of extensive studies, most of which are summarised by Kong (Kong, 1993).

To satisfy (a) and (b) simultaneously, for given spans, loads, concrete strength at transfer and in service, the positions of the tendons in the cross section may not be the same for (a) and (b). This means two different arrangements of tendons are required, achieved either by deflecting tendons, as shown in Figure 6.6, or by debonding tendons, using sheathes or similar to break the bond with concrete, as shown in Figure 6.6. Although debonded tendons still have the same position in the cross section their 'effective' position, i.e. centroid of pretensioning force, is different. The distance to the debond/deflect point depends on the service bending moment distribution, and the relative strengths f_{ck} and $f_{ck}(t)$, but for the case of uniformly distributed loading and say $f_{ck}(t) = ⅔ f_{ck}$ the distance is about ¼ span. The exact distance can be back-calculated from the bending moments at transfer and in service, or the designer may choose to debond/deflect at ¼ span and arrange the tendons to suit. The

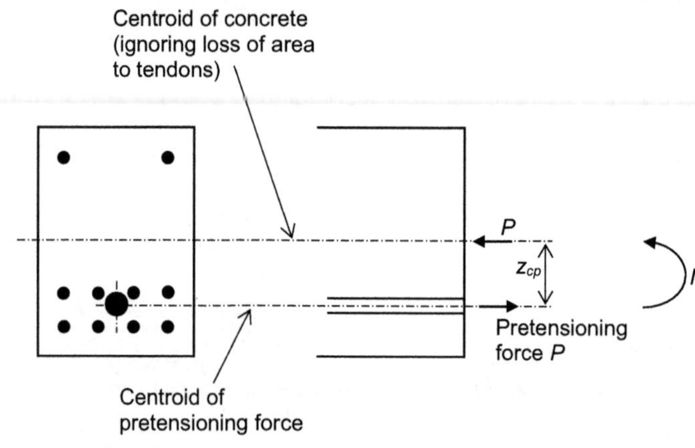

Figure 6.5 The internal couple mechanism.

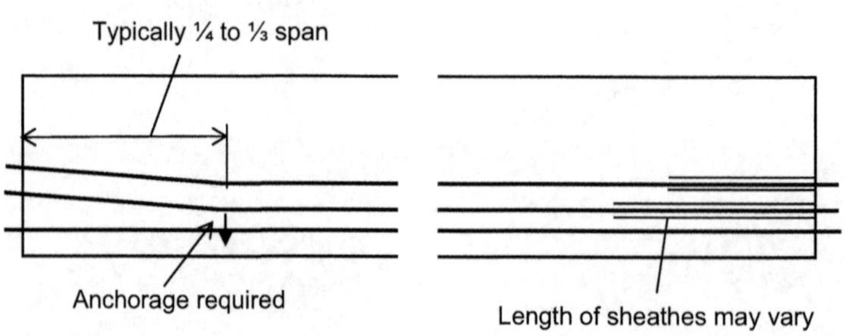

Figure 6.6 Use of deflected and debonded tendons to reduce prestress at transfer closer to the ends of elements.

length of the sheath should be one transmission length l_{pt} before the theoretical debonding point to allow the prestress to develop. This procedure is explained in Section 8.1 by using the Magnel diagram. Debonding may also take place at two points from the ends (as shown in Figure 6.6) to further optimise the moments of resistance with the service moments. Deflecting tendons may also be carried out at two points, but this is not common as two anchorage devices are required.

Debonding is often the preferred method as it is relatively cheap to provide, whereas deflected tendons require anchorages for holding down restraint and other tendons (above the deflected tendons) may be in the way. If (a) and (b) cannot be satisfied simultaneously, even by increasing $f_{ck}(t)$, the imposed bending stresses must be reduced, i.e. the design service moment of resistance M_{sR} is less than above. The ultimate moment of resistance M_{Rd} and shear capacity $V_{Rd,c}$ (flexurally uncracked section) do not control the positions of the tendons – their values are calculated *per se*, and are not usually the limiting criteria (see Chapters 9 and 10).

This is the case with hollow core floor units (hcu) (Figure 2.1) or T beams for beam and block flooring (Figure 2.5) where deflecting or debonding tendons is either not possible (e.g. slip formed or extruded hcu) or not practised by the manufacturer (e.g. T beams). However, double-tee floor units (Figure 2.19) may have deflected or debonded tendons readily incorporated to increase M_{sR} in these less efficient units compared with hcu of similar depth.

The positions of the tendons, e.g. strands or wires in hcu, are first determined by the manufacturer to satisfy (a) at transfer and then to back-calculate the maximum possible in-service values of M_{sR} and M_{Rd}. This gives what is defined as the 'highest strand pattern' where both transfer stresses (top and bottom of the section) are at the permitted levels. Also, the balance between the limiting concrete stresses at transfer and in service is dictated by the maturity of the concrete and the need to de-tension the reinforcement within 12–24 hours after casting.

The minimum clear spacing s between tendons is given in BS EN 1992-1, Figure 8.14. Vertically $s_v \geq d_g$ or 2Φ, and horizontally $s_h \geq d_g + 5$ mm, 2Φ or 20 mm, where d_g = maximum size of coarse aggregate, typically 10 or 14 mm in precast elements, and Φ = tendon diameter. Therefore, typical centre-centre distances for 12.5 mm strands are $s \geq 3\Phi = 37.5$ mm, and $s \geq 28$ mm for 9.3 mm strands.

The transfer stress, expressed in the usual manner as the characteristic strength $f_{ck}(t)$ (where (t) symbolises early strength <28 days) either as a function of the final concrete strength f_{ck} according to clause 3.1.2 (5) and (6) or is measured by the manufacturer converting cube strength to $f_{ck}(t)$ according to BS EN 1992-1-1, Table 3.1. The limiting compressive strength at the bottom fibre at transfer is according to clause 5.10.2.2 (5) $\leq 0.6\ f_{ck}(t)$, and more importantly at the top fibre by the tensile strength $f_{ctm}(t)$, where defined by clause 3.1.2 (6) and (9) as

$$f_{ctm}(t) = f_{ctm}\ f_{cm}(t) / f_{cm} \tag{6.1}$$

For example, $f_{ck} = 45$ N/mm², $f_{cm} = 45 + 8 = 53$ N/mm², $f_{ctm} = 0.3\ f_{ck}^{2/3} = 0.3 \times 45^{2/3} = 3.80$ N/mm², $f_{ck}(t) = 30$ N/mm², $f_{cm} = 30 + 8 = 38$ N/mm², then $f_{ctm}(t) = 3.80 \times 38/53 = 2.72$ N/mm². Note that $f_{ctm}(t)$ is a lower strength than where f_{ctm} is determined for the same strength at 28 days, i.e. $f_{ctm} = 0.3 \times 30^{2/3} = 2.90$ N/mm², 7% greater than $f_{ctm}(t)$. This is because in maturing concrete at say 20 hours, the products of hydration have not developed sufficiently to establish the same tensile resistance as concrete at 28 days of age. For $f_{ck} = 45$ N/mm² (the typical strength in prestressed elements) $f_{ck}(t)$ should be about 28–30 N/mm² (35–40 N/mm² cube strength). Table 4.1 shows the actual relationships between $f_{ck}(t)$ and f_{ck} at UK producers.

6.2 PRETENSIONING AND LOSS OF PRESTRESS

6.2.1 Initial prestress

A typical hollow core slab production factory is shown in Figure 4.4. Steel reinforcement, 'tendons' such as wire or helical strand of total area A_p, is stretched between jacking equipment at either end of long steel beds, about 100–150 m long, after which concrete is cast around the tendons. The tendons are positioned eccentrically relative to the centroid of the section to produce the desired pretensioning stresses shown in Figure 6.1c. The initial prestress σ_{pi} (which is set by the manufacturer) is around 70%–75% of the ultimate strength f_{pk}=1670 to 1770 N/mm^2 for wire and f_{pk}=1750–1860 N/mm^2 for strand. The many different types of tendons available (See Table 5.8) simplify to 9.3 and 12.5 mm (occasionally 10.9 mm) diameter 7-wire helical strand, or 5 and 7 mm diameter plain or indented wire.

6.2.2 Loss of prestress at transfer and in service

The tendons cannot sustain the initial stress for the following reasons:

i. During tensioning, the tendons immediately relax; about 0.5% to 1% of the initial stress is lost; some of this may be due to slight slippage of the tendons in the grips. Over the long term, the tendons lose a further 2%. The relaxation classification is given as ρ_{1000}=2.5% for relaxation *class 2* according to BS EN 1992-1-1, clauses 3.3.2 (6) and (7). The 1000 hours relaxation test value is provided by the manufacturer or as given in BS 5896 (BS 5896 2012). The initial relaxation loss at the time of transfer is according to Code Exp. 3.28 to 3.30. The initial relaxation loss is about 0.5% and is subtracted from the initial prestress before further losses are calculated. The final relaxation loss is about 2%.

ii. After the concrete has hardened around the reinforcement and the bars are released from the jacking equipment, the force in the bars is transferred to the concrete by bond. The concrete shortens elastically – this may be calculated knowing the elastic modulus of the concrete at this point in time at transfer. This is called 'elastic shortening' because the reinforcement is obliged to shorten by the same amount as the concrete. The stress in the concrete reduces too, by about 4%–6% depending on the total prestressing force. Losses (i) and (ii) are called 'transfer losses'.

iii. Desiccation of the concrete follows, causing a long-term shrinkage loss. This is the product of the shrinkage per unit length (about 400×10^{-6} for indoor and 250×10^{-6} for outdoor exposure) and modulus of elasticity of the tendons E_p=195 kN/mm^2 (strand) and 205 kN/mm^2 (wire). This gives a shrinkage loss of between 55 and 70 N/mm^2, about 4.0%–5.5%.

iv. Finally, creep strains are allowed for using a specific creep strain (i.e. creep per unit length per unit of stress) of about 2.0 for indoor curing and loading at 90 days in the U.K. Creep affects the tendons in the same manner as elastic shortening because its effect is measured at the centroid of the bars, but is also influenced by the shape of the element, called notional dimension (or depth or thickness) h_o (area-to-perimeter ratio) and f_{ck}. Hence, the creep loss is about 1.5 times the elastic shortening loss, 7%–9%.

Total losses range from about η=18% to 25% for minimum to maximum levels of prestress. There are five stages of prestress

i. initial prestress=σ_{pi}

ii. after immediate relaxation of tendons before the tendons are cut or released σ_r

iii. after initial relaxation and elastic shortening losses=σ_{pm0}

iv. after short-term losses at installation $=\sigma_{pmi}$
v. design effective prestress after all losses $=\sigma_{po}$

Note BS EN 1992-1-1 uses $\sigma_{p\infty}$ for the final prestress, and some texts use *f* for these stresses. The final force in the tendons $P_{po}=A_p\ \sigma_{po}$.

Table 6.1 gives a range of maximum strand or wire patterns for 1200 mm wide hcu, 2400 mm wide composite double-tee slabs and 92–136 mm wide T beams (in beam and block flooring). Data show the ratio of bottom $\sigma_b(t)$ and top $\sigma_t(t)$ stresses at transfer divided by their respective limiting stresses as $\sigma_c=0.6\ f_{ck}(t)=18.0$ N/mm² and $\sigma_{ct}=f_{ctm}(t)=-2.72$ N/mm² (−2.54 N/mm² for T beams for the reasons explained above), respectively. Stresses at transfer are calculated at a distance l_{pt} from the end of the elements, where the self-weight is for a span of 30×depth. They include immediate relaxation and elastic shortening losses as calculated in Section 6.2.3. Concrete strength is $f_{ck}=45$ (55 for T beams) N/mm² and $f_{ck}(t)=30$ N/mm². The initial prestressing force for strand is $\eta=0.7$, and 0.75 for T beams using 5 mm wire, of the characteristic strength $p_{yk}=1770$ N/mm². Modulus of elasticity for strand is 195 kN/mm², and for wire, it is 205 kN/mm².

The results show that transfer stresses are controlled by the limiting stress at the bottom of elements, where the average ratio for $\sigma_b(t)/\sigma_c$ is 0.86 (excluding the 150 mm depth hcu in which the number of strands is controlled by the number of webs in the section, i.e. 12×9.3 mm diameter). Apart from two of the hcus the ratio $\sigma_t(t)/\sigma_{ct}$ at the top is low, in six cases $\sigma_t(t)$ being compressive, thanks to the effect of self-weight at a distance from the support of l_t. The low stress in the top is an advantage where the elements may be subjected to vibration and impact during transportation, particularly in longer, more highly stressed elements where bearers are positioned at some distance from their ends. Note that the strands in the double-tee slabs are debonded at ¼ span such that the number of strands at mid-span is greater, e.g. 28 strands in the 500 mm depth unit. Tendons in hcu and T beams are not debonded.

Table 6.1 Examples of maximum prestress at transfer for non-composite hollow core units, composite double-tee slabs and T beams (in beam and block flooring)

Type	*Depth (mm)*	*Strand or wire diameters*		*Tendon force at transfer (kN)*	*Height to centroid of tendons (mm)*	*Eccentricity to tendons (mm)*	*Ratio of prestress at transfer/limits*	
		9.3 mm	*12.5 mm*				*At bottom* $\sigma_b(t)/\sigma_c$	*At top* $\sigma_t(t)/\sigma_{ct}$
Hollow core units	150	12	0	735.6	39.7	34.6	0.65	−0.04
	250	7	7	1167.3	40.7	82.3	0.87	0.77
	400[a]	4	10	1364.3	53.3	142.2	0.79	0.88
Double-tee units	300	0	12	1316.6	133.3	79.3	0.82	0.11
	500	0	20	2178.2	220.0	131.7	0.91	−0.26
	700	0	24	2608.3	275.0	204.1	0.89	−0.16
	900[b]	0	28	3042.5	328.6	271.1	0.87	−0.08
T beams		5 mm	7 mm					
(in beam and block)	150	4	0	96.4	40.0	21.9	0.99	−0.21
	150[c]	9	0	217.9	42.9	27.4	0.88	0.14
	225	8	0	196.7	67.9	31.7	0.76	−0.30

Note: double-tee slabs are 2400 mm wide when comparing with 1200 mm wide hcu. The strands in double-tee slabs are debonded up to ¼ span.

[a] Includes 4 no. 5 mm wires at the top.
[b] Max span = 28×depth.
[c] 150 mm depth wide T beam.

6.2.3 Immediate relaxation of tendons and elastic shortening at the time of installation

The full analysis is as follows (references to BS EN 1992-1-1 clauses are on the left side of the page)

Depth of unit $= h$
Cover to tendons $= C$ of diameter Φ
Area of each tendon A_{pi}
Height to centroid of each tendon $y_{si} = C + \Phi/2$
Height to centroid of group of tendons (if all tendons are equally stressed, otherwise replace A_{pi} with force in tendon) $y_s = \Sigma\, y_{si}\, A_{pi} / \Sigma\, A_{pi}$
Height to centroid to concrete $= y_b$
Gross concrete area (exclude infill) A_c
Eccentricity of tendons $z_{cp} = y_b - y_s$
Second moment of area I_c
Section modulus at bottom $Z_b = I_c / y_b$
Section modulus at top $Z_t = I_c / (h - y_b)$
Section modulus at centroid of tendons $Z_z = I_c / z_{cp}$
Outer perimeter of section $= u$
Notional dimension $h_o = 2A_c / u$

5.10.3 Losses at transfer

3.3.2(7) Immediate relaxation of tendon at t hours

Exp. 3.29 μ = ratio of initial prestress η (typically 0.7)

Initial prestress after friction and wedge slippage $\sigma_{pi} = \eta\, f_{pk}$ (6.2)

Initial prestress force before release $P_{pi} = \sigma_r\, A_p$

Exp. 3.29 For Class 2. $$\Delta\sigma_{pr} = \sigma_{pi} \times 0.66 \times 2.5\, e^{9.1 \times \mu} \left(\frac{t}{1000}\right)^{0.75(1-\mu)} \times 10^{-5} \quad (6.3)$$

Prestress at release $\sigma_r = \sigma_{pi} - \Delta\sigma_{pr}$

Prestress force at release $P_r = \sigma_r\, A_p$

5.10.4 Instantaneous deformation due to elastic shortening determined after relaxation loss.

Concrete stress at level of tendons $\sigma_c = P_r / A_c + P_r\, z_{cp} / Z_z$ {$- M_{self} / Z_z$ if the self-weight is considered for losses at mid-span in the stockyard} (6.4)

Exp. 5.44 Loss $\Delta\sigma_{el} = \sigma_c\, E_p / E_{cm}(t)$

5.10.3(2) After initial losses $\sigma_{pm0} = \sigma_{pi} - \Delta\sigma_{pr} - \Delta\sigma_{el} \le k_T\, f_{pk} = 0.75\, f_{pk}$ (6.5)

Prestress force at transfer $P_{pm0} = \sigma_{pm0}\, A_p$

$R_{tr} = \sigma_{pm0} / \sigma_{pi}$

No partial factor for possible variations in prestress is applied to P_{pm0} according to the UK National Annex to BS EN 1992-1-1 (NA to BS EN 1992-1-1, 2004), clause 5.10.9 (1)P, $r_{sup} = 1.0$ (while EN 1992-1-1 gives $r_{sup} = 1.05$). Some texts (e.g. Bhatt, 2011) apply $\gamma_{sup} = 1.05$

from EN 1992-1-1 to P_{pm0}. Furthermore, it is common for manufacturers to directly measure prestress either by elongation, vibrating wire or cross-bow tension meters, allowing $r_{sup}=1.0$ to EN 1992-1-1 itself.

Check maximum surface stress at transfer

$$\text{At bottom} \quad \sigma_b(t) = P_{pm0}/A_c + P_{pm0}z_{cp}/Z_b \tag{6.6a}$$

$$5.10.2.2(5) \quad \text{Limit } \sigma_b(t) \le \sigma_c = 0.6 f_{ck}(t)$$

Note if $\sigma_b(t)>0.45\ f_{ck}(t)$ is considered for the calculation of camber up to the date where $\sigma_b(t)=0.45\ f_{ck}$, which is approximately 3 days, and also just at the ends of the unit where tensile stress due to self-weight is small. In these circumstances non-linear creep is negligible.

$$\text{At top} \quad \sigma_t(t) = P_{pm0}/A_c - P_{pm0}z_{cp}/Z_t \tag{6.7a}$$

$$\text{Limit } \sigma_t(t) \ge \sigma_{ct} = -f_{ctm}(t)(-\text{ve for tension})$$

The positive (compression) stress due to the self-weight of the slab at the end of the basic transmission zone may be deducted from the above stress, unless of course the slab is held down during transfer by weight packs, thereby nullifying the self-weight stress as the unit cambers upwards. Without the weight pack, Eqs. 6.6a and 6.7a become

$$\sigma_b(t) = P_{pm0}\ /\ A_c + P_{pm0}z_{cp}\ /\ Z_b + M_{sw}\ /\ Z_{b,co} \tag{6.6b}$$

$$\sigma_t(t) = P_{pm0}\ /\ A_c - P_{pm0}z_{cp}\ /\ Z_t + M_{sw}\ /\ Z_{t,co} \tag{6.7b}$$

where $M_{sw} = w_0\ L\ l_{pt}\ /\ 2 - w_0 l_{pt}^2\ /2$

w_0=self-weight
$Z_{b,co}$ and $Z_{t,co}$=compound section modulus at the bottom and top fibre (see Section 6.2.4)

The basic transmission length is l_{pt}={0.19 (strand) or 0.25 (wire)} $\sigma_{pm0}\ \Phi/f_{bpt}$, where the bond strength f_{bpt}={3.2 (strand) or 2.7 (wire)}×0.7 $f_{ctm}(t)/\gamma_c$. Note that the prestress in the tendons σ_{pm0} has to be calculated first to determine l_{pt}. See Section 10.2.2 for full details.

6.2.4 Immediate upward camber due to prestress at transfer

The prestress is assumed to give constant curvature over the full length of the element. This is not strictly true where tendons are debonded or deflected, but the differences are small. Note that as follows δ_1 is based on the actual length of the element whereas δ_2 is for the effective span, i.e. actual – ½ bearing lengths, but the differences are small, e.g. for the 10 m hcu in Example 6.1 the difference is 0.5 mm or 2% of camber at transfer.

Immediate camber

$$\delta_1 = -P_{pm0}z_{cp}L^2\ /\ 8E_{cm}(t)I_{c,co} \tag{6.8a}$$

Strictly, the camber should allow for the development of prestress within the basic transmission length l_{pt} at each end of the unit, then

$$\delta_1 = -\left[1 - 4/3\ (l_{pt}\ /\ L)^2\right]P_{pm0}z_{cp}L^2\ /\ 8E_{cm}(t)I_{c,co} \tag{6.8b}$$

For typical values of l_{pt} and L, say l_{pt}=500 mm and L=6000 mm, the reduction in δ_1 is 0.93%, about 0.15–0.3 mm in most cases, and is therefore ignored given that variations in $E_{cm}(t)$ can be of the order of ±25% of calculated values.

Deflection due to self-weight

$$\delta_2 = 5w_oL^4 / 384E_{cm}(t)I_{c,co} \tag{6.9}$$

Resultant deflection at transfer=$\delta_1+\delta_2$

Note that the second moment of area I_c may take the compound value $I_{c,co}$ using the transformed area of tendons based on a modular ratio, without creep effects, $m=E_p/E_{cm}$. Some designers ignore this; the difference is typically 2%–3% reduction in deflection.

$$\text{Area of compound section } A_{c,co} = A_c + (m-1)A_p$$

$$y_{b,co} = \left(A_c y_b + (m-1)A_p y_s\right) / A_{c,co}$$

$$I_{c,co} = I_c + A_c\left(y_{b,co} - y_b\right)^2 + \Sigma(m-1)A_p\left(y_s - y_{b,co}\right)^2 \text{ per layer of tendons}$$

$$Z_{b,co} = I_{c,co} / y_{b,co}$$

$$Z_{t,co} = I_{c,co} / \left(h - y_{b,co}\right)$$

Example 6.1

Determine the top and bottom stresses at, and near to the ends of the 250 mm depth prestressed hollow core unit shown in cross section in Figure 6.7 at the point of transfer of prestress. Check the deflection at mid-span due to prestress at transfer and self-weight. The initial prestressing force may be taken as η=0.7 of characteristic strength of 'standard' 7-wire helical strand of f_{pk}=1770 N/mm². Manufacturer's data gives relaxation Class 2 detensioned at 20 hours after 50°C mean temperature curing.

The hcu will be simply supported over a clear span of 9.7 m, with 100 mm bearing lengths onto 300 mm wide beams. The actual length of the unit=9.7+2×0.1=9.9 m. Geometric and material data given by the manufacturer are as follows:

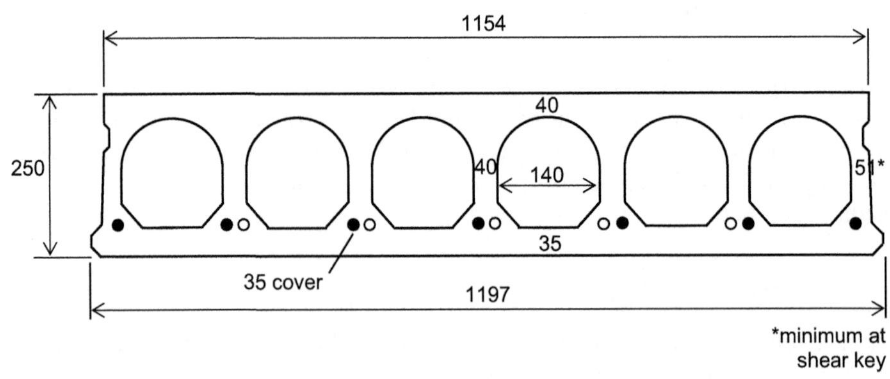

Figure 6.7 Cross-sectional hollow core floor unit to Examples 6.1 and 7.1. Strands represented by solid dots (12.5 mm dia.) and open dots (9.3 mm dia.)

Area = 168×10^3 mm²; $I_c = 1270 \times 10^6$ mm⁴; $y_b = 123$ mm; $f_{ck} = 45$ N/mm²; $f_{ck}(t) = 30$ N/mm²; $f_{pk} = 1770$ N/mm²; $E_p = 195$ kN/mm²; $A_p = 93$ and 52 mm² per 12.5 and 9.3 mm dia. strand; cover to strand = 35 mm. Concrete density = 24.5 kN/m³. Gravel aggregates. Cement CEM I grade 52.5R.

Distance to end of basic transmission length $l_{pt} = 0.19\ \sigma_{pm0}\ \Phi/(3.2 \times 0.7\ f_{ctm}(t)/1.5)$

Solution

Material properties

$f_{cm} = 45 + 8 = 53$ N/mm²; $f_{cm}(t) = 30 + 8 = 38$ N/mm²

$E_{cm} = 22 \times (53/10)^{0.3} = 36283$ N/mm²; $E_{cm}(t) = 36283 \times (38/53)^{0.3} = 32837$ N/mm²

$f_{ctm} = 0.3 \times 45^{2/3} = 3.80$ N/mm²; $f_{ctm}(t) = 3.80 \times 38/53 = 2.72$ N/mm²

$A_p = 7 \times 93 + 5 \times 52 = 911$ mm²

Section properties

Self-weight of unit = $168 \times 10^3 \times 24.5 \times 10^{-6} = 4.12$ kN/m.

$Z_b = 1270 \times 10^6/123 = 10.324 \times 10^6$ mm³

$Z_t = 1270 \times 10^6/(250 - 123) = 10.000 \times 10^6$ mm³

$y_s = (7 \times 93 \times 41.25 + 5 \times 52 \times 39.65)/911 = 40.8$ mm

$z_{cp} = 123.0 - 40.8 = 82.2$ mm; $Z_z = 1270 \times 10^6/82.2 = 15.450 \times 10^6$ mm³

Section properties of compound section with transformed area of tendons

$m - 1 = (195000/36283) - 1 = 4.37$

$A_{c,co} = 168000 + 4.37 \times 911 = 171986$ mm²; $y_{b,co} = 121.1$ mm; $I_{c,co} = 1296 \times 10^6$ mm⁴

$Z_{b,co} = 10.701 \times 10^6$ mm³; $Z_{t,co} = 10.056 \times 10^6$ mm³; $Z_{z,co} = 1296 \times 10^6/82.2 = 15.766 \times 10^6$ mm³.

Compound values are used only for calculating stresses due to loading (not for prestress), and for M_{sR} (Chapter 7) and deflections (Chapter 11).

Prestress

Eq. 6.2 Initial prestress $\sigma_{pi} = 0.7 \times 1770 = 1239.0$ N/mm²

Initial $P_{pi} = 1239.0 \times 911 = 1128729$ N

Eq. 6.3 $\Delta\sigma_{pr} = 1239.0 \times 0.66 \times 2.5 \times e^{(9.1 \times 0.7)} \times (20/1000)^{(0.75 \times (1-0.7))} \times 10^{-5} = 4.95$ N/mm²

$P_{mo} = (1239.0 - 4.95) \times 911 = 1124218$ N

$\sigma_{cp} = (1124218/168000) + (1124218 \times 82.2/15.45 \times 10^6) = 12.67$ N/mm²

Elastic shortening loss $\Delta\sigma_{el} = 195000 \times 12.67/32837 = 75.26$ N/mm²

Eq. 6.5 $\sigma_{pm0} = (1239.0 - 4.95 - 75.26) = 1158.8$ N/mm² $< 0.75 \times 1770 = 1328$ N/mm² OK

$P_{pm0} = 1158.8 \times 911 = 1055654$ N

Check transfer stresses the support

Eq. 6.6a $\sigma_b(t)=(1055654/168000)+(1055654\times82.2/10.324\times10^6)=14.69$ N/mm$^2<0.6\times30=18.0$ OK

Eq. 6.7a $\sigma_t(t)=(1055654/168000)-(1055654\times82.2/10.000\times10^6)=-2.39$ N/mm$^2>-2.72$ OK

If the above stresses are violated, the prestress at the end of the transmission length may be considered, where $l_{pt}=0.19\times1158.8\times11.17/4.064=605$ mm, where the mean diameter of the 12 strands is 11.17 mm and $f_{bpt}=3.2\times0.7\times2.72/1.5=4.064$ N/mm^2. The bending moment due to self-weight $M_{sw}=4.12\times9.9\times0.605/2-4.12\times0.605^2/2=11.58$ kNm. Then

Eq. 6.6b $\sigma_b(t)=14.69-11.58\times10^6/10.701\times10^6=13.61$ N/mm^2

Eq. 6.7b $\sigma_t(t)=-2.39+11.58\times10^6/10.056\times10^6=-1.24$ N/mm^2

Immediate camber due to prestress

Eq. 6.8a $\delta_1=-1055654\times82.2\times9900^2/(8\times32837\times1296\times10^6)=-25.0$ mm

Eq. 6.8b $\delta_1=[1-(4/3)(605/9900)^2]\times-25.0=-24.9$ mm. Ignored

Deflection due to self-weight

Eq. 6.9 $\delta_2=5\times4.12\times9900^4/(384\times32837\times1296\times10^6)=+12.1$ mm

Net deflection, or upward camber $=-25.0+12.1=-12.9$ mm.

The effects of creep and an increase in E_{cm}, between transfer and installation (at say 28 days), will cause this to increase, as shown in Section 11.2.1.

6.3 PRETENSIONING FORCE AND TENDON ECCENTRICITY IN BEAMS

6.3.1 Types of prestressed beams

The design of prestressed beams is more versatile than for floor slabs because several rows of tendons may be used, typically between 4 and 10 (in beams of depth 400–800 mm, respectively), although the tendon's positions are restricted to a predetermined pattern by an array of holes in the jacking heads, which is usually a permanent fixture at a precasting work. Figure 6.8 shows the end template for an inverted-tee beam with quite a large eccentricity from the centre on the tendons to the centroid of the section, resulting in an upward camber as shown in Figure 2.7. Upward camber is often controlled by adding two tendons (which need not be fully stressed) at the top of the beams. Top tendons also help to avoid cracking during transportation and handling.

Rectangular section beams are more efficient in terms of capacity per cross-sectional area, but inverted-tee and I beams hold the advantage of allowing the floor slab to be recessed in the upper part of the beam.

Figure 6.9a shows a full array of possible tendon positions in an inverted-tee beam and an example of a typical tendon layout. Note the symmetry. The tendons are placed at all the corners; a 40–50 mm centroidal cover distance is used in most cases. If debonded tendons are required to satisfy the transfer conditions and the service or ultimate design moments simultaneously, the procedure for calculating the required pretensioning forces and eccentricities at positions along the span is as follows. In theory, there may be several

Figure 6.8 Preparation of mould and rebars for prestressed concrete inverted-tee beam.

positions along the span where checks are required, but it is usually to satisfy transfer conditions at the support and at ¼ or ⅓ span, and the design moments at mid-span and at ¼ or ⅓ span minus the transmission length (or ultimate anchorage length). Figure 6.9b shows a typical arrangement of tendons, together with the necessary tendons to anchor the shear stirrups (in the upstand) and boot links (in the boot). Note how the tendons are staggered in the successive rows to avoid local concentrations of prestress (see also Figure 6.20).

6.3.2 Calculation for pretensioning force and eccentricity

It is desirable that the permissible stress at transfer $f_{ctm}(t)$ after the initial losses (due to elastic shortening) is made equal (or as close as possible) to 0.45 f_{ck} after all losses. In most beam designs the initial and final losses are about 8% and 25%, respectively, meaning that the ratio $f_{ck}(t)/f_{ck}$ should be at least 0.55. The transfer strength for $f_{ck}=50$ is often at about 35 N/mm^2, and therefore transfer stresses will (nearly) always govern for parallel, unbonded tendons. To overcome this problem some tendons are debonded, starting with 2 no. and increasing in multiples of two until transfer and service conditions are satisfied. It is also wise to restrict the top fibre stress to something less than $f_{ctm}(t)$ say ½ to ⅔ of this value, while accepting that there will be a small loss in moment capacity. This factor of safety is based on the experience gained from transporting and handling long and/or highly stressed prestressed beams, and the need to avoid flexural cracking at the sides of the beams for the sake of durability. During transportation, beams are supported

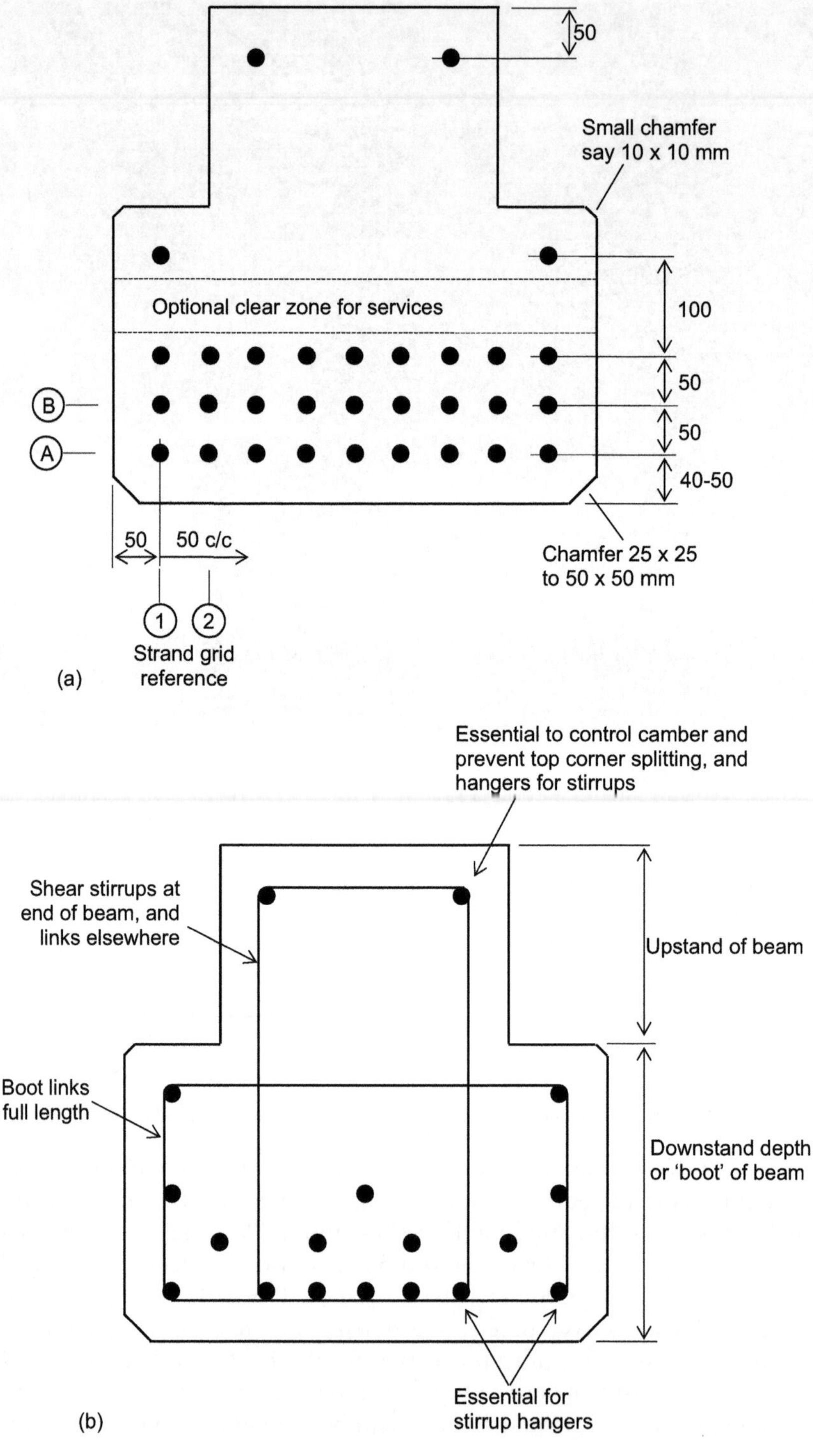

Figure 6.9 (a) Array of pretensioning tendons (usually strands) in inverted-tee beams. (b) Typical strand and link arrangement.

on bearers at about 500–750 mm from the ends, and the negative bending moment due to self-weight of the cantilevered ends causes tension in the top in addition to the negative prestress.

The pretensioning force that takes place at the moment of detensioning includes the immediate relaxation $\Delta\sigma_{pr}$ of the tendons plus most of the elastic shortening loss $\Delta\sigma_{el}$. The loss ratio at transfer is defined here as $\xi=(\Delta\sigma_{pr}+\Delta\sigma_{el})/\sigma_{pi}$ calculated according to Section 6.2.3, where $\sigma_{pi}=\eta\, f_{pk}=P_{pi}/A_p$. The loss ratio $\Delta\sigma_{pr}/\sigma_{pi}$ for Class 2 strand stressed to 70% and detensioned after 20 hours, is fixed as $0.66\times 2.5\times e^{9.1\times 0.7}\times(20/1000)^{0.75\times(1-0.7)}\times 10^{-5}=0.004$. The total loss ratio is first estimated – a figure of 0.07 is a good starting point. The force at transfer is therefore $P_{pi}\,(1-\xi)$ and the bottom and top fibre stresses are

$$\sigma_b(t)=P_{pi}(1-\xi)\,\left(1/A_c+z_{cp}/Z_b\right)<+\,0.6f_{ck}(t) \tag{6.10}$$

$$\sigma_t(t)=P_{pi}(1-\xi)\,\left(1/A_c-z_{cp}/Z_t\right)>-0.5f_{ctm}(t)^{*} \tag{6.11}$$

*it is the designer's choice to limit the tensile stress to $-0.5\,f_{ctm}(t)$.

Manipulation of the simultaneous equations Eqs. 6.10 and 6.11 will give optimum values for the initial prestressing force P_{pi} and the eccentricity z_{cp}, as follows:

$$P_{pi}=\frac{A_c}{2(1-\xi)}\left[0.6f_{ck}(t)-0.5f_{ctm}(t)+\left\{(0.6f_{ck}(t)+0.5f_{ctm}(t)\right\}\frac{1-\alpha}{1+\alpha})\right] \tag{6.12}$$

where $\alpha=Z_t/Z_b$ and ξ=initial prestress loss ratio (expressed as a decimal fraction)

The number of tendons required is

$$N=P_{pi}/\eta A_p f_{pk} \tag{6.13}$$

where η=degree of prestress.

The eccentricity zcp is given by

$$z_{cp}=\left[0.6f_{ck}(t)\ +\ 0.5f_{ctm}(t)\right]/\left[(1-\xi)P_{pi}\beta\right] \tag{6.14}$$

where $\beta=\ 1/Z_b+\ 1/Z_t$ and P_{pi} is obtained from Eq. 6.12.

Given P_{pi} and z_{cp} the actual bottom and top fibre stresses may be calculated from Eqs. 6.10 and 6.11 from which the actual value of ξ is found (as opposed to the guessed at value before Eq. 6.10), and if this differs by more than 2%–3% iteration should take place. The elastic shortening loss ratio is given by

$$\Delta\sigma_{el}/\sigma_{pi}=\sigma_c E_p/E_{cm}(t)\eta f_{pk} \tag{6.15}$$

where $\sigma_c=P_r/A_c+P_r z_{cp}/Z_z$ (6.16)

$Z_z=I_c/z_{cp}$ at the centroid of the tendons

and the prestress after initial relaxation loss $P_r=\left(\sigma_{pi}-\Delta\sigma_{pr}\right)A_p$

Finally the loss ratio $\xi=\left(\Delta\sigma_{pr}+\Delta\sigma_{el}\right)/\sigma_{pi}$ (6.17)

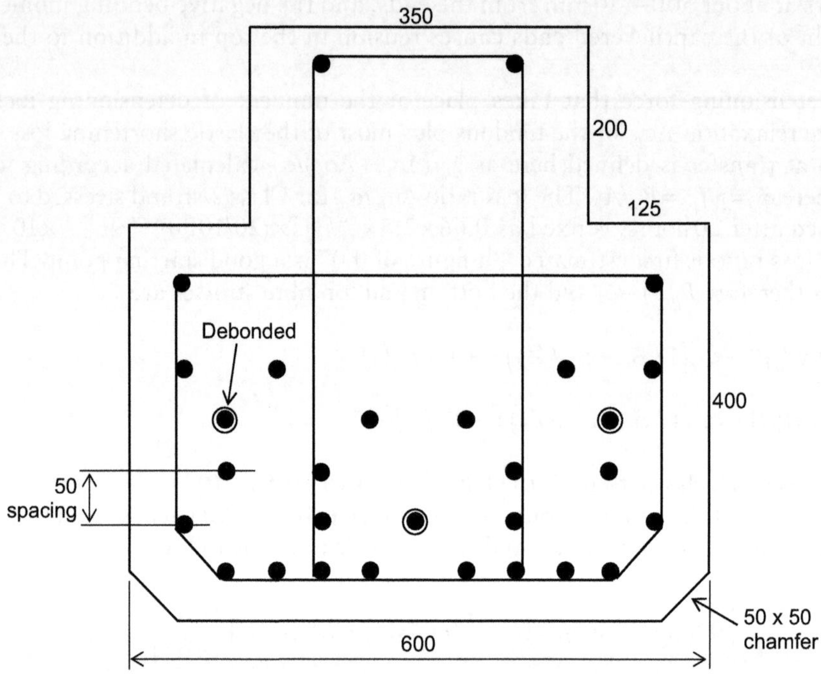

Figure 6.10 Cross section and final strand arrangement for the inverted-tee beam in Examples 6.2, 7.10, 7.11, 9.5, 10.6, 12.8, 12.9 and 12.12. The three debonded strands for Example 7.11 only are shown in circular sheathes.

Example 6.2

Calculate the initial prestressing requirements for an inverted-tee beam shown in Figure 6.10. Use $f_{ck}=45$ N/mm², $E_{cm}=36283$ N/mm², $f_{ck}(t)=30$ N/mm², $E_{cm}(t)=32837$ N/mm², diameter of strands = 12.5 mm, $A_p=93$ mm², $f_{pk}=1770$ kN/mm², $\eta=0.7$, $E_p=195000$ N/mm². Assume $\xi=0.07$ initial losses, including an initial relaxation loss ratio of 0.004.

Solution

Geometric data

$A=307.5\times10^3$ mm², $y_b=269.78$ mm, $I_c=8151.5\times10^6$ mm⁴, $Z_b=30.215\times10^6$ mm³, $Z_t=24.685\times10^6$ mm³.

Then $\alpha=24.685/30.215=0.817$

$\beta=(1/30.215\times10^6)+(1/24.685\times10^6)=7.36\times10^{-8}\text{mm}^{-3}$

$\sigma_{pi}=0.7\times1770=1239$ N/mm²

Prestressing force per strand $=93\times1239\times10^{-3}=115.23$ kN

$f_{ctm}(t)=f_{ctm}\,f_{cm}(t)/f_{cm}=3.80\times(30+8)/(45+8)=2.72$ N/mm²

Limiting transfer stresses are $\sigma_c\leq+0.6\times30=18.0$ N/mm² and $\sigma_{ct}\geq-0.5\,f_{ctm}(t)=-1.36$ N/mm²

Eq. 6.12 $$P_{pi}=\frac{307.5\times10^3}{2\times(1-0.07)}\times\left[16.64+\left(19.36\times\frac{0.183}{1.817}\right)\right]\times10^{-3}=3073.3\text{ kN}$$

Eq. 6.13 $N=3073.3\times10^3/115.23=26.67$ rounded down to 26 to prevent the possibility of overstress.

Eq. 6.14 $z_{cp}=19.36/(0.93\times3073.3\times10^3\times7.36\times10^{-8})=92.0\,\text{mm}$

The actual prestressing force $P_{pi}=26\times115.23=2995.9$ kN, and the first estimate after initial losses $=0.93\,P_{pi}=2786$ kN. The maximum fibre stresses at transfer are

Eq. 6.10 $\sigma_b(t)=2786\,(1/307.5+92/30215)=+17.54\,\text{N/mm}^2<+18.0\,\text{N/mm}^2$

Eq. 6.11 $\sigma_t(t)=2786\,(1/307.5-92/24685)=-1.33\,\text{N/mm}^2>-1.36\,\text{N/mm}^2$

At the level of the strands $Z_z=I_c/z_{cp}=8151.5\times10^6/92=88.61\times10^6\,\text{mm}^3$

Eq. 6.16 $\sigma_c=2786\times(1/307.5+92/88610)=+11.95\,\text{N/mm}^2$

As the initial relaxation loss is given as 0.004

Eq. 6.15 $\xi=[0.004+(11.95\times195000)]/(32840\times0.7\times1770)=0.062$ (original assumption of 0.07 is OK).

Having established $N=26$ strands, $A_p=2418\,\text{mm}^2$, at $z_{cp}=92\,\text{mm}$, the strands are arranged in the beam as shown in Figure 6.10 as follows:

No. strands in each row	*Distance y_i from bottom (mm)*	*$\Sigma\,N_i\,y_i$ (mm)*
2	550	1100
2	350	700
4	250	1000
2	200	400
4	150	600
4	100	400
8	50	400
$\Sigma=26$		$\Sigma=4600$

Then $y_s=4600/26=176.92\,\text{mm}$

The actual eccentricity $z_{cp}=y_b-y_s=269.78-176.92=92.86\,\text{mm}$, closest possible to 92 mm in the available array of tendon positions. The complete design is as follows:

Losses $\Delta\sigma_{pr}+\Delta\sigma_{el}$

$\Delta\sigma_{pr}=1239.0\times0.66\times2.5\times e^{9.1\times0.7}\times(20/1000)^{0.75\times(1-0.7)}\times10^{-5}=4.95\,\text{N/mm}^2$

$P_r=(1239.0-4.95)\times2418\times10^{-3}=2983.9$ kN

$\sigma_c=(2983.9/307.5)+(2983.9\times92.86/87782)=12.86\,\text{N/mm}^2$

$\Delta\sigma_{el}=195000\times12.86/32837=76.37\,\text{N/mm}^2$

$\sigma_{pm0}=1157.7\,\text{N/mm}^2<0.75\times1770=1328\,\text{N/mm}^2$ OK

The final $\xi=(1239-1157.7)/1239=0.0656$

Transfer stress check

$P_{pm0}=1157.7\times2418=2799.2$ kN

$\sigma_b(t) = (2799.2/307.5) + (2799.2 \times 92.86/30215) = 17.71\ \text{N/mm}^2 < 18.00\ \text{N/mm}^2$

$\sigma_t(t) = (2799.2/307.5) - (2799.2 \times 92.86/24685) = -1.43\ \text{N/mm}^2 > -2.72\ \text{N/mm}^2$

6.4 BURSTING, SPALLING AND SPLITTING STRESSES

6.4.1 Types of tensile stresses and cracking

The phenomenon of horizontal cracking immediately after transfer, or sometimes shortly afterwards, either along the line of pretensioned tendons or in narrow webs above the line of tendons, is known as bursting, spalling and splitting. Examples of this are shown in Figures 6.4 and 6.11, for the case of spalling in the narrow webs of hcu, and a T beam just above the top row of two rows of 3 no. wires. In the latter, the cracks were not reported until the beams had been delivered to site, i.e. the cracks did not form at the time of transfer of prestress, the usual point in time.

The three types of cracks, i.e. bursting, spalling and splitting, are generally referred to as 'splitting', the latter two colloquially known as 'crocodile cracks' (with reference to its mouth!). However, they are individually caused by different phenomena due to the transfer of prestress from the line of the pretensioning tendons to the body of the section, where the gradient of the vector, the rapid build-up of prestress within the transmission length (actual l_t, not the design value l_{pt2}) and the change in cross section above the tendons are the most important parameters. Bursting and spalling are related to the distribution of prestressing forces over the cross section. Splitting is due to bond action and the Hoyer effect due to a change in the diameter of tendons over the transmission length.

Localised compressive stresses develop in the concrete close to the point of transfer, spreading out to cause vertical and lateral tensile stresses σ_{sp} in a linear stress distribution, somewhere just beyond the end of the transmission zone. Where σ_{sp} exceeds the tensile strength of the concrete f_{ct}, sometimes less than 16 hours of age, a horizontal crack propagates rapidly, usually from the end of the element. f_{ct} is an actual axial tensile strength, not a design value, recommended to be on the basis of tests. A figure of about 1.2 f_{ctm} is a good estimate, i.e. 3.5 N/mm^2 for $f_{ck}(t) = 30$ N/mm^2.

Figure 6.11 Splitting propagating from the transmission zone in prestressed concrete T beams (courtesy Bison Precast Ltd., UK).

The immediate upward camber of elements such as hcu or slip-formed T beams, pretensioned with large eccentric forces, is thought to contribute to the severity of the cracks, particularly where the circular saw used to cut the concrete and tendons induces an upward reaction as the curvature of the element snags on the saw. Figure 6.12 illustrates the effect. To eliminate this weight packs may be placed close to the ends of the elements to nullify the camber and end curvature. See Section 4.21.

Bart Hendrikx (Hendrikx, 2011) summarised the above in a presentation at the IPHA (International Prestressed Hollowcore Association) Technical Seminar, in which the three forms of cracking were identified and discussed. Figures 6.13–6.18 are adopted from this presentation and show 2-d regions of internal compression and tension in a wide element, where the lateral stresses between tendons tend to cancel each out or do not have the space to develop due to proximity of other tendons. Although the stress field in a T beam may be 3-d, lateral splitting only occurs in wide beams where the prestress is concentrated at one of two positions across the width.

The presentation also referred to the calculation to estimate the spalling stress given in BS EN 1168 (BS EN 1168, 2005) clause 4.3.3.2.1, based on the earlier work of Den Uijl (Den Uijl, 1985) as well as the information presented in the CEB-FIB Model Code 90 (CEB-FIB, 1990). See Section 6.4.5.

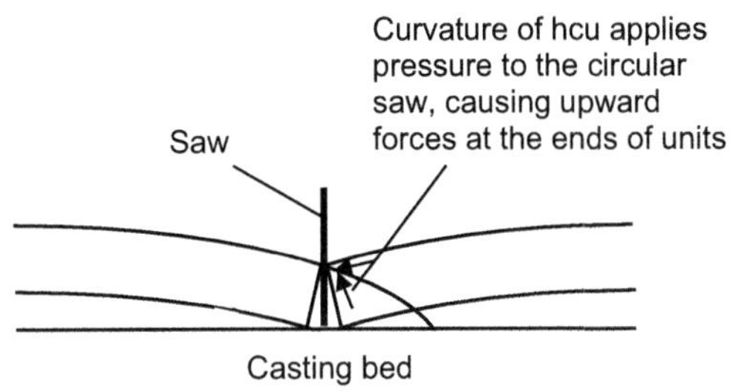

Figure 6.12 illustration of upward camber and curvature increasing the risk of spalling cracks at the ends of elements.

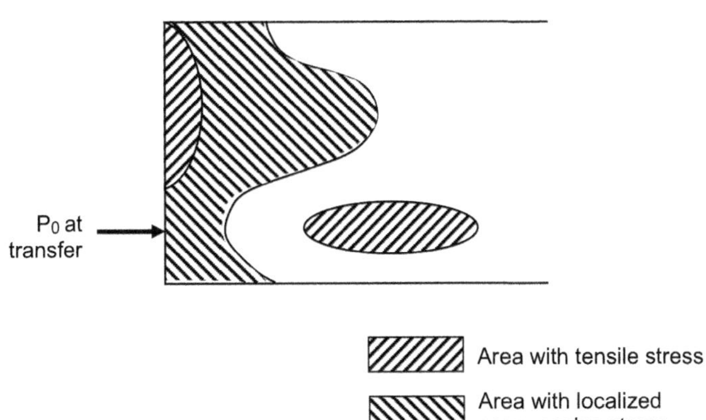

Figure 6.13 Localised compressive and tensile stresses close to ends of prestressed elements (after Hendrikx 2011).

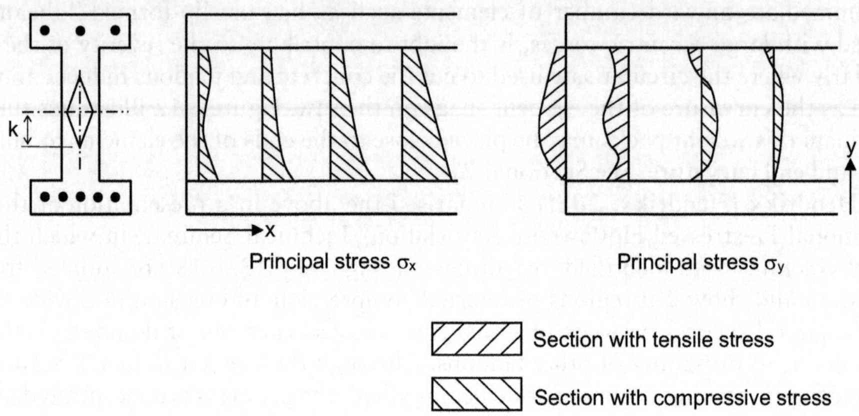

Figure 6.14 Localised compressive and tensile stresses close to ends of prestressed elements (after Hendrikx 2011).

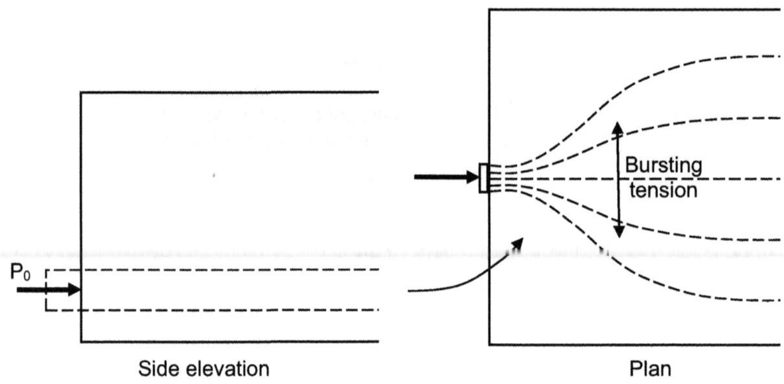

Figure 6.15 Bursting stresses created due to spreading of localised pretensioning force (after Hendrikx 2011).

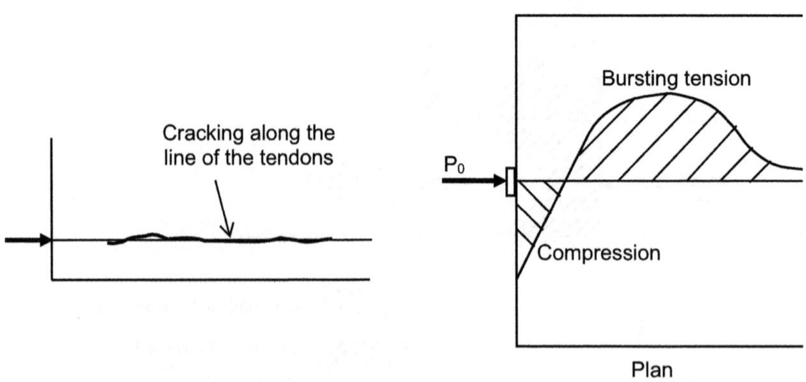

Figure 6.16 Position of bursting cracks due to the distribution of bursting tensile stresses (after Hendrikx 2011).

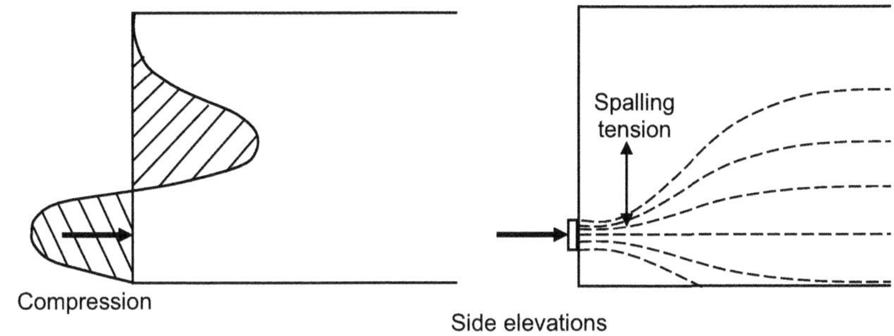

Figure 6.17 Spalling stresses created due to spreading of eccentric pretensioning force (after Hendrikx 2011).

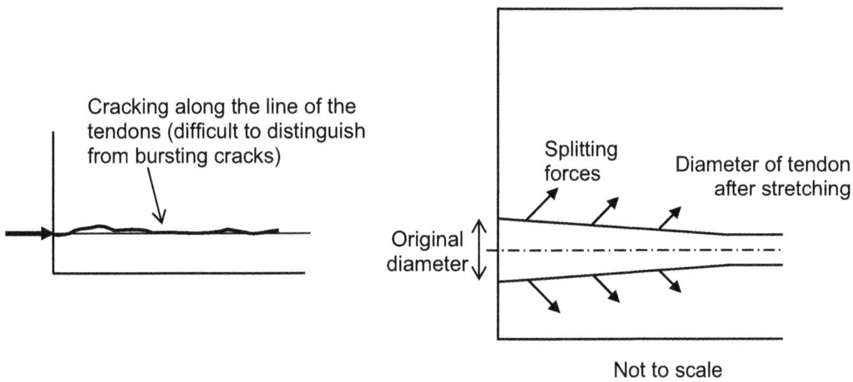

Figure 6.18 Position of splitting cracks due to radial tensile stress around tendons (after Hendrikx 2011).

6.4.2 Bursting stresses

Bursting stresses are shown in Figure 6.15 due to the spreading of prestress causing curved compressive stress trajectories which generate lateral tensile stress. The same effect is found under partially loaded areas such as bearing plates, as shown in BS EN 1992-1-1, Fig. 6.29 using a slope of trajectory of 2:1. Exp. 6.58 and 6.59 are used to determine the ratio of the lateral bursting force to normal force, typically about 0.2 for small bearing areas. Directly beneath the loaded area, the concrete is confined, and for the same reason, Figure 6.16 shows that the maximum tensile stress occurs some distance from the end of the element. A possible crack pattern occurs along the line of the tendons, developing first at transfer but increasing in length over a short time.

6.4.3 Spalling stresses

'Spalling' is the term used in BS EN 1168 to describe horizontal cracks due to a similar mechanism as bursting cracks due to the spreading of prestress into the section – except spalling cracks occur higher up the section and are most severe in narrow webs. Here 'spalling' should not be confused with the term 'spalling' to describe the surface spalling near corroding rebars or flaking due to expansive forces such as sulfates, etc. Spalling stresses are shown in Figure 6.17 due to the spreading of eccentric prestressing forces, which causes additional curving of the compressive stress trajectories, with the largest gradient being above the line of the pretensioning force. These stresses can also be increased due to internal thermal gradients during the hydration of cement in curing, particularly in the outer webs. Spalling stresses are greater than bursting stresses and occur at the end of the element, which is less confined than in the latter. The cracks are also wider than the latter. Spalling is therefore the most common of the three types, as shown in Figure 6.4, particularly in hcu with circular cores where cracking is at mid-depth, or with deep parallel sided cores where cracking occurs at about ¼ of the depth, particularly if the minimum width of the webs is less than 35 mm. It can also be a problem in highly stressed hcu of depth 400–500 mm. A calculation model is given in Section 6.4.5.

6.4.4 Splitting stresses

Splitting forces and stresses are shown in Figure 6.18. Unlike bursting and spalling these are related to the bond and dilation of tendons, particularly around helical strands where short transmission lengths (thanks to the cement matrix engaging in the tiny crevices between the six outer spiral wires) increase the gradient of the dilation. The change in diameter of the tendon due to pretensioning, known as 'Hoyer' effect, together with circumferential bond stresses, cause radial tensile stress around the tendon. Splitting cracks initiate at the ends of the elements along the line of the tendons, but are difficult to distinguish from bursting cracks as the two effects are taking place at the same time. The two main parameters controlling splitting are concrete cover and transmission length.

A calculation model is given by the FIP Recommendations (FIP, 1988) to check the radial tensile stress around tendons as a function of cover. Cover relates to any free edge, including the internal hollow cores, as shown in Figure 6.19. A frictional bond mechanism is used. Bursting stresses, which are smaller than splitting and typically 1 N/mm², are combined with splitting stresses of about 3 N/mm² using a lower and conservative coefficient of friction of $\mu = 0.32$ (Walraven's research, T. U. Delft) instead of 0.37 used elsewhere for bond analysis. See Figure 6.20a.

If the initial stress $\sigma_{po} = \eta f_{pu}$ (6.18)

Allowing 5% losses at transfer $P_o = 0.95\sigma_{po} A_p$ (6.19)

Bond stress $\tau = P_o / \pi \Phi l_t$ (6.20)

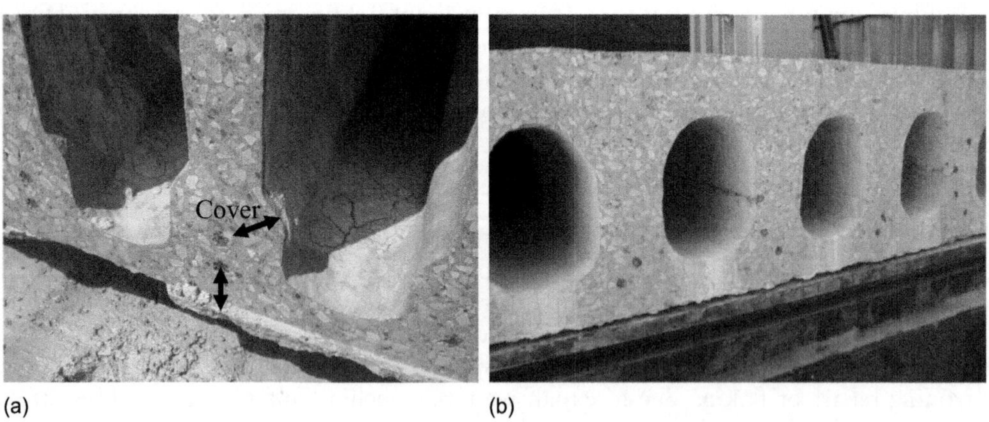

Figure 6.19 Positions of (a) strands and (b) wires in some hollow core units.

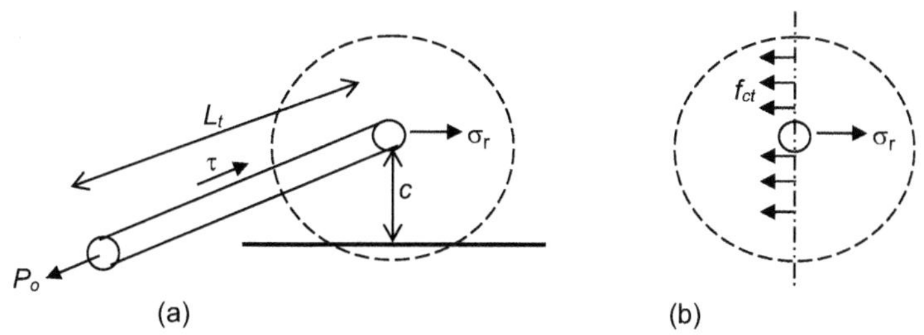

Figure 6.20 Relationship between average bond stress, circumferential stress, splitting stress and cover.

where the transmission length (according to FIP,1988) $l_t = 10\Phi\sqrt{\left(\sigma_{po}/f_{cu}(t)\right)}$ (6.21)

Radial stress $\sigma_r = \pi / \mu\Phi$ (6.22)

Splitting occurs when the average circumferential stress exceeds the tensile strength of the concrete, as shown in Figure 6.20b.

Then as $F_r = \sigma_r\Phi$ (6.23)

and $F_t = f_{ct}2C$ (6.24)

Minimum cover $C \geq \sigma_r\Phi /\ 2f_{ct}$ (6.25)

where the mean tensile strength (FIP, 1988) $f_{ct} = 0.87\left(0.05f_{cu}(t)+1\right)$ (6.26)

Example 6.3

Determine the minimum cover to satisfy splitting stress criteria for a 12.5 mm helical strand, of area A_p=93 mm², stressed to 70% with f_{pu}=1770 N/mm². Allow 5% losses at transfer with a characteristic cylinder strength $f_{ck}(t)$=28 N/mm². The factory target cube strength would be $f_c(t)$=35+5=40 N/mm².

Solution

For grade C28/35 concrete $f_{cu}(t)$=35 N/mm²

Eq. 6.18 $\sigma_{po}=0.7\times1770=1239$ N/mm²

Eq. 6.19 $P_o=0.95\times1239\times93=109.5$ kN

Eq. 6.21 $l_t=10\times12.5\ \sqrt{(1239/35)}=743$ mm

Eq. 6.20 $\tau=109.5\times10^3/\pi\times12.5\times743=3.75$ N/mm²

Eq. 6.22 $\sigma_r=3.75/0.32=11.72$ N/mm²

Eq. 6.26 $f_{ct}=0.87\ (0.05\times35+1)=2.39$ N/mm²

Eq. 6.25 Then $C\geq11.72\times12.5/(2\times2.39)=$ **30.7 mm**

This is the design value for the cover to all surfaces. The typical cover to the soffit used for 12.5 mm strands is 35–40 mm.

Repeating according to BS EN 1992-1-1, Exp 8.16 (assuming gradual release of prestress and good bond conditions) using σ_{pm0}=0.95×1239=1177 N/mm^2 and f_{bpt}= 3.2×0.7×0.3×28$^{2/3}$/1.5=4.13 N/mm^2. Then l_{pt}=0.19×12.5×1177/4.13=677 mm. f_{ct}=f_{ctm}= 0.3×28$^{2/3}$=2.76 N/mm^2. Then τ=3.75×(743/677)=4.11 N/mm^2, σ_r=12.86 N/mm^2 and C≥12.86×12.5/(2×2.76)=**29.1 mm**

It may be possible that the strands close to the bottom of the hollow cores would have this amount of cover. To check whether splitting stresses would actually be violated, using factory measured values for $f_c(t)$=40 N/mm^2 and the realistic empirical value for $f_{ct}=0.7\sqrt{f_c}(t)$.

Eq. 6.21 l_t=10×12.5 √(1239/40)=696 mm

Eq. 6.20 τ=109.5×10^3/π×12.5×696=4.00 N/mm^2

Eq. 6.22 σ_r=4.00/0.32=12.50 N/mm^2

Eq. 6.26 f_{ct}=0.7×√40=4.42 N/mm^2

Eq. 6.25 C≥12.50×12.5/(2×4.42)=**17.6 mm**

It is therefore possible that splitting stresses would not occur where the cover is at least 18 mm.

6.4.5 MODEL CALCULATIONS FOR SPALLING STRESS IN HOLLOW CORE FLOOR UNITS

Although splitting and spalling can occur in all types of prestressed slabs and beams that have particularly narrow webs in relation to the overall width and magnitude of prestress in the bottom flange, the most comprehensive body of work relates to hcu, pioneered by Den Uijl at T. U. Delft (Den Uijl, 1985).

Spalling may be avoided by following some fairly comprehensive design rules adopted in BS EN 1168, clause 4.3.3.2.1 from the work of Den Uijl. The research was applicable to hcu less than 400 mm deep, but this is not mentioned in BS EN 1168. Extensive research has been carried out on the transmission of prestress in hcu. Adopting approximations using strut-and-tie or truss models and more detailed finite element models, BS EN 1168 published the following expression for the maximum vertical spalling stress σ_{sp} in the web of an I section. The I section is transformed to have the same area and second moment of area as close as possible to the actual profile of the hcu. Hendrikx (Hendrikx, 2011) mentions that the expression for σ_{sp} is also applicable to rectangular sections, provided that the relative eccentricity α_e and web width b_w are the same, where the equivalent depth of a rectangle d_{ef} is given as

$$d_{ef} = e_o \Big/ \left[\left((e_o - k) / d \right) + \tfrac{1}{6} \right] \tag{6.27}$$

where

$$k = Z_b / A_c \geq 0 \tag{6.28}$$

is for the I section, known as the core radius (or 'kernel') as shown in Figure 6.14.

e_o=eccentricity to tendons
d=depth of I section

The figure of $^1/_6$ is based on $Z_b/A_c=d_{ef}/6$ for a rectangular section (and is checked in Example 6.4).

BS EN 1168 offers the option to either carry out a fracture mechanics design or to calculate the spalling stress using the following empirical expression:

$$\sigma_{sp} = \frac{P_{pm0}}{b_w\, e_o} \times \frac{15\alpha_e^{2.3} + 0.07}{1 + \left(\frac{l_{pt1}}{e_o}\right)^{1.5} (1.3\alpha_e + 0.1)} \quad \text{note}\left(l_{pt1}/e_o\right)^{1.5} \tag{6.29}$$

and $\sigma_{sp} \le f_{ct}$

where f_{ct}=tensile strength of concrete at transfer on the basis of tests

$$\alpha_e = (e_o - k)\ /\ h \tag{6.30}$$

$$l_{pt1} = 0.8 l_{pt} \text{ (according to BS EN 1992-1-1, Exp. 8.16 and 8.17)} \tag{6.31}$$

If the tendons are fairly uniformly distributed across the width of the hcu, then the total prestress distributed over the total cross section may be used; otherwise, individual webs must be checked.

P_{pm0}=prestressing force just after release (i.e. after immediate relaxation and shortening losses) in individual webs

where b_w=width of the corresponding web, or the total force in the slab if Σb_w=total width of webs or the width of a solid slab

e_o = eccentricity of individual tendons or total tendons likewise

h = depth of hcu or slab

Figure 6.21 offers a design graph in terms of $\sigma_{sp}/(P_{pm0}/b_w\, e_o)$ vs α_e for a range of values for l_{pt1}/e_o. While this may not be the most convenient form of presentation, in that the reader has

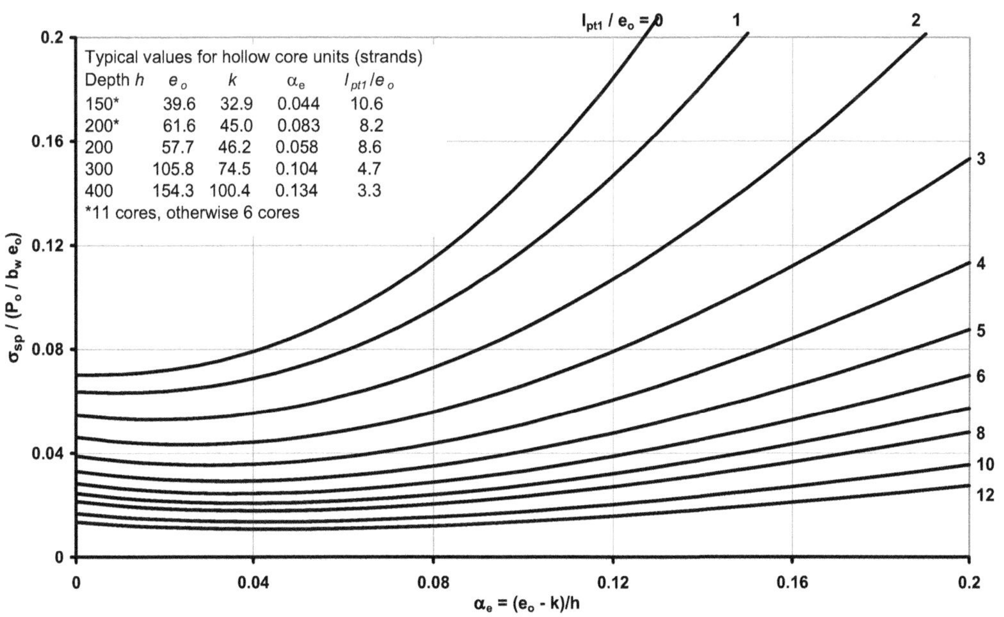

Figure 6.21 Spalling stresses according to BS EN 1168.

to do a lot of work to determine σ_{sp}, it does show the dependency of l_{pt1} on the final result – a small error in something that is quite difficult to predict can lead to large variations in (the y-axis) $\sigma_{sp}/(P_{pm0}/b_w\, e_o)$.

Designers of hcu sections show that the resulting value of σ_{sp} is less than the tensile strength of concrete at transfer f_{ct} on the basis of tests. As an initial estimate, a typical value for f_{ct} is about 3.5 N/mm^2 when $f_{ck}(t)=30$ N/mm^2. Spalling is not easily predicted by simply calculating σ_{sp}. It is significantly affected by mix design (cement type, fine aggregates), detensioning rate, prestress level (when prestress at the bottom fibre is greater than about 14 N/mm^2), bond slip and transmission length, and the environmental conditions during curing.

The effect of changes in transmission length is dramatic. For example, if the partial safety factor ($\gamma_c=1.5$) for bond stress is omitted, on the grounds that spalling is not an ultimate limit state, then l_{pt1}/e_o decreases by 1.5. For the case of the 400 mm depth hcu with 6 cores and $b_w=40$ mm, using the prestress active in an individual web σ_{sp} (derived from Eq. 6.29)=2.4 N/mm^2, but if γ_c is omitted σ_{sp} increases to 3.4 N/mm^2, by a factor of 1.42. Taking $f_{ct}\approx 3.5$ N/mm^2 the results show that splitting may only be a problem in deeper units.

Example 6.4

Calculate the spalling stress σ_{sp} according to BS EN 1168 for the webs in the 250 mm depth hcu shown in cross section in Figure 6.7 based on (i) the full width hcu and (ii) one web. Use the section properties, material data, area of strands $A_p=911$ mm^2 and prestress after transfer as $\sigma_{pm0}=1158.8$ N/mm^2 from Example 6.1. Assume the release of prestress is gradual, and bond conditions are good (see BS EN 1992-1-1, clause 8.10.2.2).

Solution

i. for the full width hcu

$Z_b=1270\times10^6/123=10.325\times10^6$ mm^3

Eq. 6.28 $k=10.325\times10^6/168000=61.5$ mm

Height to centroid of strands $y_s=40.8$ mm

$e_o=123.0-40.8=82.2$ mm

Eq. 6.30 $\alpha_e=(82.2-61.5)/250=0.083$

$f_{ctm}=0.3\times45^{2/3}=3.80$ N/mm^2

$f_{ctm}(t)=3.80\times(30+8)/(45+8)=2.72$ N/mm^2

$f_{bpt}=3.2\times0.7\times2.72/1.5=4.06$ N/mm^2

Average diameter of strands$=(7\times12.5+5\times9.3)/12=11.17$ mm

Eq. 6.31 $l_{pt1}=0.8\times0.19\times1158.8\times11.17/4.06=484$ mm

Then $l_{pt1}/e_o=5.89$

For the full width unit Ppm0=1158.8×911=1055654 N

$b_w=5\times40+2\times51=302$ mm

$P_{pm0}/b_w\, e_o=1055654/302\times82.2=42.52$ N/mm^2

Eq. 6.29 $\sigma_{sp}=42.52\times(15\times0.083^{2.3}+0.07)/[1+\{5.89^{1.5}\times(1.3\times0.083+0.1)\}]=$ **1.27 N/mm²**

ii. determine the maximum σ_{sp} in 1 no. b_w=40mm web, due to the prestress of 1 no. 12.5 and 1 no. 9.3mm diameter strands. Average diameter=10.9mm.
For the I section with one core k=62.5mm

y_s=40.4mm, e_o=82.6mm and $\alpha_e=(82.6-62.5)/250=0.0804$

P_{pm0} (for 2 strands)=1158.8×(93+52)=168026N

$P_{pm0}/b_w\ e_o=168026/40\times82.6=50.85$

$l_{pt1}=0.8\times0.19\times1158.8\times10.9/4.06=472$mm, then $l_{pt1}/e_o=5.72$

$\sigma_{sp}=50.85\times(15\times0.0804^{2.3}+0.07)/[1+\{5.72^{1.5}\times(1.3\times0.0804+0.1)\}]=$ **1.55 N/mm²**

This is 1.22 times greater than the average for the full width unit. It is unknown whether spalling would occur in an individual web at a higher stress than in all the webs or how much the prestress spreads across the unit at transfer.

If the equivalent depth d_{ef} of a rectangular section is substituted for the I section, then

Eq. 6.27 $d_{ef}=(82.6\ (0.0804+{}^1/_6)=334$mm

$k={}^1/_6\times334=55.7$mm

Eq. 6.30 $\alpha_e=(82.6-55.7)/334=0.0805$ compared with 0.0804 for the I section.

REFERENCES

BS EN 1168. 2005. Precast Concrete Products - Prestressed concrete hollow core units, +A3:2011, BSI, London, UK.

BS EN 1992-1-1. 2004. Eurocode 2: Design of concrete structures. Part 1-1: General rules and rules for buildings, +A1:2014, BSI, London, UK.

BS 5896. 2012. High tensile steel wire and strand for the prestressing of concrete, BSI, London, UK.

CEB-FIB. 1990. CEB-FIB Model Code 90, Thomas Telford, London, UK, 460 p.

Den Uijl, J. A. 1985. Bond properties of strands in connection with transmission zone cracks, *Betonwerk und Fertigteil-Technik, Journal of the Federal German Association of the Concrete and Precast Building Components Industry*, Vol. 51, No. 1, pp. 28–36.

FIP. 1988. Precast Prestressed Hollow Core Floors, Thomas Telford, London, UK, 31 p.

Hendrikx, B. 2011. Spalling in EN 1168, IPHA Technical Seminar, Prestressed Hollowcore floors - the floors for the future, RWTH-IMB Achen, Germany. October 26-27, 2011.

Kong, P. Y. L. 1993. Transfer of prestress by pretensioned wire tendons, MSc thesis, School of Civil Engineering, Curtin University of Technology, Australia. https://espace.library.curtin.edu.au

NA to BS EN 1992-1-1. 2004. UK National Annex to Eurocode 2: Design of concrete structures. Part 1-1: General rules and rules for buildings, BSI, London, UK.

Chapter 7

Flexural design in service

7.1 FLEXURAL BEHAVIOUR OF PRESTRESSED ELEMENTS

7.1.1 Flexural tests on prestressed hollow core floor units

The flexural behaviour of precast prestressed concrete elements, such as slabs and beams in bending, is quite predictable, at least up to the cracking moment of resistance and less so at the ultimate capacity. Figure 7.1 shows the results of mid-span bending moment vs. mid-span deflection on six full scale tests carried out on a 1200 mm width × 200 mm depth prestressed hollow core floor units (hcu) subjected to six-point bending (four point loads at 1/5th span spacing to produce constant bending moment near to mid-span) over a simply supported span of 5.0 m. Figure 7.2. To compare the experimental results with predicted values, given in Tables 7.1 and 7.2 as the cracking moment of resistance, and

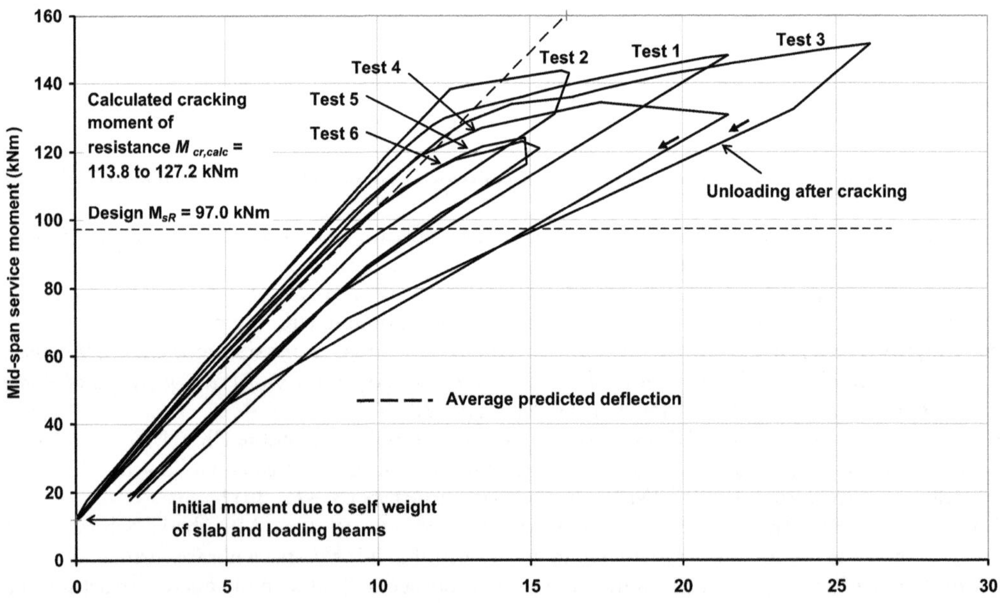

Figure 7.1 Experimental and predicted bending moment vs. mid-span deflection in full scale tests carried out on 1200 × 200 mm depth prestressed hollow core floor units. (With permission by Bison Precast, UK.)

DOI: 10.1201/9781003319450-7

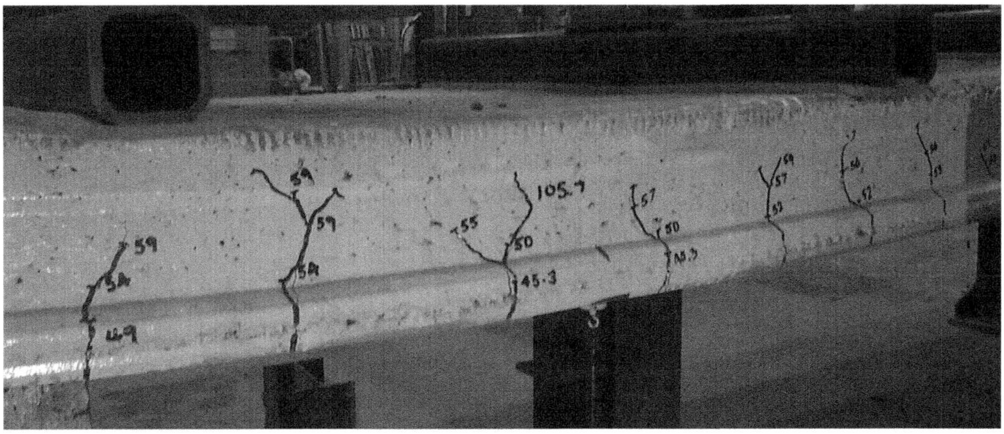

Figure 7.2 Flexural cracking in the soffit in the mid-span region of a prestressed hollow core floor unit subjected to four applied loads at 1/5th span. The numbers represent to total applied load *P* (kN), then bending moment $M=0.6$ PL, where effective span $L\approx5.0$ m. (With permission by Bison Precast, UK.)

Table 7.1 Calculation for cracking moment of resistance in full scale tests carried out on 1200 × 200 mm depth prestressed hollow core floor units

Slab ref	*Prestress in bottom of unit* σ_b *(N/mm²)*	*Compound second moment of area* I_{cco} *(mm⁴ × 10⁶)*	*Compound height to centroid of concrete* $y_{b,co}$ *(mm)*	*Compound section modulus at bottom* $Z_{b,co}$ *(mm³ × 10⁶)*	*Calculated cracking moment*[a] $M_{cr,calc}$ *(kNm)*	*Imposed moment at elastic limit*[a] M_s *(kNm)*	*Ratio* $M_s/M_{cr,crack}$
1	11.90	695.4	97.2	7.15	115.2	117.8	1.02
2	12.05	665.5	96.6	6.89	111.4	126.2	1.13
3	10.78	685.0	97.4	7.03	105.2	121.9	1.16
4	11.33	652.1	96.7	6.75	104.1	106.2	1.02
5	10.97	655.6	97.1	6.75	101.8	109.4	1.07
6	10.79	673.6	97.9	6.88	102.8	105.7	1.03
Mean	**11.30**	**671.2**	**97.1**	**6.91**	**106.7**	**114.5**	**1.07**
Unit	*9.96*	*684.3*	*97.0*	*7.06*	*97.0*[b]		

For all tests, measured mean cube strength $f_{c,cube}=59.1$ N/mm², f_{ck} taken as $0.8\times59.1=47.3$ N/mm², $f_{ctm}=3.92$ N/mm² and $f_{ct}=1.5\ f_{ctm}=5.88$ N/mm².

Compound $I_{c,co}$, $y_{b,co}$ and $Z_{b,co}$ includes transformed area of tendons (see Section 6.2.4).

[a] Excludes bending moment due to self-weight and loading equipment = 12.0 kNm.
[b] M_{cr} includes long-term losses and is based on f_{ctm}, not f_{ct}.
Unit, nominal section properties of designed hollow core unit.

the ratio of the bending moment/deflection at mid-span, respectively, the following are estimated

i. the cracking tensile strength $f_{ct}=1.5\ f_{ctm}$, i.e. $0.45f_{ck}^{2/3}$, where f_{ck} is determined from the results of cube tests at the date of testing
ii. the elastic modulus of the same concrete as $22\ (f_{ck}+8)^{0.3}$ (see Eq. 5.6)

Table 7.2 Deflections and flexural stiffness at the elastic limit in full scale tests carried out on 1200 × 200 mm depth prestressed hollow core floor units

Slab ref	*Calculated cracking moment*[a] $M_{cr,calc}$ *(kNm)*	*Mid-span deflection*[b] δ_{calc} *(mm)*	*Stiffness* $K_{calc}=M_{cr,calc}/\delta_{calc}$ *(kNm/mm)*	*Imposed moment at elastic limit*[a] M_s *(kNm)*	*Mid-span deflection in test* δ_{test} *(mm)*	*Stiffness* K_{test} *= slope of M/δ graph (kNm/mm)*	*Ratio* K_{test}/K_{calc}
1	115.2	12.24	9.41	117.8	12.20	9.66	1.03
2	111.4	12.29	9.07	126.2	12.4	10.17	1.12
3	105.2	11.25	9.35	121.9	14.4	8.44	0.90
4	104.1	11.84	8.79	106.2	11.3	9.40	1.07
5	101.8	11.40	8.93	109.4	13.5	8.12	0.91
6	102.8	11.21	9.17	105.7	12.6	8.42	0.92
Mean	**106.7**	**11.7**	**9.12**	**114.5**	**12.7**	**9.04**	**0.99**

[a] Excludes bending moment due to self-weight and loading equipment = 12.0 kNm.
[b] Based on the distance between deflection gauges at supports = effective span – 150 mm.

iii. the second moment of area I_c, height to the centroid the cross section y_b and eccentricity to the centre of the group of strands are calculated from the average measurements of the hcu at its ends, with the assumption that the measurements at mid-span will not differ too much, and with certain assumptions regarding the shape of the hollow cores
iv. the prestress at mid-span is calculated from factory information regarding the initial pretensioning force and losses of prestress at the date of testing due to strand relaxation and elastic shortening only (no creep loss).
v. the self-weight of the hcu was measured using a balancing load cell.

Using the above, Table 7.1 shows that the average ratio between the imposed bending moment at the elastic limit in the tests $M_{s,test}$ and the net calculated cracking moment of resistance $M_{cr,calc}$ (excluding the self-weight of the hcu and the loading equipment accounting for ≈ 12 kNm) is 1.07. The range is from 1.02 to 1.16, with Tests 2 and 3 under-predicting the onset of cracking. The ratios are for an assumed tensile cracking strength of $f_{ct}=1.5\ f_{ctm}=5.88$ N/mm². To provide parity, then $f_{ct}\approx 1.79\ f_{ctm}$. This is not that much greater than the value of 1.65 f_{ctm} proposed in Precast Eurocode 2: Worked Examples (Narayanan, 2008) for the cracking resistance in double-tee units.

Table 7.2 shows that the average ratio between the measured and predicted flexural stiffness, expressed here as mid-span bending moment/deflection was only 1%, at least up to the elastic limit identified from the graphs in Figure 7.1, and up to $M_{cr,calc}=(\sigma_b+f_{ct})\ Z_{b,co}$, where $Z_{b,co}=I_{c,co}/y_{b,co}$ (derived in Section 6.2.4). This fortuitous result has ratios varying from 0.90 to 1.12 with a standard deviation of 0.09. However, the overall picture is that the flexural behaviour of these units, and the serviceability limit state of bending and deflection (at least without creep effects) can be confidently predicted. The ultimate behaviour of these tests to failure is shown and discussed in Chapter 9.

7.1.2 Flexural tests on T beams

Similar flexural tests have also been carried out by the authors on prestressed T beams used in beam and block flooring. The test involved a pair of 92 mm wide × 150 mm depth beams, infilled with 440 mm wide concrete blocks, as shown in Figure 7.3a, simply

(a)

(b)

Figure 7.3 (a) General arrangement of full scale flexural testing of prestressed T beams (b) Beam subjected to uniform loading using concrete blocks giving a mid-span bending moment of 9.66 kNm per beam, equal to 1.32 and 1.83 times the calculated ultimate and service moments of resistance, respectively. (With permission by Litecast Ltd. UK.)

supported over an effective span of 4.4 m. The beams are usually critical at the design service load, and therefore the test bending moment, based on BS 8110, Part 2, Section 9 (BS 8110-2, 1985), is $M_{test}=1.125\ M_{sR}$, where M_{sR} is calculated for the same concrete strength $f_{ck}=53.5$ N/mm^2 as measured by the average of 6 cubes at the date of testing. Then $M_{sR}=1.125\times4.92=5.54$ kNm, equating to a UDL$=8\times5.54/4.40^2=2.29$ kN/m. The test results were compared with

i. $M_{sR}=(\sigma_b+f_{ctm})\ Z_{b,co}$ using losses calculated at the date of testing (12 days after casting) as 13% and the measured cross section of the beam and cover to wires
ii. cracking moment of resistance $M_{cr}=(\sigma_b+f_{ct})\ Z_{b,co}$, where f_{ct} is taken as 1.5 f_{ctm}
iii. ultimate M_{Rd} (based on the equilibrium of forces and compatibility of prestrain and ultimate strain as defined in Section 9.3.1)
iv. flexural stiffness of the beam $K=M/\delta_{test}$, where M is the imposed mid-span moment and δ_{test} the measured mid-span deflection at the same load.

The test was carried out in two loading and unloading cycles to $M_{test}=5.54$ kNm, followed by monotonic loading to failure. The bending moment due to the self-weight of the two beams and row of infill blocks is 1.51 kNm per beam.

Figure 7.3b shows the loading arrangement where up to 166 no. 18.2 kg concrete blocks were used as kentledge. Dial gauges were used to measure the mid-span deflection, and Figure 7.4 shows the mid-span bending moment vs. deflection up to the onset of cracking, shown in Figure 7.5, together with deflections during unloading, and the average predicted deflections (note the plot for Beam 2 is displaced on the graph for clarity).

Table 7.3 shows that the average ratio between the bending moment at $M_{s,test}$ and the predicted (calculated) cracking moment of resistance M_{cr} is 1.015. Therefore, the assumed tensile cracking strength of $f_{ct}=1.5\ f_{ctm}=1.5\times4.17=6.25$ N/mm^2 is valid. The average ratio between $M_{s,test}$ and M_{sR} is 1.13, showing that the calculated prestress $\sigma_b=14.74$ N/mm^2 and $f_{ctm}=4.17$ N/mm^2 are both reliable values. The ratio based on the design M_{sR}, including long-term losses and using the design $f_{ck}=55$ N/mm^2, is 1.20, giving greater confidence to the results. Ultimate moments are discussed in Section 9.1.

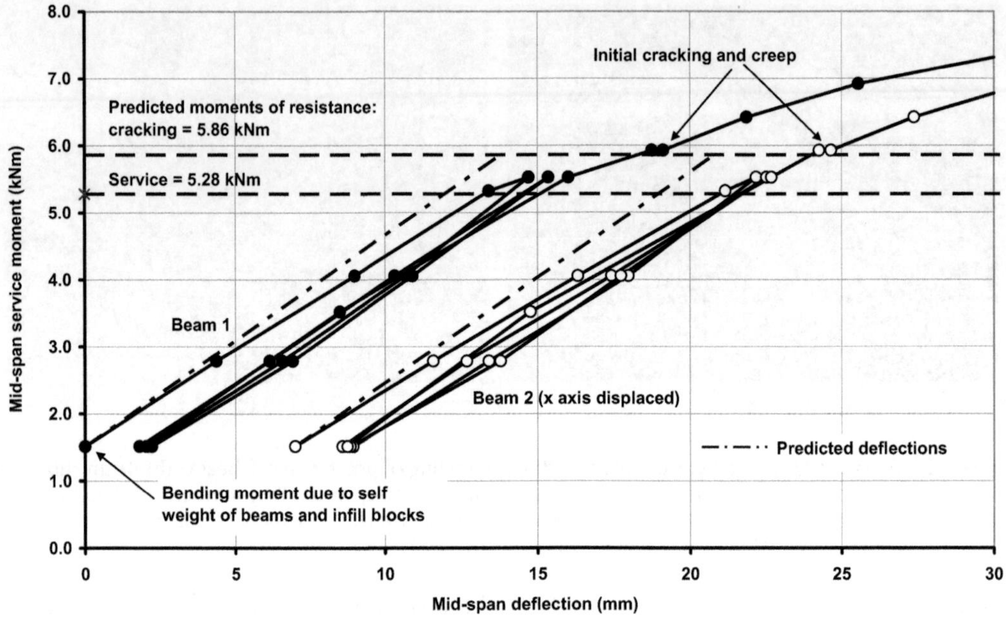

Figure 7.4 Experimental and predicted bending moment vs. mid-span deflection in full scale tests carried out on a pair of 92 mm width × 150 mm depth prestressed T beams. (With permission by Litecast Ltd. UK.)

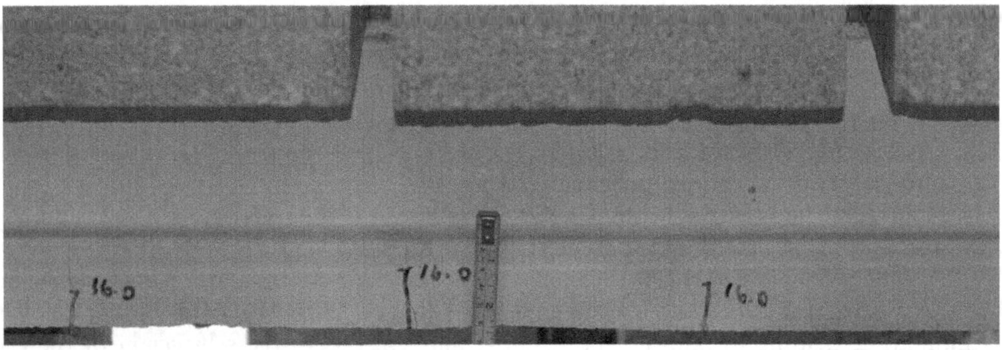

Figure 7.5 Initial flexural cracking at an applied load of 16.0 kN in the soffit in the mid-span region of prestressed T beams, giving a mid-span bending moment of 5.93 kNm per beam, equal to 1.12 times the calculated service moment of resistance. Further cracking to a height of 80 mm when the bending moment was 9.07 kNm is drawn onto this photograph. (With permission by Litecast Ltd. UK.)

Table 7.4 shows that the average ratio between the measured and predicted flexural stiffness, expressed here as mid-span bending moment/deflection (predicted $K = \partial M/\partial\delta = 0.32$ kNm/mm) was only 84% in the first load cycle, at least up to the elastic limit identified from the graphs in Figure 7.4, and 91% in the second and third load cycles. There are no creep effects in these results, which may otherwise reduce K, but in the author's experience, the value for E_{cm} (37.95 kN/mm^2 in these tests) given in BS 1992-1-1 (BS EN 1992-1-1, 2004) is higher than found in compression tests, and greater than those given by Neville (Neville, 1999) and in BS 8110, Part 2 (BS 8110-2, 1985) by at least 10%.

Table 7.3 Summary of test moments, and predicted & design moments of resistance for full scale tests carried out on 92 × 150 mm depth prestressed T beams

		Test moments (kNm)	*Predicted moments (kNm)*	*Ratio*
Beam 1	Service	5.93	5.28	1.12
	Cracking	5.93	5.86	1.01
	Ultimate	9.51	8.61	1.10
Beam 2	Service	5.93	5.23	1.13
	Cracking	5.93	5.82	1.02
	Ultimate	9.66	8.55	1.13
		Test moments (kNm)	*Design moments (kNm)*	*Ratio*
Both beams	Service	5.93	4.94	1.20
	Ultimate	9.58	6.89	1.39

Table 7.4 Deflections and flexural stiffness at the elastic limit in full scale tests carried out on 150 mm depth prestressed T beams

	Load cycle	*Deflection due to imposed load (mm)*	*Test stiffness K (kNm/mm)*	*Predicted stiffness K (kNm/mm)*	*Ratio of K*
Beam 1	First	14.7	0.274	0.320	0.86
	Second	15.4	0.297	0.320	0.93
	third	16.0	0.288	0.320	0.90
Beam 2	First	15.2	0.265	0.319	0.83
	Second	15.5	0.289	0.319	0.90
	Third	15.7	0.289	0.319	0.90

Calculated deflection at $M_{s,test}$ = 12.6 mm using E_{cm} = 37948 N/mm² (with no creep effects) and the compound second moment of area = 17.00 × 10⁶ mm⁴.

To make the predicted deflections agree with the experimental deflections, the value of E_{cm} may be reduced by 10% to 34.5 kN/mm². Now the flexural stiffness for beam 1 is K = 0.287 kNm/mm, and overall the ratio of predicted/experimental K = 0.975.

An alternative theory to the differences in flexural stiffness is that it is thought that hairline cracks in the top of the beams during the handling stage at transfer (where beams are cast upside down) may have reduced K in the first load cycle. For example, consider 5 no. cracks of 0.1 mm crack width. The additional deflection due to crack closure is determined from beam curvature ($1/r$) equal to the strain gradient through the depth of the beam ε/h = (0.5/4400)/150 = 0.76×10^{-6}/mm. The deflection due to this curvature is due to crack closure $\delta_{cc} = 0.76 \times 10^{-6} \times 4400^2/8 \approx 1.8$ mm. The difference between the actual deflection at the test load and the predicted deflection in the two beams is 2.12 and 2.60 mm, so the additional 1.8 mm could in part account for the lower K values in the first load cycle.

Finally, BS 8110, Part 2, clause 9.5.4 states that the recovery of deflections after the second load cycle should be at least equal to that after the first loading and be at least 85% for 'Class 2' prestressed concrete, i.e. allowing permitted tensile stress in the soffit. The test values are 87.7% and 86.2% for Beam 1, and 89.6% and 88.5% for Beam 2. All values are greater than 85% and close enough in the two load cycles to be considered satisfactory. As with the hcu in Section 7.1.1, the overall picture is that bending moments and deflections (at least without creep effects) can be confidently predicted.

7.1.3 Lateral distribution and flexural tests on T beam floor slab

Section 2.2.10 introduced the concept of lateral load distribution of isolated point loads and/or longitudinal wall loads (parallel to span) between several T beams in beam and block flooring, i.e. a point load acting on top of a screeded floor may be distributed between several T beams. Full scale tests were carried out by the authors, as shown in Figures 7.6 and 7.7, comprising 5 no. 150 mm depth beams and 4 rows of 440 mm wide infill blocks, to determine lateral load spread and compare mid-span deflections with calculated values. 225 mm depth beams were used in another test but are not discussed in detail here. The screed is 50 mm thickness *insitu* concrete structural topping with f_{ckT} = 32 N/mm^2 containing A142 mesh (6 mm bars at 200 mm spacing). A sheet of 0.3 mm thickness polythene was placed between the beams and topping to prevent *direct* composite action with the beams. In reality, the topping may or may not have a membrane, but is likely to contribute to the flexural strength and stiffness of the beams, as shown in the moment v deflection results. Figure 7.7 shows the completed floor and arrangement for the centralised point load.

The main application is for a domestic garage floor subjected to a service point load of 10 kN with a footprint 150×150 mm. Two tests were carried out for each depth of beam for (i) a centralised point load positioned above the central Beam 3 and (ii) offset point load above the intermediate Beam 2. The effective span of the beams was 3.50 m. The tests were carried out 56 days after casting when the beams had characteristic compressive cube strength of 90 N/mm^2, converted to f_{ck}=75 N/mm^2, i.e. 36% greater than the design strength. However, the increase in E_{cm}=41500 N/mm^2 (9% greater than the design value) may account in part

Figure 7.6 General arrangement of beam and infill block flooring in lateral point load tests prior to casting the *insitu* topping. Beam 1 is at the edges, Beam 3 is central (beneath the point load) and Beam 2 in between. (With permission by Litecast Ltd. UK.)

Figure 7.7 Completed floor slab for 150 mm deep beams with 50 mm topping, showing the position of centralised point load. (With permission by Litecast Ltd. UK.)

for the increased flexural stiffness, but would not affect the proportion of the point load carried by each of the 5 beams.

The results given here are for 150 mm depth beams that were pretensioned using 4 no. 5 mm wires with a service M_{sR}=5.55 kNm and cracking M_{cr}=6.14 kNm (using the same calculation as in Section 7.1.2).

The test was carried out in one loading and unloading cycle to a point load of 10 kN, equivalent to M_{test}=4.02 kNm, followed by monotonic loading to failure. The bending moment due to the self-weight of the beams, infill blocks and topping is 2.10 kNm per interior beam and 1.23 kNm for the edge beams.

Figure 7.8 shows the total mid-span bending moment vs. deflection for the central Beam 3, intermediate Beam 2 and edge Beam 1. The ratio between the measured and predicted flexural stiffness, expressed as mid-span bending moment/deflection (predicted $K=\partial M/\partial\delta=0.83$ kNm/mm) was 141% in the first load cycle and 111% in the second loading. The increased stiffness is thought to be due to increased E_{cm} and composite action with the topping. The recovery of deflection after the first load cycle was 88% and 83% in Beams 2 and 3, respectively. The floor slab was also tested for failure under centralised point load of 41 kN. This gave a bending moment in Beam 3 of 16.1 kNm, 1.94 times the calculated M_{Rd}. The high moment of resistance is also thought to be due to composite action as the composite M_{Rd}=13.4 kNm, assuming full interaction.

The deflections of the beams and their calculated flexural stiffness as measured for each beam enable the proportion of the total imposed load β to be determined for each beam. Variations in β up to the service load are shown in Figure 7.9. Values of β obtained from a theoretical analysis of the flexural and torsion stiffness of the beams and topping are

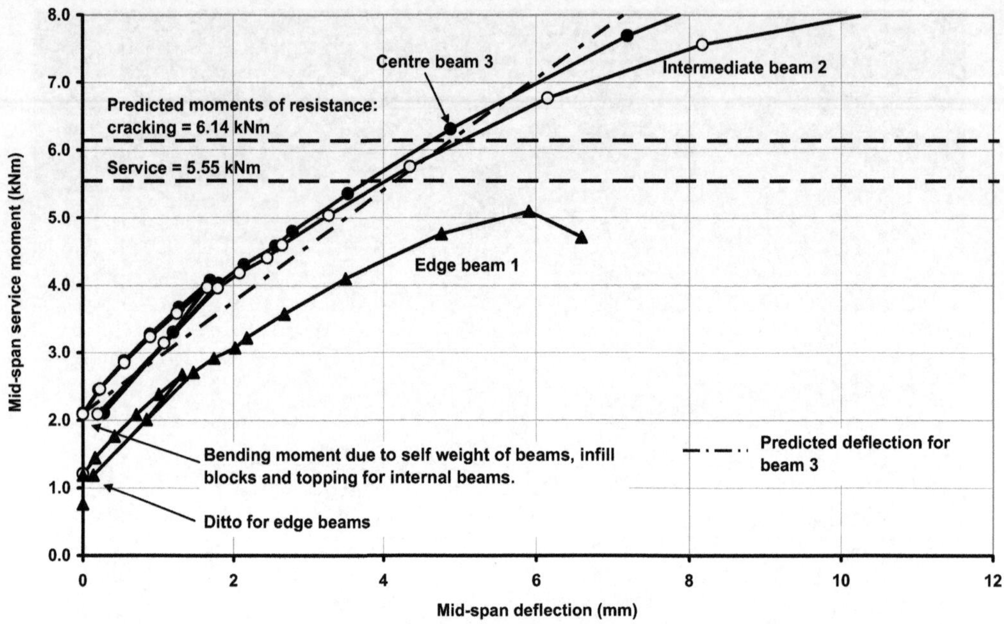

Figure 7.8 Experimental and predicted bending moment vs. mid-span deflection in full scale lateral load distribution tests on five 150 mm depth prestressed T beams subjected to a centralised point load.

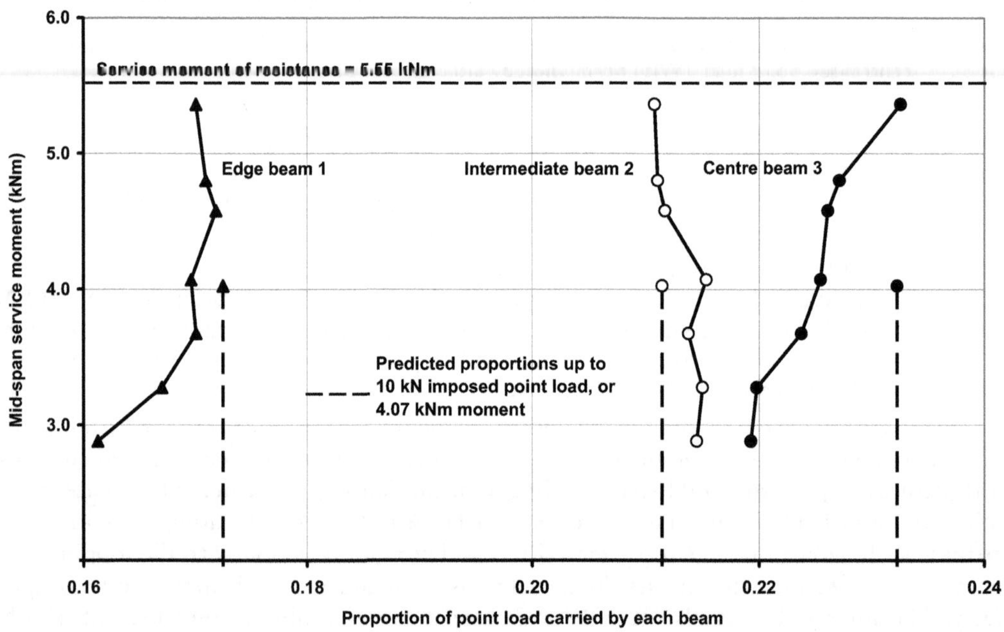

Figure 7.9 Proportion of central point load carried by each 150 mm depth T beam up to service moment of resistance.

presented in Table 7.5 (in italics) and compared with test values. For the central point load β=0.215–0.225 for Beams 2 and 3, but for the offset point load β=0.28–0.27 for Beams 1 and 2, suggesting that the point load may be distributed over 1/0.28=3.6 beams. The recommendation is that point loads may be distributed over three beams.

Table 7.5 Values for the experimental and theoretical proportion of the total imposed point load carried by each 150 mm depth prestressed T beam

Beam depth (mm)	*Position of point load*	*Imposed load of 10 kN*	*Beam 1*	*Beam 2*	*Beam 3*
150	Central over beam 3	Test	0.168[a]	0.22[a]	0.225
		Theory	*0.17*	*0.21*	*0.23*
	Offset over beam 2	Test	0.28	0.27	0.21
225	Central over beam 3	Test	0.137[a]	0.234[a]	0.254
		Theory	*0.11*	*0.24*	*0.30*
	Offset over beam 2	Test	0.24	0.30	0.23

[a] Average of two beams.

Further tests were also carried out by moving the position of the point load close to the support, at 440 mm from the support (approx l_{pt} from the end of the beam, see Section 10.2.2 for the derivation of l_{pt}). Here the load distribution factors β were close to 0.4. The recommendation is that point loads close to the support may be distributed over two beams.

7.2 FLEXURAL CAPACITY

7.2.1 Combined prestress and imposed bending stresses

The flexural capacity of precast prestressed concrete is no different from any other type of prestressed concrete. The quality control of factory cast concrete helps to explain the excellent correlation between test results and theory found in precast units, shown for example in Figures 7.1 and 7.4. The major difference in behaviour found in precast units is due more to the complex geometry found in voided units such as hcu, which may have rapidly reducing web thickness near the neutral axis, or double-tee units in which the stresses in the narrow webs dominate both the service and ultimate moments of resistance.

Subjected to compressive prestress σ_b and a service bending moment M_s the net concrete stress in the tension face should not exceed the mean tensile strength of concrete f_{ctm}, defined in Eq. 5.5. Referring to Figures 6.1c and 7.10, where the pretensioning force is applied close to the bottom of a simply supported element, if the pretensioning force after all four major losses, defined in Section 6.2.2 as tendon relation, elastic shortening, creep and shrinkage, is P_{po}, then the axial stress = force/cross-sectional area P_{po}/A_c and the flexural stress = M_p/Z_b, where M_p is the internal couple $P_{po}\, z_{cp}$, and Z_b is the section modulus at the tension face. The total prestress at the bottom of the section is

$$\sigma_b = P_{po} / A_c + P_{po} z_{cp} / Z_b \tag{7.1}$$

and at the top is

$$\sigma_t = P_{po} / A_c - P_{po} z_{cp} / Z_t \tag{7.2}$$

When subjected to a characteristic service moment M_s, at the bottom fibre (taking compression as positive) the resulting stresses are

$$+\sigma_b - M_s / Z_{b,co} \geq -f_{ctm} \tag{7.3a}$$

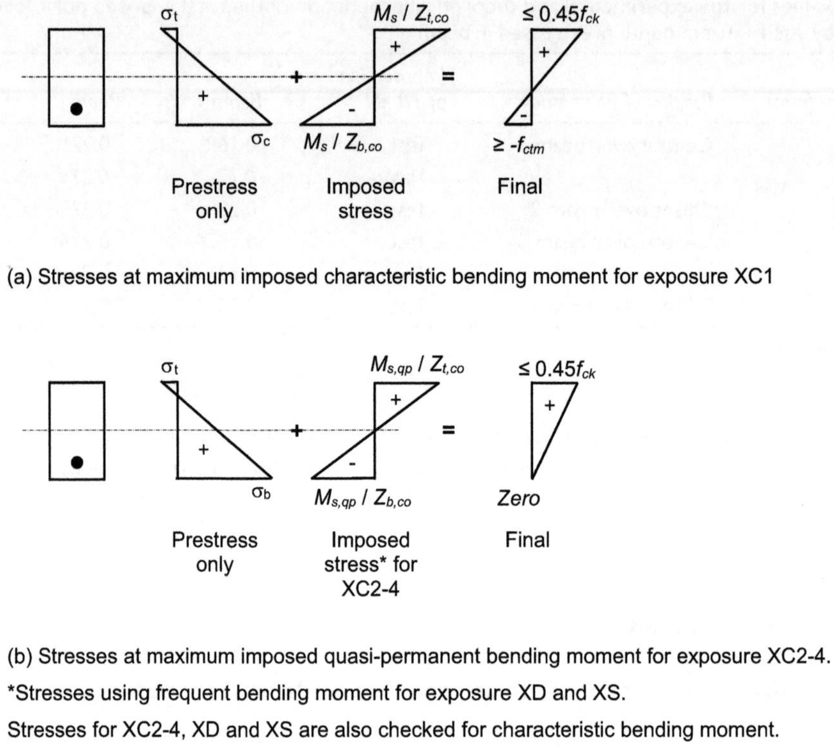

Figure 7.10 Prestress, imposed stresses and final limiting stresses. (a) Stresses at maximum imposed characteristic bending moment for exposure XC1. (b) Stresses at maximum imposed quasi-permanent bending moment for exposure XC2-4. *Stresses using frequent bending moment for exposure XD and XS. Stresses for XC2-4, XD and XS are also checked for characteristic bending moment.

Note that the compound section modulus $Z_{b,co}$ (see Section 6.2.4) may be used, although most designers ignore the small benefit from this.

The limit of f_{ctm} is based on a requirement in BS EN 1992-1-1, clause 7.3.2 (4) to prevent sudden rupture under the characteristic combination of loads, thereby restricting the tensile stress to $\sigma_{ct,p}=f_{ctm}$ in clause 7.3.2 (2). Some publications, e.g. Precast Eurocode 2: Worked Examples (Narayanan, 2008) refer to a 'minimum steel requirement' rather than limiting the stress to f_{ctm}. However, it is now shown that the minimum area of tendons $A_{p,min}$ according to clause 7.3.2 (2) & (3) and Exp. 7.1 is not practical, as follows:

$$\xi_1 A_{p,min}\Delta\sigma_p = k_c k\, f_{ctm} A_{ct} \tag{7.4}$$

This is a long and tedious calculation, but for the 250 mm depth hollow core unit used in Example 6.1, with strands at a mean distance of 40.8 mm from the soffit, the values are: $k_c=0.58$ (with $k_1=⅔$ for tension in the strands), $k=1$, $f_{ctm}=3.80$ N/mm^2, $A_{ct}\approx 70850$ mm^2 (approximate because the height $h_{c,ef}=2.5\times 40.8=102$ mm is inside the cores) and (from BS EN 1992-1-1, Table 6.2) $\xi_1=\sqrt{0.6}=0.775$ for strands. The dominant variable is $\Delta\sigma_p$, which is defined (Clause 7.3.2 (3)) as the 'stress variation in tendons from the state of zero strain of the concrete at the same level', which is understood to be the increase in tensile stress in the tendons after the strain in the concrete there was zero at the serviceability limit, i.e. $(M_{sR}/Z_z-\sigma_{cp}-0)\,E_p/(E_{cm}/(1+\varphi(t,t_0))$, where σ_{cp}=stress in concrete at level of tendons after all losses = 10.19 N/mm². M_{sR}/Z_z = 11.35 N/mm². See Example 7.1 for the derivation of the long-term creep coefficient $\varphi(t,t_0)$. Then $\Delta\sigma_p = (11.35 - 10.19 - 0) \times 195000/[36283/(1+2.065)] = 19$ N/mm². Thus $A_{p,min} = k_c\ k\ f_{ctm}\ A_{ct}/\xi_1\ \Delta\sigma_p = 0.58 \times 1 \times 3.80 \times 70850/0.775 \times 19 \approx 10600$ mm²! The area of strands in this example is 911 mm². Repeating this exercise for a 150 mm depth T beam (for beam and block flooring) $k_c = 1.02$, therefore use 1.0. $A_{p,min} = 1.0 \times 1 \times 4.21 \times 6517/0.837 \times 80 \approx 408$ mm²! The area of strands in the tension zone is 59 mm².

Other publications (e.g. Gilbert et al. (2017)) limit the tension at the bottom to 0.5 f_{ctm}, although this is used to demonstrate the procedures to calculate the upper limit of prestress rather than a design recommendation. Gilbert et al. also noted that if cracking is allowed under the characteristic (i.e. full) service load, then this limit is not specified at all, although limiting values at transfer are enforced to control the upper limit of prestress. Elements that are designed to crack under the full service load are proportioned so that no tension exists in the concrete under the sustained (dead) load. In this case, the flexurally cracked second moment of area is used in deflection calculations. It is not usual for designers of precast prestressed elements in the UK to take this option.

The limitation of f_{ctm} is appropriate only for durability exposure class XC1 (internal exposure low humidity or permanently wet) according to BS EN 1992-1-1, Table 7.1N (superceded by Table A.4 in the National Annex) when M_s is the characteristic combination of loading $g_k + q_k$. Some designers have mistakenly interpreted the definition of service loading as a reversible 'frequent' design load $g_k + \psi_1 q_k$ based on definitions given in BS EN 1990 (BS EN 1990, 2002). The frequent load is defined as a load that is exceeded for 1% of the 'reference period' of a building structure, whereas the quasi-permanent load will exceed itself for 50% of that period. The reference period is defined as the period of time for assessing statically variable actions; in other words, it may be for 1 year for a temporary structure or 50 years for the life of a concrete building.

For durability exposure classes XC2 to XC4 the limiting tension is zero (called 'decompression' in Table 7.1N) and $M_{s,qp}$ is for the quasi-permanent load $g_k + \psi_2\ q_k$. Then

$$+\sigma_b - M_{s,qp} / Z_{b,co} \geq 0 \tag{7.3b}$$

For durability XD1/XS1 or greater, the limiting tension is also zero and M_s is for the frequent load $g_k + \psi_1\ q_k$. Then

$$+\sigma_b - M_{s,freq} / Z_{b,co} \geq 0 \tag{7.3c}$$

The crack width w_k for prestressed concrete is limited $w_{max} \leq 0.2$ mm (see calculation in Section 11.3.1).

The loading, tensile stress and crack width limitations are summarised in Table 3.3. Note that for durability exposure class > XC1, the characteristic combination is checked (Eq. 7.3a) as well as decompression (Eq. 7.3b or 7.3c), so two calculations are required. This is dealt with in Examples 7.1 and 7.4.

After cracking, tension stiffening of the concrete (due to the elasticity of the reinforcement) allows reduced tensile stress in this region, but when the tensile stress reaches the narrow part of the web, cracks extend rapidly through the section and the flexural stiffness of the section reduces to a far greater extent than in a rectangular section. Figure 7.1 shows this behaviour in a flexural test carried out on a 200 mm depth hcu. The serviceability limiting state must be checked to prevent this type of behaviour, hence the need for clause 7.3.2 (4).

A second reason why the service condition is calculated in this manner is that the ratio of the ultimate moment of resistance M_{Rd} to the serviceability moment of resistance M_{sR} is usually about 1.5–1.8. Thus, with the use of the partial load factors ($\gamma_G = 1.25$ or 1.35 and

γ_Q=1.50 or 1.5 ψ_0), the serviceability condition will always be critical, provided the amount of prestress is sufficient. Finally, the problem of cracking in unreinforced zones is particularly important with regard to the uncracked shear resistance. It is therefore necessary to ensure that tensile stresses are not exceeded.

At the top fibre, the final stresses are

$$+\sigma_t + M_s / Z_{t,co} \leq 0.45 f_{ck} \tag{7.5}$$

Note that $+\sigma_t$ is most likely to be negative in the vast majority of prestressed elements.

Compressive stresses due to service loads and prestress are limited according to BS EN 1992-1-1, clauses 7.2 (2) and 7.2 (3) as discussed in Section 3.1.2 for different exposure classes. Clause 7.2 (3) specifies a limiting compressive stress of 0.45 f_{ck} under quasi-permanent loads to avoid non-linear creep in the calculation of prestress losses and deflections according to clause 3.1.4 (4).

Non-linear creep coefficient is given as

$$\varphi_k(\infty,t_0) = \varphi(\infty,t_0)\, e^{[1.5\,(\sigma_c/fcm(t_0)-0.45)]} > \varphi(\infty,t_0) \tag{7.6}$$

where σ_c is the compressive stress, i.e. $\sigma_t + M_s/Z_{t,co}$

$\varphi(\infty,t_0)$ is the long-term creep coefficient

For non-linear creep to exist, the term $[\sigma_c/f_{cm}(t_0)-0.45]$ must be positive. Then $\sigma_c \leq 0.45$ $f_{cm}(t_0)$. For example if f_{ck}=45 N/mm^2, $\sigma_c \leq 0.45\times(45+8)$=23.85 N/mm^2, or 0.53 f_{ck}. If f_{ck}=55 N/mm^2, $\sigma_c \leq 0.515$ f_{ck}, and so on.

For a wide range of solid units of depths ranging from 80 to 150 mm, and hcu from 150 to 500 mm, when the design of the element is limited by the tensile stress f_{ctm}, the stress in compression under the characteristic service load is between σ_c=0.18 and 0.39 f_{ck}, with the greatest value being recorded for depths of 200–300 mm in hcu with the highest strand pattern. For double-tee units of depths from 300 to 900 mm σ_c=0.15–0.22 f_{ck}, with values increasing for greater depths and for the highest strand patterns. For thermal slabs of widths 400–1200 mm and depth 250 mm σ_c=0.04–0.31 f_{ck}. For T beams (in beam and block flooring) of depth 150 and 225 mm σ_c=0.52 f_{ck} and 0.38 f_{ck}, respectively. Therefore, the presence of non-linear creep is avoided if all other limitations mentioned above are satisfied.

The marked difference in σ_c/f_{ck} between T beams (*viz.* 0.52) and all other elements is due to the shape of the beams and the much smaller ratio of $Z_{b,co}/Z_{t,co}$=0.68 and 0.78, respectively, compared with 0.95 for hcu, 2.00–2.44 for double-tees (although the cross section is designed with composite action in mind), and 1.22–2.22 for thermal slabs (notionally similar in cross section to double-tees).

Checking the compressive stress of 0.45 f_{ck} under quasi-permanent loads is rarely critical for exposure XC1, unless the imposed load is very small and self-weight and dead loads dominate. In design, it is often, but not always, convenient to limit the compressive stress to 0.45 f_{ck} for all situations. If necessary, in the case of certain T beams of depth 150 mm, the compressive stress need only be limited to 0.51 f_{ck}, as shown in Example 7.5, where non-linear creep is not activated.

7.2.2 Serviceability limit state of flexure

M_{sR} is calculated by limiting the flexural compressive and tensile stresses in the concrete both in the factory transfer and handling conditions and in service. Figures 6.1 and 7.10 show the stress conditions at these stages for applied sagging moments – the diagrams may be inverted for cantilever units subject to hogging moments.

To optimise the design the limiting stresses at transfer should be equally critical as those at the limiting serviceability state of stress, and that the top and bottom surface stresses should attain maximum values simultaneously, i.e. manipulating Eqs. 7.3 and 7.5, then $(0.45\,f_{ck}-\sigma_t)\,Z_{t,co}=(\sigma_b+f_{ctm})\,Z_{b,co}$. In practice, this is impossible in a (nearly) symmetrical rectangular section such as a hcu, but can be better achieved in T beams or composite double-tee slabs.

7.2.3 Serviceability limit state of flexure – calculation model

The full analysis is as follows (BS EN 1992-1-1 clause references are at left-hand side of the page)

Depth of unit $=h$	Notional dimension $=h_o$
Cover to tendons $=C$	Height to centroid of strands $y_s=C+\Phi/2$
Height to centroid to concrete $=y_b$	Gross concrete area (exclude infill) A_c
Eccentricity of tendons $z_{cp}=y_b-y_s$	Second moment of area I_c
Section modulus at bottom $Z_b=I_c/y_b$	Section modulus at top $Z_t=I_c/(h-y_b)$
Z at tendon centroid $Z_z=I_c/z_{cp}$	

Compound values $A_{c,co}$, $y_{b,co}$, $I_{c,co}$, $Z_{b,co}$ and $Z_{t,co}$ are given in Section 6.2.4.

7.2.4 Serviceability limit state of bending

7.2.4.1 Short-term losses

Short-term losses can be calculated from transfer at time t_o to installation at time t_i using relative humidity RH=70% with all faces exposed, and to the long-term 500000 hours (57 years) using service RH_s=50% (indoor) or 70% (outdoor) with the exposed sides only exposed (other sides protected by finishes, grouting, etc.)

5.10.6(1a) Loss due to creep to installation. See BS EN 1992-1-1, Annex B.1. Although the strength of concrete at 1 day will be the transfer strength $f_{ck}(t)$ N/mm², after a few days it will reach the 28 days strength f_{ck} N/mm², and so the mean strength f_{cm} is taken for the strength factors in this calculation.

Notional depth during this period is for all faces exposed (ignoring the perimeter of cores in hcu)

Exp. B.6 $$h_o = 2A_c / u \quad (7.7a)$$

Exp. B.1 Creep coefficient $$\varphi(t_i,t_0) = \varphi_{RH}\beta(f_{cm})\beta(t_0)\beta_c(t_i,t_0) \quad (7.8)$$

For t_i=installation, t_0=transfer age (days) and RH=70%

Exp. B.3b/B.8c Relative humidity factor $$\varphi_{RH} = \left[1+(1-\mathrm{RH}/100)/(0.1h_o^{1/3})\alpha_1\right]\alpha_2 \quad (7.9)$$

where $\alpha_1 = (35/f_{cm})^{0.7}$ and $\alpha_2 = (35/f_{cm})^{0.2}$

Exp. B.4 Strength factor $$\beta(f_{cm}) = 16.8/\sqrt{f_{cm}} \quad (7.10)$$

Exp. B.5 Age at release loading factor $\beta(t_0) = 1/\left(0.1 + t_{oT}^{0.2}\right)$ (7.11)

Exp. B9 Using $t_{0T} = t_T\left(\dfrac{9}{2 + t_T^{1.2}} + 1\right)^{\alpha} \geq 0.5$ day (7.12)

where α=1, 0, –1 for Class R, N, S cement

Exp. B10 Equivalent age at transfer $t_T = t_0 e^{-[4000/(273+T)-13.65]}$ (7.13)

where T=mean curing temperature °C during curing time in days, taken as 50°C

Exp. B.7 $\beta_c(t_i, t_0) = \left[(t_i - t_0)/(\beta_H + t_i - t_0)\right]^{0.3}$ where to is transfer age (7.14)

Exp. B.8b RH factor $\beta_H(\text{days}) = 1.5\left[1 + (0.012\ \text{RH})^{18}\right] h_o + 250\alpha_3$ (7.15)

where $\alpha_3 = (35\ /\ f_{cm})^{0.5}$

After initial losses $\sigma_c = P_{pm0}\ /\ A_c + P_{pm0} z_{cp}\ /\ Z_z$ (7.16a)

But within the span, particularly at mid-span, a negative bending stress due to self-weight may be subtracted from Eq. 7.16a as follows:

$\sigma_c = P_{pm0}\ /\ A_c + P_{pm0} z_{cp}\ /\ Z_z - M_{self}\ /\ Z_z$ (7.16b)

This reduces creep losses and increases the final prestress σ_b and σ_t. See also Eq. 7.22b.

Exp. 5.46 $$\Delta\sigma_{pci} = \frac{\dfrac{E_p}{E_{cm}} \times \phi(t, t_o)\sigma_c}{1 + \left[\dfrac{E_p}{E_{cm}} \times \dfrac{A_p}{A_c}\left(1 + \dfrac{A_c \times z_{cp}^2}{I_c}\right)\left(1 + 0.8\,\phi(t, t_o)\right)\right]}$$ (7.17)

Prestress after short-term creep losses $\sigma_{pmi}=\sigma_{pi}-\Delta\sigma_{pci}$
Prestress force at installation $P_{pmi}=\sigma_{pmi} A_p$

7.2.4.2 Long-term creep losses

5.10.6(1a) Loss due to creep to t=500000 hours; service RH$_s$=50%.

Notional depth is for bottom only $h_o=2A_c/b$ (7.7b)

Exp. B.2 Creep coefficient $\varphi(t,t_0)=\varphi_{RH}\ \beta(f_{cm})\ \beta(t_0)\ \beta_c(t,t_0)$ (7.18)

Exp. B.3b/B.8c $\varphi_{RH}=[1+(1-RH_s/100)/(0.1\ h_o^{1/3})\ \alpha_1]\ \alpha_2$ (7.19)

Exp. B.7 $\beta_c(t,t_0)=[(20833-t_0)/(\beta_H+20833-t_0)]^{0.3}$ (7.20)

Exp. B.8b RH factor β_H (days)=1.5 $[1+(0.012\ RH_s)^{18}]\ h_o+250\ \alpha_3$ (7.21)

$\sigma_c = P_{pm0}\ /\ A_c + P_{pm0} z_{cp}\ /\ Z_z$ (7.22a)

Within the span Eq. 7.22a is modified to

$$\sigma_c = P_{pm0} / A_c + P_{pm0} z_{cp} / Z_z - (M_{self} + M_G + y_2 M_Q{}^*) / Z_z \quad (7.22b)$$

i.e. self-weight, dead loads and quasi-permanent live loads.

*Some designers ignore the term $\psi_2 M_Q$ and only use dead loads to reduce creep losses.

Note: at cantilever supports (Section 13.2.3 and Example 13.1) and continuous supports (Section 13.6.4 and Example 13.7) the term $+(M_{self}+M_G+\psi_2 M_Q)/Z_z$ is additive, increases creep losses and reduces the final prestress σ_b and σ_t.

Exp. 5.46

$$\Delta\sigma_{p,c} = \frac{\frac{E_p}{E_{cm}} \times \phi(t,t_o)\ \sigma_{c,q-p}}{1 + \left[\frac{E_p}{E_{cm}} \frac{A_p}{A_c}\left(1 + \frac{A_c z_{cp}^2}{I_c}\right)\left(1 + 0.8\ \phi(t,t_o)\right)\right]} \quad (7.23)$$

7.2.4.3 Shrinkage losses

5.10.6(1a) Loss due to shrinkage from t_s=transfer age to t=500,000 hours. See BS EN 1992-1, Annex B.2

RH during the service period of shrinkage=RH_s

Notional depth is for bottom only $h_o = 2A_c/b$ (7.25)

Exp. B.12 RH factor $\beta_{RH} = 1.55\ [1 - (RH_s/100)^3]$ (7.26)

Type of cement (use rapid hardening)=Class R ($\alpha_{ds1}=6$, $\alpha_{ds2}=0.11$)

Exp. B.11 $\varepsilon_{cd,o} = 0.85 \times (220 + 110 \times 6)\ e^{-0.11\ f_{cm}/10}\ \beta_{RH}$ (7.27)

Table 3.3 Size coefficient

$k_n = 1.0 - 0.0015\ (h_o - 100)$ for $100 \le h_o < 200$ mm

$0.85 - 0.001\ (h_o - 200)$ for $200 \le h_o < 300$ mm

$0.75 - 0.0005\ (h_o - 300)$ for $300 \le h_o < 500$ mm (7.28)

Exp. 3.10 Age factor $\beta_{ds}(t,t_s) = (20833 - t_s) / \left(20833 - t_s + 0.04 h_o^{3/2}\right)$ (7.29)

Exp. 3.9 Drying shrinkage strain $\varepsilon_{cd} = \beta_{ds}(t,t_s)\ k_n\ \varepsilon_{cd,o}$ (7.30)

10.3.1.2 (3) Autogenous shrinkage strain is taken as zero if heat curing is used.

Exp. 3.8 Total shrinkage strain $\varepsilon_{cs} = \varepsilon_{cd}$ (7.31)

Creep coefficient $\varphi(t,t_0) = \varphi_{RH}\ \beta(f_{cm})\ \beta(t_0)\ \beta_H$ using values from above

Exp. 5.46

$$\Delta\sigma_{p,s} = \frac{E_p\, \sigma_{cs}}{1 + \left[\frac{E_p}{E_{cm}} \times \frac{A_p}{A_c}\left(1 + \frac{A_c z_{cp}^2}{I_c}\right)\left(1 + 0.8\varphi(t,t_o)\right)\right]} \quad (7.32)$$

7.2.4.4 Tendon relaxation losses

5.10.6(1b) Long-term loss due to tendon relaxation is based on σ_{pm0}

Exp. 3.29 $\quad \mu = \sigma_{pm0}/f_{pk}$ (7.33)

Exp. 3.29 $\quad \Delta\sigma_{pr} = \sigma_{pm0} 0.66 \rho_{1000} e^{9.1\mu} \left(\frac{500000}{1000}\right)^{0.75(1-\mu)}$ (7.34a)

where tendon factor for Class 2 relaxation=0.66 and per function=9.1

Exp. 5.46 $$\Delta\sigma_{p,r} = \frac{0.8\,\Delta\sigma_{pr}}{1+\left[\frac{E_p}{E_{cm}}\frac{A_p}{A_c}\left(1+\frac{A_c z_{cp}^2}{I_c}\right)\left(1+0.8\,\varphi(t,t_0)\right)\right]} \quad (7.34b)$$

Prestress after final losses $\sigma_{po}=\sigma_{pi}-(\Delta\sigma_{pr}+\Delta\sigma_{el}+\Delta\sigma_{p,c}+\Delta\sigma_{p,s}+\Delta\sigma_{p,r})$ (7.35)

Final prestress force $P_{po}=\sigma_{po}\,A_p$

$\therefore$ Total loss factor $R_s=1-\sigma_{po}/\sigma_{pi}$ (7.36)

No partial factor for variations in prestress and losses is applied to P_{po} according to the UK National Annex to BS EN 1992-1-1 (NA to BS EN 1992-1-1, 2004), clause 5.10.9(1)P, $r_{inf}=1.0$ (whilst EN 1992-1-1 gives $r_{inf}=0.95$). Some texts (e.g. Bhatt, 2011) apply $\gamma_{inf}=0.95$ from EN 1992-1-1 to P_{po}.

Check maximum surface stress in service

At bottom $\sigma_b=P_{po}/A_c+P_{po}\,z_{cp}/Z_b$ (7.1)

7.2 (3) Limit $\sigma_b \le \sigma_{c,p}=0.45\,f_{ck}$

At top $\sigma_t=P_{po}/A_c-P_{po}\,z_{cp}/Z_t$ (7.2)

7.3.2 (4) Limit $\sigma_t \ge -\sigma_{ct,p}=-f_{ctm}$ (or zero according to BS EN 1992-1-1, Table 7.1N).

7.2.4.5 Service moment of resistance M_{sR}

To calculate M_{sR}, the section is considered uncracked and the net cross-sectional area A_c and second moment of area I_c are used to compute maximum fibre stresses σ_b and σ_t at the bottom and top of the section. M_{sR} is given by the lesser of

$M_{sR,b}=(\sigma_b+f_{ctm})\,Z_{b,co}$ (for exposure class XC1) (7.37a)

and $M_{sR,b}=\sigma_b\,Z_{b,co}$ (for exposure class>XC1) (7.37b)

or $M_{sR,t}=(0.45\,f_{ck}-\sigma_t)\,Z_{t,co}$ (7.38a)

and $M_{sR,t}=(0.60\,f_{ck}-\sigma_t)\,Z_{t,co}$ (7.38b)

At the bottom fibre $M_{sR,b}$ in Eq. 7.37a is the capacity to be checked against the characteristic service load g_k+q_k. $M_{sR,b}$ in Eq. 7.37b is checked against the quasi-permanent load $g_k+\psi_2 q_k$

for exposure classes XC2 to XC4, and the frequent load $g_k+\psi_1 q_k$ for XD and XS. At the top fibre $M_{sR,t}$ in Eq. 7.38a is checked against the quasi-permanent load for all classes of durability to avoid non-linear creep as discussed above. However, quasi-permanent loads and non-linear creep are for long-term conditions, and therefore the short-term conditions for exposure class XC should also be checked using $M_{sR,t}$ Eq. 7.38b for characteristic loads. This situation is not specified in BS EN 1992-1, clause 7.2 (2), but is adopted and proposed in Precast Eurocode 2: Worked Examples (Narayanan, 2008) for exposure XC1. Finally, if the top surface is exposed to long-term to classes XD, XF or XS $M_{sR,t}$ Eq. 7.38b is checked against the characteristic service load (clause 7.2 (2)). Other publications (e.g. Gilbert et al. (2017)) limit the compression in this situation to $0.5\,f_{ck}$.

Double tee slabs present a special case. Because of its cross section, the centroid of the unit lies close to the top flange, and therefore the section modulus Z_t to the top fibre is very large, typically 2 to 3 times Z_b. Consequently, as the top fibre does not give a limiting value to M_{sR} the influence of f_{ck} is very small, as given in Eq. 7.38. As the controlling influence in Eq. 7.37 is σ_b, the stress at transfer becomes very important. It is therefore necessary with double tee units to try to achieve the maximum possible transfer strength, say about $f_{ck}(t)=35$ N/mm^2, or to debond or deflect tendons in order to reduce the demand on $f_{ck}(t)$.

Example 7.1

Continue Example 6.1 to calculate M_{sR} at the support and mid-span for the 250 mm depth prestressed hollow core unit shown in cross section in Figure 6.7. Exposure class is XC1. The effective span of hcu = 9.8 m. Self-weight of grouted unit = 3.63 kN/m^2, floor screed = 1.80 kN/m^2, finishes = 1.50 kN/m^2 and services & ceiling = 0.50 kN/m^2. Imposed live load = 5.0 kN/m^2. Geometric and material data given by the manufacturer are as follows:

$A_c=168\times10^3$ mm^2; $I_c=1270\times10^6$ mm^4; $y_b=123$ mm; $f_{ck}=45$ N/mm^2; $f_{ck}(t)=30$ N/mm^2; $f_{pk}=1770$ N/mm^2; $E_p=195$ kN/mm^2; $A_p=93$ and 52 mm^2 per 12.5 and 9.3 mm dia. strand; cover to strand = 35 mm. Concrete density = 24.5 kN/m^3. Gravel aggregates. Cement CEM I grade 52.5R. Area of longitudinal infilled joints between hcu = 9970 mm^2. Compound section properties (from Example 6.1) are $Z_{b,co}=10.701\times10^6$ mm^3; $Z_{t,co}=10.056\times10^6$ mm^3; $Z_{z,co}=15.766\times10^6$ mm^3.

Is the critical fibre stress at the top or bottom of the unit? Check M_{sR} against the characteristic combination of the design service moment M_s. Plot graphs of bending moment distributions for M_{sR} and M_s, taking the transmission length for the build-up of prestress (from Example 6.1) as $l_{pt}=605$ mm.

Solution

Material properties relevant to service M_{sR}

$f_{cm}=45+8=53$ N/mm^2, $E_{cm}=22\times(53/10)^{0.3}=36283$ N/mm^2, $f_{ctm}=0.3\times45^{2/3}=3.80$ N/mm^2.

Section properties. See Example 6.1. Note that compound section properties are used only for calculating stresses due to loading (not for prestress), and for M_{sR}.

Prestress

$$A_p=7\times93+5\times52=911\text{ mm}^2$$

Initial prestress $\sigma_{pi}=0.7\times1770=1239.0$ N/mm^2

From Example 6.1 $\Delta\sigma_{pr}=4.95$ N/mm^2, then $P_{pm0}=(1239.0-4.95)\times911=1124218$ N and $\sigma_{cp}=12.67$ N/mm^2

At mid-span, bending moment due to self-weight of ungrouted unit $M_{s0}=4.12\times9.8^2/8=49.46$ kNm

Due to M_{s0}. σ_{cp0}=−49.46/15.766=−3.14 N/mm^2
Then at mid-span σ_{cp}=12.67−3.14=9.53 N/mm^2
Elastic shortening loss $\Delta\sigma_{el}$=195000×12.67/32837=75.26 N/mm^2. At mid-span =(9.53/12.67)×75.26=56.63 N/mm^2
From Example 6.1 σ_{pm0}=1158.8 N/mm^2 and at mid-span σ_{pm0}=1239.0−4.95−56.63= 1177.4 N/mm^2<0.75×1770=1328 N/mm^2 OK
The loss of prestress at transfer at the support is 100−(100×1158.8/1239)=6.47% (this is carried forward to Example 11.1)
P_{pm0}=1158.8×911=1055654 N and at mid-span=1177.4×911=1072607 N
Maturity of concrete during curing

Mean temperature during 20 hours of curing=50°C

Eq. 7.13 $t_T=(20/24)\ e^{-[4000/(273+50))-13.65]}=2.96$ days

Factor for Cement Class R=1

Eq. 7.12 $t_{oT}=2.96\times[9/(2+2.96^{1.2})+1]^{1}=7.65$ days

Long time losses to life using RH$_s$ = 50% with bottom only exposed

Eq. 7.25 h_o=2×area/bottom=2×168000/1200=280.0 mm

$\alpha_1=(35/53)^{0.7}=0.75$. $\alpha_2=(35/53)^{0.2}=0.92$. $\alpha_3=(35/53)^{0.5}=0.81$

Eq. 7.19 $\varphi_{RH}=[1+(1-(50/100)\times0.75/(0.1\times280.0^{1/3}))]\times0.92=1.446$

Eq. 7.10 $\beta(f_{cm})=16.8/\sqrt{53}=2.308$

Eq. 7.11 $\beta(t_0)=1/(0.1+7.65^{0.2})=0.624$

Eq. 7.21 $\beta_H=((1.5\times(1+(0.012\times50)^{18})\times280.0)+(250\times0.81))=623$ days

Eq. 7.20 $\beta_c(t_i,t_0)=[(20833-1)/(623+20833-1)]^{0.3}=0.991$

Eq. 7.18 $\varphi(t,t_0)=\varphi_{RH}\ \beta(f_{cm})\ \beta(t_0)\ \beta_c(t,t_0)=1.446\times2.308\times0.624\times0.991=2.065$

Eq. 7.22a σ_c=(1055654/168000)+(1055654×82.2/15.450×10^6)=11.90 N/mm^2

At mid-span, bending moment due to self-weight, infill in gaps (9970×24×10^{-6}=0.24 kN/m), screed and finishes is

M_{s1}=(4.12+0.24+1.2×1.8+1.2×1.5)×9.8^2/8=99.88 kNm

Due to M_{s0} σ_{c0}=−99.88/15.766=−6.33 N/mm^2
Then at mid-span

Eq. 7.22b σ_c=(1072607/1055654)×11.90−6.33=12.09-6.33=5.76 N/mm^2

Denominator in Exp. 5.46=1.146

Loss of prestress due to creep $\Delta\sigma_{p,c}$=195000×2.065×11.90/(36283×1.146)=115.2 N/mm^2. At mid-span=55.74 N/mm^2

Concrete shrinkage

Eq. 7.28 k_n=0.85–0.001 (280–200)=0.77

Eq. 7.26 β_{RH}=1.55×(1– (50/100)^3)=1.356

Eq. 7.29 $\beta_{ds}(t,t_s)$=(20833–1)/[(20833–1)+0.04×280.0$^{1.5}$]=0.991

For cement grade 52.5R, class R, α_{ds1}=6 and α_{ds2}=–0.11

Eq. 7.31 $\varepsilon_{cs}=\beta_{ds}(t,t_s)\ k_n\ \varepsilon_{cd,o}$=0.991×0.77×0.85×(220+110×6)×e$^{-0.11\times53/10}$ ×1.356×10^{-6}=0.000432

$\varphi(t,t_0)$=2.065

Eq. 7.32 $\Delta\sigma_{p,s}$=0.000432×195000/1.146=73.51 N/mm^2

Strand relaxation

Eq. 7.33 μ=1158.8/1770=0.655. At mid-span=1170.6/1770=0.661

For strand relaxation Class 2, the power function in Exp. 3.29=9.1

Eq. 7.34a σ_{pr}=1158.8×0.66×2.5×e$^{(0.665\times9.1)}$×[(500000/1000)$^{(0.75\times(1-0.665))}$]×10^{-5}=36.97 N/mm^2. At mid-span σ_{pr}=39.36 N/mm^2

Eq. 7.34b $\Delta\sigma_{p,r}$=0.8×36.97/1.146=25.80 N/mm^2. At mid-span $\Delta\sigma_{p,r}$=27.47 N/mm^2 (slightly greater due to smaller initial losses)

Final prestress

Eq. 7.35 σ_{po}=1158.8–115.2–73.5–25.8=944.3 N/mm^2. At mid-span σ_{po}=1020.7 N/mm^2

P_{po}=944.3×911=860218 N. At mid-span=929835 N

The reduction in elastic shortening and creep losses at mid-span increases the prestress there by 8.1%

R_s=944.3/1239=0.762. At mid-span ratio R_s=0.824

The loss of prestress at transfer at the support is 100–(100×944.3/1239)=23.8% (this is carried forward to Example 11.1).

Final prestress at mid-span

Eq. 7.1 σ_b=(929835/168000)+(929835×82.2/10.324×10^6)=12.94 N/mm^2<0.45×45=20.25 N/mm^2

Eq. 7.2 σ_t=(929835/168000)−(929835×82.2/10.000×10^6)=−2.11 N/mm²>−3.80 N/mm²

M_{sR} at mid-span is the lesser of

Eq. 7.37 At the bottom fibre. $M_{sR,b}$=(12.94+3.80)×10.701=**179.12 kNm**

Eq. 7.38 At the top fibre. $M_{sR,t}$=(20.25−2.11)×10.056=224.84 kNm>179.12 kNm

Clause 7.2 (3). $M_{sR,t}$ at the top need only be checked against quasi-permanent service load.

Final prestress at support

σ_b=0.925×12.94=11.97 N/mm²

σ_t=0.925×−2.11=−1.95 N/mm²

M_{sR} at the support is the lesser of

$M_{sR,b}$=(11.97+3.80)×10.701=**168.75 kNm**

$M_{sR,t}$=(20.25−1.95)×10.056=223.25 kNm>168.75 kNm

The bottom fibre is critical in both cases.

The distribution of M_{sR} between support and mid-span is taken as parabolic to mimic the parabolic distribution of bending stress due to self-weight and dead loads.

$M_{sR,b}$ at the ends of the unit, where the prestress σ_b is zero is $M_{sR,b}=f_{ctm}$ $Z_{b,co}$=3.80×10.701=**40.63 kNm**

Between the end of the span and distance $x=l_{pt}$, the increase in $\sigma_{b,x}$ is linear, as

$$\sigma_{b,x}=(x/l_{pt})\ \sigma_b \tag{7.39}$$

where σ_{po} is the prestress based on the parabolic distribution between the support and mid-span. At $x=l_{pt}$, $\sigma_{b,x}$=12.19 N/mm²

$$\text{Then } M_{sR,x}=(\sigma_{b,x}+f_{ctm})\ Z_{b,co} \tag{7.40}$$

At $x=l_{pt}$, $M_{sR,x}$=171.9 kNm

The characteristic combination of the design service UDL and moment per 1.20 m wide unit is

w_s=[4.12+0.24+1.2×(1.80+1.50+0.50+5.00)]=14.92 kN/m

M_s=14.92×9.8²/8=179.1 kNm=M_{sR} at mid-span exactly!

Figure 7.11 shows the complete bending moment diagrams for the characteristic design moment M_s and M_{sR}. The design is completed for ultimate moments, shear forces and deflections in Examples 9.1, 10.1 and 11.1.

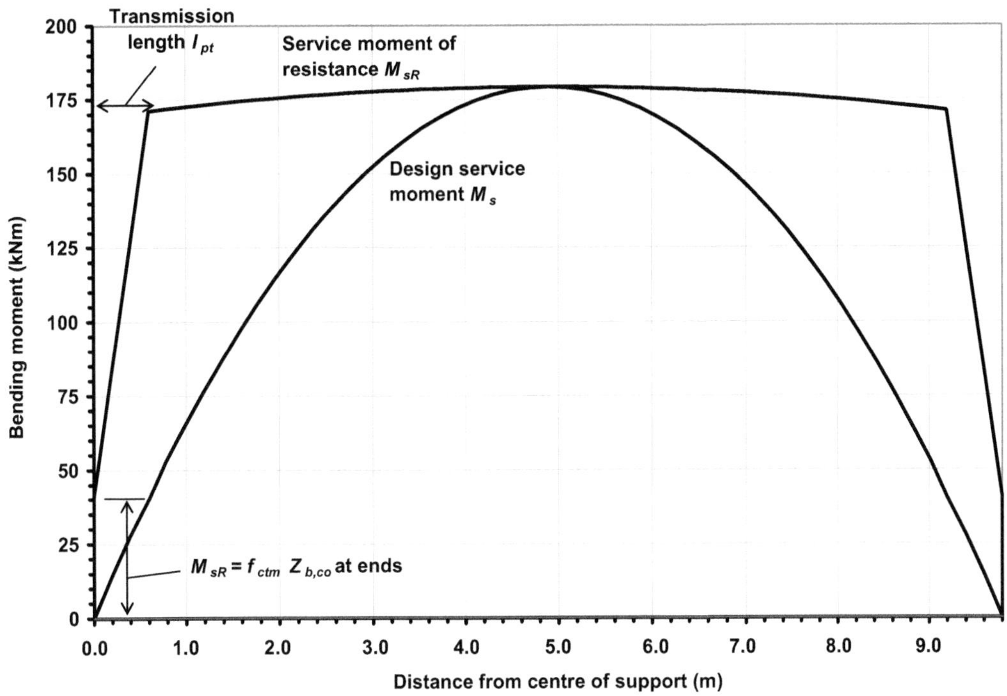

Figure 7.11 Bending moment diagrams for design service and moment of resistance for the hcu in Example 7.1.

Example 7.2

Find the required compressive f_{ck} in Example 7.1 that would equate the service moment based on the top and bottom limiting service stress conditions, thus optimising the strength of concrete.

Solution

Solving $M_{sR}=(\sigma_b+f_{ctm})\ Z_{b,co}=(0.45\ f_{ck}-\sigma_t)\ Z_{t,co}$

Then $(12.94+f_{ctm})\ 10.701\times10^6=(0.45\ f_{ck}-2.11)\ 10.056\times10^6$

Then $138.47+10.701\times0.3\ f_{ck}^{2/3}=4.527\ f_{ck}+21.22$

$117.25=4.527\ f_{ck}-3.21\ f_{ck}^{2/3}$

Hence f_{ck} = **33.2 N/mm²**

and $M_{sR,b}=(12.94+0.3\times33.2^{2/3})\times10.701=171.5$ kNm.

The result for f_{ck} is just greater than the transfer strength ($f_{ck}(t)=30$) suggesting an impractical solution. (This result further demonstrates that increasing f_{ck} to say 50 N/mm² would have little effect on the value of M_{sR}, adding 2.3%.) It is therefore necessary to modify the section properties $Z_{b,co}$ and $Z_{t,co}$ to obtain comparability, as follows.

Example 7.3

Find the required values of $Z_{b,co}$ and $Z_{t,co}$ in Example 7.1 necessary to equate the value of $M_{sR,b}$ and $M_{sR,t}$ obtained from limiting stresses. Calculate the new value of M_{sR}. Study the cross section and check whether the new values of $Z_{b,co}$ and $Z_{t,co}$ can be achieved practically. From Example 6.1 $y_{b,co}=121.1$ mm and $I_{c,co}=1296\times10^6\text{mm}^4$.

Solution

Solving $M_{sR}=(\sigma_b+f_{ctm})\ Z_{b,co}=(0.45\ f_{ck}-\sigma_t)\ Z_{t,co}$

Then $(12.94+3.80)\ Z_{b,co}=(20.25-2.11)\ Z_{t,co}$

Then $Z_{b,co}/Z_{t,co}=22.36/16.74=1.336$, i.e. $y_t/y_b=1.336$

also $y_b+y_t=250$ mm

Solving $y_b=107$ mm and $y_t=143$ mm

Then $Z_{b,co}=1296\times10^6/107=\mathbf{12.112\times10^6 mm^3}$ and $Z_{t,co}=\mathbf{9.063\times10^6 mm^3}$

and $M_{sR}=(12.94+3.80)\times12.112=$ **202.75 kNm** >179.12 kNm in Example 7.1.

To achieve this condition, the geometric centroid must be lowered by 121.1 – 107 = 14 mm. To achieve this, the voids must be repositioned or modified in shape. It is not possible to raise the position of the voids by at least twice this distance (the core area is about 44% of the gross) by making the thickness of the top flange over the top of the cores too shallow $h_{ft}=40-28=12$ mm.

Example 7.4

Repeat Example 7.1 for the durability exposure class XC3, and check whether the characteristic or quasi-permanent design service moment is critical. The floor slab is to be for offices. M_{sR} from Example 7.1 is 179.12 kNm.

Solution

$M_{sR,b}=(12.94+0)\times10.701=$ **138.47 kNm**

$M_{sR,t}=(20.25-2.11)\times10.056=224.84\text{ kNm}>134.6\text{ kNm}$

Table 2.1. For office use $\psi_2=0.3$

The quasi-permanent combination of the design service moment

$M_{s,qp}=[4.12+0.24+1.2\times(1.8+1.5+0.5+0.3\times5.0)]\times9.8^2/8=128.69\text{ kNm}$

<138.47 kNm by a ratio 0.93, therefore the quasi-permanent combination is not critical.

Example 7.5

a. Complete the calculation for M_{sR} for the 150 mm depth T beams shown in cross section in Figure 7.12 where previous in-house calculations (following the procedures in Example 7.1) found the long-term loss of prestress at mid-span as 22.3%. The initial prestressing force is $\eta=0.75$ of characteristic strength of indented wire of $f_{pk}=1770\text{ N/mm}^2$. Exposure class is XC1.
b. Calculate the maximum allowable simply supported effective span for the pair of beams infilled using 440×100 mm depth concrete blocks of density 1950 kg/m³. Use self-weight of floor screed = 1.50 kN/m², finishes = 1.0 kN/m² and services & ceiling = 0.5 kN/m². Imposed live load = 1.5 kN/m². Self-weight of infill *insitu* concrete between beams and between beam and blocks = 0.117 kN/m.

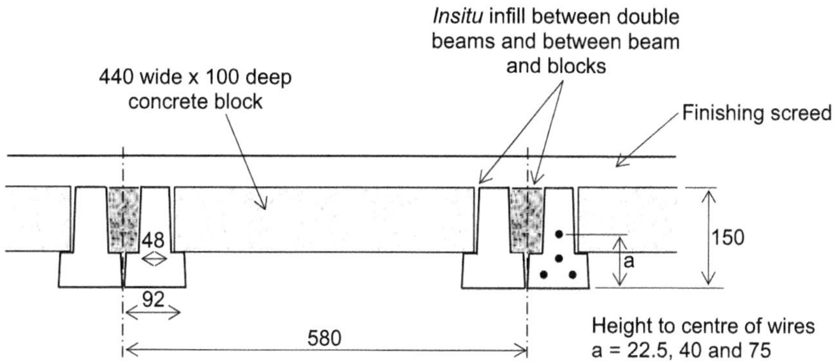

Figure 7.12 Cross section of pairs of (known as 'double') prestressed T beams and infill blocks used in Example 7.5.

c. Using the limiting stress at the top fibre according to BS EN 1992-1, clause 7.2 (3) calculate the maximum allowable simply supported effective span for the same beams as (b). The flooring is to be used for domestic purposes.

Geometric and material data given by the manufacturer are as follows:
$A_c=90\times10^3$ mm^2; $I_c=17.03\times10^6$ mm^4; $y_b=61.9$ mm; Compound $I_{c,co}=17.34\times10^6$ mm^4; $y_{b,co}=61.1$ mm; $f_{ck}=55$ N/mm^2; $A_p=19.63$ mm^2 per 5.0 mm dia. wire; Axis height to wires $=22.5$, 40 and 75 mm. Concrete density $=23.6$ kN/m^3.

Solution

Material properties

$f_{cm}=55+8=63$ N/mm^2

f_{ctm} (for $f_{ck}>50$ N/mm^2) $=2.12\log_e(1+63/10)=4.214$ N/mm^2

$A_p=4\times19.63=78.5$ mm^2

Section properties

Self-weight of beam $=9000\times23.6\times10^{-6}=0.212$ kN/m

Self-weight of infill blocks per beam $=0.440\times0.100\times(1950/102)/2=0.420$ kN/m

Self-weight of infill concrete per beam $=0.117/2=0.059$ kN/m

$Z_b=17.03\times10^6/61.9=0.275\times10^6$ mm^3

$Z_t=17.03\times10^6/(150-61.9)=0.193\times10^6$ mm^3

$y_s=(2\times22.5+40+75)/4=40.0$ mm

$z_{cp}=61.9-40.0=21.9$ mm

Section properties of compound section with transformed area of tendons

$Z_{b,co}=17.34\times10^6/61.1=0.284\times10^6$ mm^3; $Z_{t,co}$ $17.34\times10^6/88.9=0.195\times10^6$ mm^3.

Prestress

Initial prestress σ_{pi}=0.75×1770=1327.5 N/mm^2

Initial P_{pi}=1327.5×78.54×10^{-3}=104262 N

At mid-span R_s=(1−0.223*)=0.777

Final P_{po}=0.777×104262=81011 N

and σ_{po}=0.777×1327.5=1031.5 N/mm^2

*Note that the calculation for loss of prestress at mid-span (i.e. 22.3%) requires prior knowledge of the effective span in order to subtract the elastic shortening and creep losses due to self-weight and dead loads from those at the support. In this example, the second part of the question, i.e. the effective span, had previously been determined.

Solution to (a). Final prestress at mid-span

σ_b=(81011/9000)+(81011×21.9/0.275×10^6)=15.45 N/mm^2<0.45×55=24.75 N/mm^2

σ_t=(81011/9000)−(81011×21.9/0.193×10^6)=−0.19 N/mm^2>−4.214 N/mm^2

M_{sR} at mid-span is the lesser of

At the bottom fibre. $M_{sR,b}$=(15.45+4.21)×0.284=5.58 kNm

If the limiting compressive is taken as σ_c=0.6 f_{ck}=33.0 N/mm^2, then non-linear creep will be activated as σ_c/f_{cm}=33.0/63=0.52>0.45. To avoid this limit σ_c to 0.45 f_{cm}=28.35 N/mm^2.

At the top fibre. $M_{sR,t}$=(28.35−0.19)×0.195=**5.56 kNm**<5.58 kNm.

This is a good example of the optimisation of this beam at the serviceability limit state.

Solution to (b). Maximum allowable effective span

Characteristic combination of the design service UDL

Self-weight of beam, blocks and infill per beam=0.212+0.420+0.059=0.691 kN/m

Width of two beams and infill block (see Figure 7.12)=440+48+92=580 mm. Average spacing of beams=580/2 beams=290 mm

Self-weight of screed, imposed dead and live loads=0.290×(1.50+1.00+0.50+1.50)= 1.305 kN/m

Total UDL per beam w_s=0.691+1.305=1.996 kN/m

Then effective span $L=\sqrt{}(8\ M_{sR}/w_s)=\sqrt{}(8\times5.56/1.996)$=**4.721 m**

Solution to (c). Maximum allowable effective span at the top fibre for quasi-permanent load

Limiting stress 0.45 f_{ck}=0.45×55=24.75 N/mm^2

Then $M_{sR,t}$=(24.75− 0.19)×0.195=4.86 kNm.

Table 2.1. For domestic use ψ_2=0.3

Quasi-permanent imposed UDL per beam $w_{s,qp}$=0.691+[0.290×(1.50+1.00+0.50+ 0.3×1.50)]=1.692 kN/m

Then effective span $L_{qp}=\sqrt{}(8\ M_{sR}/w_{s,qp})=\sqrt{}(8\times4.86/1.692)$=**4.795 m**>4.721 m. L_{qp} is not critical.

Example 7.6

Repeat Example 7.5 for the addition of an internal partition wall positioned over one of the pairs of beams over the full span of the beams. The wall, including two coats of plaster is 130 mm thickness and has a characteristic dead load of 3.0 kN/m. It may be assumed that, because the wall is built onto the screed, the wall load may be distributed over four beams.

Previous in-house calculations found the long-term loss of prestress at mid-span, including for the effect of the wall dead load on the loss of prestress due to creep, as 22.0%.

Solution to (a). Final prestress at mid-span

At mid-span $R_s=(1-0.22)=0.780$ (a difference of 0.4% in Example 7.5); therefore, use the same prestress using $R_s=0.777$ and $M_{sR}=5.56$ kNm.

Solution to (b). Maximum allowable effective span

The wall displaces floor finishes and live loads over a width of 130 mm, giving an equivalent wall load $=3.0-0.13\times(1.0+1.5)=2.675$ kN/m. It is optional whether designers take advantage of this*. However, continuing with the equivalent load distributed over four beams

Wall load per beam $=2.675/4=0.669$ kN/m

Total UDL per beam $w_s=1.996+0.669=2.665$ kN/m

$L=\sqrt{(8\ M_{sR}/w_s)}=\sqrt{(8\times5.56/2.665)}=$ **4.085 m**

*Taking 3 kN/m for the wall load, $L=4.024$ m

Solution to (c). Maximum allowable effective span at the top fibre for quasi-permanent load

From Example 7.5. $M_{sR,t}=4.86$ kNm.

Equivalent quasi-permanent wall load $=3.0-0.13\times(1.0+0.3\times1.5)=2.812$ kN/m/4 no. beams $=0.702$ kN/m per beam

Quasi-permanent imposed UDL per beam $w_{s,qp}=1.692+0.702=2.395$ kN/m

$L_{qp}=\sqrt{(8\ M_{sR}/w_{s,qp})}=\sqrt{(8\times4.86/2.395)}=$ **4.029 m** <4.085 m. L_{qp} is critical by a difference of 1.3% showing that the differences between the two conditions for limiting stresses at the top of the beam are not significant. *Taking 3 kN/m for the wall load the difference is 0.8%.

Example 7.7

Repeat Example 7.6 replacing the internal partition as a cross wall, perpendicular to and at mid-span. Previous in-house calculations found the long-term loss of prestress at mid-span, including for the effect of the wall dead load on the loss of prestress due to creep, as 23.0%.

Solution to (a). Final prestress at mid-span

At mid-span $R_s=(1-0.23)=0.770$ (a difference of 0.9% in Example 7.5) therefore use the same prestress using $R_s=0.777$ and $M_{sR}=5.56$ kNm.

Solution to (b). Maximum allowable effective span

Equivalent cross wall load $=2.675$ kN/m

Point load per beam $=0.290\times2.675=0.776$ kN

Maximum bending moment per beam $M_s=1.996\,L^2/8+0.776\ L/4\leq5.56$ kNm

Solving the quadratic gives $L=$ **4.345 m**

Solution to (c). Maximum allowable effective span at the top fibre for quasi-permanent load

From Example 7.5. $M_{sR,t}=4.86$ kNm.

From Example 7.6. Equivalent quasi-permanent cross wall load $=2.812$ kN/m

Point load per beam $=0.290\times2.812=0.815$ kN

$M_{s,qp}=1.692\,L^2/8+0.815\ L/4\leq4.86$ kNm

Solving the quadratic gives $L_{qp}=$ **4.335 m** ≈4.345 m due to the relatively small point load.

7.3 FLEXURAL CAPACITY AT HOLES AND NOTCHES IN SLABS

Holes and notches are permitted in wide flooring units, such as solid units and hcu, voided units, double tees and thermal slabs, although in the latter two, they are only formed through the top flange (not in webs). In this respect, the voids reduce M_{sR} by about 5%–10% due to changes in cross-sectional properties, as the number of tendons is not reduced. For the case of a 1200 mm wide×250 mm depth thermal slab (Figure 7.13) containing a 500×500 mm square hole in the top flange M_{sR} reduces by 7.5% – in fact with the dispersion of prestress being uniform over a greater distance than the length of the hole, it is possible that the hole may have no effect on the moment of resistance, although this would not be the same for longer holes. Example 7.9 shows this in more detail. The design of the top flange has to be carefully checked, particularly for large holes up to ½ the width of the flange at the ultimate limit state.

This is not the case in solid and hcu, where both the cross-sectional properties and the pretensioning forces are reduced. Figures 7.14 and 7.15 show examples of holes and notches in some hcu. Figure 7.16 shows a corner notch being formed using a vacuum pump (holes mare formed likewise), but may also be formed manually using a trowel.

Figure 7.17 shows the positions and maximum recommended dimensions of voids in hcu produced by the Partek Company in Belgium. Table 7.6 shows the limits on geometry applicable to Figure 7.17. Surprisingly, the design of hcu at voids is not included in BS EN 1168 (BS EN 1168, 2005) but is included in the forthcoming FIB Manual on prestressed hcu, which also includes information of very wide holes.

Holes in the floor are dealt with in one of two ways depending on size. Small holes of less than about 600 mm in size may be formed in the precast unit (except in 600 mm wide hcu where the maximum size is 300 mm) during the manufacturing stage and before the concrete has hardened. The rough edges are beneficial to the shear key with *insitu* concrete infill.

The maximum size of hole which may be formed in the units depends on the size of the voids in the slab and how many tendons may be removed without jeopardising the strength of the unit. For example, in a certain 200 mm depth extruded hcu (e.g. *Spiroll* type) the diameter of the void is 150 mm and therefore a hole formed through the void may be of equal size. In a 200 mm depth slip-formed hcu (e.g. *Roth* type) the diameter of a hole is limited to the width of the void which is only 100 mm. Small holes, up to about 150 mm diameter, can be core-drilled on site. The holes should pass through the hollow core, and the designers must be consulted if tendons are accidentally removed. This can impose quite severe restrictions on the M&E services, but a way of increasing freedom is to provide additional tendons in the units and to specify permitted 'zones' in which a limited

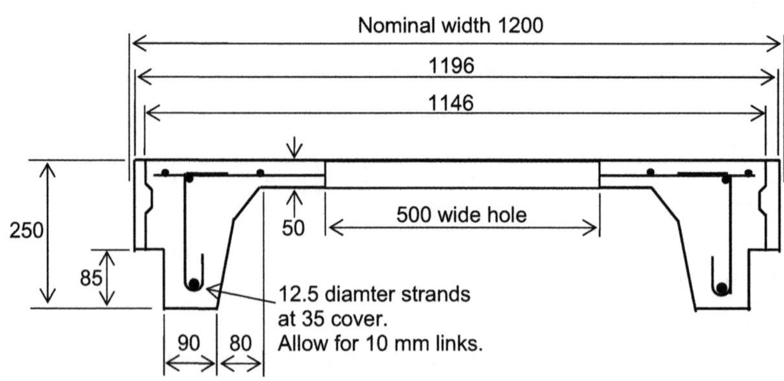

Figure 7.13 Cross section of thermal slab containing hole in top flange as used in Example 7.9.

Figure 7.14 Example of a hole in formed in a hollow core unit using a vacuum aspirator (see also Figure 4.18).

Figure 7.15 Example of side notch in a hollow core unit, the position has been marked automatically by a plotter.

Figure 7.16 Removing concrete by vacuum aspirator to form a corner notch. This unit also has a sloping end, presumably to assist fitting it between the flanges of a steel beam, or similar.

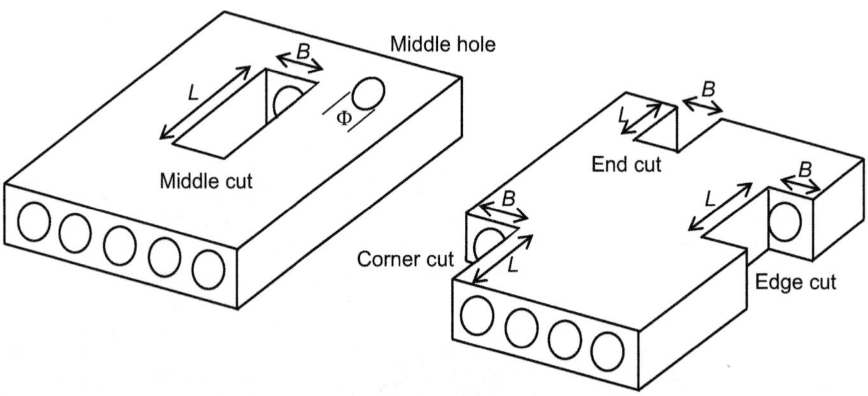

Figure 7.17 Positions and permitted sizes of holes and notches in hollow core units (see Table 7.6).

number of holes at a predetermined spacing and size may be site-drilled. Some manufacturers impose a maximum length of narrow voids (without further reference to the design office) as 2.0 m.

Larger voids that are wider than the width of the precast units are 'trimmed' using transverse supports such as steel angles (i.e. perpendicular to the span of the floor) as shown in Figure 7.18. A typical minimum section size for the steel trimmer is 100×100×8 mm equal angle. The angles carry point loads to the edges of the adjacent units. The maximum practical size of hole is about 2.4 m×2.4 m for normal office loading, but is less for superimposed floor loads greater than about 7.5 kN/m^2. There is some additional margin of safety

Table 7.6 Recommended limits imposed on cut-out geometry (see Figure 7.17)

Depth of slab (mm)	*Corner cut-out L×B (mm)*	*Edge cut-out L×B (mm)*	*End cut-out L×B (mm)*	*Middle cut-out L×B (mm)*	*Middle hole diameter, Φ (mm)*
150	600×400	600×400	1000×400	1000×400	80
200	600×380	600×400	1000×380	1000×400	130
265	600×260	600×400	1000×260	1000×400	130
320	600×260	600×400	1000×260	1000×400	170
400	600×26	600×400	1000×260	1000×400	170

Figure 7.18 Steel trimmer angles used to form large voids around hollow core units.

in the design of trimmer angles because the grouted longitudinal joints between the slabs will transfer some of the superimposed loading to adjacent slabs. The steel angles must be properly fire-protected either by precasting concrete to the soffit and sides of the angle or by spraying them with a fire-protective coating.

To calculate the prestress and moment of resistance (and shear capacity in Section 10.2.4) for units containing one or two holes in the span of the unit, and/or a notch removed from one or both ends of the basic (i.e. full width) unit, the reduction in cross-sectional properties are calculated based on the ratio of the width of the void w_v to the average width of the unit (b_{ave}=(1197+1154)/2=1176 mm), i.e. pro-rata w_v/b_{ave}. Although this sounds like a crude approximation, the results are very close to a detailed calculation of the gross area and area of hollow cores removed at the void. The dimensions are specified from the effective support, i.e. centre of bearing, not the end of the unit. Some manufacturers imposed 'cutting zones', i.e. dimensions from the edges of units where cuts can be made, usually at the edges of the hollow cores, trying to avoid cutting through the centres of hollow cores, and at least

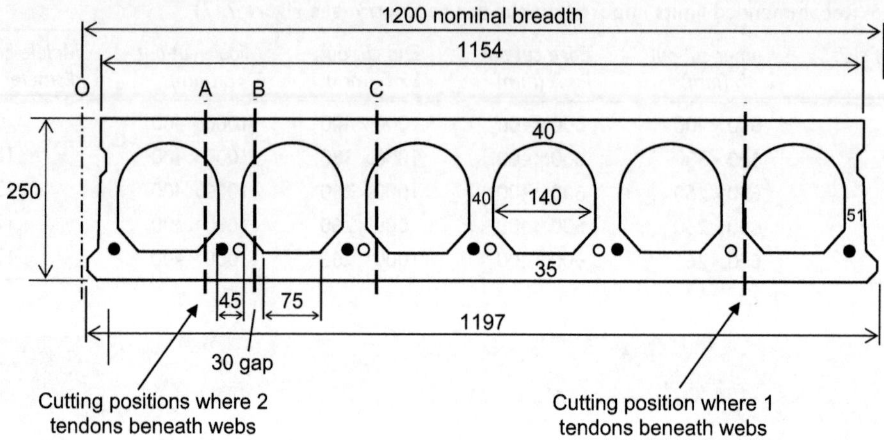

Figure 7.19 Preferred cutting positions for the hollow core floor unit to Example 7.1. Width of void A to B = 85 to 105 mm, B to C = 180 mm and A to C = 265 to 285 mm. Width of edge notch O to A = 200 mm, O to B = 285 to 305 mm, where O is at the edge of the nominal breadth of 1200 mm.

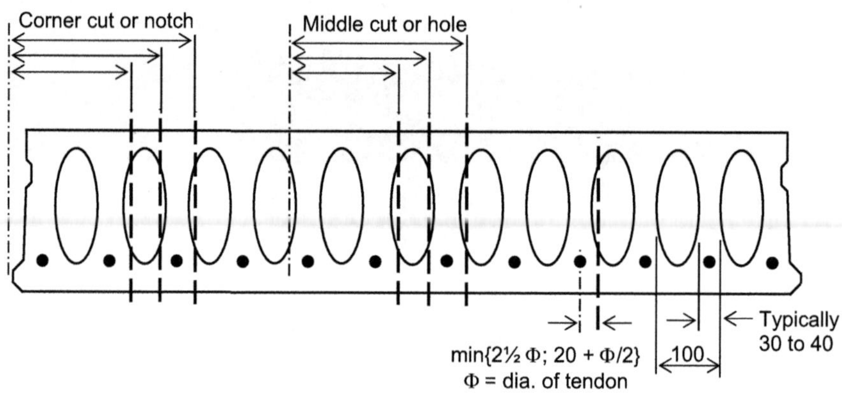

Figure 7.20 Preferred cutting positions for the hollow core floor unit with 11 no. hollow cores, as close to the edge of the core as the side cover to the tendons allows.

20 mm from the edges of tendons. In double-tee or thermal slabs, the limitations are only between points about 50 mm from the edges of the webs.

For example, in the hcu used in Example 7.1 (see Figures 6.7 and 7.19), the cutting positions are set at 20 mm clear from the edge of the tendons, thus the width A to B $w_v = 45 + 2 \times 20 = 85$ mm minimum. The preferred width of holes in the body of the hcu is from A to C $w_v = 85 + 180 = 265$ mm, which would involve a design where 4 no. strands are removed. It may be possible to increase the hole to $w_v = 265 + 180 - 85 = 360$ mm, also with 4 no. strands removed. Edge holes and notches are taken down from flooring layouts and measured from the nominal width of the hcu at point O (not at the actual edge). Thus the first edge hole from O to A is $w_v = (1200 - 4 \times 180 - 85)/2 = 198$ mm (say 200 mm). O to B is $w_v = 285$ mm, and so on. The hcu shown in Figure 7.20 have 11 no. hollow cores at 100 mm spacing is much more versatile as the cutting positions may be at the edges or several cores, at least up to the permitted width of the void.

Consider the hcu in Figure 7.19 containing a hole from A to C of width $w_v = 265$ mm, in which 2 no. webs and 4 no. tendons are removed. Based on the simplified pro-rata rule

Table 7.7 Section properties, prestress and moments of resistance for hollow core unit in Figure 7.19 with 265 mm wide hole

	Basic unit without hole	*Based on pro-rata assumption at hole* $(1-w_v/b_{ave})$	*Actual values at hole*	*Difference*
Section properties				
A_c (mm²)	168000	130127	126848	2.58%
$I_c \times 10^6$ (mm⁴)	1270.0	983.7	975.4	0.84%
y_b (mm)	123.0	123.0	122.3	0.58%
$Z_b \times 10^6$ (mm³)	10.325	7.998	7.976	0.26%
$Z_t \times 10^6$ (mm³)	9.999	7.745	7.638	1.40%
Compound values				
$I_{c,co} \times 10^6$ (mm⁴)	1296.3	1001.6	993.1	0.86%
$Z_{b,co} \times 10^6$ (mm³)	10.705	8.256	8.236	0.25%
$Z_{t,co} \times 10^6$ (mm³)	10.056	7.784	7.674	1.43%
Prestress at the hole				
σ_b (N/mm²)	12.94	11.42	11.49	−0.59%
σ_t (N/mm²)	−2.11	−1.86	−1.76	5.29%
Moments of resistance				
Service M_{sR} (kNm)	179.12	125.63	125.88	−0.20%
Ultimate M_{Rd} (kNm)	247.32	175.04	174.03	0%

Area of strands at hole $A_p = 3\times52+5\times93=621$ mm².

the ratio $w_v/b_{ave}=265/1176=0.225$, i.e. the remaining ratio is $1-0.225=0.775$. Thus the remaining cross-sectional properties are $A_c=0.775\times168000=130200\,\text{mm}^2$, $I_c=0.775\times 1270\times10^6=984.25\times10^6\,\text{mm}^4$, etc. The section properties, prestress and moments of resistance are listed in Table 7.7 together with values derived using detailed calculation of the cross section. Note that the important parameters $I_{c,co}$ and $Z_{b,co}$ differ by less than 0.86% and M_{sR} by only 0.2%, showing that the pro-rata assumption is applicable. The ultimate moment of resistance M_{Rd} is not affected by the pro-rata assumption (see Section 9.3.4). Any local stress concentrations at corners are ignored.

If the centre of this hole is located at a distance $x=6.0$ m from the left end support and has a length of 600 mm, Table 7.7 gives $M_{sR}=125.63$ kNm at the centre of the hole. The M_{sR} at a distance of l_{pt} from the ends of the hole, i.e. $x=6.0-0.6/2-0.605=5.095$ m and 6.905 m are the same values as in the bending moment graph in Figure 7.11, i.e. $M_{sR}=179$ kNm to 157 kNm, respectively. Therefore, the hcu fails by at least 31 kNm – if the hcu was a single unit acting alone. However, in service the units are grouted together after installation, known as a 'slab panel', and due to the lateral distribution of section properties over adjacent units, provided that there are no holes in the adjacent units, the net prestress may be averaged over the width of 3 no. units, or 2 no. units if the hcu is at a free edge. After subtracting the bending stress due to self-weight $\sigma_{s,0}=M_{s,0}/Z_{b,co}$ for each of the three units, the net prestress is

$$\sigma_{b,net}=[2\ (\sigma_b-\sigma_{s,0})\ (\text{full unit})+(\sigma_b-\sigma_{s,0})\ (\text{with hole})]/3 \tag{7.41}$$

The section modulus $Z_{b,co}$ may also be averaged over three units, as

$$Z_{b,co,net} = [2\ Z_{b,co}\ (\text{full}) + Z_{b,co}\ (\text{with hole})]/3 \tag{7.42}$$

$$\text{Then } M_{sR,net} = M_{s,0} + (\sigma_{b,net} + f_{ctm})\ Z_{b,co,net} \tag{7.43a}$$

$$\text{or } M_{sR,net} = M_{s,0} + \sigma_{b,net}\ Z_{b,co,net} \text{ for exposure class} > \text{XC1} \tag{7.43b}$$

The basis for this is that if point and line loads can be laterally distributed between adjacent units in the floor panel, and a hole is considered structurally as a 'dead weight', then the dead weight may be distributed too.

Figure 7.21 summarises the positions where design capacities are calculated near holes and notches. For service moments the full M_{sR} is at a one basic transmission length l_{pt} from the edge of the voids, and for ultimate moments the full M_{Rd} (Chapter 9) is at a one anchorage length l_{bpd}. Shear capacity $V_{Rd,c}$ (Chapter 10) is calculated at a distance

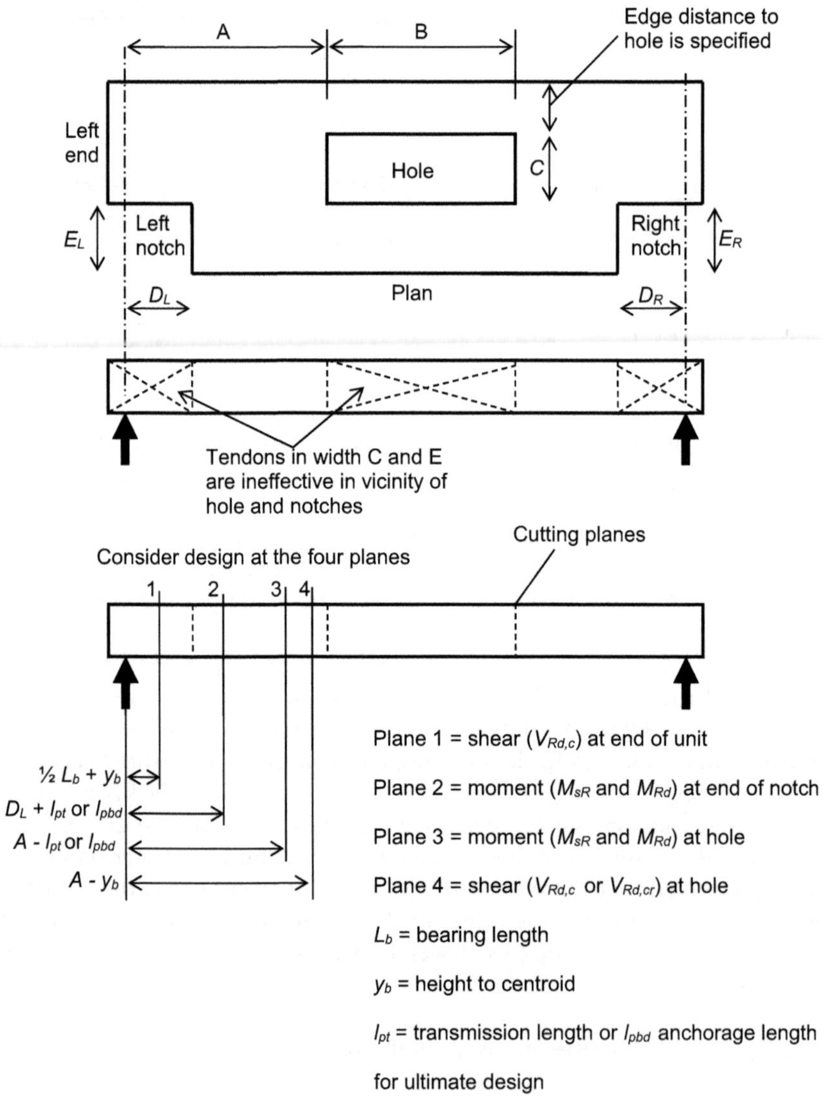

Figure 7.21 Definitions of holes and end notches, and design positions near to voids.

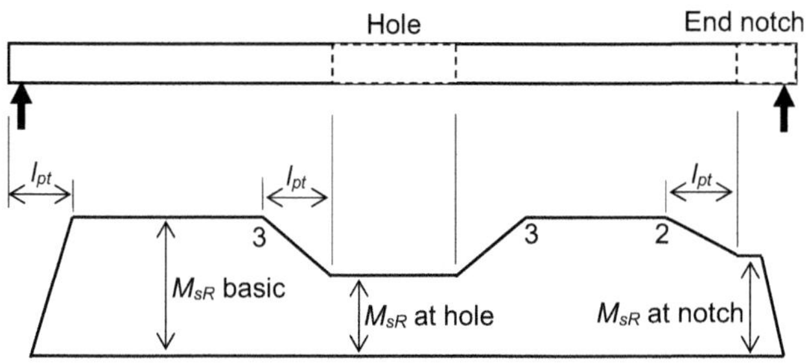

Figure 7.22 Development of prestress and M_{sR} at end notches and holes.

equal to the height of the centroid y_b from the edge of the void. Note that Figure 7.21 is reversible, such that the left end notch is considered in the same manner, and if the hole is closer to the left end of the span, the conditions are checked at that end. Figure 7.22 shows the development of prestress and M_{sR} at and between end notches and holes.

Example 7.8

The hcu shown in Figure 6.7 contains a hole of width w_v=265 mm between planes A and C as shown in Figure 7.19. The hole is located at a distance x=6.0 m from the left end support, and has a length of 600 mm. This hcu is grouted between two adjacent units at installation. Using the information given in Table 7.7 for the section properties and prestress of the three hcu, calculate the net $M_{sR,net}$ for the units and check the design moment M_s at the centre of the hole using the UDL given in Example 7.1.

Solution

The design moment due to self-weight (including infilled joints) at x=6.0 m from the support is $M_{s,0}=w_{s,0}L\ x/2-w_{s,0}\ x^2/2$ where (from Example 7.1) $w_{s0}=1.2\times3.633=4.36$ kN/m.

Then $M_{s,0}=4.36\times9.8\times6.0/2-4.36\times6.0^2/2=49.70$ kNm

In the hcu with the hole, $\sigma_{s,0}=49.70/8.256=6.02$ N/mm^2

In the basic units, $\sigma_{s,0}=49.70/10.705=4.64$ N/mm^2

From Table 7.7, Eq. 7.41. Prestress at the bottom $\sigma_{b,net}=[2\times(12.94-4.64)+(11.42-6.02)]/3=7.333$ N/mm^2

From Table 7.7, Eq. 7.42. $Z_{b,co,net}=(2\times10.325+7.998)/3=9.889\times10^6$mm^3

$M_{sR,net}=49.70+[(7.333+3.80)\times9.889]=$ **159.80 kNm**

The design UDL (from Example 7.1) $w_s=1.2\times(3.633+1.8+1.5+0.5+5.0)=14.92$ kN/m.

$M_s=14.92\times9.8\times6.0/2-14.92\times6.0^2/2=$ **170.09 kNm**>159.80 kNm, the slab panel fails by 10.3 kNm. In fact at the start of the hole, where x=5.7 m from the support M_s is slightly greater, the panel fails by 14.1 kNm, or by 8.7%.

Figure 7.23 shows the complete bending moment diagrams for the characteristic design moment M_s and M_{sR}.

The hcu in Figure 7.19 also contains an end notch from O to B of width w_v=285 mm and 600 mm length, in which 2 no. webs and 3 no. tendons are removed. The section properties and prestress at end notches are not laterally distributed between adjacent units. The resulting M_{sR} at the notch is M_{sR}=125.7 kNm at x=0.6 m, increasing to M_{sR}=171.1 kNm at $x+l_{pt}$=1.205 m. These values are also shown in Figure 7.23.

If there are two corner notches, their width is totalled as one notch width, and the greater length is considered, i.e. two notches 50 wide×100 mm long and 75×75 mm equates to a single notch 125 mm wide×100 long. The single notch does not have to be located at corners, it may be cut anywhere within the width. If two holes are present, they are considered individually, even if they are side by side – their collective reductions in capacity are subtracted from the basic unit. Holes cut exclusively through the hollow core

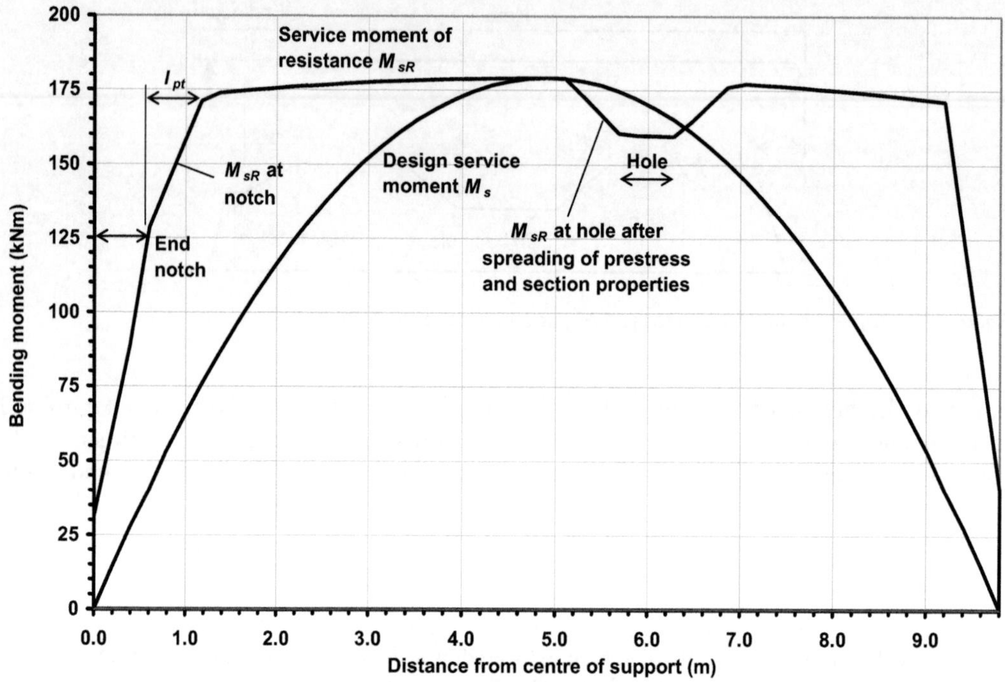

Figure 7.23 Bending moment diagrams for design service and moment of resistance for the hcu in Example 7.8.

are not considered in the structural appraisal. The reduction in self-weight due to the hole and notches is generally ignored in design. The reduced flexural stiffness at holes is not considered in the calculation of camber and deflection.

An alternative method to the design of slabs containing holes, as practised by some design engineers, is to consider the width of the hole as an effective 'dead weight' acting on a full width unit. Here the self-weight and imposed dead and live loads acting over the width of the hole are added to the loads acting on the full width unit based on the section properties of the full width unit. This is effectively reverse logic to the above where the net section properties are used. The procedure can only be used for UDL's, not point or line loads, and only where the tendons are uniformly distributed across the width of the unit such that the loss of resistance at the hole is roughly proportional to the width of the hole. To check this solution, consider that, for the 265 mm width of hole in Example 7.8, the self-weight and imposed dead and live loads are 3.633 and 8.80 kN/m^2, respectively. After grouting, the imposed load is distributed over the width of three units such that the equivalent width of the hole is 265/3=88 mm. The equivalent 'dead weight' is 3.633 (1.2+0.265)+8.80 (1.2+0.088)=16.66 kN/m and equivalent M_s=194.6 kNm>M_{sR}=178.9 kNm at the start of the hole for the full width unit by 8.8%. For this particular example, this is almost exactly the same as shown at the end of Example 7.8.

Example 7.9

The 1200 mm width×250 mm depth thermal slab shown in Figure 7.13 contains a hole of 500 mm width in the top flange. The hole is located at mid-span of a simply supported span of 6.0 m. This thermal slab is grouted between two adjacent slabs at installation. Using the information given in Table 7.8 for the section properties and prestress of the three slabs, calculate the net $M_{sR,net}$ for the slabs. The self-weight of the thermal slab (including infilled joints) is 2.75 kN/m. The slab is to be used at the ground floor using exposure class XC3.

Table 7.8 Section properties, prestress and moments of resistance for thermal slab unit in Figure 7.13 with 500 mm wide hole in top flange

	Basic unit without hole	*Values at hole without section distribution*	*Values at hole with section distribution over 3 no. units*
Section properties			
A_c (mm²)	102864	77864	
$I_c \times 10^6$ (mm⁴)	496.1	404.5	
y_b (mm)	173.9	157.4	
$Z_b \times 10^6$ (mm³)	2.854	2.569	
$Z_t \times 10^6$ (mm³)	6.515	4.369	
Compound values			
$I_{c,co} \times 10^6$ (mm⁴)	510.3	415.4	
$Z_{b,co} \times 10^6$ (mm³)	2.953	2.659	2.855
$Z_{t,co} \times 10^6$ (mm³)	6.611	4.429	
Prestress at the hole			
σ_b (N/mm²)	10.67	11.01	6.44
σ_t (N/mm²)	–2.02	–2.61	
Moments of resistance			
Service M_{sR} (kNm)	31.50	29.28	30.75
Ultimate M_{Rd} (kNm)	57.29	56.43	57.00

Further information – the slab is pretensioned using 2 no. 12.5 mm strands at 35 mm bottom cover, stressed to 70%×1770 N/mm². f_{ck}=45 N/mm² and $f_{ck}(t)$=30 N/mm².

Solution

The design moment at the hole due to self-weight $M_{s,0}=2.75\times6.0^2/8=12.375$ kNm

In the slab with the hole, $\sigma_{s,0}=12.375/2.659=4.65$ N/mm²

In the basic slabs, $\sigma_{s,0}=12.375/2.953=4.19$ N/mm²

From Table 7.7, Eq. 7.41. Prestress at the bottom $\sigma_{b,net}=[2\times(10.67-4.19)+(11.01-4.65)]/3=6.44$ N/mm²

From Table 7.7, Eq. 7.42. $Z_{b,co,net}=(2\times2.953+2.659)/3=2.855\times10^6$ mm³

$M_{sR,net}=M_{s,0}+[(\sigma_{b,net}+f_{ctm})\times Z_{b,co,net}]$ where $f_{ctm}=0$ for exposure XC3

$M_{sR,net}=12.375+[(6.44+0)\times2.855]=$ **30.75 kNm** a reduction of only 2.4% from the basic slab $M_{sR}=31.50$ kNm.

7.4 FLEXURAL DESIGN OF PRESTRESSED MAIN BEAMS

The design procedure is identical to the design of prestressed floor units given in Section 7.2 with the additional consideration of satisfying transfer, as well as working, stress conditions. This is because strands can be either deflected or debonded. There is much more freedom in selecting the strand pattern than in floor units as the design of the beam can be optimised (= economy of strands) by choosing a pattern that will simultaneously satisfy transfer at the ends of the beam and working loads at the point of maximum imposed bending moment.

Tables 9.9 and 9.10 give the minimum value of M_{sR} for a range of typical sizes for prestressed inverted-tee beams using f_{ck}=45 and 50 N/mm², respectively. The different strengths are used to demonstrate the potential increases in M_{sR}. Note that the transfer strength in both cases remains $f_{ck}(t)=30$ N/mm². In compiling the tables, the tendons were arranged for case so that the prestress at transfer was in the range $\sigma_b(t)=14.5$ to 15.0 N/mm², values that included a reduction for the self-weight at transfer. Figure 2.35 gives an ultimate beam load *vs.* clear span for a range of 600 mm wide inverted-tee beams using $f_{ck}=45$ N/mm². The tendons were arranged so that $\sigma_b(t)=14.2$ to 14.8 N/mm², also including the top and bottom fibre stresses due to self-weight at transfer. For a typical arrangement of a 6 m span floor slab and internal beam for office loading, the imposed ultimate beam load is about 80 kN/m. Therefore, according to Figure 2.35 the clear span/beam depth ratio is about 14 irrespective of depth.

Example 7.10

Complete the design of the inverted-tee beam in Example 6.2, as shown in Figure 6.10 to determine the critical service moment of resistance. Exposure class is XC1. The flooring is to be used for offices. The beam is simply supported between 300 mm wide columns at 9.0 m centres, and is supported on to 200 mm length corbels. The self-weight of floor slab = 18.0 kN/m, other dead loads = 10.0 kN/m and imposed live load = 30.0 kN/m. Geometric and material data given by the manufacturer are as follows:

$A_c=307.5\times10^3$ mm², $y_b=269.8$ mm, $I_c=8151.5\times10^6$ mm⁴, $Z_b=30.215\times10^6$ mm³, $Z_t=24.685\times10^6$ mm³, $Z_{cp}=87.782\times10^6$ mm³, $f_{ck}=45$ N/mm², $E_{cm}=36283$ N/mm², $A_p=26\times93=2418$ mm² and $z_{cp}=92.86$ mm. Concrete density = 25.0 kN/m³. Gravel aggregates. Cement CEM I grade 52.5R.

Self-weight of beam $=307.5\times10^3\times25\times10^{-6}=7.688$ kN/m

Section properties of compound section with tendons

$m-1=(195000/36283)-1=4.37$

$A_{c,co}=307500+4.37\times2418=318077$ mm², $y_{b,co}=266.7$ mm

$I_{c,co}=8449.7\times10^6$ mm⁴, $Z_{t,co}=25.351\times10^6$ mm³, $Z_{b,co}=31.683\times10^6$ mm³, $Z_{cp,co}=8449.7\times10^6/92.86=90.99\times10^6$ mm³

Solution

The effective span of beam $=9.0-0.3-2\times(0.2+0.2)/2=8.50$ m.

Prestress

From Example 6.2. Initial prestress $\sigma_{pi}=0.7\times1770=1239.0$ N/mm², $\Delta\sigma_{pr}=4.95$ N/mm², $\Delta\sigma_{el}=76.37$ N/mm², $P_{pm0}=2983929$ N and $\sigma_{cp}=12.86$ N/mm²

At mid-span, bending moment due to self-weight $M_{s0}=7.688\times8.50^2/8=69.43$ kNm

Due to M_{s0}. $\sigma_{cp0}=-69.43/90.99=-0.76$ N/mm²

At mid-span $\sigma_{cp}=12.86-0.76=12.10$ N/mm²

$\Delta\sigma_{el}$ at mid-span $=(12.10/12.86)\times76.37=71.84$ N/mm²

From Example 6.2 $\sigma_{pm0}=1157.7$ N/mm² and at mid-span $\sigma_{pm0}=1239.0-4.95-71.84=1162.2$ N/mm² $<0.75\times1770=1328$ N/mm² OK

The loss of prestress at transfer at the support is $100-(100\times1157.7/1239)=6.56\%$ (this is carried forward to Example 11.2)

$P_{pm0}=1157.7\times2418=2799263$ N and at mid-span $P_{pm0}=1162.2\times2418=2810219$ N

(Note that $\sigma_{pm0}=1157.7$ N/mm² is carried forward to Example 10.6 for the calculation of the flexurally uncracked capacity $V_{Rd,c}$.)

Long-term creep losses to life using RH = 50%

h_o = 2 × area/sides and bottom exposed = 2 × 307.5/1400 = 439 mm

$\varphi_{RH} = [1 + \{(1 - (50/100)) \times 0.75/(0.1 \times 439^{1/3})\}] \times 0.92 = 1.373$

$\beta(t_0) = 1/(0.1 + 7.65^{0.2}) = 0.624$

where $t_{oT} = 2.96 \times [9/(2 + 2.96^{1.2}) + 1]^1 = 7.65$ days, factor for CEM 52.5R = 1

$t_T = (20/24)\ e^{-[4000/(273+50)) - 13.65]} = 2.96$ days, $T = 50°C$ mean curing temp.

$\beta(f_{cm}) = 16.8/\sqrt{53} = 2.308$

$\beta_H = (1.5 \times (1 + (0.012 \times 50)^{18}) \times 439.3) + (250 \times 0.81) = 862$ days

$\beta_c\ (t,t_0) = ((20833 - 7.6)/(862 + 20833 - 7.6)^{0.3} = 0.988$

$\varphi(t,t_0) = \varphi_{RH}\ \beta\ (f_{cm})\ \beta\ (t_0)\ \beta_c\ (t,t_0) = 1.373 \times 0.624 \times 2.308 \times 0.988 = 1.954$

$\sigma_{cp} = (2799263/307500) + (2799263 \times 92.86/87.782 \times 10^6) = 12.06$ N/mm²

Denominator in BS EN 1992-1-1, Exp. 5.46 = 1.144
Creep loss $\Delta\sigma_{p,c} = 195000 \times 1.954 \times 12.06/(36283 \times 1.144) = 110.79$ N/mm²
At mid-span, bending moment due to self-weight of beam, floor slabs and other dead loads $M_{s0} = (7.688 + 18.0 + 10.0) \times 8.50^2/8 = 322.3$ kNm
Due to M_{s0}. $\sigma_{cp0} = -322.3/90.99 = -3.54$ N/mm²
At mid-span $\sigma_{cp} = [(2810219/2799263) \times 12.06] - 3.54 = 8.57$ N/mm²
and at mid-span $\Delta\sigma_{p,c} = (8.57/12.06) \times 110.79 = 78.70$ N/mm²

Concrete shrinkage

k_n factor for effective thickness = 0.72

$\beta_{RH} = 1.55 \times (1 - (50/100)^3) = 1.356$

$\beta_{ds}\ (t,t_s) = (20833 - 1)/[(20833 - 1) + 0.04 \times 439.3^{1.5}] = 0.983$

$\varepsilon_{cd} = \beta_{ds}(t,t_s)\ k_n\ \varepsilon_{cd,o} = 0.72 \times 0.85 \times (220 + 110 \times 6)\ e^{-0.11 \times 53/10} \times 1.356 \times 0.983 \times 10^{-6} = 0.000398$

$\Delta\sigma_{p,s} = 0.000398 \times 195000/1.144 = 67.86$ N/mm²

Strand relaxation

$\mu = 1157.7/1770 = 0.654$ and at mid-span $\mu = 1162.2/1770 = 0.657$

$\Delta\sigma_{pr} = 1157.7 \times 0.66 \times 2.5 \times e^{0.654 \times 9.1} \times (500000/1000)^{0.75 \times (1 - 0.654)} \times 10^{-5} = 36.83$ N/mm² and at mid-span $\Delta\sigma_{pr} = 37.4$ N/mm²

$\Delta\sigma_{p,r} = 0.8 \times 36.83/1.144 = 25.77$ N/mm² and at mid-span $\Delta\sigma_{p,r} = 26.16$ N/mm²

$\sigma_{po} = 1157.7 - 110.79 - 67.86 - 25.77 = 953.3$ N/mm² and at mid-span
$\sigma_{po} = 1162.2 - 78.70 - 67.86 - 26.16 = 989.5$ N/mm²

Final force $P_{po} = 953.3 \times 2418 = 2304984$ N. At mid-span = 2392582 N
Residual losses ratio in service $R_s = 953.3/1239 = 0.769$ (23.1% losses). At mid-span ratio $R_s = 0.799$. R_s is smaller at mid-span by a ratio of 0.963, or 3.8%.

The reduction in elastic shortening and creep losses at mid-span increased the prestress there by 3.8%. Note that this is much less than 8.1% in Example 7.1 for the hcu, due mainly to relatively higher level of the strands y_b/h, i.e. 176.9/600 = 0.29 compared with 40.8/250 = 0.16 for the hcu.

Final prestress at mid-span

$\sigma_b = (2392582/307500) + (2392582 \times 92.86/30.215 \times 10^6) = 15.13$ N/mm$^2 < 0.45 \times 45 = 20.25$ N/mm^2

$\sigma_t = (2392582/307500) - (2392582 \times 92.86/24.685 \times 10^6) = -1.22$ N/mm$^2 > -3.80$ N/mm^2

Service moment of resistance

M_{sR} at mid-span is the lesser of

At the bottom fibre. $M_{sR,b} = (15.13 + 3.80) \times 31.683 = 599.74$ kNm

At the top fibre. $M_{sR,t} = (20.25 – 1.22) \times 25.351 =$ **544.28 kNm** < 599.74 kNm

Clause 7.2 (3). $M_{sR,t}$ at the top need only be checked against quasi-permanent service load (using $\psi_2 = 0.3$ for offices) $w_{s,qp} = (7.688 + 18.0 + 10.0 + 0.3 \times 30.0) = 44.69$ kN/m. $M_{s,qp} = 44.69 \times 8.50^2/8 = 403.59$ kNm $\ll 544.28$ kNm, clearly not critical.

Therefore, as per the discussion regarding non-linear creep in Eq. 7.6 and illustrated in Example 7.5, the limiting compressive stress is $0.45\, f_{cm}(t_0) = 0.45 \times (45 + 8) = 23.85$ N/mm^2.

Then $M_{sR,t} = (23.85 - 1.22) \times 25.351 = 635.55$ kNm > 599.74 kNm. Therefore, the critical $M_{sR} =$ **599.74 kNm**

The characteristic design load $w_s = (7.688 + 18.0 + 10.0 + 30.0) = 65.69$ kN/m.

Then $M_s = 65.69 \times 8.50^2/8 =$ **593.24 kNm** < 599.74 kNm

Final prestress at support

$\sigma_b = (2304984/307500) + (2304984 \times 92.86/30215) = 14.58$ N/mm$^2 < 0.45 \times 45 = 20.25$ N/mm^2

$\sigma_t = ((2304984/307500) + (2304984 \times 92.86/24685) = –1.17$ N/mm$^2 > –3.80$ N/mm^2

M_{sR} at the support is the lesser of

$M_{sR,b}$ (bottom stress) $= (14.58 + 3.80) \times 31.683 =$ **582.2 kNm**

$M_{sR,t}$ (top stress) $= (23.85 - 1.17) \times 25.351 = 634.4$ kNm

The distribution of M_{sR} between support and mid-span is taken as parabolic to mimic the parabolic distribution of bending stress due to self-weight and dead loads.

M_{sR} at the ends of the unit $M_{sR,b} = f_{ctm}\, Z_{b,co} = 3.80 \times 31.683 =$ **120.4 kNm**

Between the end of the span and $x = l_{pt}$ the increase in $\sigma_{b,x}$ is per Eq. 7.39. At $x = l_{pt} = 0.680$ m (see Section 10.2.2 for the derivation of l_{pt}) $\sigma_{b,x} = 14.57$ N/mm^2. From Eq. 7.40, $M_{sR,x} = (\sigma_{b,x} + f_{ctm})\, Z_{b,co} = 581.9$ kNm.

Figure 7.24 shows the complete bending moment diagrams for the characteristic design moment M_s and M_{sR}. The design is completed for ultimate moment, shear forces and deflections in Examples 9.5, 10.6 and 11.2.

If the exposure class is XC3, $M_{sR,b} = (15.13 + 0) \times 31.683 =$ **479.48 kNm** and, as before, $M_{sR,t} = 544.28$ kNm > 479.48 kNm. M_{sR} is also greater than $M_{s,qp} = 403.59$ kNm by a margin of 1.19, clearly not critical compared with exposure class XC1.

Example 7.11

Repeat Example 7.10 by using debonded strands up to a distance of ¼ span to determine the maximum moment of resistance for the beam.

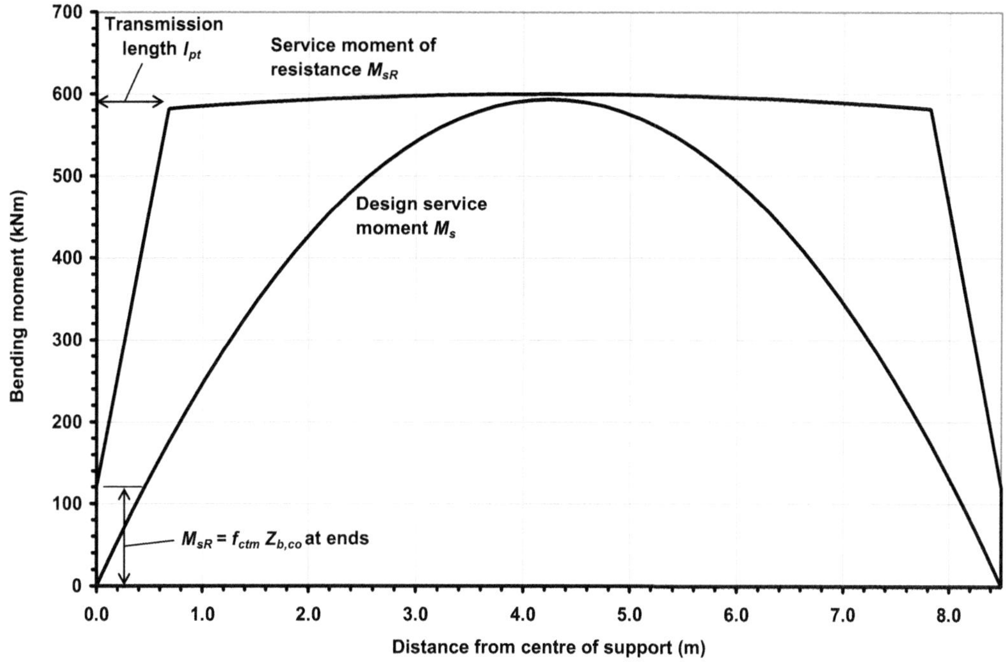

Figure 7.24 Bending moment diagrams for design service and moment of resistance for the inverted-tee beam in Example 7.10.

Solution

The allowable prestress at the bottom of the beam at the point where the strands are debonded is 0.6 f_{ck}(t) = 18.00 N/mm^2.

The bending stress due to the self-weight of the beam at x = ¼ span $\sigma_{s0,x} = M_{s0,x}/Z_{b,co}$, where $M_{s0,x}$ = 0.75* x mid-span moment of 69.43 = 52.07 kNm. Then $\sigma_{s0,x}$ = 52.07/31.683 = 1.64 N/mm^2. ($Z_{b,co}$ is from Example 7.10.) *For UDL only the bending moment at ¼ span is ¾ of the moment at mid-span.

Therefore, the maximum permitted prestress in the bottom of the beam = 18.00 + 1.64 = 19.64 N/mm^2.

Due to complex relationships between initial prestress, losses and eccentricity, it is not possible to back calculate the required arrangement of strand, and therefore the final solution is a forward calculation based on the above assumptions.

Using 29 no. strands, trial the following

No. strands in each row	*Distance y_i from bottom (mm)*	*$\Sigma N_i y_i$ (mm)*
2	550	1100
2	350	700
4	250	1000
4 (add 2)	200	800
4	150	600
5 (add 1)	100	500
8	50	400
Σ = 29		Σ = 5100

Then y_s=5100/29=175.86 mm
The actual eccentricity $z_{cp}=y_b-y_s$=269.78−175.86=93.92 mm
A_p=29×93=2697 mm²
Z_{cp}=8151.5×10⁶/93.92=86.79×10⁶ mm³

Compound section properties increase to $I_{c,co}$=8464.5×10⁶mm⁴, $Z_{t,co}$=25.367×10⁶mm³, $Z_{b,co}$=31.784×10⁶mm³, $Z_{cp,co}$=8464.5×10⁶/93.92=90.12×10⁶mm³

From Example 6.2. σ_{pi}=0.7×1770=1239.0 N/mm², $\Delta\sigma_{pr}$=4.95 N/mm²
P_r=(1239.0−4.95)×2697×10⁻³=3328229 N
At the support σ_{cp}=(3328229/307500)+(3328229×93.92/86.79×10⁶)=14.43 N/mm²
$\Delta\sigma_{el}$=195000×14.43/32837=85.66 N/mm²
σ_{pm0}=1239.0 − 4.95−85.66=1148.4 N/mm²
P_{pm0} =1148.4×2697=3097193 N

Transfer stress check at end of transmission length l_{pt}

Exp. 8.16 l_{pt}=0.19×1148.4×12.5/4.06=671 mm
Exp. 8.15 where f_{bpt}=3.2×1.0×0.7×2.72/1.5=4.06 N/mm²
Moment due to self-weight at $x=l_{pt}$ from end. $M_{s0,x}$=7.688×8.50×0.671/2−7.688×0.671²/2=20.19 kNm
Due to $M_{s0,x}$. σ_{bx}=−20.19/31.784=−0.64 N/mm²
σ_{tx}=20.19/25.367=0.80 N/mm²
At end of transmission l_{pt}. σ_b(t)=(3097193/307500)+(3097193×93.92/30.215×10⁶)−0.64=19.06 N/mm²>18.00 N/mm². Debonding required.
σ_t(t)=(3097193/307500)−(3097193×93.92/24.685×10⁶) +0.80=−0.92 N/mm²>−0.5 f_{ctm}(t)=−0.5×2.72 (from Example 6.2)=−1.36 N/mm². Debonding not required.
Later, check the service stresses at ¼ span when the full service load is determined from M_{sR}.
At mid-span.
Moment due to self-weight M_{s0}=69.43 kNm
Due to M_{s0}. σ_{cp0}=−69.43/90.12=−0.77 N/mm²
At mid-span σ_{cp}=14.43−0.77=13.65 N/mm²
$\Delta\sigma_{el}$=195000×13.65/32837=81.09 N/mm²
σ_{pm0}=1239.0−4.95−81.09=1153.0 N/mm²<0.75×1770=1328 N/mm² OK
P_{pm0} =1153.0×2697=3109531 N

Long-term creep losses to life using RH=50%

From Example 7.10. $\varphi(t,t_0)$=1.954
Due to M_{s0} for self and dead loads. σ_{cp0}=−322.3/90.12=−3.58 N/mm²
σ_{cp}=(3109531/307500)+(3109531×93.92/86.79×10⁶)−3.58=9.90 N/mm²
Denominator in Exp. 5.46=1.161
Creep loss $\Delta\sigma_{p,c}$=195000×1.954×9.90/(36283×1.161)=89.55 N/mm²

Concrete shrinkage

From Example 7.10. ε_{cd}=0.000398
$\Delta\sigma_{p,s}$=0.000398×195000/1.161=66.84 N/mm²

Strand relaxation

μ=1153.0/1770=0.651, σ_{pr}=36.25 N/mm² and $\Delta\sigma_{p,r}$=24.98 N/mm²

σ_{po}=1153.0−89.55−66.84−24.98=971.6 N/mm²
Final force P_{po}=971.6×2697=2620639 N
Residual losses ratio in service R_s=971.6/1239=0.784 (21.6% losses)

Final prestress at mid-span

σ_b=(2620639/307500)+(2620639×93.92/30.215×10⁶)=16.67 N/mm²<0.45×45=20.25 N/mm²

σ_t=(2620639/307500)−(2620639×93.92/24.685×10⁶)=−1.45 N/mm²>−3.80 N/mm²

Service moment of resistance

M_{sR} at mid-span is the lesser of

At the bottom fibre. $M_{sR,b}$=(16.67+3.80)×31.784=650.38 kNm

As per Example 7.10, the limiting compressive stress at the top is 0.45 $f_{cm}(t_0)$=23.85 N/mm².

$M_{sR,t}$=(23.85–1.45)×25.367=**641.73 kNm**<650.38 kNm.

To back-calculate the service live load. w_s=8×641.73/8.50² =71.05 kN/m

Then q_k=71.05−(7.688+18.0+10.0)=**35.37 kN/m**

Also at the top fibre. $M_{sR,t}$=(20.25−1.45)×25.367=550.46 kNm

Clause 7.2 (3). $M_{sR,t}$ is checked against $M_{s,qp}$ where $w_{s,qp}$=7.688+18.0+10.0+0.3×35.37=46.30 kN/m

Then $M_{s,qp}$=46.30×8.50²/8=418.13 kNm<550.46 kNm, clearly not critical.

Finally, check the service moment at ¼ span against M_{sR}=582.2 kNm from Example 7.10 having 26 no. strands.

M_s=0.75×71.05×8.50²/8=481.2 kNm<582.2 kNm. OK.

The distance to the end of the debonding sheathes=$L/4-l_{pt}$=8500/4−671=1454 mm.

Figure 6.10 shows the positions of the 3 no. debonded strands.

REFERENCES

Bhatt, P. 2011. *Prestressed Concrete Design to Eurocodes*, Routledge, Oxford, UK, 596 p.

BS 8110-2. 1985. British Standards Institute, BS8110, Structural Use of Concrete - Part 2: Code of Practice for special circumstances, BSI, London, UK.

BS EN 1168. 2005. Precast Concrete Products - Prestressed concrete hollow core units, +A3:2011, BSI, London, UK.

BS EN 1990. 2002. Eurocode 0: Basis of structural design, +A1:2005, BSI, London, UK.

BS EN 1992-1-1. 2004. Eurocode 2: Design of concrete structures - Part 1-1: General rules and rules for buildings, +A1:2014, BSI, London, UK.

Gilbert, R. I., Mickleborough, N. C. and Ranzi, G. 2017. *Design of Prestressed Concrete to Eurocode 2*, 2nd ed., CRC, Taylor & Francis Group, Boca Raton, FL, 665 p.

NA to BS EN 1992-1-1. 2004. UK National Annex for Eurocode 2: Design of concrete structures - Part 1-1: General rules and rules for buildings, +A1:2014, BSI, London, UK.

Narayanan, R. S. 2008. *Precast Eurocode 2: Worked Examples*, British Precast Concrete Federation, Leicester, UK, 77 p.

Neville, A. 1999. *Properties of Concrete*, 4th ed., Longman, London, UK, 844 p.

Chapter 8

Design for bending moments using the Magnel diagrams

8.1 BACKGROUND TO THE MAGNEL DIAGRAMS

In the design of prestressed elements, chiefly beams and slabs, the limiting stresses in the top and bottom fibres of the section are checked against prestress at transfer and in-service plus the bending stresses due to the minimum $M_{s,min}$ and maximum $M_{s,max}$ service moments at critical sections along the span of the element. In Chapter 6, Figure 6.1c shows the requirements at transfer at section A, where it is explained in Section 6.1 that this plane is critical at the end of the basic transmission length l_{pt} (measured from the end of the element). Figures 6.1c and 7.10 show the requirements in service at section B, i.e. at the point of maximum service moment. In the worked examples in Chapters 6 and 7, the prestressing data are given, and the task was to determine the service moment of resistance M_{sR} based on the top and bottom fibre stresses σ_t and σ_b, and to check the prestress at transfer $\sigma_t(t)$ and $\sigma_b(t)$. Thus, there are four fibre stresses to determine.

This is generally the approach of in-house precast concrete manufacturers who hold a database of predetermined arrays of tendons for a range of prestressed beams and slabs, typically in incremental steps of 50 mm for depth and/or width. In other cases, the design procedure is carried out in reverse, in which σ_t and σ_b etc. are determined from the given values of $M_{s,min}$ and $M_{s,max}$ and the limiting stresses shown in Figure 6.1c. The goal of this procedure is to determine the initial pretensioning force $P_{pi}=\sigma_{pi} A_p$ (Eq. 6.2) and eccentricity z_{cp} of the tendons at critical sections along the span.

If a single value for P_{pi} and z_{cp} is found at planes A and B, the number of tendons is constant along the length of the element, as found in Example 7.10, and no debonding or deflecting of tendons is required. However, Example 7.11 was devised to show where different values for P_{pi} and z_{cp} are required at the positions of $M_{s,max}$ and $M_{s,min}$, with the changeover (end of debonding) taking place at ¼ span. In this case, there are eight fibre stresses to determine – four at transfer at l_{pt} and ¼ span, and four in service at ¼ span and mid-span (or at the position of $M_{s,max}$). It is not always desirable to use the ¼ span position, although this is most commonly used, as it can also be at about $^1/_5$ span.

Calladine (2006) introduces this subject as follows

> The engineer normally assumes that the concrete material is linear-elastic and initially stress free; and thus at every cross-section the longitudinal component of stress always varies linearly with depth in the beam. Consequently, the extreme stresses within a cross-section at any given time can occur only at the top and bottom surfaces of the beam. The stress limits at the extreme bending moments therefore yield in principle four *inequalities* that must be satisfied at every cross-section: two at the top and two at the bottom. If the shape of the cross-section has been (tentatively) decided upon, these four inequalities involve two variables which the engineer must eventually fix: the

DOI: 10.1201/9781003319450-8

prestressing force and the *eccentricity* of the resultant prestress force within the cross-section. As is well known, G. Magnel (Magnel 1948) devised a two-dimensional plot or diagram on which these four inequalities could be displayed simultaneously as four straight lines.

Gustaaf Magnel (1889–1955) was a professor at Ghent University in Belgium, known for his expertise in reinforced concrete and prestressed concrete. Known as the 'Magnel' diagram, the equalities of top and bottom fibre stresses shown in Figure 6.1c may be plotted as straight lines in terms of $1/P_{pi}$ and z_{cp} (keeping the notation of Eurocodes, although most texts show these as $1/P_0$ and e. In some texts the axes are switched with $1/P_0$ as the abscissa.) Adapting and re-arranging Eqs. 5.6, 5.7a and 5.7b at transfer, and Eqs. 6.1, 6.2 and 6.3a,b,c in service, leads to the four equations as $1/P_{pi}=C+m\ z_{cp}$, where C is a constant and gives the intersection of the straight line where $z_{cp}=0$, and m is the gradient of the straight line. Figure 8.1 shows the construction of a typical Magnel diagram, in which the arrows on the straight lines give the permissible zone where the critical values of $1/P_{pi}$ and z_{cp} are found. (Arrows pointing *upwards* means that $1/P_{pi}$ should be greater than the equation of the straight line.) The shaded trapezoidal zone is therefore the 'design zone'. If one zone satisfies the stresses at $M_{s,max}$ and $M_{s,min}$ then one set of values for P_{pi} and z_{cp} is required, and all tendons may run uninterrupted along the entire span.

In Figure 8.1, point A gives the lowest value of P_{pi} (and hence the cheapest solution with the least number of tendons) but the greatest eccentricity z_{cp}, which may not be achievable in certain beams and slabs, particularly where splitting and spalling stresses are a problem. Point B gives the greatest P_{pi} and lowest z_{cp}. The best solution may be somewhere between the two, particularly as the final eccentricity cannot be guaranteed to be exactly at the chosen point as tendons (in beams) are typically positioned in an array of 50 mm increments.

The Magnel diagram may also contain a vertical straight line indicating the maximum achievable value for $z_{cp,max}$. This is found by some prior knowledge of the total number of tendons required, and where the tendons can or must be positioned within the cross section. For example, in beams, most of the tendons will be positioned in the bottom 2 or 3 rows at 50 mm apart, with 2 no. tendons at the top of the beam. For example in a 600×600 mm wide beam, 11 no. tendons will be positioned at 50 mm from the bottom, 11 no. at 100 mm and 2 no. at 550 mm. Then the centroid $y_s=114.6$ mm and $z_{cp,max}=300-114.6=185.4$ mm. This vertical straight line may/not cut across the design zone.

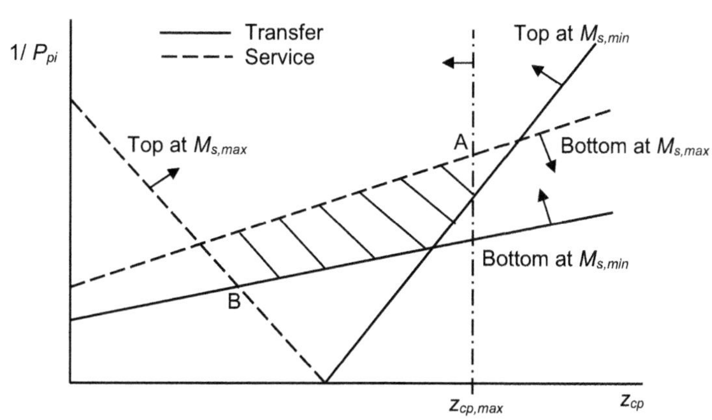

Figure 8.1 Construction of the Magnel diagram for a single solution to P_{pi} and z_{cp} at transfer and in service.

Figure 8.2 shows a possible situation where there is no single design zone. In this case, $M_{s,max}$ is greater and this drags down the dashed line representing the stresses at the bottom of the section to below the solid line for the stresses at transfer. In this case, an intermediate design point is required, at say ¼ span.

The Magnel diagrams look easy to construct, but care has to be taken when setting up the equations of equalities as will be shown in the worked examples, especially when dealing with the units of $1/P_{pi}$. It is also not easy to visualise the various states of stress at the sections, particularly as the y-axis is $1/P_{pi}$ rather than P_{pi} itself, and it is difficult to visualise where an intermediate point (e.g. ¼ span) is introduced. For these reasons, the author prefers the alternative version of the diagram, described in Section 8.4, in which the required prestress σ_t and σ_b are established as a set of stress 'boxes' at $M_{s,min}$, $M_{s,qt}$ at ¼ span and $M_{s,max}$. This method is also much easier to teach to students.

8.2 CALCULATION OF EQUATIONS FOR MAGNEL DIAGRAM

Let

P_{pi}=initial pretensioning force
R_{tr}=ratio of losses at transfer, typically 0.93–0.95
$R_{tr}\ P_{pi}$=pretensioning force at transfer
R_s=ratio of losses in service, typically 0.78–0.82
$R_s\ P_{pi}$=pretensioning force in service

8.2.1 Transfer inequalities: $1/P_{pi} = C + m\ z_{cp}$ equations

8.2.1.1 Transfer stress at the top fibre at l_{pt} from the support

Referring to Figure 6.1c and Eqs. 5.7a and 5.7b

$$R_{tr}\ P_{pi}\ (1/A_c - z_{cp}/Z_t) + M_{sw}/Z_{t,co} \geq -f_{ctm}(t) \quad (8.1a)$$

where $M_{sw} = w_0\ L\ l_{pt}\ /\ 2 - w_0 l_{pt}^2\ /\ 2$ at l_{pt} from the support
or $M_{sw}=0.75\ w_0 L^2/8$ at ¼ span

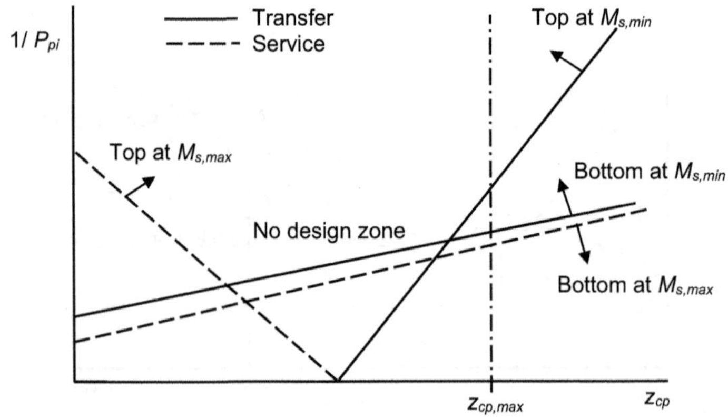

Figure 8.2 Construction of the Magnel diagram where there is no single solution to P_{pi} and z_{cp}.

w_0=self-weight of element

Note when calculating bending stresses, the compound section modulus $Z_{t,co}$ is used (see Section 6.2.4). Because this is not known at the start of the calculation, the basic section modulus may be multiplied by 1.05.

$$\text{Then } R_{tr}\, P_{pi}\, (1/A_c - z_{cp}/Z_t) \geq (-f_{ctm}(t) - M_{sw}/Z_{t,co}) \quad (8.2)$$

$$P_{pi} \geq (-f_{ctm}(t) - M_{sw}/Z_{t,co})/[R_{tr}\, (1/A_c - z_{cp}/Z_t)] \quad (8.3)$$

$$\text{and } \; 1/P_{pi} \leq R_{tr}\, (1/A_c - z_{cp}/Z_t)/(-f_{ctm}(t) - M_{sw}/Z_{t,co}) \quad (8.4)$$

However, note that the denominator $(-f_{ctm}(t) - M_{sw}/Z_{t,co})$ will be negative, then the equality $1/P_{pi} \leq$ changes to $1/P_{pi} \geq$.

$$\text{Then } 1/P_{pi} \geq R_{tr}\, (1/A_c - z_{cp}/Z_t)/(-f_{ctm}(t) - M_{sw}/Z_{t,co}) \quad (8.5)$$

At l_{pt} from the support this straight line at the top fibre is termed...(1t)
At ¼ span from the support ditto...(2t)

Example 8.1

To help visualise the diagram and position lines 1t and 2t, consider a 300 mm wide × 400 mm depth beam, using R_{tr}=0.93, $f_{ck}(t)$=30 N/mm², $f_{ctm}(t)$=2.72 N/mm² and M_{sw}=4.86 kNm. Section properties are: A_c=120×10³mm², Z_b=Z_t=8.00×10⁶mm³, $Z_{b,co}$=$Z_{t,co}$=1.05 Z=8.40×10⁶mm³. Maximum eccentricity is based on 5 no. tendons at 50 mm and 5 no. at 100 mm from the soffit, giving $z_{cp,max}$=200−75=125 mm.

Eq. 8.5 Denominator $(-f_{ctm}(t) - M_{sw}/Z_{t,co}) = -2.72 - (4.86/8.40) = -3.30$ N/mm².
$C = [0.93\,(1/120\times10^3)]/-3.30 = -2.349\times10^{-6}$
$m = -[0.93/8.00\times10^6]/-3.30 = 0.0352\times10^{-6}$
Then $1/P_{pi} \geq -2.349\times10^{-6} + 0.0352\times10^{-6}\, z_{cp}$ (units in 1/N)

Converting to kN and expressing as $1/P_{pi}\times10^3$

$10^3/P_{pi} \geq -2.349 + 0.0352\, z_{cp}$
where $10^3/P_{pi} = 0$, $z_{cp} = +2.349/0.0352 = 66.7$ mm
where $z_{cp,max} = 125$ mm, $1/P_{pi} \geq -2.349 + 0.0352\times125 = 2.055\times10^{-3}$ kN^{-1}

8.2.1.2 Transfer stress at the bottom fibre at l_{pt} from the support

Referring to Figure 6.1c and Eq. 6.6

$$R_{tr}\, P_{pi}\, (1/A_c + z_{cp}/Z_b) - M_{sw}/Z_{b,co} \leq 0.6\, f_{ck}(t) \quad (8.6)$$

$$\text{then } R_{tr}\, P_{pi}\, (1/A_c + z_{cp}/Z_b) \leq (0.6\, f_{ck}(t) + M_{sw}/Z_{b,co}) \quad (8.7)$$

$$P_{pi} \leq (0.6\, f_{ck}(t) + M_{sw}/Z_{b,co})/[R_{tr}\, (1/A_c + z_{cp}/Z_b)] \quad (8.8)$$

$$\text{and } \; 1/P_{pi} \geq R_{tr}\, (1/A_c + z_{cp}/Z_b)/(0.6\, f_{ck}(t) + M_{sw}/Z_{b,co}) \quad (8.9)$$

At l_{pt} from the support, this straight line at the bottom fibre is termed...(1b)

At ¼ span from the support ditto…(2b)

Continuing Example 8.1

Eq. 8.9 Denominator $(0.6\ f_{ck}(t)+M_{sw}/Z_{b,co})=18.00+(4.86/8.40)=18.58$ N/mm².
$C=[0.93\ (1/120\times10^3)]/18.58=0.4171\times10^{-6}$
$m=-[0.93/8.00\times10^6]/18.58=0.00626\times10^{-6}$
Then $1/P_{pi}\geq0.4171\times10^{-6}+0.00626\times10^{-6}\ z_{cp}$ (units in 1/N)

Converting to kN and expressing as $1/P_{pi}\times10^3$

$10^3/P_{pi}\geq0.4171+0.00626\ z_{cp}$
where $10^3/P_{pi}=0$, $z_{cp}=-0.4171/0.00626=-66.7$ mm
where $z_{cp,max}=125$ mm, $1/P_{pi}\geq0.4171+0.00626\times125=1.199\times10^{-3}$ kN^{-1}
The straight lines for Exp. 1b and 1t are shown in Figure 8.3.

8.2.2 Service inequalities

8.2.2.1 Service stress at the top fibre at ¼ and mid-span**

Referring to Figure 6.1c and Eqs. 5.7a and 5.7b

$$R_s\ P_{pi}\ (1/A_c-z_{cp}/Z_t)+M_{s,max}{}^*/Z_{t,co}\leq0.45\ f_{cm} \qquad (8.10)$$

$$\text{Then } R_s\ P_{pi}\ (1/A_c-z_{cp}/Z_t)\leq(0.45\ f_{cm}-M_{s,max}/Z_{t,co}) \qquad (8.11)$$

$$P_{pi}\leq(0.45\ f_{cm}-M_{s,max}/Z_{t,co})/[R_s\ (1/A_c-z_{cp}/Z_t)] \qquad (8.12)$$

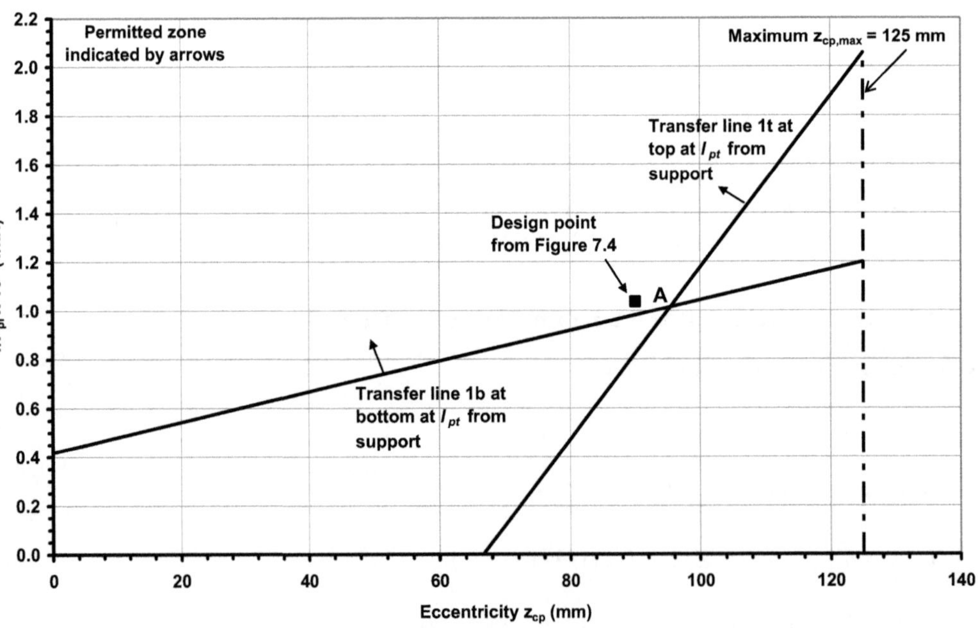

Figure 8.3 Straight lines 1t (top fibre) and 2b (bottom fibre) in the Magnel diagram for the transfer stresses in Example 8.1 at l_{pt} from the end of beam.

and $1/P_{pi} \geq R_s\ (1/A_c - z_{cp}/Z_t)/(0.45\ f_{cm} - M_{s,max}/Z_{t,co})$ (8.13)

At ¼ span this straight line at the top fibre is termed...(3t)
At mid-span** ditto...(4t)

*$M_{s,max}$ is either the frequent or quasi-permanent moment for exposure classes>XC1.
** at mid-span or the point of maximum moment

Continuing Example 8.1

Using $R_s = 0.80$, $f_{ck} = 45$ N/mm², $f_{cm} = 53$ N/mm², $f_{ctm} = 3.80$ N/mm² and $M_{s,max} = 157.5$ kNm at mid-span.

Eq. 8.13 Denominator $0.45\ f_{cm} - M_{s,max}/Z_{t,co} = 23.85 - (157.5/8.40) = +5.10$ N/mm².
$C = [0.80\ (1/120 \times 10^3)]/5.10 = 1.3072 \times 10^{-6}$
$m = -[0.80/8.00 \times 10^6]/5.10 = -0.0196 \times 10^{-6}$
Then $1/P_{pi} \geq 1.3072 \times 10^{-6} - 0.0196 \times 10^{-6}\ z_{cp}$ (units in 1/N)

Converting to kN and expressing as $1/P_{pi} \times 10^3$

$10^3/P_{pi} \geq 1.3072 - 0.0196\ z_{cp}$
where $10^3/P_{pi} = 0$, $z_{cp} = -1.3072/-0.0196 = 66.7$ mm
where $z_{cp,max} = 125$ mm, $1/P_{pi} \geq 1.3072 - 0.0196 \times 125 = -1.144 \times 10^{-3}$ kN^{-1}

8.2.2.2 Service stress at the bottom fibre

Referring to Figure 6.1c and Eq. 6.6

$R_s\ P_{pi}\ (1/A_c + z_{cp}/Z_b) - M_{s,max}{}^*/Z_{b,co} \geq -f_{ctm}{}^*$ (8.14)

then $R_s\ P_{pi}\ (1/A_c + z_{cp}/Z_b) \geq (-f_{ctm} + M_{s,max}/Z_{b,co})$ (8.15)

$P_{pi} \geq (-f_{ctm} + M_{s,max}/Z_{b,co})/[R_s\ (1/A_c + z_{cp}/Z_b)]$ (8.16)

and $1/P_{pi} \leq R_s\ (1/A_c + z_{cp}/Z_b)/(-f_{ctm} + M_{s,max}/Z_{b,co})$ (8.17)

Note that this is the only term where $1/P_{pi}$ is less than the equation.

At ¼ span this straight line at the bottom fibre is termed...(3b)
At mid-span ditto...(4b)

*$f_{ctm} = 0$ and $M_{s,max}$ is either the frequent or quasi-permanent moment for exposure classes>XC1

Completing Example 8.1

Eq. 8.17 Denominator $-f_{ctm} + M_{s,max}/Z_{b,co} = -3.80 + (157.5/8.40) = 14.95$ N/mm².

$C = [0.80\ (1/120 \times 10^3)]/14.95 = 0.4458 \times 10^{-6}$
$m = [0.80/8.00 \times 10^6]/14.95 = 0.00669 \times 10^{-6}$
Then $1/P_{pi} \leq 0.4458 \times 10^{-6} + 0.00669 \times 10^{-6}\ z_{cp}$ (units in 1/N)

Converting to kN and expressing as $1/P_{pi}\times10^3$

$10^3/P_{pi}\leq0.4458+0.00669\ z_{cp}$
where $10^3/P_{pi}=0$, $z_{cp}=-0.4458/0.00669=-66.7\,\text{mm}$
where $z_{cp,max}=125\,\text{mm}$, $1/P_{pi}\geq0.4458+0.00669\times125=1.282\times10^{-3}\ \text{kN}^{-1}$

The straight lines for Exp. 4t and 4b are shown in Figure 8.4. The (shaded) design zone is bound by the four lines, limited chiefly by the service stress at the bottom fibre at mid-span (line 4b) and the transfer stress (lines 1t and 1b). The least value of P_{pi}, and hence the least number of tendons, is at point B where $z_{cp}=97.9\,\text{mm}$ and $P_{pi}=908.7$ kN. If the initial pretensioning force per 9.3 mm diameter strand is $0.7\times1770\times52\times10^{-3}=64.4$ kN, the number of strands is $N\geq14.1$. Rounding up to $N=15$ gives $z_{cp}=88.1\,\text{mm}$ and $P_{pi}=966.4$ kN. Using an array of tendons at 50 mm increments, this is best achieved using 5, 4, 4 and 2 strands at 50, 100, 150 and 200 mm above the soffit, giving $z_{cp}=90.0\,\text{mm}$. This final 'design point' is shown in Figure 8.4, and also shown in Figure 8.3 satisfying transfer stresses close to the intersection of lines 1t and 1b at point A.

The calculation to check the losses of prestress at transfer and in service, *viz.* R_{tr} and R_s, and the assumption that $Z_{b,co}=1.05\ Z_b$, etc. together with the final stress inequalities, may be carried out as per the design procedures given in Chapters 6 and 7. For this beam, it is found that $R_{tr}=0.936$ (0.93 assumed), $R_s=0.787$ (0.80), $Z_{b,co}/Z_b=1.04$ (1.05) and $Z_{t,co}/Z_t=1.02$ (1.05). Inserting these values into the Magnel diagram point B reduces z_{cp} by 0.9 mm and increases P_{pi} by 3% to 941 kN, or $N=14.6$ strands, the major contributor to this difference being R_s (accounting for 2% of the increase). It may therefore be prudent to slightly overestimate the losses in service. However, none of these corrections changed the final design input data in this example.

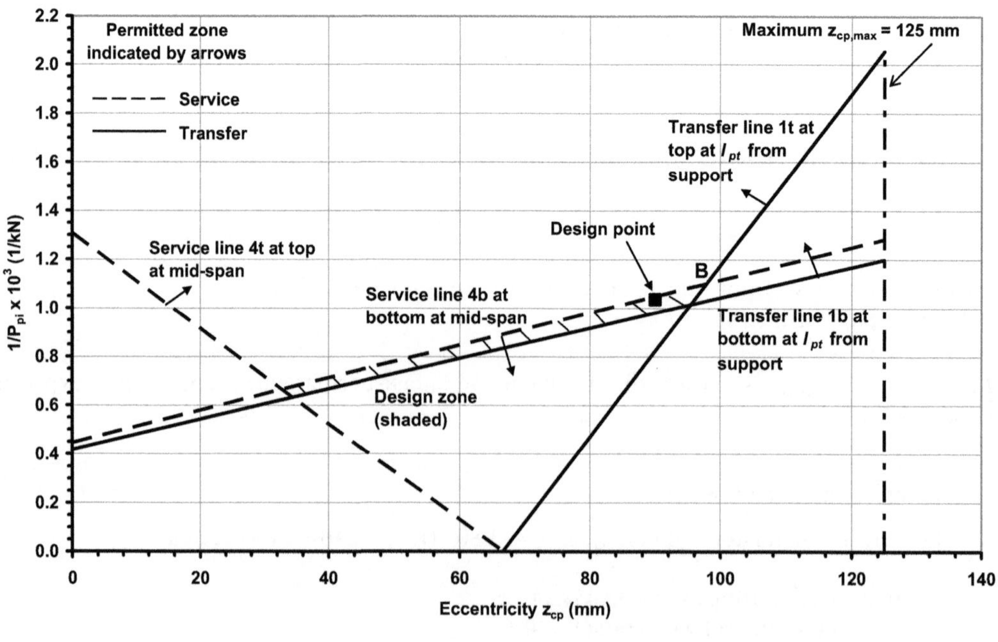

Figure 8.4 Straight lines 4t (top fibre) and 4b (bottom fibre) at mid-span, together with 1t and 2b at transfer, in the Magnel diagram for the service stresses in Example 8.1.

8.3 CALLADINE'S IMPROVEMENT OF THE MAGNEL DIAGRAM

8.3.1 Background to Calladine's work

One of the main problems in using the traditional Magnel diagram, shown in Figures 8.2 and 8.4, is the very narrow design zone (shaded area) due to the similar gradient of lines 1b (transfer) and 4b (service), particularly in rectangular beams where $Z_b = Z_t$.

Calladine (Calladine, 2006) proposed an improvement by defining a 'new' dimensionless ordinate $P_{pi}\ z_{cp}/Z_t\ \sigma_{max}$ (previously $1/P_{pi}$), where σ_{max} is the maximum compressive stress allowed, i.e. 0.6 $f_{ck}(t)$ at transfer and 0.45 f_{cm} in service, and defining a 'new' dimensionless abscissa $p = P_{pi}/A\ \sigma_{max}$ (previously z_{cp}). In this book, the dimensionless ordinate is given as $y = P_{pi}\ z_{cp}/Z_t\ \sigma_{max}$. Although the arithmetic required to derive the four inequalities is quite tedious, the final equations are simple, but not particularly easy to understand. In Calladine's paper, the allowable tensile stresses f_{ctm} and $f_{ctm}(t)$ were both set to zero, and the transfer condition was not included, making the equations appear even more simple. Note that in deriving P_{pi} from the new diagram, Calladine did not use the residual prestress ratios R_{tr} and R_s, nor the compound section modulus $Z_{t,co}$, which gives a small, but sometimes not insignificant, reduction in P_{pi}. However, the result, shown in Figure 8.5, gives a design zone from which it is easy to find a design point. The derivation of the four equalities is as follows, given in the same sequence as before. This is followed by Example 8.2, a new version but a repeat of the beam in Example 8.1.

In Figure 8.5 point A gives the lowest value of P_{pi} but the greatest eccentricity z_{cp}, which may not be achievable in certain beams and slabs. The limiting maximum eccentricity $z_{cp,max}$ may or may not be critical, usually it is not. Point B gives the greatest P_{pi} and lowest z_{cp}.

Calladine's dimensionless notations are

$$Q = Z_b/Z_t \tag{8.18}$$

$$m^- = M_{min}/Z_t\ \sigma_{max} \text{ at transfer} \tag{8.19a}$$

$$m^+ = M_{max}/Z_t\ \sigma_{max} \text{ in service} \tag{8.19b}$$

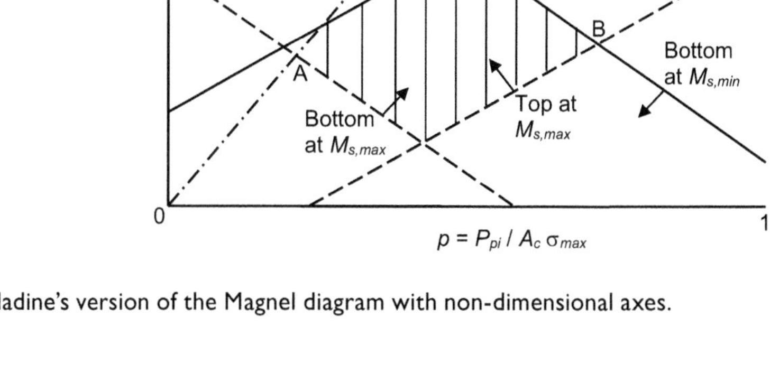

Figure 8.5 Calladine's version of the Magnel diagram with non-dimensional axes.

$$p = P_{pi}/A\ \sigma_{max} \text{ at transfer} \quad (8.20a)$$

$$p = P_{pi}/A\ \sigma_{max} \text{ in service} \quad (8.20b)$$

To use Calladine's method, a single value of σ_{max} has to be chosen to satisfy the four inequalities at the same time. In the following analysis, this is chosen as the maximum compression at the top of the section, i.e. $\sigma_{max} = 0.45\ f_{cm}$. Also $Q = Z_{b,co}/Z_{t,co}$ because both compound values are assumed to be factored by the same amount, in this case 1.05.

8.3.2 Transfer inequalities

8.3.2.1 Transfer stress at the top fibre

Referring to Figure 6.1c and Eqs. 5.7a and 5.7b, and from Eq. 8.1a

$$R_{tr}\ P_{pi}/A_c - R_{tr}\ P_{pi}\ z_{cp}/Z_t + M_{min}/Z_{t,co} \geq -f_{ctm}(t) \quad (8.1b)$$

where M_{min} is the moment M_{sw} at the end of the transfer length

Then, dividing through by 0.45 f_{cm}

$$R_{tr}\ p - R_{tr}\ y \geq -m^- - f_{ctm}(t)/0.45\ f_{cm} \quad (8.21)$$

$$\text{Finally} \quad y \leq (m^- + f_{ctm}(t)/0.45\ f_{cm})/R_{tr} + p \quad (8.22)$$

The gradient of the line in Callendine's diagram is +1 and the intercept with the y-axis is $(m^+ + f_{ctm}(t)/0.45\ f_{cm})/R_{tr}$. Using the same notation as in Figure 8.4, the line is called 1t (1 for near to support and t for the top of the section).

8.3.2.2 Transfer stress at the bottom fibre

Referring to Figure 6.1c and Eq. 6.6

$$R_{tr}\ P_{pi}/A_c + R_{tr}\ P_{pi}\ z_{cp}/Z_b - M_{sw}/Z_{b,co} \leq 0.6\ f_{ck}(t) \quad (8.6)$$

Then, multiplying through by $Q/0.45\ f_{cm}$

$$R_{tr}\ Q\ p + R_{tr}\ y - m- \leq Q\ 0.6\ f_{ck}(t)/0.45\ f_{cm} \quad (8.23)$$

where $m- = M_{min}/Z_{t,co}\ \sigma_{max}$

Multiplying by Q converts Z_b to Z_t and converts $Z_{b,co}$ to $Z_{t,co}$

$$\text{Finally} \quad y \leq (m^- + Q\ 0.6\ f_{ck}(t)/0.45\ f_{cm})/R_{tr} - Q\ p \quad (8.24)$$

The gradient of the line 1b is -Q and the intercept with the y-axis is $(m^+ + Q\ 0.6\ f_{ck}(t)/0.45\ f_{cm})/R_{tr}$.

8.3.3 Service inequalities

8.3.3.1 Service stress at the top fibre

Referring to Figure 6.1c and Eqs. 5.7a and 5.7b

$$R_s\, P_{pi}/A_c - R_s\, P_{pi}\, z_{cp}/Z_t + M_{s,max}/Z_{t,co} \leq 0.45\, f_{cm} \tag{8.10}$$

Then, dividing through by 0.45 f_{cm}

$$R_s\, p - R_s\, y + m^+ \leq 1 \tag{8.25}$$

where $p = P_{pi}/A_c\, \sigma_{max}$
and $y = P_{pi}\, z_{cp}/Z_t\, \sigma_{max}$
and $m^+ = M_{max}/Z_{t,co}\, \sigma_{max}$

x by –1 and rearrange

$$R_s\, y \geq (m^+ - 1) + R_s\, p \tag{8.26}$$

$$\text{Finally} \quad y \geq (m^+ - 1)/R_s + p \tag{8.27}$$

The gradient of the line is +1 and the intercept with the y-axis is $(m^+ - 1)/R_s$. Using the same notation as in Figure 8.4, the line is called 4t (4 for mid-span and t for top of section).

8.3.3.2 Service stress at the bottom fibre

Referring to Figure 6.1c and Eq. 6.6

$$R_s\, P_{pi}/A_c + R_s\, P_{pi}\, z_{cp}/Z_b - M_{s,max}/Z_{b,co} \geq -f_{ctm} \tag{8.14}$$

Then, multiplying through by Q/0.45 f_{cm}

$$R_s\, Q\, p + R_s\, y - m^+ \geq -\, Q\, f_{ctm}/0.45\, f_{cm} \tag{8.28}$$

where $m^+ = M_{max}/Z_{t,co}\, \sigma_{max}$

$$\text{Finally} \quad y \geq (m^+ - Q\, f_{ctm}/0.45\, f_{cm})/R_s - Q\, p \tag{8.29}$$

The gradient of the line 4b is $-Q$ and the intercept with the y-axis is $(m^+ - Q\, f_{ctm}/0.45\, f_{cm})/R_s$.

Of course the common denominator σ_{max} can take any value because it is present in all four equations. If it is omitted or set equal to unity, then y and p take the form of prestresses $P_{pi}\, z_{cp}/Z_t$ and P_{pi}/A_c, respectively, and the Magnel diagram shown in Figure 8.6 will take the same form, but with larger scales.

The gradient of Calladine's diagram is $z_{cp}\, A_c/Z_t$. Therefore, if the maximum eccentricity $z_{cp,max}$ is known a straight line may be plotted with the gradient of $z_{cp,max}\, A_c/Z_t$ below which lies the permissible design zone.

Example 8.2 - continuing Example 8.1

Draw Calladine's Magnel diagram for the beam in Example 8.1.
M_{min} is M_{sw}=4.86 kNm
M_{max} is $M_{s,max}$=157.50 kNm
σ_{max}=0.45 (45+8)=23.85 N/mm^2
m^-=4.86×10^6/(8.40×10^6×23.85)=0.02426
m^+=157.50×10^6/(8.40×10^6×23.85)=0.7862

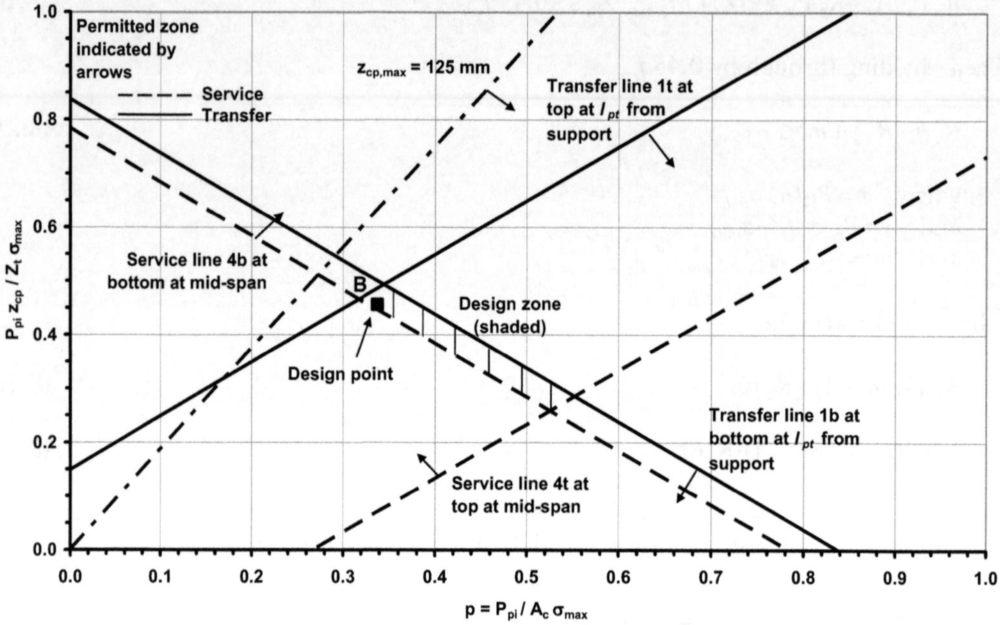

Figure 8.6 Straight lines 1t, 1b, 4t and 4b in Calladine's version of the Magnel diagram for Example 8.2.

$Q = 1.0$
$R_{tr} = 0.93$ and $R_s = 0.80$

Line 1t for transfer at the top of beam

Eq. 8.22 $y \le (m^- + f_{ctm}(t)/0.45\ f_{cm})/R_{tr} + p$
Intercept $(m^- + f_{ctm}(t)/0.45\ f_{cm})/R_{tr} = (0.02426 + 2.72/23.85)/0.93 = 0.1488$
Then $y \le 0.1488 + 1\ p$ ...(1t)

Line 1b for transfer at the bottom of beam

Eq. 8.24 $y \le (m^- + Q\ 0.6\ f_{ck}(\mathrm{t})/0.45\ f_{cm})/R_{tr} - Q\ p$
Intercept $(m^- + Q\ 0.6\ f_{ck}(\mathrm{t})/0.45\ f_{cm})/R_{tr} = (0.02426 + 1.0 \times 18.0/23.85)/0.93 = 0.8376$
Then $y \le 0.8376 - 1\ p$ ...(1b)

Line 4t for service at the top of beam

Eq. 8.27 $y \ge (m^+ - 1)/R_s + p$
Intercept $(m^+ - 1)/R_s = (0.7862 - 1)/0.80 = -0.2673$
Then $y \ge -0.2673 + 1\ p$ ...(4t)

Line 4b for service at the bottom of beam

Eq. 8.29 $y \ge (m^+ - Q\ f_{ctm}/0.45\ f_{cm})/R_s - Q\ p$
Intercept $(m^+ - Q\ f_{ctm}/0.45\ f_{cm})/R_s = (0.7862 - 1.0 \times 3.80/23.85)/0.80 = 0.7838$
Then $y \ge 0.7838 - 1\ p$ ...(4b)

The equation of the $z_{cp,max}$ line is $y = 125 \times 120000/8.0 \times 106 = 1.875\ p$.

The four straight lines and the plot for $z_{cp,max}$ for are shown in Figure 8.6. The (shaded) design zone is bound by the four lines, limited chiefly by the service stress at the bottom fibre at mid-span (line 4b) and the transfer stress (lines 1t and 1b). The least value of P_{pi}, and hence the least number of tendons, is at point B. Then from Exp. 1t=Exp. 4b, $0.1488+p=0.7838-p$. Hence $p=0.3175$ and $P_{pi}=0.3175\times120000\times23.85/103=908.7$ kN (same as in Example 8.1). This gives 14.1 strands, which is rounded to 15 no. From Exp. 1t, $y=0.1488+0.3175=0.4663$. Then $z_{cp}=0.4663\times8.0\times10^6\times23.85/908700=97.9$ mm (same as in Example 8.1). For the same reasons that were given in Example 8.1, the final design point shown in Figure 8.6 has a slightly greater P_{pi} for 15 no. strands and lower $z_{cp}=90.0$ mm.

Example 8.3

Draw the traditional Magnel diagram (i.e. $1/P_{pi}$ vs z_{cp}) for the 600×600 mm wide inverted tee beam shown in Figure 8.9. The loads, span, and materials are the same as in Examples 6.2 and 7.11, except that here the minimum bending moment due to self-weight is taken at the end of the transmission length, assumed as 600 mm. Then $M_{sw}=7.688\times8.50\times0.60/2-7.688\times0.60^2/2=18.22$ kNm.

Solution

First consider the transfer condition and construct lines 1t and 1b near to the support and lines 2t and 2b at ¼ span. Using $R_{tr}=0.93$, $f_{ck}(t)=30$ N/mm², $f_{ctm}(t)=2.72$ N/mm², $M_{sw}=18.22$ kNm near to the support, and $M_{sw,qt}=52.07$ kNm at ¼ span. Section properties are: $A_c=307.5\times10^3$ mm², $Z_b=30.215\times10^6$ mm³, $Z_t=24.685\times10^6$ mm³, $Z_{b,co}=31.726\times10^6$ mm³ and $Z_{t,co}=25.920\times10^6$ mm³. Maximum eccentricity is based on 11 no. tendons at 50 mm and 11 no. at 100 mm from the soffit and 2 no. at 50 mm from the top, giving $z_{cp,max}=155.2$ mm.

Construct line 1t at the top at transfer near to the support

Eq. 8.5 Denominator $(-f_{ctm}(t)-M_{sw}/Z_{t,co})=-2.72-(18.22/25.920)=-3.424$ N/mm².
$C=[0.93\ (1/307.5\times10^3)]/-3.424=-0.8832\times10^{-6}$
$m=-[0.93/24.685\times10^6]/-3.424=0.0110\times10^{-6}$
Then $1/P_{pi}\geq-0.8832\times10^{-6}+0.0110\times10^{-6}\,z_{cp}$ (units in 1/N)

Converting to kN and expressing as $1/P_{pi}\times10^3$

$10^3/P_{pi}\geq-0.8832+0.0110\,z_{cp}$ ...(1t)
where $10^3/P_{pi}=0$, $z_{cp}=+0.8832/0.0110=80.3$ mm
where $z_{cp,max}=155.2$ mm, $1/P_{pi}\geq-0.8832+0.0110\times155.2=0.824\times10^{-3}$ kN^{-1}

Construct line 2t at the top at transfer at ¼ span

Eq. 8.5 Denominator $(-f_{ctm}(t)-M_{sw,qt}/Z_{t,co})=-2.72-(52.07/25.920)=-4.730$ N/mm².
Then following the procedure above, and converting to kN and expressing as $1/P_{pi}\times10^3$
$10^3/P_{pi}\geq-0.6394+0.00796\,z_{cp}$ ...(2t)
where $10^3/P_{pi}=0$, $z_{cp}=+0.6394/0.00796=80.3$ mm
where $z_{cp,max}=155.2$ mm, $1/P_{pi}\geq-0.6394+0.00796\times155.2=0.597\times10^{-3}$ kN^{-1}

Construct line 1b at the bottom at transfer near to the support

Eq. 8.9 Denominator $(0.6\,f_{ck}(t)+M_{sw}/Z_{b,co})=(18.00+18.22/31.726)=18.574$ N/mm².
Then $10^3/P_{pi}\geq0.1628+0.00166\,z_{cp}$ ...(1b)
where $10^3/P_{pi}=0$, $z_{cp}=-0.1628/0.00166=-98.3$ mm
where $z_{cp,max}=155.2$ mm, $1/P_{pi}\geq0.1628+0.00166\times155.2=0.420\times10^{-3}$ kN^{-1}
Note where $z_{cp}=0$, $1/P_{pi}\geq0.1628\times10^{-3}$ kN^{-1}
$-0.8832+0.0110\,z_{cp}=0.1628+0.00166\,z_{cp}$

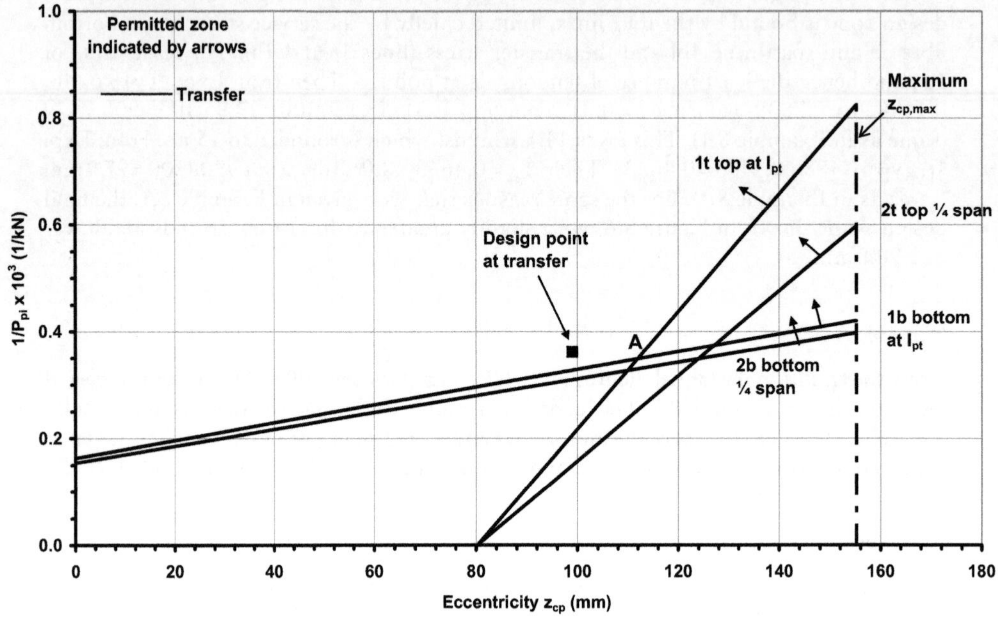

Figure 8.7 Magnel diagram for transfer stresses in the top and bottom fibre at l_{pt} from the end of beam and at ¼ span in Example 8.3.

Construct line 2b at the bottom at transfer at ¼ span

Eq. 8.9 Denominator $(0.6\,f_{ck}(t)+M_{sw,qt}/Z_{b,co})=18.00+(52.07/31.726)=19.641$ N/mm². Then $10^3/P_{pi}\geq 0.1540+0.00157\,z_{cp}$ ...(2b)
where $10^3/P_{pi}=0$, $z_{cp}=-0.1540/0.00157=-98.3$ mm
where $z_{cp,max}=155.2$ mm, $1/P_{pi}\geq 0.1540+0.00157\times 155.2=0.397\times 10^{-3}$ kN^{-1}
Note that the transfer condition at mid-span is not critical as $M_s/Z_{b,co}=-69.4\times 10^6/31.7\,25\times 10^6=-2.19$ N/mm² which is greater than $f_{ctm}(t)=-2.72$ N//mm² so the prestressing requirement is zero.

The straight lines for Exp. 1t, 1b, 2t and 2b are shown in Figure 8.7. As expected the critical point A is near to the support where lines 1t and 1b intersect. Then from Exp. 1t=Exp. 1b, $-0.8832+0.0110\,z_{cp}=0.1628+0.00166\,z_{cp}$. Hence $z_{cp}=111.9$ mm and $10^3/P_{pi}\geq 0.1628+0.00166\times 111.9=0.3483$, hence $P_{pi}\leq 2871.0$ kN. If the initial pretensioning force per 12.5 mm diameter strand is $0.7\times 1770\times 93\times 10^{-3}=115.23$ kN, the number of strands is $N\leq 24.92$. Keeping symmetry and rounding down to $N=24$ no. gives $P_{pi}=2765.4$ kN and $z_{cp}=113.1$ mm. The final arrangement of strands is given after the service requirements are determined.

Comparing with the results from Example 6.2, where the transfer condition was calculated at the support so that $M_{sw}=0$, and the limiting tensile stress at transfer was limited to $50\%\times f_{ctm}(t)=1.36$ N/mm², using the Magnel diagram, the number of strands required is 26.67 (the same as in Example 6.2) and the eccentricity for 26 strands is $z_{cp}=92$ mm (the same as in Example 6.2).

At the service condition construct lines 3t and 3b at ¼ span and lines 4t and 4b at mid-span. Use $R_s=0.80$, $f_{ck}=45$ N/mm², $f_{cm}=53$ N/mm², $f_{ctm}=3.80$ N/mm², $M_{s,qt}=481.30$ kNm at ¼ span and $M_{s,max}=641.73$ kNm at mid-span. Exposure classification XC1, then $f_{ctm}=3.80$ N/mm² (note Example 8.4 checks the same beam for exposure XC3).

Construct line 3t at the top in service at ¼ span

Eq. 8.13 Denominator $0.45\,f_{cm}-M_{s,qt}/Z_{t,co}=23.85-(481.30/25.920)=+5.281$ N/mm².
$C=[0.80\,(1/307.5\times 10^3)]/5.281=0.4926\times 10^{-6}$

$m=-[0.80/24.685\times10^6]/5.281=-0.00614\times10^{-6}$
Then $1/P_{pi}\geq0.4926\times10^{-6}-0.00614\times10^{-6}\ z_{cp}$ (units in 1/N)
Converting to kN and expressing as $1/P_{pi}\times10^3$
$10^3/P_{pi}\geq0.4926-0.00614\ z_{cp}$ ...(3t)
where $10^3/P_{pi}=0,\ z_{cp}=-0.4926/-0.00614=80.3$ mm
where $z_{cp,max}=155.2$ mm, $1/P_{pi}\geq0.4926-0.00614\times155.2=-0.460\times10^{-3}$ kN^{-1}
When $z_{cp}=0,\ 10^3/P_{pi}=0.4926$

Construct line 3b at the bottom in service at ¼ span
Eq. 8.17 Denominator $-f_{ctm}+M_{s,qt}/Z_{b,co}=-3.80+(481.30/31.726)=11.375$ N/mm^2.
Then following the procedure above, and converting to kN and expressing as $1/P_{pi}\times10^3$
$10^3/P_{pi}\leq0.2287+0.00233\ z_{cp}$ ...(3b)
where $10^3/P_{pi}=0,\ z_{cp}=-0.2287/0.00233=-98.3$ mm
where $z_{cp,max}=155.2$ mm, $1/P_{pi}\geq0.2287+0.00233\times155.2=0.590\times10^{-3}$ kN^{-1}

Construct line 4t at the top in service at mid-span
Eq. 8.13 Denominator $0.45\ f_{cm}-M_{s,max}/Z_{t,co}=23.85-(641.73/25.920)=-0.909$ N/mm^2.
Then $10^3/P_{pi}\geq-2.8631+0.03567\ z_{cp}$ ...(4t)
where $10^3/P_{pi}=0,\ z_{cp}=2.8631/0.03567=80.3$ mm
where $z_{cp,max}=155.2$ mm, $1/P_{pi}\geq-2.8631+0.03567\times155.2=2.672\times10^{-3}$ kN^{-1}

Construct line 4b at the bottom in service at mid-span
Eq. 8.17 Denominator $-f_{ctm}+M_{s,max}/Z_{b,co}=-3.80+(641.73/31.726)=16.432$ N/mm^2.
Then $10^3/P_{pi}\leq0.1583+0.00161\ z_{cp}$ ...(4b)
where $10^3/P_{pi}=0,\ z_{cp}=-0.1583/0.00161=-98.3$ mm
where $z_{cp,max}=155.2$ mm, $1/P_{pi}\geq0.1583+0.00161\times155.2=0.408\times10^{-3}$ kN^{-1}
The straight lines for Exp. 3t, 3b, 4t and 4b are shown in Figure 8.8. The ¼ span position is not critical, and considering the mid-span position alone, the design zone extends from point A where, by the equality Exp. 4t=Exp. 4b, $z_{cp}=88.7$ mm up to $z_{cp,max}=155.2$ mm where P_{pi} is the smallest but no doubt impractical value. However, the transfer condition at top fibre and at ¼ span, given by line 2t, limits the design zone at point B where by the equality Exp. 2t=Exp. 4b, $-0.6394+0.00796\ z_{cp}=0.1583+0.00161\ z_{cp}$ Hence $z_{cp}=125.6$ mm and $10^3/P_{pi}\leq0.1583+0.00161\times125.6=0.361$, hence $P_{pi}\geq2772.9$ kN requiring $N\geq24.1$ strands. The eccentricity of 125.6 mm is rather too large to achieve for a reasonable distribution of strands (it can be achieved by placing 20 no. strands in the bottom three rows, but this may lead to problems with splitting or in controlling camber), therefore try $N=27$ no. strands giving $P_{pi}=3111.1$ kN and $z_{cp}\geq101.2$ mm.

To achieve the number of strands and eccentricity, having established $N=27$ no. strands at $z_{cp}\geq101.2$ mm, the strands are arranged in the beam as shown in Figure 8.9 as follows:

Then $y_s=4500/27=166.7$ mm

No. strands in each row N_i	*Distance* y_i *from bottom (mm)*	$\Sigma\ N_i\ y_i$ *(mm)*
2	550	1100
2	350	700
2	250	500
2	200	400
6 (−2)	150	900
5 (−1)	100	500
8	50	400
Σ=27		Σ=4500

Debonded strands are given in brackets.

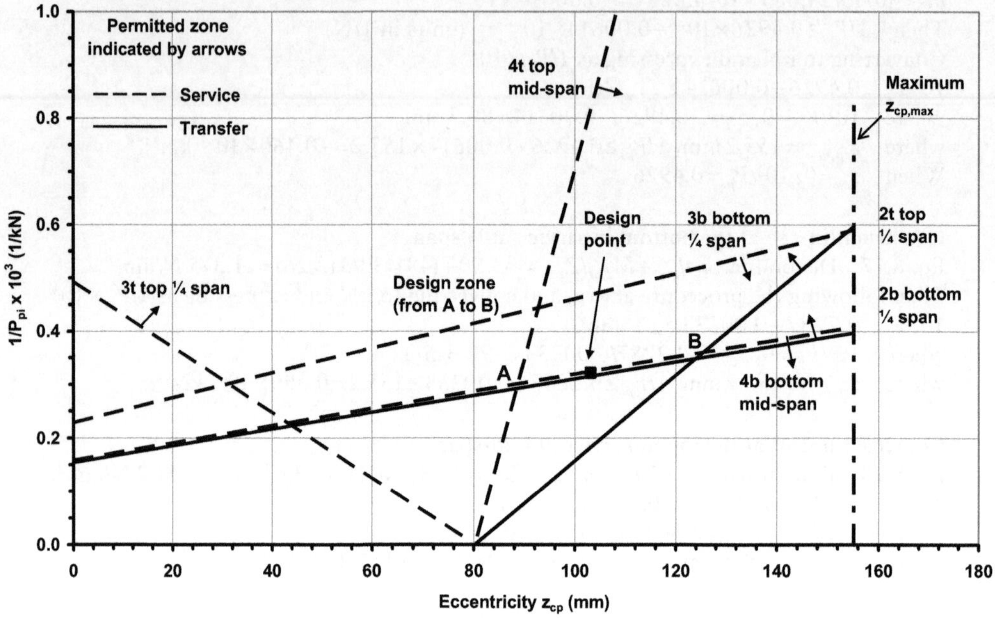

Figure 8.8 Magnel diagram for service stresses (dashed lines) in the top and bottom fibre at ¼ span and at mid-span, and for transfer at ¼ span in Example 8.3.

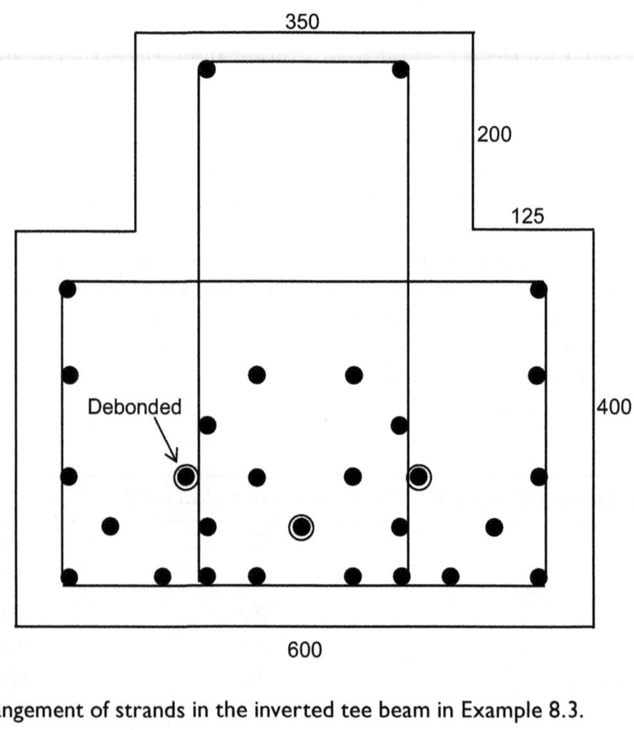

Figure 8.9 Final arrangement of strands in the inverted tee beam in Example 8.3.

The actual eccentricity $z_{cp}=y_b-y_s=269.78-166.7=103.1\,\text{mm}$, closest possible to 101.2 mm in the available array of tendon positions.

At the support $z_{cp}=y_b-y_s=269.78-4100/24=99.0\,\text{mm}$.

The design points are shown in Figures 8.7 and 8.8.

The final prestress and bending stresses are

At transfer $\sigma_t(t)+M_{sw}/Z_{t,co}=-1.95+0.70=-1.24>-2.72$ N/mm^2

$\sigma_b(t)-M_{sw}/Z_{b,co}=+16.79-0.57=+16.22<+18.00$ N/mm^2

In service $\sigma_t+M_{s,max}/Z_{t,co}=-2.30+24.76=+22.46<+23.85$ N/mm^2

$\sigma_b-M_{s,max}/Z_{b,co}=+16.59-20.23=-3.64>-3.80$ N/mm^2

A full analysis of the beam in service, based on the procedures given in Chapter 7, give $R_s=0.790$, and hence $\sigma_b-M_{s,max}/Z_{b,co}=+16.38-20.23=-3.85<-3.80$ N/mm^2 (not OK). However, the compound $Z_{b,co}=31.822\times10^6\text{mm}^3$ is such that the actual moment of resistance $M_{sR}=(16.38+3.80)\times31.822=641.94\approx641.73$ kNm. The actual M_{sR} based on the top fibre stresses is 663 kNm and clearly not critical.

Example 8.4

Repeat Example 8.3 for the service condition only for exposure classification XC3 and a quasi-permanent factor $\psi_2=0.6$ (e.g. congregation). Use $R_s=0.80$, $f_{ck}=45$ N/mm^2, $f_{cm}=53$ N/mm^2, $f_{ctm}=3.80$ N/mm^2, $M_{s,qt}=385.47$ kNm at ¼ span and $M_{s,max}=513.96$ kNm at mid-span, i.e. 80.1% of the moments in Example 8.3. For XC3 f_{ctm}=zero.

Solution

Construct line 3t at the top in service at ¼ span

Eq. 8.13 Denominator $0.45\,f_{cm}-M_{s,qt}/Z_{t,co}=23.85-(385.47/25.920)=+8.978$ N/mm^2.

$10^3/P_{pi}\geq0.2896-0.00361\,z_{cp}$ ...(3t)

where $10^3/P_{pi}=0$, $z_{cp}=-0.2896/-0.00361=80.3\,\text{mm}$

where $z_{cp,max}=155.2\,\text{mm}$, $1/P_{pi}\geq0.2896-0.00361\times155.2=-0.270\times10^{-3}$ kN^{-1}

Construct line 3b at the bottom in service at ¼ span

Eq. 8.17 Denominator $-f_{ctm}+M_{s,qt}/Z_{b,co}=0+(385.47/31.726)=12.15$ N/mm^2.

$10^3/P_{pi}\leq0.2141+0.00218\,z_{cp}$ ...(3b)

where $10^3/P_{pi}=0$, $z_{cp}=-0.2141/0.00218=-98.3\,\text{mm}$

where $z_{cp,max}=155.2\,\text{mm}$, $1/P_{pi}\geq0.2141+0.00218\times155.2=0.552\times10^{-3}$ kN^{-1}

Construct line 4t at the top in service at mid-span

Eq. 8.13 Denominator $0.45\,f_{cm}-M_{s,max}/Z_{t,co}=23.85-(513.96/25.920)=4.021$ N/mm^2.

$10^3/P_{pi}\geq0.6470-0.00806\,z_{cp}$ ...(4t)

where $10^3/P_{pi}=0$, $z_{cp}=-0.6470/-0.00806=80.3\,\text{mm}$

where $z_{cp,max}=155.2\,\text{mm}$, $1/P_{pi}\geq0.6470-0.00806\times155.2=-0.604\times10^{-3}$ kN^{-1}

Construct line 4b at the bottom in service at mid-span

Eq. 8.17 Denominator $-f_{ctm}+M_{s,max}/Z_{b,co}=0+(513.96/31.726)=16.200$ N/mm^2.

$10^3/P_{pi}\leq0.1606+0.00163\,z_{cp}$ ...(4b)

where $10^3/P_{pi}=0$, $z_{cp}=-0.1606/0.00163=-98.3\,\text{mm}$

where $z_{cp,max}=155.2\,\text{mm}$, $1/P_{pi}\geq0.1606+0.00163\times155.2=0.414\times10^{-3}$ kN^{-1}

The straight lines for Exp. 3t, 3b, 4t and 4b are shown in Figure 8.10. There is a huge shift in the position of lines 3t and 4t due to the reductions in M_s, but hardly any change in the position of the important lines 3b and 4b. Compared with Example 8.3, the equations for $1/P_{pi}$ in Exp. 4t changes by only 1.4% due to the compensating effects of $f_{ctm}=3.80$ N/mm^2 to zero, and M_s reducing by 20%. Of course, there is no change in the transfer lines 2t and 2b, so the resulting design is the same in this example as in Example 8.3. In fact, the ratio of the final $M_s/M_{sR}=0.98$. However, if the quasi-permanent factor $\psi_2=1.0$ the beam fails with $M_s/M_{sR}=1.23$.

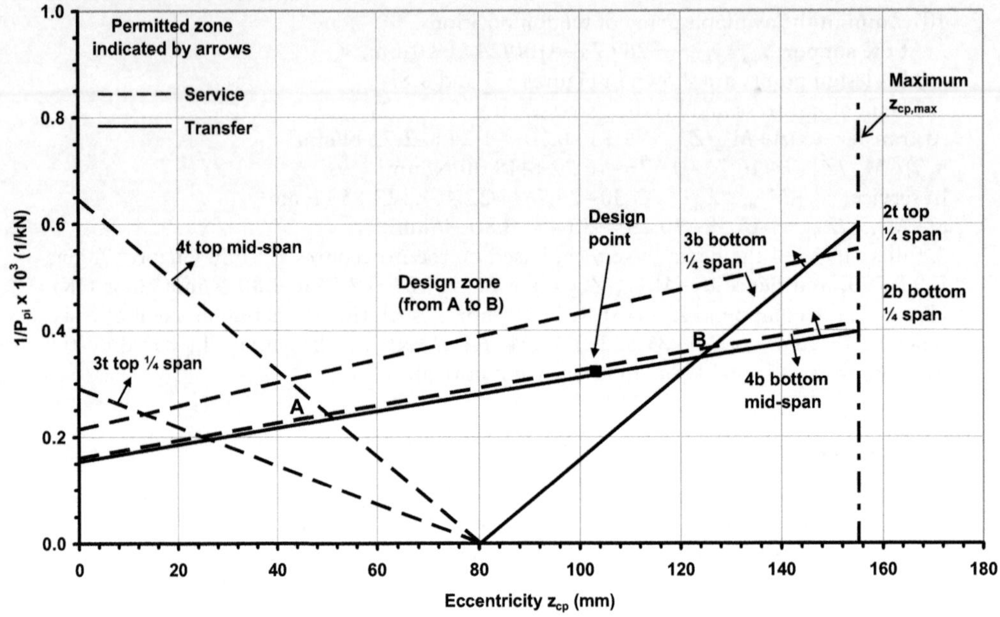

Figure 8.10 Magnel diagram for service stresses (dashed lines) in the top and bottom fibre at ¼ span and at mid-span, and for transfer at ¼ span for exposure classification XC3 in Example 8.4.

8.4 ALTERNATIVE MAGNEL DIAGRAM USING TOP AND BOTTOM FIBRE STRESSES

8.4.1 Background to the Magnel diagram

An alternative and very popular alternative to the traditional Magnel diagram is shown in Figure 8.11. Here the requirements for the initial prestress (before losses) σ_{ti} at the top fibre and σ_{bi} at the bottom fibre are plotted as a set of rectangles for the transfer and mid-span service conditions. If the rectangles overlap, there is a common pretensioning requirement for both transfer and service, highlighted as the 'overlapping zone' in Figure 8.11. So-called 'N-lines', actually P_{pi} lines with $N = P_{pi}/\eta\, f_{pk}\, A_p$ per tendon are then drawn through the overlapping zone for a selected number of tendons (say 20 to 30 no. for the beams studied in Example 8.3, or fewer in number of hollow core units, etc.) and the critical point is found in the bottom left-hand corner of the overlapping zone. So-called 'e-lines' for the eccentricity may also be drawn as shown in Figure 8.11 for chosen values of z_{cp}. e-lines are not so important as z_{cp} may easily be calculated. The popularity of this method is because the user can see the required top and bottom fibre stresses, as the rectangular stress zones move across the diagram with increasing bending moments or reducing limiting stresses.

If there is no common overlapping zone between the transfer and service conditions, a third rectangle is constructed at say ¼ span, as shown in Figure 8.12. Now there are two overlapping zones for (i) near to support to ¼ span, and (ii) ¼ span to mid-span. Note that the N-lines are common to both zones and need only be constructed once.

The four inequalities are based on the same equations 5.6, 5.7a and 5.7b, and the transfer and service conditions shown in Figures 8.11–8.14 are as follows. The limiting stress requirements are shown in Figure 8.15. Let σ_{ti} and σ_{bi} now be the initial prestress required in order to find the initial prestressing force P_{pi} and eccentricity z_{cp}

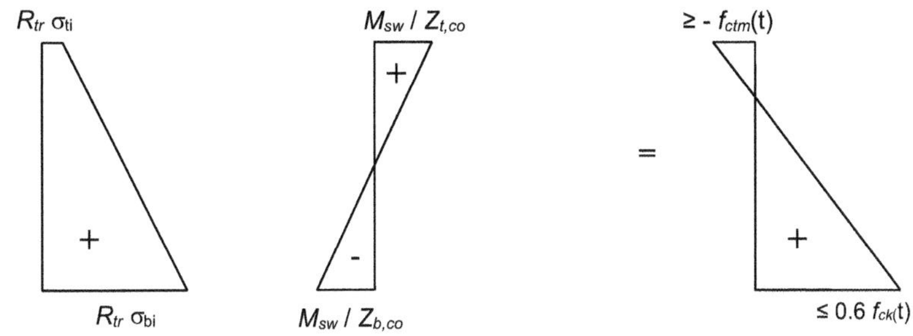

Figure 8.11 Limits of prestress at transfer near to the support, with R_{tr} losses.

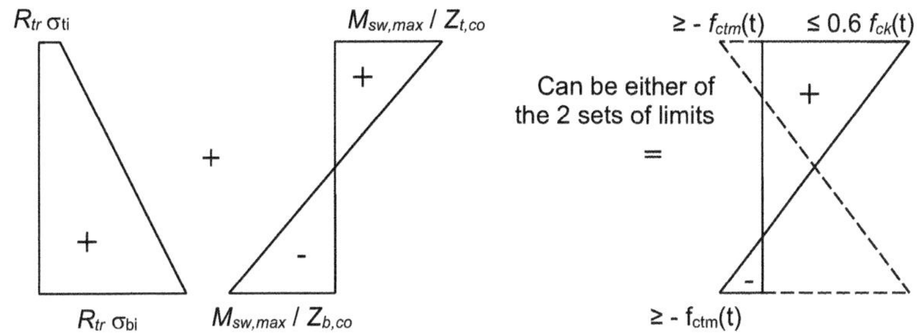

Figure 8.12 Limits of prestress at transfer at mid-span (or ¼ span if needed).

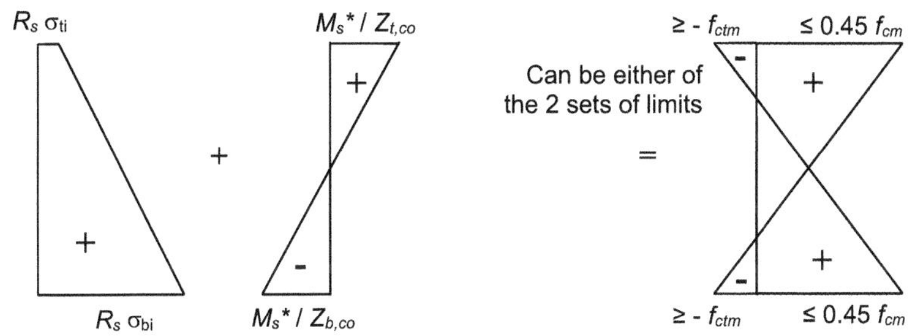

Figure 8.13 Limits of prestress in service near to the support, with R_s losses.

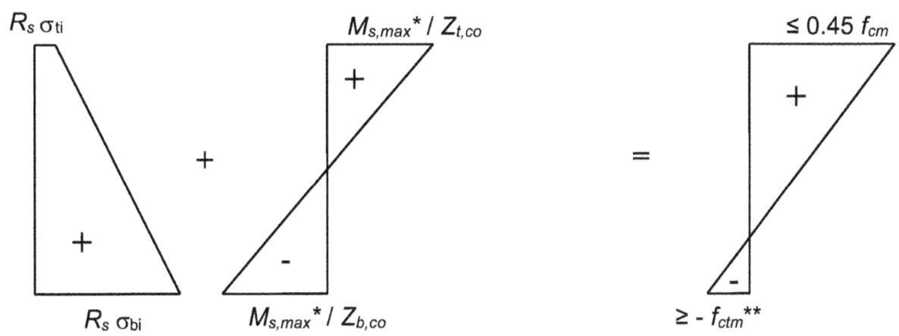

Figure 8.14 Limits of prestress in service after all losses at the position of maximum service bending moment. *Quasi-permanent moment for durability classes XC2 to XC4, and frequent moment for durability classes XD and XS (see Section 7.2.1). ** Zero tension for durability classes greater than XC1

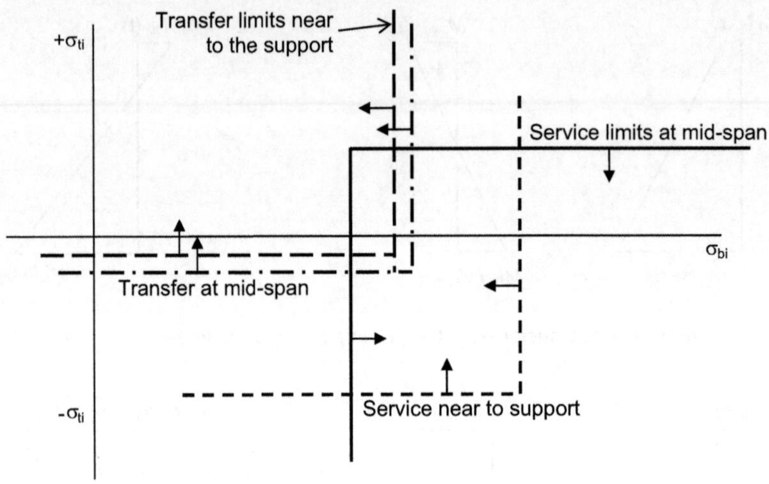

Figure 8.15 Identifying the critical limits for σ_{bi} and σ_{ti} using the Magnel diagram.

8.4.2 Transfer inequalities

8.4.2.1 Transfer stress at the top fibre at l_{pt} from the support

Eq. 8.1 $R_{tr}\ P_{pi}\ (1/A_c - z_{cp}/Z_t) + M_{sw}/Z_{t,co} \geq -f_{ctm}(t)$
Fig. 8.11 $R_{tr}\ \sigma_{ti} \geq -f_{ctm}(t) - M_{sw}/Z_{t,co}$

$\sigma_{ti} \geq (-f_{ctm}(t) - M_{sw}/Z_{t,co})/R_{tr}$ ... (1t) (8.30)

A further limit exists where $R_{tr}\ \sigma_{ti} \leq 0.6\ f_{ck}(t) - M_{sw}/Z_{t,co}$ (8.31)

At l_{pt} from the support, these straight lines at the top fibre are termed...(1t)
At ¼ span from the support ditto...(2t)

It is interesting to check the transfer conditions at mid-span, where the moment due to self-weight is $M_{sw,max}$. Intuitively, this is not critical as the flexural stresses due to the greater bending moment are counter-active to the prestress, but the differences can be quite small.

Fig. 8.12 $R_{tr}\ \sigma_{ti} \geq -f_{ctm}(t) - M_{sw,max}/Z_{t,co}$ (8.32)

Fig. 8.12 $R_{tr}\ \sigma_{ti} \leq 0.6\ f_{ck}(t) - M_{sw,max}/Z_{t,co}$ (8.33)

8.4.2.2 Transfer stress at the bottom fibre

Eq. 8.6 $R_{tr}\ P_{pi}\ (1/A_c + z_{cp}/Z_b) - M_{sw}/Z_{b,co} \leq 0.6\ f_{ck}(t)$
Fig. 8.11 $R_{tr}\ \sigma_{bi} \leq 0.6\ f_{ck}(t) + M_{sw}/Z_{b,co}$
$\sigma_{bi} \leq (0.6\ f_{ck}(t) + M_{sw}/Z_{b,co})/R_{tr}$ ... (1b) (8.34)

A further limit exists where $R_{tr}\ \sigma_{bi} \geq -f_{ctm}(t) + M_{sw}/Z_{b,co}$ (8.35)

At l_{pt} and ¼ span from the support, these are termed...(1b) and (2b), respectively.

Similarly checking the transfer conditions at mid-span

Fig. 8.12 $R_{tr}\ \sigma_{bi} \leq 0.6\ f_{ck}(t) + M_{sw,max}/Z_{b,co}$ (8.36)

Fig. 8.12 $R_{tr}\ \sigma_{bi} \geq -f_{ctm}(t) + M_{sw,max}/Z_{b,co}$ (8.37)

8.4.3 Service inequalities

For completeness, Figure 8.13 shows the limiting stresses for the service condition near to the support, but this is clearly not a critical condition as shown in Figure 8.15. The focus is therefore on mid-span as follows:

8.4.3.1 Service stress at the top fibre

Eq. 8.10 $R_s\ P_{pi}\ (1/A_c - z_{cp}/Z_t) + M_{s,max}/Z_{t,co} \leq 0.45\ f_{cm}$
Fig. 8.14 $R_s\ \sigma_{ti} \leq 0.45\ f_{cm} - M_{s,max}/Z_{t,co}$

$\sigma_{ti} \leq (0.45\ f_{cm} - M_{s,max}/Z_{t,co})/R_s$ ... (3t) (8.38)

A further limit exists where $R_s\ \sigma_{ti} \geq -f_{ctm} - M_{s,max}/Z_{t,co}$ (8.39)

At ¼ span and mid-span, these are termed...(3t) and (4t), respectively.

8.4.3.2 Service stress at the bottom fibre

Eq. 8.14 $R_s\ P_{pi}\ (1/A_c + z_{cp}/Z_b) - M_{s,max}/Z_{b,co} \geq -f_{ctm}$
Fig. 8.14 $R_s\ \sigma_{bi} \geq -f_{ctm} + M_{s,max}/Z_{b,co}$

$\sigma_{bi} \geq (-f_{ctm} + M_{s,max}/Z_{b,co})/R_s$ ... (3b) (8.40)

A further limit exists where $R_s\ \sigma_{bi} \leq 0.45\ f_{cm} + M_{s,max}/Z_{b,co}$ (8.41)

At ¼ span and mid-span, this is termed...(3b) and (4b), respectively.

8.4.3.3 Critical design zone and stresses

The limiting stresses are plotted in the Magnel diagram as shown in Figure 8.15, where a critical overlapping zone, i.e. the design zone, is bounded by the transfer stresses near to the support and service stresses at mid-span, *viz.* lines 1t, 1b, 4t and 4b. If these lines do not overlap lines 2t and 2b, and 3t and 3b are added for the transfer and service conditions at ¼ span (or wherever the designer so chooses, such as $^1/_5$ span for example.

The optimum design point, with the lowest number of tendons, is shown by the graphical solution of using P-lines in Section 8.5 to occur at the bottom left corner of the overlapping design zone, point A. Anywhere in the design zone is permitted. Point A gives the least values for σ_{bi} and σ_{ti}, from which the lowest value of the initial prestress P_{pi} and the eccentricity z_{cp} may be determined by solving the following two equations derived from Eq. 6.6 and 6.7a.

At bottom $\sigma_{bi} = P_{pi}/A_c + P_{pi}\ z_{cp}/Z_b$ (8.42)

At top $\sigma_{ti} = P_{pi}/A_c - P_{pi}\ z_{cp}/Z_t$ (8.43)

From Eq. 8.42 $z_{cp} = \sigma_{bi}\ Z_b/P_{pi} - Z_b/A_c$ (8.44)

Substituting Eq. 8.44 into Eq. 8.43

$$\sigma_{ti}=P_{pi}/A_c\,(1+1/\alpha)-\sigma_{bi}/\alpha \tag{8.45}$$

$$\text{Then } P_{pi}=A_c\,(\sigma_{ti}+\sigma_{bi}/\alpha)/(1+1/\alpha) \tag{8.46}$$

where $\alpha=Z_t/Z_b$

8.5 CONSTRUCTION OF 'P-LINES' AND 'E-LINES'

8.5.1 Construction of 'P-lines'

A graphical solution to the above is both visual and informative and gives a good impression of the relationships between P_{pi} and the limiting stresses. The overlapping design zone is extracted from Figure 8.15 into Figure 8.16. The critical design point lies within this zone and is found by constructing so-called 'P-lines', i.e. plots of surface stresses σ_{ti} and σ_{bi} for values of P_{pi} based on an increasing integer number of tendons N. Referring to Exp. 6.2, $P_{pi}=N\,\eta\,f_{pk}\,A_{pi}$ where A_{pi} is the area of one tendon, for example for 10 no. 12.5 mm diameter strands $P_{pi}=10\times0.7\times1770\times93\times10^{-3}=1152.3$ kN. (The term 'N-line' is sometimes used as P_{pi} is directly proportional to N.) The equation of the P-line is found by elimination z_{cp} from the same Eqs. 8.42 and 8.43.

$$\text{From Eq. 8.53 } z_{cp}=Z_t/A_c-\sigma_{ti}\,Z_t/P_{pi} \tag{8.47}$$

Setting Eq. 8.44=Eq. 8.47 gives the equation of the P-line

$$\sigma_{ti}=P_{pi}\,(1+1/\alpha)/A_c-(1/\alpha)\,\sigma_{bi} \tag{8.48}$$

where $\alpha=Z_t/Z_b$

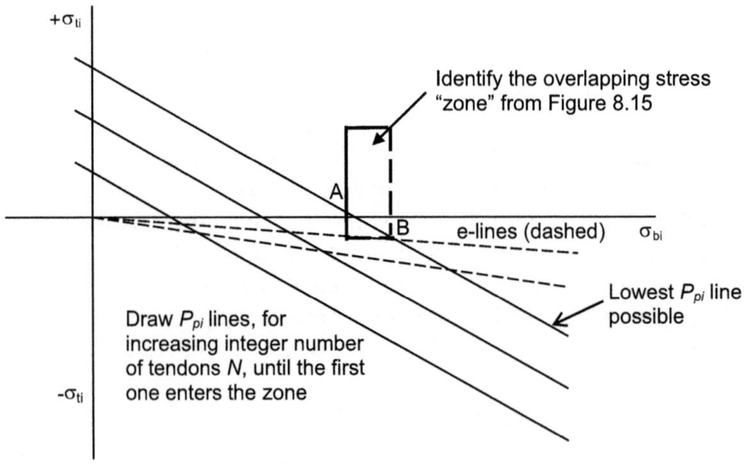

Figure 8.16 Identifying the overlapping stress zone between the support and maximum moment, and drawing the 'P-lines' using P_{pi} and 'e-lines' in the Magnel diagram.

This is a sloping line of the form $y=C-mx$. For example, in the inverted tee beam in Example 8.3, $\alpha=24.685/30.215=0.817$. If $P_{pi}/A_c=3111.1\times10^3/307.5\times10^3=10.12$ N/mm², the equation of the P-line for 27 no. strands is $\sigma_{ti}=22.51-1.224\ \sigma_{bi}$ (N/mm² units). This P-line intersects the x-axis at $\sigma_{bi}=22.51/1.224=18.39$ N/mm².

When plotting P-lines the gradient of the line ($-1/\alpha$) is constant, but the intersection with the y-axis increases with increasing number of tendons, e.g. the equation for 28 no. strands is simply $\sigma_{ti}=(28/27)\times22.51-1.224\ \sigma_{bi}=23.34-1.224\ \sigma_{bi}$ and so on. The user therefore uses trial and error to find the lowest number of tendons that enable its P-line to pass through the overlapping zone, shown as A-B in Figure 8.16. The lowest number of tendons is therefore found at the bottom left corner of the overlapping design zone, in fact at the intersection of lines 4b (service, mid-span) and 1t (transfer, support) as is the case with all other variations of Magnel diagrams. A design point is usually selected mid-way along the P-line within the overlapping zone(s) giving the designer some leeway to arrange the tendons to suit.

8.5.2 Construction of 'e-lines'

A graphical illustration of the relationship between the eccentricity z_{cp} and the limiting stresses is the use of so-called 'e-lines', as shown in Figure 8.16. There is a linear relationship between σ_{ti} and σ_{bi} as follows.

From Eq. 8.42 $P_{pi}=\sigma_{bi}/(1/A_c+z_{cp}/Z_b)$ (8.44)

From Eq. 8.43 $P_{pi}=\sigma_{ti}/(1/A_c-z_{cp}/Z_t)$ (8.49)

Setting Eq. 8.44 = Eq. 8.49 gives the equation of the e-line

$$\sigma_{ti}=\sigma_{bi}\ (1/A_c-z_{cp}/Z_t)/(1/A_c+z_{cp}/Z_b) \tag{8.50}$$

This is a sloping line of the form $y=mx$. For example in the inverted tee beam in Example 8.3, $Z_t=24.685\times10^6$, $Z_b=30.215\times10^6$, $A_c=307.5\times10^3$, trialling $z_{cp}=100$ mm, the equation of the e-line is $\sigma_{ti}=\sigma_{bi}\ (-0.799\times10^{-6})/(6.561\times10^{-6})=-0.122\ \sigma_{bi}$.

Example 8.5- continuing Example 8.1

Draw the Magnel diagram using limiting stresses, P-lines and e-lines for the beam in Example 8.1. Use $M_{sw}=4.86$ kNm, $M_{sw,max}=13.50$ kNm (for transfer), $M_{s,min}=56.70$ kNm and $M_{s,max}=157.5$ kNm (for service).

Solution

Limiting stress for transfer at the top of beam (lines 1t)
At l_{pt} from the support

Eq. 8.30 $\sigma_{ti}\geq(-f_{ctm}(t)-M_{sw}/Z_{t,co})/R_{tr}$
$\sigma_{ti}\geq(-2.72-4.86/8.40)/0.93=-3.54$ N/mm²
Eq. 8.31 $\sigma_{ti}\leq(0.6\ f_{ck}(t)-M_{sw}/Z_{t,co})/R_{tr}$
$\sigma_{ti}\leq(18.00-4.86/8.40)/0.93=+18.73$ N/mm²

At mid-span

Eq. 8.32 $R_{tr}\ \sigma_{ti}\geq-f_{ctm}(t)-M_{sw,max}/Z_{t,co}$
$\sigma_{ti}\geq(-2.72-13.50/8.40)/0.93=-4.65$ N/mm²

Eq. 8.33 $R_{tr}\ \sigma_{ti} \le 0.6\ f_{ck}(t) - M_{sw,max}/Z_{t,co}$
$\sigma_{ti} \le (18.00-13.50/8.40)/0.93=+17.62$ N/mm^2
Limiting stress for transfer at the bottom of beam (lines 1b)
At l_{pt} from the support

Eq. 8.34 $\sigma_{bi} \le (0.6\ f_{ck}(t)+M_{sw}/Z_{b,co})/R_{tr}$
$\sigma_{bi} \le (18.00+4.86/8.40)/0.93=+19.98$ N/mm^2
Eq. 8.35 $\sigma_{bi} \ge (-f_{ctm}(t)+M_{sw}/Z_{b,co})/R_{tr}$
$\sigma_{bi} \ge (-2.72+4.86/8.40)/0.93=-2.30$ N/mm^2

The limiting stresses near to the support are shown as solid lines in Figure 8.17.
At mid-span
Eq. 8.36 $\sigma_{bi} \le (0.6\ f_{ck}(t)+M_{sw,max}/Z_{b,co})/R_{tr}$
$\sigma_{bi} \le (18.00+13.50/8.40)/0.93=+21.08$ N/mm^2
Eq. 8.37 $\sigma_{bi} \ge (-f_{ctm}(t)+M_{sw,max}/Z_{b,co})/R_{tr}$
$\sigma_{bi} \le (-2.72+13.50/8.40)/0.93=-1.20$ N/mm^2

The mid-span limiting stresses are shown as dashed-dot lines in Figure 8.17.
Limiting stress for service at the top of beam (lines 4t) at mid-span

Eq. 8.38 $\sigma_{ti} \le (0.45\ f_{cm} - M_{s,max}/Z_{t,co})/R_s$
$\sigma_{ti} \le (23.85-157.5/8.40)/0.80=+6.37$ N/mm^2
Eq. 8.39 $\sigma_{ti} \ge (-f_{ctm} - M_{s,max}/Z_{t,co})/R_s$
$\sigma_{ti} \le (-3.80-157.5/8.40)/0.80=-28.18$ N/mm^2

Limiting stress for service at the bottom of beam (line 4b) at mid-span

Eq. 8.40 $\sigma_{bi} \ge (-f_{ctm}+M_{s,max}/Z_{b,co})/R_s$
$\sigma_{bi} \ge (-3.80+157.5/8.40)/0.80=+18.69$ N/mm^2

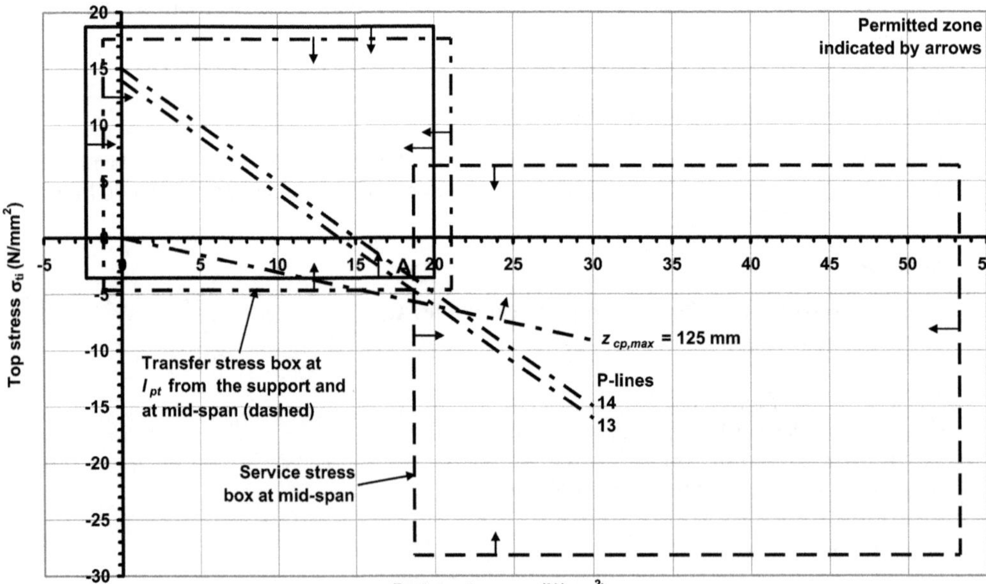

Figure 8.17 Permissible stress zones near to support and at mid-span, 'P-lines' and 'e-line' for $z_{cp,max}$ in the Magnel diagram for Example 8.5.

Eq. 8.41 $\sigma_{bi} \leq (0.45\, f_{cm} + M_{s,max}/Z_{b,co})/R_s$
$\sigma_{bi} \leq (23.85+157.5/8.40)/0.80 = +53.25$ N/mm^2

The mid-span limiting stresses are shown as dashed lines in Figure 8.17.

As expected, Figure 8.17 shows that the critical limiting transfer stresses are at the position near to the support, and the critical service stresses are at mid-span. Out of 12 conditions checked, only four are critical, and these are transferred to the overlapping design zone in Figure 8.18.

As discussed above, the optimum design point is point A where $\sigma_{bi}=+18.69$ N/mm^2 and $\sigma_{ti}=-3.55$ N/mm^2. The necessary initial pretensioning force P_{pi} is

Eq. 8.46 $P_{pi} = A_c\, (\sigma_{ti} + \sigma_{bi}/\alpha)/(1+1/\alpha)$
where $\alpha = Z_t/Z_b$ $_{=8.0/8.0\ =1.0}$
$P_{pi} = 120\times10^3\, (-3.55+18.69/1.0)/(1+1/1.0)/10^3 = 908.7$ kN

This of course exactly the same as in Example 8.1. If the initial pretensioning force per 9.3 mm diameter strand is $0.7\times1770\times52\times10^{-3}=64.4$ kN, the number of strands is $N\geq14.1$. Rounding up to $N=15$ no. gives $P_{pi}=966.4$ kN. The corresponding eccentricity is

Eq. 8.44 $z_{cp} = (18.69\times8.0\times10^6/966.4\times10^3) - (8.0\times10^6/120\times10^3) = 88.1$ mm and is exactly the same as in Example 8.1.

Using the same array of tendons as in Example 8.1 gives $z_{cp}=90.0$ mm. The final initial prestress values are $+18.69<\sigma_{bi}=+18.93<+19.98$ N/mm^2, and $-3.55<\sigma_{ti}=-2.82<+6.38$ N/mm^2. This 'design point' is shown in Figure 8.18.

Example 8.6

Draw the Magnel diagram using limiting stresses, P-lines and e-lines for the 600×600 mm wide inverted tee beam shown in Figure 8.9. The loads, span, and materials are the same

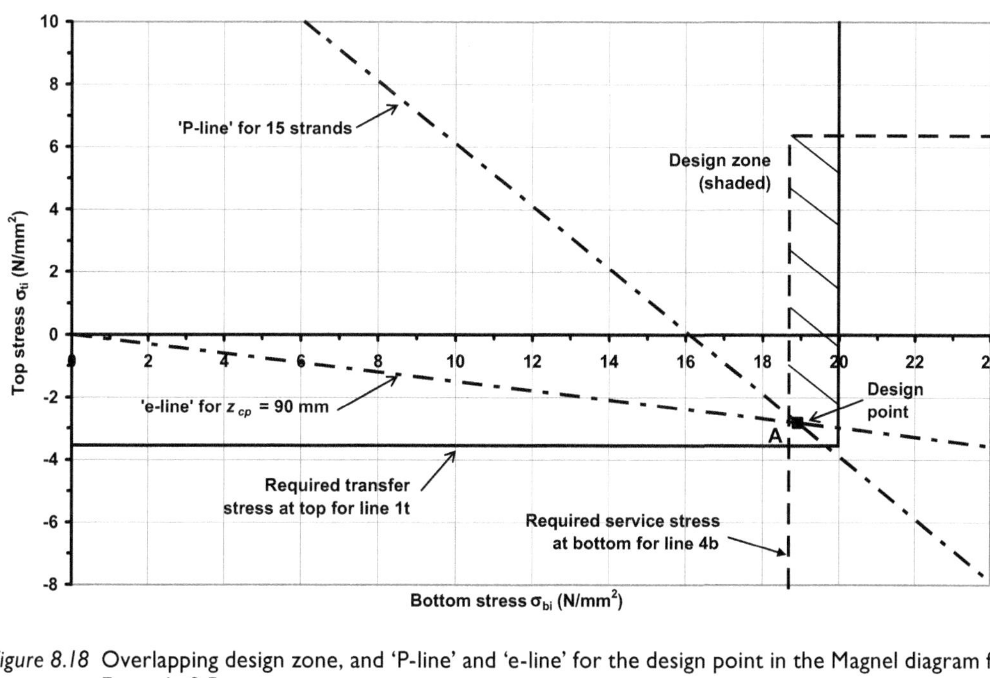

Figure 8.18 Overlapping design zone, and 'P-line' and 'e-line' for the design point in the Magnel diagram for Example 8.5.

as in Example 8.3. Use M_{sw}=18.22 kNm, $M_{sw,qt}$=52.07 kNm, $M_{s,qt}$=481.30 kNm at ¼ span and $M_{s,max}$=641.73 kNm at mid-span.

Solution

Only the critical limiting stresses identified in Example 8.5 will be calculated.
Limiting stress for transfer at the top of beam (line 1t) at l_{pt} from the support

Eq. 8.30 $\sigma_{ti} \geq (-f_{ctm}(t) - M_{sw}/Z_{t,co})/R_{tr}$
$\sigma_{ti} \geq (-2.72 - 18.22/25.920)/0.93 = -3.68$ N/mm²

Limiting stress for transfer at the bottom of beam (line 1b) at l_{pt} from the support

Eq. 8.34 $\sigma_{bi} \leq (0.6\, f_{ck}(t) + M_{sw}/Z_{b,co})/R_{tr}$
$\sigma_{bi} \leq (18.00 + 18.22/31.726)/0.93 = +19.97$ N/mm²

Limiting stress for service at the top of beam (lines 4t) at mid-span

Eq. 8.38 $\sigma_{ti} \leq (0.45\, f_{cm} - M_{s,max}/Z_{t,co})/R_s$
$\sigma_{ti} \leq (23.85 - 641.73/25.920)/0.80 = -1.14$ N/mm²

Limiting stress for service at the bottom of beam (line 4b) at mid-span

Eq. 8.40 $\sigma_{bi} \geq (-f_{ctm} + M_{s,max}/Z_{b,co})/R_s$
$\sigma_{bi} \geq (-3.80 + 641.73/31.726)/0.80 = +20.54$ N/mm²

Now $\sigma_{bi} \geq +20.54$ N/mm² at mid-span is greater than $\sigma_{bi} \leq +19.97$ N/mm² at transfer, so there is no single solution. Quarter-span limiting stresses are therefore required – transfer

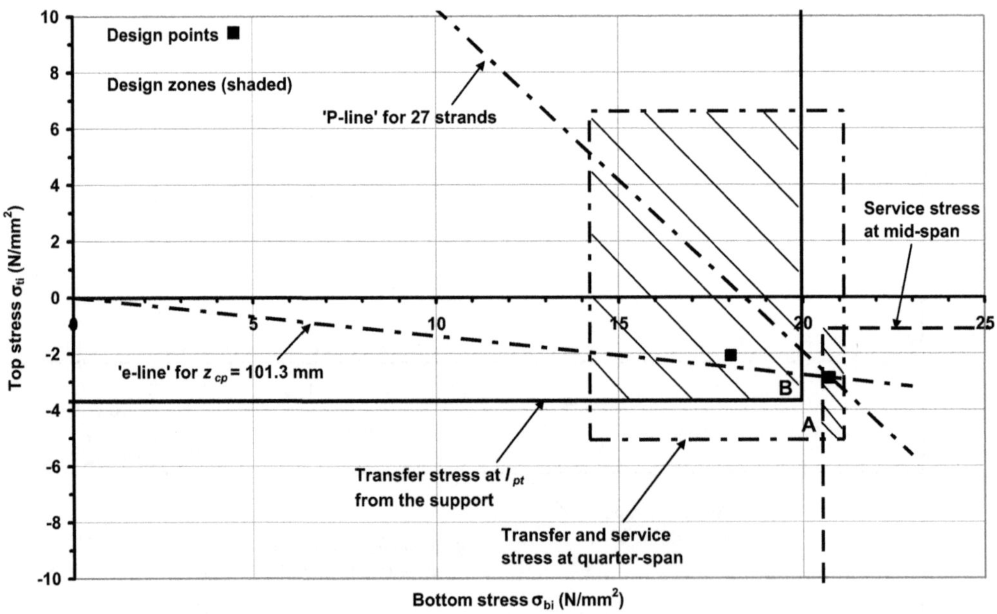

Figure 8.19 Overlapping design zones between the support & ¼ span and between ¼ span & mid-span, and 'P-line' and 'e-line' at the design point in the Magnel diagram for Example 8.6.

stresses at ¼ span must overlap with service stresses at mid-span, and service stresses at ¼ span must overlap with transfer stresses near to the support.

Limiting stress for transfer at the top of beam (line 2t) at ¼ span

Eq. 8.30 $\sigma_{ti} \geq (-f_{ctm}(t) - M_{sw,qt}/Z_{t,co})/R_{tr}$

$\sigma_{ti} \geq (-2.72 - 52.07/25.920)/0.93 = -5.09$ N/mm²

Limiting stress for transfer at the bottom of beam (line 2b) at ¼ span

Eq. 8.34 $\sigma_{bi} \leq (0.6\, f_{ck}(t) + M_{sw,qt}/Z_{b,co})/R_{tr}$
$\sigma_{bi} \leq (18.00 + 52.07/31.726)/0.93 = +21.12$ N/mm²

This is greater than $\sigma_{bi} \geq +20.54$ N/mm² at mid-span, therefore OK.

Limiting stress for service at the top of beam (line 3t) at ¼ span

Eq. 8.38 $\sigma_{ti} \leq (0.45\, f_{cm} - M_{s,qt}/Z_{t,co})/R_s$
$\sigma_{ti} \leq (23.85 - 481.30/25.920)/0.80 = +6.60$ N/mm²

Limiting stress for service at the bottom of beam (line 3b) at ¼ span

Eq. 8.40 $\sigma_{bi} \geq (-f_{ctm} + M_{s,qt}/Z_{b,co})/R_s$
$\sigma_{bi} \geq (-3.80 + 481.30/31.726)/0.80 = +14.21$ N/mm²

This is less than $\sigma_{bi} \leq +19.97$ N/mm² at l_{pt} from the support, therefore OK.

The limiting stresses between (i) the support and ¼ span and (ii) ¼ span and mid-span are shown in Figure 8.19, each with its own overlapping design zone. The optimum design point for (ii) is determined first at point A where $\sigma_{bi} = +20.54$ N/mm² and $\sigma_{ti} = -5.09$ N/mm². The necessary initial pretensioning force P_{pi} is

Eq. 8.46 $P_{pi} = A_c\,(\sigma_{ti} + \sigma_{bi}/\alpha)/(1 + 1/\alpha)$
where $\alpha = Z_t/Z_b = 24.685/30.215 = 0.817$.
$P_{pi} = 307.5 \times 10^3\,(-5.09 + 20.54/0.817)/(1 + 1/0.817)/10^3 = 2772.9$ kN

This is of course exactly the same as in Example 8.3. If P_{pi} per strand is 115.23 kN, the number of strands is $N \geq 24.1$. As explained in Example 8.3 the eccentricity is too large so that $N = 27$ no. strands are used giving $P_{pi} = 3111.1$ kN. Using the same array of tendons as in Example 8.3 gives $z_{cp} = 101.3$ mm. The final initial prestress values are $+20.54 < \sigma_{bi} = +20.74 < +21.12$ N/mm², and $-5.09 < \sigma_{ti} = -2.88 < -1.14$ N/mm². This 'design point' is shown in Figure 8.19.

Between the support and ¼ span, the next lowest number of strands is at point B where $\sigma_{bi} = +19.97$ N/mm² and $\sigma_{ti} = -3.68$ N/mm². The necessary initial pretensioning force $P_{pi} = 2871.0$ kN requiring $N = 24.9$ strands so that $N = 24$ no. strands are used giving $P_{pi} = 2765.4$ kN. Using the same array of tendons as in Example 8.3 gives $z_{cp} = 99.0$ mm. The final initial prestress values are $+14.21 < \sigma_{bi} = +18.05 < +19.97$ N/mm², and $-3.68 < \sigma_{ti} = -2.09 < +6.60$ N/mm². This 'design point' is shown in Figure 8.19.

REFERENCES

Calladine, C. R. 2006, An improvement on Magnel's diagram, and an application, *Proceedings of the Institution of Civil Engineers - Structures and Buildings*, Vol. 159, No. 3, pp. 145–152.

Magnel, G. 1948. *Prestressed Concrete*, Concrete Publications, London, UK.

Chapter 9

Ultimate bending strength

9.1 ULTIMATE FLEXURAL STRENGTH OF PRESTRESSED CONCRETE ELEMENTS

9.1.1 Ultimate flexural tests on prestressed concrete hollow core floor units

The ultimate strength of precast prestressed concrete (psc) elements, such as slabs and beams in bending, is generally under predicted, by as much as 15% in the greater pretensioned units to 25% in the least pretensioned units. Compared with design values, M_{Rd}, where the major factor of safety $\gamma_s = 1.15$ is for the tendons, there is considerable reserve flexural strength in the elements more than calculations predict. The factor of safety for concrete $\gamma_c = 1.5$ has a greater influence in shallower slabs, rather than deeper beams, due to the smaller depth to the neutral axis and hence greater lever arm.

Figure 9.1 shows the latter stages of a long span bending test on a 1200 mm width × 320 mm depth hollow core unit (hcu) where the upward camber has been nullified and the net mid-span deflection (below a straight line drawn between the top of the supports, and not including camber) is about 40 mm, or span/275. The hcu could not be taken to total failure due to testing limitations, where the net deflection was expected to be about 60 mm.

Figure 9.2 shows the ultimate failure at mid-span in a 1200 mm wide × 200 mm depth hcu subjected to a central transverse line load at mid-span. The prestress in the bottom of the hcu was 'medium', i.e. around 12 N/mm^2, and there was considerable flexural cracking, developing into diagonal flexural-shear cracks, and deflection prior to local compressive failure in the top flange. The ultimate failure moment and design moment is not known.

Figure 9.3 shows the ultimate behaviour of the same tests shown (up to service loads) in Figure 7.1 and described in Section 7.1. To compare the experimental results with predicted ultimate values, given in Table 9.1, the following are assumed

i. The prestrain after losses is based on the assumptions relating to loss of prestress in Chapter 7, Section 7.1.1
ii. The depth to the neutral axis X is first based on the full width of the unit. If $0.8\ X > h_f$ (depth of top flange) it is recalculated based on a T section where the width of the rib is equal to the total width of the webs, i.e. no account is taken of the real circular profile of the top of the hollow core, however the effect on the final moment of resistance is known to be small
iii. The ultimate strain in the wires is limited to 0.02 even though the strain in the tests may exceed this. It is thought that this was not critical as calculated strains were in the region $0.015 < 0.02$, and the ratio between test/calculated moments averaged 1.09.

DOI: 10.1201/9781003319450-9

Figure 9.1 Flexural testing of a hollow core floor unit.

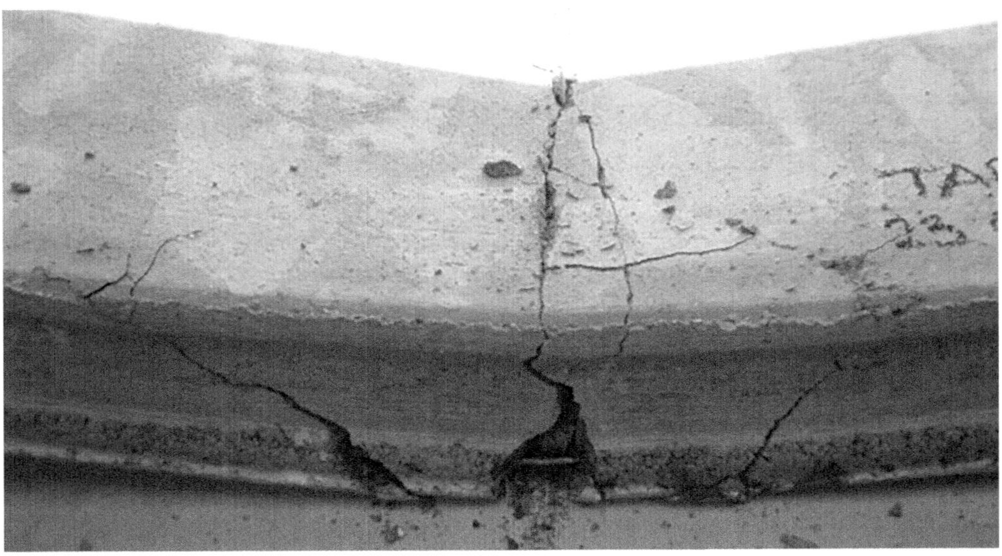

Figure 9.2 Ultimate failure in a hollow core floor unit (not the same unit as in Figure 9.1).

The main conclusions from the test was that the experimental test moment M_u ranged from 176 to 201 kNm, compared with a design value based on the nominal section and material properties of the hcu of M_{Rd}=151.9 kNm, an average ratio of 1.24. This ratio should have been even greater because the average height to the centroid of the wires was 57.0 mm

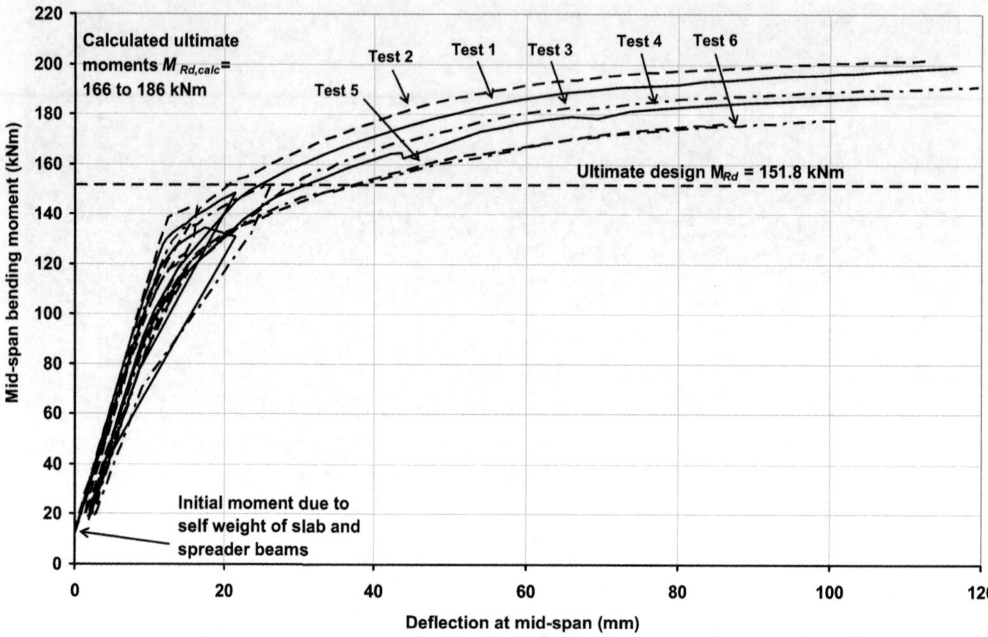

Figure 9.3 Experimental and predicted ultimate moment vs mid-span deflection in full scale tests carried out on 1200 × 200 mm depth prestressed hollow core floor units. (With permission by Bison Precast Ltd., UK.)

Table 9.1 Calculation for ultimate moment of resistance in full scale tests carried out on 1200 × 200 mm depth prestressed hollow core floor units

Slab ref	*Effective depth to wires (mm)*	*Ultimate stress in wires (N/mm²)*	*Lever arm (mm)*	*Calculated $M_{Rd,calc}$ (kNm)*	*Experimental M_u (kNm)*	*Ratio to design M_u/M_{Rd}*	*Ratio to calculated $M_u/M_{Rd,calc}$*
1	151.1	1606	137.0	186.3	198.7	1.31	1.07
2	147.7	1603	133.6	181.3	201.6	1.33	1.11
3	140.1	1596	126.1	170.4	190.9	1.26	1.12
4	139.8	1595	125.8	169.8	186.6	1.23	1.10
5	137.0	1592	123.1	165.9	176.0	1.16	1.06
6	137.6	1593	123.6	166.7	177.7	1.17	1.07
Mean	142.2	1597	128.2	173.4	188.6	1.24	1.09
Unit	*154.5*	*1330*	*134.8*	*151.8*	-	-	-

For all tests, measured mean cube strength $f_{ck,cube}$=59.1 N/mm², f_{ck} taken as 0.8×59.1=47.3 N/mm²

Experimental M_u includes bending moment due to self weight and loading equipment=12.0 kNm.

Unit=nominal section and material properties of designed hollow core unit.

compared with the design value of 45.5 mm. If the wires in the nominal unit are raised to 57.0 mm the above ratio increases to 1.31.

The calculated moments of resistance $M_{Rd,calc}$ varied from 166 to 186 kNm, with the lowest values calculated for the hcu with the highest position of wires, i.e. Tests 5 and 6 where y_s=61 and 62 mm, respectively. However this effect was reflected in M_u, which had the lowest values of 176.0 and 177.7 for Tests 5 and 6. Therefore the average ratio of

$M_u/M_{Rd,calc}=1.09$, with the lowest and therefore best predicted ratio of 1.06 paradoxically being for Test 5. The standard deviation for $M_u/M_{Rd,calc}$ is only 0.025, which shows that, in spite of the variations in section properties and height of wires, the set of results show good consistency, with a confident prediction.

9.1.2 Ultimate flexural tests on T beams

Similar flexural tests have also been carried out by the authors on psc T beams used in beam and block flooring as described in Section 7.1.2. The beams have a narrow upstand, 42 mm wide at the top increasing to 48 mm width at the bottom of the upstand. The upstand is restrained laterally by the grouted infill blocks and can therefore generate its full compressive capacity. However, the compressive strength of the upstand is limited due to its narrow width and under large curvature and ultimate strains may fail by crushing and/or lateral spalling over a length of about 150 mm, as shown in Figure 9.4.

Figure 9.5 shows the ultimate and failure stages of a bending test carried out on the same beams shown in Figure 7.3a giving the ultimate failure shown in Figure 7.3b. Unlike the tests carried out on hcu where hydraulic loading was used, 18 kg blocks kentledge was added to the beam, such that the mode of failure was uncontrolled and sudden, preceded only by flexural cracking (vertical cracks) in the mid-span region. The horizontal splitting cracks in the upstand did not occur until the beam failed. Spalling extended to the bottom of the upstand, 100 mm from the top of the beam, even though the depth to the neutral axis was calculated as 63 mm.

Figure 9.5 shows the beams exhibited excellent ductility, where the mid-span deflection at failure was more than 4.2 times the deflection at the onset of cracking. To compare the experimental results with predicted ultimate values, given in Table 9.2, the same assumptions are used as given in Section 9.1.1.

The main conclusions from the test was that the experimental test moment M_u in beams 1 and 2 was 9.51 and 9.66 kNm, respectively, compared with a design value based on the nominal section and material properties of the beam of $M_{Rd}=6.89$ kNm, an average ratio of 1.39. As the height to the centroid of the wires in the test beams was ±1 mm of the nominal values, there is no effect of the position of the wires.

The calculated moments of resistance $M_{Rd,calc}$ in beams 1 and 2 was 8.61 and 8.55 kNm, respectively. The ratio for the two beams was $M_u/M_{Rd,calc}=1.10$ and 1.13, showing good consistency. The average value of 1.117 is close to that for the hcu of 1.09. Overall

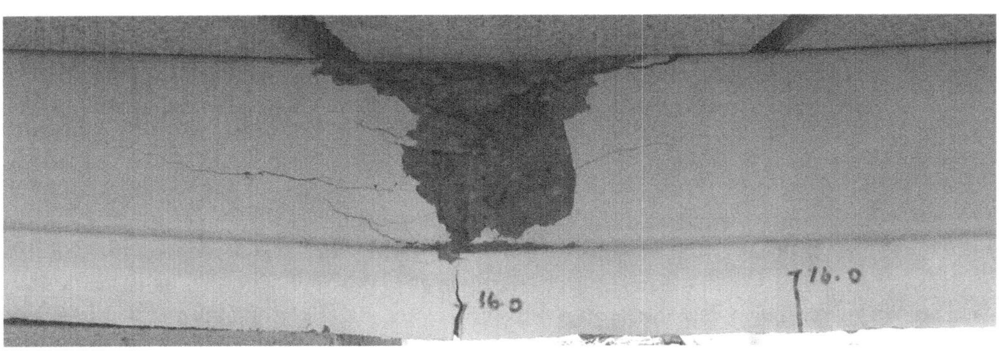

Figure 9.4 Ultimate failure, by compressive crushing, splitting and spalling in the 42 mm width upstand of a 150 mm depth prestressed T beam.

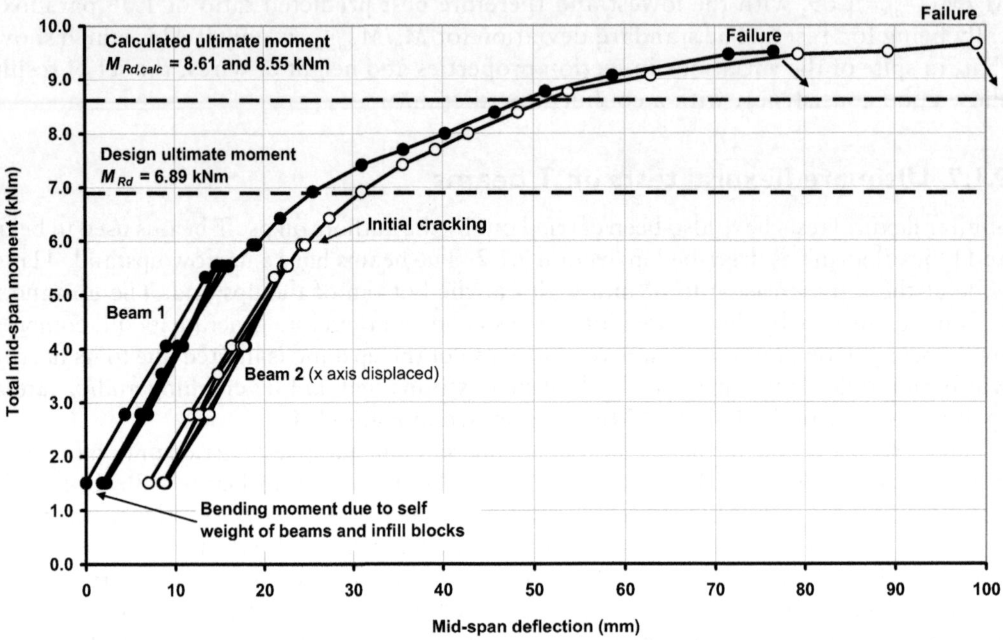

Figure 9.5 Experimental, predicted and design ultimate moment vs mid-span deflection in full scale tests carried out on a pair of 92 mm width × 150 mm depth prestressed T beams. (With permission by Litecast Ltd., UK.)

Table 9.2 Calculation for ultimate moment of resistance for full scale tests carried out on 92 × 150 mm depth prestressed T beams

Beam ref	*Effective depth to wires (mm)*	*Ultimate stress in wires (N/mm²)*	*Lever arm (mm)*	*Calculated $M_{Rd,calc}$ (kNm)*	*Experimental M_u (kNm)*	*Ratio to design M_u/M_{Rd}*	*Ratio to calculated $M_u/M_{Rd,calc}$*
1	116.5	1600	91.4	8.61	9.51	1.38	1.10
2	115.8	1599	90.7	8.55	9.66	1.40	1.13
Mean	116.2	1600	91.1	8.58	9.58	1.39	1.12
Beam	*115.8*	*1381*	*84.7*	*6.89*	-	-	-

For all tests, f_{ck} taken as 53.5 N/mm².

Beam = nominal section and material properties of designed T beam.

the results show good consistency, with confident predictions of the ultimate moment of resistance.

Deflections of flexurally cracked beams, where the total moment $M_k > M_{cr}$, may be calculated based on the partially cracked section properties. M_k is the total bending moment at a particular loading station, for example in Figure 9.5 where $M_k = 8.0$ kNm the average measured deflection for the two beams is $\delta = 38.1$ mm. M_{cr} is the calculated cracking moment defined in Section 7.1.1, i.e. using 1.5 f_{ctm} for the allowable tension prior to cracking. The average cracking moment of resistance for the two beams is $M_{cr} = 5.84$ kNm.

Generally the mid-span deflection for UDL is given as $\delta = 5w\,L^4/384\,K_{ef}$, where K_{ef} is the effective flexural rigidity defined below. But if the mid-span moment $M_k = w\,L^2/8$ then

$$\delta = 40M_kL^2/384K_{ef} \tag{9.1}$$

where L=effective span=4400 mm

BS EN 1992-1, clause 7.4.3. and Exp. 7.18 & 7.19 (BS EN 1992-1-1, 2004) the effective rigidity is given as

$$K_{ef} = K_2 + \left[(K_1 - K_2)\ (M_{cr}/M_k)^2\right] \le K_1 \tag{9.2}$$

where

$$K_1 = E_{cm}\ I_{c,co} \text{ is the uncracked compound flexural rigidity} \tag{9.3}$$

$$K_2 = E_{cm}\ I_{cr} \text{ is the flexural rigidity for the fully cracked section} \tag{9.4}$$

The modified value of E_{cm} from Section 7.1.2, to bring the calculated deflections in line with the experimental deflections, is E_{cm}=34500 N/mm^2. The mean value of $I_{c,co}$ for the two beams is 16.98×10^6 mm^4. Then $K_1 = 34500\times16.98\times10^6 = 0.586\times10^{12}$ Nmm^2.

For the flexurally cracked section $K_2 = E_{cm}\,I_{cr}$, where I_{cr} is calculated as follows

Transformed area of tendons=$m\,A_p$
Width of beam upstand=b_u
Depth to the cracked centroidal axis x_c=first moment of area from the top of the beam/area of the beam plus the transformed area of tendons

$$x_c = \left(b_u x_c^2/2 + m\,A_p d\right)\ /\ \left(b_u x_c + m\,A_p\right) \tag{9.5}$$

The quadratic $b_u\,x_c^2/2 + m\,A_p\,x_c - m\,A_p\,d = 0$ is solved for x_c

$$\text{The 2nd moment of area } I_{cr} = b_u x_c^3/3 + m\,A_p(d - x_c)^2 \tag{9.6}$$

The average values for the two beams are: b_u=41 mm, d=116.2 mm and A_p=58.9 mm. Then x_c=36.8 mm and $I_{cr}=2.886\times10^6$ mm^4. Note that this is only 17% of $I_{c,co}$, showing a huge decrease when the section is fully cracked close to failure. For the fully cracked section $K_2 = 34500\times2.886\times10^6 = 0.100\times10^{12}$ Nmm^2.

However in the partially cracked region where $M_{cr} \le M_k = 8.00$ kNm $\le M_u$

Eq. 9.2 $K_{ef} = [0.586 + (0.586 - 0.100)\ (5.84/8.00)^2]\times10^{12} = 0.360\times10^{12}$ Nmm^2.

The calculated mi-span deflection is

Eq. 9.1 $\delta = 40\times(8.00 - 1.51)\times10^6\times4400^2/384\times0.360\times10^{12} = 36.3$ mm compared with a measured deflection of 38.1 mm, i.e. 96% of the measured deflection.

Values for deflections at the other load stations in the range $M_{cr} \le M_k \le M_u$ are given in Table 9.3. The average ratio of the calculated to the measured deflections is 0.95, showing further confident predictions are possible, although the accuracy is lost towards final failure where $M_k \ge 9.07$ kNm, or 95% of M_u.

Table 9.3 Comparison of predicted deflections and measured deflections after cracking in full scale tests carried out on 92 × 150 mm depth prestressed T beams

Total moment M_k (kNm)	*Moment factor $(M_{cr}/M_k)^2$*	*Effective $K_{ef} \times 10^{12}$ (Nmm²)*	*Calculated deflections (mm)*	*Measured deflections (mm)*	*Ratio of calculated/ measured deflections*
5.93	0.971	0.572	15.6	18.4	0.85
6.42	0.827	0.502	19.7	21.1	0.94
6.92	0.713	0.446	24.4	24.8	0.99
7.41	0.621	0.402	29.6	29.8	1.00
7.69	0.576	0.380	32.8	33.9	0.97
8.00	0.535	0.360	36.3	38.1	0.96
8.38	0.486	0.336	41.3	43.6	0.95
8.77	0.443	0.315	46.5	49.0	0.95
9.07	0.415	0.301	50.6	57.3	0.88
Average					**0.95**

For both beams the average M_{cr}=5.84 kNm, uncracked K_I=0.586×10¹² Nmm².

9.2 ULTIMATE LIMIT STATE OF FLEXURAL STRENGTH

9.2.1 Background to ultimate moment of resistance and design moments

In calculating the ultimate moment of resistance of a psc section M_{Rd}, the same procedure as for statically reinforced concrete is used, with one important exception – the ultimate strength of the tendons varies depending on the level of prestress and the effective depth to the tendons. To calculate M_{Rd} three relationships are combined

i. compatibility of thru-depth strains
ii. equilibrium of forces in the concrete versus the tendons
iii. constitutive σ v ε relationship of the tendons

to determine the ultimate stress in the tendons f_p and the lever arm z from the centroid of the concrete in compression to the tendons in tension. Then, moment of resistance=stress×area×lever arm

$$M_{Rd} = f_p A_p z \tag{9.7}$$

where A_p is the total area of tendons in the tensile zone of the section, i.e. discounting tendons in the top of the section an any that lie about 0.4d from the top.

The material partial safety factors to be applied as per usual: γ_s=1.15 for steel tendons and γ_c=1.5 for concrete in flexure. When using bonded tendons (as opposed to unbounded in post-tensioned concrete design) M_{Rd} is limited by one the following

i. ultimate compressive strength of concrete, 0.567 f_{ck}
ii. the design tensile stress in the tendons, f_p

M_{Rd} is checked against the ultimate design moment M_{Ed} as defined in Section 3.1.3. Ultimate uniform, line and point loads w_{Ed}, P_{Ed}, etc. are obtained based on BS EN 1990 (BS EN 1990, 2002) for the greater of Exp. 6.10a or 6.10b. There is a closer balance between these two expressions for beams than such as hcu due to the additional self weight of the beam, which may about 15%–20% to the dead loads. Point loads in beams are usually due to wall loads for example transferred from floor slabs – they are not considered as short line loads due to lateral distribution of wall loads over the width of three (or four for composite slabs) units. Line loads on beams derive from localised cross-walls acting on floor slabs.

In solid slabs or beams it is not usual for M_{Rd} to be critical at the expense of the service moment of resistance M_{sR} where the area of tendons A_p is relatively small, say less than 0.2% of the cross-sectional area of concrete. This is because M_{sR} is a function of the prestress σ_b plus the tensile strength f_{ctm}, so even for small area of tendons M_{sR} is already $f_{ctm}\ Z_{b,co}$ even where σ_b is relatively small, say 8 N/mm². By contrast $M_{Rd}=f_p\ A_p\ z$, so a low A_p leads to a low M_{Rd}, which becomes critical.

For hcu, thanks to 70 years of evolution leading to the efficient design of the cross-section by manufacturers, the balance between M_{sR} and M_{Rd} is much closer. Where the exposure category is more severe than XC1 and $f_{ctm}=0$, the balance between M_{sR} and M_{Rd} is the opposite because imposed live loads are considerably reduced as they are multiplied by the quasi-permanent factor ψ_2 typically 0.3, or the frequent factor ψ_1 typically 0.5, as explained in Section 3.1.2.

9.2.2 Ratio of ultimate to service moments of resistance, and the critical limit state for hollow core units

Taking for example the hcu in Figure 6.7 used in Example 7.1 for exposure XC1, Table 9.4 shows how M_{sR} and M_{Rd} increase with increasing area Ap. In this exercise the effective span is adjusted for strand patterns that are different to that used in Example 7.1, i.e. 9.80 m. For the hcu the least number possible is 7 no.×9.3 mm diameter strands (having 6 no. cores) and this leads to a ratio $M_{Rd}/M_{sR}=1.14$. This is lower than the mean load factor for ultimate/service loads, which is $w_{Ed}/w_s=20.15/14.92=1.35$ in Example 7.1. Therefore the ultimate limit state of strength is reached before the serviceability limit state of stress, in this case f_{ctm} at the bottom fibre. Note that w_{Ed}/w_s is never less than 1.30 in this example. For the maximum strand pattern of 7 no.×9.3 mm plus 7 no. 12.5 mm diameter strands the ratio $M_{Rd}/M_{sR}=1.36>1.35$ so the unit is critical at the serviceability limit state. Cross-over occurs where $\rho=A_p/A_c=0.39\%$ area of strands where $M_{Rd}/M_{sR}=w_{Ed}/w_s$.

Table 9.4 Service and ultimate moment of resistance ratios for 1200 mm wide × 250 mm depth prestressed hollow core floor unit pretensioned with different amounts of strands.

Number of strands 9.3 mm	12.5 mm	*Area of strands* (mm²)	*Percentage of concrete area* (%)	*Final bottom prestress* (N/mm²)	*Service* M_{sR} (kNm)	*Ultimate* M_{Rd} (kNm)	*Moment ratio* M_{Rd}/M_{sR}	*Load ratio* w_{Ed}/w_s	*Critical limiting state*
7	0	364	0.22	5.50	97.4	110.7	1.14	1.35	Ultimate
5	2	446	0.27	6.66	110.0	134.0	1.22	1.35	Ultimate
2	5	569	0.34	8.28	127.6	168.0	1.32	1.35	Ultimate
0	7	651	0.39	9.49	140.7	189.9	1.35	1.35	Both
2	7	755	0.45	10.89	156.3	214.4	1.37	1.35	Service
5	7	911	0.54	12.94	179.1	246.3	1.37	1.35	Service
7	7	1015	0.60	14.26	194.1	264.6	1.36	1.35	Service

See Figure 6.7 for cross-section and details of pretensioning strands.

Exposure category=XC1. $f_{ctm}=3.80$ N/mm².

Table 9.5 Service and ultimate moment of resistance ratios for 1200 mm wide × 250 mm depth prestressed hollow core floor unit pretensioned with different amounts of strands for exposure category XC3

Number of strands 9.3 mm	*12.5 mm*	*Area of strands (mm²)*	*Percentage of concrete area (%)*	*Final bottom prestress (N/mm²)*	*Service M_{sR} (kNm)*	*Ultimate M_{Rd} (kNm)*	*Moment ratio M_{Rd}/M_{sR}*	*Load ratio w_{Ed}/w_s*	*Critical limiting state*
7	0	364	0.22	5.50	57.6	110.7	1.92	1.88	Service
5	2	446	0.27	6.66	70.0	134.0	1.91	1.88	Service
2	5	569	0.34	8.37	88.4	168.0	1.90	1.88	Service
0	7	651	0.39	9.49	100.5	189.9	1.89	1.88	Service
2	7	755	0.45	10.91	116.0	214.4	1.85	1.88	Ultimate
5	7	911	0.54	12.96	138.7	246.3	1.78	1.88	Ultimate
7	7	1015	0.60	14.27	153.4	264.6	1.73	1.88	Ultimate

See Figure 6.7 for cross-section and details of pretensioning strands.

Exposure category=XC3. f_{ctm}=0 N/mm². Use of floor=offices with ψ_2=0.3 for imposed live load.

It is expected that the rate of increase in M_{Rd} should mirror the rate of increase in A_p over the range, and this is true for $\rho<0.45\%$. Above this value the depth to the neutral axis (as explained in Section 9.3.1) is below the top flange and lies between the webs, making the moment of resistance less efficient – a 3% to 4% drop off in M_{Rd} is found.

For a notionally identical 400 mm depth hcu this exercise was carried out using the same strand patterns as before (except 2 no. 7 mm wires in the top of the hcu are added). As before M_{Rd} is critical up to 0.43% area of strands, with the ratio $M_{Rd}/M_{sR}=1.03$ to 1.36, virtually the same result as in Table 9.4 for 250 mm depth hcu.

Continuing with the 250 mm depth hcu, Table 9.5 shows how M_{sR} and $M_{Rd\ increase}$ for exposure XC3, where the situation is reversed due to the lower values of M_{sR} (because $f_{ctm}=0$) and the much larger load ratio $w_{Ed}/w_s=1.88$ (because the imposed live load is multiplied by $\psi_2=0.3$). For the lowest area of strands $M_{Rd}/M_{sR}=1.92>1.88$, and service governs. For the highest area of strands $M_{Rd}/M_{sR}=1.73<1.88$, and the ultimate limit state governs. For external exposure it is important that serviceability governs the long-term design condition.

The results are summarised by plotting graphs of the normalised ratio $\eta=M_{Rd}/M_{sR}$ versus w_{Ed}/w_s in Figure 9.6. The service moment is critical where $\eta\geq1$. It is shown that this occurs around $\rho\geq0.4\%$ for hcu's for exposure XC1, where there is little difference for depths of 250 and 400 mm. By coincidence this also occurs at $\rho\leq0.4\%$ for hcu's for exposure XC3, except that the trend is reversed with the ultimate moment being critical for $\rho\geq0.4\%$.

Service and ultimate moments of resistance, together with ultimate shear capacity, are also summarised for a wide range of hcu in Table 9.1, from 150 to 500 mm depth.

9.2.3 Ratio of ultimate to service moments of resistance, and the critical limit state for solid floor units and beams

Taking for example a 1200 mm width×250 mm depth solid slab, having the same cross section as the hcu in Figure 6.7 without the cores, for exposure XC1, Table 9.6 shows how M_{sR} and M_{Rd} increase with increasing area A_p. For the lowest number of strands 8 no. 9.3 mm diameter strands the ratio $M_{Rd}/M_{sR}=1.19$. This is lower than the mean load factor for ultimate/service loads, which is $w_{Ed}/w_s=1.33$ such that ultimate limit state is reached before the service stress f_{ctm}. For the maximum strand pattern of 12 no.×9.3 mm plus 4 no. 12.5 mm diameter strands (with 4 no. 7 mm diameter wires in the top) the ratio

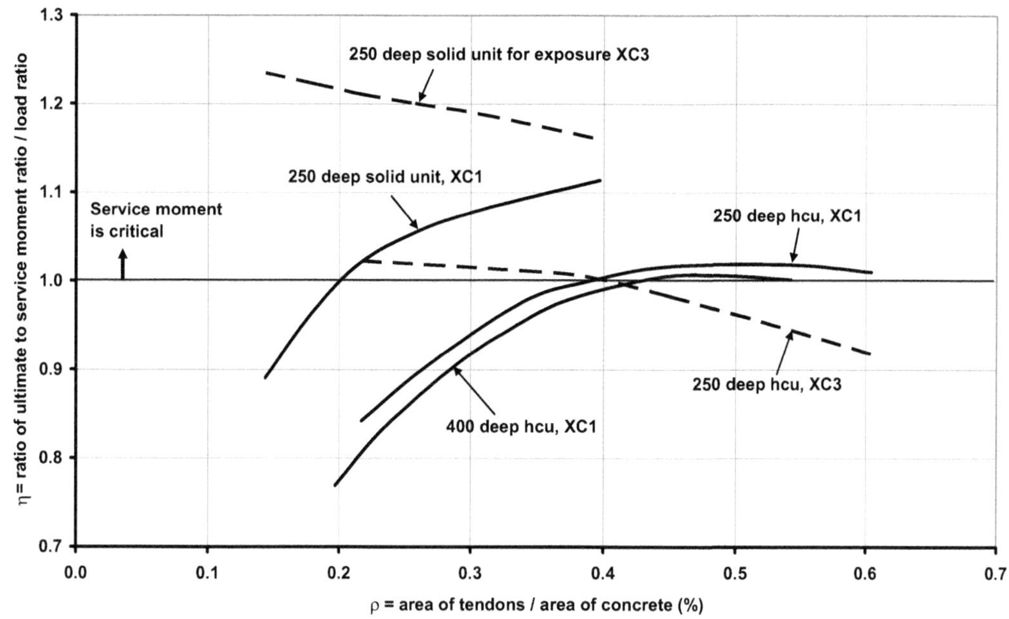

Figure 9.6 Relationships between the ratio of ultimate to service moments of resistance and ultimate to service design loads verses the ratio of the area of tendons to the area of concrete in prestressed solid and hollow core units.

Table 9.6 Service and ultimate moment of resistance ratios for 1200 mm wide × 250 mm depth prestressed solid unit pretensioned with different amounts of strands for exposure category XC1

Number of strands 9.3 mm	*12.5 mm*	*Area of strands (mm²)*	*Percentage of concrete area (%)*	*Final bottom prestress (N/mm²)*	*Service M_{sR} (kNm)*	*Ultimate M_{Rd} (kNm)*	*Moment ratio M_{Rd}/M_{sR}*	*Load ratio w_{Ed}/w_s*	*Critical limiting state*
8	0	416	0.14	4.76	106.0	125.8	1.19	1.33	Ultimate
10	0	520	0.18	5.93	120.9	155.4	1.29	1.33	Ultimate
12	0	624	0.22	7.08	135.6	184.3	1.36	1.33	Service
10[a]	2	745	0.26	7.83	145.4	204.3	1.41	1.33	Service
12[a]	2	887	0.31	8.85	158.7	228.4	1.44	1.33	Service
12[a]	4	1150	0.40	10.57	181.3	269.3	1.48	1.33	Service

See Figure 6.7 for cross-section without cores and details of pretensioning strands.

Exposure category = XC1. f_{ctm} = 3,80 N/mm².

[a] 4 no. 5 mm wires at 30 mm cover in the top of the unit.

$M_{Rd}/M_{sR} = 1.48 > 1.33$ so the unit is critical at the serviceability limit state of stress. Cross-over occurs at $\rho = 0.20\%$ – note that is half the value for hcu, as the service moment of resistance is much less efficient than for hcu. A key factor here is that the ratio of M_{Rd} to A_p is almost the same for the 250 mm depth solid unit as for the 250 mm depth hcu – the ratio varies from 0.30 to 0.23 kNm per mm² strands, with the averages for the solid and hcu being 0.28 and 0.29 kNm per mm², respectively.

Figure 9.6 also shows plots of η vs ρ for solid units, where the service moment for exposure XC1 is critical for $\rho \geq 0.20\%$, and is always critical for exposure XC3. This is because the effect of using $f_{ctm}=0$ in a solid unit has a much greater influence on M_{sR} than in a hcu because the prestress in the bottom fibre is smaller due to its larger cross-sectional area, in fact 1.72 times that of a hcu.

9.3 DESIGN PROCEDURES FOR ULTIMATE LIMIT STATE OF FLEXURAL STRENGTH

9.3.1 Ultimate force equilibrium and strain compatibility

As shown in Section 9.1 the ultimate failure of psc beams and slabs is considerably ductile, but in spite of sections being under reinforced flexurally due to the small areas of reinforcement (compared to comparable r.c. sections) the final mode of failure is sudden and often governed by compressive failure of the concrete as shown in Figures 9.2 and 9.5. It should be noted that the cross-line load in Figure 9.2 is not representative of uniformly distributed loading, and will have contributed to the localised compressive failure.

The depth of the (strain responsive) neutral axis X is obtained by considering the equilibrium of the section. If we think back to a statically reinforced section, where the σ_s v ε_s curve for rebar is a flat plateau, then referring to Figure 9.7 the ultimate forces in the concrete F_c and steel bars F_s are

$$F_c = 0.567 f_{ck} b 0.8X \tag{9.8}$$

$$F_s = \sigma_s A_s \tag{9.9}$$

Then

$$X = \left(A_s / 0.453 f_{ck} b\right) \sigma_s \tag{9.10}$$

Considering the linear strain distribution in Figure 9.7

$$X = \left[\varepsilon_{cu3} / \left(\varepsilon_{cu3} + \varepsilon_s\right)\right] d \tag{9.11}$$

Combining Eq. 9.10 and 9.11 gives

$$\sigma_s = \left[\varepsilon_{cu3} / \left(\varepsilon_{cu3} + \varepsilon_s\right)\right] 0.453 f_{ck} b \ d / A_s \tag{9.12}$$

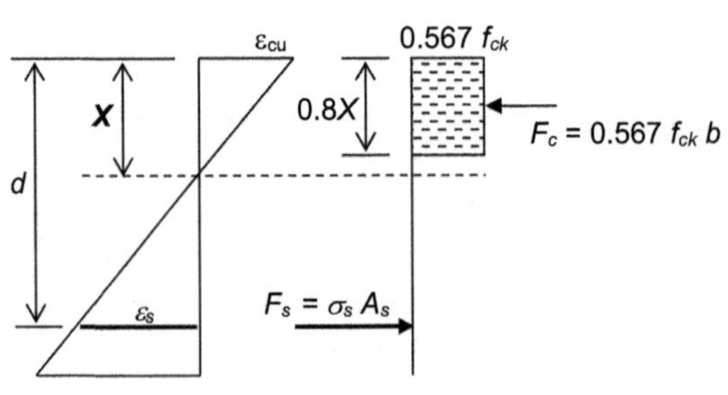

Figure 9.7 Force equilibrium and compatibility of strains in reinforced concrete flexural design.

For known values of f_{ck}, b, d, A_s and ε_{cu3}, σ_s is inversely proportional to ε_s, and the design point is where this curve intersects the plateau of the constitutive σ_s v ε_s plot for the rebars. Of course we know that $\sigma_s = 0.87\, f_{yk}$, and therefore it is not necessary to carry out this operation, but it serves as an introduction for the procedures required for pretensioned tendons where the constitutive σ_p v ε_p plot is not a plateau, but a strain hardening slope as shown in Figure 9.8.

The ultimate tensile strength of the tendons depends on the final prestress σ_{po} in the tendons after all losses and initial prestress levels have been considered. See Eq. 7.35. In most hcu production (and sections where the tendons are all near to the bottom) the ratio $\sigma_{po}/f_{pk} = 0.53$ to 0.6, depending on the area of tendons. For double tee units where most of the tendons are positioned further from the bottom $\sigma_{po}/f_{pk} = 0.45$ to 0.5. The stress history for pretensioned tendons is shown in Figure 9.9 where the effect of prestrain after losses ε_{po} (A),

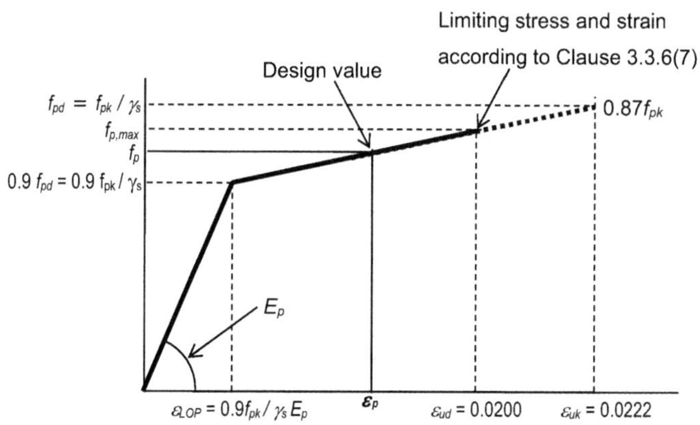

Figure 9.8 Stress v strain relationship for pretensioning tendons according to BS EN 1992-1-1, Fig. 3.10.

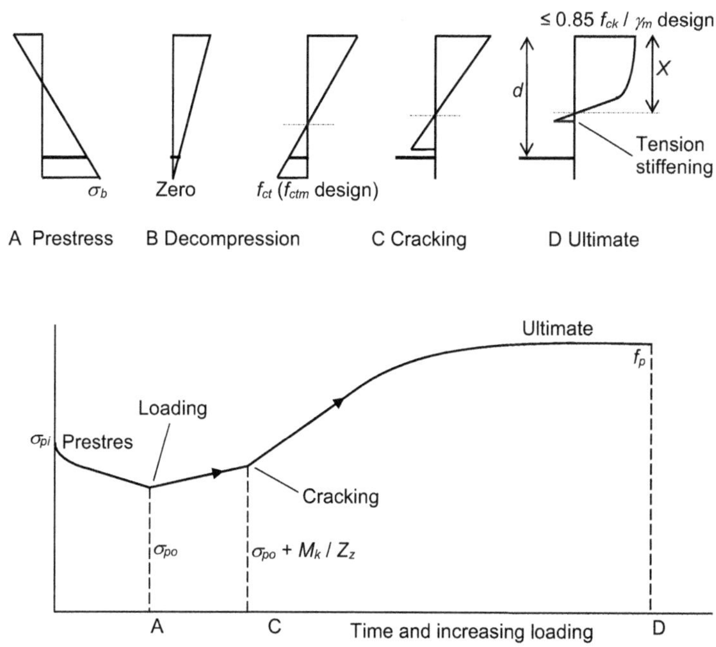

Figure 9.9 Stress history of pretensioned tendons.

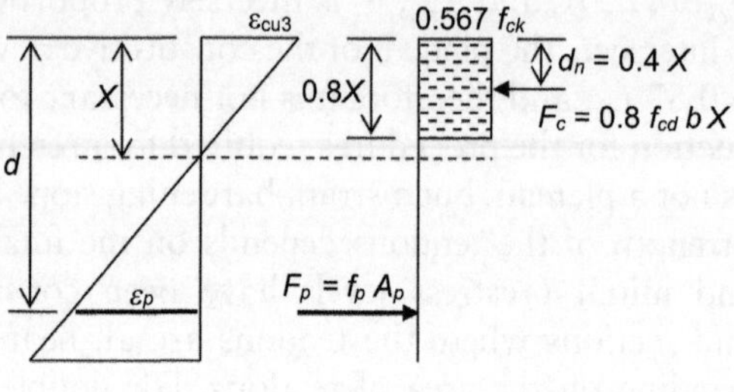

Figure 9.10 **Force equilibrium and compatibility of strains in prestressed concrete flexural design.**

decompression $M_k = \sigma_b\ Z_{b,co}$ (B), cracking $M_k > f_{ct}\ Z_{b,co}$ (C) and ultimate strength $M_k = f_p\ A_p\ z$ (assuming tendons fail before concrete) (D) are noted, where M_k is the applied bending moment at each stage.

Referring to Figure 9.10, values for X and f_p may be obtained from three considerations

i. force equilibrium $F_p = F_c$ where

$$F_p = f_p A_{pT} \tag{9.13}$$

$$F_c = f_{cd} 0.8b\ X \tag{9.14}$$

Then

$$f_p A_{pT} = f_{cd} b 0.8X \tag{9.15}$$

where $f_{cd} = 0.85\ f_{ck}/\gamma_c$ and b=width of compression zone, which for a hcu is the actual width of the unit plus the *insitu* infill in the longitudinal joints=1200 mm. Note that although the strength of the infill may be less than the precast, say $f_{cki} = 25$ N/mm² compared with $f_{ck} = 45$ N/mm², the infill is confined, and as the infill is typically only 50 mm wide (compared with 1150 mm for the unit) the effective width is taken as 1200 mm.

BS EN 1992-1-1, clause 3.1.7. For $50 < f_{ck} \leq 90$ N/mm², the stress block depth factor reduces from 0.8 to $\lambda = 0.8 - (f_{ck} - 50)/400$, and the design strength factor from 1 to $\eta = 1.0 - (f_{ck} - 50)/200$.

ii. strain compatibility of the ultimate strain in the concrete ε_{cu3} and the strain ε_p in the tendons. Referring to Figure 9.11, after the final losses in service, the ultimate compatibility strain ε_p' is given as

$$\varepsilon_p / \varepsilon_{cu3} = (d - X) / X$$

or

$$\varepsilon_p / \varepsilon_{cu3} = (d / X - 1) \tag{9.16}$$

The total ultimate strain is

$$\varepsilon_p = \varepsilon_{po} + \varepsilon_{cu3}(d / X - 1) \tag{9.17}$$

where $\varepsilon_{po} = \sigma_{po}/E_p$

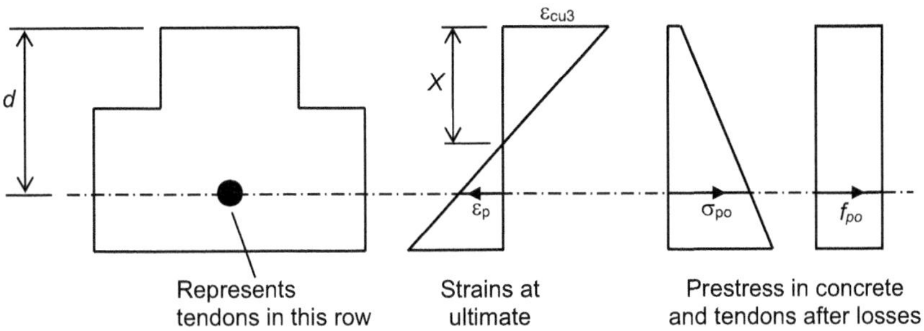

Figure 9.11 Strains in concrete and tendons at ultimate limit state.

iii. the stress v strain relationship shown in Figure 9.8 (From EN1992-1-1, Fig. 3.10), solving for X as follows

Fig. 3.4, Table 3.1 $\varepsilon_{cu3}=0.0035$ for $f_{ck}\leq 50$ N/mm^2

Otherwise $\varepsilon_{cu3}=0.0026+0.035\ [(90-f_{ck})/100]^4$ for $f_{ck}\leq 90$ N/mm^2

3.3.6(7), Fig. 3.10 $\varepsilon_{LOP}=0.9\ f_{pk}/(\gamma_s E_p)$ at which stress$=0.9\ f_{pk}/\gamma_s$ (9.18)

3.3.6(7), Fig. 3.10 $\varepsilon_{uk}=0.0222$ at which design stress $f_{pd}=f_{pk}/\gamma_s$

5.10.9 Pre-strain due to prestress after losses $\varepsilon_{po}=r_{sup}\ \sigma_{po}/E_p$ but $r_{sup}=1.0$

Effective depth of tendons in tension zone $d = h - y_{sT}$ (9.19)

Let X=depth to neutral axis, and first assuming that $0.8\ X<h_{ft}$, the depth of the top flange in such as a hcu (Figure 9.12), or the top flange of a composite beam.

Refer to Figure 9.8. If $\varepsilon_{LOP}\leq\varepsilon_p\leq\varepsilon_{uk}$ in the inclined branch of this figure

$$f_p = 0.9f_{pd} + 0.1f_{pd}\left[\left(\varepsilon_p - \varepsilon_{LOP}\right)/\left(\varepsilon_{uk} - \varepsilon_{LOP}\right)\right] \tag{9.20}$$

but where the strain is limited $\varepsilon_p\leq\varepsilon_{ud}=0.9\ \varepsilon_{uk}=0.02$

Combining Eqs. 9.15, 9.17, 9.20 gives the following quadratic equation

$$0.8f_{cd}b\left(\varepsilon_{uk} - \varepsilon_{LOP}\right)X^2 - \left[0.9\left(\varepsilon_{uk} - \varepsilon_{LOP}\right) + 0.1\left(\varepsilon_{po} - \varepsilon_{cu3} - \varepsilon_{LOP}\right)\right]A_{pT}f_{pd}X - 0.1\varepsilon_{cu3}d\ A_{pT}f_p = 0 \tag{9.21}$$

Solving yields X

Depth to centroid of concrete area $d_n=0.5\times 0.8X=0.4X$

If the compression zone lies beneath the level of the top flange, i.e. $0.8X>h_{ft}$ the compression force is the sum of the force in the top flange above the cores and webs between the cores as shown in Figure 9.12.

Thus Eq. 9.15 is amended to

$$f_pA_{pT} = b_ch_{ft}f_{cd} + 0.8f_{cd}\left(b - b_c\right)X \tag{9.22}$$

where b_c is the total width of cores, such that Eq. 9.21 is amended to

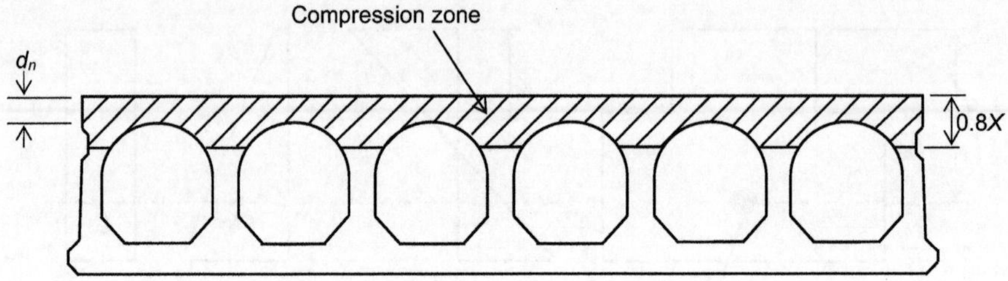

Figure 9.12 Compressive stress zones in top flange and above hollow cores.

$$0.8f_{cd}(b-b_c)(\varepsilon_{uk}-\varepsilon_{LOP})X^2-\{[0.9(\varepsilon_{uk}-\varepsilon_{LOP})+0.1(\varepsilon_{po}-\varepsilon_{cu3}-\varepsilon_{LOP})]A_{pT}f_{pd}-f_{cd}b_c h_{ft}(\varepsilon_{uk}-\varepsilon_{LOP})\}$$
$$X-0.1\varepsilon_{cu3}d\ A_{pT}f_{pd}=0 \quad (9.23)$$

Solving yields $X > h_{ft}/0.8$

Depth to centroid of concrete area is derived from the area of the top flange above the hollow cores $b_c\ h_{ft}$ and the area of the webs between the cores $(b-b_c)$ 0.8 X, as follows

$$d_n=\left[b_c h_{ft}^2/2+(b-b_c)(0.8X)^2/2\right]/\left[b_c h_{ft}+(b-b_c)\ 0.8X\right] \quad (9.24)$$

Lever arm $z=d-d_n$ (9.25)

Eq. 9.17 $\varepsilon_p=\varepsilon_{po}+0.0035\ (d/X-1)\le\varepsilon_{ud}=0.02$ (9.26)

f_p is found from Eq. 9.22, but

$$f_{p,max}\le 0.9f_{pd}+\ 0.1f_{pd}\left[(\varepsilon_{ud}-\varepsilon_{LOP})/(\varepsilon_{uk}-\varepsilon_{LOP})\right] \quad (9.27)$$

For strands with E_p=195 kN/mm^2, and wires with E_p=205 kN/mm^2, $f_{p,max}$=0.857 f_{pk}.

Finally the ultimate moment of resistance is given as

$$M_{Rd}=f_p A_{pT}z \quad (9.28)$$

9.3.2 Tabulated method for rectangular sections

A simplified tabulated method for rectangular sections is to obtain f_p/f_{pk} and X/d from tables of data derived from the equilibrium of ultimate forces and the compatibility of ultimate strains

$$F_p=A_{pT}f_p=A_{pT}f_{pk}(f_p/f_{pk}) \quad (9.29)$$

$$F_c=f_{cd}b0.8X=\ 0.8f_{cd}bd(X/d) \quad (9.30)$$

If $\varepsilon_p\le\varepsilon_{LOP}$ (elastic branch of σ-ε curve), from Eq. 9.17 x E_p/f_{pk}

$$f_p/f_{pk}=f_{po}/f_{pk}+\varepsilon_{cu3}E_p(d/X-1)/f_{pk} \quad (9.31)$$

Then

$$X / d = \left[A_{pT} f_{pk} / 0.8 f_{cd} bd\right] \left[f_{po} / f_{pk} + \varepsilon_{cu3} E_p \left(d / X - 1\right) / f_{pk}\right] \tag{9.32}$$

Solving the quadratic gives X/d, $d_n=0.4\ X$, f_p/f_{pk} from which M_{Rd} is determined.

If the final strain $\varepsilon_p=\varepsilon_{po}+\varepsilon_{cu3}\ (d/X-1)>\varepsilon_{LOP}$ Eq. 9.32 is not valid and E_p is replaced by the functions in Eq. 9.21. The results for f_p/f_{pk} and X/d are given in Table 9.7 in terms of three values of final prestress $f_{po}/f_{pk}=0.4$, 0.5 and 0.6. To check whether $\varepsilon_p \leq \varepsilon_{LOP}$ ratios of $\varepsilon_p/\varepsilon_{LOP}$ are given in Table 9.8, with elastic values in italics being in the minority. Finally $M_{Rd}/f_{ck}\ bd^2$ is given for all cases.

For example, for the hcu used in Example 6.1 and shown in cross-section in Figure 6.7, $f_{pk}\ A_{pT}/f_{ck}\ bd=1770\times 911/45\times 1200\times 209.2=0.143$ and $f_{po}/f_{pk}=1020.7/1770=0.576$.

By linear interpolation $f_p/f_{pk}=0.831$, then $f_p=1470$ N/mm²; $X/d=0.259$, then $X=0.259\times 209.2=54$ mm, and $0.8X=43$ mm which is just below the bottom of the top flange of $h_{ft}=40$ mm, therefore the section may be taken as 'rectangular'. Lever arm $z=209.2-0.4\times 54=187.6$ mm. Then $M_{Rd}=1470\times 911\times 187.6\times 10^{-6}=251\ 2$ kNm. Alternatively $M_{Rd}/f_{ck}\ bd^2=0.1056$, then $M_{Rd}=0.1056\times 45\times 1200\times 209.2^2\times 10^{-6}=$**250.5 kNm.** Rigorous analysis in Example 9.1 shows that $M_{Rd}=247.3$ kNm, a difference of 1.2% which appears to be due to the compressive block in the rigorous analysis being T-shape (as $0.8X>h_{ft}$) rather than 'rectangular'.

Table 9.7 Values of f_p/f_{pk} and X/d for prestressed design of rectangular sections

	f_p/f_{pk}			X/d		
$f_{pk}\ A_{pT}/f_{ck}\ bd$	*For values of* f_{po}/f_{pk}			*For values of* f_{po}/f_{pk}		
	0.6	0.5	0.4	0.6	0.5	0.4
0.10	0.86	0.85	0.85	0.19	0.19	0.19
0.15	0.83	0.82	0.82	0.27	0.27	0.27
0.20	0.81	0.80	0.80	0.36	0.35	0.35
0.25	0.80	0.79	0.79	0.44	0.44	0.43
0.30	0.79	0.79	0.78	0.52	0.52	0.52
0.35	0.79	0.78	0.78	0.61	0.60	0.60
0.40	0.78	0.78	0.77	0.69	0.69	0.68

Data calculated using $f_{pk}=1770$ N/mm². If this is changed to say 1500 N/mm² the differences are 0.2%.

Table 9.8 Values of $\varepsilon_p/\varepsilon_{LOP}$ and $M_{Rd}/f_{ck}\ bd^2$ for prestressed design of rectangular sections

	$\varepsilon_p/\varepsilon_{LOP}$			$M_{Rd}/f_{ck}\ bd^2$		
$f_{pk}\ A_{pT}/f_{ck}\ bd$	*For values of* f_{po}/f_{pk}			*For values of* f_{po}/f_{pk}		
	0.6	0.5	0.4	0.6	0.5	0.4
0.10	2.82	2.76	2.65	0.079	0.079	0.079
0.15	2.08	1.96	1.84	0.110	0.110	0.109
0.20	1.65	1.53	1.42	0.139	0.138	0.137
0.25	1.39	1.27	1.15	0.164	0.164	0.163
0.30	1.21	1.09	0.97	0.188	0.187	0.186
0.35	1.09	0.96	0.84	0.208	0.207	0.207
0.40	0.99	0.86	0.74	0.226	0.226	0.225

Values of $\varepsilon_p<\varepsilon_{LOP}$ in italics.

Example 9.1

Calculate M_{Rd} for the hcu used in Example 6.1 and as shown in cross-section in Figure 6.7. Draw the ultimate bending moment distributions along the span, including the effects of anchorage at the supports. Effective span of hcu = 9.8 m. Self weight of hcu including infill = 3.63 kN/m², floor screed = 1.80 kN/m², finishes = 1.50 kN/m² and services & ceiling = 0.50 kN/m². Imposed live load = 5.0 kN/m². Total width of cores $b_c = 6\times140 = 840$ mm. The f_p v ε_p relationship for strand with $f_{pk} = 1770$ N/mm² is shown in Figure 9.13.

Is the unit critical at the serviceability or ultimate limit state?

Solution

$f_{cd} = 0.85\, f_{ck}/1.5 = 0.567\times45 = 25.5$ N/mm²

$A_{pT} = 911$ mm² (all strands are in tension zone)

$y_{sT} = y_s = 40.8$ mm

Eq. 9.19 $d = 250 - 40.8 = 209.2$ mm

$\varepsilon_{po} = 1020.7/195000 = 0.005234$

Eq. 9.17 $\varepsilon_p = 0.005234 + 0.0035\ ((209.2/X) - 1)$ Eq. 1

Eq. 9.13 $F_p = f_p\times911$

Eq. 9.14 $F_c = 25.50\times0.8\times1200\ X = 24480\ X$

Then for equilibrium: $X/f_p = 911/24480 = 0.03721$ Eq. 2

Eq. 9.18 $f_{LOP} = 0.9\times1770/1.15 = 1385$ N/mm²

$f_{pd} - f_{LOP} = 1770/1.15 - 1385 = 154$ N/mm²

Eq. 9.18 $\varepsilon_{LOP} = 1385/195000 = 0.007104$

Eq. 9.20 If $\varepsilon_p > 0.007104$, then $f_p = 1385 + [154\times(\varepsilon_p - 0.007104)/(0.0222 - 0.007104)]$ Eq. 3

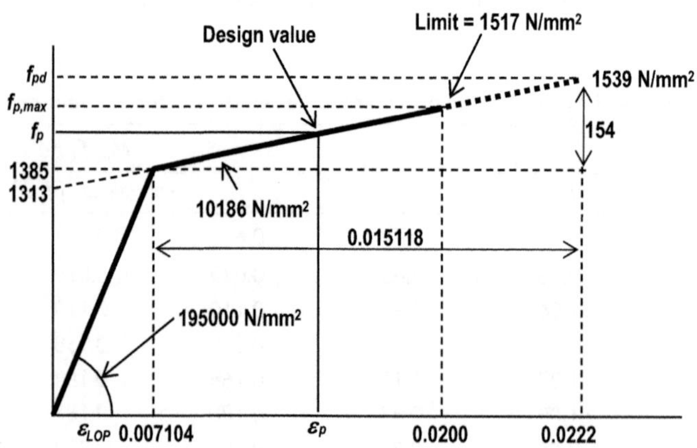

Figure 9.13 Constitutive stress *vs* strain relationship for helical strand f_{pk} = 1770 N/mm2 based on Fig. 3.10, BS EN 1992-1-1. See also Figure 9.8.

Combining Eqs. 1–3 gives: $369.6\ X^2-18298\ X-102669=0$

Then $X=54.6$ mm, but as $0.8\ X=43.7>h_{ft}=40$ mm the compression block enters the zone of the webs and the stress block is a 'T-shape', where the total width of the cores $b_c=840$ mm and the width of the webs $=1200-840=360$ mm.

The equations 1–3 now give: $110.9\ X^2-5363\ X-102669=0$

Then $X=63.1$ mm, $d_n=21.8$ mm and $z=209.2-21.8=187.4$ mm

From Eq. 1. $\varepsilon_p=0.013346$

From Eq. 3. $f_p=1449$ N/mm² < limit at $\varepsilon_s=0.02=1516$ N/mm²

Eq. 9.28 $M_{Rd}=1449\times911\times187.4\times10^{-6}=$ **247.3 kNm**

The design service dead load $=3.63+1.80+1.5+0.50=7.43$ kN/m²

The design ultimate UDL $=\max\{1.35\times7.43+0.7\times1.5\times5.00;\ 1.25\times7.43+1.5\times5.00\}=16.79$ kN/m², where $\psi_0=0.7$ for office loading

M_{Ed} per unit $=1.20\times16.79\times9.80^2/8=$ **241.8 kNm** < 247.3 kNm

Figure 9.14 shows the bending moment distributions for ultimate moments M_{Ed} and moments of resistance M_{Rd}. Close to the supports where the bond strength between the tendons and concrete is developing over an ultimate anchorage length l_{bpd} (see Section 9.5) the value of M_{Rd} is increasing bi-linearly from zero at the support (as shown in BS EN 1992-1-1, Fig. 8.17). In fact the bi-linear profile is about 95%–98% of being a linear distribution as the gradient of the development profile reduces by only 9% at $x=l_{pt2}$.

Referring to clause 8.10.2.3 (4) the anchorage length l_{bpd} to full $M_{Rd}=726+0.19\times11.2\times(1449-944.3)/2.13=1230$ mm, where $l_{pt2}=726$ mm (see Example 10.1), $\alpha_2=0.19$, average diameter of strands $=11.2$ mm, σ_{p0} at the support $=944.3$ N/mm² (see Example 7.1) and $f_{bpd}=1.2\times0.7\times3.80/1.5=2.13$ N/mm². Strictly the anchorage

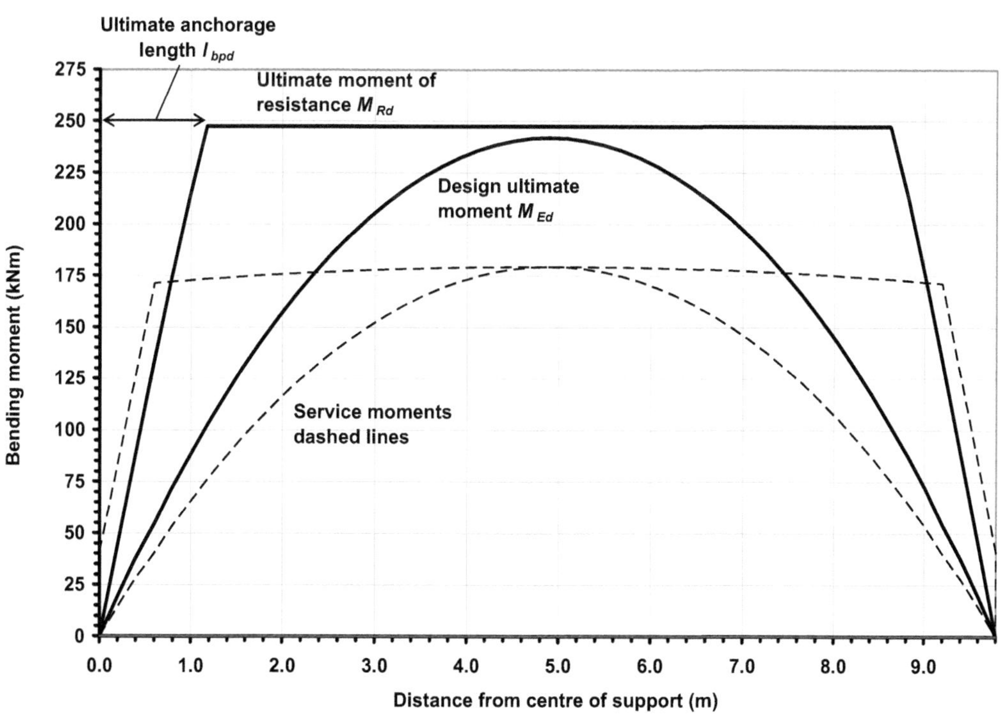

Figure 9.14 Ultimate bending moment distribution for hcu in Example 9.1.

length begins at the cut end of the hcu, e.g. at 50 mm behind the centre of support for a 100 mm bearing length.

To check whether the unit is critical at ultimate, the ratio M_{Rd}/M_{sR}=247.3/179.1=1.38. This ratio is greater than the ratio of the M_{Ed} to M_s using γ_G=1.25 and γ_Q=1.5 where, in this example M_{Ed}/M_s=241.8/179.1=1.35. Thus the hcu is not critical at the ultimate limit state.

9.3.3 Iterative trial and error method

M_{Rd} may be calculated using an iterative trial and error method by first trialing a value for f_p, called f_{p1} and substituting into Eqs. 9.15 and 9.17 to obtain values for X and ε_p and finally into Eq. 9.20 to obtain a new value for f_p, known as f_{p2}. If f_{p2} is within±5 N/mm² of f_{p1} convergence is assumed. Then $M_{Rd}=f_{p2}\,A_{sT}\,z$, where $z=d$ -0.4X. Otherwise iteration is required repeating the above and starting with f_{p2}, and so on. A good starting point is f_{p1}=1400 N/mm², roughly in the middle of the inclined branch of Figure 9.13. Eq. 9.20 for the inclined branch in Figure 9.13 is f_p=1385+[154×(ε_p−0.007104)/0.015118] N/mm². It is also reached using the stiffness of the inclined branch as f_p=1313+10186 ε_p.

Example 9.2

Calculate M_{Rd} using an iterative trial and error method for the hcu in Example 9.1, starting with f_{p1}=1400 N/mm².

Solution

Eq. 9.15 X_1=1400×911/25.5×0.8×1200=52.1 mm

Eq. 9.17 ε_{p1}=0.005234+0.0035×(209.2/52.1−1)=0.015789

Eq. 9.20 f_{p2}=1385+154×[(0.015789−0.007104)/0.015118]=1473.8 N/mm²

or f_{p2}=1313+10186×0.015789=1473.8 N/mm²

Repeat with f_{p2}=1473.8 N/mm². Then X_2=54.8 mm, ε_{p2}=0.015085 and f_{p3}=1466.6 N/mm²

Repeat with f_{p3}=1466.6 N/mm². Then X_3=54.6 mm, ε_{p3}=0.015150 and f_{p4}=1467.3 N/mm². Converges. Then z=209.2−0.4×54.6=187.4 mm

Eq. 9.28 M_{Rd}=1467.3×911×187.4×10⁻⁶=**250.5 kNm**

9.3.4 Ultimate flexural design at holes and notches

The formation and design for holes and notches in slabs is given in Section 7.4. The positions of holes and notches and limiting sizes for ultimate design are the same as those shown in Figures 6.16–6.20.

M_{Rd} for voids is calculated using the procedure in Section 9.3.1 using the residual width of the top flange after the void has been removed, and the remaining area of tendons after those at the void have been cut. The effective depth d of the tendons is recalculated based on the remaining area of tendons and their individual effective depths, unless of course the tendons are all at the same level in which case d does not change. See Example 9.3. M_{Rd} is calculated and is constant for the length of the void. Either end of the void M_{Rd} increases bilinearly to M_{Rd} for the full capacity of the unit over a distance equal to the ultimate anchorage length l_{bpd}. Voids formed entirely through hollow cores are ignored.

Lateral distribution of ultimate strains, and hence M_{Rd}, at holes in floor slabs is carried out in the same manner as in service (Section 7.4) after the units have been grouted together after installation, known as a 'slab panel', provided that there are no holes in the adjacent units. The net area of tendons $A_{pT,net}$ may be averaged over the width of 3 no. units, or 2 no. units if the hcu is at a free edge.

$$A_{pT,net} = \left[2A_{pT}\left(\text{full}\right) + A_{pT}\left(\text{with hole}\right)\right]/3 \tag{9.33}$$

$A_{pT,net}$ is input into the procedures in Section 9.3.1.

Example 9.3

Repeat Example 9.1 for the hcu with a hole and end notch as defined in Example 7.8. Using the information given in Example 9.1 for A_{pT} for the two adjacent units without holes calculate the net $M_{Rd,net}$ for the units and check the design ultimate moment M_{Ed} at the centre of the hole using the UDL given in Example 9.1.

Solution at the hole

$A_{pT}=3\times52+5\times93=621\,\text{mm}^2$

A_{pT} in adjacent units without holes$=911\,\text{mm}^2$

$A_{pT,net}=(2\times911+621)/3=814\,\text{mm}^2$

$\varepsilon_{po}=1024.2/195000=0.005252$

$\varepsilon_p=0.005252+0.0035\ ((209.2/X)-1)$ Eq. 1

Remaining $b=1200-265=935\,\text{mm}$

$F_c=25.50\times0.8\times935\ X=19074\ X$

$F_p=f_p\times814$

Then for equilibrium: $X/f_p=814/19074=0.04268$ Eq. 2

If $\varepsilon_p>0.007104$, then $f_p=1385+[154\times(\varepsilon_p-0.007104)/(0.0222-0.007104)]$ Eq. 3

Combining Eqs. 1–3 gives: $287.9\ X^2-16358\ X\ -91750=0$

Then $X=61.9\,\text{mm}$, but as $0.8\ X=49.6>h_{ft}=40\,\text{mm}$ the compression block enters the zone of the webs and the stress block is a 'T-shape', where the total width of the cores $b_c=840-140-45=655\,\text{mm}$ and the width of the webs$=935-655=280\,\text{mm}$ (or $360-2\times40=280\,\text{mm}$).

The equations 1–3 now give: $86.2\ X^2-6273\ X-91750=0$

Then $X=85.2\,\text{mm}$, $d_n=25.9\,\text{mm}$ and $z=209.2-25.9=183.2\,\text{mm}$

From Eq. 1. $\varepsilon_p=0.010342$

From Eq. 3. $f_p=1418\ \text{N/mm}^2<\text{limit}\ 1516\ \text{N/mm}^2$

Eq. 9.28 For unit with hole $M_{Rd}=1418\times814\times183.2\times10^{-6}=$**211.6 kNm**

Without load spreading, for the unit with the hole $M_{Rd}=175.0$ kNm.

The ultimate design moment at $x=6.0\,\text{m}$ from the support is $M_{Ed}=w_{Ed}\,L\,x/2-w_{Ed}\,x^2/2$ where $w_{Ed}=1.2\times16.78=20.14$ kN/m.

Then $M_{Ed}=20.14\times9.8\times6.0/2-20.14\times6.0^2/2=$**229.6 kNm**$>211.6$ kNm, the slab panel fails by 18.0 kNm. In fact, by plotting M_{Ed} and $M_{Rd,net}$ at 50 intervals along the span the greatest difference, at $x=5.68\,\text{m}$ from the left support, is 23.5 kNm.

Figure 9.15 shows the bending moment distributions for ultimate moments M_{Ed} and moments of resistance M_{Rd} at the hole and end notch.

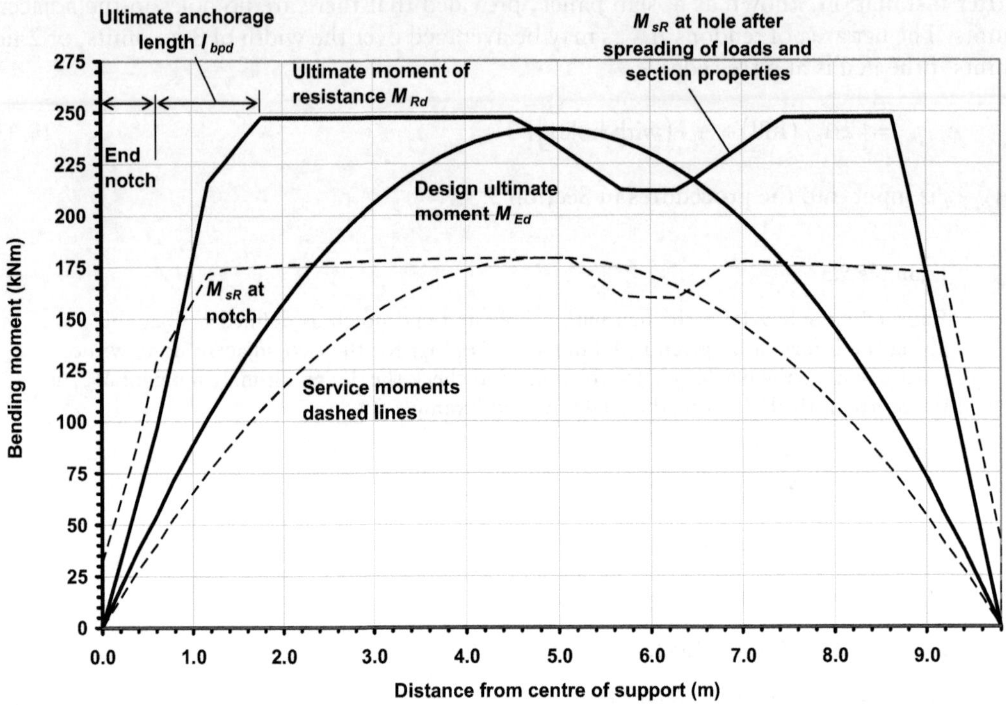

Figure 9.15 Ultimate bending moment distribution for hcu with hole and notch in Example 9.3.

The hcu in Figure 9.15 also contains an end notch from O to B of width w_v=285 mm and 600 mm length, where A_{pT}=673 mm². M_{Rd} at end notches are not laterally distributed between adjacent units. The resulting M_{Rd} at the end of the notch is M_{Rd}=96.8 kNm at x=0.6 m, increasing to M_{sR}=247.3 kNm at $x+l_{bpd}$=1.753 m.

9.3.5 Ultimate capacity for concentrated loads

Concentrated point and line loads, anywhere on the floor slab, are according to BS EN 1168, clause 4.3.3.2.5 (BS EN 1168, 2005). The first decision is whether the hcus are grouted to form a complete floor. If this is not the case (isolated units) the limiting tension is $f_{ctk,0.05}$ (= 0.7 f_{ctm}) and the capacities are at the serviceability limit state q_k and F_k. If the units are grouted to permit transverse load distribution the limiting tension f_t=f_{ctd} (= 0.7 f_{ctm}/1.5) and the capacities are at the ultimate limit state q_d and F_d.

BS EN 1168 does not explain how to deal with multiple loads in close proximity, nor diagonally crossing loads, but these should be resolved into loads parallel with the longitudinal span as far as the load is covering a width of one core plus one web. For multiple loads in close proximity, the shear force diagram may be drawn such that the design shear force should not exceed the capacities in Eqs. 9.34–9.42.

Guidance is given in BS EN 1168:2005 for the capacity of concentrated loads in psc hcu and completed slabs, such as

i. linear line loads q acting parallel with the span of the floor units not on an edge of a floor area
ii. linear line load q ditto on an edge of a floor area
iii. point loads F anywhere in the floor area.

The design is based on the resistance of the floor units to transverse bending, and as the units have no transverse reinforcement the tensile stresses caused by the bending moment are limited by the service tensile strength $f_{ctk,0.05}=0.7\,f_{ctm}$ or the ultimate strength f_{ctd} depending on the whether the loads are distributed laterally across several units, or not, respectively. Lateral load distribution of concentrated loads is not included in this book as it does not affect the behaviour of psc elements, but is covered by Elliott (Elliott, 2017) based on the guidance in BS EN 1168, Annex C.

BS EN 1168 clause 4.4.3.2.5 states 'If the elements are designed assuming no load distribution, which means that all loads acting on an element should be resisted by that element, the limiting value of the tensile stress is $f_{ctk,0.05}$ in the serviceability limit state. In this case for elements without topping, in the serviceability limit state, the capacities for concentrated loads q_k and F_k are calculated in the absence of particular justifications as follows', i.e. in Sections 9.3.4.1–9.3.4.3. This means if lateral load spreading cannot be carried out, for example at a trimmer angle around a large void, there is no moment redistribution between adjacent units and the service stress and loads govern. The clause does not state whether the 'topping' is a structural topping or not, but it is implied that it is as the topping would permit moment redistribution.

Further

> If the elements are designed by assuming load distribution according to the elastic theory, which means that a part of the loads acting on one element are distributed to adjacent elements, the limiting value of the tensile stress is f_{ctd} in the ultimate limit state. The capacities for concentrated loads in this case, in the ultimate limit state, may be derived for the same equation, but in which q_k, F_k and $f_{ctk,0.05}$ shall be replaced by (ultimate values) q_d, F_d and f_{ctd}.

In this book the notation is q_{Ed} and F_{Ed}. Punching shear capacity for point loads and hcu with three supported edges are given in Chapter 10.

It is therefore obvious, that because $f_{ctd}=f_{ctk,0.05}/1.5$, that unless there is a special situation, such as a concentrated load next to a trimmer angle, the service load is not critical. This is shown in Example 9.4. There is a paradox here – greater concentrated load capacities are therefore achieved when the loads are not laterally distributed between adjacent units?

The ultimate design concentrated load is the greater of Exp. 6.10a or 6.10b in BS EN 1990 (BS EN 1990, 2002), see Section 3.1.3. If two separate loads are present, and they overlap by a distance less than the depth of unit, the reactions are combined, or are otherwise considered individually. Line loads of length l are spread across the width b of the units at 45° as shown in Figure 9.16. If the start of the load $x<b$ the spreading length is reduced as shown in Figure 9.16.

9.3.5.1 Linear line loads not on an edge of the floor area

When the load is not at an edge the bending resistance is a function of the section modulus at the bottom of the unit. For the design of single units without lateral load distribution, the service line load capacity is

$$q_k = 20W_{lb}f_{ctk,0.05} \,/\, (l + 2b)(\text{N/mm units}) \qquad (9.34)$$

For the design of units with lateral load distribution, the ultimate line load resistance is

$$q_d = 20W_{lb}f_{ctd} \,/\, (l + 2b)(\text{N/mm units}) \qquad (9.35)$$

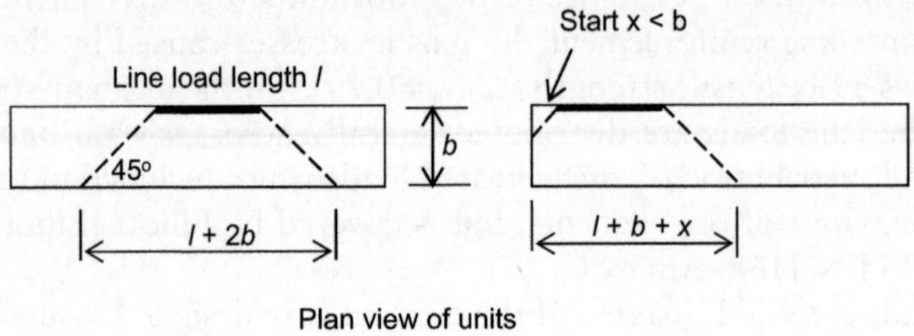

Figure 9.16 Lateral load spread for line loads acting on hollow core units at ultimate limit state

where

$$W_{lb} = I_{zz} / y_b \left(\text{mm}^3/\text{mm run}\right) \tag{9.36}$$

is section modulus at bottom in transverse direction per unit length.

$$I_{zz} = 1 \times \left[h_{ft}^3/12 + h_{fb}^3/12 + h_{ft}\left(h - h_{ft}/2 - y_b\right)^2 + h_{fb}\left(y_b - h_{fb}/2\right)^2\right] \left(\text{mm}^4/\text{mm run}\right) \tag{9.37}$$

l = length of the line load
b = width of unit

The section modulus W is calculated from the second moment of area taking a longitudinal cut through the unit, and considering the top and bottom flanges as rectangles 1 mm wide. For example, if h = 250 mm and the top and bottom flanges have thickness h_{ft} = 40 mm, h_{fb} = 35 mm (See Figure 6.7), then I_l = 856430 mm^4/mm length of unit, W_{lb} = 856430/123.0 = 6963 mm^3/mm and W_{lt} = 856430/127.0 = 6744 mm^3/mm, then the minimum W_l = 6744 mm^3/mm. If $f_{ctk,0.05}$ = 0.7 × 3.80 = 2.66 N/mm^2, length of line load l = 1000 mm and b = 1200 mm, then for an innermost line load the service capacity is

Eq. 9.34 $q_k = 20 \times 6393 \times 2.66/3400 \leq 108.8$ N/mm, or 108.8 kN/m service load for an isolated unit.

If the start of the line load is less than b from the end of the unit, the load spread term (l+2b) is reduced, e.g. if the same line load starts at 600 mm from the end of the span, (l+2b) = 600 + 1000 + 1200 = 2800 mm, and q_k = 132 kN/m. This is greater than above because some of the line load is distributed directly into the closest support. If the floor slab is grouted, part of the load will be distributed across the longitudinal joints and the ultimate capacity is now

Eq. 9.35 $q_d = 20 \times 6963 \times 1.77/3400 = 72.5$ kN/m

If this is a dead load the service capacity q_k = 72.5/1.35 = 53.7 kN/m. Line load capacities at the edge of a floor area, given later in Eqs. 9.38 and 9.39, are half of the above.

9.3.5.2 Linear line loads on an edge of the floor area

When the load is at an edge the bending resistance is a function of the section modulus at the top of the unit. The load capacity is as Section 9.3.4.1 with the following expressions

$$q_k = 10 W_{lt} f_{ctk,0.05} / (l + 2b) \tag{9.38}$$

$$q_{Ed} = 10W_{lt}f_{ctd}/(l + 2b) \tag{9.39}$$

where

$$W_{lt} = I_{zz} / y_t \left(\text{mm}^3/\text{mm run}\right) \tag{9.40}$$

is section modulus at top in transverse direction per unit length.

9.3.5.3 Service and ultimate capacities of point load anywhere in a hollow core floor area

Global capacity to point loads anywhere on a floor area are given in Eqs. 9.41 and 9.42. The size of the footprint is not given, but at least 300 mm is prudent. This is not the same as a punching shear capacity, but is derived in the manner in which full-scale test behaviour shows longitudinal splitting running through the entire span. For the design of single hcu without lateral load distribution, the total service point load F_k is applied to one unit. The service capacity is

$$F_k = 3W_l f_{ctk,0.05} \left(\text{N units}\right) \tag{9.41}$$

The ultimate resistance is

$$F_{Rd} = 3W_l f_{ctd} \left(\text{N units}\right) \tag{9.42}$$

where W_l = minimum of W_{lb} or W_{lt} (mm³/mm run)

Example 9.4

Calculate the ultimate capacity for a linear line load q_{Ed} and point load F_{Ed} for the hcu shown in Figure 6.7 where all loads are distributed over adjacent units and (i) is not at the edge of a floor area, (ii) as (i) but is present at the edge of a floor area, and (iii) F_{Ed}. The line load is from x = 2.0 to 4.0 m from the centre of the support.

For completeness calculate the service load capacities without lateral load distribution.

Use y_b = 123.0 mm, h_{ft} = 40 mm, h_{fb} = 35 mm, f_{ctm} = 3.80 N/mm², $f_{ctk,0.05}$ = 0.7 × 3.80 = 2.66 N/mm² and f_{ctd} = 2.66/1.5 = 1.77 N/mm².

Solution

$I_{zz} = 1 \times [40^3/12 + 35^3/12 + 40\ (250 - 20 - 123.0)^2 + 35\ (123.0 - 17.5)^2] = 856430$ mm⁴/mm run

i. line load not at the edge of a floor area
   W_{lb} = 856430/123 = 6963 mm3/mm
   q_d = 20 × 6963 × 1.77/(2000 + 2 × 1200) = 56.0 N/mm or kN/m
ii. line load at the edge of a floor area
   W_{lt} = 856430/127 = 6743 mm3/mm
   q_d = 10 × 6743 × 1.77/4400 = 27.1 N/mm or kN/m
iii. point load
   Minimum $W_l = W_{lt}$ = 6743 mm³/mm
   $F_{Rd} = 3 \times 6743 \times 1.77 \times 10^{-3} = 35.8$ kN

Service capacities q_k = 1.5 × 56.0 = 84.0 kN/m; q_k = 1.5 × 27.1 = 40.7 kN/m; F_k = 1.5 × 35.8 = 53.7 kN.

9.4 ULTIMATE FLEXURAL DESIGN OF PRESTRESSED BEAMS

9.4.1 Inverted-tee beams

The ultimate limit state for the flexural design of psc beams follows the procedures adopted for psc floor units given in Section 9.3, but with one major addition. Because the position of many of the tendons in beams is nearer to the neutral axis than is the case in floor units, the strains in each of the tendons should be calculated to determine whether they attain their maximum value shown in the f_p vs ε_p diagram in Figure 9.8 and Figure 9.13 for the case of $f_{pk}=1770$ N/mm². These diagrams are based on the constitutive relationship given in BS EN 1992-1-1, Fig. 3.10 where at the LOP (limit of proportionality) $\varepsilon_{LOP}=0.782\ f_{pk}/E_p$ and so for $f_p \le 0.782\ f_{pk}$

$$f_p = E_p \varepsilon_p \tag{9.43}$$

For $0.782\ f_{pk} < f_p \le f_{p,max}$ where at the limiting strain $\varepsilon_{ud}=0.02$

$$f_{p,max} = 0.782 f_{pk} + \{(0.02 - \varepsilon_{LOP})/(0.0222 - \varepsilon_{LOP})\}\ 0.087 f_{pk} \tag{9.44}$$

$$f_p = f_p' + E_p' e_p \tag{9.45}$$

where the intercept $f_p'=0.782\ f_{pk}-E_p'\ 0.782\ f_{pk}/E_p$

$$f_p' = 0.782 f_{pk}\left(1 - E_p' / E_p\right) \tag{9.46}$$

where the inclined branch Young's modulus

$$E_p' = 0.087 f_{pk}/\left(0.0222 - 0.782 f_{pk} / E_p\right) \tag{9.47}$$

For the case of $f_{pk}=1770$ N/mm², $E_p'=154/(0.022222-0.007104)=10186$ N/mm², $f_{LOP}=0.782\times1770=1385$ N/mm², $f_p'=1385\ (1-10186/195000)=1313$ N/mm² and therefore

Eq. 9.45 $f_p=1313+10186\ \varepsilon_p \le f_{p,max}$

Eq. 9.44 $f_{p,max}=1385+\{(0.02-0.007104)/(0.0222-0.007104)\}\ 154=1516$ N/mm².

In some cases (in shallow beams) it is possible that no tendon will attain its yield value and therefore the assumption that $f_p=f_{p,max}$ is not valid at all. In other cases (deeper beams and composite beams with a wide top flange) some of the tendons will attain $f_{p,max}$. Because of having to satisfy the fairly stringent serviceability stress limits, most psc rectangular and inverted-tee beams are over reinforced at ultimate – this is why composite psc beams are efficient in enabling most of the tendons to attain yield value, see Section 12.5.4. The basic analytical procedure is given below where it is assumed that the stress distribution (after losses) and a strand pattern to satisfy the service condition are known.

Firstly, the tendons in the top of the beam and at the top of the boot of inverted-tee beams are ignored. This leaves the number of tendons as N_T of total area A_{pT}. A good starting point for the first estimate of the neutral axis depth X and the ultimate stress f_p is to take all remaining tendons stressed to about 0.7 f_{pk}. Then, from the equilibrium of concrete F_c and steel F_p forces, knowing X the strain in the tendons in the bottom row is found, at the next row up and so on until the next row down from the top of the boot is reached. If the upstand is particularly tall, say equal to half the total depth of the beam, the tendons in the top of the boot may contribute a little to the total force F_p. Referring to Figure 9.11, if the distance from the top of the beam to the tendons is d, the ultimate strain in the tendon at d is given by

$$\varepsilon_p = \varepsilon_{po} + \varepsilon_{cu3}\left(d / X - 1\right) \tag{9.48}$$

where the pre-strain $\varepsilon_{po}=f_{po}/E_p$ is the final stress in the tendon after losses. The small strain due to the decompression of concrete f_g may be added to ε_{po}, but the effect is small. Referring to Figures 9.8 and 9.11 the stress in the tendon from Eq. 9.47 is $f_p=f_p'+E_p'\ \varepsilon_p$

Knowing the values of f_p in each row of tendons the force equilibrium gives

$$F_c = 0.567 f_{ck} b 0.8X \text{ in the concrete} \tag{9.49}$$

$$F_p = \Sigma f_p A_{pT} \text{ for each row of tendons below the top of the boot} \tag{9.50}$$

Hence X is found and re-substituted into Eq. 9.48 for iteration. The analysis is clearly long and tedious and involves having to calculate ε_p and f_p at each level, although a simple spreadsheet soon delivers the solution after about three (intelligent) iterations. An approximate method is to consider that the average strain exists at the centroid of the tendons in the tension zone, i.e. $d=d_T$ where d_T is the effective depth to these tendons. Eqs. 9.48–9.50 are still valid. The ultimate moment of resistance M_{Rd} is given as

$$M_{Rd} = \Sigma f_p A_{pT}\left(d_T - 0.4X\right) \tag{9.51a}$$

Note that in inverted-tee beams the compression zone comprises the upstand, of width b_w and depth h_s, plus the part of the boot of width b. Eq. 9.39 is modified to

$$F_c = 0.567 f_{ck} b_w h_s + 0.567 f_{ck} b\left(0.8X - h_s\right) \tag{9.52a}$$

Alternatively

$$F_c = 0.567 f_{ck}\left(b\ 0.8X - \left(b - b_w\right)h_s\right) \tag{9.52b}$$

$$M_{Rd} = \Sigma f_p A_{pT}\left(d_T - d_n\right) \tag{9.51b}$$

where d_n is the centroid of the inverted-T shape compressive zone.

Tables 9.9 and 9.10 give the minimum value of M_{Rd} for a range of typical sizes for psc inverted-tee beams.

Example 9.5

Calculate M_{Rd} for the beam shown in Figure 9.17 with the chosen strand pattern from Example 7.10. Draw the ultimate bending moment distributions along the span, including the effects of bond anchorage at the supports. The self weight of the beam=7.688 kN/m, the floor slab=18.0 kN/m, other dead loads=10.0 kN/m and imposed live load=30.0 kN/m. Use of beam is for offices, where $\psi_0=0.7$.

Is the beam critical at the serviceability or ultimate limit state?

Table 9.9 Ultimate moments of resistance for 500 mm wide grade C45/55 prestressed inverted-tee beams.

Soffit width (mm)	Boot depth (mm)	Upstand depth (mm)	Upstand width (mm)	Number of strands N_T	P_{pi} (kN)	z_{cp} (mm)	M_{sR} (kNm)	M_{Rd} (kNm)
500	200	150	300	11	1267	58.9	155.0	260.1
500	300	150	300	15	1728	71.9	258.3	420.4
500	400	150	300	18	2074	89.4	383.1	623.1
500	500	150	300	22	2535	104.1	544.4	884.1
500	600	150	300	26	2996	122.0	743.6	1200.5
500	200	200	300	12	1383	66.7	198.4	340.7
500	300	200	300	16	1844	77.7	311.1	513.8
500	400	200	300	20	2305	94.2	457.8	733.9
500	500	200	300	24	2765	109.4	631.5	1001.2
500	600	200	300	27	3111	131.5	831.3	1323.2
500	200	250	300	14	1613	67.9	240.5	384.5
500	300	250	300	17	1959	91.7	379.3	654.3
500	400	250	300	21	2420	105.3	534.5	842.5
500	500	250	300	25	2881	118.5	716.1	1123.6
500	600	250	300	29	3342	135.0	932.8	1458.0

All sections use f_{ck}=45 N/mm², $f_{ck}(t)$=30 N/mm², f_{ctm}=3.02 N/mm² for exposure XC1, f_{pk}=1770 N/mm², A_p=93 mm² per strand. Prestress at transfer $\sigma_b(t)$=14.5 to 15.0 N/mm².

Table 9.10 Ultimate moments of resistance for 600 mm wide grade C50/60 prestressed inverted-tee beams

Soffit width (mm)	Boot depth (mm)	Upstand depth (mm)	Upstand width (mm)	Number of strands N_T	P_{pi} (kN)	z_{cp} (mm)	M_{sR} (kNm)	M_{Rd} (kNm)
600	200	150	350	13	1498	57.1	185.8	325.6
600	300	150	350	18	2074	70.3	314.9	514.3
600	400	150	350	22	2535	85.7	469.6	771.9
600	500	150	350	27	3111	102.1	676.2	1110.8
600	600	150	350	31	3572	121.9	910.4	1495.0
600	200	200	350	15	1728	63.7	245.7	420.4
600	300	200	350	19	2189	80.5	382.5	646.7
600	400	200	350	24	2765	92.7	558.8	903.8
600	500	200	350	28	3226	112.6	768.2	1233.1
600	600	200	350	33	3802	127.2	1022.1	1637.5
600	200	250	350	16	1844	76.1	310.0	485.8
600	300	250	350	21	2420	85.2	464.5	785.9
600	400	250	350	26	2996	102.2	664.3	1063.7
600	500	250	350	30	3457	118.0	878.1	1385.2
600	600	250	350	34	3918	139.0	1140.4	1816.8

All sections use f_{ck}=45 N/mm², $f_{ck}(t)$=30 N/mm², f_{ctm}=3.02 N/mm² for exposure XC1, f_{pk}=1770 N/mm², A_p=93 mm² per strand. Prestress at transfer $\sigma_b(t)$=14.5 to 15.0 N/mm².

Solution

First estimate of X is found by considering N_T=22 strands, A_{pT}=2046 mm² stressed to 70%×1770 N/mm². F_p=2046×1239×10⁻³=2535 kN. Assume first the neutral axis (NA) depth is in the upstand, X=2535×10³/0.567×45×350×0.8=355 mm>200 mm, then NA is in the boot as follows

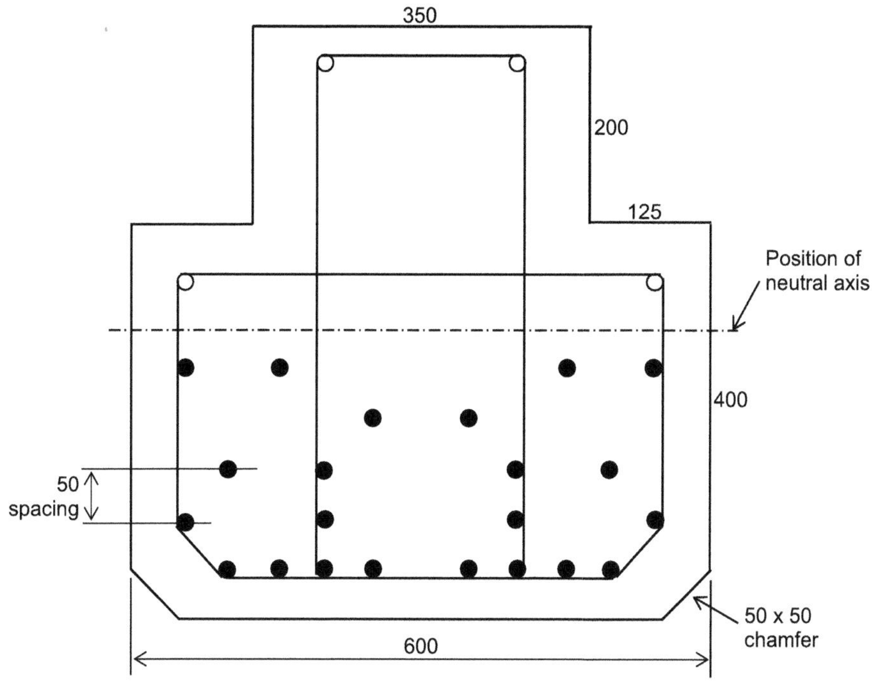

Figure 9.17 Cross-section and strand arrangement for the ultimate flexural design of the same inverted-tee beam in Example 7.10. The four strands in the top of the beam (open symbols) are excluded in the design.

$F_{c1} = 0.567 \times 45 \times 350 \times 200 \times 10^{-3} = 1784.8$ kN.

Then $X = [200 + \{(2535\text{-}1784.8) \times 10^3/0.567 \times 45 \times 600\}]/0.8 = 311$ mm. This iterates to $X = 315.8$ mm

Prestrain at mid-span $\varepsilon_{po} = 989.5/195000 = 0.005074$

Considering strands in bottom row 1, $\varepsilon_{p1} = 0.005074 + 0.0035 \times [(550/311)\ \text{-}1] = 0.007652$

$f_{p1} = 1313 + 10186 \times 0.007652 = 1391$ N/mm$^2 < 1516$, and so on gives $X = 319.1$ mm, such that $\varepsilon_{p1} = 0.007600$ and $f_{p1} = 1390$ N/mm^2, and so on until the equilibrium converges to $X = 318.9$ mm.

Then $\varepsilon_{p1} = 0.005074 + 0.0035 \times [(550/318.9) - 1] = 0.007611 > \varepsilon_{LOP} = 0.007104$

$f_{p1} = 1313 + 10186 \times 0.007611 = 1390$ N/mm^2 and in row 2 $\varepsilon_{p2} = 0.007062 < \varepsilon_{LOP}$, $f_{p2} = 1377$ N/mm^2 and so on as shown in Table 9.11.

From Table 9.11, $y_s = 315314 \times 10^3/2628466 = 120.0$ mm

$d = 600 - 120.0 = 480.0$ mm

$F_p = 2628.5$ kN

F_{c1} in the upstand $= 1784.8$ kN

$X = [200 + \{(2628.5 - 1784.8) \times 10^3/0.567 \times 45 \times 600\}]/0.8 = 318.8$ mm, converges

Centroid of concrete $d_n = 140.9$ mm

$z = 480.0 - 140.9 = 339.1$ mm

$f_p = 2628.5 \times 10^3/2046 = 1285$ N/mm^2

$M_{Rd} = 1285 \times 2046 \times 339.1 \times 10^{-6} =$ **891.5 kNm**

Table 9.11 Final total strains, stresses and forces in the rows of strands in Example 9.5

Row	No.	Height y_i (mm)	Area per strand (mm^2)	Area per row A_{pT} (mm^2)	Strain ε_p	Stress f_p (N/mm^2)	Force f_p A_p (N)	Force×y_i×10^{-3} (kNmm)
1	8	50	93	744	0.007611	1390.4	1034451	51723
2	4	100	93	372	0.007062	1377.2	512305	51230
3	4	150	93	372	0.006514	1270.1	472494	70874
4	2	200	93	186	0.005965	1163.1	216342	43268
5	4	250	93	372	0.005416	1056.1	392873	98218
Totals	22			2046			2628466	315314

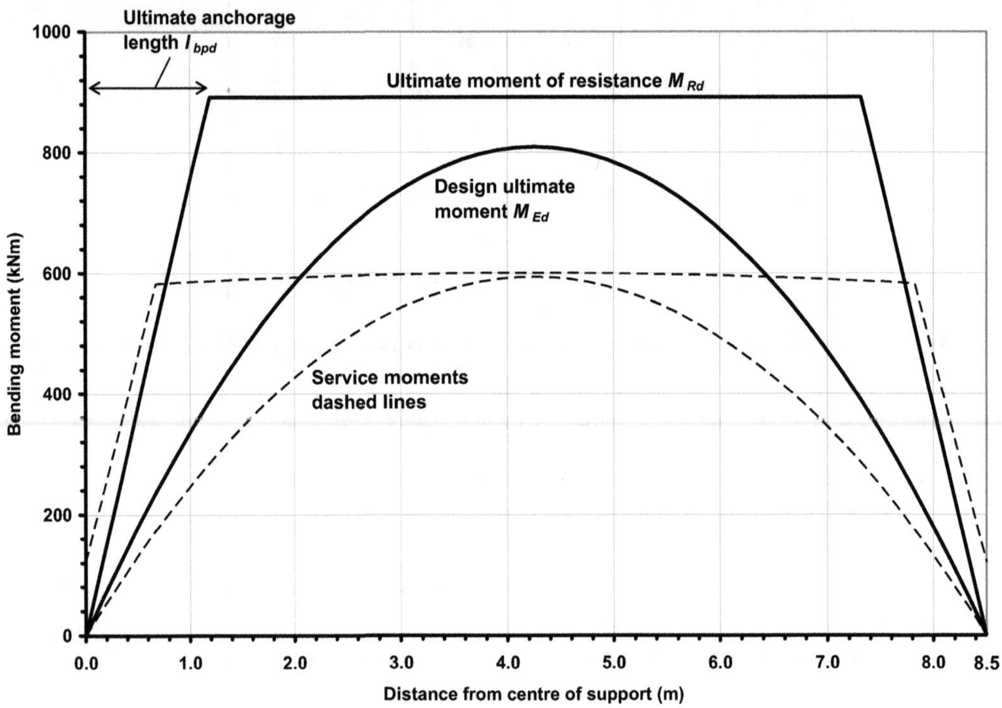

Figure 9.18 Ultimate bending moment distribution for the inverted tee beam in Example 9.5.

Figure 9.18 shows the bending moment distributions for ultimate moments M_{Ed} and moments of resistance M_{Rd}. Close to the supports where the bond strength between the tendons and concrete is developing over an ultimate anchorage length l_{bpd}=1.180 m (see Section 9.5) the value of M_{Rd} is increasing bi-linearly from zero at the support.

The ultimate design load w_{Ed}=max{1.35×(7.688+18.0+10.0)+1.05×30.0; 1.25*×(7.688+18.0+10.0)+1.5×30.0}=89.56 kN/m. M_{Ed}=89.56×8.5^2/8=808.9 kNm. (*Note this factor is actually 0.925×1.35=1.24875, hence the slight difference in the answer 89.56 kN/m.)

To check whether the unit is critical at ultimate, the ratio M_{Rd}/M_{sR}=891.5/599.7=1.49. This ratio is greater than the ratio of the M_{Ed} to M_s using γ_G=1.25 and γ_Q=1.5 where, in this example M_{Ed}/M_s=808.9/593.2=1.36. Thus the beam is not critical at the ultimate limit state.

Table 9.12 Ultimate moments of resistance for 150 to 225 mm depth prestressed T beams for use in beam and block flooring

Depth of beam (mm)	*Soffit width (mm)*	*Upstand width (mm)*	*Upstand depth (mm)*	*Number of wires N_T (& total number)*	*P_{pi} (kN)*	*z_{cp} (mm)*	*M_{sR} (kNm)*	*M_{Rd} (kNm)*
150	92	42	100	3 (4)	104.3	21.9	5.56	7.38
150	92	42	100	3 (3)	78.2	27.7	4.96	6.88
150*	215	165	100	7 (9)	208.0	27.4	11.00	18.60
180*	120	76	100	5 (6)	138.7	37.0	9.31	16.81
225**	136	80	100	6 (8)	208.52	31.7	16.61	24.22
225**	136	80	100	5 (6)	156.39	43.9	15.34	21.49

All sections use f_{ck}=55 N/mm^2 except *50 N/mm^2. Transfer $f_{ck}(t)$=32 N/mm^2, except *30 N/mm^2 and **25 N/mm^2. f_{ctm}=4.21 N/mm^2 except *4.07 N/mm^2. Exposure XC1, f_{pk}= 1770 N/mm^2, A_p= 19.6 mm^2 per wire. Prestress at transfer $\sigma_b(t)$= 14.5 to 16.0 N/mm^2.

9.4.2 T beams in beam and block flooring

The same procedure is used as for inverted-tee beams, for example in Example 9.5. The number of wires used in 150 mm depth beams is generally 4 to 9 no. for the narrow (bottom width 92 mm) and wide beams (bottom width 165 mm), respectively. Of these 3 to 7 no. wires will be used in the M_{Rd} calculations. The neglected wires are usually present to control cracking and camber at transfer. The number of wires used in 225 mm depth beams is generally 7 or 8 no., of which 6 no. will be used in the M_{Rd} calculations.

The upstand of beams is usually tapered, becoming narrower at the top of the beam, at a slope of 6/100=0.06, such that the compressive stress block is trapezoidal. When the depth to the neutral axis is known the actual width of the stress block may be determined, as shown in the following example. In such shallow beams the strain in the second row of wires may fluctuate about the limit of proportionality, i.e. ε_p is close to ε_{LOP}, and it is impossible to escape the iterative loop as the calculation progresses. Of course contains a partial safety factor γ_s=1.15, so this may not be happening in reality. It does mean that the wire in row 2 is notionally still elastic at the ultimate limit state, which should not be allowed, but as the test results in Section 9.1.2 show, it is likely this wire will also be close to yielding when ultimate failure occurs.

Table 9.12 gives values of M_{Rd} for a range of typical sizes for psc T beams. The bending stress ratio $K=M_{Rd}/f_{ck}\ b\ d^2$ varies from 0.15 to 0.22, the greater values being for the shallower beams, but showing a well balanced design with X/d=0.39 to 0.63, respectively.

Example 9.6

Calculate M_{Rd} for the 150 mm depth T beams shown in cross-section in Figure 7.12 (and Figure 5.29), where from Example 7.5 σ_{po}=1031.5 N/mm^2. The width of the trapezoidal upstand is 42 mm with a slope of 6/100=0.06. Note that for wire E_p=205 kN/mm^2.

Solution

Note that for concrete f_{ck}=55 N/mm^2, BS EN 1992-1-1, clause 3.1.7 states

i. stress block factor is $\lambda=0.8-(f_{ck}-50)/400=0.8-(55-50)/400=0.7875$
ii. design strength factor is $\eta=1.0-(f_{ck}-50)/200=1.0-(55-50)/200=0.975$
iii. Table 3.1 $\varepsilon_{cu3}=2.6+35\ [(90-f_{ck})/100]^4=0.00313$.

The beam has 2 no. wires at d_1=150-22.5=127.5 mm from the top of the beam, 1 no. wire at d_2=110 and 1 no. wire at d_3=75 mm.

First estimate of X is found by considering N_T=4 no. wires, A_{pT}=4×19.63=78.5 mm² stressed to 75%×1770=1327.5 N/mm². F_p=78.4×1327.5×10⁻³=104.2 kN. Assume first the neutral axis (NA) depth is in the upstand, X=104.2×10³/0.567×0.975×55×42×0.7875=103.7 mm>d_3=75 mm to the wire in row 3. Therefore repeat the above using N_T=3 no. wires, A_{pT}=58.9 mm², F_p=78.2 kN, and X=77.8 mm. The depth of the compression block is 0.7875 X=61.3 mm. As the upstand is trapezoidal the average width of the stress block is 42+(0.06×61.3/2)=43.8 mm, rounded to b=44 mm.

Carrying out iterations to determine f_p in each row of wires, the f_p v ε_p relationships for f_{pk}=1770 N/mm² are

ε_{LOP}=1385/205000=0.006757

$$E_p' = 154 / (0.022222 - 0.006757) = 9952 \text{ N/mm}^2$$

$$f_p' = 1385 \ (1 - 9952 / 205000) = 1318 \text{ N/mm}^2$$

Then from Eq. 9.45 f_p=1318+9952 ε_p≤1517 N/mm²

Also ε_{po}=1031.5/205000=0.005032

Commencing the iteration using X=77.8 mm, the strain in row 1 is

ε_{p1}=0.005032+0.00313 (127.5/77.8−1)=0.007028>ε_{LOP}

f_{p1}=1318+9952×0.007028=1388 N/mm²

F_{p1}=1388×39.26=54503 N

The strain in row 2 is ε_{p2}=0.005032+0.00313 (110.0/77.8−1)=0.006325<ε_{LOP}

In the elastic range f_{p2}=205000×0.006325=1297 N/mm²

F_{p2}=1297×19.63=25458 N

Total F_p=79961 N and a new X=76.0 mm

Repeating the above gives a new X=76.4 mm. Converges at X=76.3 mm

Then d_n=0.5×0.7875×76.3=30.0 mm

Lever arm z_1=127.5−30.0=97.5 mm; z_2=110.0−30.0=80.0 mm

M_{Rd}=(1389×39.26×97.5+1315×19.63×80.0)×10⁻⁶=**7.38 kNm**

If the above calculation is carried out using the closed equation Eq. 9.21, then for all three wires equally: X=77.5 mm, z=91.1 mm, ε_p=0.006811>ε_{LOP} by 0.8%, f_p=1386 N/mm² and M_{Rd}=7.43 kNm, just 0.8% greater than in Example 9.6.

If partial safety factors γ_c=1.5 and γ_s=1.15 are removed from the iterative calculations in Example 9.6, the depth to the neutral axis reduces to X=60.0 mm (as γ_c=1.5 has been removed). Then ε_{p1}=0.008542 (20% greater than in design), ε_{p2}=0.007651 (19% greater), f_{p1}=1602 N/mm² (15% greater), f_{p2}=1564 N/mm² (19% greater), and $M_{Rd,calc}$=9.19 kNm (25% greater than in Example 9.6).

9.5 ANCHORAGE LENGTH FOR ULTIMATE LIMIT STATE

Referring to EN 1992-1-1, clause 8.10.2.3.

Exp. 8.21 $$l_{bpd} = l_{pt2} + \alpha_2 \Phi\left(\sigma_{pd} - \sigma_{po}\right) / f_{bpd} \quad (9.53)$$

α_2=0.19 for 7-wire strand; 0.25 circular wire
σ_{pd}=ultimate stress f_p
σ_{po}=prestress after all losses

Exp. 8.20 $f_{bpd}=\eta_{p2}\ \eta_1\ f_{ctd}$

8.10.2.3 (1) η_{p2}=1.2 for 7-wire strand; 1.4 for wire

8.10.2.2 η_1=1 for good bond, 0.7 for poor bond, depending on casting depth

In Example 9.5, and from Example 7.10 where σ_{po}=953.3 N/mm^2 at the support, the anchorage length to full M_{Rd} is

Eq. 9.53 $l_{bpd}=812+0.19\times12.5\times(1267-953.3)/2.13=1162$ mm

where $f_{bpd}=1.2\times1.0\times0.7\times0.3\times45^{2/3}/1.5=2.13$ N/mm^2.

See Example 10.6 for the value of l_{pt2}=812 mm.

REFERENCES

BS EN 1168. 2005. Precast Concrete Products - Prestressed concrete hollow core units, +A3:2011, BSI, London, UK.
BS EN 1990. 2002. Eurocode 0: Basis of structural design, +A1:2005, BSI, London, UK.
BS EN 1992-1-1. 2004. Eurocode 2: Design of concrete structures - Part 1-1: General rules and rules for buildings, +A1:2014, BSI, London, UK.
Elliott, K. S. 2017. *Precast Concrete Structures*, 2nd ed., CRC, Taylor & Francis, London, UK, 694 p.

Chapter 10

Ultimate shear strength and torsion, and transmission length

10.1 ULTIMATE SHEAR CAPACITY OF PRESTRESSED ELEMENTS

10.1.1 Ultimate shear tests on prestressed hollow core and solid units

In this book, the resistance to shear is known as shear 'capacity' to distinguish it from moments of 'resistance'.

The ultimate shear capacity of prestressed concrete (psc) elements, particularly those without shear reinforcement such as hollow core units (hcu) and solid units, is less predictable than for moments of resistance (Sections 7.1.1. and 9.1.1). In many cases, the experimental shear capacity V_u for a set of three nominally identical units can vary by as much as ±25%, particularly for hcu where the widths of individual webs may vary and lead to differing failure loads and cracking patterns. A much more reliable method is to cut the hcu into strips and test individual webs. The other issue is the interaction between bending and shear, where the distance from the centre of support to the applied load in a shear test, typically $2.5h$, where h is the depth of the unit, has a major influence on behaviour. It is not surprising that a huge effort has been given to testing and analyzing the shear behaviour of psc elements compared with flexural behaviour, which is known to be predictable.

A further reason for the variations in V_u is that without shear reinforcement the mode of failure is brittle and sudden, and in the vast majority of elements, the load at the first major onset of cracking is also the failure load. Examples of shear failures in hcu are shown in Figures 10.1–10.3. Shear-tension failures initiate a short distance from the edge of the bearing and propagate through the webs, where the total width of the webs is typically 0.25–0.35 times the width of the unit, at an angle of 35°–40° to the horizontal, usually travelling towards the position of the applied load.

If shear tests are conducted using hydraulic loading jacks, the failure is more controlled than under kentledge due to the sudden reduction in hydraulic pressure as deformation takes place. Therefore, as shown in the experimental data in Figures 10.4 and 10.5 it is possible to control the unloading stages. This does not happen in reality. As such shear failures lack ductility, and for this reason, the design capacity $V_{Rd,c}$ contains several partial safety factors. Compared with design values, $V_{Rd,c}$, there is considerable reserve of shear capacity in the elements, but not that much more than calculations may predict.

The tests were carried out by the author according to Annex J of European Product Standard BS EN 1168:2005 +A3:2011 (BS EN 1168, 2005). The test specimens involved 1200 mm wide×150 to 400 mm depth hcu and 100 and 150 mm depth solid units. Concrete strength was around f_{ck}=47 N/mm^2 and $f_{ck,cube}$=60 N/mm^2 (compared with design grade C45/55) and flexural test strength $f_{ctm,fl}$=5.5 N/mm^2. The distance from the centre of support to the applied cross-line load was $2.5h$ and $1.5h$ for the hcu and solid units, respectively,

 DOI: 10.1201/9781003319450-10

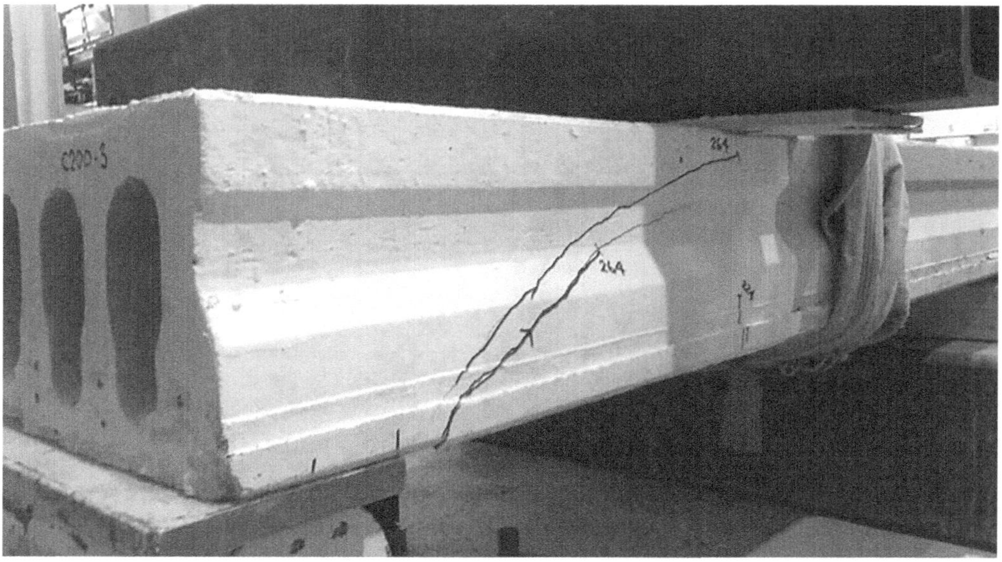

Figure 10.1 Shear-tension failure in 1200 mm width × 200 mm depth prestressed hollow core unit for test results in Figures 9.4 and 9.5. (With permission by Coltman Ltd., UK.)

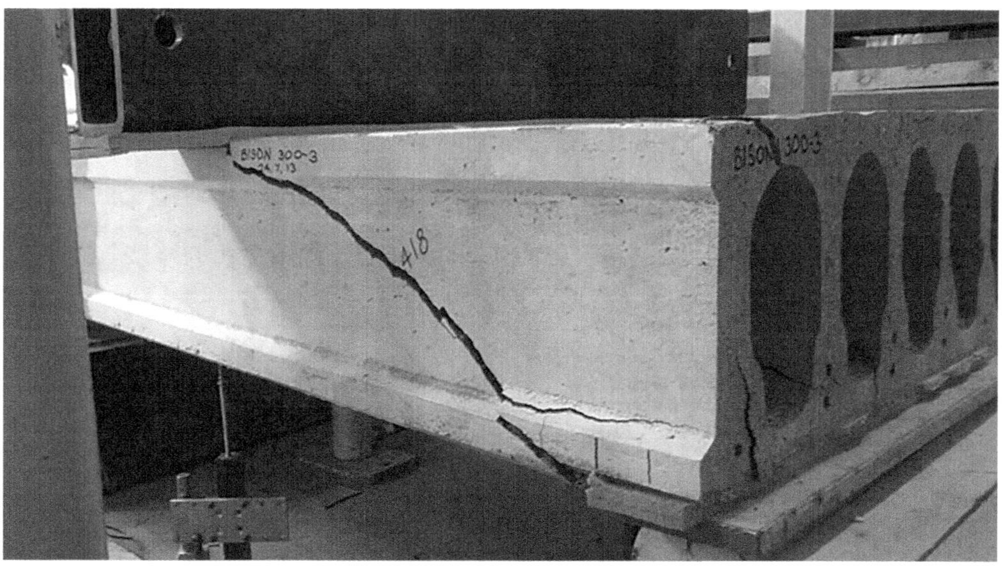

Figure 10.2 Shear-tension and bond failure in a 1200 mm width×300 mm depth prestressed hollow core unit for test results in Figure 10.6. (With permission by Bison Precast Ltd., UK.)

over a span of 4.0 m. Loading was carried out in two cycles up to a shear reaction of 0.7 $V_{Rd,c}$, as shown in Figure 10.6, where the recovery of deflection was at least 94%.

The main conclusion from the tests was that the behaviour of the hcu was different from that of the solid units because of the relatively low moment of resistance of the solid units compared with their shear capacity. The experimental test shear capacity V_u of the hcu ranged from 198 to 528 kN, for 150 to 400 mm depth, compared with design values based on the nominal section and material properties of the hcu of $V_{Rd,c}$=99 to 239 kN, respectively, an average ratio of 2.06 to 1.92, respectively. Corresponding ratios $V_u/V_{Rd,c}$ for 100 and 150 mm depth solid units were 1.12 and 1.06.

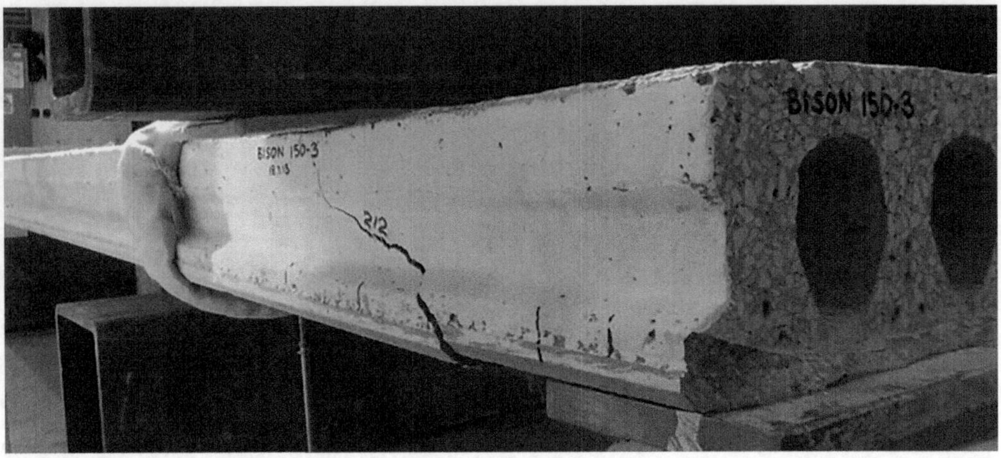

Figure 10.3 Shear-tension failure, tending towards flexural cracking, in a 1200 mm width × 150 mm depth prestressed hollow core unit. (With permission by Bison Precast Ltd., UK.)

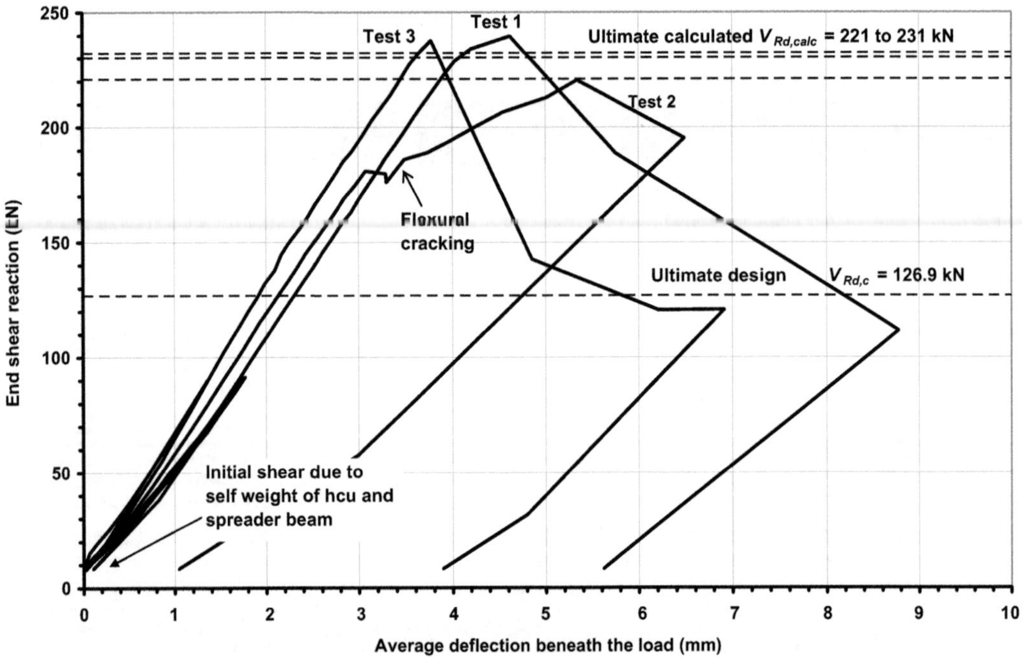

Figure 10.4 Experimental and predicted shear force vs. deflection beneath the load in full scale tests carried out on 1200 × 200 mm depth prestressed hollow core floor units. (With permission by Coltman Ltd., UK.)

The calculated shear capacities $V_{Rd,calc}$ of the hcu ranged from 187 to 469 kN, for 150 to 400 mm depth. The average ratio $V_u/V_{Rd,calc}$ was 1.13–1.03, respectively, although this ratio was as high as 1.17 in some 250 and 300 mm depth hcu. Figure 10.7 presents the ratios $V_u/V_{Rd,calc}$ in all tests. According to BS EN 1168, Annex J, the ratio of the failure test value to the respective calculated value is ≥0.95 per tests, and ≥1.0 for the mean of three specimens. The results show that all tests (except one at a ratio of 0.94) achieved these criteria.

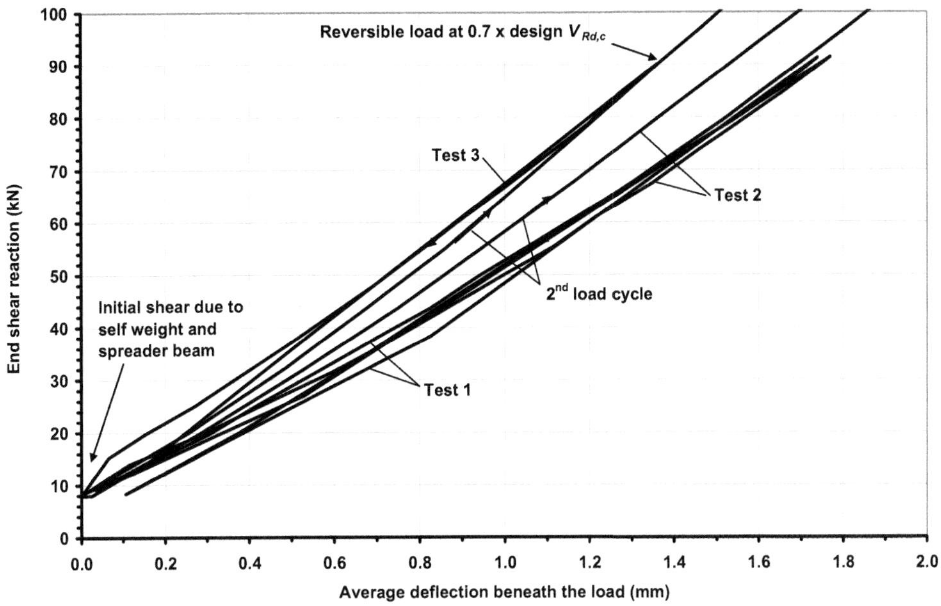

Figure 10.5 Detail of test data in Figure 10.4 up to 2 mm deflection showing recovery of test units after one cycle of load. (With permission by Coltman Ltd., UK.)

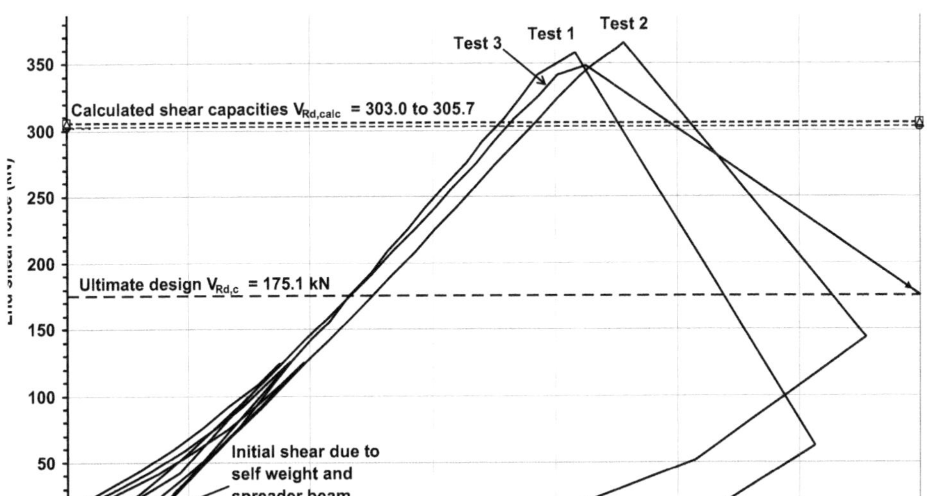

Figure 10.6 Experimental and predicted shear force vs. deflection beneath the load in full scale tests carried out on 1200×300 mm depth prestressed hollow core floor units. (With permission of Bison Precast Ltd., UK.)

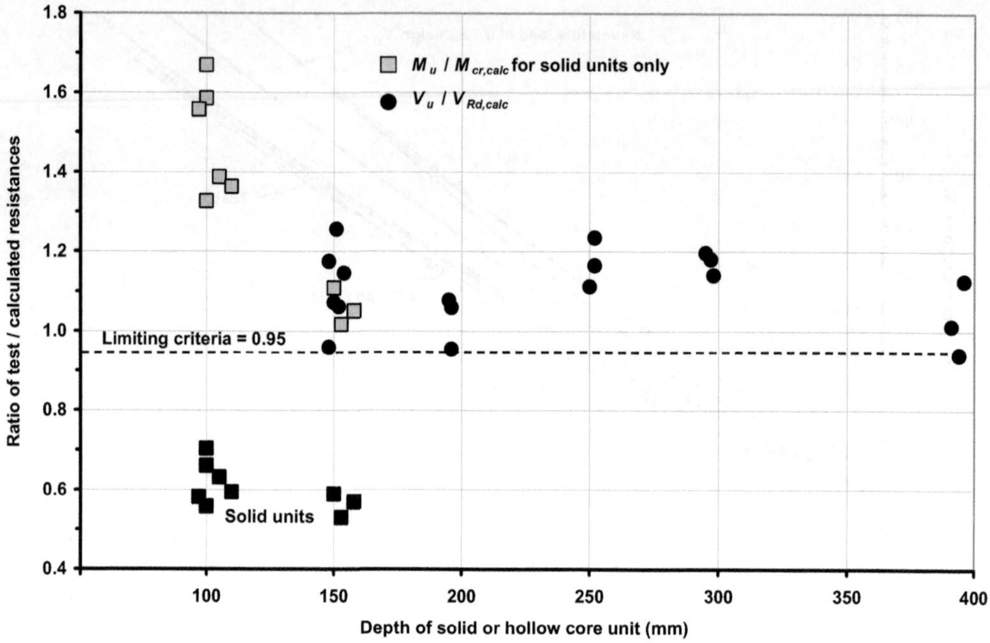

Figure 10.7 Ratios of experimental shear capacity V_u vs calculated shear capacity $V_{Rd,calc}$ for hollow core and solid units, and test/calculated moment ratios M_u vs $M_{cr,calc}$ for solid units only. (With permission of Coltman Ltd., UK and Bison Precast Ltd., UK.)

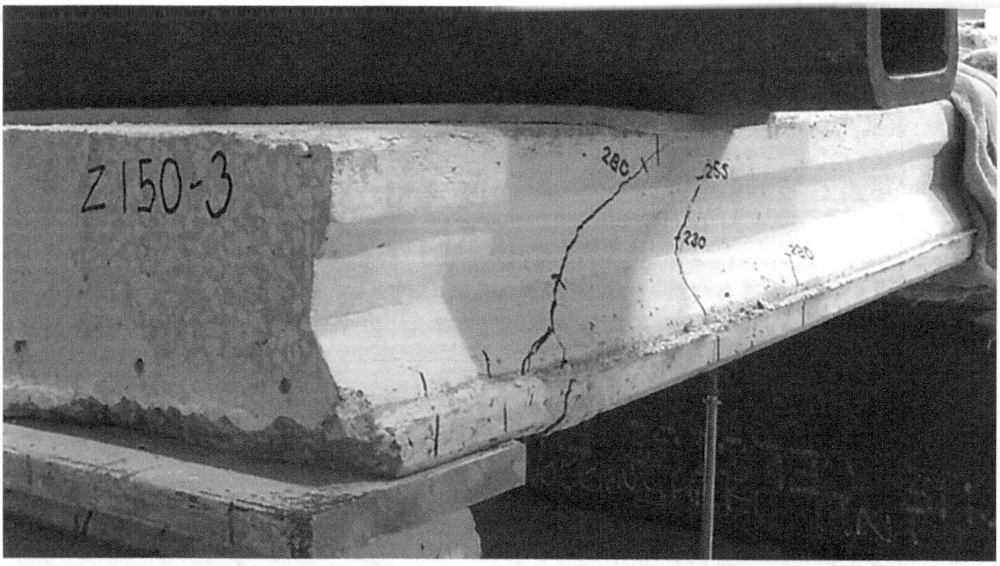

Figure 10.8 Flexural shear failure preceded by flexural cracking in a 1200 mm width × 150 mm depth solid unit. (With permission by Coltman Ltd., UK.)

The 100 and 150 mm depth solid units did not achieve the predicted shear capacity $V_{Rd,calc}$ due to prior flexural cracking as shown in Figure 10.8, although both flexural and shear cracking were observed. For this reason, the solid units are compared with their cracking moments of resistance $M_{cr,calc}$ at the position of the applied load, also shown in Figure 10.7 (grey symbols). The ratio of $M_u/M_{cr,calc}$ was in the range 1.36–1.60 for 100 mm depth units (due to higher prestress in the latter) and 1.06 for 150 mm depth units.

10.1.2 Ultimate shear tests on prestressed T beams in beam and block flooring

T beams (Figure 2.5) in beam and block flooring are mostly structurally isolated from other beams in the flooring system, i.e. known as 'singles' they support infill blocks on both sides (Figure 2.30) and therefore the shear capacity V_u of this single beam is the statically determinate capacity. Where two or three beams, or 'doubles' or 'triples' are used, there is some possibility of shear redistribution thanks to the *insitu* infill grout placed between the beams, and more than 2 (or 3) times V_u. However, the shear capacity $V_{Rd,c}$ of the single beam is used in design.

Shear tests were carried out by the author on 150×92 mm and 225×136 mm depth×width beams. The tests were carried out using a single point load positioned at a distance of 1.5 times the depth of the beam, or about 2.0 and 1.9 times the effective depth d to the pretensioning wires in the bottom of the 150 and 225 mm depth beams, respectively, over nominal effective span of 4.0 m. See Figures 10.9 and 10.10. Concrete strength was around $f_{ck}=54$ N/mm^2 and $f_{ck,cube}=77$ N/mm^2 (compared with design grade C55/67) and flexural test strength $f_{ctm,fl}=5.87$ N/mm^2. The transfer strength at 20 hours was not measured and is reliably assumed to be $f_{ck}(t)=30$ N/mm^2, with $f_{ctm}(t)=2.63$ N/mm^2. Loading was carried out in two cycles up to a shear reaction of 0.7 $V_{Rd,c}$, as shown in Figure 10.11a and b, where the recovery of deflection averaged 86% and 83% for 150 and 225 mm depth beams, respectively.

The failure of the 150 mm depth beams was by shear-tension at the bottom of the 48 mm width upstand, where the applied load at the first major diagonal crack or cracks was also the failure load, as shown in Figure 10.9. Flexural cracks did not develop into shear cracks, probably because the ratio of the bending stress to the bending resistance at this point was about 0.91. The experimental test shear capacity V_u ranged from 42.8 to 49.6 kN (a variation of ±9%) compared with design values based on the nominal section, design material properties and a bearing distance from the end of the beam of 150 mm, of $V_{Rd,c}=14.0$ kN; an average ratio of 3.34. The calculated shear capacities $V_{Rd,calc}$ ranged from 25.3 to 26.0 kN. The average ratio $V_u/V_{Rd,calc}$ was 1.83, clearly ≥1.0 for the mean of three specimens.

Figure 10.9 Shear-tension failure in 150 mm depth T beam. (With permission by Litecast Ltd., UK.)

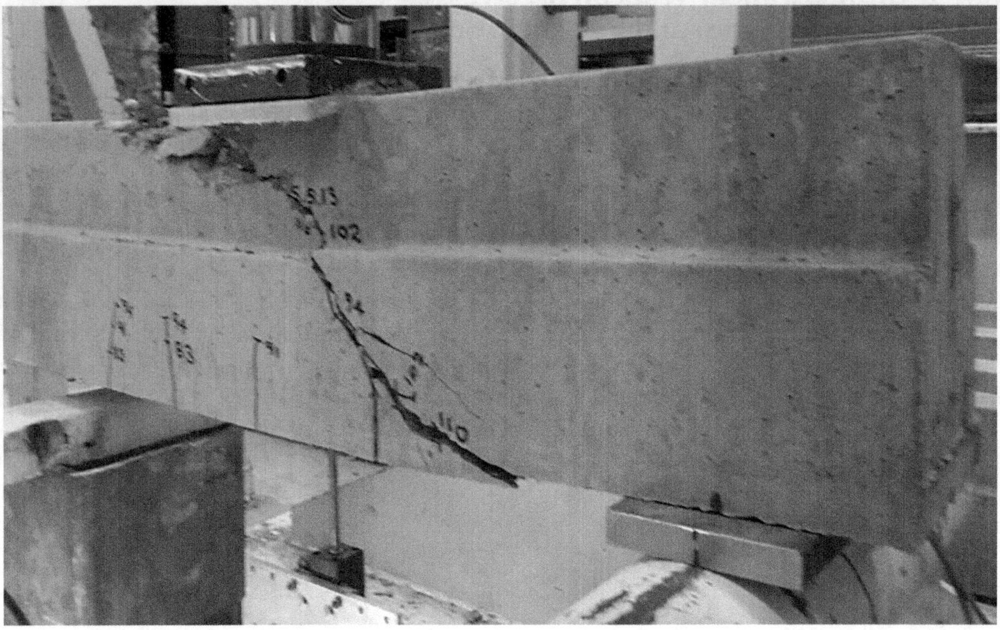

Figure 10.10 Flexural-shear and shear-tension failure in 225 mm depth T beam. (With permission by Litecast Ltd., UK.)

The failure of the 225 mm depth beams was caused by shear-tension initiated by flexural cracking at the bottom of the beam, probably because the ratio of the bending stress to the bending resistance at this point was about 1.13. The first flexural crack developed into a shear crack. Therefore, the first major diagonal crack or cracks were not necessarily at the failure load, as shown in Figure 10.10. Shear failure then occurred at the bottom of the upstand, where σ_{cp} is not only due to prestress but also by a contribution from a compressive bending stress. The experimental V_u ranged from 119.4 to 122.0 kN (a variation of ±1%) compared with $V_{Rd,c}$=35.1 kN; an average ratio of 3.45. $V_{Rd,calc}$ ranged from 62.8 to 63.8 kN. The average ratio $V_u/V_{Rd,calc}$ was 1.92. If the contribution from the bending stress at the bottom of the upstand, which averaged +5.6 N/mm^2 in the three tests, is added into the calculation the average ratio $V_u/V_{Rd,calc}$ decreases to 1.46.

The ratios $V_u/V_{Rd,calc}$=1.83 and 1.46 are considerably greater than would be expected from such shear tests; typically 1.1 to 1.3. This may in part be explained by the close clear shear distance a_v, from the edge of the bearing to the edge of the loading platen, as shown in BS EN 1992-1-1, Fig. 6.4 (BS EN 1992-1-1, 2004) 'loads near supports', being only 1.13d and 1.31d for the 150 and 225 mm depth beams, respectively. The ultimate load reduction factor $\beta=a_v/2d$, given in BS EN 1992-1-1, clause 6.2.2 (6), may now be inverted to give an increased shear capacity $V_{Rd,calc}$ for loads near to supports. As the β factor in clause 6.2.2 (6) is not entirely appropriate to the calculation of psc $V_{Rd,calc}$, as it does not affect the axial prestress, it may be applied to the principal tensile strength of concrete in the plane of cracking.

Therefore increasing the tensile strength of the concrete for the 150 mm depth beams by $1/\beta=2d/a_v$=1.78 has the effect of reducing the ratio $V_u/V_{Rd,calc}$ to 1.20. Similarly for 225 mm depth beams the factor $1/\beta$=1.53 reduces the ratio $V_u/V_{Rd,calc}$ to 1.12. Both are now within the anticipated range and give confidence to the shear behaviour of these beams.

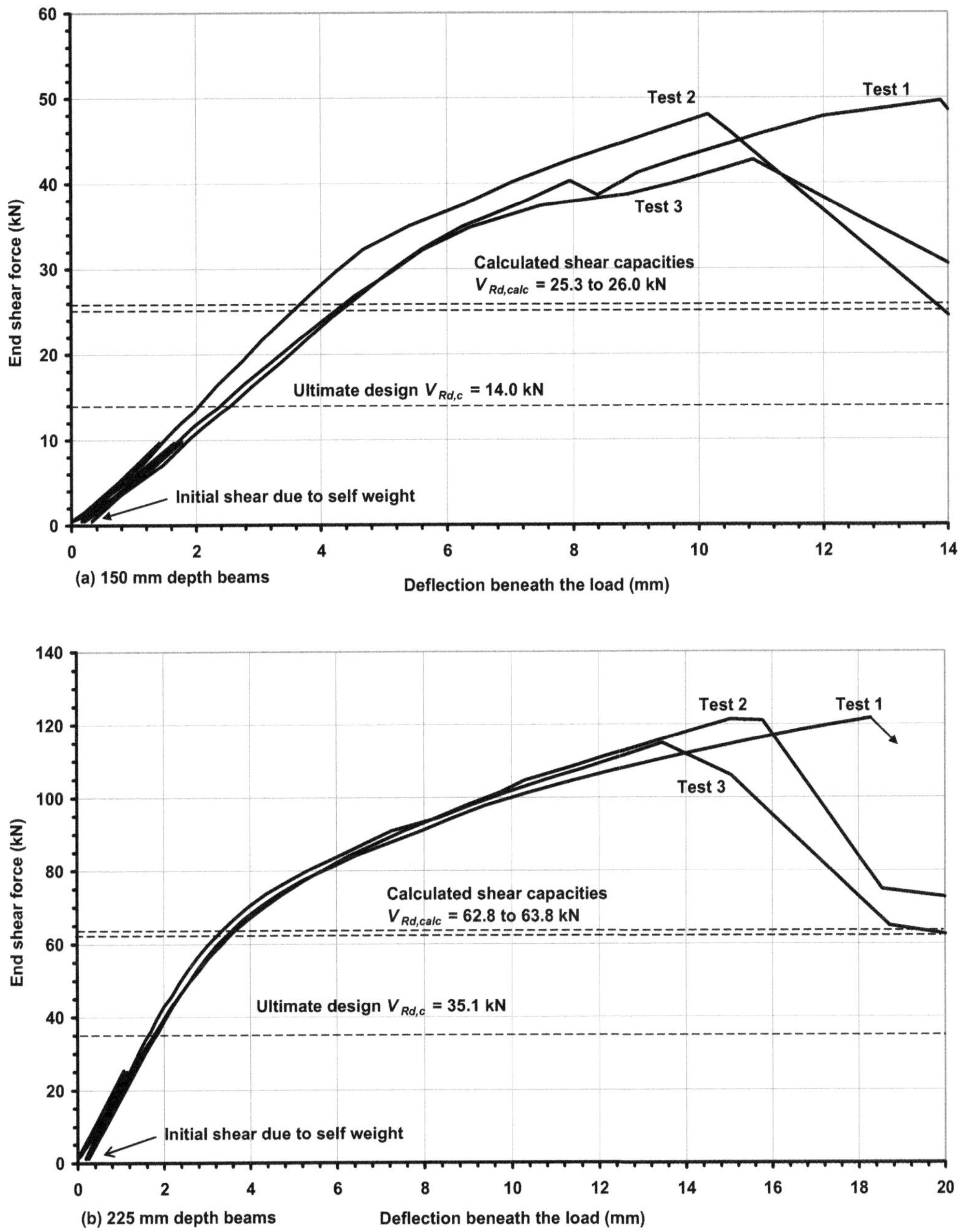

Figure 10.11 Experimental and predicted shear force vs deflection beneath the load in full scale tests carried out on (a) 92 mm width × 150 mm depth, and (b) 136 mm width × 225 mm depth prestressed T beams. (With permission by Litecast Ltd., UK.)

10.2 ULTIMATE LIMIT STATE OF SHEAR

10.2.1 Background to ultimate shear capacity and design shear forces

Calculating the ultimate shear capacity of a precast concrete section involves the equilibrium of the imposed ultimate shear stress v_{Ed} with the principal tensile strength f_{ctd} of the concrete

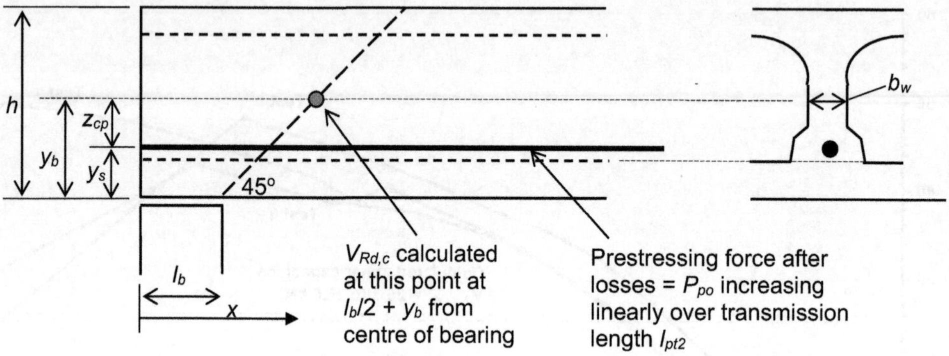

Figure 10.12 Principles and definitions used in the calculation of the flexurally uncracked shear capacity $V_{Rd,c}$ in hollow core units, similarly in solid units (without hollow cores) and beams.

combined with the precompression σ_{cp} due to prestress acting at the centroidal axis of the section. The critical design point is where an inclined plane, taken as 45° from the edge of bearing as shown in Figure 10.12, meets the intersection of the centroidal axis, providing that localised forces and stresses at end connections are dealt with and do not interrupt the distribution of shear. Shear capacity is determined only at the ultimate state – unlike in flexure there are no limiting service conditions for shear. However, there is a paradox here because some of the design equations derive from an elastic analysis.

Ultimate shear capacity is the sum of the various actions of the concrete (= aggregate interlock), prestress (= precompression), longitudinal tendons (= dowel action) and shear reinforcement (= stirrup action). However, many psc elements, such as extruded/slipformed hcu and solid units, have no shear reinforcement due to their manufacturing methods, with the only reinforcement being longitudinal pretensioning strands or wires. The shear capacity of these units therefore depends on the shear resistance of the concrete in combination with precompression and dowel action of the tendons alone.

The calculation of shear capacity carries numerous partial safety factors

i. tensile strength γ_c=1.5, then f_{ctd}=0.7 f_{ctm}/1.5
ii. precompression $\gamma_{p,fav}$=0.9
iii. strength due to aggregate interlock $C_{Rd,c}$/1.5
iv. transmission length 1.2, then l_{pt2}=1.2 l_{pt}
v. bond strength γ_c=1.5

There are good reasons for this, as shown by the sudden failures due to a single crack in Figures 10.1–10.3. The position and inclination of this crack, typically 35°–40° (depending on the position of the shear load) has a major influence of the ultimate shear capacity.

Shear capacity in prestressed sections is calculated for two conditions: the flexurally uncracked section and the flexurally cracked section in flexure, defined in Figure 10.13.

The uncracked ultimate shear resistance $V_{Rd,c}$ (BS EN 1992-1-1, clause 6.2.2, Exp. 6.4) is greater than the ultimate cracked value $V_{Rd,cr}$ (BS EN 1992-1-1, Exp. 6.2) because the full section properties are considered and a small amount of diagonal tension in the concrete is permitted. (Note that the Code's notation is $V_{Rd,c}$ for both - this book uses $V_{Rd,cr}$ for the flexurally cracked capacity.)

The ratio of $V_{Rd,c}/V_{Rd,cr}$ to BS EN 1992-1-1 is typically 1.3 to 2.0 (the latter being for maximum prestress). This is advantageous because the maximum design shear force V_{Ed}, which

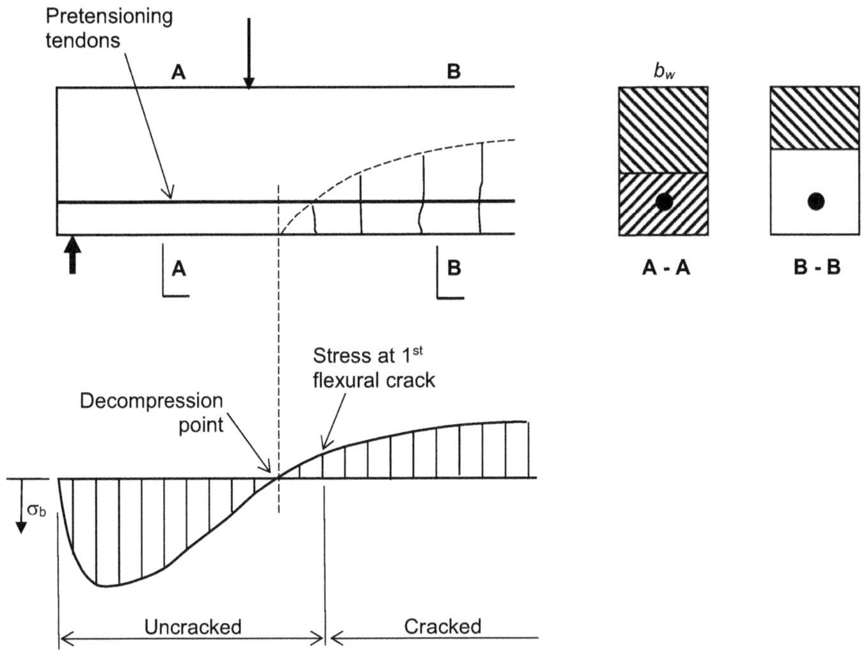

Figure 10.13 Principles of shear resistance for prestressed concrete elements.

usually occurs where the flexural moments are small may be equated with $V_{Rd,c}$ – the greater shear capacity. The position of the cracked section depends on the loading conditions, but for uniformly distributed loads, it is usually far enough into the span so as to render a check against $V_{Rd,cr}$ as unwarranted – however $V_{Rd,cr}$ should always be checked.

Two modes of shear failure are possible: flexural shear and shear tension. The occurrence of each is not merely a function of the shear span ratio, a_v/d where a_v is the distance to the primary shear force and d is the effective depth to the tendons in the shear plane, although the latter do usually occur when $a_v/d<2$ to 3.

Flexural shear occurs if a flexural crack develops into a shear crack, as shown in the failure of the solid unit in Figure 10.8. It is usually the first (or second) flexural crack which causes the shear failure. Failure occurs when the shear force exceeds the shear compression capacity and a single flexural crack initiates shear failure.

Shear tension failures are also possible where diagonal cracks are found close to the ends of the units propagating through the unprestressed and unreinforced regions of the slab (see Figures 10.1–10.3). Figure 6.14 shows the development of prestress in a simple rectangular section. According to BS EN 1992-1-1, Fig. 8.16 the length over which prestress gradually disperses to become a linear distribution across the depth of the section is given by $l_{disp}=\sqrt{(l_{pt}^2+d^2)}$. For the 250 mm depth hcu used in Example 10.1 $l_{disp}=640$ mm, or $3.1d$. Stresses in hcu are further complicated by the cross-sectional shape of the unit. It is further assumed that the extremity of the support bearing pressure spreads at an angle of about 45° from the inner bearing point. Walraven and Mercx (Walraven and Mercx, 1983) studied the bearing capacity (i.e. $V_{Rd,c}$) of hcu in considerable detail and suggested that an increment equal to half the depth of the slab is required for a shear tension crack to develop. The direct transmission of forces, by strut and tie action, does not occur in psc members which rely on the development of bond to satisfy the truss analogy.

10.2.2 Shear capacity in the flexurally uncracked region, $V_{Rd,c}$

The term 'uncracked' refers to flexural cracking, and where this exists $V_{Rd,c}$ must not be used. The ultimate shear capacity is given in BS EN 1992-1-1, clause as 6.2.2 (1) as

Exp. 6.4 $V_{Rd,c} = (I_c\, b_w / S)\, \sqrt{(f_{ctd}^2 + \alpha_l\, \gamma_{p,fav}\, \sigma_{cp}\, f_{ctd})}$ (10.1a)

For a solid rectangular section, this may be replaced by

$V_{Rd,c} = ⅔\, b_w\, h\, \sqrt{(f_{ctd}^2 + \alpha_l\, \gamma_{p,fav}\, \sigma_{cp}\, f_{ctd})}$ (10.1b)

3.1.6.2 (P) $f_{ctd} = 0.7\, f_{ctm} / 1.5$

6.2.2 (2) $\alpha_l = l_x / l_{pt2}$ (See Eq. 10.4) (10.2)

6.2.2 (2) $\sigma_{cp} = P_{po} / A_c$ = compressive stress at the centroidal axis due to prestress after all losses (10.3)

2.4.2.2 (1) $\gamma_{p,fav} = 0.9$ as this is a favourable effect

For flanged sections b_w is taken as the web width (= narrower part). In flanged sections where the centroidal axis is in the flange Eq. 10.1 should be applied at the junction of the web-flange, i.e. σ_{cp} is calculated there. For cross sections where the width varies over the height, the maximum principal stress may occur on an axis other than the centroidal axis. In such a case, the minimum value of the shear resistance should be found by calculating $V_{Rd,c}$ at various axes in the cross section. This is particularly relevant to I beams and some types of hcu with tall oval shape cores, where the principal shear tension stress occurs below the centroidal axis. See Section 10.2.5.

The critical shear plane may occur in the prestress development zone where σ_{cp} is not fully developed. It is known that prestressing forces develop somewhere between linearly and parabolically, although BS EN 1992-1-1, Figs. 8.16 and 8.17 (combined here in Figure 10.14) adopt a linear development of stress in service and bi-linear at ultimate. Therefore, a reduced value $\alpha_l\, \sigma_{cp}$ is used up to l_{pt2}, where $\alpha_l = l_x / l_{pt2}$. Known as the 'critical distance' to the shear point l_x is measured from the very end of the unit (not the centre or edge of bearing) to a shear plane drawn at 45° from the inner edge of the bearing, then $l_x = l_b + y_b$. l_{pt2} is the transmission length as follows

Exp. 8.18 Design transmission length $l_{pt2} = 1.2\, l_{pt}$ (10.4)

Exp. 8.16 Basic transmission length $l_{pt} = \alpha_1\, \alpha_2\, \sigma_{pm0}\, \Phi / f_{bpt}$ (10.5)

Transmission length coefficients are

$\alpha_1 = 1.0$ for gradual release
$\alpha_2 = 0.25$ for indented wire, $\alpha_2 = 0.19$ for 3-wire and 7-wire strand
Φ = diameter of strand or wire, or averaged if different diameters are used of the same type and same f_{pk}

The coefficient of $\alpha_2 = 0.25$ for wire comes from the basic bond length equation given in BS EN 1992-1-1, Exp. 8.3 for a circular cross section. The coefficient of $\alpha_2 = 0.19$ for strand derives from the ratio of the actual cross-sectional area of the strand to the apparent area based on the nominal diameter, e.g. for 12.5 mm diameter strand the ratio is

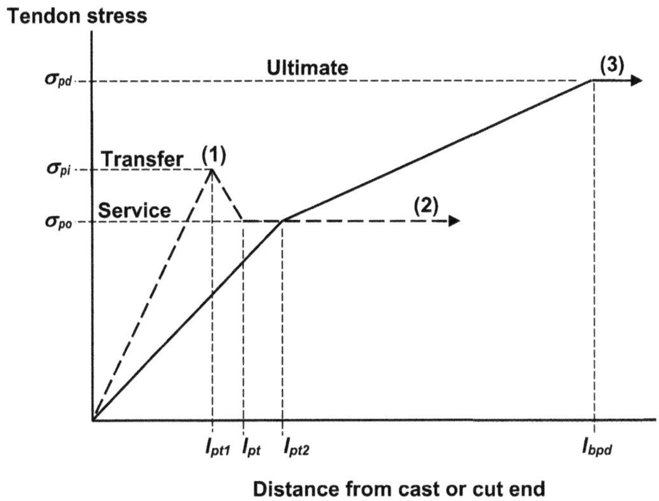

Figure 10.14 Development of stresses in the transmission zone l_{pt2} and anchorage zone l_{bpd} of prestressed members at (1) release of tendons, (2) after all losses and (3) ultimate (based on BS EN 1992-1-1, Figs. 8.16 and 8.17).

$93/(\pi\ 12.5^2/4)=0.758$. Then $\alpha_2=0.758\times0.25=0.189$. For 9.3 mm strands it is $52/(\pi\ 9.3^2/4)=0.765$, and $\alpha_2=0.765\times0.25=0.191$.

Exp. 8.15 Bond strength $f_{bpt}=\eta_1\ \eta_2\ f_{ctd}(t)$ (10.6)

where $\eta_1=1.0$ for gradual release

$\eta_2=2.7$ for indented wires, and 3.2 for 3-wire and 7-wire helical strand

The coefficient $\eta_2=2.7$ comes from the same coefficient as for high tensile ribbed bars of 2.25 (BS EN 1992-1-1, Exp 8.2) but multiplied by 1.2, i.e. $1.2\times2.25=2.7$. This gives a reduced value for the basic transmission length l_{pt} which is then compensated by the same factor 1.2 in Exp. 8.18. The coefficient for strand $\eta_2=3.2$ reflects the beneficial profile of helical strand composed of individual wires.

8.10.2.2. Design tensile strength at transfer $f_{ctd}(t)=0.7\ f_{ctm}(t)/1.5$

3.1 & 3.4 $f_{ctm}(t)=f_{ctm}\ f_{cm}(t)/f_{cm}$

Tests (Elliott, 2014) have shown that the Code's value for α_l in Exp. 6.4 may be multiplied by 1.5 for hcu manufactured using the long line manufacturing method in which the units are cut to length, rather than cast from a bare end. This comes from the results of full scale shear tests together with measurements of l_{pt} in hcu pretensioned using 5 and 7 mm wire. No results are available for strand. However, a new bond-slip relationship for the determination of transfer length l_{pt} in psc beams using strain gauges on strand has been proposed by Park and Cho (Park and Cho, 2014) followed by a finite element analysis by Oh et al. (Oh et al., 2014). The results indicate that f_{ck}, σ_{pm0} and Φ as well as concrete cover (not included in the Code's equations) are the important parameters in l_{pt}.

When the tendon is released at transfer, frictional bond stresses develop around the tendon. Equilibrium is reached when the total bond force equals the prestressing force in the tendon, and the tendon stops slipping. The length of tendon necessary to reach the state of equilibrium

Table 10.1 Ultimate shear capacity for a range of prestressed concrete hollow core floor units

Depth (mm)	*Number and width (mm) of cores*	*Self-weight (kN/m²)*	*Service moment of resistance M_{sR} (kNm)*	*Ultimate moment of resistance M_{Rd} (kNm)*	*Ultimate shear capacity $V_{Rd,c}$ (kN)*
150	11×60	2.69	67.5	100.1	120.5
200	6×140	3.33	123.7	179.0	101.1
	11×60	3.33	120.8	175.7	158.5
250	6×140	3.71	183.9	258.1	131.1
	11×60	4.03	180.9	256.8	200.1
300	6×135	4.22	249.0	342.3	174.0
	11×60	4.75	246.8	341.7	238.2
350	6×135	4.62	326.9	448.0	212.0
400	6×135	5.03	407.9	551.2	247.0
450	6×135	5.43	507.6	677.8	283.0
500	6×135	5.84	598.7	792.7	317.1

Cover or average cover to pretensioning strands = 30 mm (9.3 mm dia.) and 35 mm (12.5 mm dia.).

Final prestress at bottom (excluding self-weight) σ_b = 12.5 to 13.0 N/mm² depending on the availability of positions of tendons. Self-weight includes infilled joints between units. Exposure = XC1. f_{ck} = 45 N/mm², f_{ctm} = 3.80 N/mm².

is l_{pt} in BS EN 1992-1-1, clause 8.10.2.2. Typical values for l_{pt} are 40–80 diameters, and so 70 diameters are frequently used in design for strand, and 100 for plain or indented wire. BS EN 1992-1-1 imposes an upper bound $l_{pt2}=1.2\ l_{pt}$ for ultimate design, e.g. for $V_{Rd,c}$. When checking service stress limits, such as for slippage or bursting stresses, a lower bound is used $l_{pt1}=0.8\ l_{pt}$.

Elliott and Jolly, Section 5.2.11 (Elliott and Jolly, 2013) showed comparisons of transmission length to diameter ratio between BS EN 1992-1-1 and previous research data that are nearly all below the requirements in the code. The discrepancies are even greater when the 1.2 l_{pt} factor is added. Results from experimental measurements of transmission length showed that there is scope to considerably reduce l_{pt} in hcu, by about 30% if their unique manufacturing procedure of cutting units from long-line extruded or slip-formed castings is understood. Code values for l_{pt} are based on the development of prestress from a cast end, rather than interrupting the bond by cutting the tendons.

Table 10.1 gives the ultimate shear capacity $V_{Rd,c}$ for a wide range of hcu, from 150 to 500 mm depth. The data show that, for the same generic type of hollow cores, in this case 6 no. per 1200 mm width unit $V_{Rd,c}$ is roughly proportional to depth, typically $V_{Rd,c}=550h$ to $600h$ (h in metres).

Example 10.1

Calculate the uncracked shear capacity $V_{Rd,c}$ at the face of the support for the 250 mm depth prestressed hcu shown in Figure 6.7 and in Examples 6.1 and 7.1. Nominal bearing length $l_b=100$ mm. Effective span of hcu = 9.8 m. Self-weight of grouted unit = 3.63 kN/m², floor screed = 1.80 kN/m², finishes = 1.50 kN/m² and services & ceiling = 0.50 kN/m². Imposed live load = 5.0 kN/m². The relevant section properties are $I_c=1270\times10^6$ mm⁴, width of webs (ignoring the infill between hcu) $b_w=302$ mm, and first moment of area $S_c=6.783\times10^6$ mm³. Repeat the calculation for $V_{Rd,c}$ at a distance from the support greater than the transmission length, i.e. where $\alpha_l\geq1$. Check the ultimate shear force V_{Ed} at the face of the support using the loads given above. Draw the shear force diagrams for the ultimate design shear force V_{Ed} and capacities $V_{Rd,c}$ and $V_{Rd,cr}$.

Solution

$f_{ctd}(t)=0.7\times2.72/1.5=1.27$ N/mm²

Eq. 10.6 $\quad f_{bpt}=3.2\times1.27=4.06$ N/mm²

Average diameter = 11.17 mm
σ_{pm0} = 1158.8 N/mm² (from Example 6.1)

Eq. 10.4 $l_{pt2} = 1.2 \times 0.19 \times 1158.8 \times 11.17/4.06 = 726$ mm

$P_{po} = 860.2$ kN (from Example 7.1)
Prestress at NA (including $\gamma_{p,fav}$) $\gamma_{p,fav}$

Eq. 10.3 $\sigma_{cp} = 0.9 \times 860.2 \times 10^3/168000 = 4.61$ N/mm²

$f_{ctd} = 0.7 \times 3.80/1.5 = 1.77$ N/mm²
$l_x = l_b + y_b = 100 + 123.0 = 223.0$ mm

Eq. 10.2 $\alpha_l = 223.0/726 = 0.307$

$I_c = 1270 \times 10^6$ mm⁴, $b_w = 302$ mm, $S_c = 6.783 \times 10^6$ mm³

Eq. 10.1a $V_{Rd,c} = (1270 \times 10^6 \times 302/6.783 \times 10^6) \times [\sqrt{(1.77^2 + 0.307 \times 4.61 \times 1.77)}] \times 10^{-3}$ = **134.3 kN**

Ultimate shear force

$w_{Ed} = 1.25 \times 1.2 \times (3.63 + 3.80) + 1.5 \times 1.2 \times 5.0 = 20.14$ kN/m
Shear span $= L - l_b - 2y_b = 9800 - 100 - 2 \times 123 = 9454$ mm
$V_{Ed} = 20.14 \times 9.454/2 =$ **95.2 kN** < 134.3 kN OK

Where $\alpha_l \geq 1$, $V_{Rd,c} = (1270 \times 10^6 \times 302/6.783 \times 10^6) \times [\sqrt{(1.77^2 + 4.61 \times 1.77)}] \times 10^{-3} =$ **190.0 kN**. This information is useful only to complete the shear capacity diagram, and other than for checking hcu with holes or notches, it has no practical use.

Figure 10.15 shows the shear force diagrams for the ultimate design shear force V_{Ed} and capacity $V_{Rd,c}$.

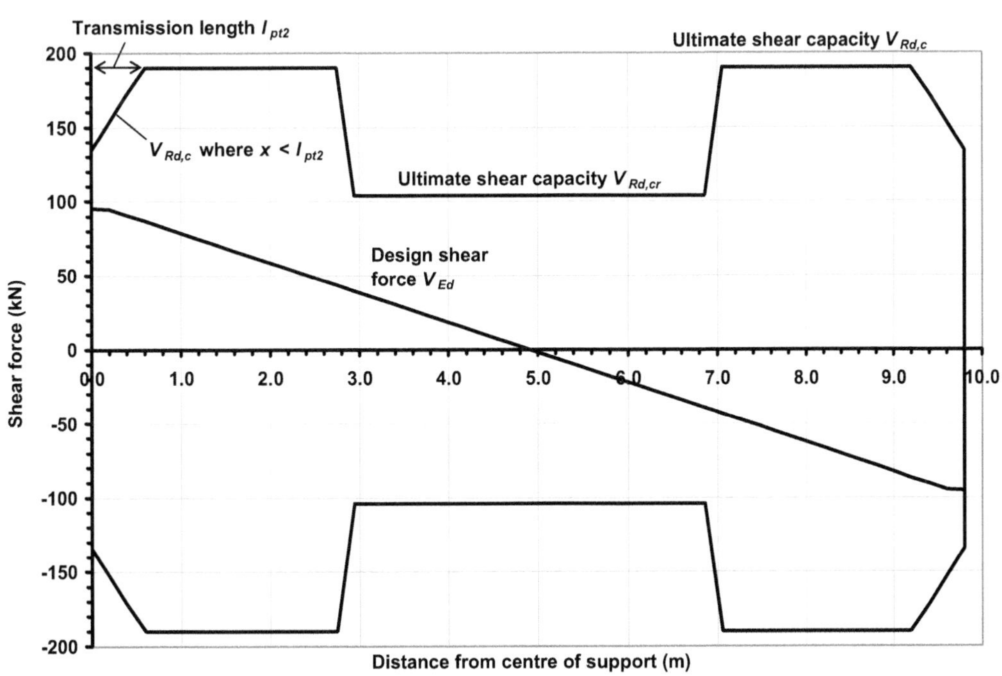

Figure 10.15 Ultimate shear force and capacity distribution for hcu in Examples 10.1 and 10.2.

10.2.3 Shear capacity in the flexurally cracked region $V_{Rd,cr}$

The shear capacity $V_{Rd,c}$ extends to a point where the service moment M_s<cracking moment $M_c=(\sigma_b+f_{ctk,0.05}/\gamma_c)\ Z_{b,co}$ according to clause 6.2.2 (2). See Figure 10.13. In the flexurally cracked region, shear resistance is a function of both the concrete and dowel action, as in reinforced concrete, plus a contribution of the level of compressive stress 0.15 σ_{cp} causing shear friction in the compressive zone. $V_{Rd,cr}$ is calculated using the following (semi-empirical) equations

Exp. 6.2a $\quad V_{Rd,cr}=[C_{Rd,c}\ k\ (100\ \rho_1\ f_{ck})^{1/3}+0.15\ \sigma_{cp}]\ b_w\ d \quad (10.7)$

6.2.2 (1) $\quad$ Depth factor $k=1+\sqrt{(200/d)}\leq 2.0 \quad (10.8)$

Steel ratio extends beyond the shear plane $\rho_1=A_p/(b_w\ d)\leq 0.02 \quad (10.9)$

$C_{Rd,c}=0.18/1.5=0.12$

Exp. 6.2b $\quad$ Minimum $V_{Rd,cr}=[v_{min}+0.15\ \sigma_{cp}]\ b_w\ d \quad (10.10)$

Exp. 6.3. $\quad v_{min}=0.035\ k^{3/2}\ \sqrt{f_{ck}} \quad (10.11)$

The critical value for $V_{Rd,cr}$ is the greater of Eqs. 10.7 and 10.10.

Example 10.2

Calculate $V_{Rd,cr}$ for the hcu in Examples 7.1 and 10.1. Determine the cracking moment of resistance where the flexurally cracked $V_{Rd,cr}$ is used. Complete the shear force diagram showing where $V_{Rd,cr}$ is critical.

Solution

$k\ (\leq 2.0)=1+\sqrt{(200/209.2)}=1.98<2.0$
$\rho_1\ (\leq 0.02)=911/(302\times 209.2)=0.0144$
Maximum $V_{Rd,cr}=[(0.18/1.5)\times 1.98\times(100\times 0.0144\times 45)^{1/3}+0.15\times 4.61)]\times 302\times 209.2\times 10^{-3}$
$=103.9$ kN

$v_{min}=0.035\times 1.98^{3/2}\ \sqrt{45}=0.653$ N/mm^2

Minimum $V_{Rd,cr}=[0.653+0.15\times 4.61]\times 302\times 209.2\times 10^{-3}=84.9$ kN
Critical $V_{Rd,cr}=$ **103.9 kN**
Distance to the flexurally cracked section

$Z_{bc,co}=10.705\times 10^6$ mm^3

$\sigma_b=11.97$ N/mm^2 at the support (from Example 7.1)

Use $V_{Rd,cr}$ where $M_s>M_c=10.705\times 10^6\times(11.97+1.77)\times 10^{-6}=$ **147.1 kNm**
From Example 7.1 and Figure 7.11 this occurs at $x=2.74$ m from the support.
Figure 10.15 shows the complete shear force diagrams for the ultimate design shear force V_{Ed} and capacities $V_{Rd,c}$ and $V_{Rd,cr}$. Note that the shear force diagram V_{Ed} is curtailed close to the support where $x<l_b/2+y_b$.

10.2.4 Ultimate shear capacity at holes and notches

The formation, sizes and positions of holes and notches in slabs are the same as those given in Section 7.4 and in Figures 7.20 and 7.21.

$V_{Rd,c}$ and $V_{Rd,cr}$ for voids are calculated using the procedure in Sections 10.2.2 and 10.2.3 using the residual width of the webs, and first and second moments of area after the void has been removed, and the remaining area of tendons after those at the void have been cut. The capacities are calculated as constant for the length of the void, and at either end of the void, they increase linearly to $V_{Rd,c}$ and/or $V_{Rd,cr}$ for the full capacity of the unit over a distance equal to the height to the centroid y_b (Figure 7.21). Voids formed entirely through hollow cores are ignored. Lateral distribution of ultimate shear capacities at holes in slabs is not carried out as the shear forces are carried by the webs in each unit individually. There is certainly no lateral distribution of section properties at notches at the supports.

Example 10.3

Repeat Examples 10.1 and 10.2 and calculate the appropriate value for $V_{Rd,c}$ or $V_{Rd,cr}$ for the hcu with a hole and end notch as defined in Example 7.8. Check the ultimate shear force V_{Ed} at the hole and face of the support using the loads given in Example 10.1. The final pretensioning force at the hole and notch is P_{po}=597.8 and 638.0 kN, respectively.

Solution at the hole

Figure 10.15 shows that the hole, at 6.0 m from the left support, is in the region of flexurally cracked shear.

k (≤ 2.0)=1+√(200/209.2)=1.98<2.0
A_{pT}=3×52+5×93=621 mm²
b_w=302−2×40=222 mm (two webs are lost to the hole)
ρ_1 (≤ 0.02)=621/(222×209.2)=0.0175

Maximum $V_{Rd,cr}=[(0.18/1.5)\times1.98\times(100\times0.0175\times45)^{1/3}+0.15\times4.61)]\times222\times209.2\times1$
0^{-3}=76.1 kN

P_{po}=597.8 kN
A_c=130127 mm² (from Table 7.7)

Prestress at NA (including $\gamma_{p,fav}$) $\gamma_{p,fav}\,\sigma_{cp}=0.9\times597.8\times10^3/130127=4.13$ N/mm²

$v_{min}=0.035\times1.98^{3/2}\sqrt{45}=0.653$ N/mm²
Minimum $V_{Rd,cr}=[0.653+0.15\times4.13]\times222\times209.2\times10^{-3}=59.1$ kN
Critical $V_{Rd,cr}$=**76.1 kN**

Ultimate shear force at the hole

w_{Ed}=20.14 kN/m
End reaction=20.14×9.80/2=98.69 kN
V_{Ed} at the hole=absolute {98.69−201.4×6.0}=**22.2 kN**<76.1 kN OK

Solution at the notch
The notch is in a flexurally uncracked position.
The transmission length is l_{pt2}=726 mm, α_l=0.307, P_{po}=638.0 kN and A_c=127269 mm²
Prestress at NA (including $\gamma_{p,fav}$) $\gamma_{p,fav}\,\sigma_{cp}=0.9\times638.0\times10^3/127269=4.51$ N/mm²

$I_c=962\times10^6$ mm⁴, b_w=222 mm, $S_c=5.139\times10^6$ mm³
$V_{Rd,c}=(962\times10^6\times222/5.139\times10^6)\times[\sqrt{(1.77^2+0.307\times4.51\times1.77)}]\times10^{-3}$=**98.4 kN**
Ultimate shear force V_{Ed}=**95.2 kN**<98.4 kN OK
Figure 10.16 shows the complete shear force diagrams for the ultimate design shear force V_{Ed} and capacities $V_{Rd,c}$ and $V_{Rd,cr}$.

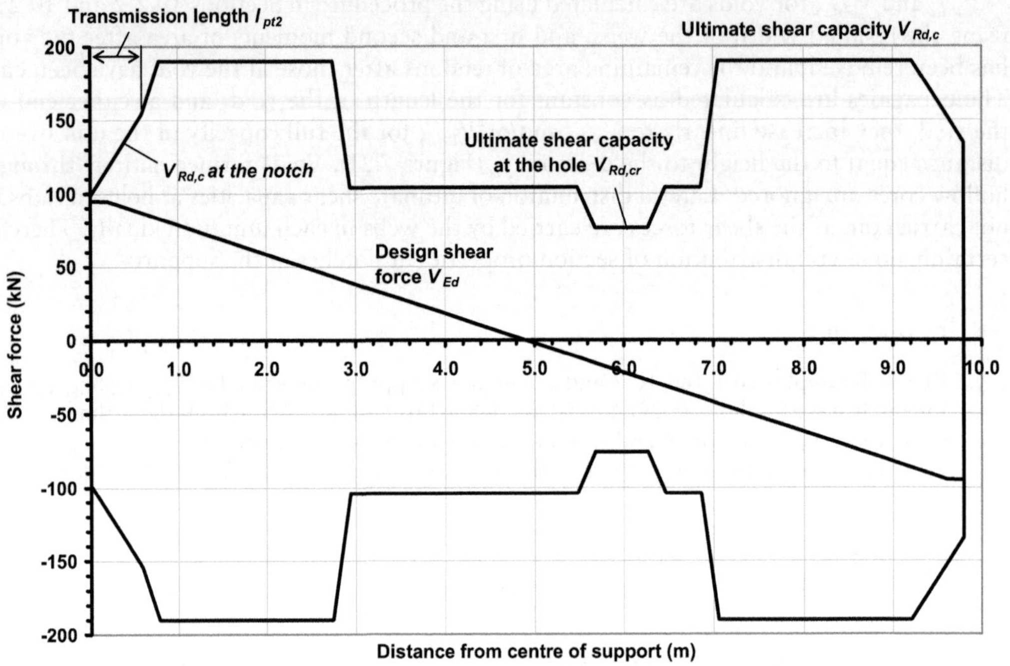

Figure 10.16 Ultimate shear force and capacity distribution for hcu with hole and notch in Example 10.3.

10.2.5 Shear capacity of hollow core units due to combined stress in the webs

The calculation for $V_{Rd,c}$ has always been carried out on a 45° shear plane at the centroidal axis, where it is understood that shear cracks will first propagate. This may not be the case in parallel-sided webs, e.g. oval or pear-shaped cores as in Figures 2.1 and 6.7, where the combined effects of the transfer of prestress and the principal shear stress may cause the shear failure to initiate at a point below the centroidal axis. BS EN 1992-1-1 recognises this in clause 6.2.2 (2)

> For cross-sections where the (web) width varies over the height, the maximum principal stress may occur on an axis other than the centroidal axis. In such a case the minimum value of the shear resistance should be found by calculating $V_{Rd,c}$ at various axes in the cross-section.

BS EN 1168, clause 4.3.3.2.2.1 presents Eqs. 10.12–10.15 to calculate $V_{Rd,c}$ at various points up the web along a 35° shear plane as shown in Figure 10.17. Then $V_{Rd,c}(y)$ at distance l_x from the end of the unit and height y from the bottom of the section, where $l_x \geq l_b + 0.5h$, is given according to clause 4.3.3.2.2.2 as

$$V_{Rd,c}(y) = \frac{I_c b_w(y)}{S_y}\left[\sqrt{f_{ctd}^2 + \sigma_{cp}(y) f_{ctd}} - \tau_{cp}(y)\right] \tag{10.12}$$

where the compressive stress ay height y due to prestress minus bending at distance l_x is

$$\tau_{cp}(y) = P_t(l_x)\left(\frac{1}{A_c} + \frac{(Y_c - y)z_{cp}}{I_C}\right) - \frac{M_{Ed}(Y_c - y)}{I_c} \tag{10.13}$$

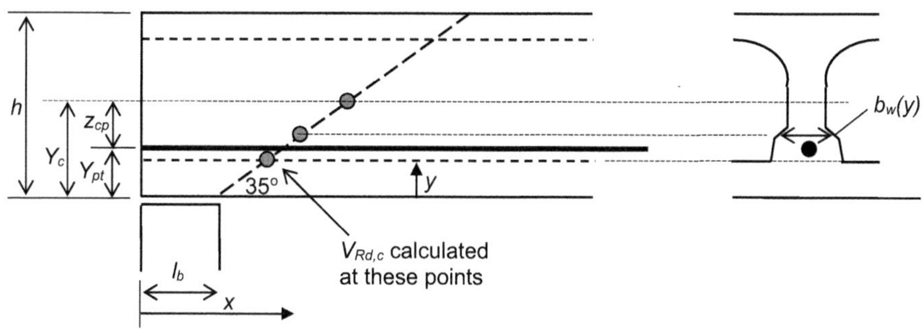

Figure 10.17 Principles and definitions used in the calculation of $V_{Rd,c}(y)$ in hollow core units after Yang (1994).

The bending stress term may be ignored if $l_x-l_b/2$ is small. BS EN 1168 does not mention the favourable effect of prestress $\gamma_{p,fav}=0.9$ to be applied to $P_t(l_x)$ in Eq. 10.13. The concrete shear stress $\tau_{cp}(y)$ at height y due to transmission of prestress at l_x is

$$\tau_{cp}(y)=\frac{\partial P_t(l_x)}{\partial x}\frac{1}{b_w(y)}\left(\frac{A_c(y)}{A_c}-\frac{S_y(Y_c-y_{Pt})}{I_c}+C_{Pt}(y)\right) \tag{10.14}$$

where I_c=second moment of area about centroidal axis

S_y=first moment of area above y and about centroidal axis
A_c=net concrete section area (note that EN 1168 calls this a *fictive* area)
$A_c(y)$=concrete section area above height y
$b_w(y)$=width of web at height y
Y_c=height to centroidal axis
y_{Pt}=height to centroid of tendons (or prestensioning force)
$P_t(l_x)$=prestressing force at distance l_x such that $\partial P_t(l_x)/\partial x$=gradient of the prestressing force in the transmission zone, which assuming a linear development of stress (Figure 10.14)$=P_{po}/l_{pt2}$
$C_{pt}(y)$=a factor according to position of tendons. $C_{pt}(y)=-1$ if $y\le y_{Pt}$ otherwise=0.
Other definitions are shown in Figure 10.17.

As an alternative to Eq. 10.12 the shear capacity at the centroidal axis may be according to Eq. 10.15, but factored as follows according to clause BS EN 1168, 4.3.3.2.2.3

$$V_{Rd,c}=(0.8I_cb_w/S)\sqrt{\left(f_{ctd}^2+0.9\,\alpha_l\gamma_{p,fav}\sigma_{cp}f_{ctd}\right)} \tag{10.15}$$

where 0.9 is a reduction factor referred to the transmission length. BS EN 1168 states that $V_{Rd,c}(y)$ need not be checked at points less than $0.5h$ from the edge of the support, $l_x<l_b+0.5h$. This clause effectively rules out checking $V_{Rd,c}$ at the bottom of the webs, where the greatest principal stresses actually occur. For example in a 250 mm depth hcu with a 35 mm bottom flange, 100 mm bearing length, $l_b+0.5h=225$ mm, whereas $l_x=100+35/\tan 35°=150$ mm. In other words $V_{Rd,c}$ is not checked until the height $y=(225-100)\tan 35°=87.5$ mm, which is only about 35 mm below the centroidal axis. (For further information for other depths of hcu see Elliott and Jolly, 2013).

Although this expression is derived using a rectangular section and not the actual flanged section appropriate to hcu, the difference is accepted as being small. The familiar term

$0.67b_wh$ is replaced in BS EN 1992-1-1 by I_cb_w/S. Safety factors are built into the equations for $V_{Rd,c}$ because $\gamma_{p,fav}=0.9$ (as a favourable effect) of the compressive stress σ_{cp} at the centroidal axis is used. The value for f_{ctd} is also conservative.

Lin Yang (Yang, 1994) has proposed a design method to calculate the shear web capacity. The equation allows for changes in cross section, material properties and prestress levels. Web shear is critical at a point at the narrowest part of the web between the voids on a diagonal line inclined at 35° to the horizontal from the edge of the support. This is typically $0.72h$ from the support, where h is the depth of the slab. Comparisons with 118 tests at VTT in Finland found that a 95% confidence limit could be used in this equation if the value so obtained is decreased by (approximately) 50 kN.

Test results, starting with Yang, have shown that where hcu have narrow webs just above the top of the strands, the maximum shear stress occurs there, and there is a reduction in $V_{Rd,c}$ of between 0.7 and 1.0, depending on the shape of the web. The ratio of the minimum value of $V_{Rd,c}$ to '$V_{Rd,c}$ at the centroid' is about 0.95 in circular cores, where the web is narrow on the centroidal axis.

The vast majority of shear tests have been carried out on hcu up to 250 mm deep, where the correlation of the ultimate shear capacity $V_{Rd,c}$ has generally been within about 10%.

Referring to the 250 mm depth hollow core unit designed in Example 10.1, $V_{Rd,c}(y)$ has been recalculated according to Eqs. 10.12–10.14. The results are shown in Figure 10.18, in which $V_{Rd,c}=136.2$ kN at the centroid, and minimum $V_{Rd,c}(y)=98.6$ kN, a huge reduction of 28%. The value of $\tau_{cp}(y)$ at the centroid is +0.42 N/mm². The steps in the results occur at changes in (i) width of webs, i.e. b_w at the centroidal axis and in the lower core are 302 and 662 mm, respectively, as shown in the dashed line in Figure 10.18, and (ii) at the level of the tendons (where $C_{Pt}(y)$ changes from 0 to –1 and causes τ_{cp} to become negative) at $y=40.8$ mm.

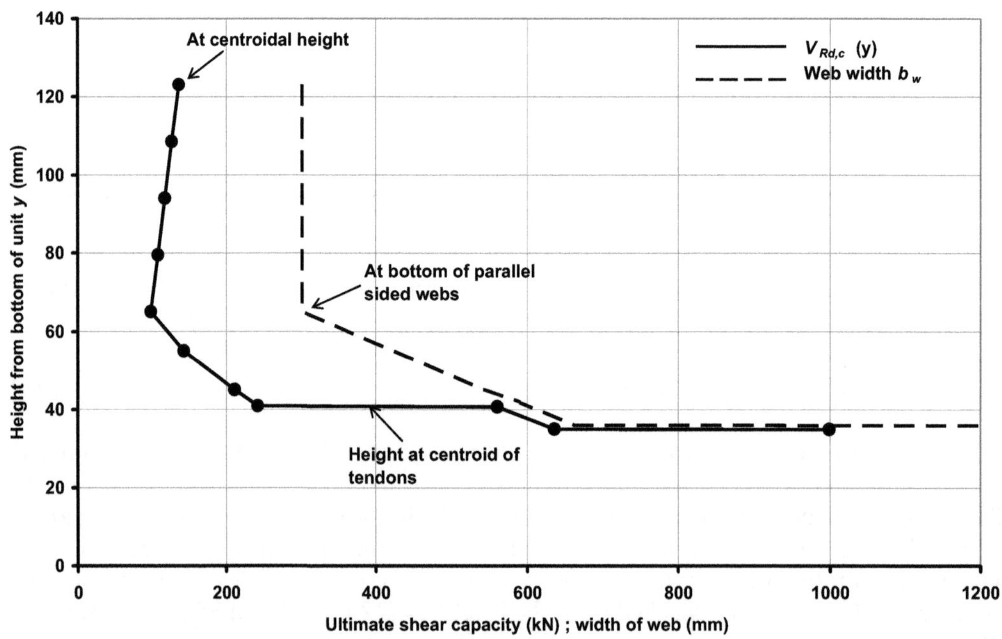

Figure 10.18 Variation in $V_{Rd,c}$ according to Eq. 10.15 in a 250 mm deep × 6 core hollow core unit of same cross section, section properties and prestress used in Exercise 10.1.

Applying the same analysis to units of similar profile but of other depths and limiting spans, and keeping $b_w = 302$ mm and the axial prestress $\sigma_{cp} \approx 5.2$ N/mm², gave the following. For $h = 150$ mm the reduction is 9%, for $h = 200$ mm it is 20%, for $h = 300$ mm it is 38%, and for $h = 350$ mm it is 45%. The minimum value of $V_{Rd,c}(y)$ always occurred at the bottom of the parallel-sided part of the webs.

The reduction in shear capacity becomes marked when $h \geq 300$ mm. In perfectly parallel-sided webs, the reductions have been shown by IPHA to be as low as 0.61, although this may have been at the very bottom of the webs and violated the $l_x < l_b + 0.5h$ rule. However, there is a clear message than in deeper units $V_{Rd,c}(y)$ will surely be less than $V_{Rd,c(c.g.)}$. ACI-318 (ACI, 2008) stated that hcu greater than 320 mm deep should have shear reinforcement when the ultimate shear force $V_u > 0.5\ V_{cw}$, i.e. without using shear reinforcement, which is impractical in production, V_{cw} is halved!

Example 10.4

Calculate $V_{Rd,c}$ at $y = 65$ mm from the bottom of the unit where the total web width is 302 mm for the hcu in Example 10.1 as shown in cross section in Figure 6.7. Section analysis has found that at this height y: $A_c(y) = 103600$ mm², $S_y = 5.735 \times 10^6$ mm³, $I_c = 1270 \times 10^6$ mm⁴, $Y_c = 123$ mm and $z_{cp} = 82.2$ mm. The shear span at this point from Example 10.1 is $9800 - 100 - 2 \times 65 = 9570$ mm. The design shear force at this point is $V_{Ed} = 20.14 \times 9.57/2 = 96.4$ kN.

Solution

l_x at $y = 65$ mm $= 100 + 65 \cot 35° = 193$ mm
$P_t(l_x) = 860.2 \times 193/726 = 228.5$ kN
$\partial P_t(l_x)/\partial x = P_t/l_{pt2} = 860.2 \times 10^3/726 = 1185$ N/mm
$Y_c - y_{Pt} = z_{cp} = 82.2$ mm
$C_{Pt}(y) = 0$ because $Y_c > y_{Pt} = 40.8$ mm
Eq. 10.14 $\tau_{cp}(y) = 1185 \times (103600/168000 - 5.735 \times 82.2/1270)/302 = 0.96$ N/mm²
M_{Ed} at $x = 193 - 100/2 = 143$ mm from centre of bearing $= w_{Ed}Lx/2 - w_{Ed}x^2/2 = 20.14 \times 9.8 \times 0.143/2 - 20.14 \times 0.143^2/2 = 13.9$ kNm
$Y_c - y = 123 - 65 = 58.0$ mm
Eq. 10.13 $\sigma_{cp}(y) = 228500\ (1/168000 + 58 \times 82.2/1270 \times 10^6) = +2.22$ N/mm² minus the bending stress $13.9 \times 10^6 \times 58/1270 \times 10^6 = -0.63$ N/mm² is $+1.59$ N/mm²
Eq. 10.12 $V_{Rd,c}(y) = (1270 \times 302/5.735) \times [\sqrt{(1.77^2 + 1.59 \times 1.77)} - 0.96] \times 10^{-3} =$ **98.9 kN**

Using the simplified Eq. 10.15 and the value of $V_{Rd,c}$ from Example 10.1 gives $V_{Rd,c} = 0.8 \times 134.3 = 107.4$ kN, some 9% greater than the minimum $V_{Rd,c}(y) = 98.6$ kN.

10.2.6 Shear capacity of hollow core units with filled cores

It is not possible to provide shear reinforcement in hcu, and therefore the prestressed concrete must do the work. (There are some production systems that enable a light mesh to be cast in the webs of very deep units, e.g. 500–750 mm, but this is not common.) A note of caution must be added here; it is not economical to increase the number of prestressing wires merely to satisfy shear. For example, increasing the area of prestressing reinforcement by a factor of 4 increases the flexural capacity by about 2.5 times and the shear capacity by only 1.2 times. The most economical method of increasing the shear capacity is to make the ends of the units solid by opening the tops of the cores and filling them with structural grade concrete, often using self-compacting concrete. A less popular alternative is to increase the depth of the slab.

The additional shear capacity with a number n (at least 2) of filled cores is according to BS EN 1168, Annex F, clause F.3 as

$$V_{Rd,t} = \tfrac{2}{3}\, n\, b_c\, d_c\, f_{ctdi} \tag{10.16}$$

where b_c=width of core (mean if cores are of different width)

d_c=depth of core (although BS EN 1168, Fig. F.1 shows this as the effective depth to the tendons)

f_{ctdi}=design tensile strength for the infill of strength f_{cki}

The ⅔ term is derived from experimental data where filled cores do not achieve their full shear capacity prior to shear failure in the surrounding webs in the psc unit.

It is important to note that because the volume of concrete above the core is already utilised in the calculation of $V_{Rd,c}$ this may not be used twice and therefore unlike Fig. F.1 the actual depth of the core is used in Eq. 10.16. If the top opened cores are wide, and option is to calculate the section properties I_c and S_c used in $V_{Rd,c}$ based on the removal of the concrete above those cores.

Some hcu have a greater shear capacity, with all things being equal, than others. The slip-formed type floor with a large number of webs has a shear capacity about 30%–40% greater than its extruded counterpart.

For example, the basic hcu designed in Example 10.1 has $V_{Rd,c}$=134.3 kN. Referring to Figure 6.7, with 6 no. filled cores using f_{cki}=25 N/mm², $V_{Rd,t}$=⅔×6×140×(250−40−35)× 1.197×10⁻³=117.3 kN, where=0.7×0.3×25^{2/3}/1.5=1.197 N/mm². Then the total $V_{Rd,c+t}$=134.3+117.3=251.6 kN. By comparison, the ultimate shear capacity of a 1200 mm width×250 mm depth solid unit, having the same material properties but with the prestress reduced by 30% in order to control transfer stresses is $V_{Rd,c}$=389 kN. A slipformed hcu with 11 no.×60 mm wide oval cores and total width of webs of 474 mm, having the same material properties and prestress as the hcu in Example 10.1, has $V_{Rd,c}$=203.7 kN (51% greater), $V_{Rd,t}$=113.2 kN and $V_{Rd,c+t}$=316.9 kN (26% greater).

10.2.7 Shear capacity of hollow core units with sloping or top notched ends

Sloping or top notched ends are often used to assist in the fixing of hcu or solid slabs onto steel, or similar, shelf angles, as shown in Figures 7.16 and 10.19. The precast unit is slewed towards one end support at an angle of about 5° to the horizontal, and the gap created by the slope or notch enables it to be positioned. The unit is pushed up against the inner face of the bearing enabling the remote end to be lowered into position.

The length of the splay is typically x=400 mm and the depth y=75–120 mm, depending on the depth of the unit, as defined in Figure 10.19. There is clearly a reduction in $V_{Rd,c}$ which is now based on the residual prestress in the shear plane x-x, obtained from striking a 45° line from the edge of the bearing. The effective depth at the shear plane=d_s. The critical depth of the section at the shear plane is h_s. If the section is cast solid at the time of manufacture, the total cross section is prestressed. The net width of the reduced section $b_s = b_t$ minus width of any vertical notch(s) or holes.

The following analysis is applicable only to non-composite units as composite units are completely filled using *insitu* concrete inclusive of the topping.

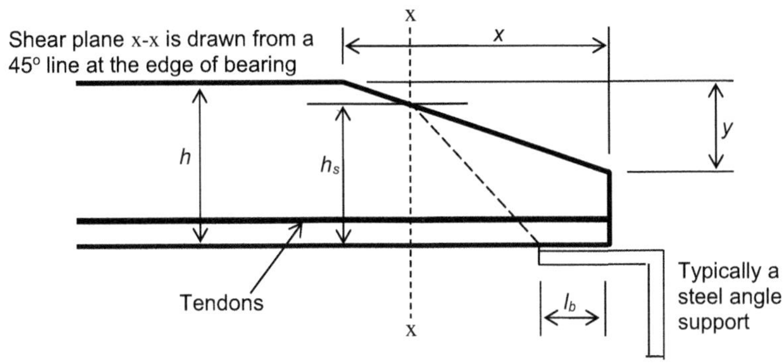

Figure 10.19 Definitions and notation for sloping ends in solid or hollow core units.

From Figure 10.19, by congruent triangles

$(h-h_s)/y=(x-h_s-l_b)/x$

Then $h_s=(hx-yx+yl_b)/(x-y)$ (10.17)

For example, if $h=250$ mm, $x=400$ mm, $y=120$ mm and $l_b=100$ mm, then $h_s=229$ mm

If length of splay $x<l_b+h$, then $h_s=h$ and $V_{Rd,c,splay}=V_{Rd,c}$

If $V_{Rd,c,splay}<V_{Rd,c}$, then $V_{Rd,c,splay}$ replaces $V_{Rd,c}$ as the critical shear capacity.

If top notched is recessed (not sloping) $h_s=h-y$

Then, following Eq. 10.1

$$V_{Rd,c,splay}=\left(I_c b_w/S\right)\surd\left(f_{ctd}^2+\alpha_1\gamma_{p,fav}\sigma_{cp}f_{ctd}\right) \tag{10.18}$$

where I_c, S, b_w and σ_{cp} are based on the section of bottom flange and webs b_w (inclusive of vertical notches and holes) of height h_s.

If the section is solid, it is effectively a rectangular section such that

$$V_{Rd,c,solid,splay}=0.67b_s h_s\surd\left(f_{ctd}^2+\alpha_1\gamma_{p,fav}\sigma_{cp}f_{ctd}\right) \tag{10.19}$$

where σ_{cp} is based on the section $b_s\ h_s$.

Example 10.5

Calculate $V_{Rd,c}$ if the basic hcu designed in Example 10.1 and shown in the cross section in Figure 6.7 contains sloping ends of length 600 mm and depth 100 mm, i.e. the bottom of the lope is level with the top of the parallel-sided webs. Also calculate $V_{Rd,c+t}$ if the remainder of the hollow cores are filled using *insitu* concrete $f_{cki}=25$ N/mm^2. The bearing length onto a steel angle is 100 mm.

Solution

Sloping end dimensions $x=600$ mm, $y=100$ mm

Eq. 10.17 $h_s=(250\times600-100\times600+100\times100)/(600-100)=200$ mm

The cross-sectional properties are calculated in the following table:

Section	Area A (mm^2)	Centroid from bottom y_b (mm)	Ay_b (mm^3)	y'(mm)	Second moment of area Ay'^2 (mm^4)	Second moment of area I_{oo} (mm^4)	First moment of area S (mm^3)
Bottom flange less webs	31324	17.5	548173	–51.5	83084300	3197676	1613241
Webs	60405	100.0	6040472	31.0	58043410	201349051	1872457
Triangles	10800	45.0	486000	–24.0	6221564	270000	259216
Totals	102529		7074645			352166001	2591455

Centroid to section of depth h_s. $y_{b,s}=7074645/102529=69.0$ mm

$P_{po}=860.2$ kN (from Example 10.1)

Prestress at NA (including $\gamma_{p,fav}$) $\gamma_{p,fav}\ \sigma_{cp}=0.9\times860.2\times10^3/102529=7.55$ N/mm²

$f_{ctd}=1.77$ N/mm²
$l_x=l_b+y_b=100+69.0=169.0$ mm
$\alpha_l=169.0/726=0.233$
$I_c=352.17\times10^6$ mm⁴, $b_w=302$ mm, $S_c=2.591\times10^6$ mm³

Eq. 10.18 $V_{Rd,c,splay}=(352.17\times10^6\times302/2.591\times10^6)\times[\sqrt{(1.77^2+0.233\times7.55\times1.77)}]\times10^{-3}=$ **102.6 kN**, a reduction of 31.7 kN from the basic unit.

With six filled cores of depth 200 – 35 = 165 mm, and using $f_{cki}=25$ N/mm², $V_{Rd,t}=2/3\times6\times140\times165\times1.197\times10^{-3}=110.6$ kN. Then the total $V_{Rd,c+t}=102.6+110.6=213.2$ kN, a reduction of 65.2 kN (or 23%) from the basic unit.

10.3 SHEAR IN PRESTRESSED BEAMS

10.3.1 Shear capacity in the flexurally uncracked region, $V_{Rd,c}$

Ultimate shear design of psc beams follows the procedures adopted for floor units given in Section 10.2 in which the uncracked $V_{Rd,c}$ and the flexurally cracked $V_{Rd,cr}$ shear resistances are calculated. In many cases, designed shear reinforcement is often quite small, and the nominal area of stirrups suffices due to the high degree of prestress in main beams and the large spans over which they operate. However, shear reinforcement, in the form of inclined bars or vertical stirrups, is required at the ends of the beam in the vicinity of the connections, and, although the design does not require it shear reinforcement is added at progressively greater spacing. The design for shear reinforcement is given in Section 10.3.3. Figure 10.20 shows an example of where shear stirrups placed in the end cage are progressively spaced further apart until the nominal area of stirrups is reached. This is not applicable to T beams for beam and block flooring, which do not have shear stirrups.

Figure 10.21 shows shear stirrups in a psc inverted-tee beam. The shape code for the stirrups in the upstand of the beam is not entirely typical, as the vertical leg of the stirrup is bent down (rather than continuous across the top of the cage) to permit the hand placement of longitudinal reinforcement on site to form a composite beam. The shear reinforcement in the bottom, or boot of the beam, will not contribute to the shear capacity and are therefore known as 'links' to resist the horizontal reaction of the floor units placed over them. See Elliott, Section 5.2.6 (Elliott, 2017) for further details and design procedures.

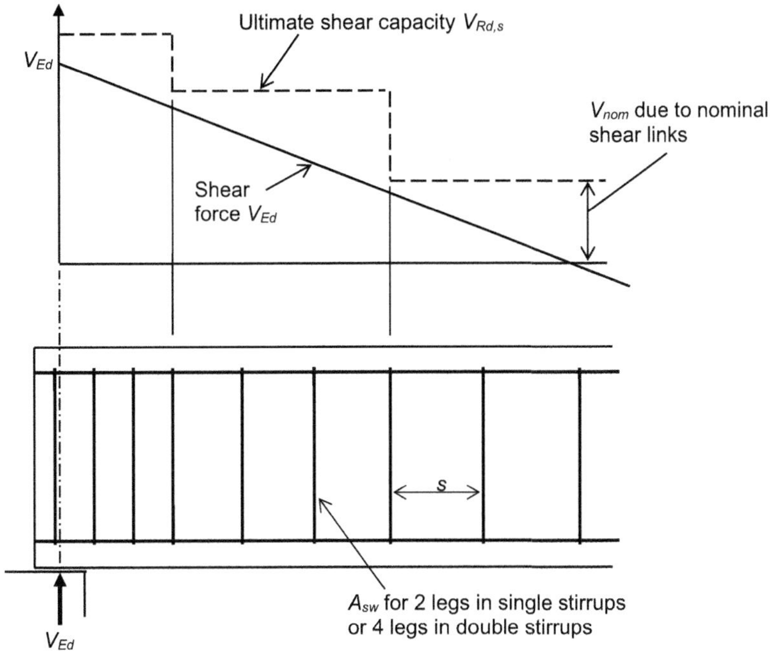

Figure 10.20 Shear reinforcement varies according to the shear force diagram.

Figure 10.21 Shear reinforcement in the end of a prestressed inverted tee beam.

Figure 10.22 Shear reinforcement in the end of the webs in a prestressed double-tee floor unit.

Figure 10.22 shows shear stirrups in the web of a psc double-tee floor unit. The pretensioning tendons have yet to be installed. The black box forms a halving joint at the end of the webs. Because the amount of prestress above the box is limited, the nib at the end of the web is designed on the basis of a static strut-and-tie. See Elliott and Jolly, Section 4.3.8 (Elliott and Jolly, 2013) and Elliott, Section 10.3 (Elliott, 2017) for further details and design procedures.

Note that because $V_{Rd,c}$ is dependent on $\alpha_l = l_x / l_{pt2}$, where the distance $l_x = l_b + y$, (Section 10.2.2) and y=height to the point of shear (which may or may not be at the centroidal height y_b from the bottom of the beam), the lowest value of $V_{Rd,c,min}$ may occur just above the centroidal axis. In rectangular sections (applies also to solid slabs) where the bearing length $l_b = 0.3h$ (h=height of beam) $V_{Rd,c,min}$ occurs at $y/h = 0.58$ where the ratio $V_{Rd,c,min} / V_{Rd,c} = 0.982$. For example, in a certain beam with b=300 mm, h=600 mm, f_{ck}=45 N/mm^2, l_{pt2}=780 mm, σ_{cp}=8.74 N/mm^2 (at centroidal height) and l_b=200 mm, the variation in $V_{Rd,c}$ over the height of the beam is shown in Figure 10.23. At y_b=300 mm $V_{Rd,c}$=415.6 kN but at y=350 mm $V_{Rd,c,min}$=408.1 kN. It is a small difference, but one which gets worse if the beam narrows above the centroidal axis, as in the case of inverted-tee beams.

In the calculation for $V_{Rd,c}$ inverted-tee beams, the shear should be considered at both the centroidal axis and at the intersection of the upstand and boot. No general guidance can be given as to which of the two positions is critical because $V_{Rd,c}$ is a function of prestress $\alpha_l\, \sigma_{cp}$ and geometry in the form of the shape function $I_c\, b_w / S_c$. Figure 10.23 also shows the variation in $V_{Rd,c}$ for the inverted-tee beam designed in Example 9.5 and shown in cross section in Figure 9.17, with $b=h$=600 mm, upstand b_w=350 mm & h_s=200 mm, f_{ck}=45 N/mm^2,

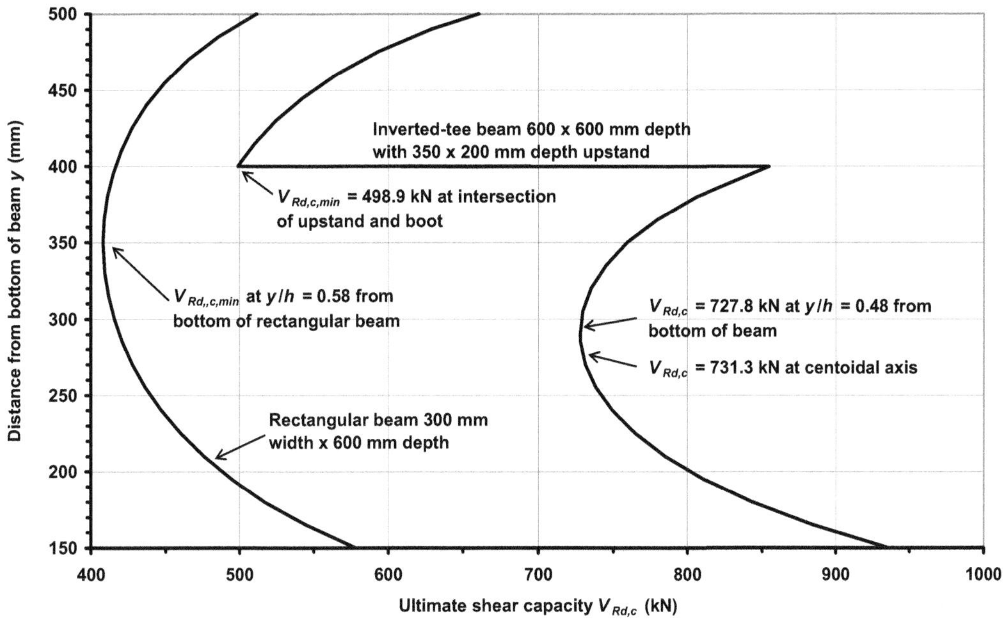

Figure 10.23 Variation in $V_{Rd,c}$ in the section of prestressed beams (left) 300 × 600 mm depth rectangular beam (right) 600 × 600 mm inverted-tee beam with 350 mm × 200 depth upstand (intersection of boot and upstand at 400 mm from bottom of beam).

l_{pt2}=812 mm, σ_{cp}=7.37 N/mm^2 (at c.g.) and l_b=200 mm. At y_b=269.8 mm $V_{Rd,c}$=727.2 kN but at the intersection of the boot/upstand where y=400 mm $V_{Rd,c,min}$=496.3 kN, this time 32% difference.

If the bending stress (introduced in Eq. 10.13) $-M_{Ed}(Y_c-y)/I_c$ is included in the expression for $V_{Rd,c}$ the shear capacity at the intersection of the web and boot increases to 605.2 kN (an increase of 106.3 kN). In this example, the M_{Ed} is calculated from the centre of bearing at x=200/2+400=500 mm, where M_{Ed}=179.2 kNm and the additional stress due to bending is +2.86 N/mm^2. The additional shear capacity due to bending is usually ignored in design.

10.3.2 Shear capacity in the flexurally cracked region $V_{Rd,cr}$

Ultimate shear design of psc beams follows the procedures adopted for prestressed floor units given in Section 10.2 in which the flexurally cracked $V_{Rd,cr}$ shear capacity is calculated. The section is considered as flexurally cracked where $M_s > M_c$, where $M_c = (\sigma_b + f_{ctd})\, Z_{b,co}$.

Shear reinforcement in the flexurally cracked region is rarely necessary because the ultimate shear resistance is a function of the ultimate flexural requirements. The shear span (M_{Ed}/V_{Ed}) for most beams in precast structures is span/4. Thus the shear force at the position of flexural decompression rarely exceeds $0.5V_{Rd,c}$, for example in Example 10.6 the shear at the point of decompression is 137 kN about 0.28 $V_{Rd,c}$. As with floor units (Section 10.2.3) a minimum value of $V_{Rd,cr}$ is calculated.

Example 10.6

Calculate $V_{Rd,c}$ and $V_{Rd,cr}$ for the inverted-tee beam in Examples 7.10 and 9.5. The bearing length may be taken as 200 mm. Check $V_{Rd,c}$ at (i) centroidal axis and (ii) intersection of the upstand and boot. Draw the shear force and shear capacity diagrams, showing the position where the section becomes flexurally cracked. Beam loads are as given in Example 9.5 where the ultimate beam load is w_{Ed}=89.56 kN/m.

Solution

Flexurally uncracked capacity $V_{Rd,c}$

i. At the centroidal axis
y_b=269.8 mm<depth of boot of 400 mm, therefore the width in shear is 600 mm
b_w=600 mm, I_c=8151.5×10^6 mm^4, y_t=600−269.8=330.2 mm
S_c=(350×200×(330.2−200/2)+600×130.2^2/2)=21.202×10^6 mm^3
$I_c\,b_w/S_c$=8151.5×10^6×600/21.202×10^6=230680 mm^2
At the support σ_{p0}=953.3 N/mm^2 and σ_{pm0}=1157.7 N/mm^2 (from Example 7.10)
$f_{ctd}(t)$=0.7×2.72/1.5=1.27 N/mm^2
f_{bpt}=3.2×1.27=4.06 N/mm^2
l_{pt2}=1.2×0.19×1157.7×12.5/4.06=812 mm
σ_{cp} including γ_p=0.9×953.3×2418/307500=6.746 N/mm^2 (from Example 7.10)
f_{ctd}=0.7×0.3×45$^{2/3}$/1.5=1.77 N/mm^2
l_x=200+269.8=469.8 mm
α_l=469.8/812=0.579
$V_{Rd,c}$=230680×√(1.77^2+0.579×6.746×1.77)×10^{-3}=**731.3 kN**

ii. At the intersection of upstand and boot
b_w=350 mm, y=200 mm, y'=230.2 mm
S_y=350×200×230.2=16.115×10^6 mm^3
$I_c\,b_w/S_y$=8151.5×10^6×350/16.115×10^6=177040 mm^2
Prestressat400mmfromthebottom=953.3×2418(1/307500+92.9×(269.8−400)/8151.5×10^6=4.08 N/mm^2
σ_{cp} at 400 mm from bottom=0.9×4.08=3.67 N/mm^2
l_x=200+400=600 mm
α_l=600/812=0.739
$V_{Rd,c}$=177040×√(1.77^2+0.739×3.67×1.77)×10^{-3}=**498.9 kN**<731.3 kN

The least of these values is used in design.

The design shear force is calculated at a distance $L_b/2+y_b$ from the centre of support. The shear span is therefore L_v=8500−2×(100+269.8)=7.760 m. V_{Ed}=89.56×7.76/2=347.5 kN<$V_{Rd,c}$=498.9 kN.

For completeness $V_{Rd,c}$ at the end of the transmission length may be calculated using α_l=1.0. Then $V_{Rd,c}$=177040×√(1.77^2+1.0×3.67×1.77)×10^{-3}=**549.6 kN.**

Flexurally cracked capacity $V_{Rd,cr}$ at intersection of upstand and boot

From Example 9.5, d=480.8 mm and A_{pT}=2046 mm^2
k=1+√(200/480.8)=1.64
v_{min}=0.035×1.64$^{1.5}$ √45=0.495 N/mm^2
$A_s/b_w\,d$=2046/(350×480.8)=0.0122
σ_{cp}=3.67 N/mm^2
$V_{Rd,cr}$=[(0.12×1.64×(100×0.0122×45)$^{1/3}$)+0.15×3.67]×350×480.8×10^{-3}=**218.7 kN**

$V_{Rd,cr}$ minimum=(0.495+0.15×3.67)×350×480.8×10^{-3}=176.0 kN<218.7 kN, use $V_{Rd,cr}$=218.7 kN.

Calculation of distance to the flexurally cracked section

This is where the service bending moment is greater than the cracking moment of resistance, $M_s>M_{cr}$ and where the flexurally cracked shear capacity $V_{Rd,cr}$ governs.

From Example 7.10. $Z_{b,co}$=31.683×10^6 mm^3, σ_b=14.58 N/mm^2
$f_{ct,0.05}$=0.7 f_{ctm}=0.7×3.80=2.66 N/mm^2

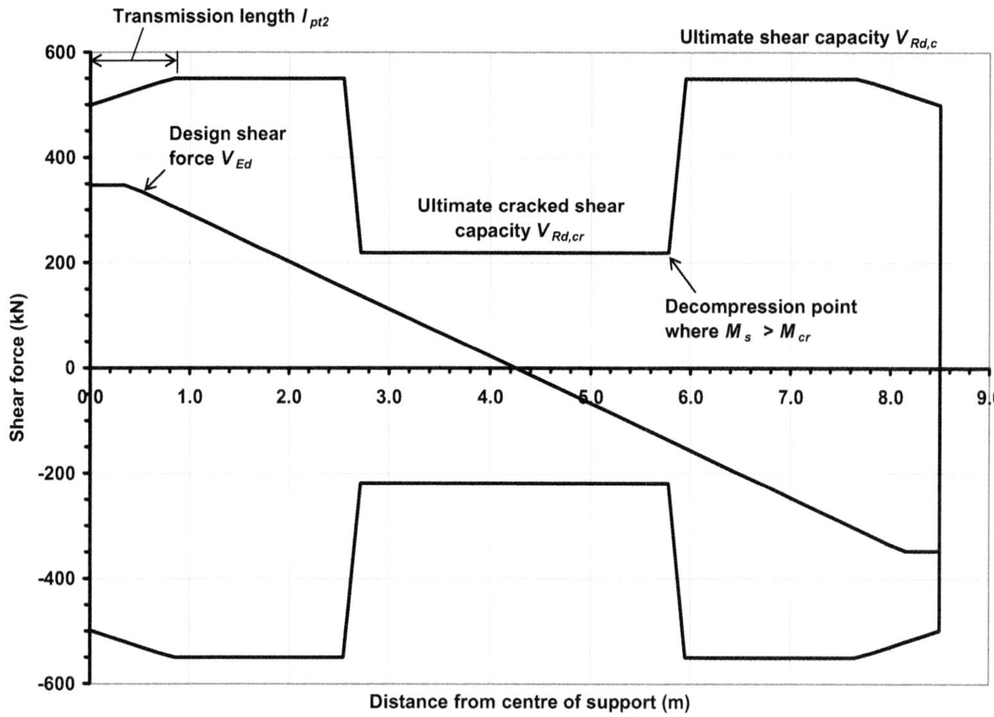

Figure 10.24 Shear force VEd and shear capacities $V_{Rd,c}$ and $V_{Rd,cr}$ for the inverted-tee beam in Example 10.6.

$M_{cr}=31.683\times(14.58+2.66/1.5)=518.0$ kNm

The service UDL $w_s=65.69$ kNm

Then at a distance x to the point from the centre of support

$M_{cr}=w_s\, Lx/2-w_s\, x^2/2$

Giving the quadratic $w_s/2\, x^2\ -w_s\, L/2\ x+M_{cr}=0$

From which $x=2.74$ m from the centre of supports

Figure 10.24 shows the complete shear force diagrams for the ultimate design shear V_{Ed} and the capacities $V_{Rd,c}$ and $V_{Rd,cr}$. The increase in $V_{Rd,c}$ is linear within the transmission length l_{pt2}. Quite often $V_{Rd,c}$ at the end of the transmission length is not calculated as it is never critical, but is shown here for completeness.

10.3.3 Prestressed beams with shear reinforcement

The design for shear follows the normal procedures for reinforced concrete sections – that is the ultimate shear capacity is the sum of the concrete resistance (= aggregate interlock+dowel action but with the additional effect of prestress) plus the ultimate (= yield stress) capacity of the shear stirrups. The approach in EN 1992-1-1 is that these variables are combined in a single function known as the variable strut inclination (VSI) method. The inclination of the shear plane may vary from $\theta=21.8°$ to 45°, or cot $\theta=1.0$ to 2.5, depending on the relationship between web width and the strength of the concrete strut. This is shown in the truss model in BS EN 1992-1-1, Fig. 6.5. Note the assumption that the lever arm between the top and bottom truss is $z=0.9d$, which is a reasonable assumption where the bending moment is small in the presence of shear, but not, for example in cantilevers where shear and bending are both large.

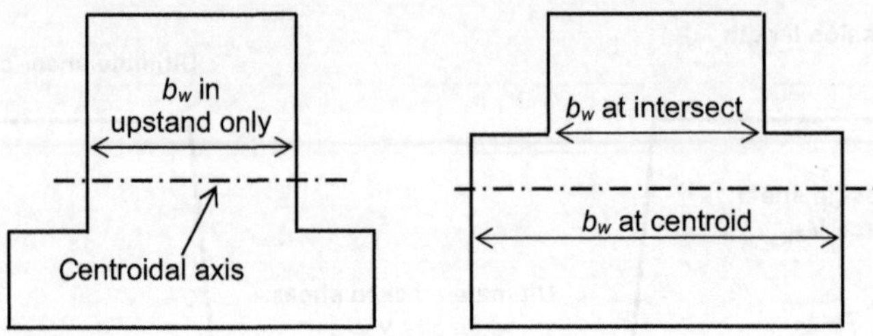

Figure 10.25 **Determination of the critical widths b_w for shear capacity in inverted-tee beams where (a) the centroidal axis is in the upstand, and (b) the centroidal axis is in the boot.**

In main inverted-tee beams, the effective width of the web b_w used in shear calculations depends on whether the centroid of the gross, flexurally uncracked, section occurs in the upstand or in the boot of the beam as shown in Figure 10.25. Where this in the upstand then b_w=upstand width. However, where it lies in the boot the critical section may lie either at (i) the centroid or (ii) at the intersection of the upstand and the boot (i.e. bottom of the upstand), and both cases must be considered as shown in Figure 10.25b.

It would be very unusual not to use shear reinforcement in beams, although some minor elements where shear failure is neither likely nor critical may rely on dowel action of the main bars fully anchored at the support, and some psc wet-cast floor slabs do not have shear reinforcement.

If $V_{Ed} > V_{Rd,c}$ designed shear reinforcement must be placed in the appropriate section. Shear stirrups will be placed in the upstand of the beam as shown in Figure 10.21. Links must also be provided in the boot of the beam to carry the bearing ledge forces, see Elliott, Section 5.2.6 (Elliott, 2017) for further details and design procedures. In the latter shear stirrups must be provided in the upstand or in the boot depending on where b_w is taken. Designed shear stirrups must be additional to those required for torsion and the minimum shear stirrups for the bearing ledge reactions (see Example 10.7).

Referring to Figure 10.26 and to BS EN 1992-1-1, clause 6.2.3, the number of stirrups crossing the shear plane=$z \cot \theta / s$. The ultimate force in each stirrup=area×stress=A_{sw} 0.87 f_{ywk}, then the ultimate resistance of the stirrups is

$$V_{Rd,s} = z\ 0.87\ f_{ywk} \cot\theta\ (A_{sw}/s) \tag{10.20}$$

Note there is no $V_{Rd,c}$ in this analysis, and $V_{Rd,c}$ is not added to $V_{Rd,s}$.

Therefore the area of designed shear reinforcement A_{sw}/s (mm²/mm run along the beam) is given by

Exp. 6.8 $$A_{sw}/s \geq V_{Rd,s}/z\ 0.87\ f_{ywk} \cot\theta \tag{10.21}$$

In the limit $V_{Ed} = V_{Rd,s}$. Then, if the diameter and hence area of the stirrups is chosen, typically in practice, then the spacing of the stirrups is given by

$$s = A_{sw}\ z\ 0.87\ f_{ywk} \cot\theta / V_{Ed} \tag{10.22}$$

The area of shear reinforcement must be appropriate to the critical shear section; in this case, it is assumed this is at the top of the boot. Clearly, the VSI $\cot\theta$ has a big influence on A_{sw}/s; it can alter it by 250%! The expression for $\cot\theta$ derives from the compressive strut, shown in Figure 10.26 and the Code's Fig. 6.5, equal to the compressive stress σ_c acting

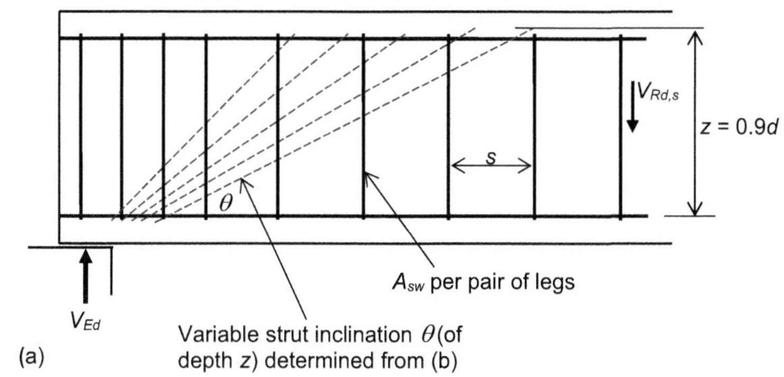

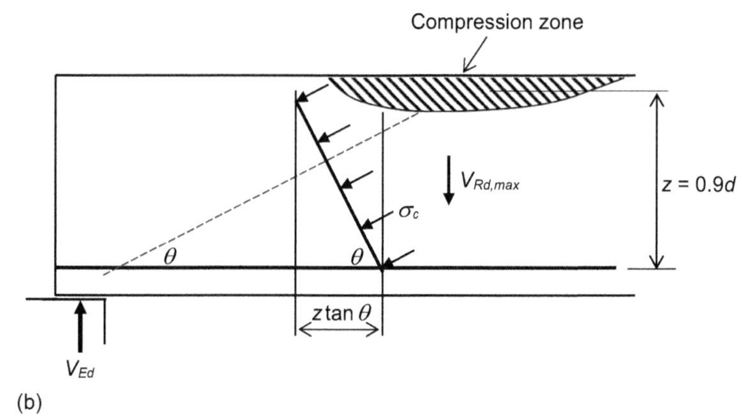

Figure 10.26 Shear design in beams (a) shear stirrups capacity $V_{Rd,s}$ (Exp 6.8, BS EN 1992-1-1) and (b) compressive strut capacity $V_{Rd,max}$ (Exp 6.9, BS EN 1992-1-1).

perpendicular to the shear plane on the width $b \times$projected length $z \cos \theta$. The vertical component of this is $b\ z \cos \theta \sin \theta$. The compressive stress is given as $\sigma_c = \alpha_{cw}\ v_1\ f_{cd}$
where

Exp. 6.11 $\alpha_{cw} = 1 + \sigma_{cp}/f_{cd}$ for $0 < \sigma_{cp} \leq 0.25\ f_{cd}$

$\alpha_{cw} = 1.25$ for $0.25\ f_{cd} < \sigma_{cp} \leq 0.5\ f_{cd}$

$\alpha_{cw} = 2.5\ (1 - \sigma_{cp}/f_{cd})$ for $0.5\ f_{cd} < \sigma_{cp} \leq 1.0\ f_{cd}$ (10.23)

According to the UK National Annex for clause 6.2.3 (3) to BS EN 1992-1-1 (NA to BS EN 1992-1-1, 2004) σ_{cp} is not calculated for a distance of $0.5d \cot \theta$ from the edge of the support, i.e. to a point at mid-height of the beam along the VSI plane, where the prestress is not yet fully developed

Note that at the end of the beam where

Exp. 6.6 shear stress parameter $v_1 = 0.6\ (1 - f_{ck}/250)$

For shear effects $f_{cd} = f_{ck}/1.5$

$V_{Rd,max} = \alpha_{cw}\ v_1\ f_{cd}\ b\ z \cos \theta \sin \theta$ (10.24)

but $\cos \theta \sin \theta = \tan \theta \cos^2\theta = \tan \theta / \sec^2\theta = \tan \theta/(1 + \tan^2\theta) = 1/(\cot \theta + \tan \theta)$

and $2 \cos \theta \sin \theta = \sin 2\theta$

Then $1/(\cot\theta+\tan\theta)=\cos\theta\sin\theta=0.5\sin 2\theta$

Exp. 6.9 $V_{Rd,max}=\alpha_{cw}\, b\, z\, v_1\, f_{cd}/(\cot\theta+\tan\theta)=\alpha_{cw}\, b\, z\, v_1\, f_{cd}\, 0.5\sin 2\theta$ (10.25)

Rearranging and in the limit replacing $V_{Rd,max}$ with V_{Ed} calculated at a distance d from edge of the support

$\theta=0.5\sin^{-1}(V_{Ed}/0.5\,\alpha_{cw}\, b\, z\, v_1\, f_{cd})$ $\therefore \cot\theta$ is determined. (10.26)

but $1.0\leq\cot\theta\leq 2.5$

For example let $V_{Ed}=700$ kN, $b=300$ mm, $\sigma_{cp}=0.2\,f_{cd}$, $z=0.9\times 500=450$ mm, $f_{ck}=32$ N/mm^2 and $f_{ywk}=500$ N/mm^2. Then $\alpha_{cw}=1.20$, $\theta=0.5\sin^{-1}(700\times 10^3/(0.5\times 1.20\times 300\times 450\times 11.16)=25.3°$. cot 25.3°$=2.108<2.5$. $A_{sw}/s=700\times 10^3/(450\times 0.87\times 500\times 2.108)=1.70$ mm^2/mm$=850$ mm^2/m per leg (if the single stirrup has two legs, or 425 mm^2/m per leg for double stirrups with four legs). Use single H10 at 90 mm spacing (872 provided).

The minimum area of links is some function of concrete tensile strength $\sqrt{f_{ck}}$ and b (or b_v in a web or upstand) as

Exp. 9.5N $A_{sw,min}/s\geq 0.08\sqrt{f_{ck}}\, b/f_{ywk}$ (10.27)

The maximum area of links is given when the concrete stress reaches 0.5 f_{cd}, as

Exp. 6.12 $A_{sw,max}/s\leq 0.5\,\alpha_{cw}\, b\, z\, v_1\, f_{cd}/0.87\, f_{ywk}$ (10.28)

But is rarely critical even in narrow web beams.

The minimum spacing of stirrups is not specified in codes, but practically is about 50 mm for small links (8–10 mm dia.) or about 70 mm (12 mm dia.). The maximum spacing is as always $0.75d$ to avoid a 45° shear plane passing between two links as

9.2.2 (6) $s\leq 0.75d$ (10.29)

Example 10.7

Calculate the requirement for shear stirrups in the prestressed inverted-tee beam shown in cross section in Figure 9.17 and designed in Example 10.6, but where the effective span is reduced to 5.50 m, the ultimate load carried by the beam is increased to 180.0 kN/m and there are dead and live point loads of 100 kN each at 1.0 m from the centre of bearing. These are chosen to ensure that the moments of resistance calculated in Examples 7.10 and 9.5 are not exceeded. Use H10 stirrups with $A_{sw}=157$ mm^2 per two-leg stirrup. Use $f_{ywk}=500$ N/mm^2. Having determined the maximum requirement check if stirrups of greater spacing may be specified. Use the end cover to the start of the stirrups as 30 mm.

Also check the minimum shear stirrups if the ultimate load (excluding self-weight of beam) from the floor slab acting on the bearing ledge is 170.5 kN/m.

Solution

$L_v=5500-2\times(100+269.8)=4.760$ m.
$V_{Ed}=180.0\times 4.76/2=428.4$ kN due to UDL.
$P_{Ed}=1.25\times 100+0.7\times 1.5\times 100=230$ kN for point loads

$V_{Ed}=180.0\times 4.76/2+230\times(5.50-1.00)/5.50=428.4+188.2=616.6$ kN>498.9 kN from Example 10.6. Shear stirrups are required.

At the opposite end $V_{Ed}=428.4+41.8=470.2<498.9$ kN. Shear stirrups are not required.

At the intersection of the upstand and boot, where b_w=350 mm, from Example 10.6 σ_{cp}=3.67 N/mm². But σ_{cp} is not required for a distance 0.5d cot θ+L_b/2=0.5×480.8×2. 5*+200/2=700 mm. *Check later. Therefore, take α_{cw}=1.

z=0.9×480.8=432.7 mm
$v_1 f_{cd}$=0.6 (1−45/250)×(45/1.5)=14.76 N/mm²

Eq. 10.26 θ=0.5 sin⁻¹ (616.6×10³/(0.5×350×432.7×14.76)=16.74°

cot 16.74°=3.32<2.5. Use cot θ=2.5

Eq. 10.21 A_{sw}/s=616.6×10³/(432.7×0.87×500×2.5)=1.31 mm²/mm=655 mm²/m per leg. Spacing required is H10 at 119.8 mm spacing. Use H10 at 115 c/c (682 provided). For 2 legs A_{sw}/s=157/115=1.365 mm²/mm.

Eq. 10.20 $V_{Rd,s}$=432.7×0.87×500×2.5×1.365×10⁻³=642.4 kN

Increase the spacing to 125 mm whence A_{sw}/s=157/125=1.256 mm²/mm.

Eq. 10.20 $V_{Rd,s}$=432.7×0.87×500×2.5×1.256×10⁻³=591.0 kN

The gradient of the shear force diagram is 180 kN/m. The shear force at the centre of bearing is V_{Ed}=180.0×5.50/2+188.2=683.2 kN. The distance to stirrups of 125 mm spacing is (683.2−591.0)×10³/180=512 mm, plus half the bearing length (L_b/2=100 mm)=612 mm from the end of the beam. Therefore, use 6 no. H10 at 115 mm from the end of the beam for a distance of 5×115+35=610 mm.

The distance from the centre of bearing where $V_{Ed}<V_{Rd,c}$=549.6 kN (from Example 10.6 at the end of the transmission length) is x=(683.2−549.6)×10³/180=742 mm. This is 742−512=230 mm beyond the end of the stirrups at 115 mm, therefore 2 no. H10 at 125 mm.

The minimum area of stirrups (these could technically be called links as they do not contribute to the shear capacity).

Eq. 10.27 $A_{sw,min}/s \geq 0.08\sqrt{45}\times350/500$=0.375 mm²/mm=188 mm²/m per leg.

Vertical reaction from the floor load acting on the bearing ledges at the top of the boot=170.5 kN/m. Then A_{sw}=0.5×170.5×10³/0.87×500=196 mm²/m per leg>188 mm²/m per leg.

The requirement is H10 at 401 mm spacing. The maximum spacing is 0.75×480.8=360 mm. In practice, the maximum spacing of links is, depending on the designer's preference, typically 300–400 mm. Use H10 at 360 c/c (218 mm²/m provided).

Eq. 10.28 $A_{sw,max}/s \leq$0.5×1.12×350×14.76/0.87×500=6.65 mm²/mm, not critical.

Figure 10.27 shows the complete shear force diagrams for the ultimate design shear V_{Ed} and the prestressed concrete capacities $V_{Rd,c}$ and $V_{Rd,cr}$ and with shear stirrups $V_{Rd,cr}$.

10.3.4 Prestressed T beams in beam and block flooring

T beams are designed and manufactured without shear reinforcement. The procedures follow those for floor units given in Section 10.2 in which the uncracked $V_{Rd,c}$ and the flexurally cracked $V_{Rd,cr}$ shear resistances are calculated. Some beams have a narrow and tapered upstand, typically 42 mm width increasing to 48 mm at the intersection with the bottom flange, and therefore the shear capacity is based on an average web width b_w=45 mm. As the centroidal axis is also just inside the upstand $V_{Rd,c}$ will be critical here, as shown in the shear tests in Figure 10.9.

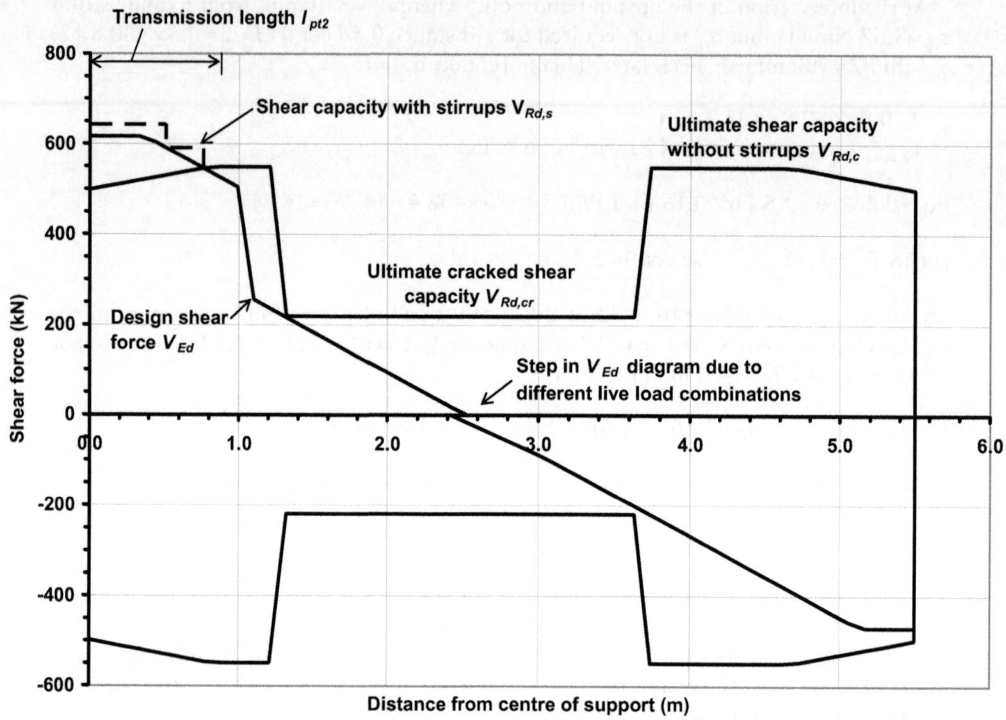

Figure 10.27 Shear force V_{Ed} and shear capacities $V_{Rd,c}$, $V_{Rd,cr}$ and $V_{Rd,s}$, dashed lines, for the inverted-tee beam in Example 10.7.

Example 10.8

Calculate $V_{Rd,c}$ and $V_{Rd,cr}$ for the 150 mm depth T beam used in beam and block flooring in Examples 7.5 and 9.6. The bearing length may be taken as 100 mm. Check $V_{Rd,c}$ at (i) centroidal axis and, if necessary at (ii) intersection of the upstand and boot. Use $f_{ck}=55$ N/mm^2 and $f_{ck}(t)=32$ N/mm^2. Area of 4 wires for prestress $A_p=78.5$ mm, and at for 3 wires at ultimate $A_{pT}=58.9$ mm^2. Prior in-house analysis has found the prestress after initial losses is $\sigma_{pm0}=1229.3$ N/mm^2 and the final prestress is $\sigma_{p0}=967.5$ N/mm^2.

Solution

Material properties

$f_{cm}=55+8=63$ N/mm^2

f_{ctm} (for $f_{ck}>50$ N/mm^2)$=2.12\log_e(1+63/10)=4.214$ N/mm^2

$f_{cm}(t)=32+8=40$ N/mm^2

$f_{ctm}(t)=(40/63)\times4.214=2.676$ N/mm^2

$f_{ctd}(t)=0.7\times2.676/1.5=1.249$ N/mm^2

$f_{bpt}=2.7\times1.249=3.37$ N/mm^2

$f_{ctd}=0.7\times4.214/1.5=1.967$ N/mm^2

Flexurally uncracked capacity $V_{Rd,c}$

From Example 7.5 $y_b=61.9$ mm>depth of bottom flange of 50 mm, therefore the width in shear is taken as the mean width (42+48)/2=45 mm (Figure 7.12) and $V_{Rd,c}$ is calculated at the centroidal axis only. $V_{Rd,c}$ is not required at the intersection of the upstand with the bottom flange, but is carried out here for interest.

(i) At the centroidal axis

b_w=45 mm, I_c=17.03×10^6 mm^4, y_t=150−61.9=88.1 mm

S_c=(42×88.1^2/2)+(2.65×88.1^2/3)=170020 mm^3

$I_c\, b_w/S_c$=17.03×10^6×45/170020=4507 mm^2

l_{pt2}=1.2×0.25×1229.3×5.0/3.37=547 mm

σ_{cp} including γ_p=0.9×967.5×78.5/9000=7.60 N/mm^2

l_x=100+61.9=161.9 mm

α_l=161.9/547=0.296

$V_{Rd,c}$=4507×√(1.967^2+0.296×7.60×1.967)×10^{-3}=**12.98 kN**

(ii) At the intersection of the upstand with the bottom flange, σ_{cp}=8.64 N/mm^2 (greater than 7.60 N/mm^2 because it is lower in the section), l_x=100+50=150 mm, α_l=150/547=0.274, b_w=48 mm and S=166656 mm^3.

Then $V_{Rd,c}$=(17.03×10^6×48/166656)×√(1.967^2+0.274×8.64×1.967)×10^{-3} =**13.43 kN**

Flexurally cracked capacity $V_{Rd,c}$
From Example 9.6, d=121.7 mm and A_{pT}=58.9 mm^2

k=1+√(200/121.7)=2.28 therefore use 2.0

v_{min}=0.035×2.0$^{1.5}$ √55=0.73 N/mm^2

$A_s/b_w\, d$=58.9/(45×121.7)=0.01076

σ_{cp}=7.60 N/mm^2

$V_{Rd,cr}$=[(0.12×2.0×(100×0.01076×55)$^{1/3}$)+0.15×7.60]×45×121.7×10^{-3}=**11.36 kN**

$V_{Rd,cr}$ minimum=(0.73+0.15×7.60)×45×121.7×10^{-3}=10.26 kN<11.36 kN therefore use $V_{Rd,cr}$=11.36 kN.

It noticed that the ratio $V_{Rd,cr}/V_{Rd,c}$=11.36/12.98=0.87 is much greater than the corresponding ratio of 0.44 for the main beam in Example 10.6. The influence of the contribution from prestress 0.15 σ_{cp} in the term $C_{Rd,c}\, k\, (100\, \rho_l\, f_{ck})^{1/3}$+0.15 σ_{cp} is a major factor in the T beam σ_{cp} is 7.60 N/mm^2 compared with 3.67 N/mm^2 in the main beam, the former being more highly stressed overall. It is unlikely that $V_{Rd,cr}$ in T beams would ever be mobilised.

10.4 DESIGN FOR CONCENTRATED LOADS IN FLOOR UNITS

This section considers the design for concentrated loads, such as punching shear from point loads and the capacity of floor units supported on three edges, where shear and torsion combine. The capacity of line loads either at the edges or within the width of psc floor elements is given in Sections 9.3.4. Whilst the design for concentrated loads in solid psc units can be found in BS EN 1992-1-1, a particular case worthy of the development of unique semi-empirical equations is prestressed hcu. These are found in BS EN 1168 clauses 4.3.3.2.4 to 4.3.3.2.6.

10.4.1 Punching shear in hollow core units

Trying to initiate a classical punching shear failure in hollow core floor units is very difficult, and the author has not been successful in several tests carried out on 200 mm depth units with 100 mm width cores, and 300 and 400 mm depth units with 140 mm width cores (similar to the profile in Figure 6.7). The failure mode is always by longitudinal cracking along the line of the cores, as shown in Figures 10.28 and 10.29, even though the footprint of the load (150 mm square), positioned at mid-width of the units, was just wider than the width of the cores. A footprint as little as w=50 to 100 mm square (e.g. as specified for computer room cabinet bearings for which unscreeded units would be inappropriate) is required to punch through the top flange directly over the core.

The mean cube strength of the hcu was measured as 57.5 N/mm^2, converted to $f_{ck}=0.8\times57.5=46$ N//mm^2, almost the same as the design strength $f_{ck}=45$ N//mm^2. The mean flexural strength was $f_{ctm}=4.37$ N/mm^2 (compared with a design value of 3.80 N/mm^2). The mean axial prestress σ_{cp} (from three test units) was 3.90, 4.95, 4.94 and 5.05 N/mm^2 for the 200 (low prestress), 200 (high prestress), 300 and 400 mm depth units, respectively. Using Eq. 10.30 the design punching strength V_{Rd} was calculated on the basis of design parameters, and the predicted strength $V_{Rd,calc}$ was based on measured parameters without partial safety factors. The effective width of the webs (see Figure 10.30) was $b_{eff}=80$ mm (2 no. webs) and 45 mm (1 no. web) for the 200 mm, and 300 and 400 mm depth units, respectively. The failure shear force V_u (i.e. point load/2) varied from 75.2 to 128.7 kN. When normalised with respect to V_{Rd} and $V_{Rd,calc}$ for each test in turn, the average ratio of V_u/V_{Rd}=1.66, 1.82, 3.04 and 2.20 for the 200 (low), 200 (high), 300 and 400 mm depth units, respectively. The corresponding average ratio of $V_u/V_{Rd,calc}$=1.10, 1.07, 2.21 and 1.59. The larger ratio for the 300 mm compared with 400 mm depth units was due to V_u being almost the same for the two sets of tests, as the failure mode was governed by longitudinal splitting of the top flange, which is the same 35 mm thickness in both profiles, but a lower value of V_{Rd} (by a factor of 1.35 reflecting the shallower depth).

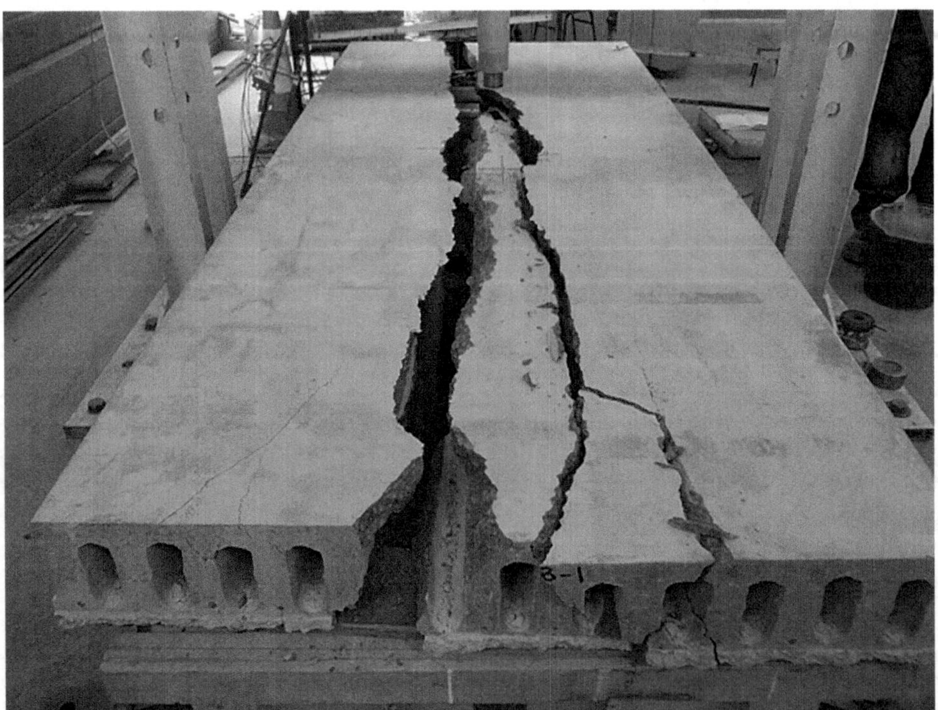

Figure 10.28 Longitudinal splitting in a punching shear test in 200 mm depth prestressed hollow core units. (With permission by Creagh Concrete Ltd., UK.)

Figure 10.29 Longitudinal splitting and radial cracks around the point load in a punching shear test in 400 mm depth prestressed hollow core units. (With permission by Creagh Concrete Ltd., UK.)

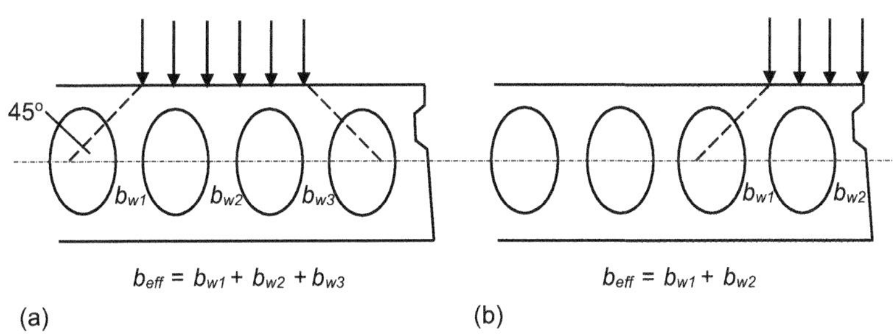

Figure 10.30 Effective width of webs under action of punching shear (a) general situation, (b) at a free edge of a floor bay. (Reproduced from BS EN 1168 Figure 5(a) and (b).)

In spite of this evidence, BS EN 1168, clause 4.3.3.2.4 gives an expression for punching shear capacity V_{Rd} in untopped hcu as

$$V_{Rd}=b_{eff}\, h\, f_{ctd}\,(1+0.3\,\alpha_l\;\sigma_{cp}/f_{ctd})\geq V_{Ed,p} \tag{10.30}$$

where $V_{Ed,p}$=reaction from the point load F_{Ed}

b_{eff}=effective widths of webs according to Figure 10.30.

h=depth of unit
$\alpha_l = l_x/l_{bpd} \leq 1$ if the point load is within a distance less than the total anchorage length (defined in Section 9.5) from the centre of support.

By rearranging Eq. 10.30, V_{Rd} is equal to the shear-tension capacity of the webs ($b_{eff}\, h\, f_{ctd}$) plus 0.3 times the effective prestress in the webs ($0.3\, \alpha_l\, \sigma_{cp}\, b_{eff}\, h$), similar to historic values. V_{Rd} is heavily influenced by the width of the footprint relative to the width of the cores, particularly if $w <$ ½ core width, e.g. in some deeper slip formed units core width=150–250 mm. Clause 4.3.3.2.4 states that if $w <$ ½ core width, h in Eq. 10.30 is replaced by h_f (top flange depth) and $b_{eff} = w$. There are several complications at free edges in the outermost web for which the best solution is to open and *insitu* infill the first core. Clause 4.3.3.2.4 states that if more than 50% of the load is acting on the outermost web (i.e. only 2 webs active) Eq. 10.30 is applicable, provided that one tendon is in that web and transverse reinforcement is present, i.e. in reality a structural topping with $A_s = F_{Ed}/0.87\, f_{yk}$ locally. The topping could be localised over the edge unit if it is possible for this unit to be (say) 50 mm shallower than the rest of the units in the floor field. If one or both of these is not satisfied, then use $0.5\, V_{Rd}$.

Consider a point dead load of 50 kN acting over a footprint 300 mm square at 1.0 m from the centre of support of a 6.0 m simply supported 250 mm depth hcu, shown in cross section in Figure 6.7. The unit is grouted in and the load is not at a free edge. The hcu contains 140 mm wide cores and 40 mm webs. Concrete $f_{ck} = 45$ N/mm², $f_{ctd} = 1.77$ N/mm², axial prestress $\gamma_{p,fav}\, \sigma_{cp} = 4.61$ N/mm², the unit contains 9.3 and 12.5 mm strands such that $l_{bpd} = 1230$ mm. If the bearing length is 100 mm, $\alpha_l = (1000 + 100/2)/1230 = 0.854$. A 300 mm wide footprint, centred over one of the cores, will be resisted by only 2 no. webs, then $b_{eff} = 80$ mm.

Eq. 10.30 $\quad V_{Rd} = 80 \times 250 \times 1.77\,(1 + 0.3 \times 0.854 \times 4.61/1.77) \times 10^{-3} = \mathbf{59.0\ kN}$

Maximum reaction from the load $V_{Ed} = 1.35 \times 50 \times (6.0 - 1.0)/6.0 = \mathbf{56.3\ kN} < 59.0$ kN.

10.4.2 Load capacity of hollow core units supported on three edges

This section has considered the local load capacities of hcu in a one-way spanning floor slab. If the edge unit is supported along one edge by a gable wall or beam, the unit is classed as supported on '*three edges*' and is subjected to shear, bending and torsion. BS EN 1168, clause 4.3.3.2.6 provides semi-empirical data for the imposed service load capacity in units of kN/m² (excludes self-weight) of such units as

$$q_k = W_t\, f_{ctk,0.05}/0.06\, l^2 \tag{10.31}$$

where W_t=elastic torsional section modulus$=\min\{2t\,(h - h_f)(b - b_w);\ b^2 h/(3 + 1.8 b/h)\}$ (10.32)

b_w=width of outermost web
$t = \min\,\{h_f;\, b_w\}$
h_f is the smaller of h_{ft} and h_{fb}

However, in the general equations for torsional moments of resistance ($h - h_f$) actually considers the actual distance between the centres of the flanges, i.e. ($h - h_{ft}/2 - h_{fb}/2$).

Continuing the same example, with $l = 6.0$ m and $b_w = 51$ mm outermost web, $W_t = 2 \times 35 \times (250 - 40/2 - 35/2) \times (1200 - 51) = 17.09 \times 10^6$ mm³ (note that $b^2 h/(3 + 1.8 b/h) = 30.9 \times 10^6$

mm^3 and is difficult to see how it applies to a hollow section). Then q_k=17.09×10^6×2.66/(0.06×6000^2)=21.05 kN/m^2. This imposed load would not exceed the service moment of resistance of the hcu considered (in Figure 6.7), and is therefore shown to be critical should the hcu happen to be supported on '*three edges*'.

10.5 SLIPPAGE OF PRETENSIONED TENDONS

Although Elliott and Jolly (2013) provide an extensive discussion of this topic, it is appropriate to include a summary here.

Slippage (sometimes called 'pull-in') ΔL_o is the shortening of tendons after transfer of prestress. It is the cumulative effect of movement of tendons relative to the concrete due to bond slip within the basic transmission length l_{pt}. It appears at the ends of prestressed elements, as shown in Figure 10.31, and gives an indication that prestress has developed correctly, particularly in the transmission zone, but is not a definitive measurement of prestress.

The subject is comprehensively reviewed, and the relationship between ΔL_o and L_t analysed in a thesis by Paul Kong (Kong, 1993). Referring to Figure 10.32

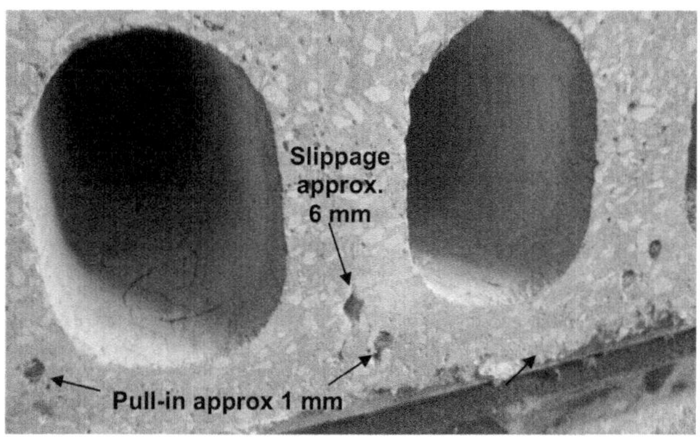

Figure 10.31 Pull-in and slippage of 5 mm diameter wires.

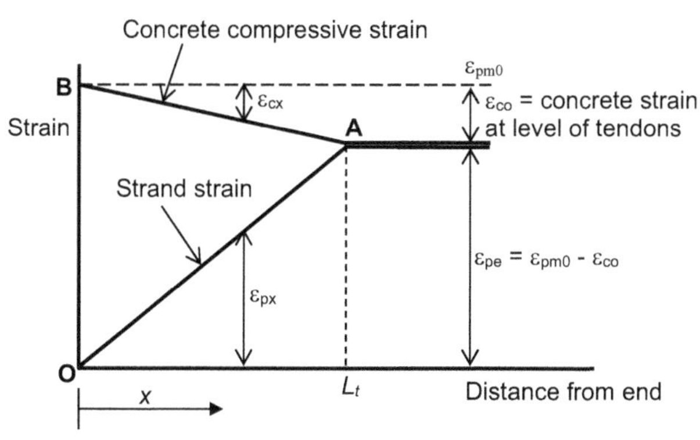

Figure 10.32 Strand and concrete strain profiles without slippage, both assumed linear according to BS EN 1992-1-1.

$\varepsilon_{pm0}-\varepsilon_{px}$=steel strain change at distance x after transfer
ε_{cx}=concrete strain at distance x after transfer
$(\varepsilon_{pm0}-\varepsilon_{px})-\varepsilon_{cx}$=steel strain change at distance x after transfer net of concrete strain

$$\Delta L_o=\int_0^{L_t} (\varepsilon_{pm0}-\varepsilon_{px}-\varepsilon_{cx})\,\delta x \text{ (the area OAB)} \tag{10.33}$$

$$\Delta L_o=\varepsilon_{pm0}\, L_t/2=\sigma_{pm0}\, l_{pt}/2\, E_p \tag{10.34}$$

where σ_{pm0}=prestress at transfer

E_p=modulus of elasticity of tendons

BS EN 1992-1-1, clause 8.10.2.1 specifies a linear increase in prestress, such that

$$\Delta L_o=\sigma_{pm0}\, l_{pt}/2\, E_p \tag{10.35}$$

For a parabolic increase, the denominator is 3, i.e. $\Delta L_o=\sigma_{pm0}\, l_{pt}/3\, E_p$. If it is linear-parabolic increase in prestress, the denominator is 3, i.e. $\Delta L_o=\sigma_{pm0}\, l_{pt}/2.5\, E_p$ and it takes the familiar form given in BS EN 13369, clause 4.3.2.3.4 (BS EN 13369, 2018)

$$\Delta L_o=0.4\, l_{pt2}\, \sigma_{pm0}/E_p \tag{10.36}$$

Note the transmission length is increased to l_{pt2} even though this is not a design issue. Slippage is routinely observed and often measured by manufacturers. In typical production, slippage is about 1–2 mm. Theoretically, it cannot be zero, but often appears to be so. Slippage of wire tends to be more consistent than with strand, particularly if there is a concentration of large diameter strands in a small area, and is quite common if a group of 3 strands are positioned in a line, rather than in a triangular formation.

The limit for slippage is a manufacturing issue, rather than a design issue, and as such is not given in BS EN 1992-1-1. However, BS EN 13369, clause 4.2.3.2.4 states that mean slippage in an element (both ends) should not exceed

$$\Delta L_o=0.4\, l_{pt2}\, \sigma_{pm0}/E_p \tag{10.37}$$

and for an individual tendon not more than 1.3 ΔL_o

For example of strand, if $\sigma_{pm0}=0.7\times1770\times0.95$ (5% loss), and l_{pt2}=(say) 600 mm, ΔL_o= $0.4\times1177\times600/195000=1.45$ mm, and 1.3 ΔL_o=1.88 mm. These are reasonable limits to work to, but this clause does not explain how, or if, the capacity of a psc element may be reassessed if is ΔL_o exceeded. This is also known as 'slippage' δ_s and therefore to avoid confusion ΔL_o is sometimes known as 'pull-in'.

If $\delta_s>\Delta L_o$ by 2 or 3 mm in a couple of tendons, the effect on l_{pt} and mechanical strength will be small, but if δ_s=5 or 6 mm in all tendons, re-analysis for residual prestress, or rejection of the unit, are possible courses of action.

The FIP Manual 'Quality assurance of hollow core slab floors', clause 3.5.3. (FIP 1992) gives a production limit based on an upper bound value for transmission length $l_{pt}=13\,\Phi\,\sqrt{\sigma_{pm0}/f_{cu}(t)}$ and slippage $\Delta L_o=0.5\,\varepsilon_{pm0}\,l_t$. For example, consider a hcu with $f_{cu}(t)$=35 N/mm^2, with 12.5 mm strands stressed to 70%×1770 N/mm^2 and 5% losses at transfer, $l_{pt}=13\times12.5\times\sqrt{1177/35}=942$ mm, and $\Delta L_{o,limit}=0.5\times0.00604\times942=2.85$ mm. This is multiplied by 1.5 for an individual tendon $\Delta L_{o,limit}$=4.3 mm, and this would be an acceptable limit.

For the relationship between slippage and residual prestress, some designers take the FIP limit to be non-contributory and re-calculate the capacity on a reduced number of tendons.

Others use a linear scale to determine residual prestress, taking the ratio of prestress with/without slippage from 1.0 where $\delta_s = \Delta L_o$ to zero where δ_{free}=free elastic shortening over half the length of the unit for the tendons having $\delta_s > \Delta L_o$. Then $\delta_{free} = 0.5\ L\ \sigma_{pm0}/E_p$.

Continuing the example above, let slippage $\delta_{s,mean} = 4.0$ mm occur in half the number of tendons in a hcu of 6.0 m length. Now $\delta_{free} = 1177 \times 3000/195000 = 18.1$ mm. The prestress in tendons with slippage is given pro-rata as $[(\delta_{free} - \delta_{s,mean})/\ \delta_{free}]\ \sigma_{pm0} = [(18.1 - 4.0)/18.1] \times 1177 = 917$ N/mm². The mean effective prestress $\sigma_{pm0,eff} = (0.5 \times 917 + 0.5 \times 1177) = 1\,047$ N/mm², and this is used to calculate losses due to elastic shortening, creep and relaxation, leading to a reduced value of M_{sr}. Comparison with the research by Evans (Evans and Robinson, 1955) who measured the loss of strain at end of the transmission zone as 0.14 $\delta_{mean}/l_{pt} = 0.14 \times 2.7/600 = 0.00063$, where $\delta_{mean} = (0.5 \times 4.0 + 0.5 \times 1.45) = 2.7$ mm. Residual strain=(1177/195000)−0.00063=0.00541, and residual stress=0.00541×195000=1055 N/mm² (giving reasonable agreement with 1047 N/mm²).

The increase in l_{pt2} for strands with slippage is given as $l_{pt2}' = \delta_s\ E_p/0.4\ \sigma_{pm0} = 4.0 \times 195000/0.4 \times 1177 = 1656$ mm. The mean length by superposition $l_{pt2}'_{,mean} = 0.5 \times 1656 + 0.5 \times 600 = 1\,128$ mm. This is used to recalculate the reduced shear capacity $V_{Rd,c}$.

Slippage will cause a reduction in camber according to the effective prestress. Changes in l_{pt2} have no influence on flexural stiffness when calculating deflections due to live loads, but would nevertheless result in greater long-term positive deflections due to reduced visco-elastic negative camber.

The results of shear tests carried out on 305, 407 and 508 mm depth prestressed hcu with strand slippage of up to 2.9 mm by Palmer and Schultz (2011) are reviewed and discussed by Elliott and Jolly, Section 5.2.10.1 (Elliott and Jolly, 2013). Comparisons with BS EN 1992-1-1 are mostly unconservative, with only six out of 24 tests having $V_u/V_{Rd,c} > 1$. Tests with the largest slippage (UMN A 16) have a mean $V_u/V_{Rd,c} \approx 0.75$. However, comparisons with Yang, on which Eq. 10.12 is based, are all conservative, with $V_{obs}/V_{EC2} = 1.0$ to 2.0. The results indicate that whilst BS EN 1992-1-1 is largely unconservative for deep units, Yang's approach has tipped the balance too far.

10.6 TORSION IN BEAMS AND SLABS

10.6.1 Equilibrium and compatibility torsion

Torsion, or a twisting bending moment, is best visualised by twisting a cardboard Smartie tube and observing the separation of the joints in the spiral. Pure torsion, without the presence of prestress, shear or bending actions, induces a spiral of tensile stresses τ_t at 45° to the longitudinal axis. To maintain equilibrium, compressive struts develop perpendicular to the tensile stress leading to maximum shear stresses τ_t in outside of webs and flanges of slabs or beams, as shown in the truss analogy in Figure 10.33. In the presence of shear stress v, the total vertical shear stress is the sum of $\tau = \tau_t + v$. When τ or $\tau_t > f_{ctm}$, torsion cracking occurs as shown in the top of a prestressed hcu in Figure 10.34. In this photograph, the crack spiralled around the sides of the hcu but was arrested by the compressive prestress in the bottom of the hcu, therefore resulting in just one crack rather than a series of spiral cracks. This is known as 'equilibrium' torsion – the internal resistance of individual elements in which the cracking $T_{Rd,c}$ and ultimate $T_{Rd,max}$ torsional moments of resistance are calculated against the ultimate design torque T_{Ed}. Equilibrium torsion is the result of eccentric forces acting on slabs and beams, for example due to edge balconies connected by steelwork brackets to the sides of solid prestressed units (not hcu), as shown in Figure 10.35, or a secondary beam connected to the side of a primary beam as shown in Figure 10.36. Torsion may also be due to unequal floor loads acting on psc inverted-tee beams, for example where a long hcu is bearing on one side of the beam, but is spanning parallel to the beam on the other side.

Gilbert et al. (2017) gave extensive account of both types of torsion and the truss analogy used for equilibrium torsion.

This section deals only with equilibrium torsion, i.e. the calculation of T_{Ed} from eccentric slab or beam loads, the resistances $T_{Rd,c}$ and ultimate $T_{Rd,max}$. Where structural elements, such as edge beams and floor slabs, are connected to each other, and the stability of the frame is provided by the torsional stiffness and strength of the connection between the two elements, this is known as 'compatibility' torsion. This form of torsion is required to

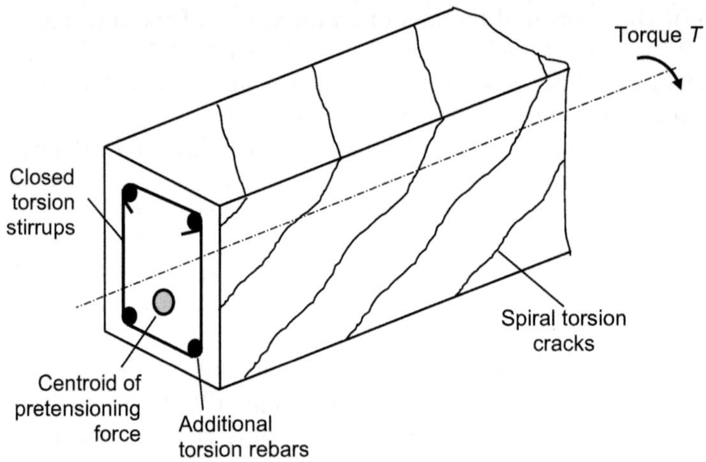

Figure 10.33 Torsional behaviour of a rectangular element subject to pure torsion.

Figure 10.34 Torsion cracking in prestressed hollow core floor unit. (Courtesy Bison Precast Ltd., UK.).

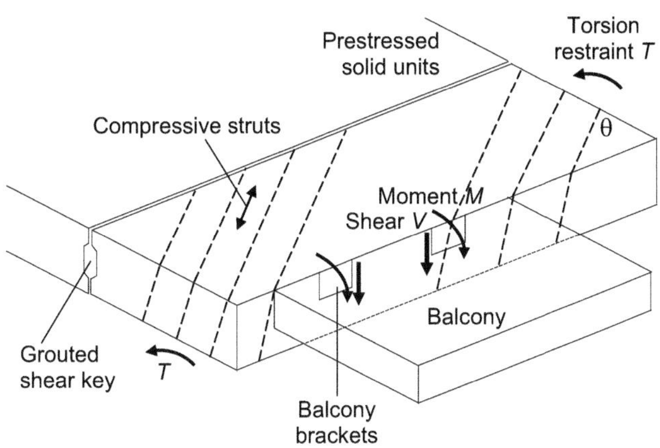

Figure 10.35 Torsion, shear and bending effects from steel balconies attached to the side of prestressed solid floor units.

Figure 10.36 Secondary beam attached to the side of a primary beam causes torsion, shear and bending in the primary beam

maintain compatibility of deformations between elements in the frame and is dealt with by Elliott and Jolly, Section 4.3.6 (Elliott and Jolly, 2013) for the particular case of the compatibility between L-shape edge beams and hcu. The applied pure torque at cracking was 30 kNm, which, using the sand-heap theory, gave a torsional stress in the boot of the beam of 1.8 N/mm^2. Calculating the same stress according to BS EN 1992-1-1, Exp. 6.26 (see Eq. 10.39 later) gave $\tau_{t,i}$=2.1 N/mm^2, about 0.6 f_{ctm} for the test beam. Based on the results of these limited tests, torsional cracking occurs at a stress lower than the mean axial tensile strength. At this point it is worth noting that the inclination of the torsion cracks in the tests was close to 45°, such that cot $\theta \approx 1$. BS EN 1992-1-1, clause 6.3.2 (2) assumes the same value for θ for torsion as well as shear, and it is more than likely that in most beams, with the exception of I beams with narrow webs, for shear cot θ will be in the region of 2.0–2.5. This means that the subsequent equations for torsional reinforcement will be conservative as cot θ=2.5 will be applied to the effects of torsion.

10.6.2 Torsional stress

The torsional stiffness of a psc element is based on elastic theory up to the cracking moment of resistance, $T_{Rd,c}$. The reinforcement is ignored. Torsion produces diagonal compressive struts between inclined ties – the equilibrium between compressive struts and the longitudinal and transverse ties is shown in the truss analogy in Figure 10.37a. The maximum compressive stress is assumed to be located close to the outer surface of the element. According to BS EN 1992-1-1, clauses 6.3.1 (3) and 6.3.2 (1) the torsional resistance of solid section may be modelled and calculated on the basis of a thin-walled closed section in which equilibrium is satisfied by closed shear flow as shown in BS EN 1992-1-1, Fig. 6.11, and simplified in Figure 10.37b. It is well known that the shear stress τ in a prismatic thin-walled section subjected to a pure torsional moment T is given as

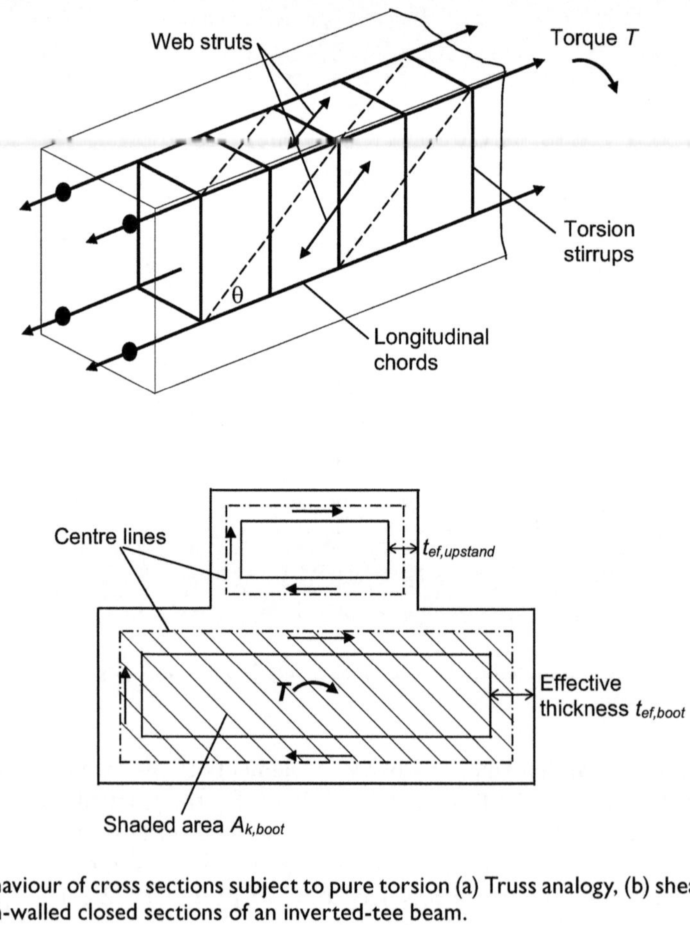

Figure 10.37 Behaviour of cross sections subject to pure torsion (a) Truss analogy, (b) shear flow in idealised thin-walled closed sections of an inverted-tee beam.

$$\tau = T/2\,At \tag{10.38}$$

where t=wall thickness

A=area enclosed within the median line of the wall thickness (e.g. for a circular tube of outer radius r and thickness t, $A=\pi\,(r-t/2)^2$

BS EN 1992-1-1, Exp. 6.26 expresses the torsional shear stress $\tau_{t,i}$ for each part i of the thin walled section as

$$\tau_{t,i} = T_{Ed}/2\,A_k\,t_{ef,i} \tag{10.39}$$

where T_{Ed}=design ultimate torsion

$t_{ef,i}$=effective thickness of the wall=A/u but not less than twice the axis distance to main bars

A_k=area enclosed by the centre lines of the connecting walls

A=total area of cross section within the outer circumference, including inner hollow areas (viz. hollow core slabs)

u=outer circumference of the cross section

The shear stress in the vertical wall of the beam in Figure 10.37b is given as

$$\tau_{t,i} = V_{Ed,i}/t_{ef,I}\,z_i \tag{10.40}$$

where $V_{Ed,i}$=design ultimate shear in the wall i

z_i=side length of the wall I between the intersection of the walls

Equating Eqs. 10.39 and 10.40 gives the shear force in each of the walls as

$$V_{Ed,i} = T_{Ed}\,z_i/2\,A_k \tag{10.41}$$

For example in a rectangular beam b=400 mm×h=600 mm, with main bars axis distance=50 mm, and subject to T_{Ed}=50 kNm, t_{ef}=240000/2000=120 mm>2×50=100 mm. A_k=280×480=134400 mm^2. z=600−2×50=500 mm. In the vertical walls $V_{Ed,i}$=50×10^6×500/(2×134400×10^3)=93.0 kN.

Gilbert et al. (2017) state that the test shows that prestress increases the torsional stiffness significantly by delaying the onset of torsional cracking, but does not greatly affect the strength. After cracking the contribution of the concrete to ultimate strength is only marginally increased by prestress, and the contribution of the stirrups is unchanged by prestress. On the contrary, Kong and Evans (1987) claim that the torsional resistance of psc is significantly higher than for a corresponding reinforced element, but also that BS 8110, Part 2 (BS 8110-2, 1985) recommends that the procedures used for reinforced concrete should be used for prestressed.

10.6.3 Design for torsion in prestressed beams

Torsion is considered for (i) temporary loads during construction, to cater for all floor slab dead load on one side of the beam, and (ii) floor and beam loads at the ultimate limit state. See Figure 10.38. BS EN 1992-1-1, clause 6.3.2 (2) states that the concrete section should be checked for torsional capacity combined with shear. Stresses due to bending moments are not explicitly mentioned, probably because they are small where the torsion and shear is greatest, and are also small close to the centroidal axis, but should be considered in

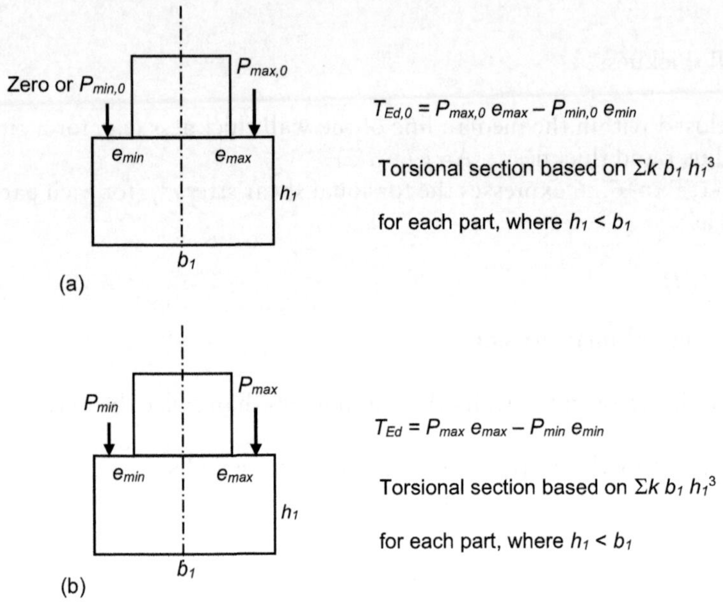

Figure 10.38 Calculating torsional moments (a) during construction (b) in service (for non-composite beams).

cantilevers and continuous beams where negative bending moments near to supports add tensile stress to the torsion stress in the top fibre. However, in most beams, the top of the beam is cast monolithic with the floor slab where the torsion is virtually eliminated, but there may be exceptions.

The design ultimate torsional moment, known as 'torque' T_{Ed} is the greater of BS EN 1990, Exp. 6.10a or 6.10b as $\{1.35\ T_G+1.5\ \psi_0\ T_Q;\ 1.25\ T_G+1.5\ T_Q\}$ where T_G and T_Q is the service torque due to dead and live loads, respectively. Note that at the construction stage $T_{Ed,0}=1.35\ T_G$ only (Figure 10.38a).

The service torque considers the imbalanced ultimate reactions P_{max} and P_{min} from floor slabs, or from otherwise eccentric loads, on either side of the beam, as shown in Figure 10.38b. At the construction stage (suffix 0) it may be possible that floor slabs are first positioned on one side of the beam only, in which

$$T_{Ed,0}=P_{max,0}\ e_{max} \tag{10.42a}$$

or if the depth, span and self-weight of slabs on either side of the beam are different, then

$$T_{Ed,0}=P_{max,0}\ e_{max}-P_{min,0}\ e_{min} \tag{10.42b}$$

Fixing tolerances are allowed for by adding 15 mm to the eccentricity of $P_{max,0}$, and subtracting 15 mm from the eccentricity of $P_{min,0}$. This is a notional allowance adopted by industry, but is in line with the deviations Δa_2 plus Δa_3 in BS EN 1992-1-1, clause 10.9.5.2. For example, in the 600 mm width inverted-tee beam shown in cross section in Figure 6.10, supporting floor units with a nominal bearing length of 100 mm, the nominal eccentricity is $e_{nom}=600/2-100/2=250$ mm. After fixing allowances $e_{max}=265$ mm and $e_{min}=235$ mm. Given that the self-weight of the floor units one either side of the beam is, say 12 kN/m, and the shear span of the beam (from Example 10.6) at 370 mm from the centres of support is 7.76 m, then $T_{Ed,0,max}=1.35\times12\times0.265\times7.76/2=16.66$ kNm.

In service, with floor slabs present on both sides of the beam,

$$T_{Ed}=P_{max}\, e_{max}-P_{min}\, e_{min} \quad (10.43)$$

Given that the service dead and live loads on one side of the beam are $w_G+w_Q=18.67+20.0$ kN/m, and on the other side are $w_G=9.33$ kN/m, and that ultimate load combination Exp. 6.10b is critical, then $P_{max}=1.25\times18.67+1.5\times20=53.33$ kN/m, and $P_{min}=11.67$ kN. Then $T_{Ed}=(53.33\times0.265-11.67\times0.235)\times7.76/2=44.2$ kNm.

The design shear force V_{Ed} is determined as per Example 10.6. based on the same shear span.

Torque for imposed dead and live loads may be small due to the connection of the slab (and topping) to the beam. Here the effective bearing shifts closer to the shear centre of the beam, and twisting of the beam is resisted by the horizontal diaphragm reaction in the slab. This is particularly found in edge beams.

The first step is to ensure that beams are torsionally under-reinforced, i.e. that the longitudinal bars and stirrups reach their yield strength before the ultimate torsional strength is reached. The effects of torsion and shear may be superimposed, assuming the same angle of inclination of the compressive strut θ, with $1\leq\cot\theta\leq2.5$. To avoid crushing the concrete struts, for a beam in combined torsion and shear, the following interaction ratio according to BS EN 1992-1-1, clause 6.3.2 (4) is given as

Exp. 6.29 $$T_{Ed}/T_{Rd,max}+V_{Ed}/V_{Rd,max}\leq1.0 \quad (10.44)$$

Exp. 6.30 $$T_{Rd,max}=2\,v\,\alpha_{cw}\,f_{cd}\,A_k\,t_{ef}/(\cot\theta+\tan\theta) \quad (10.45)$$

Exp. 6.9 $$V_{Rd,max}=v\,\alpha_{cw}\,f_{cd}\,b_w\,0.9d/(\cot\theta+\tan\theta)$$ (see note below for I beams) (10.46)

6.2.2.6 $$v=0.6\,(1-f_{ck}/250)$$

b_w=width of web at centroidal axis or at the intersection of the boot and upstand if the centroidal axis is in the boot

d=effective depth to tension reinforcement

For prestressed concrete $\alpha_{cw}=(1+\sigma_{cpx}/f_{cd})$ for $\sigma_{cpx}/f_{cd}\leq0.25$, 1.25 for $\sigma_{cpx}/f_{cd}\leq0.5$ and 2.5 $(1-\sigma_{cpx}/f_{cd})$ for $\sigma_{cpx}/f_{cd}\leq1.0$.

BS EN 1992-1-1 clause 6.3.2 (4) does not specify where is σ_{cp} taken, but it is assumed to be at the mid-height of the section in the first inclined strut, i.e. at $x=0.5\,h\cot\theta$. This point may lie within the transmission length such that $\sigma_{cpx}=\sigma_{cp}\,l_x/l_{pt2}$ (see Eq. 10.2). The concrete design strength is $f_{cd}=f_{ck}/1.5$.

If Eq. 10.44 is violated, the cross section of the member must be increased – adding more stirrups etc. is not the solution.

10.6.4 Proportion of torque per sub-section

In flanged, T, I or inverted tee beams the total torsional moment is proportioned into each of the component rectangles (sub-sections) of the cross section according to the compatibility of the angle of twist θ at the intersection of flanges/boot and webs. The full derivation is given by Kong and Evans, Section 6.8 (Kong and Evans, 1987). As the angle of twist θ at all intersections is equal, and θ is given by the general equation

$$\theta=T/G\,K \quad (10.47)$$

Table 10.2 Values for torsional constant k for rectangular sections

Max/min side h/b	*k*	*Max/min side h/b*	*k*
1.0	0.141	4.0	0.281
1.5	0.196	6.0	0.299
1.75	0.214	8.0	0.307
2.0	0.229	10.0	0.313
2.5	0.249	∞	0.333
3.0	0.263		

where K is known as St. Venant's torsional constant, and has the values given in Table 10.2.

G=shear modulus for concrete=$E_{cm}/(1+2\nu)$
ν=Poisson's ratio taken as 0.2 (BS EN 1992-1-1, clause 3.1.3 (4))

the proportion of T in each sub-section is proportional to K. This is defined in BS EN 1992-1-1, clause 6.3.1 (4) where the ultimate torque T_{Ed} is proportioned according to the uncracked torsional stiffness K of each sub-section. This is based on an elastic analysis, and therefore K is

$$K=k\,h\,b^3 \text{ where } h\geq b \tag{10.48}$$

where k=torsional constant for rectangular areas

$$\text{Thus, for sub-section } i,\ T_{Edi}=T_{Ed}\,(k\,h\,b^3)_i/\Sigma\,k\,h\,b^3 \tag{10.49}$$

Triangular sections are simplified into rectangles of mean depth, e.g. 150 mm depth bottom flange plus 50 mm sloping flange, equates to 175 mm. Different depths of upstands and boot depths are averaged. Sub-sections are idealised where $\Sigma k\,h\,b^3$ is a maximum, generally achieved if the widest rectangle is made as deep as possible. Inverted-tee beams are obviously divided into the upstand and boot, and I beams into top and bottom flanges and the web.

The torsional constant k for rectangular areas can be obtained by curve-fitting numerical data from Table 10.2 by Hearn (Hearn, 1977).

$$\text{For } h/b\leq 6,\quad k=-0.0002(h/b)^4+0.0047(h/b)^3-0.0438(h/b)^2+0.1824(h/b)+0.0023$$

$$\text{For } h/b>6,\quad k=0.28+0.0035(h/b)$$

e.g. for h/b=2, k=0.226 (but is actually 0.229 directly from the table).

The proportioning of T_{Ed} into the upstand and boot of the inverted-tee beam in Figure 6.10 (excluding the 50×50 mm chamfers at the bottom corners) is as follows:

Boot width=600 mm×depth 400 mm. Maximum side h=600 mm and minimum side b=400 mm (note that the notation for width and depth have been reversed. Then h/b=1.5 and k=0.196. Then $K=k\ h\ b^3=0.196\times600\times400^3=7526.4\times10^6$ mm^4. Similarly in the upstand h/b=1.75 and k=0.214. $K=0.196\times350\times200^3=599.2\times10^6$ mm^4. The proportion of T_{Ed} into the boot of the beam is therefore 7526.4/(7526.4+599.2)=0.926.

The fully plastic torque resisted by each section is given by Kong and Evans, Section 6.8 (Kong and Evans, 1987) as $T_{Edi}=T_{Ed}\,(h\,b^2)_i/\Sigma h\,b^2$. In the case of the inverted-tee beam, the proportion of T_{Ed} into the boot is now 96.0/(96.0+14.0)=0.873. Although this is ignored in design, it suggests some redistribution of torque away from the boot after cracking.

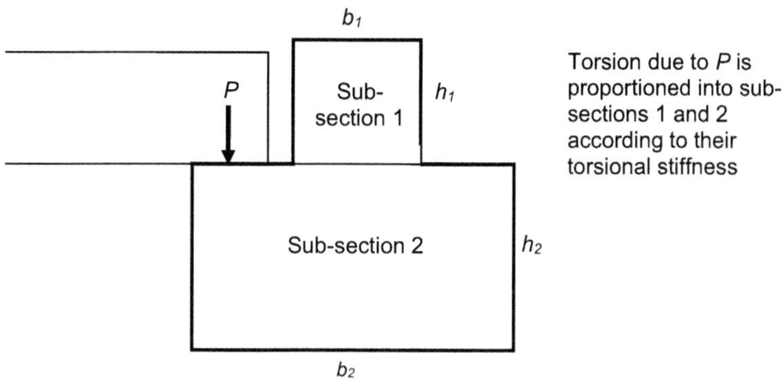

Figure 10.39 Proportioning the cross section of an inverted-tee beam into rectangular sections.

Referring to Figure 10.39

$A = bh$, or $A_1 = b_1\ h_1$ etc. for each sub-section
$u = 2(b+h)$, or $u_1 = 2(b_1 + h_1)$ etc. for each sub-section
$t_{ef} = A/u$ or $t_{ef1} = A_1/u_1$ etc.
$A_k = (b - t_{ef})\ (h - t_{ef})$
$u_k = 2\ (b + h - 2t_{ef})$ is the perimeter of area A_k

For I beams the term $V_{Ed}/V_{Rd,max}$ is for the full depth and width of the web. V_{Ed} in the top and bottom flanges is based on the summation of the horizontal shear stress $\tau \times$area of flange as $V_{Ed,fl} = 0.67\ (V_{Ed}\ A\ y'/I_{xx}\ b_f)\ b_f\ h_{fl}$, where 0.67 is the average stress in parabolic distribution, $A = bh_f$, y'=distance from centroid to centre of area A, and I_{xx}=second moment of area.

For the bottom flange $V_{Ed,fl} = 0.67 V_{Ed} b_f h_{fl}^2 \left(y_b - h_{fl}/2\right)/I$. $V_{Rd,fl,max} = \upsilon\ \alpha_{cw}\ f_{cd}\ b_{fl}\ 0.9\ (h_{fl} - 50$ mm)/(cot θ+tan θ). Note that both $V_{Ed,fl}$ and $V_{Rd,fl,max}$ are based on the full width b_f and so the ratio $V_{Ed,fl}/V_{Rd,max,fl}$ is not affected. If the flanges are shallow relative to the depth of the beam, e.g. say 1/5th, it is unlikely compressive struts will develop in the presence of deep webs.

10.6.5 Torsion reinforcement

Torsion causes deformation in two directions – perpendicular and parallel with the beam as shown in Figure 10.37a. Therefore, torsional reinforcement is required, simultaneously in both directions in the form of vertical stirrups and longitudinal bars, which are calculated separately. BS EN 1992-1-1 is vague in this as clause 6.3.2 (5) only refers to the minimum reinforcement in clause 9.2.1.1 for longitudinal reinforcement, and not the minimum shear links (which is in clause 9.2.3). However, it may be implied that if designed longitudinal bars are required, then so too are designed stirrups as torsion cracks cut across both sets of bars. The traditional approach has been to provide torsion stirrups when the minimum concrete torsion strength, typically 0.4 N/mm^2, is exceeded.

BS EN 1992-1-1, clause 6.3.2 does not specify at what magnitude of T_{Ed} *designed* torsion stirrups are required and does not provide an expression for the area of stirrups, although this may be derived later. However, clause 9.2.3 (1) shows examples of shapes of closed torsional links and clause 9.2.3 (2) states that Exp. 9.5N for minimum shear links is generally sufficient to provide minimum torsion links, see Eq. 10.27.

10.6.5.1 Additional longitudinal bars due to torsion

BS EN 1992-1-1 is vague in this respect, as clause 6.3.2 (5) states that for 'approximately rectangular solid sections' only minimum reinforcement is required according to clause 9.2.1.1, i.e. longitudinal tension reinforcement $A_{s,min}$=max{0.26 f_{ctm} b_t d/f_{yk}; 0.0013 b_t d), provided that the interaction formula satisfies

$$T_{Ed}/T_{Rd,c}+V_{Ed}/V_{Rd,c}\leq 1.0 \quad (10.50)$$

where T_{Ed} and V_{Ed} are the design ultimate torque and shear force at a section

$T_{Rd,c}$ is the torsional cracking moment of resistance from Eq. 10.39 letting $\tau_{ti}=f_{ctd}$.

$$\text{Then } T_{Rd,c}=2\,A_k\,t_{ef}\,f_{ctd} \quad (10.51)$$

Eq. 10.50 should be used with care and consideration because, if the ratio $V_{Ed}/V_{Rd,c}$ is close to 1.0, or even $V_{Ed}/V_{Rd,c}>1.0$ such that shear stirrups are provided, a small amount of torsion will violate Eq. 10.50 and additional longitudinal bars (and closed torsion stirrups (Section 10.6.5.2)) are immediately required.

Gilbert et al. (2017) include the beneficial effect of prestress to replace f_{ctd} with $f_{ctd}\sqrt{(1+10\,\sigma_{cp}/f_{ck})}$, where is the average prestress over the beam or sub-section. Gilbert also replaces 2 A_k t_{ef} with the torsional constant $J=0.33\,b^2\,h$ (where $b<h$) for solid sections.

$$\text{Then } T_{Rd,c}=J\,f_{ctd}\sqrt{(1+10\,\sigma_{cp}/f_{ck})} \quad (10.52)$$

Figure 10 does not multiply σ_{cp} tenfold, but derives from the denominator actually being the tensile strength of concrete, approximately $f_{ck}/10$. The result of this will be tested in Example 10.9.

The clause states that '$V_{Rd,c}$ follows from Expression 6.2', i.e. the flexurally cracked shear capacity $V_{Rd,cr}$ (Eq. 10.7 or 10.10). However, in a psc beam where T_{Ed} is calculated near to the support the critical shear capacity is $V_{Rd,c}$ (Eq. 10.1). Therefore, the calculation for torsion should also be made where $V_{Rd,cr}$ is critical, typically at about ¼ to ⅓ span (or not at all in some designs) where the torque will be about ½ the value at the supports (for UDL).

The clause does not elaborate on what an 'approximately rectangular solid section' is, *viz.* is the boot of the inverted-tee beam in Figure 6.10 so dominant in the total cross section that behaves as a rectangular section, or if the component parts of this beam qualify as rectangular sections? The authors consider this to be the case, but it would not apply to I beams with tapered or curved top and bottom flanges.

The clause does not explain how $T_{Ed,i}$ and $T_{Rd,c,i}$ in the sub-sections of beams such as inverted-tee or I beams are apportioned, as this clause is only intended for rectangular sections where this does not apply. The obvious approach is to proportion both $T_{Ed,i}$ and $T_{Rd,c,i}$ in each sub-section. The same logic applies to $V_{Ed,i}$ and $V_{Rd,c,i}$ in each sub-section, in order to be able to combine the terms in Eq. 10.50. Proportioning V_{Ed} into the upstand (up) and boot of an inverted-tee beam requires the shear stress in the upstand to be calculated from $\tau_{up}=V_{Ed}\,S/I_{xx}\,b_w$, and thus $V_{Ed,up}=0.67\,\tau_{up}\,b_w\,h_{up}$. For example, in the inverted-tee beam in Figure 6.10 and Example 10.6 is subject to V_{Ed}=347.5 kN, with $I_{xx}\,b_w/S$=177040 mm², then $\tau_{up}=347.5\times10^3/177040=1.96$ N/mm², and $V_{Ed,up}=0.67\times1.96\times350\times200\times10^{-3}=91.6$ kN. The remainder of the shear force carried by the boot is $V_{Ed,boot}=347.5-91.6=255.9$ kN. Although this calculation will not influence the designed area of torsion stirrups nor longitudinal bars (as they are both a function of $T_{Ed,i}$) it may tip the requirement for minimum rather than designed reinforcements.

The requirements for longitudinal reinforcement A_{sl} due to torsion (additional to the flexural requirement of top and bottom tendons) are given in clause 6.3.2 (3) as

$$\text{Exp. 6.28}\quad A_{sl}=T_{Ed}\,u_k\cot\theta/2\,A_k\,0.87\,f_{yk} \quad (10.53)$$

where cot θ is the same as in the shear calculations at the same cross section in the span.

As the area A_{sl} is based on the strength of rebars (f_{yk}=500 N/mm^2), if the reinforcement is to use some of the spare capacity of the pretensioning tendons (f_{pk}=1770 N/mm^2) near to the support their effective area is A_{sl}×500/1770. If rebars are used the area is of course A_{sl}.

The bars should be arranged so that there is at least one bar in each corner of each sub-section, with the remainder spread around the inner perimeter of the links, with spacing≤350 mm (see clause 9.2.3 (4)). The minimum number of bars is 4 (one in each corner). If the dimension of the sub-section is greater than 450 mm (i.e. 350+2×50 edge axis distance), then 1 no. extra bar is required in each mid-side. If greater than 800 mm (2×350+2×50) then 2 no. extra bars are required in each mid-side.

In compression zones, the bars may be reduced in proportion to the available compressive force. If the ultimate bending moment is M_{Ed} at the point where T_{Ed} is calculated, i.e. 0.5 d cot θ from the support, and the lever arm z between the compression zone and bottom bars is known, the compressive force $F_c = M_{Ed}/z$ may be used to reduce the area of top bars in A_{sl}. (See Example 10.9).

In tension zones, the bars should be added to, or replace the spare capacity of, other flexural reinforcement. Gilbert's approach (Gilbert et al., 2017, Example 7.2) is to determine F_c in the compressive zone above the neutral axis and to calculate A_{sl} in the bottom bars based on the remainder of the axial force in the tension zone. Alternatively, the stress in the pretensioned tendons around the torsional perimeter is small near to the supports and their spare capacity may be used to contribute to A_{sl}. Note that the tendons are already pretensioned to σ_{p0}, typically to about 60% of 0.87 f_{pk}, such that their remaining capacity is about 40%.

10.6.5.2 Additional stirrups due to torsion

Although BS EN 1992-1-1, clause 6.3.2 does not specify at what magnitude of T_{Ed} torsion stirrups are required, it is implied from clause 6.3.2 (5) that they are required when Eq. 10.50 is violated. Note that the calculated area of a torsion link is for one leg only.

The equation for area of links is not given in BS EN 1992-1-1, but may be derived from Eqs. 10.39, 10.40 and 10.20 as

$$\tau_{t,i} = V_{Ed,i}/t_{ef,i}\, z_i = T_{Ed}/2\, A_k\, t_{ef,i}$$

$$\text{Then } V_{Ed,} = T_{Ed}\, t_{ef,}\, z/2\, A_k\, t_{ef,} = T_{Ed}\, z/2\, A_k$$

$$\text{Eq. 10.20} \quad A_{sw}/s \geq V_{Ed}/z\ 0.87\ f_{ywk} \cot\theta$$

$$\text{Then } A_{sw}/s \geq T_{Ed}/2\, A_k\ 0.87\ f_{ywk} \cot\theta \tag{10.54}$$

Generally, the diameter and area of the stirrup is decided upon, and so the spacing is

$$s = A_{sw}\ (1\ \text{leg})\ 2\, A_k\ 0.87\, f_{ywk} \cot\theta / T_{Ed} \tag{10.55}$$

cot θ is the same as in shear design

9.2.3 (3) $s \leq \min\{u/8,\ b^*;\ 0.75d\}$ *the lesser dimension of the section

The shape of the closed links is shown in BS EN 1992-1-1, Fig. 9.6. The minimum diameter is the same as for shear links. If shear requirements are small, and torsion requirements are large, then the diameter of torsion links may be increased with greater spacing to give the same A_{sw}/s.

Example 10.9

Calculate the torsional strength $T_{Rd,c}$ and the requirements for torsion stirrups and additional longitudinal reinforcement for the inverted-tee beam in Example 10.6. Let the maximum ultimate torque at the end of the beam T_{Ed}=50 kNm and V_{Ed}=347.5 (Example 10.6) kN. The edge distance to the centre of the strands is 50 mm.

Solution

Material properties

$f_{cd}=45/1.5=30$ N/mm^2
$v=0.6\ (1-45/250)=0.492$
$f_{ctd}=0.7\times0.3\times45^{2/3}/1.5=1.77$ N/mm^2
$f_{yd}=500/1.15=435$ N/mm^2

The proportioning of T_{Ed} into the upstand and boot is given in Section 10.6.4, respectively, as

$T_{Ed,up}=0.074\times50=3.70$ kNm
$T_{Ed,boot}=0.926\times50=46.30$ kNm

Dimensions and effective torsional section properties for the boot are: b (lesser dimension)=400 mm, h=600 mm, A=240000 mm^2, u=2000 mm, t_{ef}=max{120; 2×50}=120 mm, $A_k=(400-120)\times(600-120)=134400$ mm^2, and $u_k=2\times(280+480)=1520$ mm.

Eq. 10.51 $T_{Rd,c}=2\times134400\times120\times1.77\times10^{-6}=57.13$ kNm

Using the equations per Gilbert et al. taking advantage of the prestress in the boot

Eq. 10.52 $T_{Rd,c}=(0.33\times400^2\times600)\times1.77\times\sqrt{(1+10\times4.34/45)}\times10^{-6}=78.63$ kNm is 38% greater than Eq. 10.51.

where σ_{cp} in the boot=0.579×(6.746/0.9)=4.34 N/mm^2 (from Example 10.6)
From Examples 10.6 and 10.7 by inspection $\cot\theta=2.5$, then $\cot\theta+\tan\theta=2.5+1/2.5=2.9$

Eq. 10.45 $T_{Rd,max}=2\times0.492\times1.14\times30\times134400\times120\times10^{-6}/2.9=187.9$ kNm

where $\alpha_{cw}=1+(4.34/30)=1.14$

V_{Ed}=255.9 kN (proportioned into the boot in Section 10.6.5.1)
$V_{Rd,c}$=731.3 kN (from Example 10.6)
Eq. 10.46 $V_{Rd,max}=1.14\times600\times(0.9\times480.8)\times0.492\times30\times10^{-3}/2.9=1512.3$ kN
Eq. 10.50 $T_{Ed}/T_{Rd,c}+V_{Ed}/V_{Rd,c}=46.30/57.13+255.9/731.3=0.810+0.350=1.160>1$, then torsion stirrups and additional longitudinal bars are required.

Using $T_{Rd,c}$ calculated from Eq. 10.52 after Gilbert et al., $T_{Ed}/T_{Rd,c}+V_{Ed}/V_{Rd,c}=46.30/78.63+0.350=0.94<1$, then no torsion links are required, with clear scope for discussion.

Eq. 10.44 $T_{Ed}/T_{Rd,max}+V_{Ed}/V_{Rd,max}=46.30/187.9+255.9/1512.3=0.246+0.169=0.415<1$, therefore the concrete section is adequate.

Corresponding values for the upstand are

Eq. 10.50 $T_{Ed}/T_{Rd,c}+V_{Ed}/V_{Rd,c}=3.70/8.86+91.6/498.8=0.418+0.184=0.602<1$

It is unlikely for the upstand to fail independently from the boot. Torsion stirrups are not required, but the same area and spacing of stirrups used in the boot will be provided for the continuity of strength.

Eq. 10.44 $T_{Ed}/T_{Rd,max}+V_{Ed}/V_{Rd,max}=3.70/28.00+91.6/848.2=0.132+0.108=0.240<1.$

Torsion stirrups in the boot.

Choose H10 stirrups (to match the shear stirrups in Example 10.7) with $A_{sw}=78.5$ mm^2 per leg.

Eq. 10.55 $s=78.5\times2\times134400\times435\times2.5/46.30\times10^6=496$ mm

Minimum spacing per clause 9.2.3 (3). $s\leq\min\{2000/8, 400, 0.75\times480.8\}=250$ mm. Use H10 closed stirrups at 250 mm spacing, detailed according to BS EN 1992-1-1, Fig. 9.6.

Additional longitudinal reinforcement.

Eq. 10.53 $A_{sl}=46.30\times10^6\times1520\times2.5/(2\times134400\times435)=1505$ mm^2

This is clearly a conservative calculation as the angle of the compressive struts for torsion will be closer to cot $\theta=1$.

The equivalent area of strands is $1505\times500/1770=425$ mm^2

If, for example, 10 bars are notionally placed around the perimeter of the beam - two in each bottom corner, two in the top of the boot, two in the upstand, and two in the mid-width of the bottom (see Figure 6.10), the area required per strand is 42 mm^2. The area of a 12.5 mm strand is 93 mm^2. However, the strands are already prestressed to $\sigma_{po}=953.3$ N/mm^2 (Example 7.10), which is $953.3/0.87\times1770=62\%$ of available capacity, leaving 38% of the area spare for the torsional requirement$=0.38\times93=35$ mm^2. Although this is less than 42 mm^2 required it is acceptable, given the assumption above and the complexity of the calculation.

Further analysis of the beam is required to find the distance from the support where Eq. 10.50 is violated, which is 960 mm from the centre of the support. Therefore, 5 no. torsion stirrups at 250 mm spacing are required, with the first stirrup at 50 mm from the end of the beam. Beyond these, use minimum links H10 at 360 mm spacing.

10.6.6 Design for torsion in prestressed hollow core units

10.6.6.1 Torsion cracking in hollow core units

Torsion cracks may occur at the corners of hcu, typically at 45° in plan view from a point about ⅓ of the width of the unit, and where the unit is supported along its longitudinal edge, or in floors with large openings, skew or uneven floor supports, etc. Figure 10.34 shows a torsion crack due to unequal level bearings. Equilibrium torsion may also be induced due to eccentric loads from trimmer angles at voids or linear line loads or point loads acting at the edges of the units which cannot be transversely distributed into the slab field because of the presence of large voids in the floor as shown in Figure 10.40. T_{Ed} is due to the eccentric load of the trimmer angles and the walls surrounding the void.

The torsional resistance may be calculated using an idealised cross section comprising of the upper and lower flanges and the outermost webs of the units, as shown in Figure 10.41. An extensive research program and carried out at VTT in Finland, known as 'Holcotors' (European Commission, 2004), found that for hcu of less than 250 mm depth that the two outermost webs contribute to the resistance; otherwise, it is one outermost web. Figures 10.42 and 10.43 show an individual hcu subjected to two point loads at opposite corners of a hcu producing pure torsion, and five hcu placed in a floor system subject to point loads that would otherwise be used to calculate torsion in an individual unit. Ludgren et al. (Lundgren et al.,

Figure 10.40 Example of torsion induced into unit A from trimmer angles at B. After the longitudinal joint C is concreted the remainder of the floor loads pass through the joint. A wall built around the void at D also imposes a load at point B.

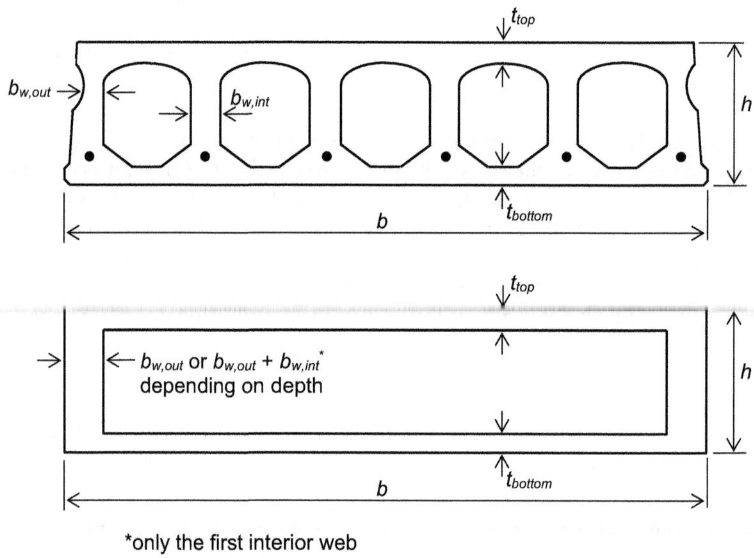

Figure 10.41 Transformation of a hollow core cross section into an equivalent tubular cross section for the calculation of torsional properties.

2004) summarised the project and concluded that 'hollow core units can carry higher amounts of the shear and torsion when placed in a floor system, and not only due to the transfer to the neighbouring elements … the boundaries with adjacent units seem to yield a more favourable stress state, or better possibilities for redistribution within the unit.' Finite element modeling found that shear-torsion interaction follows the circular plots shown in Figure 10.44 (for the units shown in Figure 10.45), rather than the linear interaction according to BS EN 1168, clause 4.3.3.2.2.4. See Section 10.6.6.5.

10.6.6.2 Torsional strength of hollow core units

The shear stress in a cross section subjected to pure torsion can be calculated as

$$\tau_t = T/W_t \leq f_{ctd} \qquad (10.56)$$

Figure 10.42 400 mm depth prestressed concrete hollow core unit subjected to pure torsion (by two point loads at opposite corners) in Holcotors project. (Courtesy VTT, Espoo, Finland.)

Figure 10.43 Torsional failure of 400 mm depth hollow core floor unit in Holcotors project. (Courtesy VTT, Espoo, Finland.)

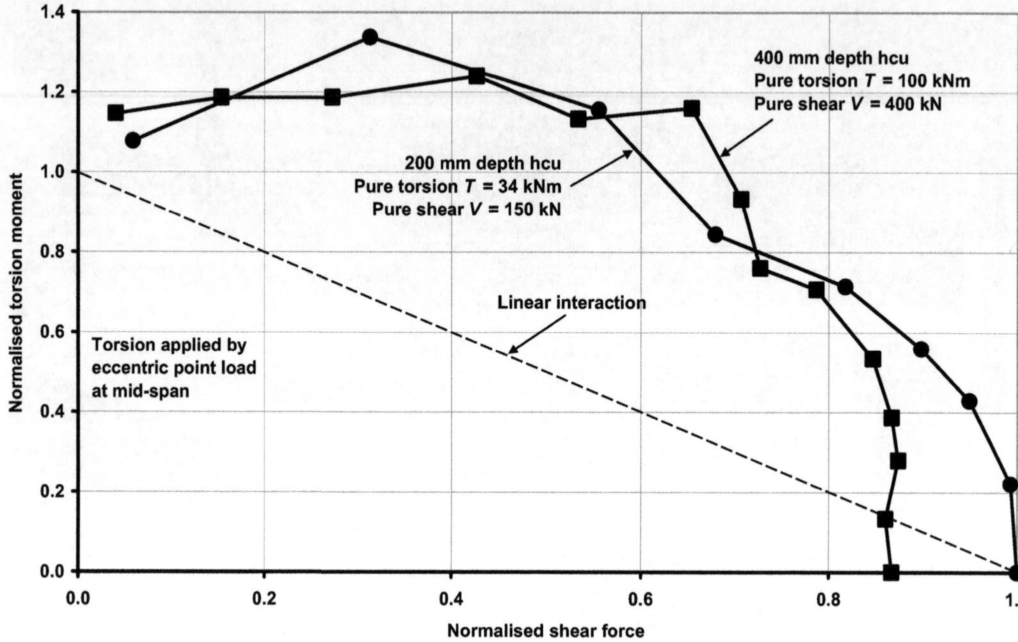

Figure 10.44 Shear-torsion interaction based on finite element results for 200 and 400 mm depth hcu in Holcotors project. (Adapted and normalised from Lindström (2005).)

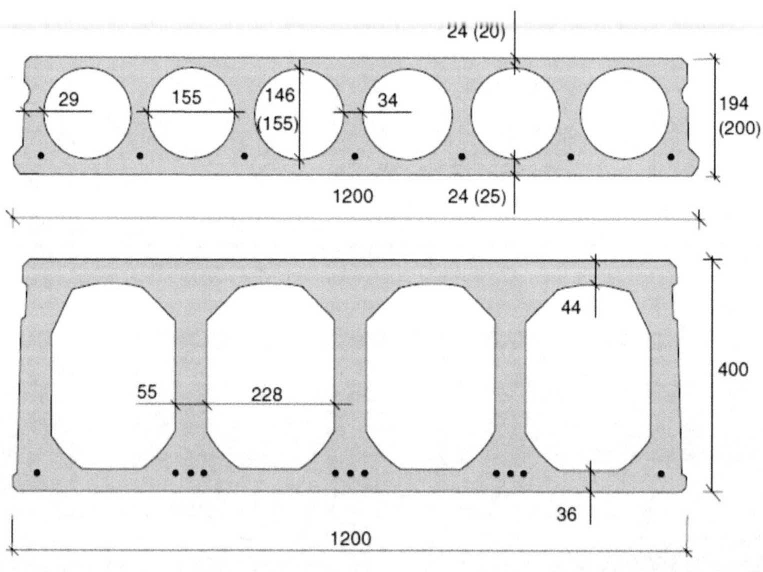

Figure 10.45 Cross sections of prestressed hollow core units tested in the Holcotors project. Values in parenthesis are nominal dimensions if different from measured. (Courtesy VTT, Espoo, Finland.)

where W_t is the torsional section modulus

The shear stress due to torsion is largest at the midpoint of the wider outer surfaces of a solid rectangular section. For thin-walled tubular cross sections, the shear flow in the walls is constant and the largest shear stress is in the wall with minimum thickness. For hcu, the torsional stiffness and resistance is very similar to a tubular thin-walled box section as idealised in Figure 10.41. Referring to this diagram W_t may be calculated as

$$W_t=2\ [h-0.5\ (h_t+h_b)]\ (b_y-b_{w,o})\ t \tag{10.57}$$

where t is the thickness of the thin-walled part for which the shear stress is calculated, i.e. for the web $t=b_{w,o}$.

h_t and h_b are the depth of the top and bottom flanges
b_y=total width at centroidal axis, typically 1150–1160 mm
$b_{w,o}$=width of the outermost web

For different thin-walled parts, the resistance due to torsion is based on

$$W_{t,top}=W_t(h_t) \tag{10.58a}$$

$$W_{t,web}=W_t(b_{w,o}) \tag{10.58b}$$

$$W_{t,bottom}=W_t(h_b) \tag{10.58c}$$

For hollow core cross sections wall thickness should be limited by

$b_{w,o}\leq 0.265\ h$ when using $W_{t,web}$
$h_f\leq 0.175\ h$ when using $W_{T,\ top}$ or $W_{T,bottom}$

For hcu subjected to torsion and normal stresses due to prestressing and imposed loading, the torsional moment of resistance $T_{Rd,c}$ is found when the principal stress reaches the concrete design tensile strength f_{ctd}. The principal stress for the top flange, where the axial stress due to prestress σ_{ct} is the least, and likely to be in tension, can be expressed as (Mohr's circle approach)

$$f_{ctm}=\sigma_{ct}/2+\sqrt{\left[(\sigma_{ct}/2)^2+t_t^2\right]} \tag{10.59}$$

where σ_{ct} is the axial concrete tensile stress (positive) in the top flange due to prestress and flexural stress due to bending moments at the end of the transmission length.

Re-arranging Eq. 10.59 gives the shear stress in the top flange $\tau_{t,top}$

$$\tau_{t,top}=f_{ctm}\ \sqrt{[1-(\sigma_{ct}/f_{ctm})]} \tag{10.60}$$

The torsional moment of resistance $T_{Rd,c}$ is calculated as

$$T_{Rd,c,top}=W_{t,top}\ f_{ctd}\ \sqrt{[1-(\sigma_{ct}/f_{ctd})]} \tag{10.61}$$

Example 10.10

Calculate the ultimate torsion moment of resistance $T_{Rd,c}$ for the webs and flanges in the 250 mm depth hcu shown in cross section in Figure 6.7, and material and section properties from Examples 7.1 and 10.1. The width at the centroidal axis b_y may be taken as 1154 mm.

Solution

Width of external web $b_{w,o}$=51 mm, $b_y-b_{w,o}$=1154−51=1103 mm

Note that $b_{w,o}$ =(51/250) h=0.204 h<0.265 h, and only the outermost webs are used in this example.

Depth of flanges h_t=40 mm and h_b=35 mm. $h-0.5\ (h_t+h_b)=250-0.5\times75=212.5$ mm

Eq. 10.58a $W_{t,web}=2\times212.5\times1103\times51=23.91\times10^6$ mm³

Eq. 10.58b $W_{t,top}=2\times212.5\times1103\times40=18.75\times10^6$ mm³

Eq. 10.58c $W_{t,bottom}=2\times212.5\times1103\times35=16.41\times10^6$ mm³

At the end of the transmission length and after losses of prestress from Example 7.1 $\sigma_{c,t}$=1.95 N/mm² (tension)

The stress due to the self-weight in the top of the hcu at the end of the transmission length l_{pt2} from the centre of bearing is required. This is at $x=726-100/2=676$ mm. Then $M_{s,x}=4.12\times9.8\times0.676/2-4.12\times0.676^2/2=12.7$ kNm. $Z_{t,co}=10.056\times10^6$ mm³. $\sigma_{t,x}=12.7/10.056=1.26$ N/mm² (compression). Calculating the stress due to self-weight gives the minimum value for $T_{Rd,c}$.

$\sigma_{c,t}=1.95-1.26=0.69$ N/mm²
$f_{ctd}=1.77$ N/mm²

Eq. 10.60 $\tau_{t,top}\leq1.77\times\sqrt{(1-0.69/1.77)}=1.38$ N/mm²

Eq. 10.61 $T_{Rd,c,top}=18.75\times1.38=25.9$ kNm

In the webs, $\sigma_{c,t}$ at the height of the centroid of the hcu is -5.12 N/mm² (compression), and the stress due to self-weight is zero. Then $\tau_{t,web}\leq1.77\times\sqrt{(1-5.12/1.77)}=3.49$ N/mm² and $T_{Rd,c,web}=23.91\times3.49=83.4$ kNm, clearly not critical. Even adding the shear stress due to vertical self-weight shear, v=0.36 N/mm² will not change the outcome.

10.6.6.3 Comparison of torsion moment of resistance with Holcotors

The following compares the calculated values of $T_{Rd,c}$ using the above procedures with the results from the pure torsion tests in the Holcotors project for 200 mm and 400 mm depth hcu shown in Figures 10.43–10.45. The test results were reported by Pajari at VTT in Finland (Pajari, 2003). Table 10.3 gives the theoretical prediction of the torsional cracking stresses and moment using Eqs. 10.59–10.61, but with f_{ctd} replaced by the measured tensile strength f_{ct}, and the actual material properties and measured dimensions from the tests. The ratio of the predicted T_{Rd} to the experimental torsion moment T_u in the tests 0.86 and 0.97 for the 200 and 400 mm depth units, respectively.

Although there are several assumptions in the calculations in Table 10.3, such as the losses in pretensioning force at the time of the test, assumed as 6 weeks after transfer, is 15%, the critical parameter in the calculations is the tensile strength f_{ct} in the top flange. If this varies by more than ±5%, the agreement with T_u changes by about 7 percentage points (i.e. the ratio 0.97 changes to 1.05 and 0.90), but is still within acceptable limits.

10.6.6.4 Shear and torsion interaction

Torsion loading on a hcu produces shear forces around the perimeter of the cross section, as shown in Figure 10.46. In units of greater depth with fewer cores, e.g. 400 mm with 4 cores, torsional shear stress exists also in the first interior web, as illustrated in Figure 10.46, the magnitude dependent on depth. These shear forces act upwards in one and downwards

Table 10.3 Comparison of predicted and experimental torsional moments at failure using data from Holcotors project

Test reference	*PT200*	*PT400*
Depth of unit (mm)	200	400
Number of strands	7	11
Cover to strands (mm)	30	30
Diameter of strands (mm)	12.5	12.5
Initial prestress (N/mm^2)	900	1000
Prestress at time of test (assumed 6 weeks) (N/mm^2)	764	847
Width of unit at centroidal axis b_y	1160	1160
Width of outer web(s) (two outer webs for 200 depth) $b_{w,o}$ (mm)	66	53
Net width $b_y - b_{w,o}$ (mm)	1094	1107
Depth of top flange h_{ft} (mm)	23.8	40.0
Depth of bottom flange h_{fb} (mm)	23.6	38.5
Net depth $h - [0.5\ (h_t + h_b)]$ (mm)	176	361
Section modulus W_{top} (× 10^6 mm^3)	9.10	30.76
Test data		
Measured tensile strength f_{ct} (N/mm^2)	4.13	4.25
Prestress tension at top of hcu σ_{ct} (N/mm^2)	0.97	2.33
Shear stress in top $\tau_{t,top} = f_{ct}\ (1 - \sqrt{(\sigma_{ct}/f_{ct})}]$ (N/mm^2)	3.61	2.86
Predicted $T_{Rd} = W_{top}\ \tau_{t,top}$ (kNm)	32.9	87.9
Torsion moment (mean of 2 tests) T_u (kNm)	38.4	90.2
Ratio of predicted to experimental T_{Rd}/T_u	**0.86**	**0.97**

in the other of the outermost webs of the unit. The calculation method given in Section 10.6.6.2 assumes a theoretical addition of torsion and shear stress in the webs. In reality, there will be a redistribution of the stresses over the whole section due to deformations and compatibility within the units, the softening behaviour of cracked concrete, and possible restraints at the boundaries of the units.

The forthcoming FIB Manual on prestressed hcu establishes the shear and torsion interaction relationship based on the results of the 'Holcotors' project (European Commission, 2004) using finite element analysis with nonlinear material properties. The manual also gives a simplified method to establish the interaction diagram using some key values for shear force and torsion moment alone.

10.6.6.5 Shear-torsion interaction formula according to BS EN 1168

BS EN 1168, clause 4.3.3.2.2.4 and Fig. 4 (adapted here in Figure 10.47) gives a calculation method taking into account the interaction between shear and torsion to determine the net shear capacity of hcu. Essentially, the net shear capacity $V_{Rd,n}$ after the shear force $V_{Ed,t}$ due to the torsional shear flow in the webs is subtracted and is determined as follows. Generally, from Eq. 10.41, $V_{Ed,i} = T_{Ed}\ z/2\ A_k$. The shear force in the outer webs $V_{Ed,out}$ of width $b_{w,o}$ is

$$V_{Ed,out} = T_{Ed}\ z/2\ A_k$$

But in the vertical webs $z = h - 0.5\ (h_f + h_{fb})$

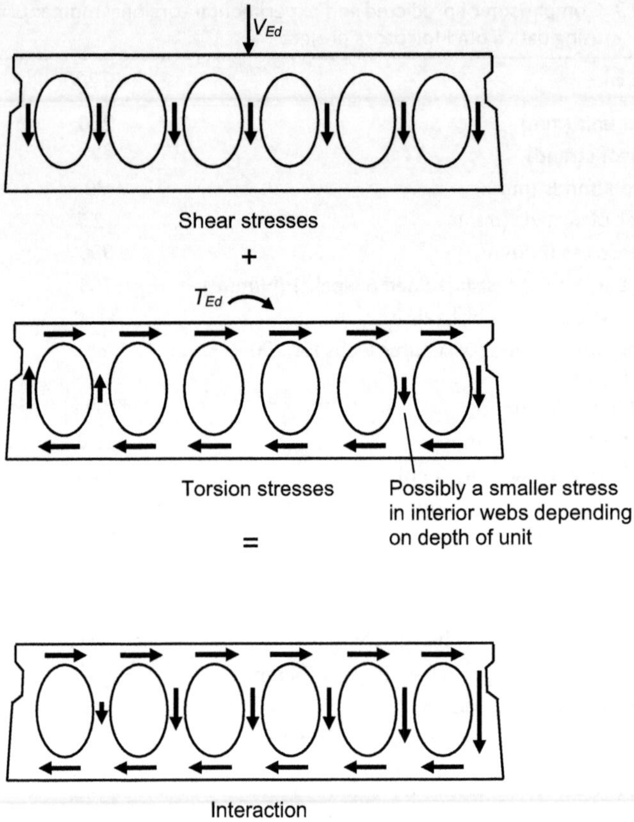

Figure 10.46 Vectors of the distribution of shear and torsion stresses, and shear-torsion interaction in hollow core units.

and A_k = the area enclosed by the median line of the outer walls of the hcu $= (b - b_{w,o})(0.5\,(h_f + h_{fb}))$

Then $V_{Ed,out} = T_{Ed}\,(h - 0.5\,(h_f + h_{fb}))/[2\,(b - b_{w,o})\,(h - 0.5\,(h_f + h_{fb})]$

Finally $\quad V_{Ed,out} = T_{Ed}/[2\,(b - b_{w,o})]$ (10.62)

Note that the overall width b is taken at the centroidal axis.

But the shear force in all webs $V_{Ed,t}$ is pro-rata the sum of the width of all webs Σb_w divided by the width of the outer web $b_{w,o}$, as

$$V_{Ed,t} = T_{Ed}\,\Sigma b_w/2\,(b - b_{w,o})\,b_{w,o}$$

This is expressed in BS EN 1168, clause 4.3.3.2.2.4 as

$$V_{Ed,t} = (T_{Ed}/2\,b_{w,o}) \times [\Sigma b_w/(b - b_{w,o})] \quad (10.63)$$

Finally $\quad V_{Rd,n} = V_{Rd,c} - V_{Ed,t}$ (10.64)

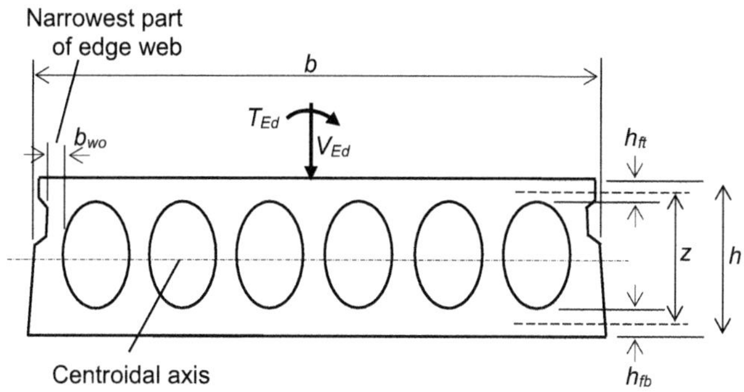

Figure 10.47 Torsion caused by eccentric shear force. (Adapted from BS EN 1168; 2005+A3 (2015), Fig. 4.)

Example 10.11

Calculate the remaining shear capacity V_{Rdn} of the 250 mm depth hcu shown in cross section in Figure 6.7 and designed in Example 10.1 if the applied ultimate torsional moment near to the support T_{Ed} is 10 kNm. The width of the outer webs is 51 mm and the total width of webs is 302 mm. The overall width at the centroidal axis may be taken as 1154 mm.

Solution

Eq. 10.63 $V_{Ed,t} = 10 \times 10^6 \times 302 \times 10^{-3} / [102 \times (1154 - 51)] = 26.9$ kN

(This may be conservative as more than the outside webs are contributing to T_{Rd})

Example 10.1. $V_{Rd,c} = 134.3$ kN

Eq. 10.64 $V_{Rd,n} = 134.3 - 26.9 =$ **107.4 kN**

Check the result using Eqs. 10.56 and 10.58 and data in Examples 10.1 and 10.10

Eq. 10.58a $W_{t,web} = 23.91 \times 10^6$ mm^3 (from Example 10.10)

Eq. 10.56 $\tau_t = 10.0/23.91 = 0.42$ N/mm^2

Shear stress in webs (from Example 10.1) $\tau_v = V_{Rdn}\ S/I_{xx}\ b_w = 107.4 \times 10^3 \times 6.783 \times 10^6 / (1270 \times 10^6 \times 302) = 1.90$ N/mm^2

Total stress in webs $= 0.42 + 1.90 = 2.32$ N/mm^2

Maximum principal stress in concrete (from Example 10.1) is $\sqrt{(1.77^2 + 0.307 \times 4.61 \times 1.77)} = 2.37$ N/mm^2

The discrepancy derives because $W_{t,web}$ is calculated for the outside web alone, whereas v is an average stress in all webs. If $W_{t,web}$ is recalculated for the *average* width of the webs, 43.1 mm, then $\tau_t = 0.42 \times 51/43.1 = 0.49$ N/mm^2 and the total stress in webs $= 0.49 + 1.90 = 2.39$ N/mm$^2 \approx 2.37$ N/mm^2.

REFERENCES

ACI-318, 2008. Building Code Requirements for Structural Concrete and Commentary, ACI Committee 318, American Concrete Institute, Farmington Hills, MI 48331, USA.

BS 8110-2, 1985. Structural Use of Concrete - Part 2. Code of Practice for special circumstances, British Standards Institute, London, UK.

BS EN 1168. 2005. Precast Concrete Products - Prestressed concrete hollow core units, +A3:2011, BSI, London, UK.

BS EN 13369. 2018, Common rules for precast concrete products, BSI, London, UK.

BS EN 1992-1-1. 2004. Eurocode 2: Design of concrete structures - Part 1-1: General rules and rules for buildings, +A1:2014, BSI, London, UK.

Elliott, K. S. 2014. Transmission length and shear capacity in prestressed concrete hollow core slabs, *Magazine of Concrete Research*, Vol. 66, No. 12, pp. 585–602.

Elliott, K. S. 2017. Precast Concrete Structures, 2nd ed., CRC, Taylor & Francis, London, UK, 694 p.

Elliott, K. S. and Jolly, C. K. 2013. Multi-Storey Precast Concrete Framed Structures, 2nd ed., John Wiley, London, UK, 750 p.

European Commission, 2004. Holcotors: Shear and torsion interaction of hollow core units, Project funded by the European Community, Contract No G6RD-CT-2001-00641, 2002 to 2004.

Evans, R. H and Robinson, G. W. 1955. Bond stress in prestressed concrete from X-ray photographs, *Proceedings of the Institution of Civil Engineers*, Part 1, Vol. 4, No. 2, pp. 212–235.

FIP. 1992. Quality Assurance of Hollow Core Slab Floors, Guide to Good Practice, Fédération Internationale de la Précontrainte, Lausanne, Switzerland.

Gilbert, R. I., Mickleborough, N. C. and Ranzi, G. 2017. Design of Prestressed Concrete to Eurocode 2, 2nd ed., CRC, Taylor & Francis Group, Boca Raton, FL, 665 p.

Hearn, E. J. 1977. Mechanics of Materials, Vol. 2, Pergammon Press, Oxford.

Kong, F. K. and Evans, R. H. 1987. Reinforced and Prestressed Concrete, 3rd ed., Van Nostrand, London, UK, 508 p.

Kong, P. 1993. Transfer of prestress by prestressed wire tendons, MEng Thesis, Curtin University, Perth, Australia.

Lin Yang. 1994. Design of prestressed hollow core slabs with reference to web shear failure, *Journal of Structural Engineering, ASCE*, Vol. 120, No. 9, pp. 2675–2696.

Lindström, G. 2005. Holtocors results - shear and torsion interaction, Presentation at IPHA Seminar, Prague.

Lundgren, K., Broo, H. and Engström, B. 2004. Analysis of hollow core floors subjected to shear and torsion, *Structural Concrete*, Vol. 5, No. 4, pp. 161–172.

NA to BS EN 1992-1-1. 2004. UK National Annex for Eurocode 2: Design of concrete structures - Part 1-1: General rules and rules for buildings, +A1:2014, BSI, London, UK.

Oh, B. H., Lim, S. N., Lee, M. K. and Yoo, S. W. 2014. Analysis and prediction of transfer length in pre-tensioned, prestressed concrete members, *ACI Structural Journal*, Vol. 111, No. 3, pp. 549–560.

Pajari, M. 2003. Shear and torsion interaction of hollow core slabs, Holcotors project, Test Report 1/2002 - Pure torsion tests on single slab units, VTT, Espoo, Finland, 53 p.

Palmer, K. D. and Schultz, A. E. 2011. Experimental investigation of the web-shear strength of deep hollow core units, *PCI Journal*, Vol. 56, No. 4, pp. 84–104.

Park, H. and Cho, J-Y. 2014. Bond-slip-strain relationship in transfer zone of pretensioned concrete elements, *ACI Structural Journal*, Vol. 111, No. 3, pp. 503–514.

Walraven, J. C. and Mercx, W. 1983. The bearing capacity of prestressed hollow core slabs, *Heron*, Vol. 28, No. 3, pp. 1–46.

Chapter 11

Serviceability limit state for deflections and cracking

11.1 DEFLECTIONS IN PRESTRESSED CONCRETE SLABS AND BEAMS

11.1.1 Background to deflection calculations in prestressed concrete

Previous UK Codes of Practice, such as CP 110 (CP 110, 1972), BS 8110 Part 1 (BS 8110-1, 1997) and BS 8110 Part 2 (BS 8110-2, 1985), had a rather casual approach to the calculation of deflections in prestressed concrete (psc) elements, such as '*where it is considered necessary to calculate deflections ...*', and '*the serviceability requirements for deflection are given ... but no numerical limits are set*'. This approach no doubt stems from the major influence of eccentric prestress causing upward camber which continues to increase due to the effects of creep, in fact historically the authors recall in-house limit for the final (long-term) downward deflection of simply supported prestressed slabs as span/1000. Long-term deflections have always been considered to be controlled by limiting the serviceability tensile stress to f_{ctm}, or similar values such as $0.45\sqrt{f_{cu}}$ in earlier codes. Elastic analysis based on the gross (or compound) concrete cross-section may be used for instantaneous and long-term deflections if the tensile stress is so limited.

BS EN 1992-1-1 (BS EN 1992-1-1, 2004) does not distinguish between reinforced concrete and psc, but states a Principle in clause 7.4.1 (1)P '*The deformation of a member or structure shall not be such that it adversely affects its proper functioning or appearance.*' and the limits that have been traditionally set for reinforced concrete members, such as limiting the final (long-term) deflection of span/250, are also applicable to prestressed concrete (psc). This is rarely critical in the vast majority of designs, apart from lightly loaded long span elements where self-weight and dead loads are greater than imposed live loads, where the tensile stress is limited to f_{ctm}. For example in the hollow core unit (hcu) shown in cross-section in Figure 6.7 and subjected to the gravity loads in Example 7.1, with $\psi_2=0.3$, it will be shown later in this chapter that the final deflection is 9.2 mm<span/250=9800/250=3 9.2 mm. For the deflection to exceed span/250, the span must be increased to 12.2 m and the imposed live load decreased from 5.0 to 0.9 kN/m^2, i.e. lightly loaded. If the durability exposure>XC1 and limiting tensile stress f_{ctm}=zero, it is impossible for the final deflection to be critical.

Calculations for the upward camber, instantaneous and long-term deflections are given in Section 11.2.2, together with the limits in BS EN 1992-1-1, clause 7.4.1 (4) & (5). Figure 11.1 shows an illustration of the imposed service bending moment vs mid-span deflection for a simply supported psc element. The deflections are linear until flexural cracking occurs, and thereafter a function of the uncracked and cracked flexural stiffness according to BS EN 1992-1-1, clause 7.4.3 (3), see Section 11.2.4. Note that the 'artificial jump' in the long-term

DOI: 10.1201/9781003319450-11

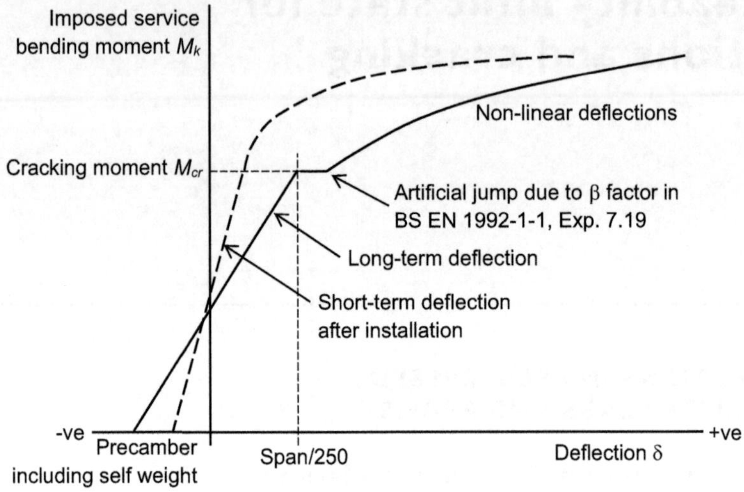

Figure 11.1 Illustration of deflections in simply supported prestressed concrete element.

deflections is due to a factor $\beta=0.5$ applied to the cracked stiffness in BS EN 1992-1-1, Exp. 7.19 where $M_k>M_{cr}$. Of course if the tensile stress is limited to f_{ctm} the deflection is *de facto* linear. Note that the characteristic service moment is denoted as M_k, rather than M_s, to distinguish it from flexural design. The service cracking moment is also denoted $M_{cr}=Z_{b,co}$ (σ_b+f_{ctm}) to distinguish it from the cracking moment M_c used in ultimate shear design (see Section 10.2.3). The long-term value of M_{cr} is less, typically 10%, than the instantaneous value because of prestress losses.

Although there is no specified limit to upward camber, which has a major influence on the depth of level screeds near to supports, the industry standard is typically span/300, although this may vary by ±50% due to differences in curing and storage conditions, e.g. the camber of a hcu stored on top of a stack on a cold sunny day may be up to 25% greater than those beneath it. BS EN 1992-1-1, clause 7.4.1 (4) suggests a limit of span/250 for upward camber incorporated into formwork, which is not quite the same camber due to prestress, but may be adopted as a check.

11.1.2 Normalised moment vs deflections in bending tests of hollow core units and T beams

Figures 7.1 and 7.4 show plots of applied bending moments vs mid-span deflections for a range of psc hcu and T beams (in beam and block flooring), respectively, up to the bending moments at the occurrence of flexural cracking. Figures 9.3 and 9.5 show the same data, but up to the ultimate moment of resistance and beyond to the end of the tests at failure. These diagrams also show the moments of resistance for each hcu or T beam, but they do not show where the deflections exceed the usual limit of span/250 for psc units at the serviceability limit state. Figures 11.2 and 11.3 show the same data, but these are normalised with respect to the mid-span deflection δ divided by the limiting deflection of the measured effective span/250. Similarly, these figures show the normalised mid-span bending moment M divided by the design service moment of resistance M_{sR} for each hcu or T beam. Where the axis of the graphs cross at $M/M_{sR}=1.0$ and $\delta/\text{span}/250=1.0$ is effectively the criterion for design. Of course, the deflections do not include long-term creep, nor the proportion of δ due to self-weight, imposed dead and imposed quasi-permanent live loading – long-term

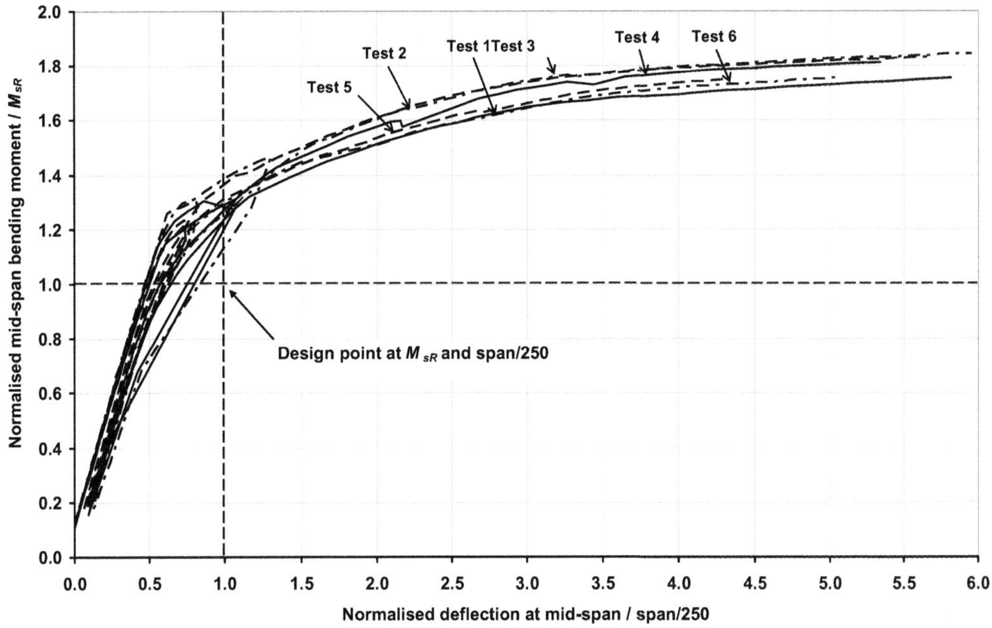

Figure 11.2 Normalised bending moment M/M_{sR} and mid-span deflections δ/span/250 for full scale tests on prestressed concrete hollow core units. (With permission by Bison Precast Ltd., UK.)

bending tests would be required for this – but (as shown in Section 11.2.2) given that the quasi-permanent factor ψ_2 for residential and offices is 0.3, and the long-term creep coefficient for imposed loading is ψ_∞=1.5–2.0, the effects of $1/\psi_2$ and $(1+\psi_\infty)$ virtually cancel out. Therefore M/M_{sR} vs δ/span/250 plots in Figures 11.2 and 11.3 give a reasonable indication of the behaviour relative to the design point. Note that the normalised plots are for span/depth of 5000/200=25.0 and 4400/150=29.3 for the hcu and T beams, respectively. Different span/depth ratios would inevitably lead to different results.

The important information in Figure 11.2 is that, in the first load cycle, at M/M_{sR}=1.0 then δ/span/250=0.5–0.7, i.e. deflection would not be critical. It is interesting to note that for the same hcu (span, section properties, materials and pretensioning) subjected to the same loading to give $M_s=M_{sR}$, the mid-span deflection (excluding self-weight of hcu) for long-term office loading with ψ_2=0.3 is 12.1 mm, i.e. δ/span/250=12.1/20=0.61, a result that lies in the centre of the experimental range. At the limiting deflection, where δ=span/250, the imposed bending moment in the tests exceeded M_{sR} by between 1.25 and 1.40. Therefore a comparison with the design example cannot be made because the deflections after cracking were non-linear. The ductility factor of structural elements is the ratio of ultimate deformation, in this case, mid-span deflections may be used, to the deformation at the limit of proportionality. In these tests, the ductility factor averaged 6.0.

The important information in Figure 11.3 is that, in the first load cycle, at M/M_{sR}=1.0 then δ/span/250=0.7–0.8. For the same T beams subjected to the same loading to give $M_s=M_{sR}$, the mid-span deflection for long-term domestic loading with ψ_2=0.3 (excluding self-weight of beam) is 13.9 mm, i.e. δ/span/250=13.9/17.6=0.79, a result that lies within the experimental range. At the limiting deflection, where δ=span/250, M_s exceeded M_{sR} by between 1.18 and 1.20. In the design example, this ratio is 1.26, close to 1.20 because the deflections were just inside the linear range. In these tests, the ductility factor averaged 5.2.

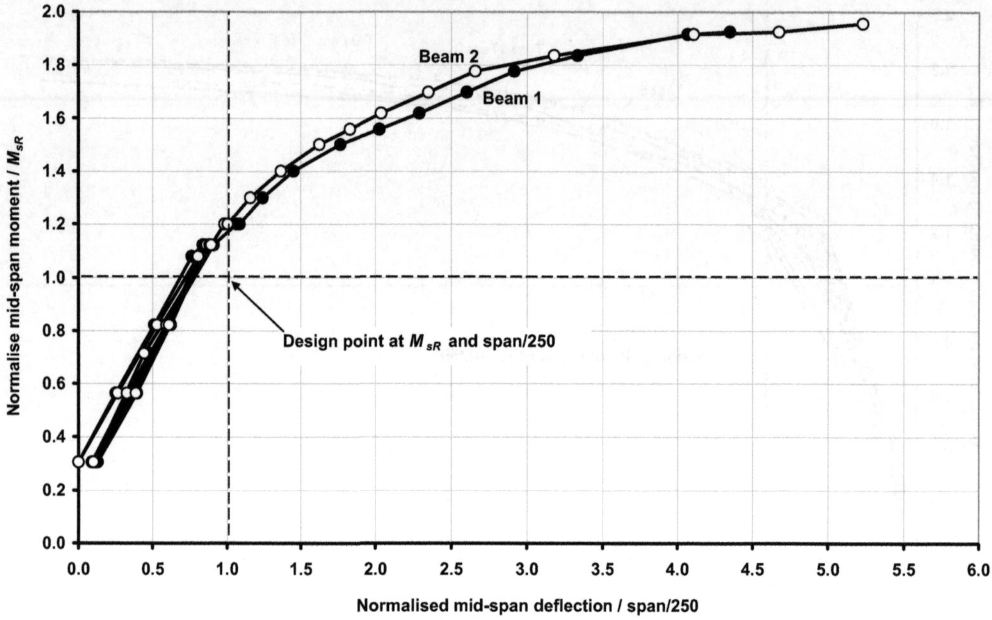

Figure 11.3 Normalised bending moment M/M_{sR} and mid-span deflections δ/span/250 for full scale tests on prestressed concrete T beams (in beam and block flooring). (With permission by Litecast Ltd., UK.)

The conclusion to the tests is that the normalised M/M_{sR} and δ/span/250 plots are similar and consistent for both hcu and T beams, e.g. at δ/span/250=2.0 then $M/M_{sR} \approx 1.6$. Although limited in terms of their span/depth range, short-term loading gives a reasonably reliable prediction of the long-term behaviour, but only where the quasi-permanent factor is $\psi_2=0.3$ for the reasons explained above. These tests would not predict deflections for example for storage where $\psi_2=0.8$.

11.2 CALCULATION OF DEFLECTIONS

11.2.1 Deflections up to the cracking moment of resistance

Deflection calculations are always carried out for psc elements – it is not sufficient to check span-effective depth ratios as in a reinforced section. This is because the strength-to-stiffness ratio of a psc section is considerably greater than in a reinforced section. The effects of tendon relaxation, creep, etc. have greater effects as the degree of prestress increases. The general method of curvature-area may be adopted in psc design. For non-deflected tendons the curvature diagram for prestress is rectangular. For debonded tendons, the curvature diagram comprises three rectangles, with a slope between them at the point of debonding, although the slope may be ignored and replaced by a step.

Net deflection is found by the superposition of upward cambers due to prestress and downward deflection due to gravity loads. In a cantilever the upward camber in the main span causes downward deflection in the cantilever span, see Figure 13.16a. To avoid this tendons may either be debonded or (less favourably) deflected to the top of the section.

Calculations are based on a flexurally uncracked stiffness $E_{cm}\, I_{c,co}$, using $E_{cm}(t)$ at transfer for initial camber due to prestress and the final value E_{cm} and appropriate creep coefficients for (i) long-term total deflections and (ii) deflections due to imposed dead and live loads after finishes are applied. In this book (ii) is hereby referred to as 'movement' to differentiate from total deflections. The compound value $I_{c,co}$ may be used, although many designers ignore the transformed section and use I_c, making a difference of about 2%–3% compared to I_c.

Young's modulus, called the 'modulus of elasticity', is according to BS EN 1992-1-1, Table 3.1, $E_{cm}=22\ [(f_{ck}+8)/10)]^{0.3}$ (kN/mm²) for gravel aggregates. This is reduced by 10% for limestone aggregates. The same equation may be used for the transfer strength. When considering the gain in the modulus of elasticity over time between the age at transfer (typically 16–20 hours) and at installation, typically 28 days, a factor may be used representing the approximate integration of E_{cm} over that period, i.e. $E_{cm}(t)/[0.5\ (E_{cm}(t)+E_{cm})]$. For example if $f_{ck}=45$ N/mm² and $f_{ck}(t)=30$ N/mm², using gravel aggregates $E_{cm}=36.283$ kN/mm² and $E_{cm}(t)=32.837$ kN/mm². The factor is 32.837/34.56=0.950. This factor may be used to adjust the short-term creep coefficients when calculating deflections at installation.

Precamber deflection comprises of four parts

a. immediate upward camber (defined as negative) after detensioning (and cutting of machine formed solid or hcu slabs) due to the immediate prestressing force P_{pm0} after the initial relaxation and elastic shortening of tendons
b. short-term value due to P_{pmi} after P_{pm0} plus short-term creep losses
c. long-term value due to P_{po} after all losses
d. self-weight deflection (downward, positive).

In-service long-term deflections are calculated in the usual manner taking into consideration the support conditions, loading arrangement and creep. Dead plus quasi-permanent live loads $G_k+\psi_2 Q_k$ are used. Final deflection, including upward camber, is limited according to BS EN 1992-1-1, clause 7.4.1 (4) to effective span/250. Movement, or the deflection after finishes are applied, is limited to span/500 where masonry partitions and/or brittle finishes are to be applied, or span/350 for non-brittle finishes. Architects may also impose a limit of 20 mm where span/350 or span/500 limit for longer spans may lead to damage of finishes.

Deflections are determined for the immediate elastic deformation, plus the visco-elastic effects of creep, which is reflected in the changing values of pretension force and the modulus of elasticity with time. The creep coefficients, with notation ψ, are obtained from the Italian Concrete Association ASSAP Manual (ASSAP, 2002) used in recent *fib* publications, and the Eurocode design guide by the Precast Concrete Federation (British Precast, 2008). Because the stress conditions in the top and bottom of the unit are different, the values for ψ determined for the effect of dead and imposed loads are not the same as the internal strain-related creep coefficients $\varphi(t,t_0)$ (long-term), $\varphi(t_i,t_0)$ (up to installation) and $\varphi(t,t_i)$ (after installation) used for calculating losses in prestress (hence the different notation).

Long-term deformations are based on a single long-term creep coefficient $\psi_\infty=2.5$ together with a concrete ageing coefficient, which allows for stress increments after initial loads, of X=0.8. Creep coefficients applied over an interval of time are determined by multiplying a 'coefficient of development' ψ' by $X\psi_\infty$. Over time intervals from $t_i=1$ day (at transfer) to 12 months these are as follows (ASSAP, 2002)

Coefficient of development ψ' at

transfer	= 0.1	15 days	= 0.3
28 days	= 0.4	2 months	= 0.5
3 months	= 0.6	6 months	= 0.7
12 months	= 0.8	∞	= 1.0

Intermediate values may be linearly interpolated up to 15 days, e.g. for 7 days $\psi'=0.0857+t_i/70=0.186$, thereafter curve fitting the data gives $\psi'=0.1579 \log_e t_i - 0.126$.

For example the creep coefficient from transfer at 1 day to installation at say 28 days is $\psi_1=(0.4-0.1)\times 2.5=0.75$ multiplied by the factor $E_{cm}(t)/[0.5\ (E_{cm}(t)+E_{cm})]$. In the previous example $\psi_1=0.75\times 0.950=0.713$. From 28 days to the long-term $\psi_{28}=(1.0-0.4)\times 0.8\times 2.5=1.20$, and so on.

Short-term and long-term deflections, and deflections after the application of finishes (known here as 'movement' to distinguish from the aforementioned deflections) are calculated at different periods of time and situations. These are summarised in Table 11.1.

11.2.2 Calculation of deflections

Let L=effective span (assume this is the same as the actual length of an element for calculating camber at transfer), P_{pm0}=prestressing force at transfer, P_{pmi}=prestressing force at installation, P_{po}=final prestressing force, z_{cp}=eccentricity, E_{cm} and $E_{cm}(t)$=modulus of elasticity at 28 days and at transfer, $I_{c,co}$=second moment of area of compound section, ψ_1=creep coefficient from transfer to installation, ψ_{28}=creep coefficient from installation (assume 28 days) to long-term, and ψ_∞=creep coefficient at long-term, w_0, w_1, w_2 and w_3=uniformly distributed loads due to self-weight of unit, self-weight of slab after grouting (or self-weight on infill blocks and infill concrete (for T beams), dead loads and imposed live loads, respectively, ψ_2=quasi-permanent live load factor.

The general term for the upward camber (often called 'pre-camber') at the mid-length (taken as mid-span for convenience) of an element is $\delta=-P\, z_{cp}\, L^2/8\, EI$, where P and E are the prevailing pretensioning forces and modulus of elasticity where the camber is to be calculated. The general term for downward deflection due to uniformly distributed gravity loads at the mid-span is $\delta=5\, w\, L^4/384\, EI$, where w is the prevailing load.

a. At transfer

Immediate camber

$$\delta_1 = -P_{pm0} z_{cp} L^2 / 8 E_{cm}(t) I_{c,co} \tag{11.1}$$

Table 11.1 Summary of instantaneous and creep deflections used in deflection calculations

	Camber due to prestress			*Self-weight*			*Screed*		*Finishes, services, partitions*		*Imposed live load*	
Deflection	*Instant*	*Creep[1]*	*Creep[2]*	*Instant*	*Creep[1]*	*Creep[2]*	*Instant*	*Creep[2]*	*Instant*	*Creep[2]*	*Instant*	*Creep[2]*
Installation	✓	✓		✓	✓							
Short-term	✓	✓		✓	✓		✓		✓		✓	
Long-term final	✓	✓	✓	✓	✓	✓	✓	✓	✓	✓	✓	✓
Movement												
Short-term after finishes									✓[a]		✓	
Long-term after finishes			✓			✓		✓	✓[a]	✓	✓	✓

Creep[1], from transfer to installation; Creep[2], from installation to long-term.

[a] Services and partitions only.

Immediate deflection due to self-weight

$$\delta_2 = 5w_0L^4/384E_{cm}(t)I_{c,co} \tag{11.2}$$

Resultant deflection at transfer $=\delta_1+\delta_2$

b. At installation

The effective creep coefficient ψ_1 between 1 and 28 days takes into account the mean change in E_{cm} as

$$\psi_1 = \left[E_{cm}(t)/0.5\left(E_{cm} + E_{cm}(t)\right)\right]\times(0.4-0.1)\times 2.5 \tag{11.3}$$

Camber at installation is due to the camber at transfer, plus further visco-elastic movement ψ_1 δ_1 minus a (optional) small reduction due to the reduction in pretensioning force from transfer P_{pm0} to installation P_{pmi}. The second term is often ignored as it makes about a 2%–3% difference in δ_3.

$$\delta_3 = (1+\psi_1)\delta_1 + \left(P_{pm0} - P_{pmi}\right)z_{cp}L^2/8E_{cm}I_{c,co} \tag{11.4}$$

Deflection at installation due to self-weight is due to the static deflection plus further visco-elastic movement after transfer

$$\delta_4 = +\ (1+\psi_1)\delta_2 \tag{11.5a}$$

For the case of T beams with infill blocks the additional static deflection for the infill blocks and infill concrete is $5\ w_1L^4/384\ E_{cm}\ I_{c,co}$, then

$$\delta_4 = +\ (1+\psi_1)\delta_2 + 5w_1L^4/384E_{cm}I_{c,co} \tag{11.5b}$$

Resultant deflection at installation $=\delta_3+\delta_4$

c. Long-term deflections

The effective creep coefficient after installation takes into account the increase in strength given by X=0.8, and modulus of elasticity after 28 days

$$\psi_{28} = \left(\psi'_\infty - \psi'_{28}\right)X\,\psi_\infty = (1.0-0.4)\times 0.8\times 2.5 = 1.20 \tag{11.6}$$

If the time to installation t_i is different from 28 days the appropriate coefficient of development ψ' at t_i is used, e.g. at 15 days the creep coefficient after installation is $(1.0-0.3)\times 0.8\times 2.5=1.40$.

Long-term camber is due to camber at installation plus further visco-elastic movement due to the pretensioning force at installation P_{pmi}, minus a (optional) reduction due to the reduction in pretensioning force from installation P_{pmi} to long-term P_{po}

$$\delta_5 = \delta_3 - \left[\psi_{28}P_{pmi} - \left(P_{pmi} - P_{po}\right)\right]z_{cp}L^2/8E_{cm}I_{c,co} \tag{11.7}$$

Long-term deflection due to self-weight and any longitudinal infill, or infill blocks, w_1 is due to the deflection at installation plus visco-elastic movement of w_1 using ψ_{28}

$$\delta_6 = +\delta_4 + \psi_{28}\ 5w_1L^4/384E_{cm}I_{c,co} \tag{11.8}$$

Static plus visco-elastic deflection due to UDL dead w_2 added after installation using ψ_{28}

$$\delta_7 = +(1+\psi_{28})\ 5w_2L^4/\ 384E_{cm}I_{c,co} \tag{11.9}$$

Static plus visco-elastic deflection due to the quasi-permanent UDL live load $\psi_2\ w_3$ using $0.8\ \psi_\infty$

$$\delta_8 = +(1+0.8\psi_\infty)\ 5\psi_2\ w_3L^4/384E_{cm}I_{c,co} \tag{11.10}$$

The total long-term downward deflection is from Eqs. 11.5 and 11.8 to 11.10 as $\delta_9=\delta_4+\delta_6+\delta_7+\delta_8$ and the final long-term deflection $\delta_{10}=\delta_9+\delta_5$

There are similar expressions for imposed point loads and line loads, e.g. point dead load P_G at distance a from the centre of support, where $a<L-a$. If $a>L-a$, replace a with $L-a$

$$\delta_{11} = P_GL^3(1+\psi_{28})\ \left(3a/L-4a^3/L^3\right)/\left(48E_{cm}I_{c,co}\right) \tag{11.11a}$$

At mid-span, $a/L=0.5$ and $\delta_{11}=P_G\ L^3\ (1+\psi_{28})/(48\ E_{cm}\ I_{c,co})$ plus ditto point live load P_Q

$$\delta_{12} = \psi_2P_QL^3(1\ +\psi_\infty)\left(3a/L-4a^3/L^3\right)/\left(48E_{cm}I_{c,co}\right) \tag{11.11b}$$

d. Long-term deflection, or movement, after installation of finishes

Long-term movement due to visco-elastic self-weight and dead load deflections after finishes are applied is due to the post-installation creep coefficient ψ_{28}, static and visco-elastic live load deflections due to $1+\psi_{28}$ (not $1+0.8\psi_\infty$), and changes in camber after installation. It appears from some experimental data on composite hollow core slabs (Elliott and Ibrahim, 2006, Ibrahim and Elliott, 2009) that changes in deflection due to imposed loads applied after about 28 days are closer to the effect of $1+\psi_{28}$ than $1+0.8\psi_\infty$. Movement is the greater of δ_{13} or δ_{14}

$$\delta_{13} = 5\ \left[(1+\psi_{28})\psi_2\ w_3\right]L^4/384E_{cm}I_{c,co} \tag{11.12}$$

or $$\delta_{14} = 5\ \left[\psi_{28}(w_1+w_2)+(1+\psi_{28})\psi_2\ w_3\right]L^4/\ 384E_{cm}I_{c,co}+(\delta_5-\delta_3)$$

i.e. $$\delta_{14} = \delta_{13}+5\left[\psi_{28}(w_1+w_2)\right]L^4/384E_{cm}I_{c,co}+(\delta_5-\delta_3) \tag{11.13}$$

plus deflections due to imposed point dead loads where, over the long-term, in Eq. 11.11a $(1+\psi_{28})$ is replaced by ψ_{28} after installation.

For imposed live point loads in Eq. 11.11b $(1+\psi_\infty)$ is replaced by $(1+\psi_{28})$

$$\delta_{15} = \delta_{11}\psi_{28}(1+\psi_{28}) \tag{11.14a}$$

$$\delta_{16} = \delta_{12}(1+\psi_{28})/(1+\psi_\infty) \tag{11.14b}$$

Summary deflections

7.4.1 (4) Final long-term deflection $=\delta_{10}+\delta_{11}+\delta_{12}\leq L/250$

7.4.1 (5) Variable long-term movement due to live load only $=\delta_{13} \leq L/500$ or $L/350$
7.4.1 (5) Overall long-term movement, due to δ_{13} plus creep of camber and dead loads $=\delta_{14} \leq L/500$ or $L/350$.

Example 11.1

Calculate the long-term deflection and the deflection after applying finishes for the 250 mm depth prestressed hcu shown in cross-section in Figure 6.7 and designed for serviceability in Example 7.1. The slab is subjected to imposed dead and live loads of 3.80 and 5.0 kN/m^2, respectively over a simply supported span of $L=9.80$ m. The self-weight of the precast unit alone is 4.12 kN/m run. After the joints are grouted on site the self-weight of the precast floor is 3.63 kN/m^2, or 4.36 kN/m. The floor slab is for use in offices. Check the Code's limits if the floor finishes are brittle.

Solution

From Example 7.1, considering the prestress at the support (not mid-span) the initial loss at transfer is 6.47% and final losses are 23.8%. Also (but not included in Example 7.1) the losses at installation at 28 days after transfer are 10.5%.

Then $P_{pi}=1239\times911\times10^{-3}=1128.7$ kN
$P_{pm0}=(1-0.0647)\times1128.7=1055.6$ kN
$P_{pmi}=(1-0.105)\times1128.7=1010.2$ kN
$P_{po}=(1-0.238)\times1128.7=860.2$ kN
$L=9800$ mm, $E_{cm}(t)=32837$ N/mm^2, $E_{cm}=36283$ N/mm^2, $z_{cp}=82.2$ mm, $I_{c,co}=1296\times10^6$ mm^4

a. Deflection at transfer
   Eq. 11.1 $\delta_1=-1055.6\times10^3\times82.2\times9800^2/(8\times32837\times1296\times10^6)=-24.5$ mm
   Eq. 11.2 $\delta_2=5\times4.12\times9800^4/(384\times32837\times1296\times10^6)=+11.6$ mm
   Net deflection at transfer $=-12.9$ mm
b. Deflection at installation
   $\psi_\infty=2.5$; ψ'_1 (at 1 day transfer) $=0.1$; ψ'_{28} (at 28 days) $=0.4$
   Eq. 11.3 $\psi_1=[32837/(0.5\times(32837+36283))]\times(0.4-0.1)\times2.5=0.71$
   $P_{pm0}-P_{pmi}=1055.6-1010.2=45.4$ kN
   Eq. 11.4 $\delta_3=1.71\times-24.5+[45.4\times10^3\times82.2\times9800^2/(8\times36283\times1296\times10^6)]=-41.9+1.0=-40.9$ mm
   Eq. 11.5a $\delta_4=1.71\times11.6=+19.8$ mm
   Net(upward)deflectionatinstallation $=-40.9+19.8=-21.1$ mm $<L/250=39.2$ mm.
   This is carried forward to the composite design in Examples 12.1 and 12.5.
c. Long-term deflections
   Eq. 11.6 $\psi_{28}=(1-0.4)\times2.00=1.20$ where $X\psi_\infty=0.8\times2.5=2.00$
   Long-term camber. $P_{pmi}-P_{po}=1010.2-860.2=150.0$ kN
   Eq. 11.7 $\delta_5=-40.9-\{[(1.20\times1010.2\times10^3)-150.0\times10^3]\times82.2\times9800^2/(8\times36283\times1296\times10^6)\}=-40.9-22.3=-63.2$ mm
   Self-weight of slab $w_1=4.12$ kN/m
   Floor imposed dead load per unit width $w_2=1.2\times3.80=4.56$ kN/m
   Offices $\psi_2=0.3$, then the quasi-permanent live load per unit width is $\psi_2 w_3=0.3\times5.00\times1.2=1.80$ kN/m
   Eqs. 11.5a, 11.8 to 11.10 $\delta_9=\delta_4+\delta_6+\delta_7+\delta_8=19.8+[(1.20\times4.36+(1+1.20)\times4.56+(1+2.00)\times1.80]\times[5\times9800^4/(384\times36283\times1296\times10^6)]=19.8+20.66\times2.554=+72.6$ mm
   Final deflection $\delta_{10}=\delta_9+\delta_5=+72.6-63.2=$ **+9.4 mm** $<9800/250=39.2$ mm
d. Long-term movement due to live load only
   Eq. 11.12 $\delta_{13}=(1+1.20)\times1.80\times2.554=+10.1$ mm

e. Long-term movement due to visco-elastic deflections of dead loads and changes in camber, and long-term movement due to live load

$\delta_{14} = 10.1 + [1.20 \times (4.36 + 4.56) \times 2.554] + (-63.2 - -40.9) = 10.1 + 27.3 - 22.3 =$ **+15.1 mm**$> +10.1$ mm $< 9800/500 = 19.6$ mm

Figure 11.4 shows distributions of camber, long-term deflections and movement along the span, together with the appropriate limits. It is usual for movement after installation to be critical, rather than long-term deflections. In this example the ratios of δ_{14}/span/500 = 0.77 and δ_{10}/span/250 = 0.24.

Example 11.2

Calculate the long-term deflection and the deflection after applying finishes for the inverted-tee beam shown in cross-section in Figure 6.10 and designed for serviceability in Example 7.10. The self-weight of the beam is 7.688 kN/m. The self-weight of floor slab = 18.0 kN/m, other imposed dead loads = 10.0 kN/m and imposed live load = 30.0 kN/m. The effective simply supported span is $L = 8.5$ m. The beam is for use in offices. Check the Code's limits if the floor finishes are brittle.

Solution

From Example 7.10, considering the prestress at the support (not mid-span) the initial loss at transfer is 6.56% and final losses are 23.06%. Also (but not included in Example 7.10) the losses at installation at 28 days after transfer are 9.92%.

Then $P_{pi} = 1239 \times 2418 \times 10^{-3} = 2995.9$ kN

$P_{pm0} = (1 - 0.0656) \times 2995.9 = 2799.3$ kN

$P_{pmi} = (1 - 0.0992) \times 2995.9 = 2698.6$ kN

$P_{po} = (1 - 0.2306) \times 2995.9 = 2305.0$ kN

$L = 8500$ mm, $E_{cm}(t) = 32837$ N/mm^2, $E_{cm} = 36283$ N/mm^2, $z_{cp} = 92.9$ mm,

$I_{c,co} = 8449.7 \times 10^6$ mm^4

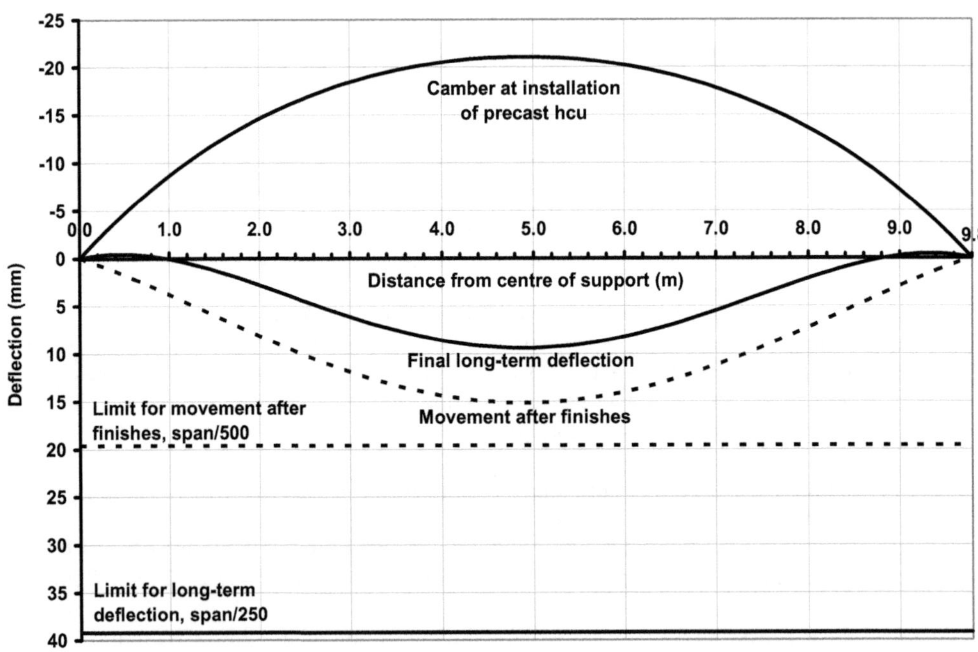

Figure 11.4 Long-term final deflections, movement after installation of finishes and camber at installation of the precast hcu for the design in Example 11.1.

a. Deflection at transfer

Eq. 11.1 δ_1=-2799.3×10^3×92.9×8500^2/(8×32837×8449.7×10^6)=−8.5 mm

Eq. 11.2 δ_2=5×7.688×8500^4/(384×32837×8449.7×10^6)=+1.9 mm

Net deflection at transfer=−6.6 mm

b. Deflection at installation

ψ_∞=2.5; ψ'_1 (at 1 day transfer)=0.1; ψ'_{28} (at 28 days)=0.4

Eq. 11.3 ψ_1=[32837/(0.5×(32837+36283))]×(0.4−0.1)×2.5=0.71

$P_{pm0}-P_{pmi}$=2799.3–2698.6=100.7 kN

Eq. 11.4 δ_3=1.71×−8.5+[100.7×10^3×92.9×8500^2/(8×36283×8449.7×10^6)]
=−14.5+0.3=−14.2 mm

Eq. 11.5a δ_4=1.71×1.9=+3.2 mm

Net (upward) deflection at installation=−11.0 mm<L/250=34.0 mm

c. Long-term deflections

Eq. 11.6 ψ_{28}=(1−0.4)×2.00=1.20 where Xψ_∞=0.8×2.5=2.00

Long-term camber. $P_{pmi}-P_{po}$=2698.6−2305.0=393.6 kN

Eq. 11.7 δ_5=−14.2−[(1.20×2698.6×10^3)−393.6×10^3]×92.9×8500^2/
(8×36283×8449.7×10^6)=−14.2–7.7=–21.9 mm

Self-weight of beam w_1=7.688 kN/m

Imposed dead load w_2=18.0+10.0=28.0 kN/m

Offices ψ_2=0.3, then ψ_2 w_3=0.3×30.0=9.0 kN/m

Eqs. 11.5a, 11.8 to 11.10 $\delta_9=\delta_4+\delta_6+\delta_7+\delta_8$=3.2+[(1.20×7.688+(1+1.20)×28.0
+(1+2.00)×9.0]×[5×8500^4/(384×36283×8449.7×10^6)]=3.2+97.8×0.2217=+
24.9 mm

Final deflection $\delta_{10}=\delta_9=\delta_5$=+24.9–21.9=**+3.0 mm**<8500/250=34.0 mm

d. Long-term movement due to live load only

Eq. 11.12 δ_{13}=(1+1.20)×9.0×0.2217=+4.4 mm

e. Long-term movement due to visco-elastic deflections of dead loads and changes in camber, and long-term movement due to live load

δ_{14}=4.4+[1.20×(7.688+28.0)×0.2217]+(–21.9−−14.2)=4.4+9.5−7.7=**+6.2 mm**>+4.4 mm<8500/500=17.0 mm

It is usual for movement after installation to be critical, rather than long-term-deflections. In this example the ratios of δ_{14}/span/500=0.37 and δ_{10}/span/250=0.09.

It is noticeable that the deflections are much less for this beam compared with those for the hcu in Example 11.1, for example the movement after installation is 6.2 mm compared with 15.1 mm, a ratio of 0.41. The main reasons for this are: $I_{c,co}$ for the beam is 6.5 times that of the hcu; the ratio of spans^2 is 0.75; and the loading ratio is 37.0/10.72 kN/m=3.45. Multiplying these ratios gives 3.45×0.75/6.5=0.40≈0.41.

Example 11.3

Calculate the long-term deflection and the deflection after applying finishes for the pairs of 150 mm depth T beams shown in cross-section in Figure 7.12 (and Figure 5.29) for the designs in Examples 7.5 and 7.7. From Example 7.5, the self-weight (beam, blocks and infill), and the imposed dead and live loads are 0.691, 0.870 and 0.435 kN/m per beam. From Example 7.7 the point dead load for the cross-wall, built onto the screed, is 0.776 kN per beam. The self-weight of the beam is 0.212 kN/m. The effective simply supported span is L=4.345 m. The beam is for use in residential housing. Check the Code's limits if the floor finishes are not brittle.

Solution

From Example 7.5, considering the prestress at the support (not mid-span) the initial loss at transfer is 7.40% and final losses are 27.12%. Also (but not included in Example 7.5) the losses at installation at 28 days after transfer are 12.19%.

Then P_{pi}=1327.5×78.54×10^{-3}=104.3 kN
P_{pm0}=(1−0.0740)×104.3=96.55 kN
P_{pmi}=(1−0.1219)×104.3=91.55 kN
P_{po}=(1−0.2712)×104.3=76.0 kN
L=4345 mm, $E_{cm}(t)$=33346 N/mm², E_{cm}=38214 N/mm², z_{cp}=21.9 mm, $I_{c,co}$=17.34×10^{6}mm^4

a. Deflection at transfer
Eq. 11.1 δ_1=−96.55×10³×21.9×4345²/(8×33346×17.34×10⁶)=−8.6 mm
Eq. 11.2 δ_2=5×0.212×4345⁴/(384×33346×17.34×10⁶)=+1.7 mm
Net deflection at transfer=−6.9 mm

b. Deflection at installation
Eq. 11.3 ψ_1=[33346/(0.5×(33346+38214))]×(0.4−0.1)×2.5=0.70
$P_{pm0}-P_{pmi}$=96.55−91.55=5.00 kN
Eq. 11.4 δ_3=1.70×−8.6+[5.00×10³×21.9×4345²/(8×38214×17.34×10⁶)]=−14.6+0.4=−14.2 mm
Self-weight of blocks and infill between the beams=0.691−0.212=0.479 kN/m. Eq. 11.5b is used for the additional self-weight of the infill blocks and infill concrete between beams.
Eq. 11.5b δ_4=1.70×1.7+[5×0.479×4345⁴/(384×38214×17.34×10⁶)]=2.9+3.3=6.2 mm
Net (upward) deflection at installation=−8.0 mm<L/250=17.4 mm

c. Long-term deflections
Long-term camber. $P_{pmi}-P_{po}$=91.55−76.0=15.55 kN
Eq. 11.7 δ_5=−14.2−[(1.20×91.55×10³)−15.55×10³]×21.9×4345²/(8×38214×17.34×10⁶)=−14.2−7.4=−21.6 mm
Self-weight of beam and blocks w_1=0.691 kN/m
Imposed dead load w_2=0.870 kN/m
Offices ψ_2=0.3, then ψ_2 w_3=0.3×0.435=0.13 kN/m
Eqs. 11.5b, 11.8 to 11.10 $\delta_9=\delta_4+\delta_6+\delta_7+\delta_8$=6.2+[(1.20×0.691+(1+1.20)×0.870+(1+2.00)×0.13]×[5×4345⁴/(384×38214×17.34×10⁶)]=6.2+3.13×7.004=+28.2 mm
The long-term deflection due to the cross-wall dead point load of 0.776 kN is
Eq. 11.11a $\delta_{11}=P_G L^3 (1+\psi_{28})/(48\ E_{cm}\ I_{c,co})$=0.776×10³×(1+1.20)×4345³/(48×38214×17.34×10⁶)=4.4 mm
Final deflection $\delta_{10}=\delta_9+\delta_5+\delta_{11}$=+28.2−21.6 +4.4=**+11.0 mm**<4345/250=17.4 mm

d. Long-term movement due to live load only
Eq. 11.12 δ_{13}=(1+1.20)×0.13×7.004=+2.0 mm

e. Long-term movement due to visco-elastic deflections of dead loads and changes in camber, and long-term movement due to live load
Eq. 11.13 δ_{14}=+2.0+[1.20×(0.691+0.870)×7.004]+(−21.6−−14.2)=2.0+13.1−7.4=+7.7 mm
Long-term movement due to visco-elastic deflection of dead point load
Eq. 11.14a $\delta_{15}=\delta_{11}\times[\psi_{28}/(1+\psi_{28})]$=4.4×1.2/2.2=2.4 mm
Total long-term movement=7.7+2.4=**10.1 mm**<4345/350=12.4 mm

It is usual for movement after installation to be critical, rather than long-term-deflections. In this example the ratios of $\delta_{14}+\delta_{15}$/span/350=0.81 and δ_{10}/span/250=0.63.

11.2.3 Short-term deflections

11.2.3.1 Total short-term deflections

Short-term deflections, typically t_i=28–90 days, or when the full design imposed dead and live loads are applied, should also be calculated. The procedures in Section 11.2 are followed, but the creep development coefficient $\psi'=0.4$ to 0.6 replaces $X\psi_\infty=2.0$, and the quasi-permanent factor ψ_2 is either removed (i.e. $\psi_2=1$) or increased, depending on t_i, although there is no information in BS EN 1990 (BS EN 1990, 2002) on what ψ_2 might be after say 90 days. The maximum value in BS EN 1990, Table NA.A1.1 (NA to BS EN 1990, 2002) is $\psi_2=0.8$ for storage areas, and this seems an appropriate value for any type of short-term loading. Of course at $t_i=28$ days $\psi_{2t}=1.0$.

The deflection at installation is still from Eqs. 11.1 to 11.5 as $\delta_3+\delta_4$

The effective creep coefficient after installation at 28 days up to say 90 days is

$$\psi_{28t}=\left(\psi_t'-\psi_{28}'\right)X\psi_\infty=(0.6-0.4)\times0.8\times2.5=0.40 \tag{11.15}$$

Short-term camber is as per Eq. 11.7, but to time t_i. The small difference in P_{pmi} between installation and t_i may be ignored

$$\delta_{16}=\delta_3-\psi_{28t}\,P_{pmi}z_{cp}L^2/8E_{cm}I_{c,co} \tag{11.16}$$

Short-term deflections δ_{17} and δ_{18} due to self-weight and dead loads, w_1 and w_2 is the same as Eqs. 11.8 and 11.9 but using ψ_{28t}. Deflection due to the quasi-permanent live load $\psi_{2t}\,w_3$ is similar to Eq. 11.10 but using ψ_{28t} and ψ_{2t}

$$\delta_{19}=+\left(1+\psi_{28t}\right)\ 5\psi_{2t}w_3L^4/384E_{cm}I_{c,co} \tag{11.17}$$

The total short-term downward deflection is $\delta_{20}=\delta_4+\delta_{17}+\delta_{18}+\delta_{19}$ and the final short-term deflection $\delta_{21}=\delta_{20}+\delta_{16}$

11.2.3.2 Short-term movement after installation

Short-term movement for self-weight and dead loads is due to ψ_{28t}, quasi-permanent static and visco-elastic live load deflections due to $1+\psi_{28t}$, and changes in camber after installation. Movement is the greater of δ_{22} or δ_{23}

$$\delta_{22}=5\left[\left(1+\psi_{28t}\right)\psi_{2t}w_3\right]L^4/384E_{cm}I_{c,co} \tag{11.18}$$

$$\delta_{23}=\delta_{22}+5\left[\psi_{28t}\left(w_1+w_2\right)\right]L^4/384E_{cm}I_{c,co}+\left(\delta_{16}-\delta_3\right) \tag{11.19}$$

Example 11.4

Repeat Example 11.1 for short-term deflections and movement after 90 days. Let the quasi-permanent factor for office loading $\psi_{2t}=0.8$ at 90 days.

Solution

From Example 11.1, deflection at installation $=19.8-40.9=-21.1$ mm

Eq. 11.15 $\psi_{28t}=(0.6-0.4)\times0.8\times2.5=0.40$

a. Short-term total deflection

Eq. 11.16 $\delta_{16}=-40.9-\ (0.4\times1010.2\times10^3\times82.2\times9800^2/(8\times36283\times1296\times10^6))=-40.9-8.5=-49.4$ mm

Imposed live load is $\psi_{2t}\,w_3=0.8\times1.2\times5.0=4.8$ kN/m

Eqs. 11.5a and 11.17 $\delta_{20}=\delta_4+\delta_{17}+\delta_{18}+\delta_{19}=19.8+[(0.40\times4.36+(1+0.40)\times4.56+(1+0.40)\times4.8]\times[5\times9800^4/(384\times36283\times1296\times10^6)]=19.8+14.85\times2.554=+57.7$ mm

Final deflection $\delta_{21}=\delta_{20}+\delta_{16}=+57.7-49.4=$ **+8.3 mm**$<9800/250=39.2$ mm

b. Short-term movement

Eq. 11.18 $\delta_{22}=(1+0.4)\times 4.8\times 2.554=+17.2$ mm due to live load only

Eq. 11.19 $\delta_{23}=17.2+[0.4\times(4.36+4.56)\times 2.554]+(-49.4-40.9)=17.2+9.1-8.5$ $=$ **+17.8 mm**$>+17.2$ mm$<9800/500=19.6$ mm

For completeness, if the imposed dead and live loads are applied at 28 days (same time as installation), $\psi_{28t}=0$, $\psi_{2t}=1.0$. Then $\delta_{16}=-40.9$ mm, $\delta_{20}=19.8+(4.56+6.0)$ $\times 2.554=+46.8$ mm and final deflection $\delta_{21}=+46.8-40.9=$ **+5.9 mm**. Short-term movement $\delta_{22}=6.0\times 2.554=$ **+15.3 mm**.

11.2.4 Deflections in partially cracked sections

11.2.4.1 Calculation of deflections

Although psc elements designed with the flexural tension limited to f_{ctm} are not flexurally cracked, it is informative to look at the behaviour of the elements when the service moment exceeds the cracking moment of resistance, M_{cr} which is also M_{sR} where the exposure classification is XC1. Deflections beyond M_{cr} are illustrated in Figure 11.1 and test data are shown in Figures 9.3 and 9.5 and the normalised data in Figures 11.2 and 11.3 for hcu and T beams, respectively.

The design procedure is given in BS EN 1992-1-1, clause 7.4.3 (3) and Exp. 7.18 & 7.19. The design is based on the rigidity of a partially cracked section, mid-way between flexurally uncracked and fully cracked. The flexurally uncracked rigidity is $E_{c,eff}\, I_{c,co}$, where $I_{c,co}$ is the compound value used in Section 11.2.2. $E_{c,eff}$ is defined in BS EN 1992-1-1, Exp. 7.20 as $E_{c,eff}=E_{cm}/(1+\varphi(\infty,t_0))$, where $\varphi(\infty,t_0)$ is the appropriate creep factor, but as explained in Section 11.2.2 creep coefficients $\psi=\psi'\, X\psi_\infty$, for defined time intervals in the loading history of psc elements, are used instead. This is because several creep coefficients are appropriate, as shown in Eqs. 11.1 to 11.10 ψ_1, ψ_{28} and $X\psi_\infty$ are used, such that an overall creep coefficient is required. Taking the calculation of deflections for the hcu in Example 11.1, the final long-term deflection due to gravity loads is $\delta_9=72.6$ mm, but as shown in Example 11.5 a single calculation using the quasi-permanent UDL gives an equivalent deflection, without creep effects, of $\delta_{eq}=27.4$ mm. Therefore the equivalent creep coefficient is $\psi_{eq}=\delta_9/\delta_{eq}=(72.6/27.4)-1=1.65$, a reasonable outcome given that the short-term and long-term creep coefficients are 0.71 & 1.20, and 2.0, respectively. The same calculation can be made for camber, based on the final prestress with around 20% losses, but without creep effects, which gives $\psi_{eq}=2.50$. The two values of $(1+\psi_{eq})$ may then be used as constants to calculate deflections for the partially cracked section, between M_{cr} and M_{Rd}.

The fully cracked rigidity is $E_{c,eff}\, I_{cr}$, where I_{cr} is the second moment of area of the cracked section, using the transformed area of the tendons mA_p, where $m=E_p/E_{c,eff}$, where $E_{c,eff}=E_{cm}/(1+\psi_{eq})$ and A_p is the area of tendons, including tendons in the top of the section.

The effective flexural rigidity is given in BS EN 1992-1-1, clause 7.4.3 (3) as

$$\text{Exp. 7.18} \qquad \alpha = \varsigma\alpha_{II} + (1-\varsigma)\alpha_{I} \qquad (11.20)$$

where α is a deformation parameter which may be considered for curvature, or rotation, but in this case may be taken as the flexural rigidity of the element for the purpose of deflections.

ζ is a tension stiffening coefficient, recognising that the section does not become instantly fully cracked, given as

$$\text{Exp. 7.19} \qquad \zeta=1-\beta\,(M_{cr}/M_k)^2 \qquad (11.21)$$

α_I and α_{II} are the uncracked and fully cracked flexural rigidities which are renamed K_1 and K_2, respectively

β=a coefficient for the duration of the load, as
β=1.0 for short-term loading
β=0.5 for long-term, sustained or repeated loading
M_k=service bending moment

$$M_{cr} = Z_{b,co}\left(\sigma_b + f_{ctm}\right) \tag{11.22}$$

$Z_{b,co}$=uncracked compound section modulus (see Section 6.2.4)
Eqs. 11.20 and 11.21 are more easily understood if they are rearranged as

$$K_{eff} = K_2 + \left[(K_1 - K_2)\beta(M_{cr}/M_k)^2\right] \le K_1 \tag{11.23}$$

i.e. K_{eff} is equal to the fully cracked rigidity K_2 plus a portion of the difference $K_1 - K_2$
where uncracked flexural rigidity $K_1 = E_{c,eff}\, I_{c,co}$

fully cracked flexural rigidity $K_2 = E_{c,eff}\, I_{cr}$
The long-term camber, where $M_k > M_{cr}$ is also a function of K_{eff} assuming that, after different creep effects, $E_{c,eff}$ may not have the same value for the concrete in the bottom of the element which is under compression due to prestress.

The increased camber, due to the final pretensioning force P_{po} is

$$\delta_{17} = -P_{po} z_{cp} L^2 / 8K_{eff} \tag{11.24}$$

The deflection due to the quasi-permanent gravity UDL wqp is

$$\delta_{18} = +5w_{qp} L^4 / 384K_{eff} \tag{11.25}$$

The final net deflection is $\delta_{19} = \delta_{17} + \delta_{18}$ (11.26)

11.2.4.2 Calculation of flexurally cracked second moment of area

For the fully flexurally cracked section, the calculation of the second moment of area I_{cr} requires the transformed area of tendons=mA_p and the width of the section=b.

Referring to Figure 11.5, the depth to centroidal axis x_c is given as the 1st moment of area about the top of the section $bx_c^2/2+mA_pd$, divided by the area of the section bx_c+mA_p as

$$x_c = bx_c^2 / 2 + mA_pd / \left(bx_c + mA_p\right) \tag{11.27}$$

Giving the quadratic equation $0.5b\, x_c^2 + mA_px_c - mA_pd = 0$ (11.28)

If x_c is within the top flange, i.e. $x_c \le h_f$ the second moment of area about the centroidal axis is

$$I_{cr} = bx_c^3 / 3 + mA_p\left(d - x_c\right)^2 \tag{11.29}$$

Otherwise if $x_c > h_f$ restart the calculation by splitting the section into the webs $b_w x_c$ and remainder of the top flange $(b - b_w)\, h_f$ as shown in Figure 11.6

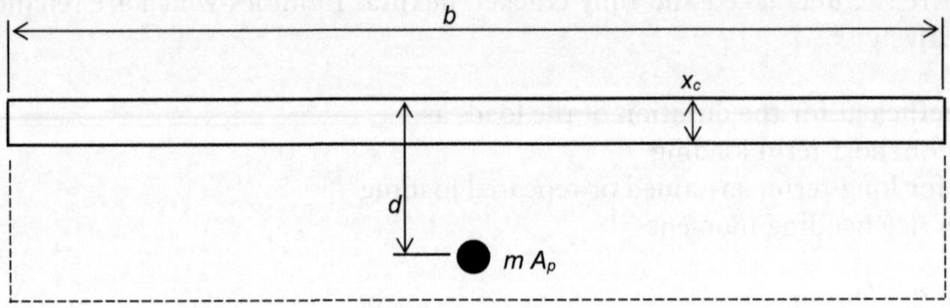

Figure 11.5 Idealised cross-section for the flexurally cracked rectangular section.

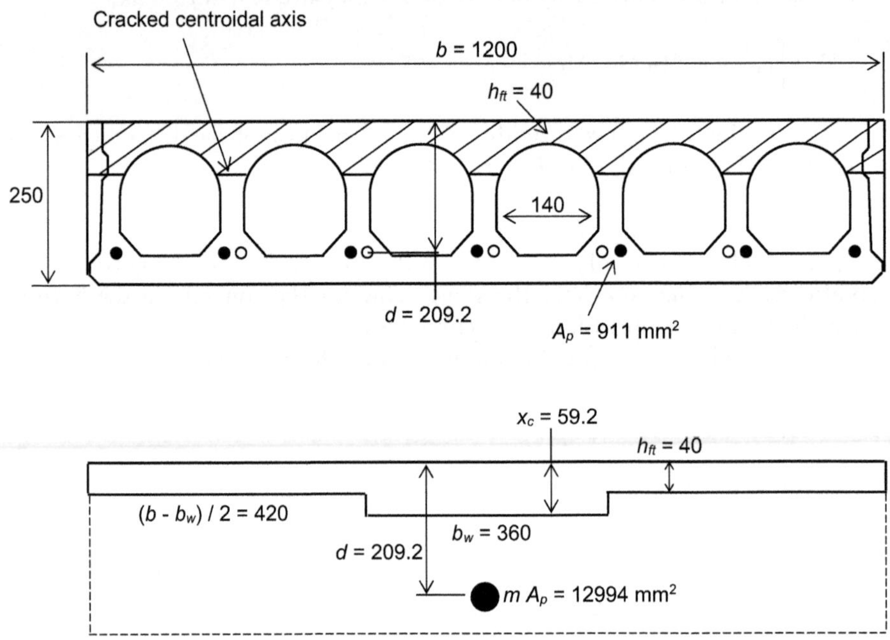

Figure 11.6 Actual and idealised cross-section for the flexurally cracked hcu for the design in Example 11.5.

$$x_c = b_w x_c^2 / 2 + (b - b_w) h_f^2 / 2 + mA_p d / (b_w x_c + (b - b_w) h_f + mA_p) \quad (11.30)$$

Giving $$0.5 b_w x_c^2 + [mA_p + (b - b_w) h_f] x_c - [mA_p d + (b - b_w) h_f^2 / 2] = 0 \quad (11.31)$$

Then $$I_{cr} = b x_c^3 / 3 - (b - b_w)(x_c - h_f)^3 / 3 + mA_p (d - x_c)^2 \quad (11.32)$$

Example 11.5

Calculate the post-cracking long-term deflections for the 250 mm depth prestressed hcu shown in cross-section in Figure 6.7 and designed in Example 11.1. Use the final long-term deflections due to gravity loads and camber at the cracking moment of resistance M_{cr} and the equivalent deflection without creep effects δ_{eq} to determine the equivalent creep factors ψ_{eq}. Plot a graph of the deflections from M_{cr} to M_{Rd} (from Example 9.1), and for completeness from the installation of the precast hcu to M_{cr} using the values in Example 11.1. For the uncracked rigidity use $I_{c.co}=1296\times10^6$mm^4. For the cracked rigidity $A_p=911$ mm^2, $b=1200$ mm, $b_w=360$ mm and $d=209.2$ mm. For the camber, $P_{po}=860.2$ kN and $z_{cp}=82.2$ mm.

Solution

Final long-term gravity load deflection at M_{cr}, $\delta_9=72.6$ mm

Quasi-permanent UDL $w_{qp}=1.2\times(3.63+3.8+0.3\times5.0)=10.72$ kN/m

$\delta_{eq}=5\times10.72\times9800^4/(384\times36283\times1296\times10^6)=27.4$ mm.

$\psi_{eq}=\delta_9/\delta_{eq}=(72.6/27.4)-1=1.654$

Equivalent long-term $E_{c,eff}=36283/(1+1.654)=13670$ N/mm^2

Modular ratio $m_{eq}=195000/13670=14.264$

Eq. 11.28 $\quad 0.5\times1200x_c^2+14.264\times911x_c-14.264\times911\times209.2=0$

$600x_c^2+12995x_c-2718628=0$

Then $x_c=57.3$ mm$>h_f=40$ mm, restart with $b-b_w=840$ mm

Eq. 11.31 $\quad 0.5\times360x_c^2+\left[14.264\times911+840\times40\right]x_c-\left[14.264\times911\times209.2+840\times40^2/2\right]=0$

$180\,x_c^2+46595\,x_c-390628=0$

Then $x_c=59.2$ mm

Eq. 11.32 $\quad I_{cr}=1200\times59.2^3/3-840\times19.2^3/3+14.264\times911\times150^2=373.4\times10^6$mm^4

Then $K_1=13670\times1296\times10^6=17.72\times10^{12}$ Nmm^2

and $K_2=13670\times373.4\times10^6=5.105\times10^{12}$ Nmm^2

Where the imposed service moment $M_k>M_{cr}$ the effective rigidity is

Eq. 11.23 $\quad K_{eff}=5.105\times10^{12}+0.5\times12.615\times10^{12}\times(179.1/M_k)^2$

Let $M_k=179.1$ kNm, then $K_{eff}=[5.105+(6.307\times1.000^2)]\times10^{12}=11.412\times10^{12}$ Nmm^2

The deflection due to the quasi-permanent gravity load is

Eq. 11.25 $\delta_{18}=5\times10.72\times9800^4/384\times11.412\times10^{12}=+112.8$ mm,

The effect of β and K_{eff} on camber has to be considered as follows

Final long-term camber after final losses, $\delta_8=-63.2$ mm

Camber due to initial pretensioning force after initial losses

$\delta_{eq}=860.2\times10^3\times82.2\times9800^2/(8\times36283\times1296\times10^6)=-18.1$ mm.

$\psi_{eq}=(63.2/18.1)-1=2.50$

Note that ψ_{eq} is greater than expected for creep factors alone, that are $1.71\times1.20=2.05$, because of the 23% loss of prestress.

$E_{c,eff}=36283/(1+2.50)=10354$ N/mm^2

and $K_{eff}=(10354/13670)\times11.412=8.643\times10^{12}$ Nmm^2

Eq. 11.24 $\quad \delta_{17}=-860.2\times10^3\times82.2\times9800^2/8\times8.643\times10^{12}=-98.2$ mm

Eq. 11.26 $\quad \delta_{19}=112.8-98.2=14.6$ mm

an increase of 5.2 mm, or 1.55 times the deflection in the uncracked section. Of course, this does not happen in reality, as shown in Figure 9.3 for 200 mm depth hcu and is due entirely to the β factor.

Let $M_k = M_{Rd} = 247.3$ kNm, then $M_{cr}/M_k = 0.724$

$$K_{eff} = [5.105 + (6.307 \times 0.724^2)] \times 10^{12} = 8.413 \times 10^{12} \text{ Nmm}^2$$

The UDL for a moment of 247.3 kNm is $w_s = 8 \times 247.3/9.8^2 = 20.60$ kN/m, of which 8.92 kN/m is dead load.

Then $w_{qp} = 8.92 + [0.3 \times (20.60 - 8.92)] = 12.42$ kN/m
Eq. 11.25 $\delta_{18} = 5 \times 12.42 \times 9800^4/384 \times 8.413 \times 10^{12} = +177.4$ mm

For camber, $K_{eff} = (10354/13670) \times 8.413 = 6.372 \times 10^{12}$ Nmm²

Eq. 11.24 $\delta_{17} = -860.2 \times 10^3 \times 82.2 \times 9800^2/8 \times 6.372 \times 10^{12} = -133.2$ mm
Eq. 11.26 $\delta_{19} = 177.4 - 133.2 =$ **44.2 mm**

11.2.4.3 Discussion of deflections and deflection profiles

Figure 11.7 shows the results from the deflection calculations in Example 11.1, up to $\delta_{10} = +9.4$ mm, and from Example 11.5 from 9.4 mm to $\delta_{19} = 44.2$ mm. The gradient of the graph from 14.6 to 44.2 mm is slightly parabolic, due to the term $\beta\ (M_{cr}/M_k)^2$, but not as much as is seen in Figure 9.3 from full scale tests, although this was short-term loading without changes in camber. Figure 11.7 therefore suggests a probable curved profile for 'actual' post-cracking deflections. Interestingly if $\beta = 1$ and $(M_{cr}/M_k)^3$ the calculated deflections follow the curved profile accurately, suggesting that for hcu, where cracking propagates through the region of the narrow webs much quicker than for a solid rectangular section, the power term in BS EN 1992-1-1, Exp. 7.19 might be revised to 3 for hcu.

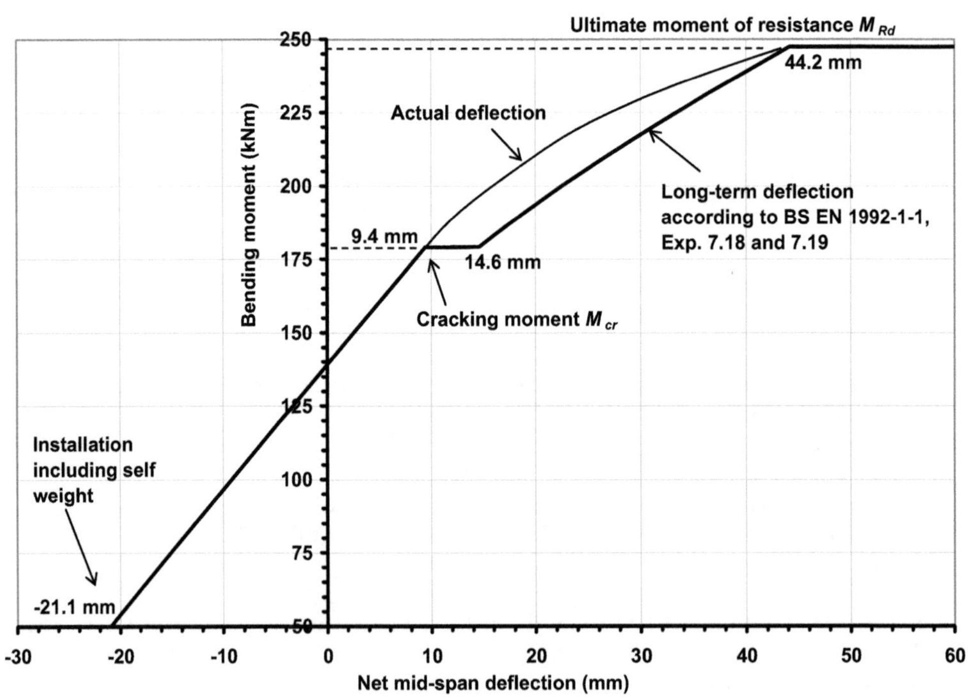

Figure 11.7 Service bending moment vs deflection for the design in Examples 11.1 and 11.5.

11.3 CRACK WIDTHS

11.3.1 Calculation of crack width

Prestressed concrete elements designed with the limiting tensile stress zero or f_{ctm} should not, by definition, require a check for crack width, and this has been the practice for many years. Unacceptable cracking may be avoided if the level of prestress does not exceed 0.75 f_{pk} in service, typically 1330 N/mm^2, which is uncommon and means that the initial prestress needs to be as high as 0.85 f_{pk}.

BS EN 1992-1-1, clause 7.1 (2) states that for the purpose of calculating crack widths, service stresses should be for the flexurally uncracked section and the flexural tensile stress $\leq f_{ctm}$. Further, clause 7.3.2 (4) states that where crack control is necessary no minimum reinforcement is required providing that, under characteristic loading, the tensile stress is less than f_{ctm}. This is always the case as the only time characteristic loading is not used, the limiting tensile stress is zero for exposure classes greater than XC1, according to BS EN 1992-1-1, Table 7.1N, replaced by the UK National Annex to BS EN 1992-1-1, Table NA.4 (NA to BS EN 1992-1-1, 2004). Table 11.2 shows the part of Table 7.1N applicable to psc with bonded tendons. Note that Table 7.1N has not been accepted in all European countries, particularly for zero tension for exposure classes XC2 to XC4 where a limiting tensile stress of $f_{ctm}/1.2$ is used for example. Although Table 7.1N suggests that crack widths are required by calculation, and should be ≤0.2 mm under the frequent load combination, it is not common practice to calculate crack widths. Two alternatives are therefore possible

1. under the frequent combination of load, the tensile stress is maintained as ≤1.65 f_{ctm}. This is proposed in Technical Report 43 (Concrete Society, 2005) albeit on post-tensioned concrete slabs. It is also used in the Eurocode design guide by the Precast Concrete Federation (British Precast, 2008) for replacing crack width calculations for the frequent load.
2. crack spacing and crack width calculations are undertaken according to BS EN 1992-1-1, clause 7.3.4.

The design crack width w_k is the product of crack spacing $s_{r,max}$ and effective tensile strain between the steel bars ε_{sm} and the concrete ε_{cm}, allowing for tension stiffening, as

Exp. 7.8 $$w_k = s_{r,max}\left(\varepsilon_{sm} - \varepsilon_{cm}\right) \quad (11.33)$$

Table 11.2 Maximum crack width (mm) for prestressed concrete elements with bonded tendons (extract from UK NA to BS EN 1992-1-1, Table NA.4)

Class designation	*Frequent load combination = 1.0 $G_k + \psi_1 Q_k$*	*Quasi-permanent load = 1.0 $G_k + \psi_2 Q_k$*
XC1	0.2	-
XC2 XC3 XC4	0.2	Decompression i.e. zero tension in soffit[a]
XD1/XS1 XD2/XS2 XS3	Decompression i.e. zero tension in soffit[a]	-
XD3	Depends on nature of aggressive agent	

[a] All wires lie at least 25 mm within concrete in compression, i.e. cover from soffit ≥ 25 mm.

Note that in calculating ε_{sm} only the additional strain, not the total strain, beyond the state of zero strain in the concrete, i.e. for psc this means after decompression.

The net strain is given as

Exp. 7.9 $$\varepsilon_{sm} - \varepsilon_{cm} = \left[\sigma_s - k_t f_{ct,eff}\left(1+\alpha_e \rho_{eff}\right)/\rho_{eff}\right]/E_p \geq 0.6\sigma_s/E_p \quad (11.34a)$$

For psc, the steel stress σ_s may be replaced by $\Delta\sigma_p$, the stress variation in the tendons from the state of zero strain in the concrete at the level of tendons nearest to soffit (Figure 11.8).

7.3.2 (2) $f_{ct,eff}=f_{ctm}$ or lower $f_{ctm}(t)$ if cracking is expected within 28 days. For long-term loading where $k_t=0.4$ Exp. 7.9 is rewritten as

Exp. 7.9 $$\varepsilon_{sm} - \varepsilon_{cm} = \left[\Delta\sigma_p - 0.4 f_{ctm}\left(1 + \alpha_e \rho_{eff}\right)/\rho_{eff}\right]/E_p \geq 0.6\Delta\sigma_p/E_p \quad (11.34b)$$

If the limiting stress in the bottom of the section is f_{ctm}, then de facto the applied bending moment is M_s. $\Delta\sigma_p$ is calculated from Figure 11.8 as

$$\Delta\sigma_p = \left[(h-x-a)/(h-x)\right] f_{ctm}\alpha_e \quad (11.35)$$

where a=axis distance=minimum cover+tendon diameter

x=depth to zero stress after decompression

$$x = \left[\left(M_s/Z_{t,co} + \sigma_t\right)/\left(M_s/Z_{t,co} + \sigma_t + f_{ctm}\right)\right]h \quad (11.36)$$

where f_{ctm} is a positive value, e.g. +3.80 N/mm^2 for f_{ck}=45 N/mm^2

$\alpha_e = E_p/E_{cm}$

$\rho_{eff} = A_p/A_{c,eff}$

7.3.2 (3) $A_{c,eff}$=area in tension zone to a height of min{$(h-x)/3$; $2.5\,(h-d)$; $0.5\,h$}

For the majority of prestressed sections that are critical where M_{sr} is based on f_{ctm}, x is very close to the bottom of the section, and if the bending stress was calculated using the frequent load decompression would not be attained and hence $\Delta\sigma_p$ would be zero.

The design crack spacing $s_{r,max}$ is given as

Exp. 7.11 $$s_{r,max} = k_3\, C + k_1\, k_2\, k_4\, \Phi/\rho_{eff} \quad (11.37)$$

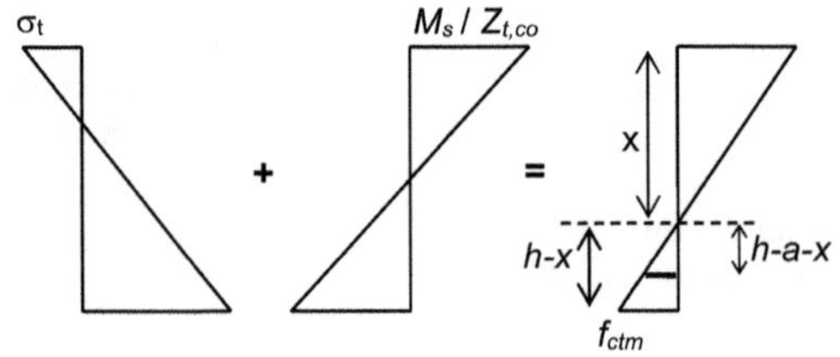

Figure 11.8 State of stress used in the calculation of $\Delta\sigma p$ and crack width in prestressed concrete.

where $k_1=1.6$, $k_2=0.5$, $k_3=3.4$, $k_4=0.425$

Φ=mean tendon diameter based on area
C=minimum cover

Finally, using clause 7.3.4 (3) check that tendon spacing<5 $(C+\phi/2)$ (11.38)

Otherwise, Exp. 7.14 $s_{r,max}=1.3\ (h-x)$ (11.39)

Example 11.6

Calculate the tensile stress in tendons after decompression $\Delta\sigma_p$ for the 250 mm depth hcu designed in Example 7.1, and check whether a crack width calculation is required, and if so calculate the crack width. Cover=35 mm to average diameter of tendons=11.17 mm. Use $\psi_1=0.5$ for office loading.

Solution

$M_s/Z_{t,co}=179.1/10.056=17.81$ N/mm².
Final prestress in top (from Example 7.1) is $\sigma_t=-2.11$ N/mm².

Eq. 11.36 $x=[(17.81-2.11)/(17.81-2.11+3.80)]\times 250=201.3$ mm

Mean axis distance to strands $a=35+11.17/2=40.6$ mm

$\alpha_e=195000/36283=5.37$

Eq. 11.35 $\Delta\sigma_p=[(250-201.3-40.6)/(250-201.3)]\times 3.80\times 5.37=3.39$ N/mm² (tension).
A very small stress as expected.

$A_{ct,eff}=1197\times(250-201.3)/3=19431$ mm²
$\rho_{eff}=A_p/A_{ct,eff}=911/19431=0.0469$
$(1+\alpha_e\ \rho_{eff})=1+5.37\times 0.0469=1.252$

Eq. 11.34b $\varepsilon_{sm}-\varepsilon_{cm}=[3.39\text{-}0.4\times 3.80\times 1.252/0.0469]/195000=-190.7\times 10^{-6}$
But at least $\varepsilon_{sm}-\varepsilon_{cm}=0.6\times 3.39/195000=10.41\times 10^{-6}$
Eq. 11.38 Distance between strands=180 mm<5×(35+11.17/2)
Eq. 11.37 $s=3.4\times 35+(1.6\times 0.5\times 0.425\times 11.17/0.0469)=200$ mm
Eq. 11.33 $w_k=10.41\times 10^{-6}\times 200=$ **0.002 mm**<0.2 mm
As expected these equations are trying to say that w_k is zero as the tensile stress is limited to f_{ctm}.

REFERENCES

ASSAP. 2002. *The Hollow Core Floor Design and Applications, Manual 1st Edition*, Association of Manufacturers of Prestressed Hollow Core Floors (ASSAP), Verona, Italy.

BS EN 1990. 2002. Eurocode 0- Basis of structural design, +A1:2005, BSI, London, UK.

BS EN 1992-1-1. 2004. Eurocode 2: Design of concrete structures - Part 1-1: General rules and rules for buildings, +A1:2014, BSI, London, UK.

BS 8110-1. 1997. British Standards Institute, BS 8110, Structural Use of Concrete - Part 1: Code of practice for design and construction, BSI, London, UK.

BS 8110-2. 1985. British Standards Institute, BS 8110, Structural Use of Concrete - Part 2: Code of practice for special circumstances, BSI, London, UK.

British Precast. 2008. *Precast Eurocode 2: Worked Examples*, British Precast Concrete Federation, Leicester, UK, 77 p.

Concrete Society. 2005. *Technical Report 43, Post-Tensioned Concrete Floors - Design Handbook*, 2nd ed., The Concrete Society, Camberley, UK.

CP 110. 1972. British Standards Institute, CP 110, The Structural Use of Concrete, BSI, London, UK.

Elliott, K. S. and Ibrahim, I. 2006. Composite construction in prestressed precast concrete floors, CONCET 2006, 9th International Conference on Concrete Engineering, Malaysia/Kuala Lumpur. Institution of Engineers Malaysia, pp. 104–116.

Ibrahim, I. and Elliott, K. S. 2009. Long-term behaviour of precast prestressed hollow core slabs with concrete toppings, 7th Asia Pacific Structural Engineering and Construction Conference (APSEC) and 2nd Asian European Civil Engineering Forum (EACEF), Kuala Lumpur, Malaysia.

NA to BS EN 1990. 2002. UK National Annex for Eurocode 0- Basis of structural design, +A1:2005, BSI, London, UK.

NA to BS EN 1992-1-1. 2004. UK National Annex for Eurocode 2: Design of concrete structures - Part 1-1: General rules and rules for buildings, +A1:2014, BSI, London, UK.

Chapter 12

Composite slabs and beams

12.1 BACKGROUND TO THE USE OF COMPOSITE SLABS AND BEAMS

12.1.1 Composite floor slabs

Non-composite precast prestressed concrete (psc) floor units, such as hollow core units (hcu) and solid plank have very efficient structural properties, in terms of moment of resistance per unit area of tendons, easily twice as efficient as reinforcing bars (although the relative cost of strands vs rebars has to be taken into account as shown in Section 2.1, Figure 2.10) and greater span/depth ratios where deflections are critical, at least 1.3–1.6 times greater than a corresponding statically reinforced slab with the same material properties and quantities. The units can also be tied together with tie steel, for example long lengths of unstressed 7 wire helical strand(s), and site bars cast into to the top of broken out hollow cores, to facilitate robustness in the event of accidental loading – this design and construction is given in the FIB Bulletin 63 (FIB, 2012) and covered by Elliott's Chapter 11 (Elliott, 2017). However, in many countries, particularly South East Asia and in earthquake-affected zones, the preferred choice is to design and construct all precast reinforced and psc floors as composite slabs, i.e. a prestressed unit made to behave as a monolithic slab by the addition of a reinforced structural topping. (Note the terminology 'topping' is different to a 'screed' which is considered as a non-composite concrete layer.) Figure 2.4 shows the preparation for a structural topping on precast hcu – a steel mesh, typically A142 (6 mm rebars at 200 mm spacing in two directions) is positioned at mid-depth of the topping, with loose tie bars between the sheets and around the edges, then the precast surface is dampened, not ponded, prior to pouring *insitu* concrete Topping, typically f_{ckT}=20 to 30 N/mm^2, with f_{ckT}=25 N/mm^2 the most common. Figure 12.1 shows power floating the structural topping in the same project – the extra effort over tamping the surface may eliminate the need for additional finishes, saving costs as well as about 1.5–1.8 kN/m^2 in self-weight.

The main benefit from composite action is to increase the bending resistance and flexural rigidity – shear and bearing resistance is barely increased. There are however a number of other reasons why a structural topping may be specified, such as

i. to improve vibration, thermal and acoustic performance of the floor
ii. to provide horizontal diaphragm action
iii. to provide horizontal stability ties in the form of steel mesh and rebars, or steel fibre reinforcement, across floors
iv. to provide a continuous and monolithic floor finish (e.g. where brittle finishes are applied)

DOI: 10.1201/9781003319450-12

Figure 12.1 Power floating a structural topping, that may avoid adding further levelling finishes.

Solid slabs and hcu may be used with or without a structural topping because the individual floor units are keyed together over the full surface area of their edges to form a floor 'slab'. This may not be the case with all the other types of floor units, such as double tee (Figure 2.19) or multi-rib units (Figure 2.21) where a structural topping (i.e. containing adequate reinforcement) must be used either for horizontal load transfer, flexural and shear strength, or simply to complete the construction.

Precast concrete units are designed compositely with a structural topping of at least 50 mm depth at the 'crown' – the highest point at mid-span in a pre-cambered unit, up to 100 mm depth. There is some debate about whether depths of up to 150 mm are suitable, due to the large restraint to thermal gradients particularly at the roof level. Toppings of more than 100 mm are counterproductive to the service moment of resistance as the additional self-weight is consumed by the larger bending stresses because the rising centroidal axis causes a reduction in $Z_{b,c,co}$, the composite compound section modulus at the bottom fibre (see Section 12.3.2.1). Double tee and thermal units are designed compositely with a structural topping, in which case the top flange depth is 50–75 mm. Solid psc units, often known as 'planks' of depths 75–200 mm and spanning 3–8 m, as shown in Figure 12.2, are popular in South East Asia where the structural topping provides wet room facilities with embedded water and/or service pipes. The common name for these is 'half-slab'. T beams (in beam and block flooring) are not generally designed compositely, although Figure 2.30c shows a successful detail in spite of the absence of a mechanical connection (e.g. dowels post-fixed into the top of the beams) should the interface at the top of the beam fail in horizontal shear.

12.2 FLEXURAL AND SHEAR BEHAVIOUR OF COMPOSITE PRESTRESSED SOLID SLABS

12.2.1 Flexural tests on composite prestressed slabs

Six full scale flexural tests were carried out on 1200 mm width × 150 mm depth composite psc solid slabs. The tests aimed to demonstrate that the failure of the composite slab is not

Figure 12.2 Prestressed concrete composite plank, often known as 'half-slab'.

due to bond failure at the interface between the precast unit and the topping, i.e. failure is by ultimate bending or large deflections.

Figure 12.3a and b shows the results of the applied bending moment vs. mid-span deflection. The composite slabs comprised 100 and 75 mm depth psc solid units with nominally 50 and 75 mm *insitu* concrete structural topping, respectively. The average cube strength of the topping was 57.5 N/mm^2, which converts to f_{ckT}=46 N/mm^2. The topping was reinforced using A142 steel mesh at 25 mm cover. The slabs were subjected to four point loads at ¼ span spacing to produce constant bending moment near to mid-span over simply supported spans of 5.0 and 3.75 m, respectively. The precast units were pretensioned using 33 and 28×5 mm wires, respectively to produce prestress in the bottom fibre of $\sigma_b \approx 9.5$ N/mm^2. The top surface of the precast units was left as slipformed, i.e. not roughened, and was dampened 30 minutes before casting the *insitu* topping using 2.5–3.0 L of water per sq. m.

The results show, as with the non-composite hcu tests in Figures 7.1 and 9.3, the behaviour of composite psc slabs in bending is predictable and consistent. The calculated (i.e. predicted) ultimate $M_{Rd,c,calc}$ (note suffix *c* for composite) and cracking $M_{cr,c,calc}$ moments of resistance are compared with the corresponding experimental values $M_{u,c}$ and $M_{c,c}$ in Table 12.1, together with the design $M_{Rd,c}$ based on nominal dimensions and material properties. The ratio of $M_{u,c}/M_{Rd,c,calc}$ varies from 1.18 to 1.35 with a mean value of 1.25, a result that is about 15% greater than for the non-composite hcu (Table 9.1) most probably because at failure the top flange is solid and uninterrupted by the hollow cores. The ratio at first cracking $M_{c,c}/M_{cr,c,calc}$ varies depending on the depth of the topping – the average for the 100+50 mm slabs was 1.24 compared with 1.52 for the 75+75 mm slabs. $M_{c,c}$ in the tests varies only by ±5%, whereas $M_{cr,c,calc}$ for the 75+75 mm slab was 20% lower because of the much larger self-weight stress acting on the shallower precast unit. The average ratio of the test/design $M_{u,c}/M_{Rd,c}$ is 1.77, greater than the maximum possible load factor γ_f of 1.5, showing how much the composite slabs have in reserve.

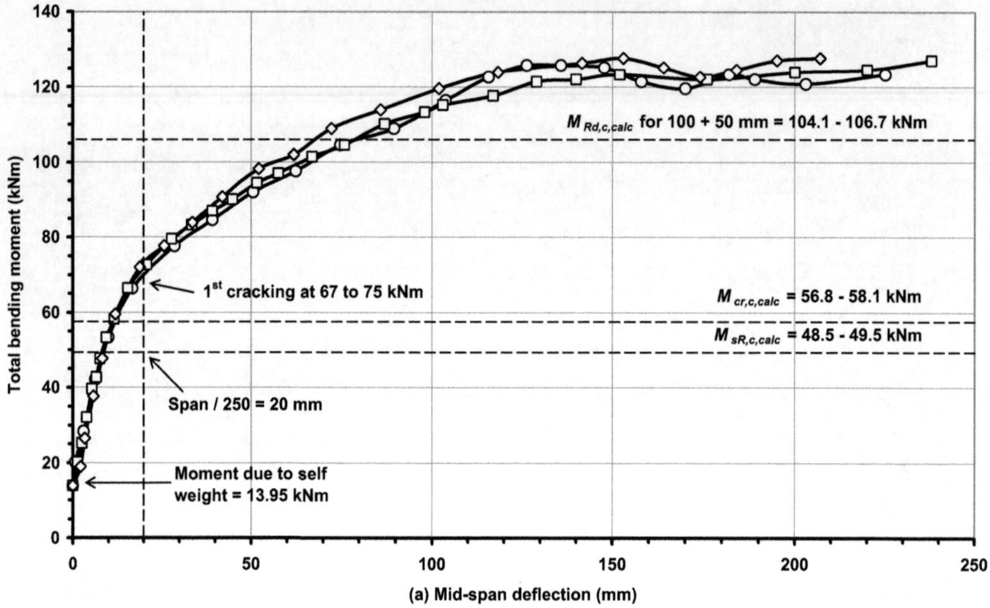

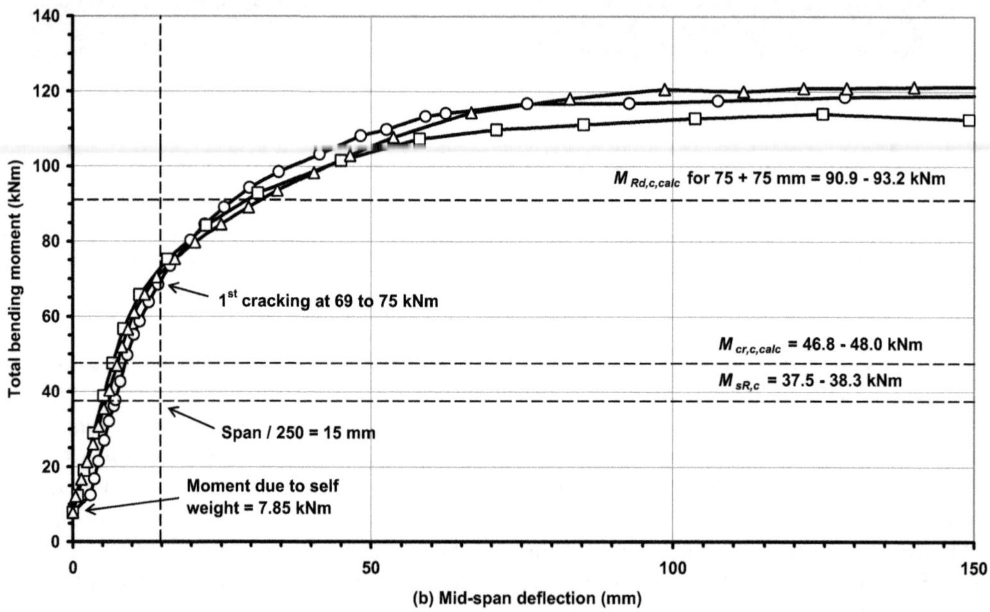

Figure 12.3 (a) and (b). Experimental and predicted service, cracking and ultimate moments of resistance vs mid-span deflection in full scale tests carried out on 1200 × 150 mm depth composite prestressed solid slabs. (a) 100 mm depth precast with 50 mm structural topping, (b) 75 mm depth precast with 75 mm structural topping. (With permission by Bison Precast Ltd., UK.)

Figure 12.4 shows an important feature for the position and direction of flexural cracks, which pass through the interface at mid-depth and develop into compressive-splitting cracks with no interface shear failure.

The flexural stiffness, i.e. mid-span moment/deflection, M/δ_c, where δ_c is the deflection at first cracking, agrees with calculated values for the 100+50 mm slabs, where the mean

Table 12.1 Comparison of the experimental and calculated ultimate moments of resistance $M_{Rd,c,calc}$ and cracking moments $M_{cr,c,calc,}$ and design $M_{Rd,c}$ in full scale flexural tests on composite in full scale tests carried out on 1200 × 150 mm depth composite prestressed solid slabs.

Test ref.	*Depth of precast[a] (mm)*	*Depth of topping (mm)*	*Moment at first cracking $M_{c,c}$ (kNm)*	*Ultimate test moment $M_{u,c}$ (kNm)*	*Ratio $M_{c,c}/M_{cr,,c,calc}$*	*Ratio $M_{u,c}/M_{Rd,calc}$*	*Ratio to design $M_{u,c}/M_{Rd,c}$*
100A	100	53	70.1	125.7	1.21	1.18	1.65
100B		54	67.0	127.0	1.18	1.22	1.66
100C		53	75.1	127.6	1.32	1.23	1.67
75A	75	77	75.4	118.9	1.57	1.28	1.87
75B		74	68.8	114.0	1.47	1.25	1.79
75C		74	71.1	123.2	1.52	1.35	1.94
Mean					1.38	1.25	1.77

The design $M_{Rd,c}$ is based on nominal cross section 1200×150 mm and design strengths f_{ck}=45 N/mm^2 for the precast and f_{ckT}=25 N/mm^2 for the *insitu* topping. Precast units are pretensioned using 5 mm diameter wires, p_{yk}= 1770 N/mm^2 initially stressed to 70%. 75 mm depth units have 18 and 10 no. at 20 and 40 mm cover, and 100 mm depth units have the same plus 5 no. at 60 mm cover.

[a] Nominal depth, was not measured.

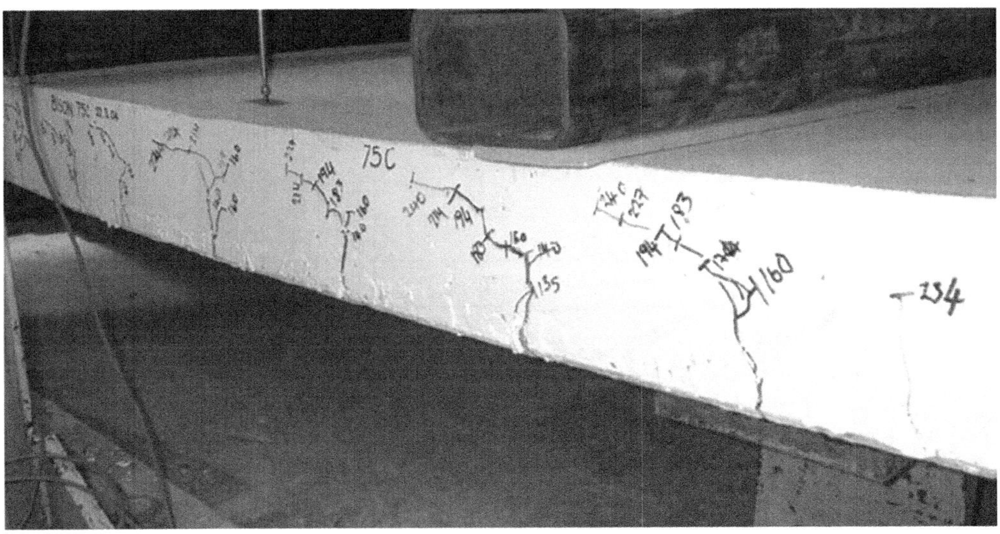

Figure 12.4 Flexural cracking in a composite 75 mm depth prestressed solid slab with 75 mm structural topping. The applied bending moment is 121 kNm, or 1.35 times the calculated $M_{Rd,c,calc}$. Note that the cracks pass through the interface at mid-depth and develop into compressive-splitting cracks with no interface shear failure. (With permission by Bison Precast Ltd., UK.)

experimental and calculated values are 4.15 vs 4.30 kNm/mm, respectively, but less so for the 75+75 mm slabs of 5.60 vs 7.50 kNm/mm, respectively. It is not known why the experimental stiffness of the latter is 26% lower than predicted, as the span/depth ratio of 3750/150=25 in not particularly small. The horizontal interface deformation between the precast unit and the topping at the ends of the slab Δ_s is less than 0.2 mm at the first cracking load, compared with Δ_s<0.12 mm for the 100+50 mm slab, but in all cases, there was no slippage at the cracking load.

Overall the results show good consistency, with the one exception as above, with confident predictions of the cracking and ultimate moments of resistance.

12.2.2 Shear tests on composite prestressed slabs

Seven shear tests were also carried out on similar specimens (4 no. 100 mm precast plus 50 mm topping, and 3 no. 75 plus 75 mm) in an attempt to produce horizontal interface shear failure (see Section 12.4). The average cube strength of the topping was 60.7 N/mm², which converts to f_{ckT}=50.5 N/mm². The slabs were subjected to a single transverse point load at a distance of 2.5h=375 mm, where h is the total depth of the composite slab. In all tests the slabs failed in flexure beneath the point load, and not in shear. As shown in Figure 12.5 horizontal interface shear failure occurred in two out of three 75+75 mm slabs, where the average total bending moment (inclusive of self-weight) of $M_{u,c}$=79.3 kNm is combined with the average applied shear force of $V_{u,c}$=203.5 kN (does not include the self-weight of the composite slab). There was no interface failure in the 100+50 mm slabs where averages for $M_{u,c}$=74.0 kNm and $V_{u,c}$=187.7 kN. It is expected that the interface shear stress will be greater in the 75+75 mm depth slabs as the interface is at the mid-depth and centroid of the composite section and the forces in the 75 mm depth topping will be greater.

Figure 12.5 also shows the calculated values for the flexurally cracked composite shear capacity $V_{Rd,cr,c,calc}$ where the ratio of the experimental total shear force (inclusive of self-weight) to $V_{Rd,cr,c,calc}$ is 0.96 and 1.00 for the 75+75 mm and 100+50 mm depth slabs, respectively. Although these ratios should be greater than unity, say 1.1–1.2, the complexity of testing and calculating the flexurally cracked shear capacity is recognised in the results.

Table 12.2 presents the calculated (i.e. predicted) horizontal interface shear strength $v_{Rdi,calc}$, without partial safety factors and the experimental horizontal interface shear stress v_{ui}, calculated according to the methods given in Section 12.4, based on BS EN 1992-1-1, clause 6.2.5 and Exp 6.24, (BS EN 1992-1-1, 2004), using a cohesion factor c=0.2. Note that the depth of the nominally 50 and 75 mm topping averaged 60 and 81 mm, respectively.

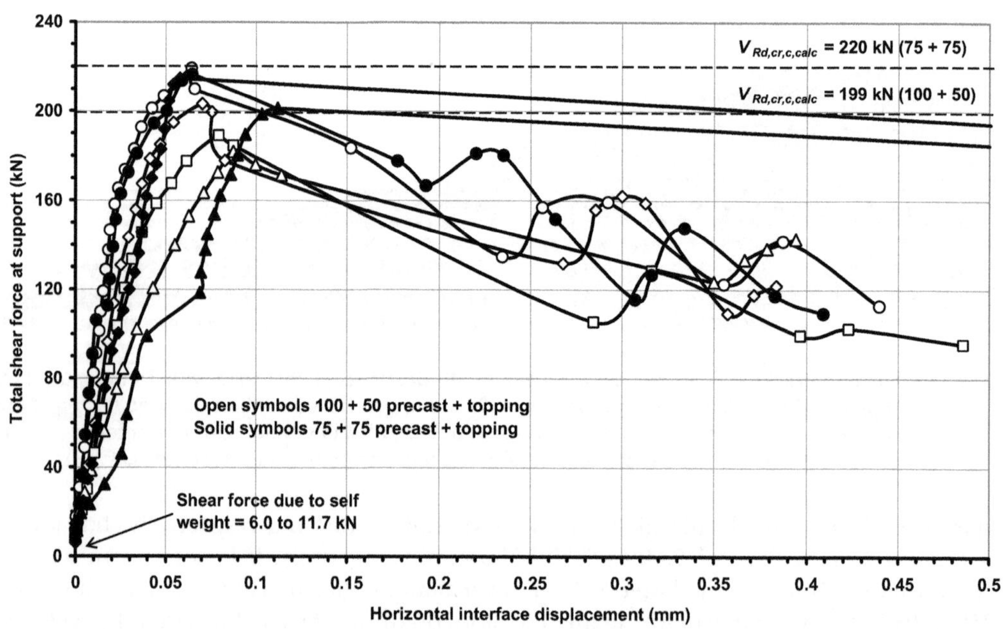

Figure 12.5 Experimental horizontal displacement of the interface between prestressed concrete solid units and structural topping in shear tests. (With permission by Bison Precast Ltd., UK.)

Table 12.2 Comparison of the experimental v_{ui} and calculated shear strength $VR_{di,calc}$ and horizontal interface slippage Δ_s in full scale composite shear tests carried out on 1200 × 150 mm depth composite prestressed solid slabs.

Test ref.	*Depth of precast unit (mm)*	*Depth of topping (mm)*	*Ultimate test applied shear force $V_{u,c}$ (kN)*	*Interface shear stress v_{ui} (N/mm²)*	*Ratio $v_{ui}/v_{Rdi,calc}$*	*Slip at maximum shear force (mm)*
100D1	100	67	207.5	1.32	2.26	0.064
100D2		59	179.5	1.22	2.09	0.079
100E2		55	191.4	1.35	2.32	0.070
100D1		60	172.4	1.12	1.92	0.104
75D1	75	80	207.8	1.49	2.64	0.065
75D2		82	205.0	1.44	2.55	No value[a]
75E2		81	208.6	1.46	2.58	0.059
75E1		81	192.6	1.36	2.41	0.112
Mean				**1.35**	**2.35**	**0.079**

[a] Due to instrument failure.

The average stress v_{ui}=1.25 and 1.44 N/mm², for the 100+50 mm and 75+75 mm slabs, respectively, all greater than $v_{Rdi,calc}=c\ 0.7f_{ctm}$=0.583 and 0.565 N/mm², respectively, by an average ratio of 2.35. This ratio is in general agreement with the figure of 2.86 based on similar work by Ibrahim (Ibrahim et al., 2016) in Section 12.4. Table 12.2 also gives the horizontal interface slippage Δ_s just prior to failure at the maximum experimental shear force. The average for the tests is Δ_s=0.08 mm. After interface failure in the three 75+75 mm slabs then $\Delta_s \approx$0.7, 30 and 32 mm, i.e. uncontrolled failure in the latter.

12.3 DESIGN OF COMPOSITE FLOORS

12.3.1 Precast floors with composite toppings

The structural capacity of precast psc floor units may be increased by adding a layer of structural reinforced concrete to the top of the unit. Providing that the topping concrete is fully anchored and bonded to the precast unit the two concretes, precast and cast *insitu*, may be designed as monolithic. The section properties of the precast unit plus the topping are used to determine the structural performance of the composite floor. A composite floor may be made using any type of precast unit, but clearly, there is more to be gained from using voided psc units, such as hcu or double-tee, which are lightweight and therefore cheaper to transport and erect than solid psc units. Figure 12.6 shows the details for the most common types of composite floors.

The minimum depth of the topping should not be less than 50 mm. When calculating the average depth of the screed allowances for camber should be made – span/300 will suffice. The grade of *insitu* concrete is usually C25 or C30, but there is no reason why higher strength cannot be used except that the increased strength of the composite floor resulting from the higher grade will not justify the additional costs of materials and quality control. See Tables 4.2 and 4.3 for the concrete data. The topping must be reinforced, but, as explained later, interface shear reinforcement between the precast and *insitu* topping is only required if the design dictates, although some designers specify minimum loops or dowels

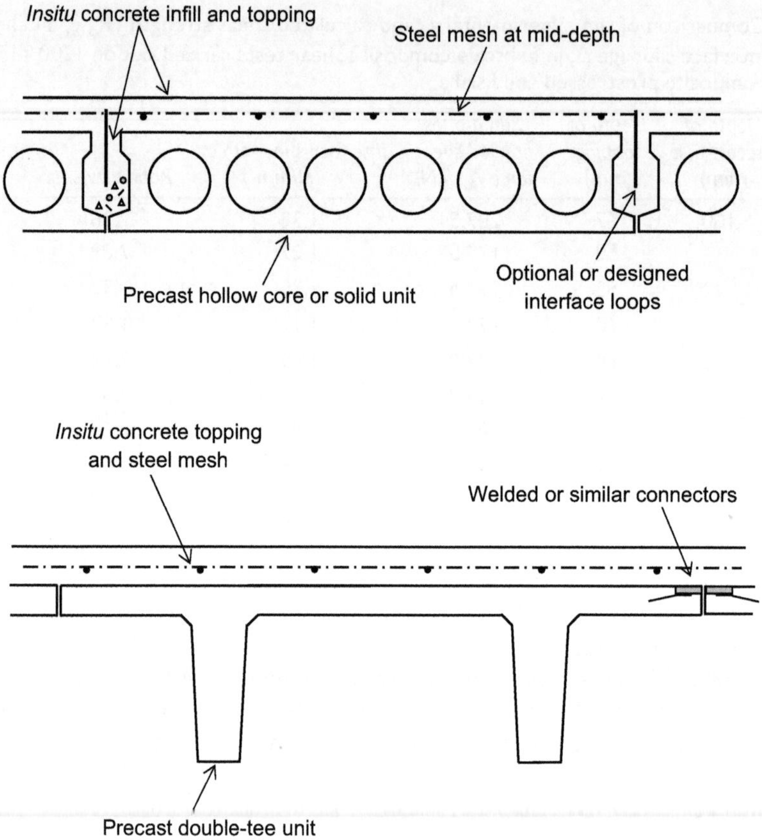

Figure 12.6 Composite floor profiles.

such as H12 at 1200 mm spacing. Mesh reinforcement of minimum area 0.13% or 0.26 $f_{ctm}/f_{yk}\times$concrete area is the preferred choice, e.g. 75 mm minimum depth topping, having a maximum depth due to camber of say 90 mm, requires $0.0013\times1000\times90=117$ mm²/m, use a square mesh A142 (142 mm²/m).

Composite floor design is carried out in two stages, before and after the *insitu* topping becomes structural. The transfer stress condition in the psc unit must have previously been satisfied. Therefore, the precast unit must carry its own weight plus the self-weight of the wet *insitu* concrete (plus a construction traffic allowance typically 1.5 kN/m²). The composite floor (= precast+hardened topping) carries imposed dead and live loads. In the final analysis, the stresses and forces resulting from the two cases (minus the construction traffic allowance which is temporary) are additive. Relative shrinkage stresses between the *insitu* topping and precast unit should be added, see Section 12.3.2.2.

Figures 2.14, 2.18, 2.22 and 2.26 show the load capacity for solid and hcu slabs, wet-cast voided floor slabs, double-tee floors and thermally insulated floor slabs, all achieved using a 75 mm thick structural topping. The precast units are all unpropped during the construction of the topping – see Section 12.3.3 to discuss the effect of propping. The benefit from the topping decreases as (i) the span increases because the self-weight of *insitu* topping nullifies the additional capacity and (ii) the unit depth increases because the section modulus of the composite section is proportionately less. The following design procedures are used.

12.3.2 Flexural analysis for composite prestressed concrete elements

12.3.2.1 Serviceability state of stress

Permissible service stresses are checked at two stages of loading – Stages 1 and 2 before and after hardening of the *insitu* topping as follows:

1. Stage 1: for the self-weight of the precast slab w_0 plus the self-weight of *insitu* Topping w_T, plus an allowance for temporary construction traffic w_c of up to 1.5 kN/m². The section properties of the precast unit alone are Z_b and Z_t for stress analysis and $Z_{b,co}$ and $Z_{t,co}$ for moment of resistance.
2. Stage 2: for imposed dead load w_2 and superimposed live load w_3. The section properties of the composite compound section $Z_{b,c,co}$ and $Z_{t,c,co}$ for stresses. The temporary construction load is removed.

Stage 1. Referring to Figure 12.7, a precast psc unit has the final prestress after losses of σ_b and σ_t according to Eqs. 6.1 and 6.2. (It is likely that σ_t will be negative=tension.) Let M_{s1} be the maximum service bending moment due to the stage 1 load. The flexural stresses in the bottom and top of the precast unit must first satisfy

$$f_{b1} = +\sigma_b - M_{s1}/Z_{b,co} \leq \mathbf{+0.45}\ \boldsymbol{f_{ck}} \text{ and} \geq -f_{ctm} \tag{12.1}$$

$$f_{t1} = +\sigma_t + M_{s1}/Z_{t,co} \geq \mathbf{-}\ \boldsymbol{f_{ctm}} \text{ and} \leq +0.45\ f_{ck} \tag{12.2}$$

(the most likely condition in **bold**) where $Z_{b,co}$ and $Z_{t,co}$ are the stage 1 section compound modulii for the precast unit alone defined in Section 6.2.4. The transfer stress condition in the precast unit must have previously been satisfied.

Stage 2. Let M_{s2} be the maximum service bending moment due to the stage 2 load. The flexural stresses in the bottom and top of the precast unit and at the top of the composite floor are derived from the composite section as

$$f_{b2} = -M_{s2}/Z_{b,c,co} \tag{12.3}$$

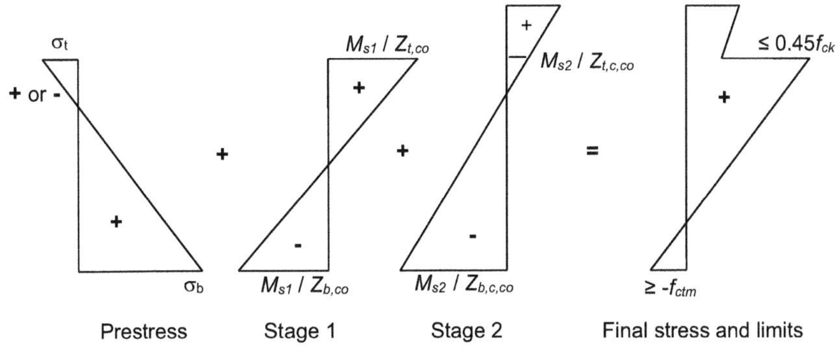

Figure 12.7 Principles of serviceability stress limitations for composite prestressed concrete elements. The limiting stress for exposure XC1 is $-f_{ctm}$, otherwise it is zero.

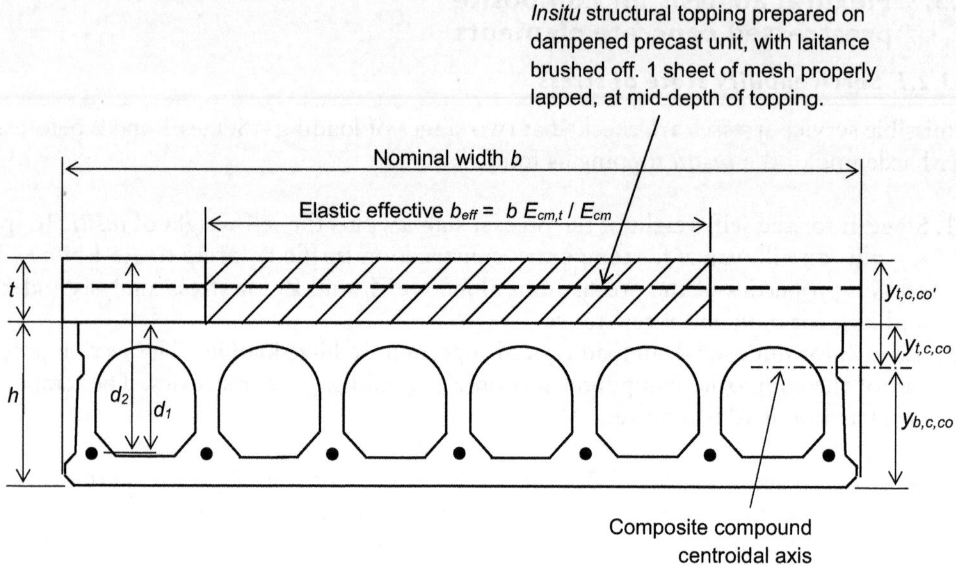

Figure 12.8 Cross section of composite hollow core floor unit with reinforced structural topping.

$$f_{t2}=+M_{s2}/Z_{t,c,co} \tag{12.4}$$

$$f'_{t2}=+M_{s2}/Z'_{t,c,co} \tag{12.5}$$

where $Z_{b,c,co}$, $Z_{t,c,co}$ and $Z'_{t,c,co}$ the stage 2 composite compound section modulii for the composite section at the bottom of the precast unit, the top of the precast unit and the top of the topping, respectively.

To calculate these values, the second moment of area $I_{c,c,co}$ of the composite section must be determined using the modular ratio approach for f_{ckT} and E_{cmT} of the *insitu* topping. This is to allow for the different elastic modulii of the precast and *insitu* concrete, typically (from Eq. 4.6) $[(25+8)/(45+8)]^{0.3}=0.87$. Referring to Figure 12.8, the width of the *insitu* topping is reduced in proportion to E_{cm} such that the effective width of the topping is

$$b_{eff}=b\ E_{cmT}/E_{cm} \tag{12.6}$$

where E_{cmT} is the elastic modulus for the Topping

Note that in this case b=full width of the precast unit, e.g. 1200 mm, not the manufactured top width, e.g. 1154 mm.

The composite compound section properties, using the transformed area of tendons in the precast unit, are as follows.

Composite compound area of concrete

$$A_{c,c,co}=A_{c,co}+b_{eff}\,t \tag{12.7}$$

where precast $A_{c,co}$ is the compound area of the precast unit

Height to the composite compound centroidal axis

$$y_{b,c,co}=[A_{c,co}\,y_{b,co}+b_{eff}\,t\,(h+0.5t)]/A_{c,c,co} \tag{12.8}$$

where $y_{b,co}$ is the height to the precast compound centroidal axis

Composite compound second moment of area

$$I_{c,c,co}=I_{c,co}+A_{c,co}\,(y_{b,c,co}-y_{b,co})^2+b_{eff}\,t^3/12+b_{eff}\,t\,(h+0.5t-y_{b,c,co})^2 \tag{12.9}$$

where $I_{c,co}$ is the compound second moment of area of the precast unit

Composite compound section modulii for the bottom and top of the precast unit

$$Z_{b,c,co}=I_{c,c,co}/y_{b,c,co} \tag{12.10}$$

$$Z_{t,c,co}=I_{c,c,co}/(h-y_{b,c,co}) \tag{12.11}$$

And for the top of the topping

$$Z'_{t,c,co}=I_{c,c,co}/\left(h+t-y_{b,c,co}\right) \tag{12.12}$$

Then, the stresses from prestressing and stage 1 loads are added to those from stage 2 loads and must satisfy

$$f_b=\sigma_b-M_{s1}/Z_{b,co}-M_{s2}/Z_{b,c,co}\geq -f_{ctm} \tag{12.13a}$$

$$f_t=\sigma_t+M_{s1}/Z_{t,co}+M_{s2}/Z_{t,c,co}\leq 0.45\,f_{ck} \tag{12.14}$$

$$f'_{t2}=+M_{s2}/Z'_{t,c,co}\leq 0.567 f_{ckT} \tag{12.15}$$

where, in Eq. 12.14, M_{s2} is the quasi-permanent load and is therefore rarely critical. Equation 12.15 is rarely critical and does not affect the design of the precast unit.

As with non-composite design in Section 7.2.1, the limit $-f_{ctm}$ in Eq. 12.13a only applies to durability exposure class XC1. For XC2 to XC4 and XD1/XS1, the limiting tension is zero and M_{s2} is replaced by $M_{s2,qp}$ and $M_{s2,freq}$ is for the quasi-permanent and frequent imposed loads, respectively. Then

$$f_b=\sigma_b-M_{s1}/Z_{b,co}-M_{s2,qp}/Z_{b,c,co}\geq 0 \tag{12.13b}$$

$$f_b=\sigma_b-M_{s1}/Z_{b,co}-M_{s2,freq}/Z_{b,c,co}\geq 0 \tag{12.13c}$$

The critical situation nearly always occurs at the bottom of the slab because the new position of the centroidal axis is often close to the top of the precast unit, as shown in Figure 12.8. Equation 12.13a can be written as

$$M_{sR2}=M_{s2}\geq (f_{ctm}+\sigma_b)\,Z_{b,c,co}-[M_{s1}\,(Z_{b,c,co}/Z_{b,co})] \tag{12.16}$$

with the same replacement of f_{ctm} with zero for exposure>XC1.

As the situation at the top of the precast unit should also be checked for completeness, Eq. 12.14 can be written as

$$M_{sR2}=M_{s2}\geq (0.45\,f_{ck}-\sigma_t)\,Z_{t,c,co}-[M_{s1}\,(Z_{t,c,co}/Z_{t,co})] \tag{12.17}$$

$$\text{Total } M_{sR,c}=\text{Minimum } M_{sR2}+M_{s1} \tag{12.18}$$

However, as explained in Section 7.2.1 under certain conditions in order to avoid non-linear creep the final compressive stress in the top may be limited to $+0.45\,f_{cm}$. Then Eq. 12.14 may be written as

$$f_t=\sigma_t+M_{s1}/Z_{t,co}+M_{s2}/Z_{t,c,co}\leq 0.45\,f_{cm} \tag{12.19}$$

The construction traffic loading needs only be considered as part of M_{s1} when checking that the service stresses do not exceed 0.45 f_{ck} and $-f_{ctm}$. It does not have to be included in the final calculation in Eq. 12.17 because it is not a permanent load. (Nor does if affect the ultimate limit state.)

It is not possible to prepare in advance standard calculation for composite floor slabs. This is because the final stresses in Eqs.12.16 and 12.17 depend on the respective magnitudes of stage 1 and stage 2 loads and moments, i.e. the same precast concrete unit may have different load-bearing capacity when used in different conditions. The following examples will show this.

Example 12.1

Calculate the stage 2 service bending moment M_{sR2} that is available if the 250 mm depth prestressed concrete hollow core unit (hcu) shown in cross section in Figure 12.9 and in Example 7.1 has a 75 mm minimum depth of structural topping. The top surface of the topping is power floated (as shown in Figure 12.1) such that the screed may be eliminated and floor finishes can be laid directly. Therefore, floor finishes, ceiling and services may be taken as 2.0 kN/m². The floor is simply supported over an effective span of 9.8 m. The pre-camber of the hcu at installation may be taken from Example 11.1 as –21 mm. Use f_{ck}=45 N/mm² and f_{ckT} for the *insitu* concrete topping=25 N/mm². Self-weight of topping (including mesh)=25 kN/m³. Exposure class is XC1. Calculate the maximum imposed service load.

Before commencing the solution, a technical note that the creep loss is smaller because the tensile bending stress due to self-weight $\{-M_{self}/Z_z\}$ in Eq. 6.16, which serves to reduce the compressive strain causing creep, now includes the self-weight of the topping. As a result, σ_b after losses increases slightly from the value in Example 7.1 from 12.94 to 12.98 N/mm². The final σ_t after losses is –2.116 N/mm².

Solution

Section properties

Elastic modulus for topping $E_{cmT}=22\times(25+8)^{0.3}=31476$ N/mm²

Self-weight of topping=25×[75+(50% camber at installation) 0.5×21]×10⁻³=2.14 kN/m²

m insitu/precast=31476/36283=0.868

b_{eff} (serviceability)=1200×0.868=1041 mm

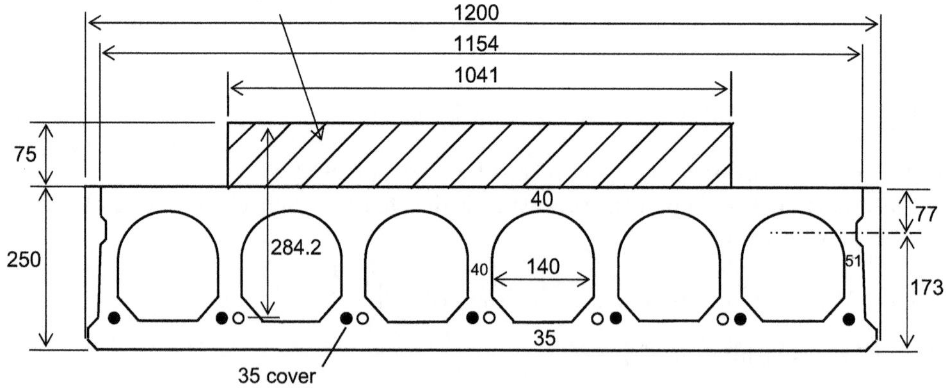

Figure 12.9 Cross section of composite hollow core floor unit in Example 12.1.

y_{bT}=250+0.5×75=287.5 mm
$A_{c,c,co}$=171985+1041×75=250060 mm^2
$y_{b,c,co}$=[(171985×121.1)+(1041×75×287.5)]/250060=173.0 mm
$I_{c,c,co}$=1296×10^6+[171985×(173.0−121.1)^2]+(1041×75^3/12)+[1041×75×(287.5−173.0)^2]=2819.9×10^6 mm^4
$Z_{b,c,co}$=2819.9×10^6/173.0=16.295×10^6 mm^3
$Z_{t,c,co}$=2819.9×10^6/(250−173.0)=36.645×10^6 mm^3
Then $Z_{b,c,co}/Z_{b,co}$=16.295/10.701=1.523
$Z'_{t,c,co}$ 2819.9×10^6/(250+75−173.0)=18.55×10^6 mm^3

Stage 1

Self-weight of the hcu plus infill w_0=4.36 kN/m
Self-weight of topping w_1=2.14×1.2=2.57 kN/m for 1.2 m wide unit
w_0+w_1=4.36+2.57=6.93 kN/m
M_{s1}=6.93×9.8^2/8=83.2 kNm
f_{b1}=83.2/10.701=−7.77 N/mm^2 tension
f_{t1}=83.2/10.056=+8.27 N/mm^2 compression

Stage 2

After topping. w_2=1.2×2.0=2.4 kN/m; w_3=1.2×5.0=6.0 kN/m; M_{s2}=100.8 kNm
f_{b2}=100.8/16.295=−6.19 N/mm^2 tension
f_{t2}=100.8/36.645=+2.75 N/mm^2 compression
Check at top of topping f_{t2}'=100.8/18.55=+5.43 N/mm^2<0.45×25=11.25 N/mm^2. OK

Final stresses

Bottom f_b=12.98+−7.77+−6.19=−0.98 N/mm^2>f_{ctm}=−3.80 N/mm^2
Top f_t=−2.116+8.27+2.75=+8.90 N/mm^2<0.45 f_{ck}=20.25 N/mm^2

The additional moment capacity available, due to the additional stress at the bottom of f_b-f_{ctm}=−0.98−(−3.80)=+2.82 N/mm^2, is $M_{sR,c}-M_2$=2.82×16.295×10^6×10^{-6}=45.9 kNm.

Eq. 12.18 $M_{sR,c}$=100.8+45.9+83.2=**229.9 kNm**

Alternatively Eqs. 12.16 & 12.18, $M_{sR,c}$=(3.80+12.98)×16.295−83.2×1.523 +83.2=146.7+83.2=**229.9 kNm**

For completeness at the top, due to an additional stress of 0.45 $f_{ctm}-f_t$=20.25−8.90=+11.35 N/mm^2, is $M_{sR,c}$ - M_2=11.35×36.645×10^6×10^{-6}=415.9 kNm, clearly not critical

The allowable imposed load=8×(100.8+45.9)/9.8^2=12.22 kN/m=12.22/1.2=**10.18 kN/m^2**>7.0 kN/m^2

Note that $M_{sR,c}$=229.9>179.1 kNm is 28% greater than for the non-composite unit in Example 7.1, but the volume of concrete used in the composite slab is 59% more, although not having a 75 mm depth screed brings this down to 9%. Therefore, providing the structural topping replaces the screed the composite slab is more efficient in terms of capacity vs. mass of concrete.

Alternative ways of looking at the additional structural benefits of a composite slab are: (i) using the original design loading the maximum effective span for this composite slab could be increased from 9.80 to 10.82 m, or (ii) using the original effective span the imposed live load could be increased from 5.00 to 8.20 kN/m^2. In both cases, serviceability would always be critical over the ultimate limit states of bending and shear.

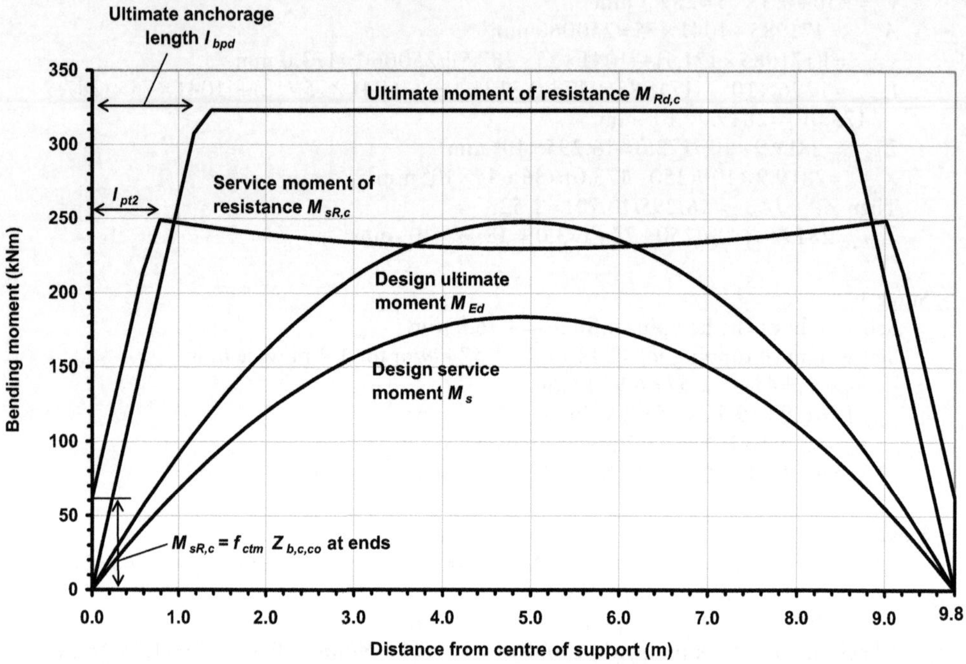

Figure 12.10 Service and ultimate design moments and the moments of resistance for the composite 250 mm depth hcu with 75 mm structural topping in Examples 12.1 and 12.2.

Figure 12.10 shows the service design moments and the moments of resistance for Example 12.1. Note the reduction in $M_{sR,c}$ as the stage 1 stresses increase towards mid-span. At the supports, where $M_s=0$, $M_{sR,c}$ is due to f_{ctm} and the compound section modulus $Z_{b,c,co}$. Then $M_{sR,c}=3.80\times16.295=61.9$ kNm. From there to $x=l_{pt2}$, the end of the transmission length, $M_{sR,c}$ increases linearly.

12.3.2.2 Relative shrinkage between insitu topping to precast unit

In calculating service stress and deflections, the effects of the relative shrinkage of the topping to that of the precast unit should be added to those resulting from loads and prestressing. For the ultimate limit state BS EN 1992-1-1, clause 2.2.2 (2) states that shrinkage (and creep) should 'generally' be taken into account in the serviceability limit state and should only be considered at ultimate where their effects are significant, for example in the verification of ultimate limit states of stability where second order effects are important, which rules out floor slabs and beams. If the topping is cast in bays, possibly a few days apart, the shrinkage effect will be considerably reduced due to the localised curvatures beneath each section of the pour. The effect of relative movement between precast and *insitu* concrete is well documented in text books and FIB documents (e.g. Bhatt, 2011, FIP Recommendations, 1982) and may be summarised as follows

Refer to Figure 12.11. Relative horizontal shrinkage (suffix $_{sh}$) between the *insitu* topping and precast unit is

$$\varepsilon_{sh}=\varepsilon_{csT}-\varepsilon_{cs} \tag{12.20}$$

where ε_{scT} is the nominal unrestrained, or 'free', shrinkage of the topping according to BS EN 1992-1-1, Table 3.2 and Exp. 3.11. This term may include the axial restraint due to steel

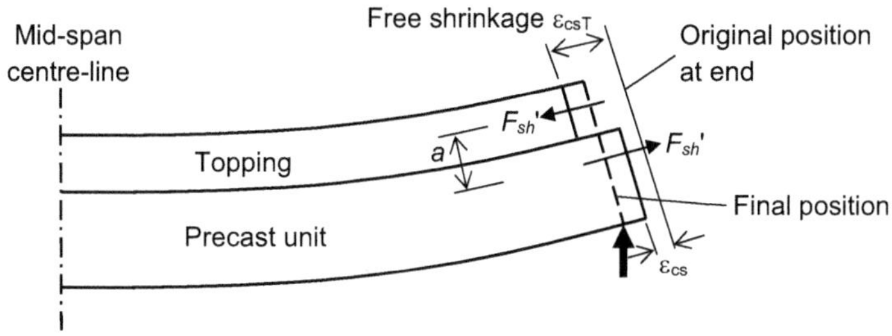

Figure 12.11 Curvature due to relative shrinkage between cast *insitu* topping and precast unit.

mesh of area A_s, or other rebars, such that ε_{scT} is divided by $(1+m\rho)$ where $\rho=A_s/A_{cT}$ and m is the long-term modular ratio of the rebars to *insitu* concrete usually taken as 15. For example using A142 mesh in 75 mm topping $(1+m\rho)=1+(15\times142/1000\times75)=1.03$.

ε_{cs} is for the precast unit as determined in Section 7.2.4.3. This term may include the eccentric restraint of the tendons, but is usually ignored as small.

$$\text{Free shrinkage stress } \sigma_{sh}=\varepsilon_{sh}\,E_{cmT} \tag{12.21}$$

$$\text{Restoring force } F_{sh}=\sigma_{sh}\,A_{cT}=\varepsilon_{sh}\,E_{cmT}\,A_{cT} \tag{12.22}$$

where $E_{cmT}\,A_{cT}$ is the axial stiffness of the *insitu* concrete topping of area A_{cT}

Lever arm from centroid of topping (depth t) and the centroid of the composite section is

$$a=h+0.5t-y_{b,c,co} \tag{12.23}$$

Restoring stress in the precast slab are $\sigma_{sh,b}$ in the bottom (compression positive) and $\sigma_{sh,t}$ in the top

$$\sigma_{sh,b}=+F_{sh}/A_{c,c,co}-F_{sh}\,a/Z_{b,c,co} \tag{12.24}$$

$$\sigma_{sh,t}=+F_{sh}/A_{c,c,co}+F_{sh}\,a/Z_{t,c,co} \tag{12.25}$$

For example, assuming that the topping is cast in a single pour, and subjecting the composite slab in Example 12.1 to a relative shrinkage strain of $\varepsilon_{sh}=528-432=96\times10^{-6}$, where from Example 7.1, $\varepsilon_{cs}=432\times10^{-6}$, and for 50% relative humidity and $f_{ck}=25$ N/mm^2 $\varepsilon_{csT}=(\varepsilon_{cd}+\varepsilon_{ca})/(1+m\rho)=\{506\times10^{-6}+[1.00\times2.5\times(25-10)\times10^{-6}]\}/1.03=528\times10^{-6}$. This is roughly in line with the recognised value between *insitu* toppings and precast elements as $\varepsilon_{sh}=100\times10^{-6}$. Continuing with $\varepsilon_{sh}=100\times10^{-6}$

The restoring force $F_{sh}=\sigma_{sh}\,A_{cT}=3.148\times75\times1200\times10^{-3}=283.3$ kN

where $\sigma_{sh}=\varepsilon_{sh}\,E_{cmT}=100\times10{-}^{6}\times31476=3.148$ N/mm^2

$$a=287.5-173.0=114.5 \text{ mm}$$

$$\sigma_{sh,b}=283.3\times10^3\times[(1/250060)-(114.5/16.295\times10^6)]=+1.13-1.99=-0.86 \text{ N/mm}^2$$

In effect, the shrinkage-induced stress reduces the short-term $M_{sR,c}$ by $0.86\times16.295\times10^6\times10^{-6}=14.0$ kNm. In the long term, the stress reduces due to the exponential creep growth factor $\Psi=1-e^{-\varphi}$, e.g. if $\varphi=2.0$, then $\Psi=0.86$. Then $M_{sR,c}$ reduces by $0.86\times14.0=12.0$ kNm. This may be reduced further by the end restraints and/or continuity at the end of slabs and beams.

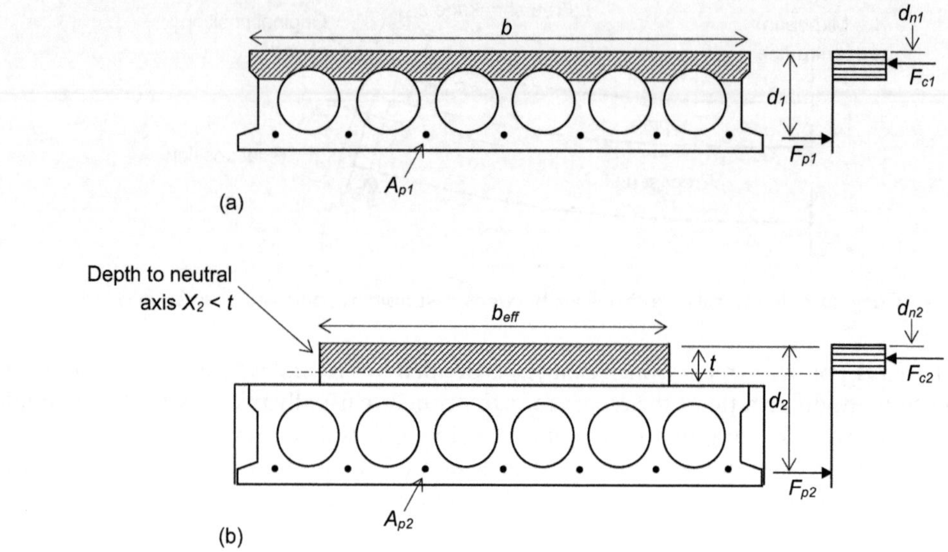

Figure 12.12 Principles for the ultimate strength for prestressed concrete composite elements (a) stage 1 in precast unit, (b) stage 2 in composite slab.

12.3.2.3 Ultimate limit state of bending

The design at ultimate limit state is also a two-stage process, with the flexural stresses resulting from the self-weight of the precast element plus the wet *insitu* concrete being carried by the precast unit alone. The lever arm in stage 1 is the same as in a non-composite design, but is now known as d_1. The method used is to calculate the area of steel A_{p1} required in stage 1 and to add the area A_{p2} required in stage 2 using an increased lever arm. See Figure 12.12a. The effect of the structural topping is to increase the lever arm to the steel reinforcement by an amount equal to the depth of the screed t, proving the depth to the neutral axis is less than t. See Figure 12.12b. In stage 2, the effective ultimate width of the topping is

$$b_{ef} = b \times f_{ckT} \ (\textit{insitu} \text{ topping}) / f_{ck} \ (\text{precast}) \tag{12.26}$$

where b is the full width of the precast unit, e.g. 1200 mm, not the manufactured width. (Note that b_{eff} is used for the elastic effective width in service.)

Some design engineers choose not to separate the design into two stages, using the composite section properties alone. This was a satisfactory approach in the past where the plateau in the stress-strain relationship meant that the ultimate stress f_p was not dependent on the design strain ε_p. However, the inclined branch of BS EN 1992-1-1, Fig. 3.10 (Figure 9.8) means that f_p is still increasing with ε_p. Adopting a two-stage approach, Figure 12.12a equilibrium in the section due to stage 1 moment M_{Ed1} is determined first, followed by the compatibility and equilibrium due to stage 2 moment M_{Ed2} as follows.

Stage 1. M_{Ed1} = due to self-weight, topping, propping (and any trimmer angle dead loads acting as a point load, see Figure 10.40).

$$K_1 = M_{Ed1} / f_{ck}(t)\ b_1\ d_1^2 \tag{12.27}$$

where b_1 = width of precast unit, d_1 = effective depth from top of precast unit

$f_{ck}(t)$ = strength of precast concrete at installation at t days < 28 days, but not less than $f_{ck}(t)$ at transfer

Exp. 3.2 $f_{cm}(t) = \exp\{s\ [1 - \sqrt{(28/t)}]\}\ f_{cm}$ (12.28)

$f_{cm} = f_{ck} + 8$ N/mm^2

$f_{ck}(t) = \exp\{s\ [1 - \sqrt{(28/t)}]\}(f_{ck} + 8) - 8$ (units of N/mm^2 only)

Cement s = 0.2 (Class R), 0.25 (Class N) or 0.38 (Class S)

$z_1 = [0.5 + \sqrt{(0.25 - K_1/1.134)}]\ d_1 \leq 0.95\ d_1$ (12.29)

$x_1 = (d_1 - z_1)/0.4$ (12.30)

If $0.8x_1 \leq h_{ft}$ the depth of the top flange

then $A_{p1} = M_{Ed1}/z_1\ f_{p1}$ (12.31)

Fig 3.10 where f_{p1} = stress in tendons at strain $\varepsilon_s = 0.02$

If $0.8\ x_1 > h_{ft}$ compression is in webs of width b_w

Restart the analysis using $K_{1w} = M_{Ed1}/f_{ck}(t)\ b_w\ d_1^2$ (12.32)

$z_1 = (0.5 + (0.25 - [K_{1w}/1.134 - \beta])\ d_1$ (12.33)

where $\beta = (b_1 - b_w)\ h_{ft}\ (d_1 - 0.5\ h_{ft})/(2\ b_w\ d_1^2)$ (12.34)

Then $A_{p2} = A_p - A_{p1}$ (12.35)

The composite $M_{Rd,c2}$ is calculated according to Section 9.3.1 but using A_{p2} instead of A_p

Total $M_{Rd,c} = M_{Rd,c2} + M_{Ed1}$ (12.36)

There is a problem in calculating standard values for M_{Rd} for specific units because the stage 1 moments and area of steel must be first known. As these are span-dependant, the superimposed moment capacity $M_{Rd,c2}$ is a function of span and stage 1 loads.

Example 12.2

Calculate the ultimate bending moment of resistance $M_{Rd,c}$ for the 250 mm depth hcu with a 75 mm depth topping shown in cross section in Figure 12.9 and designed in Example 12.1. In this example ONLY let the age at installation = 7 days. What is the maximum ultimate imposed load. Is the composite slab critical at service or at ultimate?

Solution

Effective depth in the hcu $d_1 = 209.2$ mm and $b_1 = 1154$ mm
Effective depth in composite section $d_2 = 209.2 + 75 = 284.2$ mm
Consider stage 1 moment acting on precast alone at installation at $t_i = 7$ days

Eq. 12.28 $f_{ck}(t_i) = [38\ e^{0.2\ (1 - \sqrt{(28/7)})}] - 8 = 35.4$ N/mm^2

$M_{Ed1} = 1.25 \times 83.2 = 104.0$ kNm where $\gamma_G = 1.25$ from Exp 6.10b

Eq. 12.28 $K_1=104.0\times10^6/(35.4\times1154\times209.2^2)=0.058$

Eq. 12.29 $z_1=\{0.5+\sqrt{[0.25-(0.058/1.134)]}\}\times209.2=197.9\text{ mm}<0.95d_1$

Ultimate stress f_p is not known until the stage 2 M_{Rd} is found, therefore taking the limiting strain as 0.02 then $f_{p1}=1516$ N/mm² (this assumption is checked at the end of this calculation).

$A_{p1}=104.0\times10^6/(197.9\times1516)=346\text{ mm}^2$

$A_{p2}=911-346=565\text{ mm}^2$

$\varepsilon_{po}=0.004843$

$\varepsilon_p=0.004843+0.0035\ (284.2/X-1)=0.001343+0.995/X$ Eq. 1

Effective width of topping $b_{ef}=1200\times25/45=667$ mm

$F_c=0.567\times45\times667\times0.8\ X=13600\ X$

$F_{s2}=565\ f_p$

$X/f_p=565/13600=0.0415$ Eq. 2

Then $X=61.9\text{ mm}<75$ mm depth of topping

$d_{n2}=0.4\times61.9=24.8$ mm

$z_2=284.2–24.8=259.4$ mm

From eq. 1 $\varepsilon_p=0.001343+0.995/61.9=0.017413$

From eq. 2 $f_p=61.9/0.0415=1491$ N/mm²

$M_{Rd,c}=f_p\ A_{p2}\ z_2+M_{Ed1}=1491\times565\times259.4\times10^{-6}+104.0=$ **322.5 kNm**

The design ultimate UDL is $w_{Ed}=20.65$ kNm. Then $M_{Ed}=20.65\times9.8^2/8=247.9$ kNm<322.5 kNm

Stage 2 $M_{Rd,c2}=322.5-104.0=218.5$ kNm

Ultimate imposed load $w_{Ed}=8\times218.5/9.8^2=18.2$ kN/m$=$**15.1 kN/m²**>7.0 kN/m²

Note that $M_{Rd,c}=322.5>247.3$ kNm is 30% greater than for the non-composite unit in Example 9.1, almost the same percentage increase (28%) as for the serviceability moments.

Referring to Example 12.1, the ratio of ultimate/service resistance$=322.5/229.9=1.40$. As this value is less than $\gamma_Q=1.5$ and greater than $\gamma_G=1.25$ the composite slab may be critical under either limit states. In this loading case $\gamma=1.35<1.40$, therefore the composite slab is critical at service. Using the original design loading the maximum effective span for this composite slab could be increased from 9.80 to 11.08 m, but now $M_{sR,c}$ fails by 6%, or using the original effective span the imposed live load could be increased from 5.00 to 8.44 kN/m² but now $M_{sR,c}$ fails by 2%. Therefore, in both cases, the service limit is first critical.

The assumption is that the ultimate stress at stage 1 $f_{p1}=1516$ N/mm² is not too far from the final stress $f_p=1491$ N/mm². If iteration is carried out by say letting $f_{p1}=1491$ N/mm² then f_p increases slightly to 1492 N/mm² and $M_{Rd,c}$ reduces by 1.6 kNm to 320.9 kNm.

Figure 12.10 shows the ultimate design moments and the moments of resistance for Example 12.2.

12.3.2.4 Ultimate shear capacity of composite sections

The design for ultimate shear capacity for composite hcu slabs is defined in BS EN 1168 (BS EN 1168, 2005) as a two-stage design, but is not mentioned as such in BS EN 1992-1-1. It is technically correct to deduct the elastic shear stress consumed in stage 1 before the topping hardens as

$$v_{Ed1} = V_{Ed1}\, S_c / b_w\, I_c \quad (12.37)$$

Clearly, this approach may be used for solid slabs, double-tee, thermal units, etc., but it again highlights the paradox that $V_{Rd,c,c}$ is an *ultimate* strength based on an *elastic* analysis. However, there is some ductility if the shear-tension failure mode is critical (as is the case of UDL) even in hcu with narrow webs, and so adopting this method is valid. Referring to BS EN 1168, Annex F, clause F2.2, the stage 2 shear capacity is

$$V_{Rd,c,c2} = I_{c,c}\, b_w / S_{c,c}\, [\sqrt{(f_{ctd}^2 + \alpha_l\, \sigma_{cp}\, f_{ctd})} - v_{Ed1}] \quad (12.38)$$

For hcu, where shear cracking commences in the narrowest part of the webs, the same value for $\alpha_l = l_x / l_{pt2}$ as for the non-composite design is used for the composite design (Eq. 9.2). Similarly the same applies to σ_{cp} calculated at the centroid of the precast hcu. This rule does not apply to solid slabs where the shear plane moves up to $y_{b,c}$, then $\alpha_l = (L_b/2 + y_{b,c})/l_{pt2}$ and σ_{cp} is calculated at the same level.

The total composite shear capacity now includes V_{Ed1} as

$$V_{Rd,c,c} = V_{Rd,c,c2} + V_{Ed1} \quad (12.39)$$

where V_{Ed1} = ultimate dead load acting on the unit, including self-weight of wet *insitu* topping

$S_c\ I_c$ = first and second moment of area of the non-composite section
$S_{c,c}$ = first moment of area for the composite section above the centroidal axis of the composite section

$$S_{c,c} = b_{eff}\, t\, (h + 0.5t - y_{b,c}) + b(h - y_{b,c})^2/2 \quad (12.40a)$$

where b_{eff} is the elastic effective width (not the ultimate b_{ef})

If centroidal axis is below top flange $y_{b,c} < h - h_{ft}$, subtract the area for the hollow cores

$$S_{c,c} = b_{eff}\, t\, (h + 0.5t - y_{b,c}) + b(h - y_{b,c})^2/2 - (b - b_w)(h - h_{ft} - y_{b,c})^2/2 \quad (12.40b)$$

Note that due to camber, the depth of the structural topping at the supports, where $V_{Rd,c,c}$ is calculated, may be 10–20 mm more than t, provided the topping is floated level (not following the profile of camber, which may imply variable thickness screed is being used). For example, in the present design of the 250 mm depth hcu, the camber at installation is 21 mm, such that the depth of the topping is 75+21=96 mm. This is usually ignored in design as it may not be known whether the topping is to be floated level, or will follow the camber with a uniform depth t.

6.2.2 (2) Use $V_{Rd,c,c}$ where $M_s = M_{s1} + M_{s2} <$ composite cracking moment $M_{cr,c}$ for two-stage loading as follows.

The composite section is flexurally cracked at ultimate when

$$\sigma_b - M_{s1}/Z_{b,co} - M_{s2}/Z_{b,c,co} \geq -f_{ctk,0.05}/\gamma_c = -f_{ctd}$$

$$M_{s2}=(\sigma_b-M_{s1}/Z_{b,co}+f_{ctd})\ Z_{b,c,co}$$

$$M_{cr,c}=M_{s1}+(\sigma_b-M_{s1}/Z_{b,co}+f_{ctd})\ Z_{b,c,co}$$

$$M_{cr,c}=(\sigma_b+f_{ctd})\ Z_{b,c,co}-M_{s1}\ [(Z_{b,c,co}/Z_{b,co)}-1)] \tag{12.41}$$

i.e. $M_{cr,c}$ varies along the span according to stage 1 service moment, including the effects of propping. Where $M_{s1}=0$, $M_{cr,c}=(\sigma_b+f_{ctd})\ Z_{b,c,co}$.

For the ultimate cracked shear capacity $V_{Rd,cr,c}$

Using BS EN 1168, clause F4, the same equations as non-composite units apply except that the composite effective depth $d_2=d+t$ replaces d. There is no two-stage design for $V_{Rd,cr,c}$ as Eq. 9.7 is based on the dowel action of the tendons provided, irrespective of the state of stress at stage 2, and the term 0.15 σ_{cp} is not affected by the negative bending stress due to stage 1 loads. Similarly, Eq. 10.10 is based on the shear strength of concrete using design concrete data.

Example 12.3

Calculate the ultimate shear capacity $V_{Rd,c,c}$ and $V_{Rd,cr,c}$ for the 250 mm depth hcu with a 75 mm depth topping shown in cross section in Figure 12.9 and designed in Example 12.1. Derive an expression from Eq. 12.41 where the section becomes flexurally cracked in the span and the shear capacity reduces to $V_{Rd,cr,c}$.

To remind the reader, for the non-composite section in Example 10.1 $V_{Rd,c}=(1270\times10^6\times302/6.783\times10^6)\times[\surd(1.77^2+0.307\times4.61\times1.77)]\times10^{-3}=134.3$ kN, and the shear span $l_{v1}=9.454$ m. The prestress at the support from Example 7.1 is $\sigma_b=11.97$ N/mm^2. The stage 1 load from Example 12.1 is 6.93 kN/m per unit width.

Solution

Uncracked composite shear capacity $V_{Rd,c,c}$

$S_c=6.783\times10^6$ mm^3, I_c (not compound)$=1270\times10^6$ mm^4

V_{Ed1} at shear plane$=1.25\times6.93\times9.454/2=40.94$ kN

Eq. 12.37 $v_{Ed1}=40940\times6.783\times10^6/(302\times1270\times10^6)=0.724$ N/mm^2

The height to the composite centroidal axis (not compound)

$y_{b,c}=[(168000\times123.0)+(1041\times75\times287.5)]/(168000+1041\times75)=43.11\times10^6/246075$ mm$=175.2$ mm.

Eq. 12.40b $S_{c,c}=[1041\times75\times(287.5-175.2)]+(1200\times74.8^2/2)-[(1200-302)\times(74.8-40)^2/2]$
$=11.576\times10^6$ mm^3

$I_{c,c}=1270\times10^6+[168000\times(175.2-123.0)^2]+(1041\times75^3/12)+[1041\times75\times(287.5-175.2)^2]$
$=2749.0\times10^6$ mm^4

Eq. 12.38 $V_{Rd,c,c2}=(2749.0\times10^6\times302/11.576\times10^6)\times[\surd\ (1.77^2+0.307\times4.61\times1.77)-0.724]=118.47$ kN

Eq. 12.39 $V_{Rd,cc}=118.47+40.94=$ **159.41 kN**

Cracked capacity $V_{Rd,cr,c}$

$d_2=d+t=209.2+75=284.2$ mm

$k = 1 + \sqrt{(200/284.2)} = 1.84$

$v_{c,min} = 0.035 \times 1.84^{1.5} \times \sqrt{45} = 0.585$ N/mm²

$\rho = 911/(302 \times 284.2) = 0.0106$

Maximum $V_{Rd,cr,c} = \{[(0.18/1.5) \times 1.84 \times (100 \times 0.0106 \times 45)^{1/3}] + 0.15 \times 4.61\} \times 302 \times 284.2 \times 10^{-3} = \mathbf{128.1}$ **kN**

Minimum $V_{Rd,cr,c} = (0.59 + 0.15 \times 4.61) \times 302 \times 284.2 \times 10^{-3} = 109.6$ kN < 128.1 kN
Use $V_{Rd,cr,c} = 128.1$ kN
Use $V_{Rd,c,c} = 159.41$ kN when $M_s < M_{cr,c}$
Eq. 12.41 $M_s < M_{cr,c} = (11.97 + 1.77) \times 16.295 - M_{s1}$ $[(16.295/10.701) - 1] = \mathbf{223.9 - 0.522}$ M_{s1} kNm, otherwise $V_{Rd,cr,c} = 128.1$ kN. This condition exists where $x = 4.31$ m from the centre of support where $M_s = 181.3$ kNm and $M_{s1} = 82$ kNm. The ultimate shear force at this point is 12.1 kN, clearly not critical.

12.3.3 Propping

12.3.3.1 Flexural stresses in propped composite slabs

Propping is a technique that is used to increase the service moment capacity by reversing the stage 1 stresses, particularly at mid-span. This is achieved by placing a rigidly founded support, load-bearing props, such as Peri or similar, in the desired place whilst the *insitu* concrete topping is hardening. See Figures 12.13 and 12.14. To ensure that the props are always effective, many contractors prefer to use two props rather than one – just in case

Figure 12.13 Propping of prestressed floor slab. (Courtesy Bison Precast Ltd., UK.)

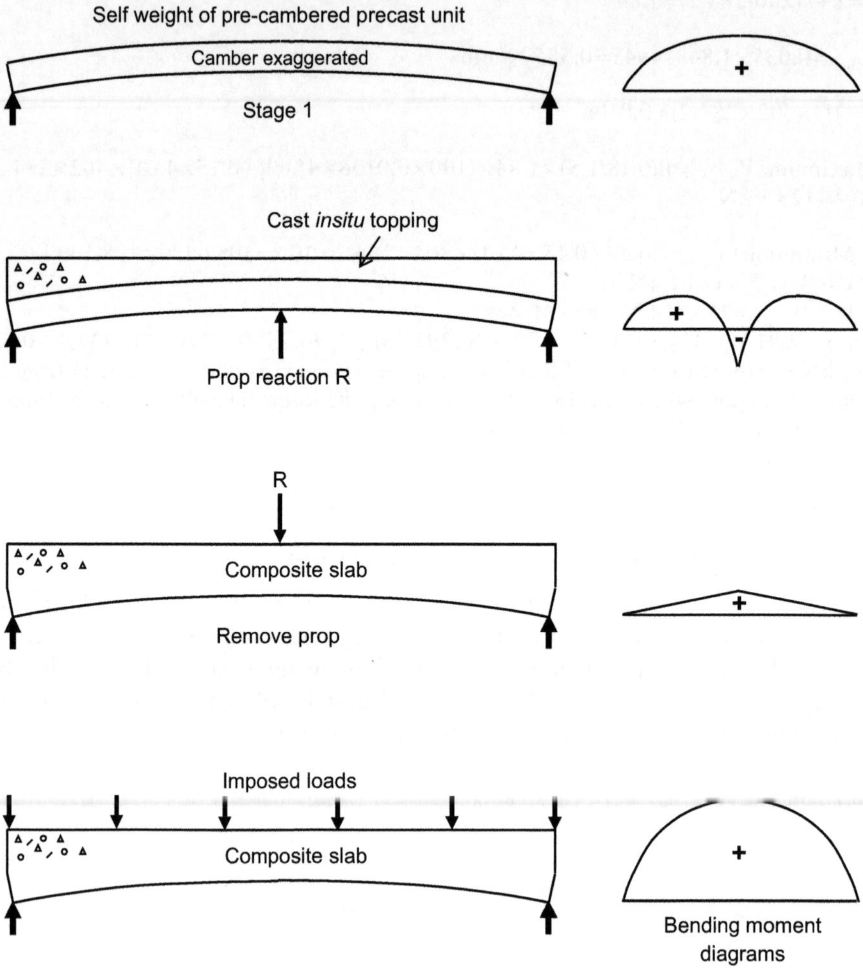

Figure 12.14 Bending moments resulting from propping.

the foundation to one of the props is 'soft'. However, the following analysis will consider a single mid-span propped floor slab. The reader can easily extend the same analysis to twin props at ⅓ and ⅔ span.

The benefit derives from the fact that the stage 1 moments due to the self-weight of the wet concrete screed are determined over a continuous double span, each of $L/2$. When the props are removed, the prop reaction R creates a new moment which is carried by the composite section. Finally, the superimposed loads are added as shown in Figure 12.14.

When rigidly secured at mid-span, the single prop nullifies the deflection due to the self-weight w_T of the wet *insitu* topping (allowing for precamber of the slab) of an effective span L, which would otherwise be $5\,w_T\,L^4/384\,E_{cm}\,I_{c,co}$. The virtual upward deflection due to a prop reaction R is $R\,L^3/48\,E_{cm}\,I_{c,co}$. When these deflections are equal $5\,w_T\,L^4/384 = R\,L^3/48$, and hence $R = 240\,w_T\,L/384 = 0.625\,w_T\,L$. Prop reactions for single and twin props are shown in Figure 12.15.

The negative moment at mid-span due to the prop $M = -0.625\,w_T\,L^2/4 = -5\,w_T\,L^2/32$. The stage 1 moment is now $M_{s1} = w_0\,L^2/8 + w_T\,L^2/8 - 5\,w_T\,L^2/32 = w_0\,L^2/8 - w_T\,L^2/32$, where w_0 is the self-weight of the precast unit. The additional Stage 2 moment following the removal of the prop is $M_{s2} = +5\,w_T\,L^2/32$.

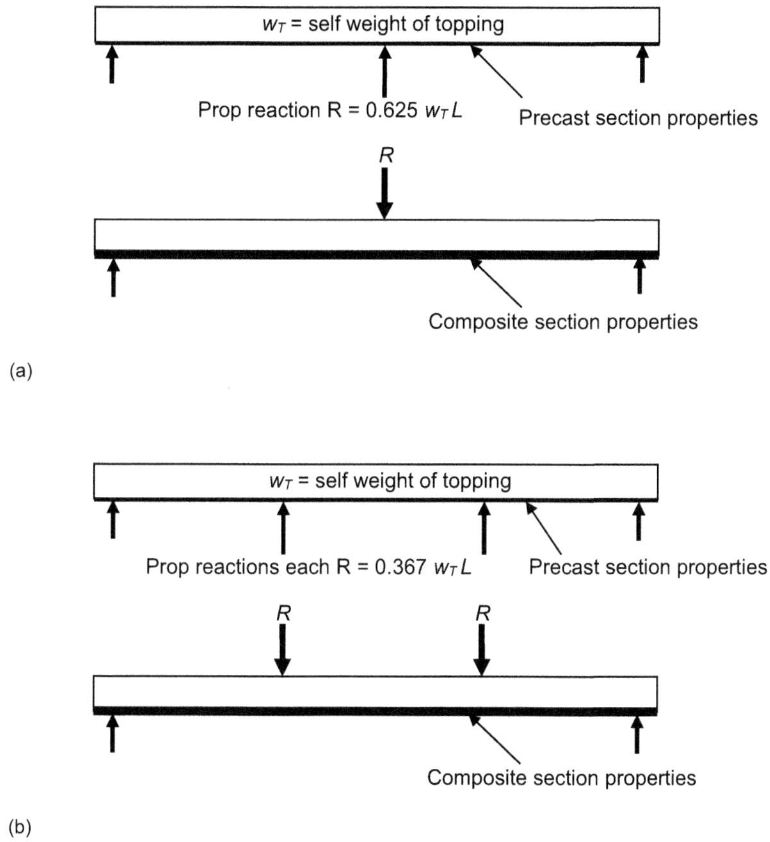

Figure 12.15 Reactions from props at (a) mid-span, and (b) two props at 1/3 span.

Eq. 12.16 is therefore modified as

$$M_{sR2} = (f_{ctm} + \sigma_b)\, Z_{b,c,co} - [M_{s1}\,(Z_{b,c,co}/Z_{b,co})] - 5\, w_T\, L^2/32 \tag{12.42a}$$

For 2 props positioned at ⅓ and ⅔ span

$$M_{sR2} = (f_{ctm} + \sigma_b)\, Z_{b,c,co} - [M_{s1}\,(Z_{b,c,co}/Z_{b,co})] - 11\, w_T\, L^2/90 \tag{12.42b}$$

with the same replacement of f_{ctm} with zero for exposure>XC1.

When the prop(s) is removed M_{s2}=+5 w_T $L^2/32$ or +11 w_T $L^2/90$, respectively, plus all imposed dead and live loads.

Finally, the total $M_{sR,c}$ from Eq. 12.8 is

$$M_{sR,c} = M_{sR2} + M_{s1} \tag{12.43}$$

but where in Eq. 12.43 $M_{s1} = (w_0 + w_T)\, L^2/8$, i.e. the negative moment due to the prop is not included twice.

12.3.3.2 Temporary situation in propped composite slabs

When checking the moment of resistance in the temporary stage M_{sR} of the precast unit alone, due to self-weight of the precast unit, the propped topping and an allowance for the propped construction traffic of say w_c=1.5 kN/m², note that the maximum bending moment is not at mid-span. If the end reaction F due to self-weight and construction is $(w_0+w_T+w_c)\ L/2$, and due to the prop reaction is $-0.625\ (w_T+w_c)\ L/2$, then

$$R=[0.5\ (w_0+w_T+w_c)-0.3125\ (w_T+w_c)]\ L \tag{12.44}$$

The distance to zero shear force, and hence the position of $M_{s1,max}$ is at $a=F/(w_0+w_T+w_c)$, then

$$a=[0.5-0.3125\ (w_T+w_c)/(w_0+w_T+w_c)]\ L \tag{12.45}$$

and

$$M_{s1,max}=Fa-(w_0+w_T+w_c)\ a^2/2 \text{ (see Example 12.4)} \tag{12.46}$$

The service moment of resistance in the temporary situation is calculated in the usual manner, i.e. $(\sigma_b+f_{ctm})\ Z_{b,co}$. (Note that in the temporary situation, f_{ctm} is used also for exposure classification greater than XC1.) The prestress σ_b may, if the designer wishes, be calculated at the time of installation, not the long-term value – the difference is usually about 10%.

The economical and practical benefits of propping wide slabs such as hcu and double-tees should be carefully considered. Propping can be quite expensive and may slow down site erection rates. Refer to Chapter 1, Section 1.2.1 on the subject of propping from a manufacturer's point of view.

Example 12.4

Repeat Example 12.1 with the hcu propped at mid span. The effective span is now increased to 11.2 m to reflect the increased service moment capacity due to propping. The increased length of slab has reduced creep losses such that σ_b=13.28 N/mm² and σ_t=−2.17 N/mm². The camber at installation has reduced from 21.1 to 19.6 mm, due to the compensating effect of self-weight. The self-weight of the topping in now w_T=2.54 kN/m.

Check the service moment of resistance in the temporary stage, due to the self-weight of the hcu, propped topping and a construction allowance of 1.5 kN/m². It is known that the prestress at installation at 28 days is σ_b=14.53 N/mm² (this is greater than the long-term value due to smaller losses of prestress at installation).

Solution

Separating the bending stresses into stages 1 and 2

Stage 1

$w_1=w_0+w_T=4.36+2.54=6.90$ kN/m

$M_{s1}=6.90\times11.2^2/8=108.25$ kNm minus moment (due to propping)
$=5\times2.54\times11.2^2/32=49.85$ kNm

$M_{s1}=108.25-49.85=58.40$ kNm

$f_{b1}=58.40/10.701=-5.45$ N/mm²

$f_{t1}=58.40/10.056=+5.81$ N/mm²

Stage 2. Prop removed

$M_{s2}=49.85+1.2\times7.0\times11.2^2/8=49.85+131.71=181.56$ kNm

$f_{b2}=181.56/16.295=-11.14$ N/mm^2

$f_{t2}=181.56/36.645=+4.95$ N/mm^2

Final stresses

$f_b=13.28+-5.45+-11.14=-3.32$ N/mm^2>-3.80 N/mm^2

$f_t=-2.17+5.81+4.95=+8.60$ N/mm^2 <<+20.25 N/mm^2

Additional moment of resistance due to propping $M_{sR}-M_s=(3.80-3.32)\times16.295\times10^6\times10^{-6}=7.82$ kNm

$M_{sR,c}=58.40+181.56+7.82=$**247.8 kNm**

$M_{s2}=131.71+7.82=139.5$ kNm

The maximum imposed service load in stage 2, $w_{s2}=8\times139.5/11.2^2=8.90$ kN/m=**7.41 kN/m²**>7.0 kN/m² design imposed load.

Alternatively, using Eqs. 12.42a and 12.43 directly

$M_{sR2}=(3.80+13.28)\times16.295-(58.40\times16.295/10.701)-49.85=139.5$ kNm

$M_{sR,c}=M_{sR2}+M_{s1}=139.5+101.25=$**247.8 kNm**

This is the situation at mid-span. However, the negative bending moment due to propping (as shown in Figure 12.14) causes a shift in the position of critical stress from mid-span $x/L=0.50$ to about 0.48 (the shift increases for greater topping depth). At this point $w_{s2}=7.38$ kN/m^2 a reduction of only 0.5% compared to the mid-span calculation, but as the depth of the topping increases relative to the depth of the precast unit, this difference is greater, e.g. if the topping depth increases to 150 mm the critical stress is at $x/L=0.43$ and difference in w_{s2} is 3.5%. It should be noted that although propping increases $M_{sR,c}$ it does not increase $M_{Rd,c}$ (except for a small increase due to lower M_{Ed1}) nor $V_{Rd,c,c}$, and as $M_{Rd,c}$ becomes critical the full extent of propping is not advantageous. For example, in this same example, the ultimate limit state is critical, i.e. $M_{Ed}=323.3$ kNm and $M_{Rd,c}=323.5$ kNm, so that $L=11.2$ m is in fact the critical span.

Figure 12.16 shows the service and ultimate design moments and the moments of resistance for Example 12.4. Note the reduction in $M_{sR,c}$ as the stage 1 stresses increase is nullified by propping.

It was noted in Example 12.1 that, under the same design loading, the maximum effective span in the unpropped composite slab was 10.82 m. The maximum effective span in the propped slab is 11.20 m – the designer and contractor have therefore to assess the benefit of propping, *viz*, additional span against the cost of propping.

Temporary condition after casting topping

The distance a to the maximum bending moment $M_{s1,max}$ in stage 1 is as follows:

Self-weight of precast, topping and construction load $w_0+w_T+w_c=4.36+2.54+(1.2\times1.5)=8.70$ kN/m

Self-weight of topping and construction $w_0+w_T+w_c=4.34$ kN/m

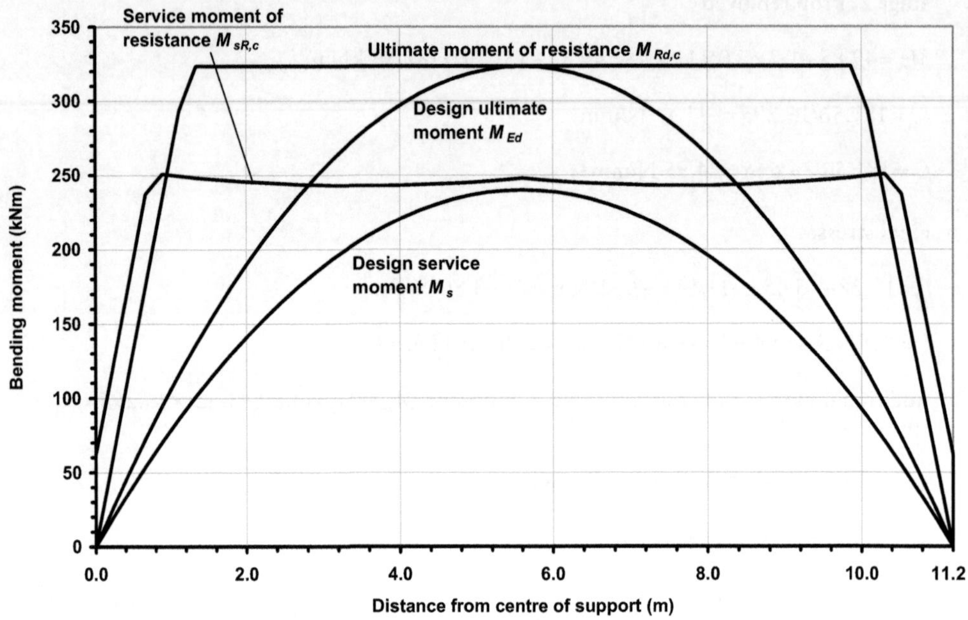

Figure 12.16 Service and ultimate design moments and the moments of resistance for the composite 250 mm depth hcu with 75 mm propped structural topping in Example 12.4.

Eq. 12.44 End reaction $F=(0.5\times8.70-0.3125\times4.34)\times11.2=33.53$ kN

Eq. 12.45 $a=33.53/8.70=3.85$ m, i.e. $x=0.344\ L$

Eq. 12.46 $M_{s1,max}=33.53\times3.85-8.70\times3.85^2/2=64.6$ kNm.

Note that at mid-span $M_{s1}=51.3$ kNm

Service moment of resistance for the hcu alone $M_{sR}=(14.53+3.80)\times10.701=196.1$ kNm>64.6 kNm. If the long-term value for σ_b were used, then $M_{sR}=182$ kNm.

12.3.4 Deflections in composite slabs

The calculation of deflections in composite slabs differs from non-composite units in the following ways

i. second moment of area after the topping has hardened is the transformed composite compound value $I_{c,c,co}$
ii. creep coefficients for visco-elastic movement are taken after the topping has hardened
iii. if relevant, the effect of propping is to eliminate stage 1 deflection due to the topping (zero deflection at the actual position of the prop and small deflections between the props and the supports).

Transfer deflections δ_1 and δ_2, and deflections at installation due to camber and self-weight δ_3 and δ_4 are as for the non-composite unit using Eqs. 10.1–10.5. After installation the static deflection due to the self-weight of topping w_T

$$\delta_5=+5\ w_T\ L^4/384\ E_{cm}\ I_{c,co} \tag{12.47}$$

where $I_{c,co}$ is for the precast unit alone.

The equivalent upward deflection due to single propping at mid-span of the wet *insitu* topping is due to the prop reaction force is $0.625\ w_T\ L$, then

$$\delta_6 = -0.625\ w_T\ L^4/48\ E_{cm}\ I_{c,co} \quad (12.48)$$

Ditto due to twin propping at ⅓ and ⅔ points due to reaction forces $0.367\ w_T\ L$, then

$$\delta_6 = -w_T\ L^4/76.84\ E_{cm}\ I_{c,co} \quad (12.49)$$

Resultant deflection at installation if propped $= \delta_3 + \delta_4 + \delta_5 + \delta_6$

In the long-term, imposed dead loads are taken as being applied immediately after installation at t_i days). The creep coefficient for the *insitu* concrete topping is $0.8\psi_\infty = 2.0$ (because *insitu* concrete is fresh at installation). This controls the long-term deflection for imposed dead and live loads after the topping has hardened. Note that E_{cm} for the precast is used, not E_{cmT} because $I_{c,c,co}$ is the transformed section based on E_{cm}.

Long-term camber is due to camber at installation, plus further visco elastic movement based on $I_{c,c,co}$ minus a reduction due to the reduction in pretensioning force from installation P_{pmi} to the long-term P_{po}

$$\delta_7 = \delta_3 - [\psi_{28}\ P_{pmi} - (P_{pmi} - P_{po})]\ z_{cp}\ L^2/8\ E_{cm}\ I_{c,c,co} \quad (12.50)$$

The long-term deflection δ_8 due to the self-weight and infill of the slab w_0 is due to the previous deflection δ_4 plus further visco elastic movement, and the static and visco elastic deflections due to topping w_T, using a net creep factor $\psi_{\infty,t} = 0.8\psi_\infty - \psi_1$ after installation at time t. But, because the former is a function of $I_{c,c,co}$ and the latter a function of $I_{c,co}$ before hardening of the topping and then $I_{c,c,co}$ afterwards, this particular creep coefficient is multiplied by the ratio of the respective second moments of area $I_{c,co}/I_{c,c,co}$

$$\psi_{\infty,t} = (0.8\psi_\infty - \psi_1)\ I_{c,co}/I_{c,c,co} \quad (12.51)$$

$$\delta_8 = \delta_4 + 5\ (\psi_{\infty,t}\ w_0 + (1+\psi_{\infty,t})\ w_T)\ L^4/384\ E_{cm}\ I_{c,co} \quad (12.52)$$

The creep coefficients ψ_1 and ψ_{28} are based on the same data used in Section 11.2.1 based on the ASSAP Manual (ASSAP 2002).

The deflection δ_9 due to imposed dead w_2 and quasi-permanent live load $\psi_2\ w_3$ is

$$\delta_9 = +\ 5\ [(1+\psi_{28})\ w_2 + (1+0.8\psi_\infty)\ \psi_2\ w_3]\ L^4/384\ E_{cm}\ I_{c,c,co} \quad (12.53)$$

There is no effect of creep due to the removal of the prop, i.e. instantaneous downward deflection, acting on the composite section is

$$\delta_{10} = +0.625\ w_T\ L^4/48\ E_{cm}\ I_{c,c,co} \text{ for a single prop} \quad (12.54a)$$

$$\delta_{10} = +w_T\ L^4/76.84\ E_{cm}\ I_{c,c,co} \text{ for twin props} \quad (12.54b)$$

$$\text{The final deflection } \delta_{11} = \delta_7 + \delta_8 + \delta_9 + (\delta_6 + \delta_{10}) \quad (12.55)$$

The long-term deflection after installation of finishes, otherwise known as 'movement', is due to imposed live load using $1+\psi_{28}$, viscoelastic self-weight and topping using $\psi_{\infty,t}$ and viscoelastic imposed dead loads using ψ_{28}, and changes in camber. Movement is the greater of δ_{12} or δ_{13}

$$\delta_{12} = 5\ [(1+\psi_{28})\ \psi_2\ w_3]\ L^4/384\ E_{cm}\ I_{c,c,co} \quad (12.56)$$

$$\delta_{13}=\delta_{12}+5\ (w_0+w_T)\ \psi_{\infty,t}\ L^4/384\ E_{cm}\ I_{c,co}+5\ w_2\ \psi_{28}\ L^4/384\ E_{cm}\ I_{c,c,co}+(\delta_7-\delta_3) \quad (12.57)$$

The visco-elastic live load deflection due to $1+\psi_{28}$ is as explained in Section 11.2.2 and based on the composite slab tests by Ibrahim (Elliott and Ibrahim 2006, Ibrahim and Elliott 2009).

Example 12.5

Calculate the long-term deflection and movement after installation for the 250 mm depth hcu with a 75 mm depth topping shown in cross section in Figure 12.9 and designed in Example 12.1. The slab is subjected to imposed dead and live loads of 2.0 kN/m^2 and 5.0 kN/m^2, respectively. Check the Code's limits if the floor finishes are brittle.

Solution

From Examples 11.1 and 12.1, $P_{pmi}=1010.2$ kN, $P_{po}=860.2$ kN, $L=9800$ mm, $E_{cm}=36283$ N/mm^2, $z_{cp}=82.2$ mm, $I_{c,co}=1296\times10^6$ mm^4 and $I_{c,c,co}=2819.9\times10^6$ mm^4. Self-weight slab and topping=4.36 and 2.57 kN/m.

Net deflection at installation (from Example 11.1) $\delta_3+\delta_4=-40.9+19.8=-21.1$ mm

Long-term final deflection

$\psi_1=0.71$; $\psi_{28}=0.8\times2.5\times(1.0-0.4)=1.20$ (from Example 11.1)

$\psi_{\infty,t}=(0.8\times2.5-0.71)\times1296\times10^6/2819.9\times10^6=0.592$

Due to long-term camber

$P_{pmi}-P_{po}=1010.2-860.2=150$ kN

$\delta_7=-40.9-[(1.20\times1010.2\times10^3)-150\times10^3]\times82.2\times9800^2/(8\times36283\times2819.9\times10^6)=-51.0$ mm

Due to self-weight and topping

$\delta_8=19.8+5\times[0.592\times4.36+(1+0.592)\times2.57]\times9800^4/(384\times36283\times1296\times10^6)$ $=+36.8$ mm

Due to imposed dead and quasi-permanent live loads

$\delta_9=5\times[(1+1.20)\times2.4+(1+2.00)\times(0.3\times6.00)]\times9800^4/(384\times36283\times2819.9\times10^6)$ $=+12.5$ mm

Final deflection $\delta_{11}=\delta_7+\delta_8+\delta_9=-51.0+36.8+12.5=$ **-1.7 mm** $<9800/250=39.2$ mm

Movement after installation

Due to visco* quasi-permanent live load

$\delta_{12}=5\times2.20^*\times1.8\times9800^4/(384\times36283\times2819.9\times10^6=+4.6$ mm. *See commentary to Eq. 10.12.

Due to visco self-weight, topping and imposed dead loads, and static and the above live load.

Visco camber $\delta_7+\delta_3=-(51.0-40.9)=-10.1$ mm

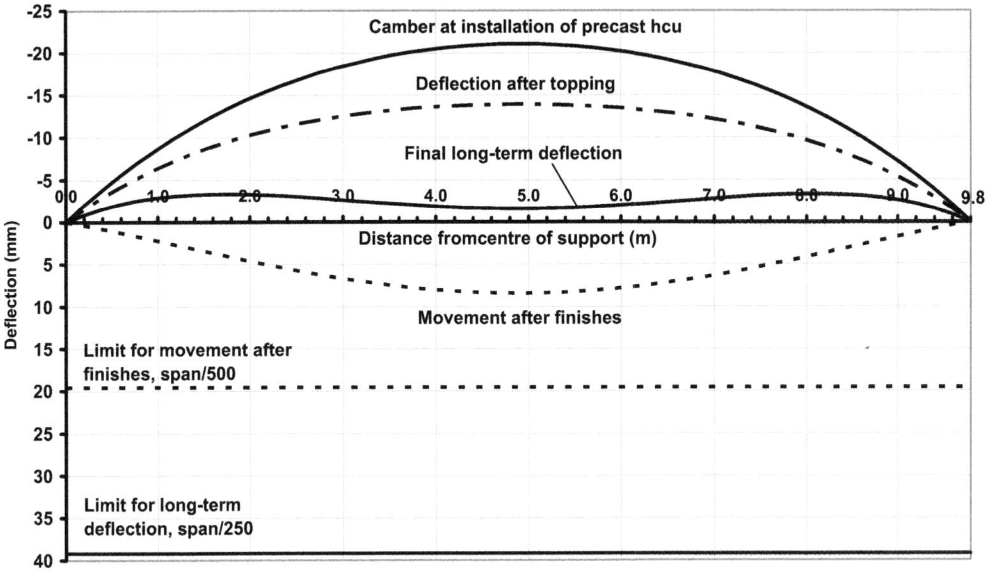

Figure 12.17 Final deflections and movement after installation of finishes for the composite 250 mm depth hcu with 75 mm structural topping in Example 12.5.

$\delta_{13}=-10.1+5\times(0.592\times6.93)\times9800^4/(384\times36283\times1296\times10^6)+5\times1.20\times2.4\times9800^4/(384\times36283\times2819.9\times10^6)=-10.1+10.5+3.4+4.6=+8.4$ mm>4.6 mm

Movement after installation=**8.4 mm**<9800/500=19.6 mm

Figure 12.17 shows the final deflections and movement after installation of finishes for Example 12.5.

It is usual for movement after installation to be critical, rather than long-term deflections. In this example, the ratios of δ_{13}/span/500=0.43 and δ_{11}/span/250=–0.04. The major influence is the increased second moment of area $I_{c,c,co}/I_{c,co}=1.43$, reducing the deflections after installation or after the topping has hardened from 72.6 mm (non-composite hcu in Example 11.1) to 49.3 mm in this example, in all 23.3 mm. Meanwhile, the camber only reduced from –63.2 to –51.0=12.2 mm, a swing of 11.1 mm, such that the final deflection reduced by the same amount to –1.7 mm. Note the large influence of the camber at installation of –40.9 mm affects both cases.

Example 12.6

Calculate the long-term deflection and movement after installation for the 250 mm depth hcu with a 75 mm Vdepth topping shown in cross section in Figure 12.9 and designed in Example 12.4, propped at mid-span. Effective span $L=11200$ mm. Self-weight of topping=2.54 kN/m.

Solution

Ratio of effective span in this example to that in Example 12.5=11.2/9.8=1.143. Then for camber is increased by $1.143^2=1.306$, and gravity load deflections by $1.143^4=1.706$

Net deflection at installation (pro-rata from Example 12.4) $\delta_3+\delta_4=1.306\times-40.9+1.706\times19.8=-53.5+33.9=-19.6$ mm

The deflection due to the topping at mid-span is eliminated by propping, then $\delta_5+\delta_6$=zero, except that at ¼ span the net deflection due to self-weight and the prop reaction is +0.32 mm.

Long-term final deflection
Due to long-term camber

δ_7=1.306×–51.0=–66.6 mm

Due to self-weight and topping,

δ_8=33.9+5×[0.592×4.36+(1+0.592)×2.54]×11200^4/(384×36283×1296×10^6)= +62.8 mm

Due to imposed dead and quasi-permanent live loads

δ_9=1.706 x +12.5=+21.3 mm

Net deflection due to addition and removal of the prop

$\delta_6+\delta_{10}$=(0.625×2.54×11200^4/48×36283×1296×10^6)×(–1+1296×10^6/2819.9×10^6)
=–11.0+5.1=–5.9 mm

Final deflection
δ_{11}=–66.6 +62.8 +21.3–5.9=**+11.6 mm**<11200/250=44.8 mm

Movement after installation and after prop removed
Due to visco quasi-permanent live load δ_{12}=1.706×4.6=7.8 mm
Due to visco self-weight, topping and imposed dead loads, and static and the above live load
Visco camber $\delta_7+\delta_3$=–(66.6–53.5)=–13.1 mm

δ_{13}=–13.1+1.706×(10.5+3.4)+7.8=18.4 mm>7.8 mm

Movement after installation=**18.4 mm**<11200/500=22.4 mm
It is usual for movement after installation to be critical, rather than long-term deflections. In this example, the ratios of δ_{13}/span/500=0.82 and δ_{11}/span/250=0.26. The former is 1.9 times that of the unpropped slab in Example 12.5, mainly due to the ratio of span^4=1.706 but also that the ratio span^2 for camber=1.306, so the visco elastic increase in camber is relatively small.

12.3.5 Dual propping of slabs over two storeys

Dual propping of slabs is occasionally carried out, where props are positioned in vertical alignment over two floors, as shown in Figure 12.18. Instead of the upper floor prop being supported through to a rigid support at the foundation, it is supported on the composite slab below. The main reason for doing this is to free up space at the ground floor to allow following trades to progress unhindered by the props. Because the composite slab below is flexible, the reaction in the upper prop will be smaller than if it were rigidly founded. The sequence is as follows:
The prop is erected at the first floor and the *insitu* topping, of self-weight w_{T1}, is added (Figure 12.18a). The prop is erected to the soffit of the floor unit at the second floor (Figure 12.18b). After about 7 days the *insitu* topping w_{T2} is added at the second floor (Figure 12.18c). When the topping has hardened the first floor slab is acting compositely (Figure 12.18c). After the prop has been removed at the first floor the reaction from the second floor prop R_2 is

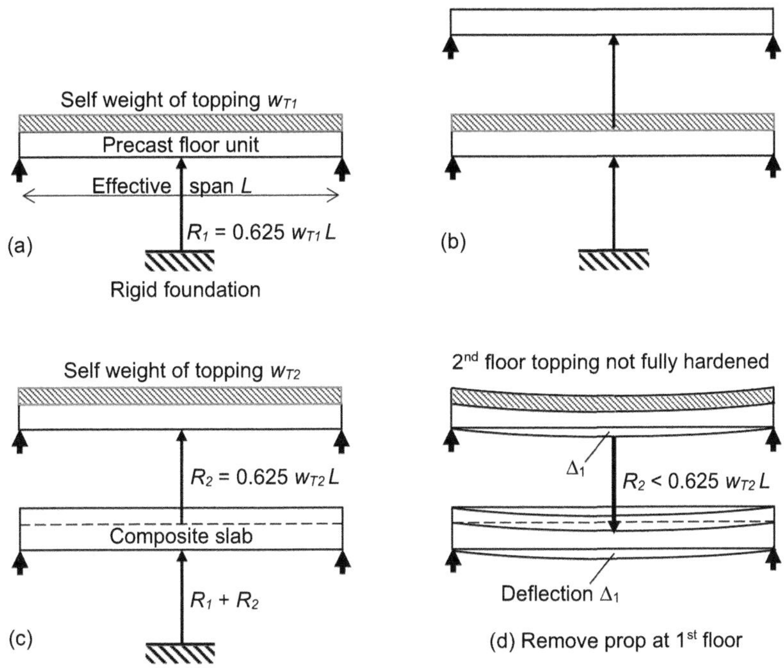

Figure 12.18 The sequence of dual propping over two nominally identical floors.

temporarily carried by the composite slab at the first floor (Figure 12.18d). Given that the prop foundation at the first floor is flexible (not rigid as at the foundation) and will deflect Δ_1 under R_2, the reaction in the second floor prop R_2 will decrease. This is because the deflection of the second floor unit due to the topping alone will be much greater than Δ_1.

Assuming initially that the first floor slab is 'fully rigid', the reaction from the second floor prop $R_2 = 0.625\ w_{T2}\ L$ causes a mid-span deflection in the first floor composite slab (general term $RL^3/48EI$) of

$$\Delta_1 = R_2\ L^3/48\ E_{cm}\ I_{c,c,co,1}$$

$$\text{Then } \Delta_1 = 0.625\ w_{T2}\ L\ L^3/48\ E_{cm}\ I_{c,c,co,1} \tag{12.58a}$$

where $I_{c,c,co,1}$ is the transformed, compound second moment of area of the composite slab at first floor

The reduction in R_2, due to the flexibility in Eq. 12.58a, is re-calculated from the equalisation of deflections due to the topping w_{T2} and the upward reaction R_2 plus the shortfall in deflection Δ_1 as

$$(5\ w_{T2}\ L^4/384\ E_{cm}\ I_{c,co,2}) = (R_2\ L^3/48\ E_{cm}\ I_{c,co,2}) + \Delta_1 \tag{12.59}$$

where $I_{c,co,2}$ is the compound second moment of area of the precast unit at second floor

$$\text{Then } R_2 = 0.625\ w_{T2}\ L - 48\ \Delta_1\ E_{cm}\ I_{c,co,2}/L^3 \tag{12.60a}$$

Substituting Eq. 12.58a into Eq. 12.60a gives

$$\text{Then } R_2 = 0.625\ w_{T2}\ L\ [1 - (I_{c,co,2}/I_{c,c,co,1})]$$

Letting $\alpha = I_{c,co,2}/I_{c,c,co,1}$

Then $R_2 = 0.625\ w_{T2}\ L\ (1-\alpha)$ (12.61a)

This reduced prop force causes a reduced deflection Δ_1 requiring iteration of R_2 until the changes in Δ_1 are small, as follows

New $\Delta_1 = [0.625\ w_{T2}\ L\ (1-\alpha)]\ L^3/48\ E_{cm}\ I_{c,c,co,1}$ (12.58b)

Substituting Eq. 12.58b into Eq. 12.60a gives

New $R_2 = 0.625\ w_{T2}\ L\ (1-\alpha+\alpha^2)$ (12.61b)

... and so on until the final prop reaction is the common power series

$R_2 = 0.625\ w_{T2}\ L\ (1-\alpha+\alpha^2-\alpha^3+\alpha^4-\alpha^5\ ...)$ (12.61c)

This is calculated in the following example.

The reaction R_2 may cause a greater bending moment in the first floor composite slab than under service loads, and should be checked, particularly for cracking at the serviceability stress. Because the limiting value for permissible tension is limited to f_{ctm} for the purpose of controlling long-term cracking and deflections, some designers choose to check the temporary condition using $f_{ctm,fl}$, typically 1.3–1.4 f_{ctm}. This applies to all conditions of exposure.

Deeper slabs, and/or reduced spans, at second floor level should be avoided as these will reduce the stiffness ratio α and lead to lower prop reactions R_2, as shown in the following example.

Let the propped composite hcu slab in Example 12.6 be dual-propped over two storeys. Using non-composite $I_{c,co,2} = 1296 \times 10^6$ mm^4 and composite $I_{c,c,co,1} = 2819.9 \times 10^6$ mm^4, the stiffness ratio $\alpha = 1296/2819.9 = 0.4596$ (assuming E_{cm} is the same for both floors). The prop reaction $R_1 = 0.625 \times 2.54 \times 11.2 = 17.78$ kN. After iteration, R_2 is given as

Eq. 12.61c. $R_2 = 17.78\ (1-0.4596+0.2112-0.0971+0.0446\ ...) = 17.78 \times 0.685 = 12.18$ kN

The mid-span deflection at the first floor due to R_2 is $\Delta_1 = 12.18 \times 10^3 \times 11200^3/48 \times 36283 \times 2819.9 \times 10^6 = 3.5$ mm. As the virtual upward deflection of the precast unit at the second floor due to the reaction in the prop would have been $\Delta_2 = 17.78 \times 10^3 \times 11200^3/48 \times 36283 \times 1296 \times 10^6 = 11.1$ mm > 3.5 mm, the prop reaction R_2 decreases. Thus the ratio of the shortfall in the upward deflection is $(11.1-3.5)/11.1 = 0.685$, confirming the calculation for R_2. The shortfall in the mid-span hogging bending moment due to the second floor prop is $(17.78-12.18) \times 11.2/4 = 15.7$ kNm.

The effect of a precast floor unit of greater depth at the second floor is a reduction in the prop reaction R_2 because the virtual deflection of the precast unit due to the self-weight of the topping (assumed to be the same as in the above example) is reduced. For example replace the 250 mm depth hcu at second floor with a generically similar 300 mm depth unit, where $I_{c,co,2} = 2123 \times 10^6$ mm^4. Then $\alpha = 2123/2819.9 = 0.753$. After iteration using Eq. 12.61c, $R_2 = 17.78 \times 0.571 = 10.14$ kN, a reduction of 25% compared with the rigid prop reaction.

12.4 INTERFACE SHEAR STRESS IN COMPOSITE SLABS

12.4.1 Horizontal interface shear stresses

Under the action of vertical flexural shear, the horizontal interface between the precast unit and *insitu* topping will be subjected to an ultimate horizontal shear stress v_{Edi}, as shown in

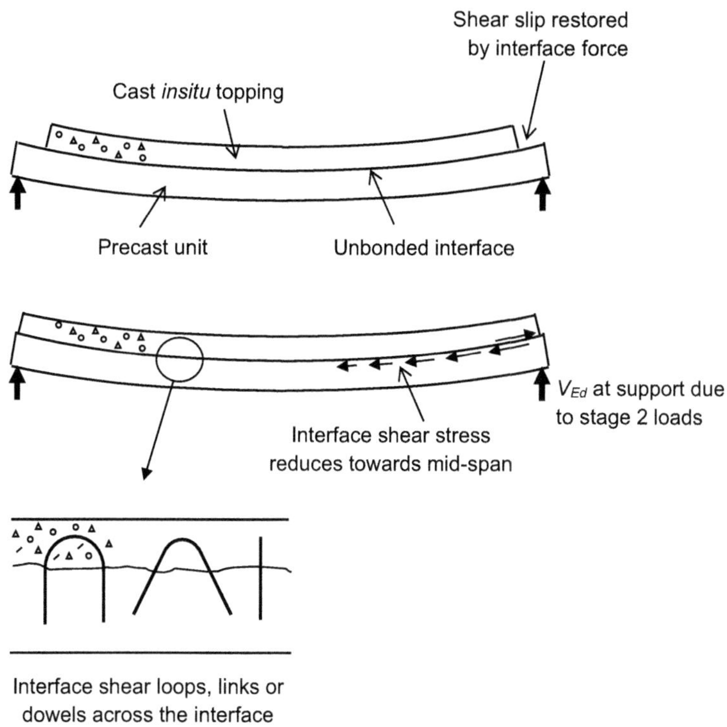

Figure 12.19 Interface shear stress and forces in composite elements.

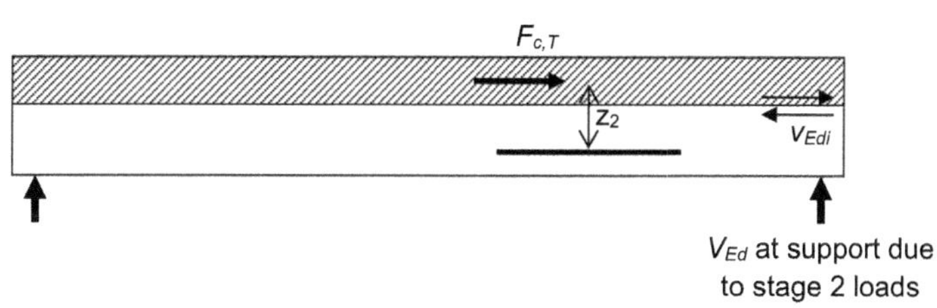

Figure 12.20 Interface shear stress based on BS EN 1992-1-1, clause 6.2.5.

Figures 12.19 and 12.20. Often termed 'shear flow', because it is measured in force per linear length, this shear derives from the equilibrium of the vertical shear at a section. It is the result of imposed, stage 2, loads present only after the *insitu* concrete topping has hardened. The distribution of interface shear is identical to the imposed shear force distribution and must therefore be checked at all critical sections. Interface shear need only be checked for the ultimate limit state. The design method is based on partly experimental evidence for the empirical terms in BS EN 1992-1-1, clause 6.2.5 (1) and Exp. 6.25, and will ensure that serviceability conditions are satisfied. BS EN 1168, Annex F2.3 for hcu refers to this clause.

The texture and surface moisture condition of the precast unit just before receiving concrete topping may affect the overall strength of the composite slabs, in particular the interface shear resistance. Ibrahim et al. (Ibrahim et al., 2016) carried out an experimental study on the shear-flexure behaviour of 24 no. composite slabs using nominally 1200×150 mm depth hcu with 75 mm depth topping. The hcu had different surface roughness (smooth and rough, as defined in BS EN 1992-1-1, clause 6.2.5 (2)) and surface moisture conditions (dry, optimum wet and ponded) Figure 12.21. Four tests were carried out for each permutation.

The smooth surface was the original extruded hcu surface while the roughened surface was obtained by scraping using a steel brush. The dry surface condition was obtained by leaving the hcu in the ambient temperature. The optimum wet, which is the preferred condition as mentioned in the FIP document (FIP Recommendations, 1982), was obtained by spraying the precast surface with 4.5 L/sq.m. of water and leaving it to dry under ambient temperature until the surface colour turned to a lighter dark grey with no excess surface water. The ponded condition was obtained by pouring water until the excess water was left on the surface with a depth of approximately 1.8 mm. The surface moisture content was recorded as 12%–13% for the dry condition, 26%–28% for optimum wet condition and 41%–43% for ponding condition.

The composite slabs were tested using a single point load at 2.5 h_s, about 562 mm where h_s is the total depth of the slab. The results for the experimental shear capacity V_u at the failure of the interface are given in Table 12.3, showing the influence of the surface conditions. The ratio for V_u for dry/optimum wet condition is 0.91 and 0.71 for the smooth and rough surface, respectively. Table 12.3 also shows the calculated interface shear stress, based on the equation $v_{ui} = V_u / b_i\, z$, where b_i is the width of the interface (in this case the measured width of the hcu, on average 1184 mm) and z is the lever arm between the tendons in the bottom of the hcu and the centre of the compression depth in the topping, on average 193 mm. For the optimum wet condition, v_{ui} is 0.75 N/mm² for the smooth, extruded surface profile. This may be compared with the value for $v_{Rdi} = c\, f_{ctdT}$ given in BS EN 1992-1-1, clause 6.2.5 (2). For an as-cast surface profile the cohesion factor $c = 0.2$, and for $f_{ckT} = 25$ N/mm², $v_{Rdi} = 0.2 \times 1.20 = 0.24$ N/mm², or 3 times less than v_{ui} in the tests. Figure 12.22a plots the values for v_{ui} for rough and smooth profiles, where the benefit of the optimum wet condition may be seen. The graph also shows design values for v_{Rdi} using $f_{ckT} = 20–40$ N/mm².

The same data are normalised in Figure 12.22b with respect to f_{ctdT} obtained from the measured cube strength of the topping, by calculating the experimental cohesion factor

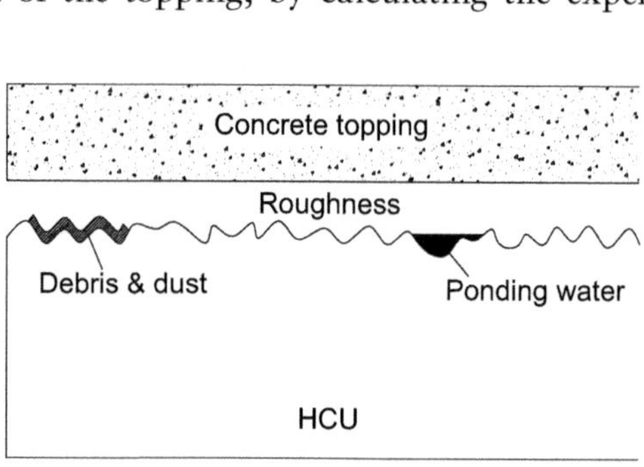

Figure 12.21 Problems related to surface roughness in concrete topping (Ibrahim et al., 2016).

Table 12.3 Comparison of experimental shear capacity V_u and interface shear resistance v_{ui} for rough and smooth surfaces between *insitu* concrete topping and prestressed hollow core floor units (based on Ibrahim et al., 2016)

	Shear capacity V_u (kN)		*Interface shear stress v_{ui} (N/mm²)*	
Surface moisture condition	*Smooth*	*Rough*	*Smooth*	*Rough*
Dry	156.33	170.65	0.69	0.75
Optimum wet	171.50	239.60	0.75	1.05
Ponded	145.98	215.50	0.64	0.95
Ratio dry/optimum wet	0.91	0.71	0.91	0.71
Ratio ponded/optimum wet	0.85	0.90	0.85	0.91

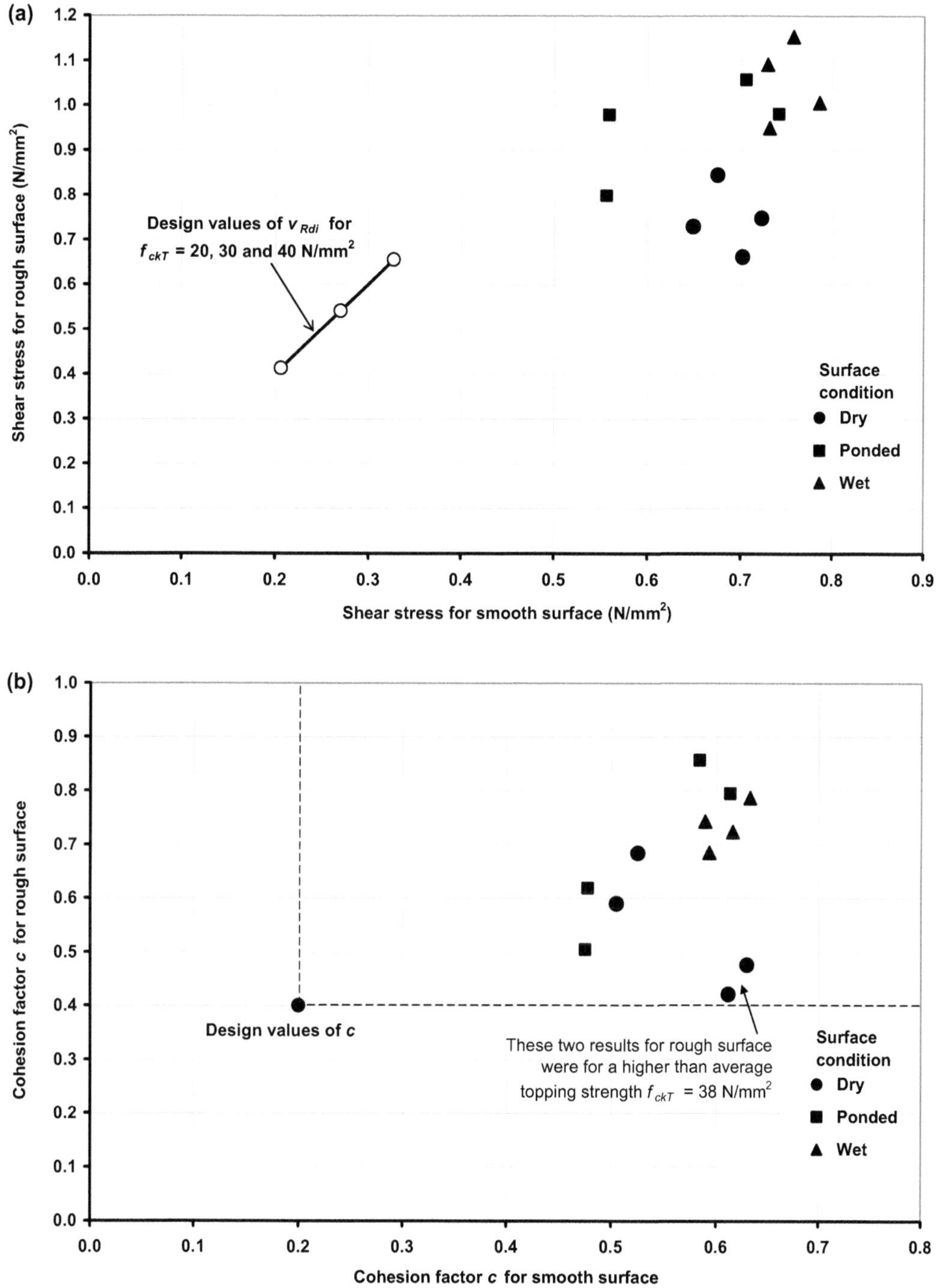

Figure 12.22 (a) Comparison of experimental and design interface shear resistance for rough and smooth surfaces between *insitu* concrete topping and prestressed hollow core floor units (based on Ibrahim et al., 2016). (b) Comparison of experimental and design values for the cohesion factor for rough and smooth surfaces between *insitu* concrete topping and prestressed hollow core floor units (based on Ibrahim et al., 2016).

$c=v_{ui}/f_{ctdT}$, for example in the optimally wet smooth surface where $f_{ckT}=26.2$ N/mm^2, then $c=0.75/1.235=0.61$. Allowing for scatter in the results the characteristic value for $c=0.57>0.2$ by a factor of 2.85. However, for the optimally wet rough surface where $f_{ckT}=33.0$ N/mm^2, then $c=1.05/1.436=0.73$ and the characteristic value $c=0.66>0.4$ by a factor of only 1.65. In all the 24 experimental values for c were greater than the design values for the smooth and rough surfaces by a ratio of 2.86 and 1.64, respectively, showing that the figure of $c=0.2$ for smooth is very conservative, and $c=0.4$, when allowing for experimental scatter and a factor of safety on the experimental results, is satisfactory. Note that the original version of clause 6.2.5 published in 2004 has $c=0.35$, in which case the above ratio would be 1.63, roughly as expected?

Providing that the *insitu* topping is fully bonded to the precast unit, full interaction is assumed, i.e. there is no relative slippage between the two concretes. The horizontal shear force $F_{c,T}$ at the interface is equal to the total force in the *insitu* topping due to stage 2 imposed loads. It is therefore necessary to have carried out a calculation for the ultimate limit state in flexure and to have determined the depth to the neutral axis X_c before this is attempted. (This is a benefit from having carried out a two-stage approach to ultimate flexure.)

The section is flexurally uncracked at the support where the interface shear is a maximum. Therefore, only stage 2 loads, after hardening of the topping, contribute to v_{Edi}. Let M_{Ed2}=maximum ultimate moment due to stage 2 loads, including dead and live point and line loads. Then

$$F_{cT}=0.567\, f_{ckT}\, b_i\, 0.8\, X_c \tag{12.62}$$

where

$$X_2=(d_2-z_2)/0.4<0.6\, d_2 \tag{12.63}$$

d_2=effective depth from top of composite section

$$z_2=d_c\,[0.5+\sqrt{(0.25-K_2/1.134)}]\le 0.95\, d_2 \tag{12.64}$$

$$\text{Stage 2 } K_2=M_{Ed,2}/f_{ckT}\, b_i\, d_2^2 \tag{12.65}$$

$$\text{If } 0.8\, X_c>\text{topping depth } t, \quad F_{c,T}=0.567\, f_{ckT}\, b_i\, t \tag{12.66a}$$

otherwise

$$F_{c,T}=0.567\, f_{ckT}\, b_i\, 0.8\, X_c \tag{12.66b}$$

$F_{c,T}$ only acts at the point of maximum bending moment – elsewhere it is less than this and may even change sign in continuous elements. Therefore, the distance over which this force is distributed along the span of the floor is taken as the distance from the maximum to the zero or minimum moment.

The average ultimate shear stresses at the interface may be calculated according to Figure 12.20 for ultimate interface force $F_{c,T}$ due to a change in moment ∂M over an incremental length Δx

$F_{c,T}=(\partial M/\partial x)\ \Delta x/z_2$ but $V_{Ed}=(\partial M/\partial x)$

then $F_{c,T}=V_{Ed}\ \Delta x/z_2$ (12.67)

Interface shear stress=force/contact area, then from Eq. 12.67

$v_{Edi}=F_{c,T}/b_i\ \Delta x=V_{Ed}/b_i\ z_2$ (12.68)

This is presented in BS EN 1992-1-1, clause 6.2.5 (1)

Exp. 6.24 $v_{Edi}=(F_{cT}/F_c)\ V_{Ed}/b_i\ z_2$ (12.69)

where $F_{c,T}$=compression in topping due to ultimate stage 2 loading

F_c=total compression in topping and precast unit due to ultimate stage 2 loading
V_{Ed}=ultimate vertical shear force at the support (not on shear plane) due to loads after the topping has hardened.
z_2=lever arm of rebars-topping due to stage 2 loads
b_i=width of interface (defined in BS EN 1992-1-1, Fig. 6.8)

Strictly, for units such as hcu and double-tee b_i should be the actual width of the unit, e.g. 1154 mm for the hcu in Figure 6.7, unless the *insitu* infill in the longitudinal joint is cast at the time of erection, the preferred sequence, and the structural topping is cast at a later date. As this sequence is often unknown at the design stage it is wiser to use the actual width, rather than the nominal width.

v_{Edi} is then distributed in accordance with the magnitude of the vertical shear V_{Ed} at any section.

The design interface shear stress v_{Edi} is first checked against the cohesive and frictional shear resistance according to BS EN 1992-1-1, clause 6.2.5 (1)

Exp. 6.25 $v_{Rdi}=c\ f_{ctd}+\mu\ \sigma_n\leq 0.5\ v\ f_{ck}/\gamma_c$ (12.70)

where σ_n=normal stress=0 (the vertical pressure of imposed load and self-weight of topping is ignored)

Exp. 6.6N $v=0.6\ (1-f_{ck}/250)$

3.1.6 (2)P Design *insitu* tensile strength of the topping or the weaker material

$$f_{ctd}=0.7\times 0.3f_{ck}^{2/3}\ /1.5$$

6.2.5 (2) c=0.025 to 0.1, μ=0.5 ('very smooth' against a steel, plastic or prepared wooden mould)

c=0.2, μ=0.6 ('smooth' after casting, slip-formed or extruded)
c=0.4, μ=0.7 ('rough',>3 mm rough ridges at 40 mm spacing by raking or similar)
c=0.5, μ=0.9 ('indented', according to the profile shown in Figure 12.23)

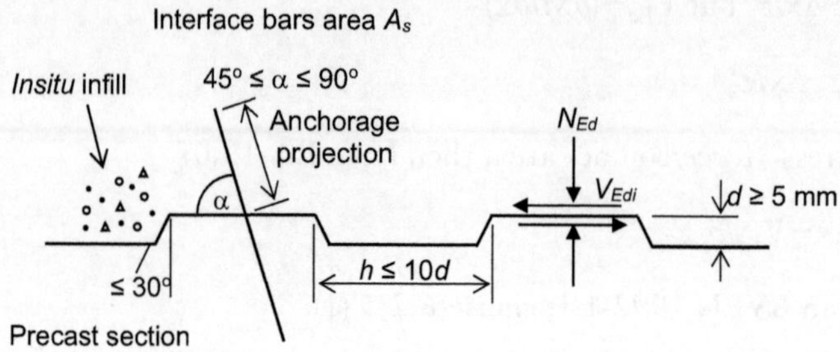

Figure 12.23. Indented profile at the edge of inverted tee beams (or other similar interfaces) (based on BS EN 1992-1-1, Fig. 6.9).

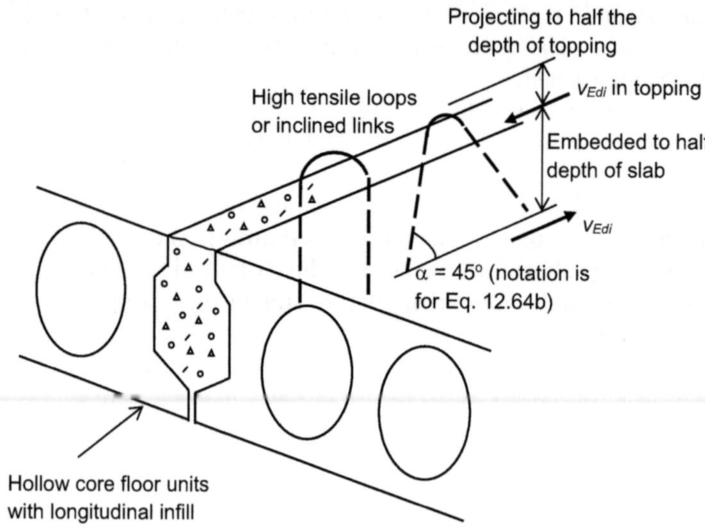

Figure 12.24 Interface shear links (or loops or dowels) in joints between hollow core floor units.

12.4.2 Interface shear reinforcement

If v_{Edi} is greater than v_{Rdi} then shear reinforcement is added (per 1 m run) projecting at right angles from the precast unit into the structural topping

Exp. 6.25 $$v_{Rdi} = c\, f_{ctd} + \mu\, \sigma_n + \mu\, A_s\, f_{yd}/A_j \le 0.5\, v\, f_{ck}/\gamma_c \tag{12.71a}$$

where A_s = area interface shear reinforcement

A_j = area of interface joint

If inclined loops are used, at an angle not less than α = 45°, and the other leg at α′ ≤ 135°, as shown in Figure 12.24, BS EN 15037-1 (BS EN 15037-1, 2008), Annex C2, gives an equation in which Eq. 12.71a is modified to

$$v_{Rdi} = c\, f_{ctd} + \mu\, \sigma_n + (\mu \sin\alpha + \mu \sin\alpha' + \cos\alpha)\, A_s\, f_{yd}/A_j \le 0.5\, v\, f_{ck}/\gamma_c \tag{12.71b}$$

i.e. the component (parallel to the surface of the interface) of the area in each leg of the loop is $\mu \sin\alpha\, A_s$ but the component of the area of the leading leg ($\cos\alpha'$) is ignored as the bar is leaning away from the direction of the interface force. For example, using $\mu = 0.6$ and 45° legs, the parallel component is $(0.6 \times 0.707 + 0.6 \times 0.707 + 0.707)\ A_s = 1.55\ A_s$, as opposed to 2.0 A_s for a 90° loop.

Exp. 6.25 is at odds with Annex C2 because the leg leaning away from the direction of the interface force has the component cos α, where in Annex C2 it does not, but the above calculation gives the same result if there was the cos α' term in Exp. 6.25 and it is ignored.

The distribution of A_s should respond to the distribution of V_{Ed} as shown in BS EN 1992-1-1, Fig. 6.10.

If the normal pressure σ_n is zero, the required area interface shear reinforcement using vertical loops or dowels is

$$A_s = (v_{Edi} - c\ f_{ctd})\ b_i/\mu\ f_{yd}\ (\text{mm}^2/\text{mm}) \qquad (12.72)$$

but not less than $A_{s,min} = 0.13\%$ or $0.26\ f_{ctmT}/f_{yk}$ of the thru-depth contact area as the bars are required for confinement. The horizontal contact area is not used as this leads to a huge area, e.g. for 1200 mm wide units $A_{s,min} = 0.13\% \times 1200 \times 1000 = 1560$ mm²/m requiring H12 loops at 145 mm spacing! The reinforcement should be adequately anchored on both sides of the interface. If loops are used, as shown in Figure 12.24, the clear space beneath the bend should be at least 5 mm+size of aggregate. The spacing of links should not be too large, 1.20–1.50 m being typical for hcu. If $v_{Edi} < v_{Rdi}$ no interface shear reinforcement is required, although some contractors choose to place H10 or H12 loops (as shown in Figure 12.24) at about 1200 mm spacing anyway. The loops should pass over the top of the bars in the structural topping and be concreted into the joints between the precast units.

Example 12.7

Calculate the interface shear reinforcement necessary to satisfy the ultimate horizontal shear force at the precast – *insitu* interface for the 250 mm depth hcu with a 75 mm depth topping shown in cross section in Figure 12.9 and designed in Example 12.2 if

i. UDL loading as Example 12.1
ii. Increased live load of 10 kN/m²

The top surface of the hcu is 'as extruded' finish. In this example use b_i as the actual width of the hcu. Use high tensile reinforcement for interface links or loops $f_y = 500$ N/mm².

Solution

i. UDL loading only

$M_{Ed1} = 1.25 \times 6.93 \times 9.8^2/8 = 104.0$ kNm

$M_{Ed2} = 247.9 - 104.0 = 139.9$ kNm (from Example 12.2)

$K_2 = 139.9 \times 10^6/(25 \times 1200 \times 284.2^2) = 0.051$

$z_2 = 268.4$ mm

$X_2 = (284.2 - 268.4)/0.4 = 39.4$ mm

thus $0.8\ X_2 <$ topping depth 75 mm, then $F_{ct}/F_c = 1.0$, i.e. ratio of flexural force in topping/total force due to stage 2 bending $= 1.00$

V_{Ed2} at the support=12.0×9.8/2=58.8 kN

where w_{Ed2}=(1.25×2.4+1.5×6.0)=12.00 kN/m

Eq. 12.69 v_{Edi}=1.00×58.8×10³/1154×268.4=0.19 N/mm²

f_{ctmT}=0.7×0.3×25^{2/3}/1.5=1.20 N/mm²

Eq. 12.70 v_{Rdi}=0.2×1.20=0.24 N/mm²<0.5×0.6×(1–25/250)×25/1.5=4.50 N/mm²

No interface shear loops required

ii. Imposed live load=10 kN/m²

w_{Ed2}=1.25×2.4+1.5×12.0=21.0 kN/m

M_{Ed2}=21.0×9.8²/8=252.1 kNm

K_2=252.1×10⁶/(25×1200×284.2²)=0.104; then z_2=255.2 mm

X_2=72.6 mm; 0.8 x_2<75 mm, then F_{ct}/F_c=1.0 and V_{Ed2}=102.9 kN

Eq. 12.69 v_{Edi}=1.00×102.9×10³/1154×255.2=0.35 N/mm²>0.24 N/mm² but<4.50 N/mm², use interface loops with two vertical legs
Smooth surface profile μ=0.6
Eq. 12.72 A_s=1000×(0.35−0.24)×1154/(0.6×0.87×500)=486 mm²/m run
Minimum area is the greater of 0.13%×1000×75=98 mm²/m or 0.26×2.56×1 000×75/500=100 mm²/m
Use H10 at 320 mm spacing (2×78.5×1000/320=490 mm²/m)
Check Eq. 12.71a, v_{Rdi}=0.2×1.20+0.6×490×435/1154×1000=0.24+0.11=0. 35 N/mm².

12.5 COMPOSITE BEAMS

12.5.1 Internal beams

Internal beams are usually symmetrically loaded, i.e. floor slabs are on both sides of the beam, and therefore the beam is symmetrical in cross section as shown in Figures 2.7 and 6.9. The limiting design criterion is often minimum depth in order to maximise headroom and reduce the drop beam, or 'downstand' depth defined in Figure 6.9. For this reason, internal beams are often pretensioned to maximise their structural performance. To minimise the downstand, part of the beam may be recessed within the depth of the floor slab, giving rise to the so-called 'inverted-tee' beam. Internal beams may be designed compositely, with the floor slab acting as a compression flange.

There is great potential for designing composite beams as continuous, as shown in Figure 12.25 where the negative moment of resistance is achieved by passing rebars through slots in columns, or as shown in Figure 12.26 placing the bars in the gap between single-story columns.

Figure 12.25 Composite beam design with negative moment continuity of rebars either through sleeves in columns.

Figure 12.26 Composite inverted-tee beam showing mesh in the topping and tie steel into the slots in hollow core floor units at North Galaxy building, Brussels, Belgium. (Courtesy Ergon nv/sa, Belgium.)

12.5.2 Composite prestressed concrete beam design

Composite action in psc beams is achieved in exactly the same manner as described in Section 12.2 for psc slabs. However, there is potential for greater enhancement in the flexural capacity of composite beams than with composite slabs because of the greatly increased section modulus at the top of the beam. This is because the neutral axis in the composite section is near to the top of the beam, and as seen in Example 12.8 the need to render the beam flexurally 'balanced' or 'under reinforced' is important. Shear design is not carried out using the composite properties where hcu alone are used, particularly at edge columns where, unlike flexural behaviour, the composite section may not be fully effective at the face of the support, although the term for $I_{c,c}\ b_w/S_{c,c}$ in Eq. 12.38 may be calculated from the width of the beam and total composite depth of construction, particularly in rectangular beams. Another reason for this is where floor slabs are notched around columns and the *insitu* infill is not active with the precast unit. However, composite design may be used at internal columns where hcu are continuous at the face of the support (see Example 12.10).

The composite section will be effective where precast-*insitu* half-slab is used. A two-stage approach is used, similar to Eq. 12.37, where the shear stress due to stage 1 $v_{Ed1}=V_{Ed1}\ S_c/b_w\ I_c$ is subtracted from the principal stress in stage 2 according to Section 12.3.2.4 and Eq. 12.38. Deflections are significantly reduced in composite beams due to the greater $I_{c,c}$ values in the composite section.

Composite action in beams is not considered where double-tee floor slabs or thermal slabs are used, or where solid psc slabs are used without a structural topping. If hcu are used without a structural topping the top flange is formed by placing tie steel into a number of top opened hollow cores, known as 'slots', typically 3–5 per 1200 mm wide (depending on the number of cores in the hcu)×500 to 600 mm in length (depending on the diameter of the tie bars, 10 or 12 mm, respectively). For example in a 11 core unit, 5 slots could be filled, that would effectively form a continuous flange even though 6 cores were blocked off at their ends. Similarly for 6 cores 3 slots could be filled.

Figure 12.27 shows a composite beam test being carried out on an inverted-tee beam and hcu. It was found that the effective width of the floor was at least 1.0 m wide in spite of the fact that the floor section includes many empty hollow cores. The resulting span-to-depth ratio for this type of construction varies from about 15:1 to 22:1 for typical floor loads (15 kN/m^2 ultimate) and floor spans (6–10 m). Figure 2.36 shows load v span graphs for 600 mm wide inverted tee beams made composite with 200 mm depth hcu and 1 m wide *insitu* infill in the cores.

The precast-*insitu* interface must be reinforced. If the interface is horizontal (e.g. above a rectangular or I beam) the options for projecting loops or dowels are as shown in Figure 12.28. If the interface is vertical (e.g. inverted-tee beam) the options are shown in Figure 12.29. Factory cast projecting bars are bent upwards for transportation and fixing and later bent down on site into the opened cores of the hcu, Figure 12.29a, or site placed through holes in the upstand, Figure 12.19b. For shallow upstands, or deeper slabs, projecting links or additional loops may project into the *insitu* infill above the beam, which should have a roughened surface profile, as shown in Figure 12.29c, where the depth of the *insitu* infill should be at least 75–100 mm (depending on the size of the beam, etc.) as it is not possible to generate the necessary compressive force in the small quantity of *insitu* concrete in this region.

Figure 12.27 Full scale testing of composite beams with hollow core floor slab. (Courtesy Bison Precast Ltd., UK.)

12.5.3 Serviceability limit state of composite beams

As with composite slab design, Section 12.2, service and ultimate stresses are checked at two stages of loading (or three if a structural topping is added to the floor) and superimposed elastically. Figure 12.30 shows either a two-stage or three-stage design: (i) precast beam alone, (ii) precast beam with *insitu* infill into the top opened cores of hcu, and (iii) stage 2 plus *insitu* structural topping. If a structural topping is used the loading conditions are as follows

Stage	*Loading*	*Section properties based on*
1	Self-weight of beam and dry floor slab	Precast beam only
2	As stage 1 plus self-weight of topping	As stage 1 plus *insitu* infill near the beam
3	Superimposed dead and live	As stage 2 plus topping

Stage I stresses exist in the precast beam only, Figure 12.30a. These are the result of prestress, elastic shortening and short-term relaxation, self-weight of the beam, precast floor units and wet concrete. Stage 2 stresses exist in the precast beam and *insitu* infill, Figure 12.30b, and are due to the self-weight of the topping, Figure 12.30c. Stage 3 stresses in the composite beam, Figure 12.30d are in addition to that mentioned earlier and are the result of superimposed, services and partition loading, differential shrinkage and the total creep relaxation

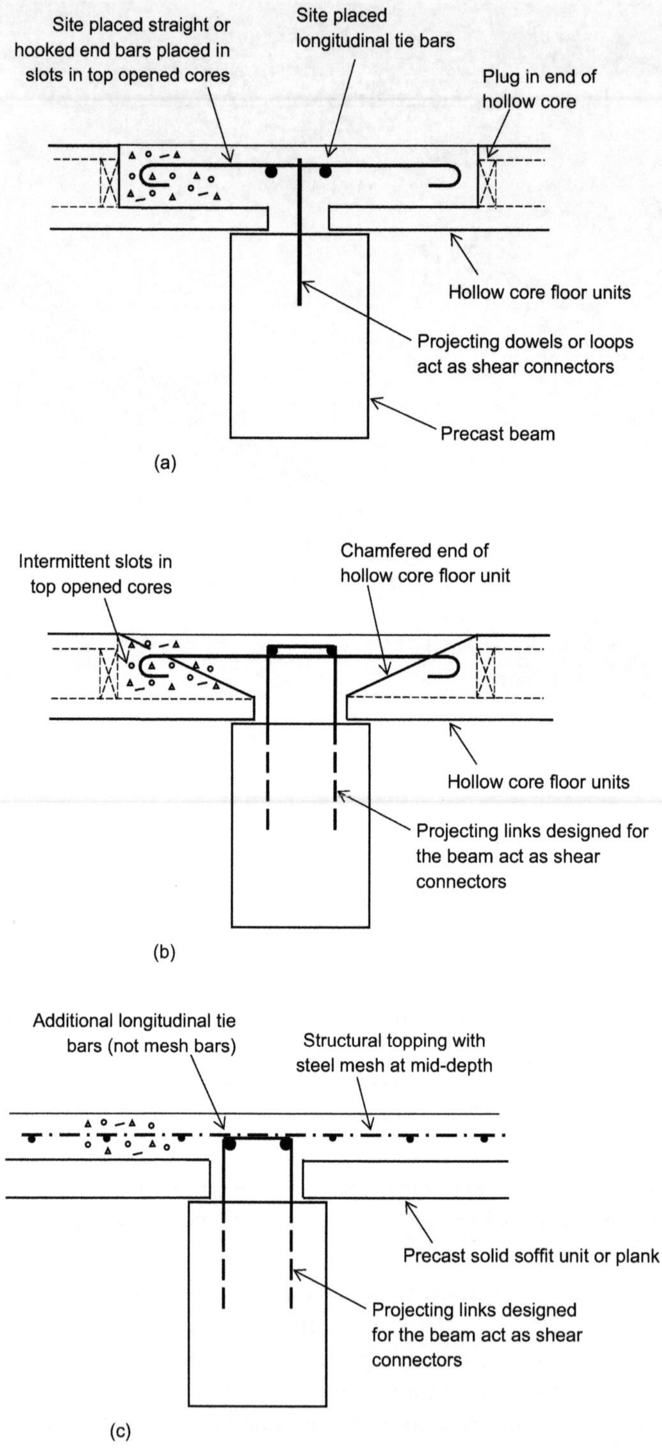

Figure 12.28 Interface shear reinforcement in composite rectangular beams using (a) single dowel and slotted top opened cores in hollow core unit (hcu), (b) projecting loop into slotted holes in chamfered hcu (a popular detail using steel beams with shear studs), (c) projecting loop into the topping of half-slab.

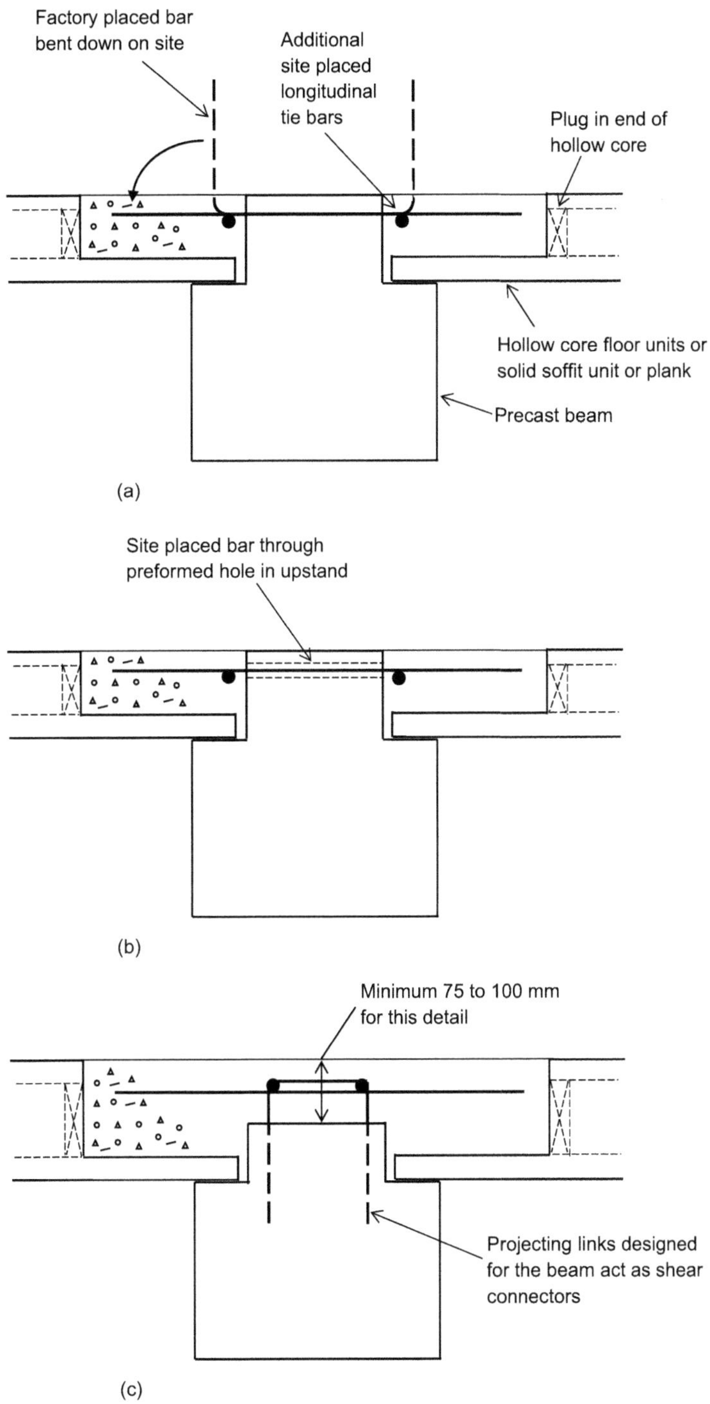

Figure 12.29 Procedures for placing interface shear reinforcement and longitudinal and transverse tie steel in inverted-tee beams (a) projecting bar cast in at the factory and bent down on site, (b) loose bar placed through sleeve in the upstand, (c) projecting links into *insitu* infill above a shallow upstand or deeper slab.

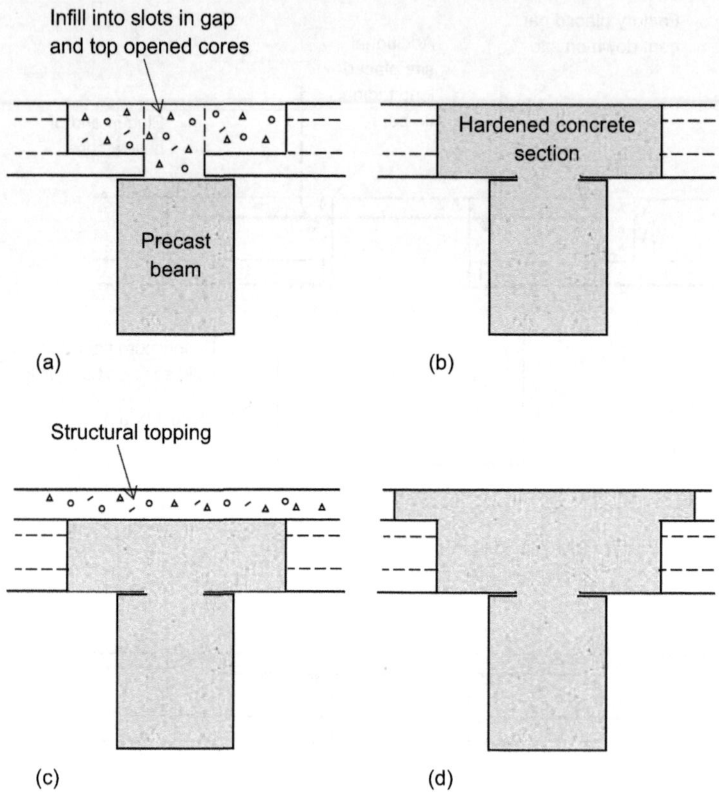

Figure 12.30 Stages in cross-sectional properties of a composite beam with precast slab and structural topping (a) stage 1 with wet *insitu* infill in hollow core units (hcu), (b) stage 2 where infill has hardened, (c) stage 3 with wet *insitu* topping and (d) final stage 3 with hardened topping.

after hardening of the *insitu* concrete. Eqs. 12.1, 12.2 and 12.13–12.15 are appropriate, except that if a structural topping is included Eqs. 12.13 and 12.14 are modified to

$$f_b=\sigma_b-M_{s1}/Z_{b,co}-M_{s2}/Z_{b,c,co}-M_{s3}/Z_{b,c3,co}>-f_{ctm} \tag{12.73a}$$

$$f_t=\sigma_t+M_{s1}/Z_{t,co}+M_{s2}/Z_{t,c,co}+M_{s3}/Z_{t,c3,co}<0.45\ f_{ck} \tag{12.74}$$

where M_{s3} and $Z_{b,c3,co}$ and $Z_{t,c3,co}$ are the design moment and section moduli for stage 3.

As before the limit $-f_{ctm}$ in Eq. 12.13a only applies to XC1. For XC2 to XC4 and XD1/XS1 the limiting tension is zero and M_{s3} is replaced by $M_{s3,qp}$ and $M_{s3,freq}$ is for the quasi-permanent and frequent imposed loads, respectively. Then

$$f_b=\sigma_b-M_{s1}/Z_{b,co}-M_{s2}/Z_{b,c,co}-M_{s3,qp}/Z_{b,c3,co}>0 \tag{12.73b}$$

$$f_b=\sigma_b-M_{s1}/Z_{b,co}-M_{s2}/Z_{b,c,co}-M_{s3,freq}/Z_{b,c3,co}>0 \tag{12.73c}$$

For 2 stages $M_{sR,c}=[(\sigma_b+f_{ctm}-M_{s1}/Z_{b,co})\ Z_{b,c,co}]+M_{s1}$ (12.75)

For 3 stages $M_{sR,c}=[(\sigma_b+f_{ctm}-M_{s1}/Z_{b,co}-M_{s2}/Z_{b,c,co})\ Z_{b,c3,co}]+M_{s1}+M_{s2}$ (12.76)

with the same replacement of f_{ctm} with zero for exposure>XC1.

The effective width of the top flange b_{eff} for the *insitu* topping or for half-slab (precast unit with structural topping) is calculated as Eq. 12.6, but b (on either side of the top of the beam of width b_w) is determined from BS EN 1992-1-1, clause 5.3.2.1 (3), Figs. 5.2 and 5.3, and Exp. 5.7, 5.7a and 5.7b as

$$\text{Exp. 5.7a} \quad b=0.2\,b_1+0.1\,l_o \leq 0.2\,l_o \tag{12.77}$$

where for simply supported beams l_o=effective span L, and b_1 is the clear spacing between beams.

For example for a rectangular beam of 400 mm width at 6.0 m column centres, and with a span $L=8.0$ m, on either side of the beam $b=0.2\times5600+0.1\times8000=1920$ mm$\leq0.2\times8000=1600$ mm. If floor slabs are on both sides total $b=400+2\times1600=3600$ mm. The rule of thumb used to be ¼ span on either side of the beam, then $b=4000$ mm.

Note that for two-stage precast design, if the upstand of the beam (e.g. inverted-tee beams) extends level with the topping, b_w is discounted as the beam is already in compression from stage 1 loads.

For hcu with top opened cores, known as 'slots' b is equal to the total length of the slots, plus any infill between the ends of the hcu, as shown in Figures 12.28 and 12.29. For example for a rectangular beam of 400 mm width supporting hcu with 100 mm bearing length and with 600 mm length slots, $b=400-2\times100+2\times600=1400$ mm. If $f_{ckT}=25$ N/mm^2 and $f_{ck}=45$ N/mm^2, $b_{eff}=1400\times31476/36283=1214$ mm (an alternative is $b_{eff}=1400\times(33/53)^{0.3}=1214$ mm). The depth of the top flange is equal to the depth of the *insitu* filled hollow cores, typically h -30 to 40 mm, where h is the depth of the hcu.

12.5.4 Ultimate limit state of flexure in composite beams

The ultimate moment of resistance $M_{Rd,c}$ is calculated as per Section 9.4.1, i.e. by using iteration of the strains, stresses and forces in the tendons in each row in the beam, until the depth to the neutral axis converges with the initial assumption. Because the compression is in the top flange of either the topping or the infill in top opened cores in hcu, Eq. 8.49 is modified to

$$F_c=0.567\,f_{ckT}\,b_{eff}\,0.8X \text{ in the topping} \tag{12.78a}$$

$$F_c=0.567\,f_{cki}\,b_{eff}\,0.8X \text{ in the infilled cores} \tag{12.78b}$$

It is also calculated in a two-stage design as per Section 12.3.2.3, except that if a structural topping is included the moment if resistance due to stages 2 and 3 are combined (the difference compared with separate calculation for the two stages is negligible), such that the subscripts 2 become 2,3 (= stages 2 and 3 combined). To calculate the total $M_{Rd,c}$ in a beam with multiple rows of tendons it is not possible to simply subtract A_{p1} from A_{pT} because the final strains, stresses and forces are summated for each row of tendons. It is therefore necessary to calculate the equivalent area of each tendon to be subtracted, as follows

Let M_{Ed1}=stage 1 ultimate design moment

$$K_1=M_{Ed1}/f_{ck}\,b_1\,d_1^2 \tag{12.79}$$

where b_1=width of precast unit, d_1=effective depth from top of precast unit

$$z_1=[0.5+\sqrt{(0.25-K_1/1.134)}]\,d_1\leq0.95\,d_1 \tag{12.80}$$

$$X_1=(d_1-z_1)/0.4 \tag{12.81}$$

If $0.8X_1\leq h_{ft}$ the depth of the top flange

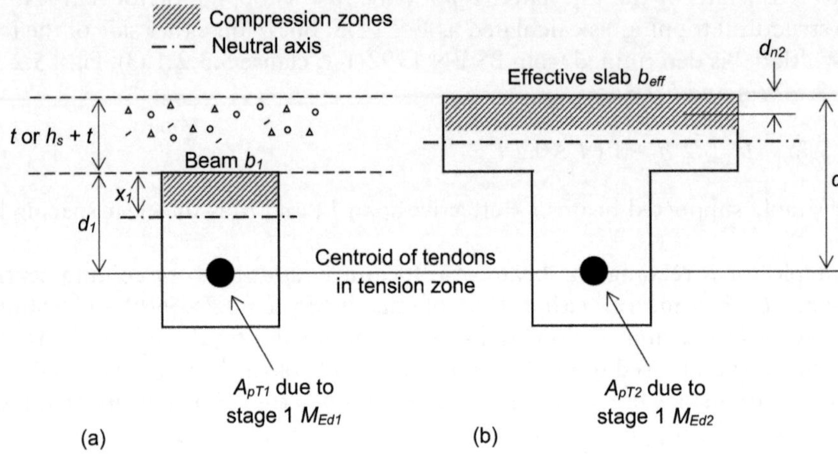

Figure 12.31 Ultimate limit state design in composite beams (a) stage 1 based on precast alone prior to hardening of the *insitu* topping, (b) stage 2 composite when topping of depth t, or topping plus infill of depth $t+h_s$, has hardened.

then $A_{p1}=M_{Ed1}/z_1\, f_{p1}$ (12.82)

Fig 3.10 where f_{p1}=stress in tendons at strain $\varepsilon_s=0.02$

The equivalent area subtracted from each tendon is $A_p'=A_{p1}/N$, where N is the total number of tendons contributing to A_{pT}. The net area per row is required in the analysis because the final compatibility strains and stresses are required per row (see Example 12.9 and Table 12.4b).

The final ultimate stress in the tendons f_p is calculated from the net area $A_{pT2}=A_{pT}-A_{p1}$

Referring to Figure 12.31, the effective depth to the tendons is $d_2=d_1+h_s+t$, where h_s and t are the depths of slab and topping, respectively. Then without topping, e.g. hcu with filled top opened cores

$$M_{Rd,c}=f_p\, A_{pT2}\, (d_2-d_{n2})+M_{Ed1} \quad (12.83a)$$

where $M_{Rd,c}$ is the total moment of resistance for stage 2 added to the stage 1 M_{Ed1}.

If a topping is added to the infill in the slabs, then

$$M_{Rd,c}=f_p\, A_{pT2}\, (d_2-d_{n2,3})+M_{Ed1} \quad (12.83b)$$

where $M_{Rd,c}$ is for stages 2 and 3 added to M_{Ed1}.

The effective width of the flange b_{ef} is based on the full depth of the *insitu* topping or the full depth of the hcu if slots in top opened cores are used. If the two-stage design is carried out b_{ef} does not include the width of the beam, which is consumed in stage 1 (see Example 12.9).

The width of the beam can be made as wide as possible, in some cases 1200 mm as shown in Figure 12.32 where the precast beam in this photograph is acting essentially as a permanent shuttering to hcu. The rebar girder acts as a bending truss to carry the self-weight of the beam and floor slabs until *insitu* concrete is placed over the top of the beam to form a composite section. To further increase the moment capacity of the beam and reduce the mid-span sagging moment, continuity reinforcement is placed in the top of the rebar girder, and passes through the column, thus producing a hogging moment of resistance equal to the sagging moment. In cases such as these the columns are usually discontinuous at the floor level, which leaves space for the continuity reinforcement.

Figure 12.32 Precast wide beam with top rebar girder to be constructed compositely with prestressed hollow core floor units.

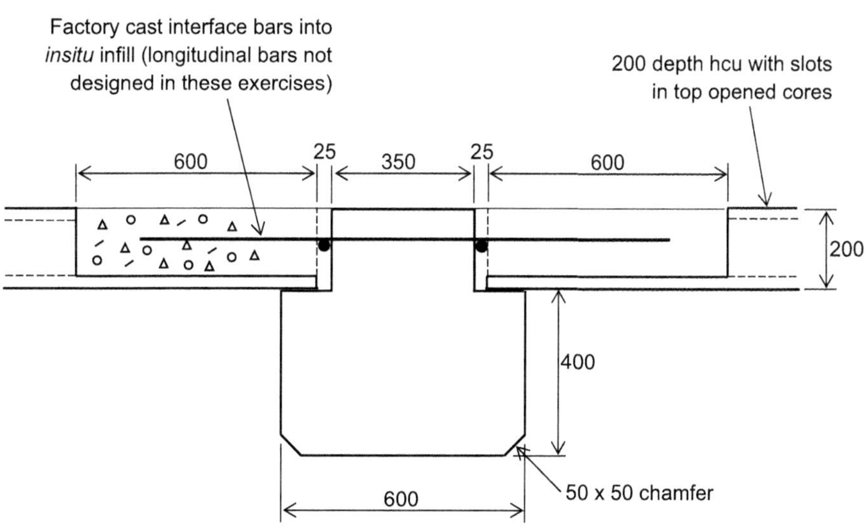

Figure 12.33 Cross section of composite inverted-tee beam with hcu and *insitu* infill in slots in top opened cores used in Examples 12.8–12.13.

Example 12.8

Calculate the serviceability moment of resistance $M_{sR,c}$ for the inverted-tee beam shown in cross section in Figure 6.10 (without debonded strands) and Figure 12.33 designed compositely using 200 mm depth hcu with a clear span from the face of the beam of 6.0

m and a bearing length of 100 mm. The length of the slots in top opened cores is 600 mm. The infill is f_{cki}=25 N/mm^2. Exposure class is XC1. Draw the service bending moment distributions along the span.

The effective span of the beam is 8.50 m. The self-weight of the beam=7.688 kN/m, floor slab=18.0 kN/m, other dead loads=10.0 kN/m and imposed live load=45.0 kN/m (increased from 30 kN/m in the non-composite beam). Geometric and material data given by the manufacturer and/or derived in Example 7.10 are as follows

$A_{c,co}$=318077 mm^2, $y_{b,co}$=266.7 mm, $I_{c,co}$=8449.7×10^6 mm^4, $Z_{b,co}$=31.683×10^6 mm^3, $Z_{t,co}$=25.351×10^6 mm^3, f_{ck}=45 N/mm^2, E_{cm}=36283 N/mm^2, A_p=26×93=2418 mm^2 and z_{cp}=92.86 mm. For the *insitu* infill f_{cki}=25 N/mm^2 and E_{cmi}=31476 N/mm^2.

Use the same pretensioning data as in Example 7.10, where the final prestress at mid-span is σ_b=15.13 N/mm^2 and σ_t=-1.22 N/mm^2.

Solution

Actual width of top *insitu* flange=2×25+2×600=1300 mm

Effective width b_{eff}=1300×31476/36283=1128 mm

Area of *insitu* flange=1128×200=225600 mm^2

Height to centre of flange=600-200/2=500 mm

Compound composite section properties.

Using $m-1$=(195000/36283) −1=4.37

$A_{c,c,co}$=318077+225600=543677mm^2

$y_{b,c,co}$=(318077×266.7+225600×500)/543677=363.5 mm

$I_{c,c,co}$=8449.7×10^6+318077×(363.5−266.7)^2+1128×200^3/12+225600×(500−363.5)^2 =16385×10^6 mm^4,

$Z_{b,c,co}$=16385×10^6/363.5=45.076×10^6 mm^3

$Z_{t,c,co}$=16385×10^6/(600-363.5)=69.278×10^6 mm^3

Stage 1 w_{s1}=7.688+18.0=25.69 kN/m

M_{s1}=25.69×8.5^2/8=232.0 kNm

f_{b1}=232.0×10^6/31.683×10^6=−7.32 N/mm^2 tension

f_{t1}=232.0×10^6/25.351×10^6=+9.15 N/mm^2 compression

Stage 2 w_{s2}=10.0+45.0=55.0 kN/m, M_{s2}=496.7 kNm

f_{b2}=496.7×10^6/45.076×10^6=−11.02 N/mm^2

f_{t2}=496.7×10^6/69.278×10^6=+7.17 N/mm^2

Check at top of *insitu* infill f_{t2}'=f_{t2}=+7.17 N/mm^2<0.45×25=11.25 N/mm^2. OK

Final stresses

Bottom f_b=15.13+−7.32+−11.02=−3.21 N/mm^2>f_{ctm}=−3.80 N/mm^2

Top f_t=−1.22+9.15+7.17=+15.10 N/mm^2<0.45 f_{ck}=20.25 N/mm^2

The additional moment capacity available, due to the additional stress at the bottom of f_b-f_{ctm}=−3.21−(−3.80)=+0.59 N/mm^2, is $M_{sR,2}-M_{s2}$=0.59×45.076×10^6×10^{-6}=26.6 kNm. By inspection the top stress is not critical as $M_{sR,c}$=1085 kNm.

Eq. 12.18 $M_{sR,c}$=496.7+26.6+232.0=**755.3 kNm**>M_s=232.0+496.7=728.7 kNm.

Note that $M_{sR,c}$=755.3>599.7 kNm is 26% greater than for the non-composite beam in Example 7.10. The additional volume of *insitu* concrete used in the slots in the hcu is effectively free of charge because the infill and tie steel is required for other purposes, e.g. horizontal diaphragm action and stability ties for robustness.

Figure 12.34 shows the service design moments M_s and the service moments of resistance $M_{sR,c}$ for Example 12.8. Note the reduction in $M_{sR,c}$ as the stage 1 stresses increase towards mid-span. At the supports, where M_s=0, $M_{sR,c}$ is due to f_{ctm} and the compound composite section modulus $Z_{b,c,co}$. Then $M_{sR,c}$=3.80×45.076=171.3 kNm. From there to $x=l_{pt2}$, the end of the transmission length, $M_{sR,c}$ increases linearly.

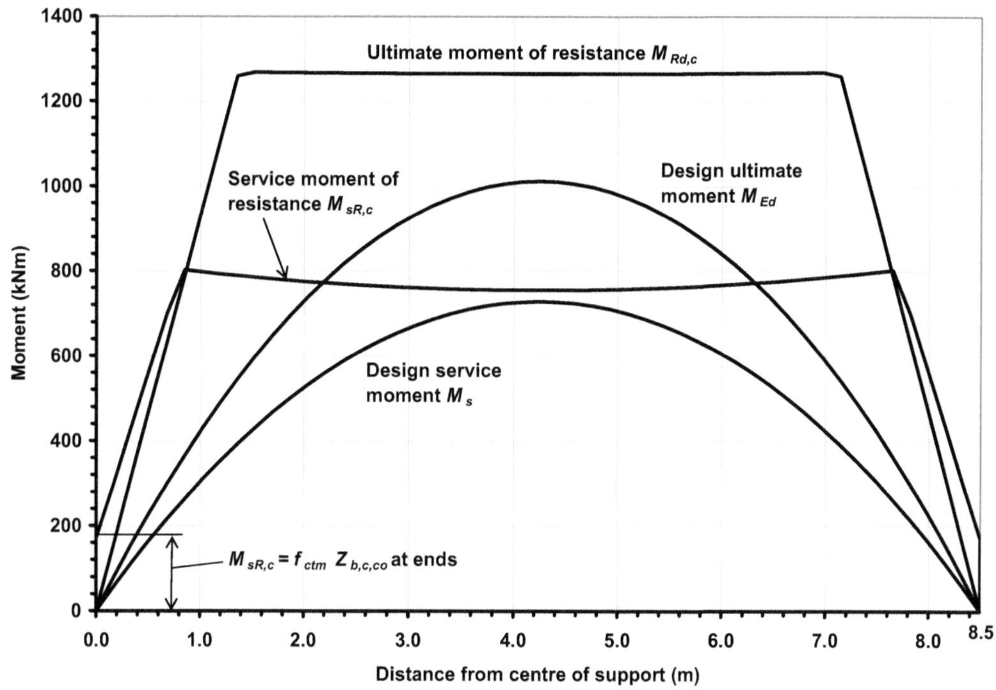

Figure 12.34 Service and ultimate bending moments for the composite inverted-tee beam in Examples 12.8 and 12.9.

Example 12.9

Calculate the ultimate moment of resistance $M_{Rd,c}$ for the inverted-tee beam in Example 12.8. The beam is to be used for offices, where $\psi_0=0.7$. The area per 12.5 mm diameter strand is 93 mm^2. Draw the ultimate bending moment distributions along the span, including the effects of bond anchorage at the supports.

Solution

In the non-composite beam in Example 9.5 the number of strands contributing to the tension force is 22, i.e. the 2 strands in the top of the beam and 2 strands in the top of the boot do not contribute as $X=319$ mm. For the composite beam the neutral axis will be higher and it is likely that X will be about half this value ≈ 160 mm as b_{eff} is found to be 2 times greater but the forces in the active strands will be about 25% greater. Therefore, the first estimate of X is found by considering an extra 2 strands (in the top of the boot) giving $N_T=24$ strands. The axis distance to these strands is $y_s=145$ mm.

For the two-stage design, stage 1 $M_{Ed1}=1.25^{*}\times 232.0=290.0$ kN (from Example 12.8). $^{*}\gamma_G=1.25$ is used as it is later shown that Exp. 610(b) in BS EN 1990 (BS EN 1990) is critical.

Eq. 12.79 $K_1=290.0\times 10^6/(45\times 350\times 455^2)=0.089$

where $d_1=600-145=455$ mm

Eq. 12.29 $z_1=\{0.5+\sqrt{[0.25-(0.089/1.134)]}\}\times 455=416.0 \text{ mm}<0.95d_1$

Refer to Example 12.2 to show that f_{p1} may be taken as 1516 N/mm^2

A_{p1}=290.0×10^6/(416.0×1516)=460 mm²

The average equivalent area is A'_p=460/24=19.2 mm²

The net area per strand=93−19.2=73.8 mm²

Then A'_{pT}=24×73.8=1771.2 mm²

Taking an initial stress of 70%×1770 N/mm², F_p=1771.2×1239×10⁻³=2196 kN.

The effective width in stage 2 is b_{ef}=(25/45)×1300=722 mm

Assume first the neutral axis (NA) depth is in the top flange, X=2196×10³/0.567×45×722×0.8=149.0 mm<200 mm, then NA is in the top flange as follows

Prestrain at mid-span ε_{po}=989.5/195000=0.005074

Considering strands in bottom row 1 at d=550 mm, ε_{p1}=0.005074+0.0035×[(550/149.0)−1]=0.014498

f_{p1}=1313+10186×0.014498=1461 N/mm²

Table 12.4a completes the calculation of the initial strains, stresses and forces using X=149.0 mm for 6 rows of strands that are below the neutral axis.

The total force in 24 strands is now F_p=2548 kN

X=2548×10³/0.567×45×722×0.8=172.8 mm such that ε_{p1}=0.012712 and f_{p1}=1442 N/mm², and so on until the equilibrium converges to X=170.5 mm.

Table 12.4b completes the calculation of the final strains, stresses and forces. Finally F_p=2513 kN with an average stress f_p=1418<1516 N/mm²

From Table 12.4b, y_s=360912/2513=143.6 mm

d=600−143.6=456.4 mm

d_n=0.4×170.5=68.2 mm

z=456.4−68.2=388.2 mm

Table 12.4a Initial strains, stresses and forces in the rows of strands in the first iteration in Example 12.9

Row	No.	Height y_i (mm)	Area per strand (mm²)	Area per row A_{pT} (mm²)	Strain ε_p	Stress f_p (N/mm²)	Force $f_p A_p$ (N)	Force×y_i×10^{-3} (kNmm)
1	8	50	73.8	591	0.014498	1461	862874	43144
2	4	100	73.8	295	0.013323	1449	427899	42790
3	4	150	73.8	295	0.012148	1437	424360	63654
4	2	200	73.8	148	0.010973	1425	210411	42082
5	4	250	73.8	295	0.009798	1413	417284	104321
6	2	350	73.8	148	0.007449	1389	205104	71786
Totals	24			1771			2547932	367777

Table 12.4b Final strains, stresses and forces in the rows of strands in the last iteration in Example 12.9

Row	No.	Height y_i (mm)	Area per strand (mm²)	Area per rowA_{pT} (mm²)	Strain ε_p	Stress f_p (N/mm²)	Force $f_p A_p$ (N)	Force×y_i×10^{-3} (kNmm)
1	8	50	73.8	591	0.012865	1444	853042	42652
2	4	100	73.8	295	0.011839	1433	423430	42343
3	4	150	73.8	295	0.010813	1423	420338	63051
4	2	200	73.8	148	0.009786	1413	208624	41725
5	4	250	73.8	295	0.008760	1402	414156	103539
6	2	350	73.8	148	0.006707	1308	193150	67603
Totals	24			1771			2512739	360912

$M_{Rd,c}$=1418×1771.2×388.2×10^{-6}+290.0=**1265.4kNm**

The ultimate design load w_{Ed}=max{1.35×(7.688+18.0+10.0)+1.05×45.0; 1.25×(7.688+18.0+10.0)+1.5×45.0}=112.1 kN/m

M_{Ed}=112.1×8.5^2/8=1012.1 kNm<1265.4 kNm.

The ratio $M_{Ed}/M_{Rd,c}$=0.80, compared with the service ratio M_s/M_{sR}=0.96, so clearly the serviceability state of stress is critical, a feature common with composite beams, particularly in shallower beams with fewer tendons.

Not included in this example but worthy of comment, the situation at the support, where M_{Ed1}=0, provides some information required for the horizontal interface shear reinforcements between the upstand of the beam and *insitu* infill in the slots. Here b_{eff}=722+350=1072 mm, A_{pT}=24×93=2322 mm^2, ε_{po}=0.004889, and following the same equations for equilibrium and compatibility of strains, the final results is d=456 mm, X=146 mm, z=397 mm, f_p=1437 N/mm^2 and $M_{Rd,c}$=1273.9 kNm. It is not overlooked that the difference between $M_{Rd,c}$ at the support and mid-span is only 8.5 kNm, less than would be expected based on the design of composite floor slabs, but it appears that the greater effective depth of strands in beams makes this particular calculation more robust.

Figure 12.34 shows the ultimate design moments M_{Ed} and the ultimate moments of resistance $M_{Rd,c}$ for Example 12.9. Close to the supports where the bond strength between the tendons and concrete is developing over an ultimate anchorage length l_{bpd}=1.375 m (see Section 9.5) the value of M_{Rd} is increasing bi-linearly from zero at the support.

12.5.5 Ultimate limit state of shear in composite beams

The ultimate shear capacities $V_{Rd,c,c}$ and $V_{Rd,cr,c}$ for composite beams are calculated as per Sections 10.2.2 and 10.2.3 for the basic equations for capacity, together with the two-stage design given in Section 12.3.2.4 for composite slab construction. Shear capacity is checked at two positions in the beam – at the centroidal axis of the composite beam and, in the case of inverted-tee beams, at the intersection (known as the 'intersection') of the upstand and boot. For I beams the critical section is in the narrow webs. Then

Eq. 12.37 $$v_{Ed1}=V_{Ed1}\, S_c/b_w\, I_c \qquad (12.37)$$

where V_{Ed1}=ultimate dead load acting on the beam, including self-weight of the precast unit and *insitu* infill

S_c=first moment of area at the narrowest point in the webs, taken as the centroidal axis of the composite beam at height $y_{b,c}$ (not at the centroid of the precast beam y_b because stage 2 shear stress is to be added at $y_{b,c}$, or at the intersection of the upstand and boot about the composite beam at height $y_{b,c}$

Eq. 12.38 $$V_{Rd,c,c2}=I_{c,c}\, b_w/S_{c,c}\, [\sqrt{(f_{ctd}^2+\alpha_l\, \sigma_{cp}\, f_{ctd})}-v_{Ed1}] \qquad (12.38)$$

Eq. 12.39 $$V_{Rd,c,c}=V_{Rd,c,c2}+V_{Ed1} \qquad (12.39)$$

where $S_{c,c}$=first moment of area at the centroidal axis of the composite beam at $y_{b,c}$, or at the intersection

σ_{cp}=prestress at the centroidal axis of the composite beam, or at the intersection

All other terms as per Section 12.3.2.4.

For the flexurally cracked section $V_{Rd,cr,c}$ the equations are as per Section 10.3.2 except that the composite effective depth, or with a topping $d+t$, replaces d. Use $V_{Rd,cr,c}$ where M_s>cracking moment of resistance $M_{cr,c}$ as

Eq. 12.41 $$M_{cr,c}=(\sigma_b+f_{ctd})\ Z_{b,c,co}-M_{s1}\ [(Z_{b,c,co}/Z_{b,co})-1)] \quad (12.41)$$

Example 12.10

Calculate $V_{Rd,c,c}$ and $V_{Rd,cr,c}$ for the composite inverted-tee beam in Example 12.8 and using beam section properties from Example 10.6. The floor slab comprises hcu that are continuous across the face of the support of the beam, with *insitu* infill in slots in top opened cores. The bearing length may be taken as 200 mm. Check the capacity at (i) centroidal axis and (ii) intersection of the upstand and boot.

Solution

Precast beam: $A_c=307.5\times10^3$ mm^2, $y_b=269.8$ mm, $I_c=8151.5\times10^6$ mm^4, $A_p=26\times93=2418$ mm^2

S_c (intersection)$=350\times200\times(400+200/2-367.2)=9.296\times10^6$ mm^3

S_c (centroid of composite beam)$=9.296\times10^6+600\times(400-367.2)^2/2=9.618\times10^6$ mm^3, where

Composite beam: $y_{b,c}=367.2$ mm, $I_{c,c}=15799.3\times10^6$ mm^4, $S_{c,c}$ (centroid)$=39.573\times10^6$ mm^3, $S_{c,c}$ (intersection)$=39.251\times10^6$ mm^3

At the support $\sigma_{po}=953.3$ N/mm^2 and $P_{po}=953.3\times2418\ \times10^{-3}=2305.0$ kN, and $l_{pt2}=812$ mm

Uncracked shear capacity $V_{Rd,c}$

i. At the centroidal axis of the composite beam
   Effective shear span $l_v=8500-200-2\times367.2=7566$ mm

$V_{Ed1}=1.25\times25.69\times7.566/2=121.5$ kN

$v_{Ed1}=121.5\times10^3\times9.618\times10^6/(8151.5\times10^6\times600)=0.24$ N/mm^2

σ_{cp} at centroid of composite beam$=0.9\times2305.0\times10^3\times[1/307500-92.86\times(367.2-269.8)/8151.5\times10^6]=4.44$ N/mm^2

$l_x=200+367.2=567.2$ mm

$\alpha_l=567.2/812=0.699$

$V_{Rd,c,c}=(15799.3\times10^6\times600/39.573\times10^6)\times[\sqrt{}(1.77^2+0.699\times4.44\times1.77)-0.24]\times10^{-3}+121.5=$ **768.2 kN**

ii. At the intersection of the upstand and boot

$l_v=8500-200-2\times400=7500$ mm

$V_{Ed1}=1.25\times25.69\times7.50/2=120.4$ kN

$v_{Ed1}=120.4\times10^3\times9.296\times10^6/(8151.5\times10^6\times350)=0.39$ N/mm^2

σ_{cp} at intersection$=0.9\times2305.0\times10^3\times[1/307500-92.86\times(400-269.8)/8151.5\times10^6]=3.67$ N/mm^2

l_x=200+400=600 mm

α_l=600/812=0.739

b_w at intersection=600 mm as the upstand is in the slab

$V_{Rd,c,c}$=(15799.3×10^6×600/39.251×10^6)×[√(1.77^2+0.739×3.67×1.77)−0.39]×10^-3+120.4=**706.2 kN**<768.2 kN

V_{Ed2}=80.0×7.50/2=300.0 kN

V_{Ed}=120.4+300.0=420.4 kN<706.2 kN.

Note that $V_{Rd,c,c}$=706.2>498.9 kN is 41% greater than for the non-composite unit in Example 10.6, a much greater percentage increase (28% and 30%) than for the serviceability and ultimate moments.

For interest, at the centroidal axis of the beam $V_{Rd,c,c}$=720.9 kN>706.2 kN but<768.2 kN, as follows

l_v=8500−200−2×269.8=7760 mm and V_{Ed1}=124.6 kN

v_{Ed1}=124.6×10^3×21.202×10^6/(8151.5×10^6×600)=0.54 N/mm^2

From Example 10.6, σ_{cp}=6.746 N/mm^2, α_l=0.579 and S_c (centroid of beam)=41.814×10^6 mm^3

$V_{Rd,c,c}$=(15799.3×10^6×600/41.814×10^6)×[√(1.77^2+0.579×6.75×1.77)−0.54]×10^-3+124.6=**720.9 kN**

Cracked shear capacity $V_{Rd,cr,c}$ at intersection of upstand and boot

From Example 12.9, d_2=455.7 mm and A_{pT}=24×93=2232 mm^2

k=1+√(200/455.7)=1.66

v_{min}=0.035×1.66^1.5 √45=0.503 N/mm^2

$A_s/b_w d$=2232/(600×455.7)=0.00816

σ_{cp}=3.67 N/mm^2

$V_{Rd,cr,c}$=[(0.12×1.66×(100×0.00816×45)^1/3)+0.15×3.67]×600×455.7×10^-3=**331.8 kN**

$V_{Rd,cr}$ minimum=(0.503+0.15×3.67)×600×455.7×10^-3=288.1 kN<331.8 kN, use $V_{Rd,cr,c}$=331.8 kN

Calculation of distance to the decompression point

This is where the service bending moment is greater than the cracking moment of resistance, $M_s>M_{cr,c}$ and where the flexurally cracked shear capacity $V_{Rd,cr,c}$ governs.

From Example 12.8, $Z_{b,co}$=31.683×10^6 mm^3, $Z_{b,c,co}$=45.076×10^6 mm^3, σ_b=15.13 N/mm^2 and f_{ctd}=1.77 N/mm^2

Eq. 12.41 $M_s<M_{cr,c}$=(15.13+1.77)×45.076−M_{s1} [(45.076/31.683)−1]=**761.9−0.423** M_{s1} kNm. This condition exists where x=3.06 m from the centre of support where M_s=671.6 kNm and M_{s1}=213.8 kNm. The ultimate shear force at this point is 133.4 kN, clearly not critical.

12.5.6 Interface shear in composite beams

The integrity of composite construction relies on the long-term continuity between the precast and *insitu* concrete. Composite beams are designed with interface shear links, dowel bars or other similar mechanical fasteners. Shear friction alone is not recommended for beams. The ultimate interface shear stress v_{Edi} is checked for the uncracked section against v_{Rdi} values in BS EN 1992-1-1, clauses 6.2.5 (1) and (2).

Exp. 6.24 $v_{Edi}=(F_{c,T}/F_c)\ V_{Ed2}/z_2\ b_i$ (12.84)

Exp. 6.25 $v_{Rdi}=c\ f_{ctd}+\mu\ A_s\ f_{yd}/A_j \leq 0.5\ \nu\ f_{cd}$ (12.85)

where $F_{c,T}$=compression in slab or slab & topping due to ultimate stage 2 loading

F_c=total compression in slab and precast beam due to ultimate stage 2 loading
V_{Ed2}=stage 2 ultimate vertical shear force at the support (not on shear plane)
z_2=lever arm from tendons to the centroid of the compression zone due to stage 2 loads
b_i=interface contact length=horizontal gap width between ends of slabs for rectangular beams or I beams with no upstand, or 2 x upstand depth for inverted tee beams, see Figure 12.29
A_s=area interface shear reinforcement
A_j=area of interface joint

c and μ are defined in Section 12.4.1

Note that for inverted-tee or I beams having an upstand, for smooth and rough surfaces where the interface plane is in the vertical sides of the upstand in flexural conditions, the term $c\ f_{ctd}$ is ignored because the end of the slab is subjected to end rotation, causing cracks in at least in the upper part of the interface. This does not apply to an indented surface (Figure 12.23) where the depth of the shear key is at least 5 mm. Neither is it the case for rectangular beams where the interface plane is horizontal.

Further, Eq. 12.85 does not include the contribution from the shear friction of the floor units acting on the bearing ledges of the beams, and in this respect it is conservative (see Example 12.10) even though shear friction is generally ignored in the interface between beams and slabs.

Nominal interface shear bars should always be provided. The minimum area is the greater of $A_{s,min}/A_j$=0.13% or 0.26 f_{ctmi}/f_{yk} of the contact area, or, if the depth of slab is greater than the upstand $A_{s,min}/A_j$=0.13% or 0.26 f_{ctmi}/f_{yk} of the slab area, where f_{ctmi} is for the *insitu* infill. If $v_{Edi}>v_{Rdi}$ typically about 0.3–0.5 N/mm², horizontal bars are provided projecting from the upstand, or U loops or links projecting from the top of the upstand and lapped to horizontal bars projecting into the infill in floor units or the topping. See Figure 12.29. The area of dowels links or loops that are perpendicular to the interface is

$$A_s=v_{Edi}\ b_i/\mu\ 0.87\ f_{yk}\ (\text{mm}^2/\text{mm}) \tag{12.86}$$

Horizontal bars act in dowel action serving to restrain lateral tensile strains. The distribution of the bars may be stepped, as shown in BS EN 1992-1-1, Fig. 6.10, usually in not more than one step between the designed and minimum areas of $A_{s,min}$. If the designed spacing is more than about 600 mm it is not worth the possible confusion of changing the spacing, and certainly the opposite ends of the same beam should not be different (without good cause). The spacing of the bars into hcu should not be more than 300 mm and preferably in alternate opened cores, e.g. 5 no. for 11 cores per 1.2 m width (try to avoid opening the first and last edge cores), 4 no. for 7 to 9 cores, and all cores opened for 6 cores.

Example 12.11

Calculate the interface shear reinforcement necessary to satisfy the ultimate horizontal shear force at the precast – *insitu* interface for the inverted-tee beam shown in cross section in Figure 12.33 and designed in Example 12.9. Use f_{cki}=25 N/mm^2 for the *insitu* infill and f_{yk}=500 N/mm^2 for the interface shear reinforcement. The upstand of the beam has vertical shear keys conforming to Figure 12.23.

Solution

Stage 1 M_{Ed1}=290 kNm (from Example 12.9)

Stage 2 M_{Ed2}=1012.1–290.0=722.1 kNm

The actual width of the *insitu* infill b=1300 mm and d_2=455.7 mm

K_2=722.1×10^6/(25×1300×456^2)=0.107; then z_2=406.7 mm

X_2=120 mm; 0.8 X_2<200 mm, then F_{cT}/F_c=1.0

V_{Ed2}=80.0×8.5/2=340.0 kN

The interface 'width' is now the depth of the upstand on both sides of the beam, b=2×200=400 mm

Eq. 12.69 v_{Edi}=1.00×340.0×10^3/400×406.7=2.09 N/mm^2>0.60 N/mm^2
where for an indented profile c=0.5 and μ=0.9

Eq. 12.70 v_{Rdi}=0.5×1.20=0.60 N/mm^2 but<4.50 N/mm^2, use interface bars (not loops)

Minimum area is the greater of 0.13%×1000×200=260 mm^2/m or 0.26×2.56×1000×200/500=266 mm^2/m in each side of the upstand

Eq. 12.72A_s=1000×(2.09-0.6)×200/(0.9×0.87×500)=760 mm^2/m run in each side of the upstand. The area required per 1200 mm width hcu is 1.2×760=912 mm^2. Trialing H12 bars (113 mm^2 per bar) the number of bars required is n=912/113=8.1. The projecting length of the bar is notionally 40 diameters=480 mm which fits comfortably into a 600 mm long slot without the need for an end hook.

The spacing of the slots in the top opened cores depends on the generic type of hcu used, i.e. for 11 cores the spacing is a multiple of 100 mm, for 9 cores it is about 120 mm and for 6 cores it is about 180 mm. For 11 and 9 cores, typically 60 to 90 mm wide, only one bar is provided per core. For 6 cores, typically 140 mm wide two bars per core are provided.

Therefore a 6 core hcu would require 9 bars with 1 or 2 bars in each alternate slot (about 80 mm spacing where 2 bars are provided). An 11 core hcu would require 9 bars in all the interior slots.

The minimum number of bars per unit is n=1.2×266/113=2.8. Therefore, provide 3 no. bars at a distance along the span from the support where

Eq. 12.85 v_{Edi}=(0.5×1.20)+[0.9×266×0.87×500/(200×1000)]=1.12 N/mm2,
Eq. 12.84 V_{Ed2}=1.12×406.7×200×10^{-3}=91.1 kN

As the gradient of the stage 2 shear force is 80.0 kN/m, this occurs at 91.1/80.0=1.14 m from mid-span, or x=8.5/2−1.14=3.11 m from the support.

In between at x=1.56 m from the support V_{Ed2}=(340.0+91.1)/2=215.6 kN, v_{Edi}=1.32 N/mm^2 and A_s=370 mm^2/m in each side of the upstand. Then n=1.2×370/113=3.9. Therefore, provide 4 no. bars per hcu from 1.56 to 3.11 m from the support.

12.5.7 Propping of composite beams

The basic concept for propping composite sections before the *insitu* concrete is added is as given in Section 12.3.3 for floors. The potential for structural enhancement is much greater for psc beams than in floors because the stresses due to the self-weight of the floor slab, *insitu* infill and topping are reduced by the use of 2 or 3 props. The props are added after the beam is positioned, before the slabs are installed, and therefore the stresses due to the self-weight of the beam are calculated for the unpropped span.

It is worthwhile propping beams of over 6 m to 10 m span with 2 or 3 props, respectively, particularly if a structural screed is being used and the ratio of the composite properties to the basic section properties exceeds 1.50. Propping increases beam erection time by about 15% to 25%, and so the delicate balance between material saving and time lost on site has to be considered. Only the precast manufacturer can give exact information on this.

Let w_0=self-weight of beam, w_1=self-weight of precast floor units plus wet concrete infill and (if present) structural topping, and w_2=imposed loads (all beam loads, e.g. kN/m). For a single prop at mid-span the reaction $R=0.625\ w_1\ L$. The negative moment$=-0.625\ w_1\ L^2/4=-0.15625\ w_1\ L^2$. The various service moments at mid-span are

$$M_{sw}=+0.125\ w_0\ L^2 \tag{12.87}$$

$$M_{s1}=(0.125-0.15625)\ w_1\ L^2=-0.03125\ w_1\ L^2 \tag{12.88}$$

$$M_{s2}=+0.15625\ w_1\ L^2+0.125\ w_2\ L^2 \tag{12.89}$$

Alternatively, the serviceability moment of resistance M_{sR2} at the bottom and top fibres may be calculated in the same manner as in Section 12.3.3.1, Eqs. 12.42 and 12.43 for a single prop at mid-span as

$$M_{sR2}=(f_{ctm}+\sigma_b)\ Z_{b,c,co}-[(M_{sw}+M_{s1})\ (Z_{b,c,co}/Z_{b,co})]-0.15625\ w_1\ L^2 \tag{12.90}$$

$$M_{sR2}=(0.45\ f_{ck}-\sigma_t)\ Z_{t,c,co}-[(M_{sw}+M_{s1})\ (Z_{t,c,co}/Z_{t,co})]-0.15625\ w_1\ L^2 \tag{12.91}$$

$$\text{Total } M_{sR,c}=\text{Minimum } M_{sR2}+(M_{sw}+M_{s1}) \tag{12.92}$$

but where in Eq. 12.92 $M_{sw}+M_{s1}=(w_0+w_1)\ L^2/8$, i.e. the negative moment due to the prop is not included twice.

If the prop is first erected beneath the beam and takes the full reaction from the self-weight, then M_{s0} is zero and

$$M_{s1}=-0.03125\ (w_0+w_1)\ L^2 \tag{12.93}$$

$$M_{s2}=+\ 0.15625\ (w_0+w_1)\ L^2+0.125\ w_2\ L^2 \tag{12.94}$$

However, if the prop is set level with the supports, and the beam is cambered upwards due to the net effect of eccentric prestress and self-weight, that part of the self-weight which brings the beam down to a horizontal level (until the beam becomes flat with zero camber) will not be resisted by the prop, and so Eq. 12.91 will not include the full w_0. Let the proportion of w_0 needed to cause zero deflection equal $\beta\ w_0$. Then, if the prestress (causing negative deflection) and creep coefficient at installation is P_{pmi} and $\varphi(t_i,t_0)$ and the prop is at mid-span

$$-P_{pmi}\ z_{cp}\ L^2\ \varphi(t_i,t_o)/8\ E_{cm}\ I_{c,co}+5\ \beta\ w_0\ L^4\ \varphi(t_i,t_0)/384\ E_{cm}\ I_{c,co}=0$$

$$\text{Then}\ \beta\, w_0 = 9.6\ P_{pmi}\ z_{cp}/L^2 \quad (12.95)$$

The resulting prestress requirements are

$$\sigma_b = (M_{sw}+M_{s1})/Z_{b,co} - M_{s2}/Z_{b,c,co} - f_{ctm}\ \text{(at bottom)} \quad (12.96)$$

$$\sigma_t = 0.45\ f_{ck} - (M_{sw}+M_{s1})/Z_{t,co} + M_{s2}/Z_{t,c,co}\ \text{(at top)} \quad (12.97)$$

However, due to the sagging moment either side of the prop the critical position in the span is usually at about $x=0.45\ L$, as shown in Example 12.12. The effect is obviously more noticeable for shallower beams with deeper slabs, and/or deeper topping, resulting in a greater prop force.

Example 12.12

Calculate the serviceability moment of resistance $M_{sR,c}$ for the composite inverted tee beam in Example 12.8, as shown in cross section in Figure 6.10 (without debonded strands) and Figure 12.33 using a prop at mid-span. The beam is propped during installing the precast floor units and casting the infill in slots in top opened cores. Calculate the increased imposed live load made possible by propping. Draw the final bending moment and show where the critical stress for the effect of the prop is located.

Solution

Stage 1

Service load due to self-weight of precast beam plus wet *insitu* infill slab=7.69 and 18.0 kN/m. Stage 1 mid-span service moment is from Eqs. 12.87 and 12.88

$M_{sw}+M_{s1}=0.125\times7.69\times8.5^2\text{-}0.03125\times18.0\times8.5^2=28.8$ kNm

Note that the maximum stage 1 moment occurs at $x=2.38$ m ($x/L=0.28$) from the support and is $M_{s1}=73.28$ kNm. However, this point is not critical when stage 2 loads are added.

Stage 2. Prop removed

Eq. 12.90 $M_{sR2}=(3.80+15.13)\times45.076-28.8\times45.076/31.683-203.3=609.1$ kNm

Eq. 12.91 $M_{sR2}=(20.25--1.22)\times69.278-28.8\times69.278/25.351-203.3=1205.3$ kNm>609.1 kNm

The stage 1 moment due to self-weight loads (without prop force)$=25.69\times8.5^2/8=232.0$ kNm,

Eq. 12.92 $M_{sR,c}=609.1+232.0=$ **841.1 kNm**

The allowable imposed load$=8\ M_{sR2}/L^2=8\times609.1/8.5^2=67.4$ kN/m, and the imposed live load 67.4 − 10.0 = **57.4 kN/m**>45.0 kNm in Example 12.8. Note that $M_{sR,c}=841.1>755.3$ kNm is 11% greater than for the unpropped composite beam in Example 12.8.

Alternatively, using the previous service loading the effective span may be increased to 9.12 m, an extra 0.62 m for the cost of installing a single prop.

Many contractors prefer to use two props, each at ⅓ span, giving greater assurance in case the foundation for a single prop is not completely rigid. As there is a small sagging

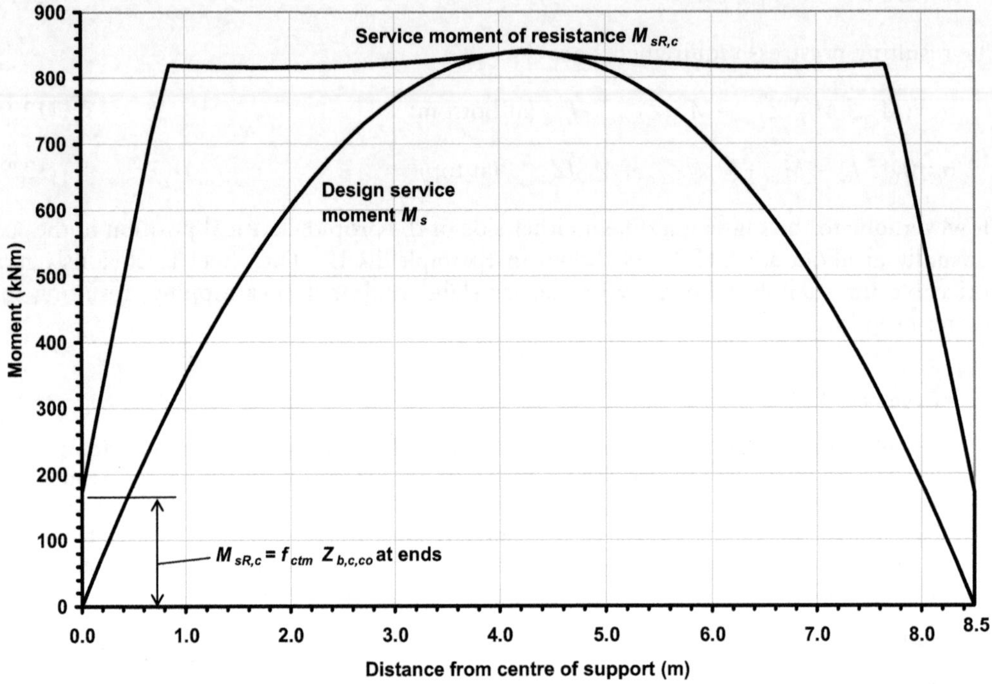

Figure 12.35 Service bending moments for the propped composite inverted-tee beam in Example 12.12.

moment between the two props $M_{sR,c}$ will be less than for a single prop. If two props were used in Example 12.12 then $M_{sR,c}$=822.4 kNm<841.1 kNm, 2% less.

Figure 12.35 shows the service design moments and the moments of resistance for Example 12.12. Note the reduction in $M_{sR,c}$ towards mid-span, seen in Figure 12.34 for the unpropped beam, is eliminated by propping, in fact $M_{sR,c}$ at mid-span is now greater than near to the support.

It is just possible to discern from the graphs of M_s and $M_{sR,c}$ that the critical position is a short distance from mid-span, in this example at x=4.08 m from the support, where $M_{sR,c}$=837.8 kNm, or 3.3 kNm less than at mid-span. To obtain this accurate figure this point in the span must be known in advance, which is impossible, or else the calculation is adjusted afterwards. In this example the author calculated $M_{sR,c}$ at intervals of L/50 (0.17 m) either side of mid-span.

The position of the critical value for $M_{sR,c}$ varies depending on the relative magnitudes of the stage 1 prop reaction, where for heavier slabs and greater spans the position of the critical value for $M_{sR,c}$ drifts further from mid-span.

Example 12.13

Repeat Example 12.12 by reducing the number of strands necessary to maintain the original mid-span service moment of 728.2 kNm in Example 12.8, i.e. imposed live load of 45 kN/m. Determine a suitable arrangement of strands.

Solution

The previous example has shown that the critical stresses are in the bottom fibre, and therefore σ_t at the top fibre is not required.

With reference to Figure 6.10 (excluding the debonded strands) remove 2 no. strands at each rows 3 and 5. Try 22 no. strands $A_p=22\times93=2046$ mm^2 with $z_{cp}=97.1$ mm, $Z_{b,co}=31.600\times10^6$ mm^3 and $Z_{b,c,co}=44.884\times10^6$ mm^3

Initial prestress $P_{pi}=1239.0\times2046\times10^{-3}=2535.0$ kN

By the analysis in previous examples, the final losses at mid-span=18.3%, then $P_{po}=(1\text{-}0.183)\times2535.0=2070.5$ kN

$\sigma_b=(2070.5\times10^3\times[(1/307500)+(97.1/30.215\times10^6)]=13.38$ N/mm^2

Stage 1

Stage 1 mid-span service moment $M_{s1}=28.81$ kNm (from Example 12.12). Then

$f_{b1}=28.81\times10^6/31.600\times10^6=-0.91$ N/mm^2

Stage 2

Prop reaction$=0.625\times18.0\times8.5=95.6$ kN

Mid-span service moment due to imposed loads of 10.0+45.0 kN/m plus prop

$M_{s2}=55.0\times8.5^2/8+95.6\times8.5/4=699.8$ kNm incl. moment due to prop force

$f_{b2}=699.8\times10^6/44.884\times10^6=-15.59$ N/mm^2

Net tension in bottom$=+13.38-0.91-15.59=-3.12$ N/mm$^2>-3.80$ N/mm^2 OK

The effect of propping is to save 26−22=4 no. strands.

REFERENCES

ASSAP. 2002. *The Hollow Core Floor Design and Applications, Manual 1st Edition*, Association of Manufacturers of Prestressed Hollow Core Floors (ASSAP), Verona, Italy.

Bhatt, P. 2011. *Prestressed Concrete Design to Eurocodes*, Routledge, Oxford, UK, 596 p.

British Precast. 2008. *Precast Eurocode 2: Worked Examples*, British Precast Concrete Federation, Leicester, UK, 77 p.

BS EN 1168. *2005*. Precast Concrete Products - Prestressed concrete hollow core units, +A3:2011, BSI, London, UK.

BS EN 1990:2002 +A1:2005. Eurocode 0 - Basis of structural design, BSI, London, UK.

BS EN 1992-1-1. 2004. Eurocode 2, Design of concrete structures - Part 1-1: General rules and rules for buildings, BSI, London, UK.

BS EN 15037-1. 2008. Precast concrete products, Beam and block floor systems, Part 1 - Beams, BSI, London, UK.

Elliott, K. S. 2017. *Precast Concrete Structures*, 2nd ed., CRC, Taylor & Francis, London, UK, 694 p.

Elliott, K. S. and Ibrahim, I. 2006. Composite construction in prestressed precast concrete floors, CONCET 2006, 9th International Conference on Concrete Engineering, Malaysia/Kuala Lumpur. Institution of Engineers Malaysia, pp. 104–116.

FIB. 2012. *Bulletin 63, Design of Precast Concrete Structures against Accidental Actions, Guide to Good Practice,* Fédération Internationale du Béton, Lausanne, 72 p.

FIP Recommendations. 1988. *Precast Prestressed Hollow Cored Floors*, FIP Commission on Prefabrication, Thomas Telford, London, UK, 31 p.

FIP Recommendations. 1982. *Shear at the Interface of Precast and Insitu Concrete*, Federation International de la Precontrainte, Wexham Springs, Slough, UK, 31 p.

Ibrahim, I. and Elliott, K. S. 2009. Long-term behaviour of precast prestressed hollow core slabs with concrete toppings, 7th Asia Pacific Structural Engineering and Construction Conference (APSEC) and 2nd Asian European Civil Engineering Forum (EACEF), Kuala Lumpur, Malaysia.

Ibrahim, I. S., Elliott, K. S., Abdullah, R. and Sarbini, N. N. 2016. Experimental study on the shear behaviour of precast concrete hollow core slabs with concrete topping, *Engineering Structures*, Vol. 125, pp. 80–90.

Chapter 13

Cantilevers and continuous slabs and beams

13.1 INTRODUCTION TO CANTILEVERS AND BALCONIES

Although prestressed concrete (psc) floor slabs with cantilevers represent a small percentage of the flooring market, used mainly internally in buildings because of the need to provide a thermal bridge in external situations or in car parks, their use to provide the restraint for separate balconies has increased considerably over the past 20 years in multi-storey residential buildings, for example as shown in Figures 13.1 and 16.19. The main function is as a residential balcony. Both solid and hollow core units (hcu) are used, often with a structural topping to cope with the negative cantilevering bending moment in the main span, known as the 'back-span'. For a typical 1.5 m span balcony, carrying imposed dead and live loads of about 1.0 and 3.0 kN/m^2, together with a balustrade, the typical back-span is 6–8 m for a 200 mm depth composite solid plank or hcu.

Figure 13.1 Steel balconies connected to the perimeter of a precast structure using solid precast floor slabs as the back-spans.

DOI: 10.1201/9781003319450-13

Balcony units are separate reinforced precast concrete or steel units with timber decking that typically have a steel member acting as the cantilever. Figure 13.2 shows a flooring layout comprising mainly of psc hollow core units (hcu) but also solid precast units providing the back-span to the cantilever balcony. To satisfy building regulations and avoid cold bridges, a thermal break is provided within the cavity zone as shown in Figures 13.3 and 16.18. There are many proprietary types depending on capacity and function. A continuous prestressed slab crossing the cavity would not satisfy these thermal requirements.

Recently, it has become popular to attach a structural steelwork balcony to the side of psc solid units, as shown in Figure 13.4. The balconies are typically 2.5–4.0 m in length (parallel with the side of the precast unit) and 1.5 m from the edge of the precast unit, which is formed with a square edge (not a profiled/keyed edge). The depth of the precast unit is at least 200 mm in most cases. A 75 mm thickness structural topping may be added to either (i) increase the main span of the floor unit supporting the balcony or (ii) increase the shear and torsion capacity of the unit supporting the balcony. Sturdy moment resisting brackets are bolted to the side of the precast unit at say 1.3–1.5 m centres, and the bay of steel beams, cross beams, bracing and steel/glass balustrade cantilevers from the side of the precast floor. The tie back is formed by one steel strap per bracket, typically 200 mm width × 12 mm thickness, anchored and grouted through the thickness of the precast unit using threaded bolts or

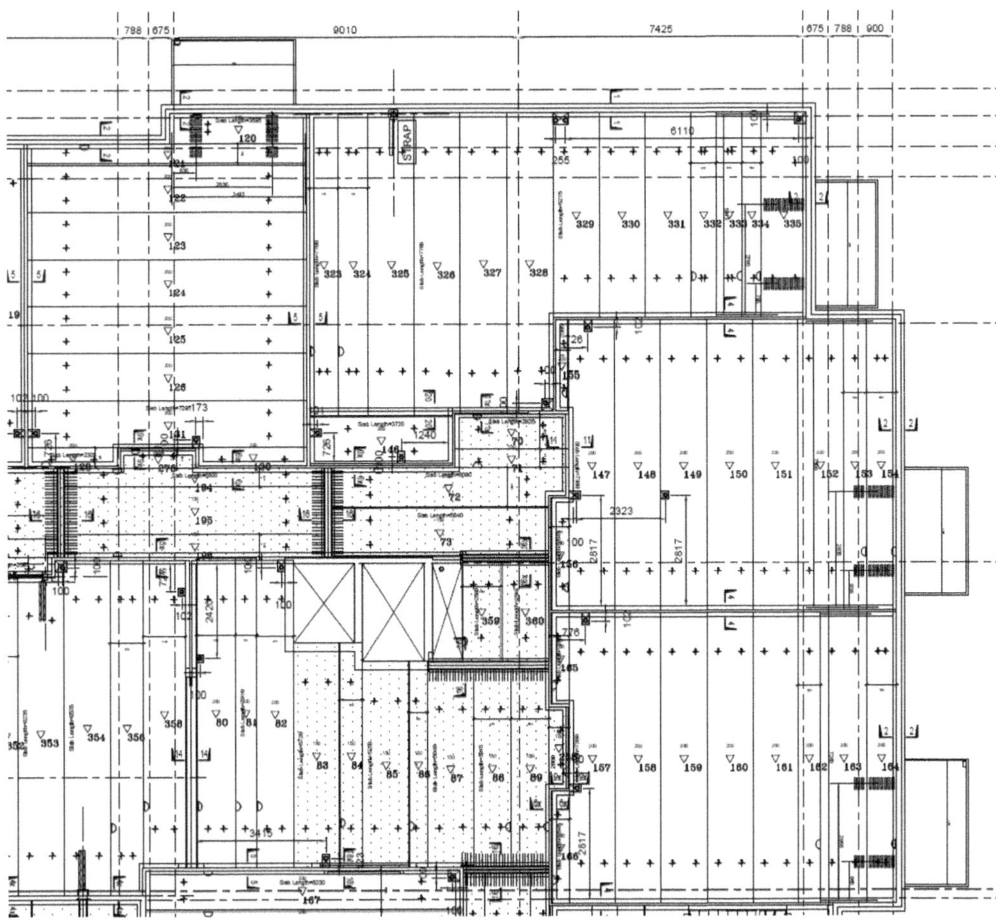

Figure 13.2 Precast hollow core flooring layout, also showing solid composite precast slabs providing the back-span for balconies in a residential building.

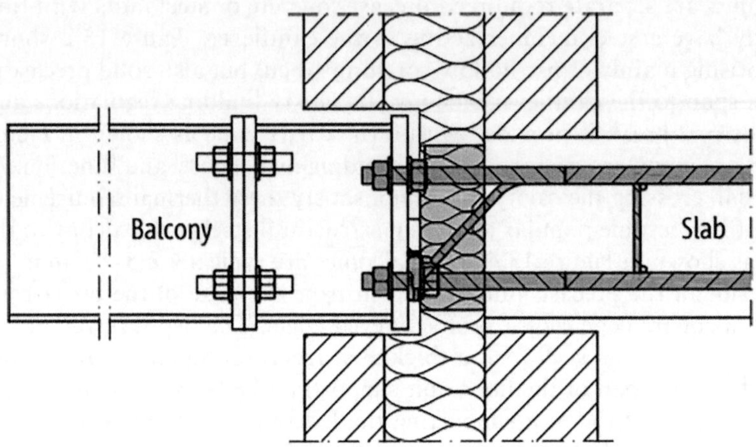

Figure 13.3 Thermal bridge between the balcony and the precast slab. (Courtesy Schoeck, UK.)

Figure 13.4 Structural steel brackets used to provide a moment-resisting connection to a steelwork balcony.

dowels. Concrete screed, typically 75 mm thickness completes the construction (if no topping is used). The back-span, this time perpendicular to the span of the floor, is achieved by structurally grouting two precast units together, making it at least 2.3 m. A third precast unit, usually a hcu of equal depth, may be used to increase the back-span if necessary. Therefore, the precast unit at the edge of the balcony is subjected to torsion, in addition to shear at the supports, and bending further from the supports, together with a line reaction in the grouted joint between the edge and adjacent unit. Figure 13.5 shows the completed solution.

Figure 13.6 shows cantilevered prestressed hcu providing a balcony to a residential building in Australia. Details of the thermal bridge are unknown. However, cantilevering slabs

Figure 13.5 The completed solution uses the bracket and tie-back straps to support the balcony off prestressed concrete solid floor units.

Figure 13.6 Cantilevered hollow core floor units in residential building. (Courtesy Simon Hughes, Precast Concepts & Hollow Core Concrete, Australia.)

Figure 13.7 Roof canopy to a supermarket in Malaysia formed by cantilevered prestressed hollow core units.

may be used if the thermal insulation is provided and supported at the end of the cantilever, for example by thermally insulated cladding or similar façade. Figure 13.7 shows a more flamboyant use of hcu cantilevers as a roof canopy at a supermarket in Malaysia – suffice to say that the imposed dead and live loads on this roof are quite small. Details of the structural ties to the supporting beam, if any, are not known.

13.2 STRUCTURAL DESIGN OF CANTILEVERS USING SOLID AND HOLLOW CORE UNITS

13.2.1 Definition of spans and loading cases

Cantilever slabs are designed as statically determinate units with a simple support (usually at one end but as shown in Figure 13.7 could be at both ends) and a continuous support over a wall or primary rectangular or I beam. These supports provide no restraint to rotation at the ends of the main span. No moments are distributed unless one end is continuous as shown in Section 13.5. There is no moment redistribution at a cantilever support. The maximum length, or effective span, of the cantilever is limited by

i. moment of resistance at the cantilever support, usually the ultimate M'_{Rd} governs as static rebars are placed in the top of the precast unit or mesh and some rebars in the structural topping
ii. overturning of the main span, i.e. the reaction at the simple support becomes negative, allowing of course for appropriate factors of safety

iii. restrictions placed on the long-term deflection, and/or deflection after finishes, at the end of the cantilever
iv. The manufacturer/designer's in-house limitation, often in the order of effective span to depth of precast unit or total depth including topping of 7–10. A common limiting ratio is 7½.

To cater for all conditions, maximum and minimum imposed dead and live loads are placed on the main span and cantilever as shown in Figure 13.8, resulting in critical bending moments and shear forces, as follows:

- Loading Case I. Maximum load in the main span with minimum load (self-weight and imposed dead loads) in the cantilever. This establishes the greatest sagging (positive) moment in the main span. It also gives the maximum shear force at the simple support as the upward reaction from the cantilever moment is the least.
- Loading Case II. Maximum load in the cantilever with minimum load in the main span. This establishes the greatest negative moment at the cantilever support, and the critical distance into the back-span for negative top reinforcement.
- Loading Case III. Maximum load in both the cantilever and main span. This establishes the greatest shear force in the main span at the face of the cantilever support (with the main span bending moments being less than in Case I).

It is usual to combine load cases I and II in bending moment diagrams, for both serviceability and ultimate limit states, as shown later by the kinks in the graphs in Figures 13.33 and 13.34.

Load case II is also used to check the stability, i.e. over-turning criteria, of the unit, including an allowance for construction traffic in the cantilever, typically 1.5 kN/m². See Section 13.2.5.

Service and ultimate bending moments, and ultimate shear forces are calculated in the usual manner for the three load cases. Let L=effective main span, L'=effective cantilever span, calculated according to BS EN 1992-1-1, clause 5.3.2.2 (1) (BS EN 1992, 2004) as

Fig. 5.4(e) $$L' = \text{clear span } l'_n + \min\{0.5l_b;\ 0.5h\} \tag{13.1}$$

where l_b=cantilever bearing length and h=depth of cantilever. The latter may be critical where the bearing is a beam of say 300 mm width.

Referring to Figure 13.8:

For Case I loading, let w_s=total self-weight, imposed dead and live service UDLs in the main span, and w'_{s1}=self-weight and dead UDL only in the cantilever

The reaction F at the simple support

$$F = w_s L/2 - \left(w'_{s1} L'^2/2\right)/L \tag{13.2}$$

The maximum sagging (positive) moment M_s in the main span is at the point of zero shear force, known as the 'shear span' of length a as

$$a = F/w_s \tag{13.3}$$

and

$$M_s = F\,a - w_s\, a^2/2 \le M_{sR} \text{ for the psc unit (Section 7.2.4.5)} \tag{13.4}$$

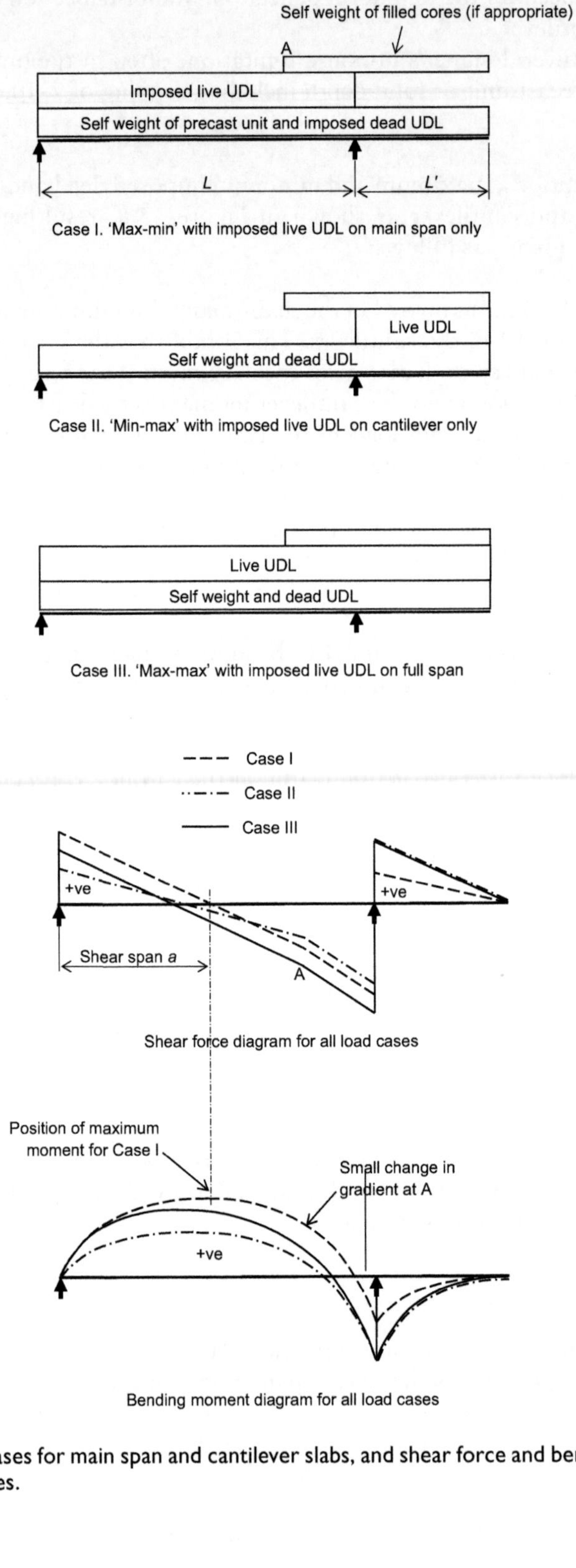

Figure 13.8 Loading cases for main span and cantilever slabs, and shear force and bending moment diagrams for all cases.

If dead point loads P' at the end of the cantilever are included Eq. 13.2 is modified to $F = w_s L/2 - \left(w'_{s1} L'^2/2 + P'L'\right)/L$. If there are point and line loads in the main span, then Eqs. 13.3 and 13.4 are not applicable and are modified to suit.

For Case II loading, let w_{s1} = self-weight and dead UDL in the main span and w'_s = total dead and imposed UDLs in the cantilever.

The maximum negative moment M'_s in the cantilever

$$M'_s = -w'_s\, L'^2/2 \tag{13.5}$$

For the ultimate limit state replace w_s and w'_s with appropriate w_{Ed} and w'_{Ed} for load combinations Exp. 6.10a or 6.10b from BS EN 1990 (BS EN 1990, 2002). For Case I loading, the support reaction is now F_{Ed}, shear span a_{Ed} and maximum positive moment M_{Ed}. For Case II loading, the maximum design negative moment in the cantilever is M'_{Ed}. The back-span calculation for a_{Ed} is only required to determine the position of zero bending moment in the main span, i.e. for the purpose of calculating the point of curtailment in the top reinforcement. Let x = distance from simple support to point of zero ultimate moment, then

$$M_{Ed} = F_{Ed}\, x - w_{Ed}\, x^2/2 = 0$$

Then $w_{Ed}\, x^2/2 = F_{Ed}\, x$

hence $x = 2\, F_{Ed}/w_{Ed}$, but as $a_{Ed} = F_{Ed}/w_{Ed}$, then $x = 2\, a_{Ed}$

$$x = 2\, F_{Ed}/w_{Ed} \tag{13.6}$$

The logic behind this is that the shape of the bending moment diagram is symmetrical about $x = a_{Ed}$.

For composite cantilever slabs, the same equations Eqs. 13.2–13.6 are used except that the UDLs are separated into stage 1 and stage 2 loads, before and after the structural topping has hardened. Strictly, the effective span of the cantilever, according to Eq. 13.1, may be slightly different for stage 1 where $h_s < l_b$, then for stage 2 where it might be that $h > l_b$, but this is generally ignored and stage 2 value is used.

Stage 1 and 2 service moments are separated only for the determination of the maximum compressive stress in the soffit of the psc unit as given in Section 13.2.3, and to check whether the cantilever is flexurally cracked or uncracked for the calculation of $V'_{Rd,c}$ or $f'_{t,max}$, respectively, as given in Section 13.2.6.

Stage 1 and 2 ultimate moments are separated for the design of top rebars in the precast unit in stage 1 – these may be in excess of the moment of resistance provided by top tendons if present. Stage 2 ultimate moments are for the design of the steel mesh and other loose rebars in the structural topping. See Section 13.2.4.

The maximum design shear force in the cantilever is for Case II loading at the start of the shear span $M'_{Ed} = L' - l_b/2 - y_b$ as

$$V'_{Ed} = w'_{Ed}\, l'_v \tag{13.7}$$

This is the case for hcu, where it is assumed that the design point is at the non-composite centroid height y_b. For solid composite slabs, the point is at the composite height $y_{b,c}$ – the difference is insignificant as the shear capacity of solid slabs is huge.

In the main span, maximum V_{Ed} is at the face of the cantilever support for Case III loading. The reaction in the main span at the cantilever support is

$$F_{Ed} = w_{Ed}L/2 + \left(w'_{Ed}\,L^2/2\right)/L \tag{13.8}$$

$$V_{Ed} = F_{Ed} - w_{Ed}\,(l_b/2 - y_b) \tag{13.9}$$

Loads for balustrades at the end of the cantilever, and sometimes at the edge of the cantilever, are specified in kN/m according to their construction. There is nothing written in BS EN 1991 for balustrade loads, although the UK National Annex to BS EN 1991-1-1 (NA to BS EN 1991-1-1, 2002) requires a minimum point load of 2.7 kN in office areas to be considered, but this is not a balustrade load. For 20 mm toughened glass balustrades with galvanised steel posts for example the self-weight is typically 1.0 kN/m. The horizontal reaction of 0.74 kN/m (or 1.5 kN/m for congregation areas) is taken from NA to BS EN 1991-1-1, Table NA.8 for parapet walls. The typical height to this load is 0.75 m, generating a constant moment in the cantilever of 0.56–1.12 kNm/m.

It is not possible to produce load vs. span charts for cantilever slabs as the effective span vs. imposed load in the main span depends on the effective span of the cantilever. However, as the main span reduces, so do the requirements for prestress leading to much smaller end rotations due to camber, which in turn reduces deflection at the end of the cantilever. For example, for an effective span of 1.5 m, it is shown in Example 13.4 that when the main span is 9.8 m the long-term deflection at the end of the cantilever is 15 mm, but if the main span is reduced to 7.0 m and the prestress is halved the deflection there reduces to 7 mm. The limiting deflections in cantilevers are discussed and given in Section 13.3.3.

Figure 13.8 shows an example of the loading cases that includes the self-weight of filled cores w_{0c}, suggesting that there are no tendons at the top of the unit, and there is no reinforced structural topping. Figures 4.32, 13.9, 13.10, and 13.14 show examples of hcu and solid units that include top rebars in opened cores or slots.

This leads to two possible conditions for the design of cantilever slabs

1. Referring to Figure 13.11, a psc unit alone with either
   i. pretensioned tendons in the top sufficient to carry the ultimate moment at cantilever support
   ii. no top tendons, or small area of tendons, with top opened cores and factory placed rebars in the top of filled cores
2. Referring to Figure 13.12, a composite psc unit with topping, with either
   i. pretensioned tendons in the top of the unit sufficient to carry stage 1 ultimate moment at the cantilever support due to self-weight and topping, plus a construction traffic allowance of 1.5 kN/m^2. Stage 2 loads, after hardening of the topping, are resisted by a steel mesh (with loose rebars as necessary) in the topping
   ii. no top tendons, or a small area of tendons, with top opened cores and factory placed rebars in the top of filled cores to carry stage 1 moment and stage 2 loads in the same manner.

Wet cast solid or hollow core (Figure 2.16) units are suitable for cantilevered slabs, either non-composite or composite, because the bottom tendons can be debonded in the region of the cantilever, unlike dry-cast extruded or slipformed units that cannot. Top tendons, for example small strands (9.3 mm diameter) or wires (5 or 7 mm diameter) may be positioned

Figure 13.9 Prestressed hollow core unit with slots to receive *insitu* concrete infill and threaded rebars into couplers to support a cantilever balcony.

Figure 13.10 Preparing the top steel for prestressed hollow core floor unit, 3 no. H12 rebars, to be placed inside top opened cores. (Courtesy Bison Precast Ltd., UK.)

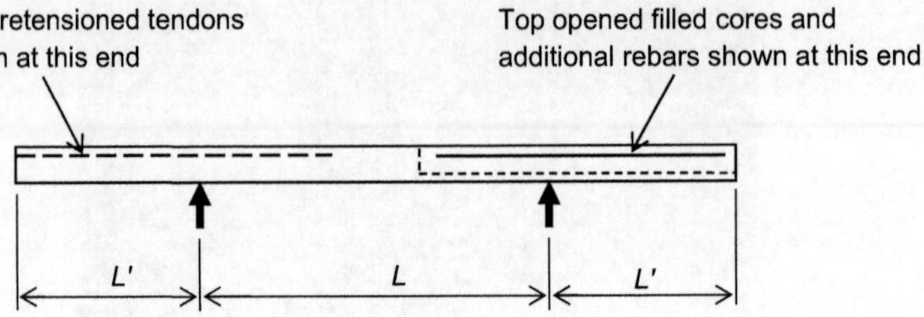

Figure 13.11 Options for reinforcing the cantilevers of non-composite slabs using (left) pretensioned tendons, which may be debonded in wet-cast units, and (right) additional rebars in filled cores.

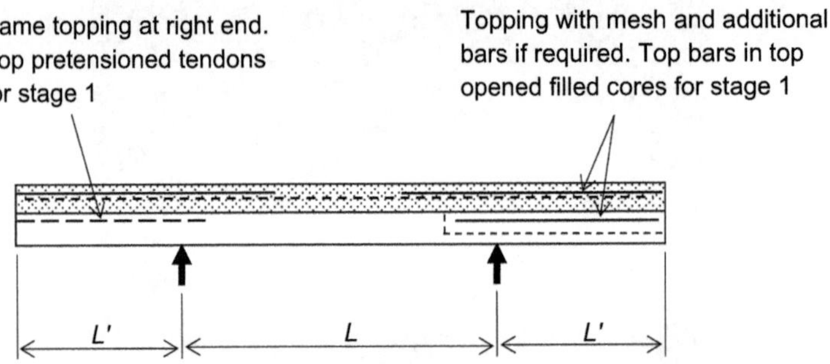

Figure 13.12 Options for reinforcing the cantilevers of composite slabs using mesh and additional rebars (if required) in the topping for stage 2 with (left) pretensioned tendons, which may be debonded in wet-cast units, and (right) additional rebars in filled cores for stage 1.

at the top of the unit and debonded in the region of the main span, thus maximising M_{sR} in the main span. This method also has the structural advantage of greater negative M'_{sR} compared with dry cast units in addition to upward camber in the cantilever. Distances to the point of debonding are calculated in a similar manner to the curtailment distances for rebars in the top opened cores in Section 13.2.2, except that the anchorage length for prestressing tendons is l_{bpd} (not l_{bd}).

13.2.2 Design of reinforcement in top opened cores and/or with top pretensioned tendons

Let M'_{Ed}=ultimate negative moment

b'=width of precast unit at soffit (minus notch width if present)
d'=effective depth to top bars or top tendons
f_{ck} of the precast unit

Referring to Figure 13.11

Case 1. Top bars in opened cores with area of top tendons A'_p

$$K' = M'_{Ed} / f_{ck} b' d'^2 \tag{13.10}$$

$$z' = d' \min\{0.95; 0.5 + \surd(0.25 - K'/1.134)\} \tag{13.11}$$

$$A'_s = M'_{Ed} / 0.87 f_{yk} z' - A'_p \tag{13.12}$$

Note that due to strain compatibility of loose rebars and top tendons, the relative yield strength of tendons $p_{yk}=1770$ N/mm^2 to rebars $f_{yk}=500$ N/mm^2 is not considered, i.e. the strength of tendons is also limited to 0.87 f_{yk}.

The number of top opened cores should be at least 2, and not more than 2 less than the number of cores. The number of bars in narrow cores, of say 60–80 mm in width, should be 1, but in wider cores, such as 100–150 mm in width, could be 1 or 2.

The curtailment and anchorage length of rebars in the top opened cores, beyond point of zero M_{Ed} in the main span (Eq. 13.6) is given as

Let f_{cki}=strength of concrete in filled cores

Although f_{cki} could be as low as 25 N/mm^2 most manufacturers prefer to use higher strength concrete, often equal to the precast f_{ck}, but not more than $f_{cki}=45$ N/mm^2. This improves the bond strength at the interface of the interior of the cores, which can be quite smooth for extruded units, but less so for slipformed units where the fluctuating piston action of the machine causes small crests at about 200 mm spacing on the inside of the cores.

8.4.4 $l_{bd}=\alpha_1\,\alpha_2\,\alpha_3\,\alpha_4\,\alpha_5\,l_{b,rqd}>\max\{0.3\,l_{b,rqd};\ 10\,\phi;\ 100\,\text{mm})$ (13.13)

Table 8.2 $\alpha_1=1$ for straight bars

$0.7\le\alpha_2=1-0.15\,(c_d-\Phi)/\Phi\le 1$ for straight bars (13.14)

$\alpha_3=\alpha_4=\alpha_5=1$

Fig. 8.3 where $c_d=\min$ {horizontal spacing/2; top cover c'}

8.4.3.(2) $l_{b,rqd}=0.25\,\sigma_{sd}\,\Phi/f_{bdi}$ (13.15)

where σ_{sd} = actual design stress = $0.87 f_{yk} A'_s\,/\,A'_{s\ provided}$ (13.16)

8.4.2 $f_{bd}=2.25\,\eta_1\,\eta_2\,f_{ctdi}$ (13.17)

Table 3.1 and 3.1.6 2(P) $f_{ctdi} = 0.7\times 0.3 f_{cki}^{2/3}\,/\,1.5$

Fig. 8.2 $\eta_1=1$ for good casting condition, except where the depth of the cores is $\ge$ 300 mm where $\eta_1=0.7$

$\eta_2=1$ for bar $\phi\le 32$ mm diameter

9.2.1.3 (2) Curtailment length beyond the point of zero $M_{Ed}=l_{bd}+d'$ (13.18)

where d' is for the 'shift' rule in this clause

Case 2. Top tendons A'_p alone

Eqs. 13.11 and 13.12 K' and z' as above

$X'=(d'-z')/0.4$ (13.19)

Referring to Section 9.3.1

Fig. 3.10 $\varepsilon_p=\varepsilon_{po}+\varepsilon_{cu3}\,(d'/X'-1)$ where $\varepsilon_p\le\varepsilon_{ud}=0.02$ (13.20)

ε_{po}=prestrain after losses at the support

$$f'_p = 0.9f_{pd} + 0.1f_{pd}\left(\varepsilon_p - \varepsilon_{Lop}\right)/\left(\varepsilon_{uk} - \varepsilon_{Lop}\right) \quad (13.21)$$

$$A'_p = M'_{Ed} / z' f'_p \quad (13.22)$$

If cantilever span L'<anchorage development length l_{bpd} for the top tendon, f_p is modified by the ratio of L'/l_{bpd}, but not greater than 1.

Referring to Sections 9.5 and 10.2.2

Exp. 8.21 $l_{bpd}=l_{pt2}+\alpha_2\ \Phi\ (\sigma_{pd}-\sigma_{po})/f_{bpd}$ (13.23)

Exp. 8.16, 8.18 $l_{pt2}=1.2\times1.0\ \alpha_2\ \Phi\ \sigma_{pm0}/f_{bpt}$ (13.24)

where Φ=diameter of top tendon. Other terms are defined in Sections 9.5 and 10.2.2.

13.2.3 Service stress due to prestress plus negative flexural stress

The bottom fibre of the cantilever is checked for the combined service stress due to prestress and the maximum cantilever moment M'_s. As given in Section 7.2.1: 'For non-linear creep to exist the term $[\sigma_c/f_{cm}(t_0)-0.45]$ must be positive. Then $\sigma_c\leq0.45\ f_{cm}(t_0)$.' Therefore, the combined compressive stress is

Combined compressive stress=prestress+flexural stress

$$f'_{b,max} = \sigma_b + M'_s / Z_{b,co} \leq 0.45 f_{cm} \quad (13.25)$$

where $f_{cm}=f_{ck}+8$ N/mm^2

At the top fibre of the cantilever, the combined tensile stress=prestress+flexural stress

$$f'_{t,max} = \sigma_t - M'_s / Z_{t,co} \geq -f_{ctm} \quad (13.26)$$

If cantilever span L'<transmission length l_{pt2} for the top tendon, σ_b and σ_t are modified by the ratio of L'/l_{pt2}, but not greater than 1.

If in Eq. 13.26 $f'_{t,max}\geq-f_{ctm}$ all section properties are based on the flexurally uncracked modulus and the shear capacity at the face of the cantilever support is $V_{Rd,c}$.

If in Eq. 13.26 $f'_{t,max}<-f_{ctm}$ the section is considered to be flexurally cracked, in which case filled cores are required and the shear capacity at the face of the cantilever support is $V_{Rd,cr+t}$ the addition of the precast unit and the filled cores, as given in Section 13.2.6.

13.2.4 Composite cantilever slabs

13.2.4.1 Ultimate design for stage 1 and 2 bending moments

The design of reinforcement in the structural topping, and if necessary in the top opened filled cores (Section 13.2.2) is a two-stage design at the ultimate limit state. The serviceability limit state is also checked for combined compressive prestress plus flexural stress (similar to Eq. 13.25), but the remainder of the design is at the ultimate limit state.

At stage 1. $M'_{Ed1} = w'^2_{Ed1} / 2$ (13.27)

where w'_{Ed1} is the self-weight of the precast unit plus the self-weight of the wet structural topping. For a depth of topping t the self-weight is taken as 25 $(t+5)$ (kN/m^2) allowing for the slope of the psc unit due to camber, typically 10 mm for $L'=1.5$ m. The self-weight may be adjusted after stage 1 calculation is completed if necessary.

The ultimate load combination used for w'_{Ed1} is the same as that used for the total loads based on Exp. 6.10a or 6.10b of BS EN 1990. Therefore, referring to Section 3.1.3, if it is found that for the total loads $\gamma_G=1.25$ and $\gamma_Q=1.5$, these values are also used for w'_{Ed1}. If $\gamma_G=1.35$, because w'_{Ed1} is a dead load, an anomaly in the total ultimate load will accrue.

If top tendons A'_p are present, Eqs. 13.11–13.13 determine whether top bars are required at stage 1. If not the precast is then checked for the combined service stresses in the top fibre according to Eq. 13.26, and top bars are then specified or not.

The top bars are considered to be at their design ultimate strength $0.87 f_{yk}$, even if the area provided $A'_{s\,prov} > A'_{s\,req}$, and therefore do not contribute further to stage 2.

At stage 2. $M'_{Ed2} = w'^2_{Ed2} / 2$ (13.28)

plus any additional point or line loads

where w'_{Ed2} is for the imposed dead and live loads based on Exp. 6.10a or 6.10b of BS EN 1990.

Let the strength of the topping $= f_{ckT}$

d'_2 = effective depth to the top bars in the topping, usually a A142 or A193 steel mesh placed at mid-depth of the topping giving top cover C' ≈ 30 mm.

It is not known how the sheets of mesh will be nested, and whether the mesh bars parallel to the span of the cantilever will be nearest to the top surface or further down. Therefore, in deeper toppings, such as $t=100$ mm, the parallel bars may possibly be at a depth from the top surface of 30 + say 12 + radius 6 = 48 mm.

$$K'_2 = M'_{Ed2} / f_{ckT} b' \, d'^2_2 \quad (13.29)$$

$$z'_2 = d'_2 \min\{0.95;\ 0.5 + \surd(0.25 - K'_2 / 1.134)\} \quad (13.30)$$

$$A'_{s2} = M'_{Ed2} / 0.87 f_{yk} z'_2 \quad (13.31)$$

A'_{s2} is provided by a steel mesh, and if necessary additional loose rebars placed in between the parallel mesh bars particularly if the size of the mesh becomes large, e.g. A393, as it may only be required for a short distance in the main span. Therefore a better option would be to use A142 mesh plus H8 mm bars at 200 mm spacing. The minimum area of mesh is the usual requirement of 0.13% or $0.26\, f_{ctm}/f_{yk}$ × concrete area.

If cantilever span L' < anchorage length l_{bd} for the mesh, f_{yk} is modified by the ratio of L'/l_{bd}, but not greater than 1. Alternatively A'_{s2} is modified by the ratio of l_{bd}/L'.

13.2.4.2 Service stress due to prestress plus stage 1 and 2 negative flexural stress

The bottom fibre of the cantilever is checked for the combined service stress due to prestress and the maximum cantilever moments M'_{s1} and M'_{s2}.

Combined compressive stress = prestress + flexural stress

$$\sigma_b + M'_{s1} / Z_{b,co} + M'_{s2} / Z_{b,c,co} \le 0.45 f_{cm} \quad (13.32)$$

Combined tensile stress

$$\sigma_t - M'_{s1} / Z_{t,co} - M'_{s2} / Z_{t,c,co} \geq -f_{ctm} \tag{13.33}$$

If cantilever span L'<transmission length l_{pt2} for any top tendons present, σ_b and σ_t are modified by the ratio of L'/l_{pt2}, but not greater than 1.

If Eq. 13.33 $<-f_{ctm}$ the section is considered to be flexurally cracked, in which case the shear capacity at the face of the cantilever support is $V_{Rd,cr,c}$.

13.2.4.3 Propped cantilevers in composite cantilever slabs

Propping the end of a cantilever while the *insitu* concrete topping is placed is fraught with danger. Let w_T=self-weight of the topping per unit width. The problem is that positive deflections and negative end rotations θ at the cantilever support in the main span due to w_T will cause uplift at the end of the cantilever. (Rotations are defined as positive clockwise if the cantilever is viewed at the right-hand end of the span.) Therefore the reaction in a previously rigidly founded prop will reduce, in turn reducing the effectiveness of the prop to nullify stage 1 stresses due to w_T. For example due to w_T in the main span $\theta=-w_T L^3/24 E_{cm} I_{c,co}$ and due to the moment $w_T L'^2/2$ in the cantilever $\theta'=+(w_T L'^2/2) L/3 E_{cm} I_{c,co}$. If the main span L is considerably greater than L', which is often 4 to 5 times, the deflection at the end of the cantilever will be negative upwards $\delta'=(\theta+\theta') L'+w_T L'^4/8 E_{cm} I_{c,co}$ (the free deflection of the cantilever, usually small).

For example, for the model 1200×250 mm prestressed hcu used in the worked examples in this chapter, let $E_{cm} I_c=36283\times1296\times10^6=47.0\times10^{12}$ Nmm². Let L=9.8 m and L'=1.5 m. Assume a 75 mm depth topping of average depth 80 mm, then $w_T=1.2\times0.08\times25=2.4$ kN/m (or 2.4 N/mm) per unit. Then due to the main span $\theta=-2.4\times9800^3/24\times47.0\times10^{12}=-0.002000$ radians. Due to the cantilever moment $\theta'=+(2.4\times1500^2/2)\times9800/3\times47.0\times10^{12}=+0.000187$ rad. Then $\delta'=(-0.002000+0.000187)\times1500=-2.72$ mm plus the free deflection $+2.4\times1500^4/8\times47.0\times10^{12}=0.03$ mm $\delta'=-2.7$ mm. To nullify this deflection, the prop reaction R' at the end of the cantilever must have a downward axial force $R'=3 E_{cm} I_c \delta'/(L'^3+L'^2L)=-14.95$ kN. In other words, the initial precompression in the prop will reduce by an equivalent amount of 14.95 kN* due to curvature when the cantilever slab is loaded by the *insitu* topping. The reaction in the prop is therefore unknown. The actual reaction R' will be lower than this because creep effects will have taken place between transfer and installation of the topping, say $\psi_1=0.5$, and this will reduce R' to≈−14.95/(1+0.5)=−10 kN. Furthermore site tolerances and variations in slab properties $E_{cm} I_{c,co}$ will lead to other variations, but the effect remains that, unless the precompression in the prop is known, the net prop reaction is unknown. *The same result is found by carrying out Hardy-Cross moment distribution or using stiffness matrix for a two-span beam of 9.8 and 1.5 m subjected to a UDL of 2.4 kN/m; −14.92 kN in this case.

To eliminate this effect, the ratio L/L' has to be approximately 2.3, clearly an impractical solution, e.g. a 2.0 m cantilever requires a main span not greater than 4.6 m, theoretically although the precompression in the prop will allow this ratio to increase.

Propping the main span at mid-span has a dramatic effect on the action of the cantilever prop because θ and θ' are reduced by factors of around 4. Repeating the above calculation for R' by adding a prop at mid-span gives a reaction after creep effects R'=−0.2 kN, effectively zero. Therefore the prop is not active and the stage 1 moment due to topping $M'_{sT}=w_T L'^2/2$ remains at 2.7 kNm. To eliminate this moment the prop reaction needs to be R'=+1.8 kN. When the topping has hardened and the prop is removed the reverse moment M'_{sT} acts on the composite section.

Eqs. 13.32 and 13.33 are modified to

$$\sigma_b + M'_{s0} / Z_{b,co} + \left(M'_{s2} + M'_{sT}\right) / Z_{b,c,co} \le 0.45 f_{cm} \tag{13.34}$$

$$\sigma_t - M'_{s0} / Z_{t,co} - \left(M'_{s2} + M'_{sT}\right) / Z_{t,c,co} \ge -f_{ctm} \tag{13.35}$$

where M'_{s0} is due to self-weight of the precast unit

13.2.5 Instability, or over-turning, criteria for cantilever slabs

The instability of a cantilever slab is essentially a check for over-turning to ensure that, with the appropriate factors of safety obtained from the NA to BS EN 1990, Table NA.A1.2 (A) (NA to BS EN 1990, 2002) for 'equilibrium', the reaction R at the simple support is not negative.

The loading is Case II, defined in Figure 13.13a, where the maximum cantilever moment M'_{Eq} is checked against the reactions in the back-span L. The suffix Eq means ultimate-equilibrium.

The design values for actions (EQU) given in Table NA.A1.2 (A) are given in Figure 13.13b and are essentially $\gamma_{Gkj,inf}$=0.9 for the (inferior) dead load in the back-span, and $\gamma_{Gkj,sup}$=1.1 and γ_Q=1.5 for the (superior) dead and live loads in the cantilever, respectively. The factor for an accompanying live load is $\gamma_Q = \psi_0$ 1.5, with each live load(s) considered as the accompanying load in turn.

The design values for actions (EQU) given in Table NA.A1.2 (B) are essentially based on the ultimate combination Eq. 6.10 in BS EN 1990. These are shown in Figure 13.13c as $\gamma_{Gkj,inf}$=1.15 in the back-span, and $\gamma_{Gkj,inf}$=1.35 and γ_Q=1.5 (and $\gamma_Q = \psi_0$ 1.5) in the cantilever, respectively.

Consider the latter from Table NA.A1.2 (B) for UDL both in the cantilever and back-span. In the cantilever

$$M'_{Eq} = \left(1.35 w_{G,sup} + 1.5 w_Q\right) L'^2 / 2 \tag{13.36}$$

In the back-span

$$F_{Eq} = 1.15 w_{G,inf} L / 2 - M'_{Eq} / L \tag{13.37}$$

When $R_{Eq} \ge 0$ the minimum back-span is given as

$$L \ge \sqrt{\left(2M'_{Eq} / 1.15 w_{G,inf}\right)} \tag{13.38}$$

Considering the cantilever slab example in Example 13.1, where the cantilever span L'=1.5 m, and the UDLs for the self-weight, filled cores and imposed dead loads are $w_{G,sup}$=3.63+1.24 5+3.80=8.675 kN/m², $w_{G,inf}$=3.63 kN/m² (neglecting the small extent of filled cores in the back-span) and w_Q=5.0 kN/m².

Eq. 13.36 $\quad M'_{Eq} = (1.35 \times 8.675 + 1.5 \times 5.0)\ 1.5^2 / 2 = 21.6$ kNm

Eq. 13.38 $\quad L \ge \sqrt{(2 \times 21.6 / 1.15 \times 3.63)} =$ **3.22 m**

Using data from Table NA.A1.2 (A) (EQU)

Eq. 13.36 $\quad M'_{Eq} = (1.10 \times 8.675 + 1.5 \times 5.0)\ 1.5^2 / 2 = 19.2$ kNm

Eq. 13.38 $\quad L \ge \sqrt{(2 \times 19.2 / 0.9 \times 3.63)} =$ **3.43 m**

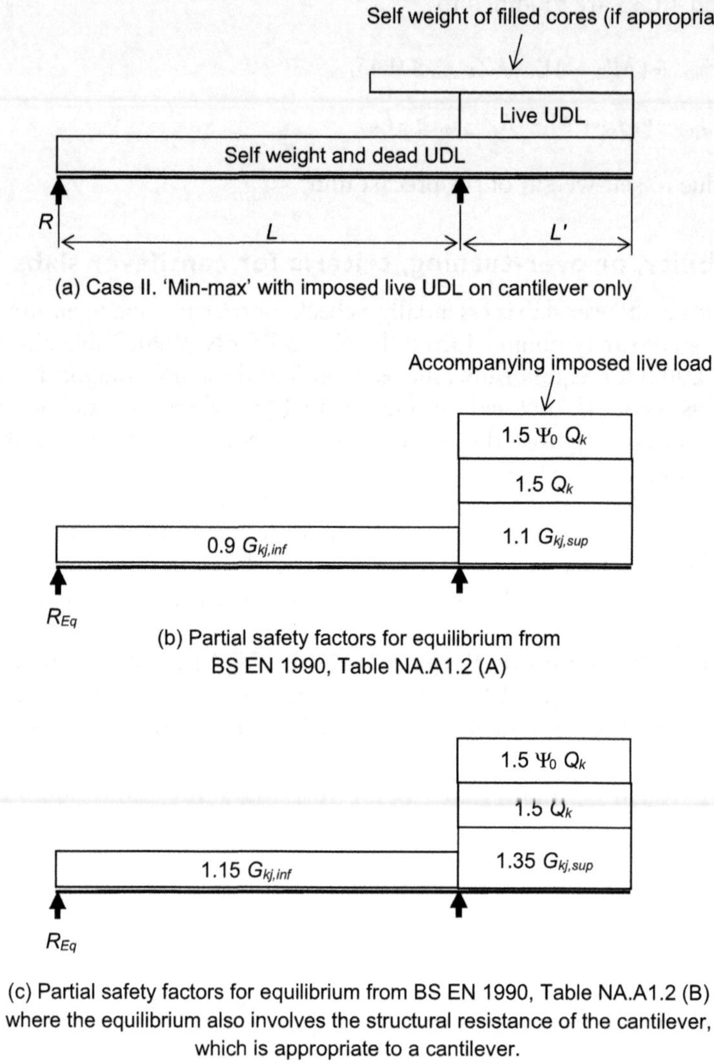

Figure 13.13 Partial safety factors for the equilibrium check against over-turning in a cantilever slab. (a) Case II. 'Min-max' with imposed live UDL on cantilever only. (b) Partial safety factors for equilibrium from BS EN 1990, Table NA.A1.2 (A). (c) Partial safety factors for equilibrium from BS EN 1990, Table NA.A1.2 (B) where the equilibrium also involves the structural resistance of the cantilever, which is appropriate to a cantilever.

This is actually less conservative than the recommendation to use Table NA.A1.2 (B) (STR). Given that a separate calculation is required Table NA.A1.2 (A) (EQU) may as well be adopted.

Example 13.1

Calculate M_{sR} at the simple support, mid-span and at the cantilevered support for the 250 mm depth psc hollow core unit shown in cross section in Figure 6.7. The section properties, material data and prestress are as given in Example 7.1. Exposure class to the bottom and top surfaces is XC1. The effective span of the main span $L = 9.8$ m and the cantilever span $L' = 1.5$ m, as shown in Figure 13.14. As there are no tendons in the

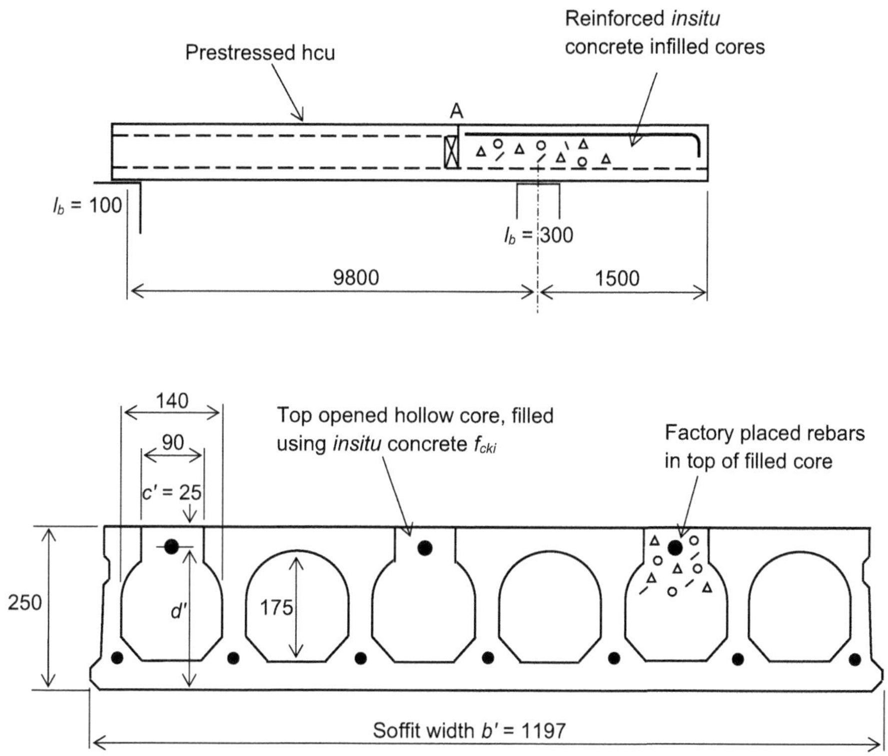

Figure 13.14 Elevation and cross section of prestressed concrete hcu designed in Examples 13.1–13.4, showing position and extent of reinforced concrete fille″d cores. Distance to point A to be determined from Example 13.2.

top of the hcu the negative moment of resistance of the cantilever is to be provided by placing rebars of strength f_{yk} = 500 N/mm^2 into the top of 140 mm wide opened cores. The slots are filled with *insitu* concrete of the same strength as the precast unit f_{cki} = 45 N/mm^2 and known density 24.5 kN/m^3. The distance to the curtailment of rebars in the main span (point A) is determined in Example 13.2 (ultimate design). Use self-weight of grouted unit = 3.63 kN/m^2, imposed dead loads = 3.30 kN/m^2, ceiling 0.5 kN/m^2 and imposed live load both in the main span and cantilever = 5.0 kN/m^2.

Using the two load combinations Case I and II, given in Figure 13.8, check M_{sR} against the characteristic combination of the design service moment M_s. Plot graphs of bending moment distributions for M_{sR} and M_s, taking l_{pt} = 605 mm (from Example 6.1).

Solution

At simple support

M_{sR} is the same as in Example 7.1, M_{sR} = **168.75 kNm**

At mid-span

M_{sR} is calculated according to Example 7.1 except that the bending moments due to self-weight and imposed dead loads now include the cantilever, inclusive of the self-weight of the filled cores.

It is therefore necessary at this early stage to decide on the number of filled cores. This depends on the number and diameter of rebars, but for a 6-core unit should be 2 or 3.

Manufacturing data give the cross-sectional area of one core as 20330 mm^2, therefore taking 3 no. cores the self-weight is $3\times20330\times24.5\times10^{-6}/1.2=1.245$ kN/m^2.

It is shown later in Example 13.2 that the ultimate cantilever moment $M'_{Ed}=-24.76$ kNm. If the lever arm z' to 12 mm diameter top bars is taken as say 0.9 d', where $d'=250-25$ cover – say 6 = 219 mm, $z'=197$ mm and $A'_s=24.76\times10^6/(197\times0.87\times500)\approx289$ mm^2, or 96 mm^2 per core. Therefore use 3 no. H12 top bars (339 mm^2). Two filled cores would require 147 mm^2 per bar, so 2 no. H16 could be used (402 mm^2), but the 3 bars give a better distribution of resistance across the width of the unit and less reliance on the interface shear strength inside the hollow cores.

The cantilever dead load $w'_0=1.2\times(3.63+1.245+3.30)=9.81$ kN/m, and $M'_{S0}=-9.81\times1.5^2/2=-11.04$ kNm. At mid-span, the moment is $99.88-11.04/2=94.36$ kNm. Now, following the procedures in Example 7.1

Due to M'_{S0} $\sigma_{cp0}=-94.36/15.766=-5.98$ N/mm^2

Then at mid-span $\sigma_{cp}=(1072607/1055654)\times11.90-5.98=6.11$ N/mm^2

Loss of prestress due to creep $\Delta\sigma_{p,c}=59.13$ N/mm^2, shrinkage $\Delta\sigma_{p,s}=73.51$ N/mm^2 and relaxation $\Delta\sigma_{p,r}=27.47$ N/mm^2

Final prestress. At transfer, at the mid-span position of the main span (not at mid-span of the overall length of unit of 11.3 m) and discounting the cantilever which does not exist, the prestress after initial losses is $\sigma_{pm0}=1177.4$ N/mm^2

$\sigma_{po}=1177.4-59.13-73.51-27.47=1017.3$ N/mm^2 (compared with 1020.7 N/mm^2 without the cantilever).

Final prestress at mid-span. $\sigma_b=12.89$ N/mm^2 and $\sigma_t=-2.10$ N/mm^2

M_{sR} at mid-span is the lesser of

$M_{sR,b}=(12.89+3.80)\times10.701=$ **178.66 kNm**

$M_{sR,t}=(20.25-2.10)\times10.056=224.76$ kNm >178.66 kNm

At cantilever support

M_{sR} is calculated as above except that the negative bending moments due to self-weight and imposed dead loads are for the cantilever alone, adding compression to the concrete stress at the centroid of the tendons, increasing creep losses.

The cantilever $M'_{S0}=-9.81\times1.5^2/2=-11.04$ kNm

Due to M'_{S0} $\sigma_{cp0}=-11.04/15.766=+0.70$ N/mm^2

$\sigma_{cp}=(1072607/1055654)\times11.90+0.70=12.79$ N/mm^2

Loss of prestress $\Delta\sigma_{p,c}=123.85$ N/mm^2, $\Delta\sigma_{p,s}=73.51$ N/mm^2 and $\Delta\sigma_{p,r}=27.03$ N/mm^2

After initial losses close to the end of the span $\sigma_{pm0}=1158.8$ N/mm^2

$\sigma_{po}=1158.8-123.85-73.51-27.03=934.4$ N/mm^2

Final prestress at the cantilever. $\sigma_b=11.84$ N/mm^2 and $\sigma_t=-1.93$ N/mm^2

$M_{sR,b}=(11.84+3.80)\times10.701=$ **167.42 kNm** and $M_{sR,t}=223.04$ kNm >167.42 kNm

Therefore the reduction is M_{sR} between mid-span and the cantilever is 11.24 kNm, or 6.3%, not a huge amount and which is in fact not critical, but it is worth carrying out the exercise to see the differences.

This above value of $M_{sR,b}$ does not take into account the mass of the reinforced filled cores. Manufacturer's data for this hcu with three empty cores gives $I_c=1395\times10^6$ mm^4 (increased from 1270×10^6 mm^4 for 6 no. cores).

There is a small increase to allow for the transformed area of the tendons and rebars in the top of the cores such that, allowing for the manufacturing operations of opening the tops of cores, the best approximation is $I_{c,co} \approx 1497 \times 10^6 \text{mm}^4$ and $Z_{b,co} \approx 12.38 \times 10^6 \text{mm}^3$ (14% increase). Then $M_{sR,b} \approx (11.84+3.80) \times 12.38 =$ **193.6 kNm**. This higher value may be ignored.

M_{sR} at the ends of the unit is the same as in Example 7.1. $M_{sR,b} =$ **40.63 kNm**

Design moments M_s and M'_s in the cantilever

Case I. Maximum loads in the main span and minimum (dead) loads in the cantilever.

Main span $w_s = 1.2 \times (3.63+3.80+5.0) = 14.92$ kN/m

Cantilever dead load $w_0 = 1.2 \times (3.63+1.245+3.80) = 10.41$ kN/m, and $M'_s = -11.71$ kNm.

Eq. 13.2 Reaction at simple support $F = 14.92 \times 9.8/2 - 11.71/9.8 = 71.91$ kN

Eq. 13.3 Shear span $a = 71.91/14.92 = 4.820$ m from simple support

Eq. 13.4 $M_s = F\,a - w_s\,a^2/2 = 71.91 \times 4.820 - 14.92 \times 4.820^2/2 = 173.3$ kNm < 178.66 kNm

Strictly the self-weight of the filled cores should be added to w_s in the main span, but only over the length they are filled. This is not known until the ultimate moment analysis, i.e. distance to zero M_{Ed} plus the anchorage length, is completed in Example 13.2, which is found to be 0.90 m from the cantilever support. Therefore F is increased by $(1.2 \times 1.245 \times 0.90^2/2)/9.8 = 0.06$ kN to 71.97 kN. The new shear span $a = 71.97/14.92 = 4.824$ m and $M_s = 71.97 \times 4.824 - 14.92 \times 4.824^2/2 =$ **173.6 kNm** < 178.66 kNm.

Case II. Maximum loads in the cantilever and minimum in the main span.

Main span $w_s = 1.2 \times (3.63+3.80) = 8.92$ kN/m

Cantilever dead load $w_0 = 10.41 + 1.2 \times 5.00 = 16.41$ kN/m, and $M'_s = -18.46$ kNm.

$F = 8.92 \times 9.8/2 - 18.46/9.8 = 41.82$ kN

$a = 41.82/8.92 = 4.689$ m

$M_s = 41.82 \times 4.689 - 8.92 \times 4.689^2/2 = 98.2$ kNm clearly not critical.

However, these data give the distance to the point of zero moment at

Eq. 13.6 $x = 2a = 9.378$ m from the simple support, or 0.422 m from the cantilever support, as shown in Figure 13.15.

For interest only at this stage in the analysis, $M_{sR,t}$ at the top of the section is based on prestress $\sigma_t = -1.93$ N/mm². The remaining tension in the top is $-3.80 - -1.93 = -1.87$ N/mm². $M_{sR,t} \approx 1.87 \times 10.056 = 18.81$ kNm > 18.46 kNm, therefore the section remains uncracked for the purpose of calculating shear capacities at either side of the cantilever support.

Also, check the total compression in the bottom fibre due to prestress and flexural compression. The total compressive stress is $\sigma_b + M'_s/Z_{b,co} = 11.84 + 18.46/10.701 = 13.56$ N/mm² $< 0.45\,f_{cm} = 0.45 \times (45+8) = 23.85$ N/mm².

Figure 13.15 shows the complete bending moment diagrams for the characteristic design moments M_s and M_{sR}. The design is completed for ultimate moments, shear forces, deflections in Examples 13.2, 13.3 and 13.4.

Example 13.2

Calculate M_{Rd} and M'_{Rd} for the basic section and for the transformed composite section with filled cores at the cantilevered support for the 250 mm depth prestressed hcu shown in cross section in Figures 6.7 and 13.14, and using the same properties as in Examples 9.1 and 13.1.

Using the two load combinations Case I and II, given in Figure 13.8, check M_{Rd} and M'_{Rd} against the ultimate design moments M_{Ed} and M'_{Ed} in the main span and cantilever, respectively. Plot graphs of bending moment distributions taking $l_{bpd} = 1230$ m (from Example 9.1) and calculating l_{bd} for the rebars in the top of the opened cores. The slab is used for offices with $\psi_0 = 0.7$.

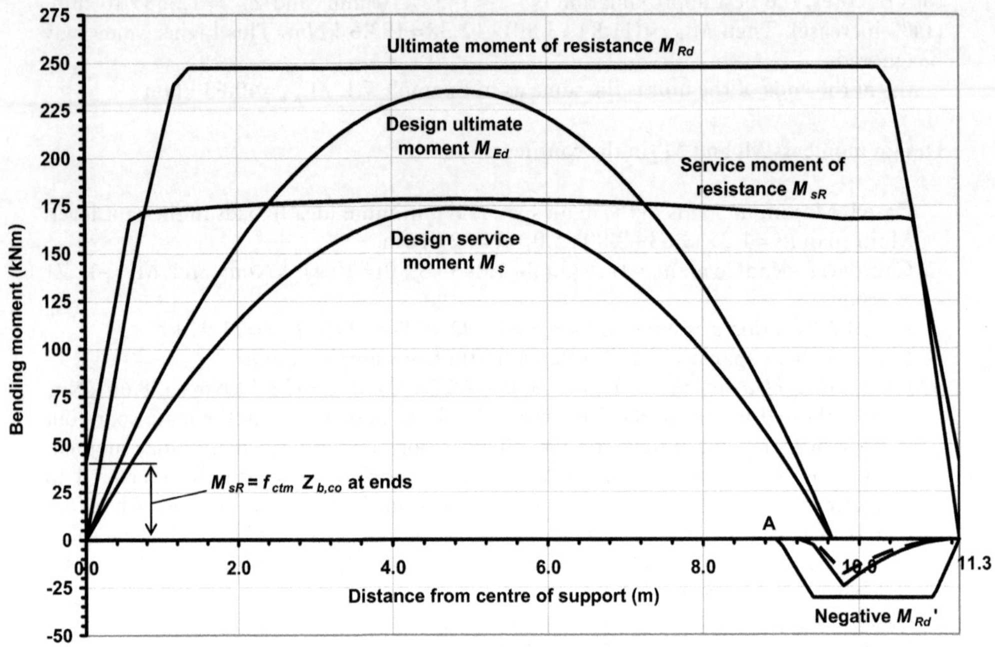

Figure 13.15 Service and ultimate design bending moments and moments of resistance for the cantilever hcu designed in Examples 13.1 and 13.2.

Solution

At mid-span

M_{Rd} is the same, within 0.01% as in Example 9.1, M_{Rd}=**247.3 kNm** (the tiny difference is due to the final prestress σ_{po}=1017.3 vs 1020.7 N/mm^2.

At cantilever support

M_{Rd} for the precast unit alone is calculated as above except that the bending stress due to self-weight and imposed dead loads is for the cantilever is subtracted to give σ_{po}=934.4 N/mm^2. Hence M_{Rd}=**246.8 kNm**

M'_{Rd} for the transformed section with filled cores is based on the following statically reinforced design. For exposure class XC1 top cover $C' \geq 20$ mm (see Table 5.4). However, for manufacturing purposes use C'=25 mm. Assuming 12 mm diameter bars d'=250−25−6=219 mm. The compression width is the actual width of the hcu b'=1197 mm (ignoring the 8×8 mm fillets in the bottom corners)

For Case II load combination, the ultimate cantilever load is the maximum of Exp. 6.10a and 6.10b of BS EN 1990. The cantilever dead UDL is 1.20×(3.63+1.245+3.80)= 10.41 kN/m and live UDL=1.20×5.00=6.00 kN/m. Then

$$w'_{Ed} = \max[1.35 \times 10.41 + 0.7 \times 1.5 \times 6.00; 1.25 \times 10.41 + 1.5 \times 6.00]$$

$$= \max\{20.35;\ 22.01\} = 22.01 \text{ kNm}$$

Note that the critical load case is from Exp. 6.10b and therefore γ_G=1.25 will also be used for the dead load in the main span.

$$M'_{Ed} = -22.01 \times 1.5^2 / 2 = -24.76 \text{ kNm}$$

Eq. 13.10 $K'=24.76\times10^6/(45\times1197\times219^2)=0.0096$
Eq. 13.11 $z'/d'=\min\{0.95;\ 0.5+\sqrt{(0.25-0.0096/1.134)}\}=\min\{0.95;\ 0.99\}$
$z'=0.95\times219=208\,mm$
Eq. 13.12 $A_s'=24.76\times10^6/(208\times0.87\times500)=274\,mm^2$

Use 3 no. H12 bars (339 mm²)

Negative M'_{Rd} is calculated as follows

$$F_s'=0.87\times500\times339\times10^{-3}=147.5\text{ kN}$$

$$X'=147.5\times10^3/(0.567\times45\times0.8\times1197)=6.0\,mm$$
$$z'=\min\{0.95\times219;\ 219-0.4\times6.0\}=208\,mm$$

$$M'_{Rd}=147.5\times208\times10^{-3}=30.7\text{ kNm}>24.76\text{ kNm}$$

Calculation of anchorage length l_{bd} at the end of the slot from the point of zero negative bending moment in the top of the section

Eq. 13.16 σ_{sd}=actual design stress$=0.87\times500\times274/339=352\text{ N/mm}^2$
Eq. 13.17 $f_{bd}=2.25\times1.0\times1.0\times(0.7\times3.80/1.5)=4.00\text{ N/mm}^2$
$\eta_1=1$ for good casting condition $h\le250\,mm$
$\eta_2=1$ for bar $\phi\le32\,mm$ diameter
Eq. 13.15 $l_{b,rqd}=0.25\times352\times12/4.00=264\,mm$
Eq. 13.13 $l_{bd}=0.84\times264=222\,mm$
Eq. 13.14 where $\alpha_2=1-0.15\,(25-12)/12=0.84$ and $\alpha_1=\alpha_3=\alpha_4=\alpha_5=1$

Curtailment length beyond zero M'_{Ed} is $l_{bd}+d'=222+219=441\,mm$

Calculation to the point of zero M'_{Ed} actually requires prior knowledge of this point to take account of the self-weight of the filled cores in the main span, but this is found to be insignificant, i.e. changes the curtailment point by only 10 mm.

The curtailment length at the end of the cantilever is taken as the minimum anchorage 10Φ plus d', as $120+219=339\,mm$.

Calculation to the point of zero M'_{Ed} is as follows
Case II main span $w_{Ed}=1.25\times8.92=11.15$ kN/m
Simple support reaction $F_{Ed}=11.15\times9.8/2-24.76/9.8=52.11$ kN
Shear span $a_{Ed}=52.11/11.15=4.674\,m$
Point to zero M'_{Ed} $x=2a_{Ed}=9.347\,m$
Curtailment point $x=9.347-0.441=8.906\,m$, rounded to 8.90 m, or 0.90 m from the cantilever support (this information is carried to Example 13.1).

For Case I load combination, the cantilever dead UDL $w'_{Ed}=1.25\times10.41=13.01$ kN/m

$$M'_{Ed}=13.01\times1.5^2/2=-14.64\text{ kNm}$$

In the main span $w_{Ed}=1.25\times8.92+1.5\times6.00=20.145$ kNm

Plus the filled cores $w_{Edt}=1.25\times1.2\times1.245=1.87$ kNm for a distance of 0.90 m from the cantilever support

Simple support reaction $F_{Ed}=20.145\times9.8/2+(1.87\times0.90^2/2)/9.8-14.64/9.8=98.71+0.08-1.49=97.30$ kN

Shear span $a_{Ed}=97.30/20.145=4.830\,m$

$M_{Ed}=97.30\times4.830-20.145\times4.830^2/2=$ **235.0 kNm**

Figure 13.15 shows the complete bending moment diagrams for the ultimate design moments M_{Ed}, M'_{Ed}, M_{Rd} and M'_{Rd} and the curtailment point A to the reinforced top opened cores.

13.2.6 Shear capacity of cantilever floor units

Refer to Section 10.2.6. After production of the hcu, and after detensioning, the cores are filled using factory placed concrete f_{cki}. The additional shear capacity of plain-filled cores, i.e. without rebars in the tension zone to act in dowel action against shear forces is

Eq. 10.16 $V_{Rd,t} = ⅔\, n\, b_c\, d_c\, f_{ctdi}$

The total shear capacity $V_{Rd,c+t} = V_{Rd,c} + V_{Rd,t}$

Example 13.3

Calculate the additional shear capacity $V_{Rd,t}$ to the basic capacities $V_{Rd,c}$ and $V_{Rd,cr}$, respectively for the reinforced concrete filled top opened cores for the 250 mm depth prestressed hcu shown in Figures 6.7 and 13.14 and designed in Examples 10.1, 13.1 and 13.2. Check the ultimate shear force V_{Ed} at the cantilever supports both in the main span and in the cantilever using the ultimate loads given in Example 13.2. The nominal bearing length at the cantilever support $l_b = 200$ mm.

Solution

At the face of the cantilever support, which is beyond the end of the transmission zone where $\alpha_l \geq 1$, the shear capacity from Example 10.1 is $V_{Rd,c} = 190.0$ kN.

Similarly, the flexurally cracked capacity in the basic unit (without filled cores) from Example 10.2 is $V_{Rd,cr} = 109.3$ kN

At the cantilever support in the main span

The filled cores are designed as plain concrete. The breadth of the core is 140 mm and depth of the core $d_c = 250 - 40 - 35 = 175$ mm.

Using $f_{cki} = 45$ N/mm², $f_{ctdi} = 0.7 \times 0.3 \times 45^{2/3}/1.5 = 1.77$ N/mm²

Eq. 10.16 $V_{Rd,t} = ⅔ \times 3 \times 140 \times 175 \times 1.77 \times 10^{-3} = 86.7$ kN

$V_{Rd,c+t} = 190.0 + 86.7 = \mathbf{276.7}$ **kN**

At the cantilever support in the cantilever

If the cantilever section at the support is flexurally uncracked, as was shown to be the case in Example 13.1, the ultimate shear capacity remains as $V_{Rd,c} = 190.0$ kN. The filled cores are similarly designed as plain concrete $V_{Rd,t} = 86.7$ kN. Then $V_{Rd,c+t} = \mathbf{276.7}$ **kN.**

At the end of the cantilever (from Example 10.1) $V_{Rd,c} = 134.3$ kN. Then $V_{Rd,c+t}$ reduces to $134.3 + 86.7 = \mathbf{221.0}$ **kN.**

If the cantilever section at the support is flexurally cracked $V_{Rd,cr}$ is based on the area A_s' and d' of the bars in top opened cores, in addition to the contribution of prestress at the centroidal axis.

$k = 1 + \sqrt{(200/219)} = 1.96 < 2.0$

$\rho_1 = 339/(302 \times 219) = 0.0051$

Maximum $V_{Rd,cr} = [(0.18/1.5) \times 1.96 \times (100 \times 0.0051 \times 45)^{1/3} + 0.15 \times 4.61)] \times 302 \times 219 \times 10^{-3} = 89.9$ kN

$v_{min} = 0.035 \times 1.96^{3/2}\sqrt{45} = 0.642$ N/mm²

Minimum $V_{Rd,cr} = [0.642 + 0.15 \times 4.61] \times 302 \times 219 \times 10^{-3} = 88.2$ kN

Critical and total $V_{Rd,cr+t} = 89.9 + 86.7 = \mathbf{176.6}$ **kN**

Design shear forces at the cantilever support

The critical loading is Case III with the maximum load in both the main span and cantilever. This gives the maximum downward reaction at the cantilever support in the main span.

Maximum ultimate UDL where filled notches $w_{Ed} = w'_{Ed} = 22.01$ kN/m over a distance of $0.9 + 1.5 = 2.4$ m. Distance from the simple support to the centre of this load is $11.3 - 2.4/2 = 10.1$ m

The same applies where no filled notches w_{Ed}=22.145 kN/m for a distance of 9.8−0.9=8.9 m from the simple support

Taking moments about the simple support, the total reaction at the cantilever support is F_{Ed}=(20.145×8.9²/2+22.01×2.4×10.1)/9.8=135.9 kN

Reaction at cantilever support in cantilever=22.01×1.5=33.0 kN

Net reaction at cantilever support in main span=135.9−33.0=102.9 kN

Shear force V_{Ed} is calculated at $l_b/2+y_b$ from the centre of support=200/2+123.0=223 mm. Then V_{Ed}=102.9−0.223×22.01=**98.0 kN**<276.7 kN

In the cantilever V_{Ed}=33.0−0.223×22.01=**28.1 kN**<176.6 kN

13.3 DEFLECTIONS DUE TO END ROTATION AND NEGATIVE MOMENTS

13.3.1 Deflection in main span in non-composite units

These are calculated in the same manner as for simply supported spans, as given in Section 11.2.1, at least in terms of the use of creep factors ψ_1, ψ_{28} and 0.8 ψ_∞. The main difference is that the end rotation due to the cantilever moment M'_s, based on self-weight and imposed dead loads, reduces the mid-span deflection in the main span. In fact, the distance to the point of maximum negative moment is at x=0.5774 L from the simple support, but the difference between this and the deflection at mid-span is only 2%.

To remind the reader, upward camber in the main span, Eq. 11.1 is $\delta_1=-P_{pm0}\ z_{cp}\ L^2/8\ E_{cm}(t)\ I_{c,co}$ and deflection due to self-weight, Eq. 11.2 is $\delta_2=+\ 5\ w_0 L^4/384\ E_{cm}(t)\ I_{c,co}$.

Then at transfer, the upward deflection at mid-span due to the self-weight of the cantilever is

$$\delta_{2a}=-w_0 L^2\ L'^2/32\ E_{cm}(t)\ I_{c,co} \tag{13.39a}$$

or $$\delta_{2a} = -M'_{s0}\ L^2 / 16E_{cm}(t)I_{c,co} \tag{13.39b}$$

where w_o is the self-weight of the unit and $M'_{s0}=w_0\ L'^2/2$ is the negative moment at the cantilever support.

Then δ_{2a} is added to $\delta_1+\delta_2$ in Section 11.2.2.

Strictly one could argue that deflections at transfer should be based on the full length of the unit $L+L'$, particularly when the short-term creep effects are considered, and that the above deflections do not occur until the units are installed, at which point in time the above deflections are now valid.

Similarly at installation the upward deflection due to self-weight is

$$\delta_{4a}=+\ (1+\psi_1)\ \delta_{2a} \tag{13.40}$$

with an additional static deflection due to the self-weight of filled cores.

Then δ_{4a} is added to $\delta_3+\delta_4$ in Section 11.2.2.

The long-term negative deflection at mid-span due to the cantilever moment M'_{s1} is due to the visco-elastic part of the self-weight (and filled cores if present) w_1 and static plus the visco-elastic part of M'_{s2} due to imposed dead loads w_2.

$$\delta_{6a} = \delta_{4a} - \left[(\psi_{28}M'_{s1}\ L^2 + (1+\psi_{28})M'_{s2}\ L^2\right] / 16E_{cm}I_{c,co} \tag{13.41}$$

and δ_{6a} is added to $\delta_5+\delta_6+\delta_7$ in Section 11.2.2.

The term 'visco-elastic' means the deflection takes place due to creep only over time after the instantaneous static deflection has taken place.

Changes in deflection, or movement at mid-span, due to visco-elastic dead loads

$$\delta_{14a} = -\left[\psi_{28} M'_{s1} + (1+\psi_{28}) M'_{s2}\right] L^2 / 16 E_{cm} I_{c,co} \quad (13.42)$$

and δ_{14a} is added to δ_{14} in Section 11.2.2.

Note that the increased second moment of area due to top bars in filled cores is ignored in the main span deflections.

13.3.2 Deflection in main span in composite slabs

For composite cantilevers, refer to Section 12.3.4 for deflections in composite elements. The reduction in deflection at mid-span at installation due to the moment $M'_{sT} = w_T L^2/2$ due to the self-weight of topping w_T is

$$\delta_{5a} = -M'_{sT} \, L^2 / 16 E_{cm} I_{c,co} \quad (13.43)$$

Then δ_{5a} is added to δ_5 in Section 12.3.4.

The reduction in long-term deflection at mid-span due to the moment M'_{s0} due to self-weight of the precast unit w_0 plus the moment M'_{sT} due to topping

$$\delta_{8a} = -\left(\psi_{\infty,t} M'_{s0} + \left(1+\psi_{\infty,t}\right) M'_{sT}\right) L^2 / 16 E_{cm} I_{c,co} \quad (13.44)$$

Eq. 12.51 $\psi_{\infty,t} = (0.8\,\psi_{\infty} - \psi_1)\, I_{c,co}/I_{c,c,co}$

The reduction at mid-span due to the moment M'_{s2} due to imposed dead loads w_2

$$\delta_{9a} = -\left(1+\psi_{28}\right) M'_{s2} \, L^2 / 16 E_{cm} I_{c,c,co} \quad (13.45)$$

Then $\delta_{8a} + \delta_{9a}$ is added to $\delta_8 + \delta_9$ in Section 12.3.4.

The reduction at mid-span in the long-term deflection after finishes, or 'movement', is due to imposed live load using $1+\psi_{28}$, visco-elastic self-weight and topping using $\psi_{\infty,t}$ and visco-elastic imposed dead loads using ψ_{28}, and changes in camber.

$$\delta_{13a} = \psi_{\infty,t}\left(M'_{s0} + M'_{sT}\right) L^2 / 16 E_{cm} I_{c,co} + \psi_{28} M'_{s2} \, L^2 / 16 E_{cm} I_{c,c,co} \quad (13.46)$$

Then δ_{13a} is added to δ_{13} in Section 12.3.4.

13.3.3 Deflection at the end of the cantilever in non-composite units

Cantilever deflections are calculated at transfer, installation of precast unit, for total long-term deflection, and movement after installation of finishes. They are due to

i. (a) effects of prestress in the upward camber and end rotation in the main span and (b) small deflections due to the internal prestressing moment in the cantilever span
ii. gravity loads (a) due to end rotation in the main span at the cantilever support, (b) upward deflection due to (iia), and (c) 'free' deflection of the cantilever from an encastre support, such as $wL^4/8\,EI$ or $PL^3/3\,EI$ for UDL and point load, respectively.

Note that +ve is a clockwise rotation if the simple support is viewed at the left end.

Table 13.1 gives a summary of rotations and deflections, showing the progression from transfer to installation, and from installation to long-term. Note that the critical loading is Case II with only dead loads in the main span causing end rotations θ_2, θ_4 and θ_6 at the cantilever support, whereas the cantilever is fully loaded causing end rotations θ_{2a}, θ_{4a} and θ_{6a}. The internal moment due to prestress in the main span causes end rotations θ_1, θ_3 and θ_5 at the cantilever support. The internal prestress in the cantilever does not cause rotation (as there is no external reaction). The term 'free' is the deflection in the cantilever from an encastre support, and is added to the deflection due to θL′. Note that, for clarity, Table 13.1 does not include the small changes in prestress included in the equations for δ_3 and δ_5.

a. At transfer

Rotation at the cantilever support θ_1 is due to prestress, and θ_2 is due to self-weight of the unit (without infill) in the main span and θ_{2a} is due to self-weight in the cantilever

$$\theta_1 = + P_{pm0}\, z_{cp}\, L/2\, E_{cm}(t)\, I_{c,co} \tag{13.47}$$

$$\theta_2 = -\, w_0 L^3/24\, E_{cm}(t)\, I_{c,co} \tag{13.48}$$

$$\theta_{2a} = + (w_0\, L'^2/2)\, L/3\, E_{cm}(t)\, I_{c,co} \tag{13.49}$$

Referring to Figure 13.16a, deflection is equal to $\theta_1\, L'$ plus the free internal moment due to prestress δ_{1a}

$$\delta_1 = \theta_1\, L' + \delta_{1a} \tag{13.50}$$

where $\delta_{1a} = P_{pm0}\, z_{cp}\, L'^2/2\, E_{cm}(t)\, I_{c,co}$ (13.51)

Table 13.1 Summary of end rotations at cantilever support θ and deflections at the end of the cantilever δ at transfer, installation and long term

Transfer		*Installation*		*Long term*	
Prestress moment in main span $PzL/2EI$	$+\theta_1$	$(1+\psi_1)\,\theta_1$	$+\theta_3$	$\theta_3+\psi_{28}\, PzL/2EI$	$+\theta_5$
Gravity UDL in main span $w_0L^3/24EI$	$-\theta_2$	$(1+\psi_1)\,\theta_2$	$-\theta_4$	$\theta_4+[\psi_{28}\, w_1 + (1+\psi_{28})\, w_2]\, L^3/24EI$	$-\theta_6$
Gravity UDL in cantilever $ML/3EI$	$+\theta_{2a}$	$(1+\psi_1)\,\theta_{2a}$	$+\theta_{4a}$	$\theta_{4a}+[\psi_{28}\, M_1 + (1+\psi_{28})\, M_2 + (1+X\psi_\infty)\, \psi_2 M_3]\, L/3EI$	$+\theta_{6a}$
$\theta_1\, L'+\delta_{1a}$	δ_1	$\theta_3\, L'$	δ_3	$\theta_5\, L'$	δ_5
Free moment due to prestress $PzL'^2/2EI$	δ_{1a}	$(1+\psi_1)\,\delta_{1a}$		$\psi_{28}\, PzL'^2/2EI$	
$(\theta_2+\theta_{2a})\, L'+\delta_{2a}$	δ_2	$(\theta_4+\theta_{4a})\, L'$	δ_4	$(\theta_6+\theta_{6a})\, L'$	δ_6
Free gravity UDL $wL'^4/8EI$	δ_{2a}	$(1+\psi_1)\,\delta_{2a}$		$+[\psi_{28}\, w_1 + (1+\psi_{28})\, w_2 + (1+X\psi_\infty)\, \psi_2 w_3]\, L'^4/8EI$	

Notation for this table. P=pretensioning force. z=eccentricity. w=prevailing gravity UDL (suffix 0, 1, 2, 3 for unit weight, self-weight including infill, dead and live loads). L=effective main span. L'=effective cantilever span. M=moment in cantilever $wL'^2/2$. E=prevailing modulus of elasticity. I=prevailing second moment of area. ψ_1=creep coefficient from transfer to installation. ψ_{28}=creep coefficient from installation (say 28 days) to long-term service. $X\psi_\infty$=creep coefficient at long-term for live loads. 'Free' means deflection in the cantilever from an encastre support.

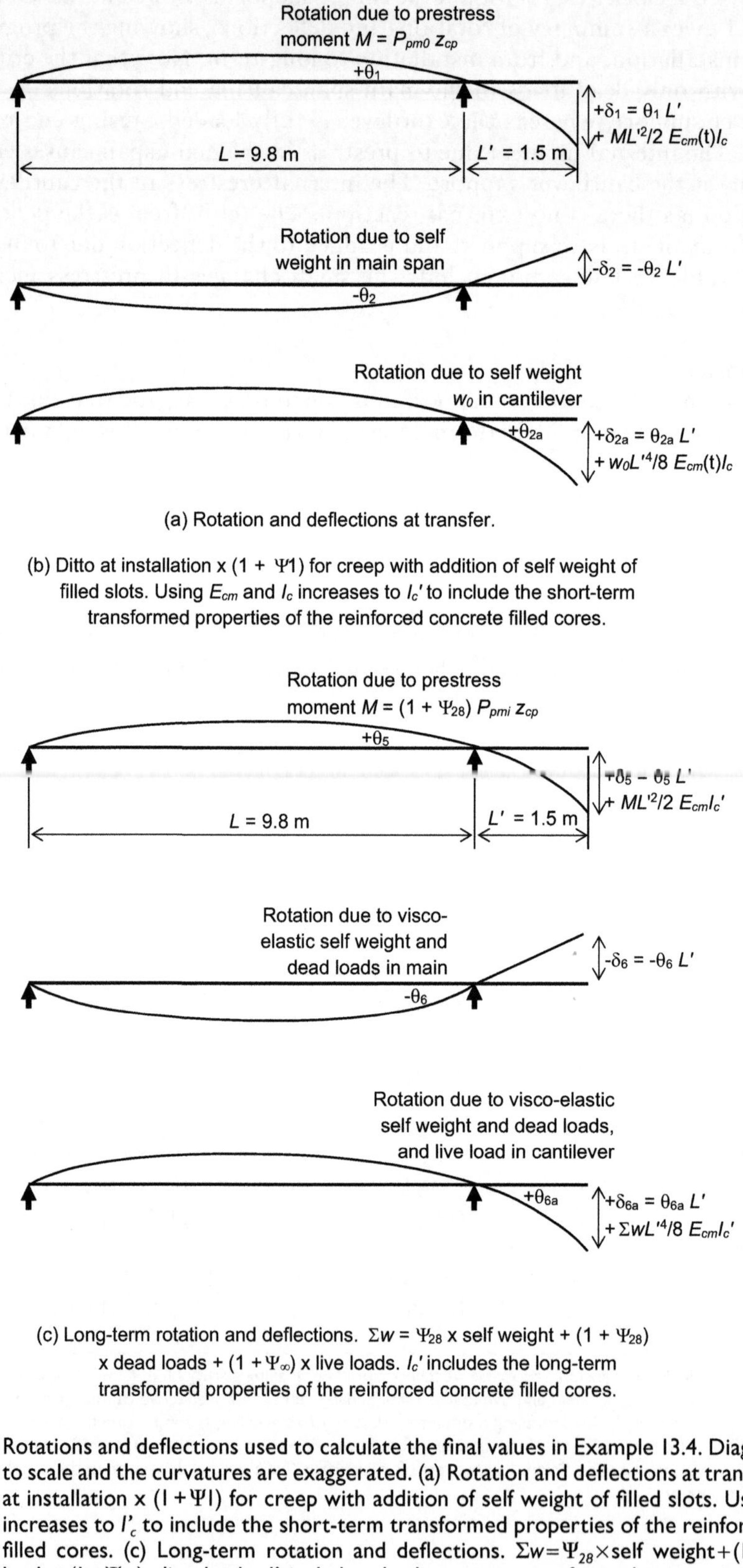

Figure 13.16 Rotations and deflections used to calculate the final values in Example 13.4. Diagrams are not to scale and the curvatures are exaggerated. (a) Rotation and deflections at transfer. (b) Ditto at installation x $(1 + \Psi 1)$ for creep with addition of self weight of filled slots. Using E_{cm} and I_c increases to I'_c to include the short-term transformed properties of the reinforced concrete filled cores. (c) Long-term rotation and deflections. $\Sigma w = \Psi_{28} \times$ self weight $+ (1 + \Psi_{28}) \times$ dead loads $+ (1 + \Psi\infty) \times$ live loads. I'_c includes the long-term transformed properties of the reinforced concrete filled cores.

Deflection due to self-weight is equal to $(\theta_2+\theta_{2a})\ L'$ plus the free deflection δ_{2a}

$$\delta_2=(\theta_2+\theta_{2a})\ L'+\delta_{2a} \tag{13.52}$$

where $\delta_{2a}=+\ w_0\ L^4/8\ E_{cm}(t)\ I_{c,co}$ (13.53)

Net deflection at transfer $=\delta_1+\delta_2$

b. Deflection at installation

The creep factor to installation is ψ_1 (Eq. 11.3). The reduction in pretensioning force from transfer to installation is $P_{pm0}-P_{pmi}$

Refer to Figure 13.16b

$$\theta_3=(1+\psi_1)\ \theta_1-(P_{pm0}-P_{pmi})\ z_{cp}\ L/2\ E_{cm}(t)\ I_{c,co} \tag{13.54}$$

$$\theta_4=(1+\psi_1)\ \theta_2 \tag{13.55}$$

$$\theta_{4a}=(1+\psi_1)\ \theta_{2a}+(w_{0c}\ L^2/2)\ L/3\ E_{cm}\ I_{c,co} \tag{13.56}$$

where w_{0c} is the self-weight of filled cores per unit width. This is only for the actual size of the core and does not include that part of the precast concrete removed to be replaced by factory placed infill.

If reinforced top opened cores are required (Section 13.2.2, Eq. 13.12) after the concrete has hardened the deflections due to prestress and self-weight in the cantilever are now based on a greater second moment of area, which, after the transformed areas of the top bars are added, is $I'_{c,co}$. Let the ratio $\alpha=I'_{c,co}\ /\ I_{c,co}$.

Referring to Figure 13.16b, the camber is equal to $\theta_3\ L'$, minus the change due to reducing prestress from transfer to installation, plus the deflection due to the free internal moment due to prestress divided by α

$$\delta_3=\theta_3 L'-\left(P_{pm0}-\ P_{pmi}\right)z_{cp}L'^2\ /\ 2E_{cm}\left(t\right)I'_{c,co}+\left(1+\psi_1\right)\delta_{1a}\ /\ \alpha \tag{13.57}$$

The deflection is equal to $(\theta_4+\theta_{4a})\ L'$ plus the free deflection of any filled cores plus the deflection due to self-weight divided by α

$$\delta_4=\left(\theta_4+\theta_{4a}\right)L'+w_{0c}L'^4\ /\ 8E_{cm}I'_{c,co}+\left(1+\psi_1\right)\delta_{2a}/\alpha \tag{13.58}$$

Net deflection at installation $=\delta_3+\delta_4$

c. Long-term deflections

The creep factors from installation to long-term $=\psi_{28}$ (Eq. 11.3) and at long-term 0.8 ψ_∞ (Eq. 11.10). The reduction in pretensioning force from installation to long-term is $P_{pmi}-P_{po}$

Refer to Figure 13.16c. Long-term rotation due to long-term camber

$$\theta_5=\theta_3+[\psi_{28}\ P_{pmi}-(P_{pmi}-P_{po})]\ z_{cp}\ L/2\ E_{cm}\ I_{c,co} \tag{13.59}$$

Long-term rotation due to long-term visco-elastic self-weight w_1 and dead load w_2 only in the main span

$$\theta_6=\theta_4-[\psi_{28}\ w_1+(1+\psi_{28})\ w_2]\ L^3/24\ E_{cm}\ I_{c,co} \tag{13.60}$$

Long-term rotation due to the cantilever moment for Case II loading

$$\theta_{6a} = \theta_{4a} + \left[\psi_{28} M'_{s1} + (1+\psi_{28}) M'_{s2} + (1+0.8\,\psi_{\infty}) M'_{s3}\right] L / 3E_{cm} I_{c,co} \tag{13.61}$$

where M'_{s1} is the moment due to self-weight (including any filled cores)

M'_{s2} the moment due to imposed dead loads
M'_{s3} the moment due to quasi-permanent live load.

The effects of long-term creep on concrete in the top opened cores increases to $I''_{c,co}$
Referring to Figure 13.16c, cantilever deflection is $\theta_5\ L'$ plus the visco elastic free internal moment $\psi_{28}\ P_{pmi}\ z_{cp}$ (less a small reduction to P_{po}) due to prestress

$$\delta_5 = \theta_5 L' + \left[\psi_{28} P_{pmi} - (P_{pmi} - P_{po})\right] z_{cp} L'^2 / 2E_{cm} I''_{c,co} \tag{13.62}$$

Cantilever deflection is $(\theta_4+\theta_{4a})\ L'$ plus the free deflections due to self-weight w_1, any filled cores w_{0c}), dead w_2 and quasi-permanent live loads $\psi_2\ w_3$

$$\delta_6 = (\theta_6 + \theta_{6a}) L' + \left[\psi_{28}\ (w_1 + w_{0c}) + (1+\psi_{28}) w_2 + (1\ + 0.8\ \psi_{\infty}) \psi_2 w_3\right] L'^4 / 8E_{cm} I''_{c,co} \tag{13.63}$$

Final deflection $\delta_7=\delta_5+\delta_6\leq$limit as follows
As the limit for the long-term deflection of a cantilever is not given in BS EN 1992-1-1, clause 7.4.1 (4), nor the limit for live load deflections or movement after finishes in clause 7.4.1 (5), it is hereby suggested that the Code's limits may be based on an apparent effective span of the cantilever L'_{ef} of *twice* the distance from the end of the cantilever to the point maximum negative deflection in the main span, at x' from the simple support. (See later Figure 13.17 in Example 13.4). Then $L'_{ef}=2\ [L'+(L-x')]$ and the deflection limit is therefore $L'_{ef}/250$. Similarly for live deflections or movement after finishes the limit is $L'_{ef}/500$, or $L'_{ef}/350$ for non-brittle finishes or partitions.

d. Long-term movement due to live load only
This is due to the component of live load in the rotation θ_{6a} plus the component of live load in the deflection δ_6. For case II loading there is no contribution from the live load in θ_6 in the main span. Then

$$\theta_8 = +[(1+\ \psi_{28}) M'_{s2}] L / 3E_{cm} I_{c,co} \tag{13.64}$$

$$\delta_8 = \theta_8 L' + (1+\psi_{28}) \psi_2 w_3 L'^4 / 8E_{cm} I''_{c,co} \tag{13.65}$$

e. Long-term rotation and movement due to changes in camber, visco-elastic deflections of dead loads and long-term movement due to live load

Change in camber $\delta_9=\delta_5-\delta_3$ (13.66)

Rotation due to the visco-elastic component of the self-weight and imposed dead load in the main span in θ_6 and in the cantilever in θ_{6a}

$$\theta_{10}=-\psi_{28}\ (w_1+w_2)\ L^3/24\ E_{cm}\ I_{c,co} \tag{13.67}$$

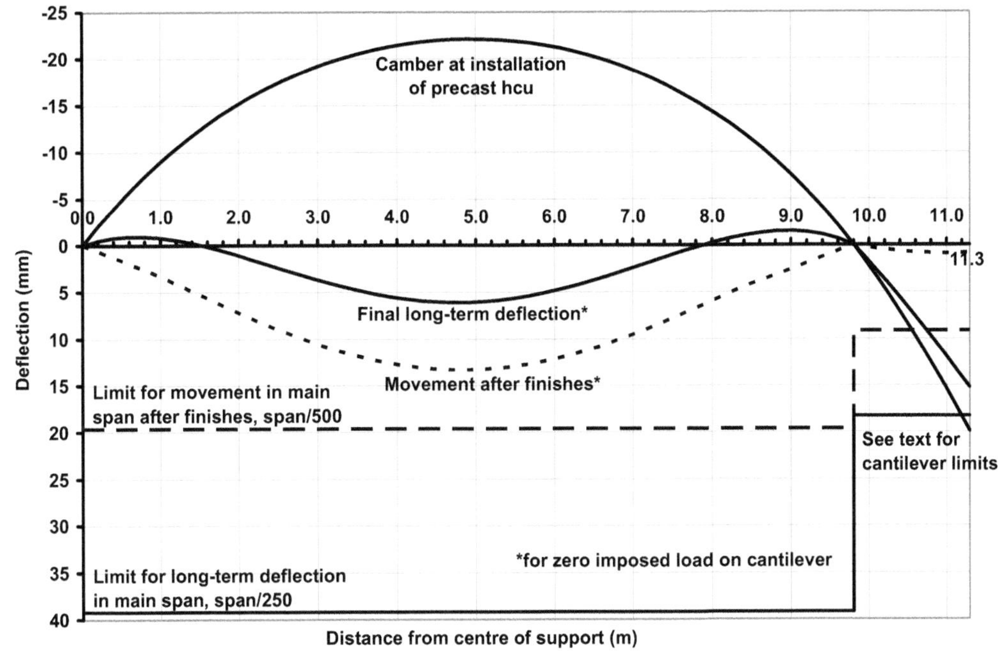

Figure 13.17 Long-term final deflections, movement after installation of finishes and camber at installation of the precast cantilevered hcu for the design in Example 13.4.

$$\theta_{10a} = +\psi_{28} M'_{s1} L / 3E_{cm} I_{c,co} \tag{13.68}$$

Cantilever deflection is $(\theta_{10}+\theta_{10a})\ L'$ and free deflections due to visco-elastic self-weight and dead loads

$$\delta_{10} = (\theta_{10} + \theta_{10a}) L' + \psi_{28} (w_1 + w_2) L'^4 / 8E_{cm} I''_{c,co} \tag{13.69}$$

Final movement $\delta_{11} = \max\{\delta_8;\ \delta_8 + \delta_9 + \delta_{10}\} \leq$ limit of $L'_{ef} / 500$ or $L'_{ef} / 350$.

13.3.4 Deflection at the end of cantilevers due to point and partial line loads

These are due to a point load P at distance a from the cantilever support, and a line load w extending to a from the cantilever support. The following are all multiplied by the appropriate ψ_{28} or $0.8\psi_\infty$ creep factors, and all divided by E_{cm} and the appropriate $I_{c,co}$ in Section 13.3.1.

Due to cantilever moments causing rotation at the cantilever support

$$M = Pa + wa^2/2, \text{ then } \theta = PaL/3 + (wa^2/2)\ L/3 \tag{13.70}$$

$$\text{Then}\quad \delta = \theta\ L' = PaL\ L'/3 + wa^2 L\ L'/6 \tag{13.71}$$

Due to free cantilever moments

$$\delta = Pa^3/3 + (Pa^2/2)\ (L' - a) + wa^4/8 + wa^3\ (L' - a)/6 \tag{13.72}$$

13.3.5 Calculations of deflections in partially cracked sections in the cantilever

The value of $I'_{c,co}$ depends whether the top cantilever section is reinforced using top opened cores and rebars and whether the section is flexurally cracked or not. For continuous top strands or wires in which case the section is, by definition in Section 13.2.4.2, uncracked, use $I'_{c,co}$. Otherwise, the flexurally cracked section is based on the partially cracked section as given in Section 11.2.4 and Eqs. 11.20–11.32. The only differences are that

i. M_{cr} and M_k refer to the cantilever values M'_{cr} and M'_k
ii. section modulus also refers to the top fibre $W_t = I'_u / (h - x'_u)$ in the cantilever
iii. A_s refers to the area of top bars or tendons A'_s
iv. d is to the top bars d'
v. b is the soffit width b'.

Note that in checking the top fibre stress the prestress σ'_t, which may be tensile and negative, is adjusted for the ratio of the cantilever span to the transmission length of tendons in the cantilever, i.e. $\sigma'_t = \sigma_t \, L'/l_{pt2} \leq 1$.

Example 13.4

Calculate the long-term deflection and the deflection after applying finishes at the end of the cantilever for the 250 mm depth prestressed hcu shown in cross section in Figures 6.7 and 13.14 and designed for serviceability in Examples 11.1 and 13.1. The loading is Case II for maximum cantilever and minimum main span loads. The floor slab is for use in offices. Check the Code's limits if the floor finishes are brittle.

Solution

From Example 11.1 at the simple support P_{pm0}=1055.6 kN, P_{pmi}=1010.2 kN and P_{po}=860.2 kN. From Example 13.1 at the cantilever support is slightly less, i.e. P_{po}=934.4×911×10^{-3}= 851.2 kN, but it is only localised in the region of the cantilever support and may be ignored.

To calculate deflections in the cantilever it is necessary to determine the end rotation in the main span θ due to prevailing prestress and/or gravity loads, from which $\delta = \theta L'$ plus the free cantilever deflections.

Although the prestress is released over the full length of the unit, $L + L'$=9.8+1.3=11.3 m, this cannot be used to calculate rotation at installation because deflections and end rotations will revert back to those determined for the main span and cantilever individually. Referring to Figure 13.16a the rotations at transfer are as follows

a. Rotation at transfer (+ve is a clockwise rotation if the simple support is viewed at the left end)
   Eq. 13.47 θ_1=+1055.6×10^3×82.2×9800/(2×32837×1296×10^6)=+0.009990 radians (rad.)
   Eq. 13.48 θ_2=−4.12×9800^3/(24×32837×1296×10^6)=−0.003792 rad.
   Eq. 13.49 θ_{2a}=+(4.12×1500^2/2)×9800/(3×32837×1296×10^6)=+0.000355 rad.
   Referring to Figure 13.16(a)
   Eq. 13.51 δ_{1a}=+ 1055.6×10^3×82.2×1500^2/(2×32837×1296×10^6)=2.3 mm
   Eq. 13.50 δ_1=[+0.009990×1500]+2.3=17.3 mm
   Deflection due to self-weight is equal to (θ_2+θ_{2a}) L' plus the free deflection
   Eq. 13.53 δ_{2a}=+4.12×1500^4/(8×32837×1296×10^6)=+0.06 mm
   Eq. 13.52 δ_2=[(−0.003792 +0.000355)×1500] +0.06=−5.2 +0.06=−5.1 mm
   Net deflection at transfer=+17.3−5.1=+12.2 mm

b. Deflection at installation

Referring to Figure 13.16(b), ψ_1=0.71 (from Example 11.1) and $P_{pm0}-P_{pmi}$=45.4 kN

Eq. 13.54 θ_3=[1.71×+0.009990]−45.4×10³×82.2×9800/(2×32837×1296×10⁶)=+0.016719 rad.

Eq. 13.55 θ_4=1.71×-0.003792=-0.006495 rad.

Eq. 13.56 θ_{4a}=1.71×+0.000355+(1.50×1500²/2)×9800/(3×36283×1296×10⁶)=+0.000725 rad.

where self-weight of the filled cores=1.2×1.245=1.50 kN/m.

The greater second moment of area based on the transformed areas of the top bars in the top opened cores is $I'_{c,co}$=1472×10⁶mm⁴. This modifies the free deflections in the cantilever by a factor of β=1472/1296=1.136.

Eq. 13.57 δ_3=[+0.016719×1500]−[45.4×10³×82.2×1500²/(2×36283×1472×10⁶)]+1.71×2.3/1.136=+25.1−0.1+3.5=+28.5 mm

Eq. 13.58 δ_4 =[(−0.006495 +0.000725)×1500]+(1.50×1500⁴/(8×36283×1472×10⁶)+1.71×0.06/1.136=−8.7+0.02+0.1=−8.6 mm

Net (downward) deflection at installation=+28.5−8.6=+19.9 mm

c. Long-term deflections

ψ_{28}=1.20 and ψ_∞=2.00 (from Example 11.1)

P_{pmi}=1010.2 kN and $P_{pmi}-P_{po}$=150.0 kN

Refer to Figure 13.16(c). Long-term rotation due to long-term camber

Eq. 13.59 θ_5=+0.016719+(1.20×1010.2×10³−150.0×10³)×82.2×9800/(2×36283×1296×10⁶)=+0.025816 rad.

Due to self-weight and dead load only in the main span

Eq. 13.60 θ_6=−0.006495−[(1.20×4.36+(1+1.20)×4.56)]×9800³/(24×36283×1296×10⁶)=−0.019222 rad.

where the self-weight of slab w_1 =1.2×3.63=4.36 kN/m and imposed dead w_2 =1.2×3.80=4.56 kN/m

Long-term rotation due to the cantilever moment for Case II loading

Eq. 13.61 θ_{6a}=+0.000725+[1.2×6.58+(1+1.20)×5.13+(1+2.00)×2.025]×10⁶×9800/(3×36283×1296×10⁶)=+0.002480 rad.

where the moment for self-weight (including filled cores) is 6.58 kNm, imposed dead loads are 5.13 kNm, and the quasi-permanent (ψ_2=0.3) moment due to live load is 1.80×1.5²/2=2.025 kNm.

The effects of long-term creep on the concrete in the top opened cores increase further to $I'_{c,co}$=1497×10⁶mm⁴.

Referring to Figure 13.16(c), cantilever deflection due to θ_5 plus the free internal moment due to prestress is

Eq. 13.62 δ_5=[+0.025816×1500]+[(1.20×1010.2×10³) -150.0×10³]×82.2×1500²/(2×36283×1497×10⁶)=38.7+1.8=+40.5 mm

Cantilever deflection due to rotation and free deflections due to self-weight, filled cores, dead and live loads

Eq.13.63 δ_6=[(−0.019222+0.002480)×1500]+[1.20×(4.36+1.50)+(1+1.20)×4.56+(1+2.00)×1.80]×1500⁴/(8×36283×1497×10⁶)=−25.2+0.2=−25.0 mm

Final deflection δ_7=40.5−25.0=**+15.5 mm**<limit of 18.4 mm

The apparent effective length is based on x'=9.0 m from the simple support as shown in Figure 13.17. L'_{ef}=2×(1.50+(9.8−9.0))=4.60 m. The deflection limit is therefore 4600/250=18.4 mm.

d. Long-term movement due to live load only

This is due to the component of live load in the rotation θ_{6a} plus the component of live load in the deflection δ_6. Then

Eq. 13.64 θ_8=(1+1.20)×2.025×10⁶×9800/(3×36283×1296×10⁶)=+0.000309 rad.

Eq. 13.65 δ_8=[+0.000309×1500]+[(1+1.20)×1.80]×1500⁴/(8×36283×1497×10⁶)=0.46+0.04=**+0.5 mm**

″e. Long-term rotation and movement due to changes in camber, visco-elastic deflections of dead loads and long-term movement due to live load

Change in camber

Eq. 13.66 $\delta_9=\delta_5-\delta_3$=40.5−28.5=12.0 mm

Rotation due to the visco-elastic component of the self-weight and imposed dead load in the main span in θ_6 and cantilever in θ_{6a}

Eq. 13.67 θ_{10}=−1.20×8.92×9800³/(24×36283×1296×10⁶)=−0.008925 rad.

Eq. 13.68 θ_{10a}=+1.20×11.71×10⁶×9800/(3×36283×1296×10⁶)=+0.000976 rad.

Cantilever deflection due to θ_{10} and free deflections due to visco-elastic self-weight and dead loads

Eq. 13.69 δ_{10}=(-0.008925+0.000976)×1500+[1.20×8.92×1500⁴/(8×36283×1497×10⁶)=-11.9+0.2=-11.7 mm

Final movement $\delta_{11}=\delta_8+\delta_9+\delta_{10}$=0.5+12.0−11.7=**+0.8 mm**<limit of L'_{ef} / 500=4600/500=9.2 mm.

Figure 13.17 shows distributions of camber, long-term deflections and movement along the span, together with the appropriate limits. Note that the deflections in the main span are for Case I loading ('max-min') and the deflections in the cantilever are for Case II loading ('min-max'). This explains why there is a discontinuity in the profile of the deflections.

Example 13.5

Repeat the calculations in Example 13.4 for the long-term deflection at the end of the cantilever only in a composite hcu slab, but replacing the dead load of the screed with a 75 mm nominal depth structural topping. Calculate whether reinforced concrete-filled cores are required for stage 1 loads, including an allowance of 1.5 kN/m² for construction traffic, and calculate the requirements for steel rebars in the topping. Use f_{ckT}=25 N/mm² for the topping. Cover to bars in the topping is 30 mm.

The section properties for the composite hcu slab, without top bars in filled cores and without the steel mesh in the topping, may be taken from Example 12.1 as $I_{c,c,co}$=2819.9×10⁶mm⁴, and for the precast unit alone as $I_{c,co}$=1296×10⁶mm⁴.

Solution

Stage 1 service stress check. Self-weight of 75 mm depth screed plus 5 mm allowance for camber=0.080×25=2.0 kN/m².

Stage 1 load. w'_{s1}=1.2×(3.63+2.0+1.5)=8.56 kN/m

M'_{s1}=8.56×1.5²/2=9.63 kNm<18.81 kNm from Example 13.1.

No additional rebars in filled cores are required. In fact, the actual moment of resistance could be based on clause 3.1.8 as $f_{ctm,fl}$=1.35 f_{ctm}=5.13 N/mm² resulting in M_{sR}=32.2 kNm, giving a factor of safety of 3.3.

Stage 2 ultimate load w'_{Ed2}=1.2×[1.25×(3.63+2.0+1.5+0.5)+1.5×5.0]=20.45 kN/m

b'=1197 mm, d'_2=250+75 − 30 − say 6=289 mm

M'_{Ed2}=−20.45×1.5²/2=− 23.00 kNm

K_2'=23.00×10⁶/(45×1197×289²)=0.005

z'_2 / d'_2=min{0.95; 0.5+√(0.25 − 0.005/1.134)}=min {0.95; 0.99}

z'_2=0.95×289=274.5 mm

$A'_{s2}=23.00\times10^6/(274.5\times0.87\times500)=193\,mm^2$ per 1.2 m width $=161\,mm^2/m$
$A'_{s2,min}=\max\{0.13\%\times1000\times75;\ 0.26\times2.56\times1000\times75/500\}=100\,mm^2/m$

Use A193 mesh ($193>161\,mm^2/m$)
Calculation for deflection at the end of the cantilever
Referring to Example 13.4, net deflection at transfer = +12.2 mm
Deflection at installation due to precast unit and topping

Eq. 13.55 $\theta_4=1.71\times-0.003792-2.40\times9800^3/(24\times36283\times1296\times10^6)=-0.008486$ rad.

where self-weight of the topping $=1.2\times2.00=2.40$ kN/m.

Eq. 13.56 $\theta_{4a}=1.71\times+0.000355+(2.40\times1500^2/2)\times9800/(3\times36283\times1296\times10^6)=+0.000795$ rad.

Due to creep rotation due to self-weight at transfer plus the instant rotation due to topping in main span

Eq. 13.57 $\delta_3=+25.1-0.1+1.71\times2.3=+28.9$ mm due to camber

Eq. 13.58 $\delta_4=[(-0.008486+0.000795)\times1500]+(2.40\times1500^4/(8\times36283\times1296\times10^6)+1.71\times0.06=-11.54+0.03+0.10=-11.4$ mm

Net (downward) deflection at installation $=+28.9-11.4=+17.5$ mm

Long-term deflections
The greater second moment of area based on the transformed areas of the mesh in the topping is $I'_{c,c,co}=2867.6\times10^6\,mm^4$.
Due to camber in the main span

Eq. 13.59 $\theta_5=+0.016719+(1.20\times1010.2\times10^3-150.0\times10^3)\times82.2\times9800/(2\times36283\times2867.6\times10^6)=+0.020831$ rad.

Due to self-weight, topping and imposed dead load only in the main span

Eq. 13.60 $\theta_6=-0.008486-[(1.20\times(4.36+2.40)+(1+1.20)\times2.40)]\times9800^3/(24\times36283\times2867.6\times10^6)=-0.013534$ rad.

where the self-weight of topping and imposed dead $w_2=1.2\times(1.50+0.50)=2.40$ kN/m
Long-term rotation due to the cantilever moment for Case II loading

Eq. 13.61 $\theta_{6a}=+0.000795+[1.2\times7.60+(1+1.20)\times2.70+(1+2.00)\times2.025]\times10^6\times9800/(3\times36283\times2867.6\times10^6)=+0.001458$ rad.

where the moment for the self-weight, topping is 7.60 kNm and imposed dead loads is 2.70 kNm

Eq. 13.62 $\delta_5=[+0.020831\times1500]+[(1.20\times1010.2\times10^3)-150.0\times10^3]\times82.2\times1500^2/(2\times36283\times2867.6\times10^6)=31.2+1.0=+32.2$ mm

Eq. 13.63 $\delta_6=[(-0.013534+0.001458)\times 1500]+[1.20\times(4.36+2.40)+(1+1.20)\times 2.40+(1+2.00)\times 1.80]\times 1500^4/(8\times 36283\times 2867.6\times 10^6)=-18.1+0.1=-18.0\,\text{mm}$

Final deflection $\delta_7=+32.2-18.0=$ **14.2 mm** < limit of 26.4 mm

By calculating and plotting the long-term deflection in the main span, the apparent effective length is found to be based on $x'=8.0\,\text{m}$ from the simple support. $L'_{ef}=2\times(1.50+(9.8-8.0))=6.60\,\text{m}$. The deflection limit is therefore 6600/250 = 26.4 mm.

Long-term movement due to live load only

This is due to the component of live load in the rotation θ_{6a} plus the component of live load in the deflection δ_6. Then

Eq. 13.64 $\theta_8=(1+1.20)\times 2.025\times 10^6\times 9800/(3\times 36283\times 2867.6\times 10^6)=+0.000140\,\text{rad.}$

Eq. 13.65 $\delta_8=[+0.000140\times 1500]+[(1+1.20)\times 1.80]\times 1500^4/(8\times 36283\times 2867.6\times 10^6)=0.2+0.02=$ **+0.2 mm**

Long-term rotation and movement due to changes in camber, visco-elastic deflections of dead loads and long-term movement due to live load

Change in camber

Eq. 13.66 $\delta_9=\delta_5-\delta_3=32.2-28.9=3.3\,\text{mm}$

Rotation due to the visco-elastic component of the self-weight, topping and imposed dead load in the main span in θ_6 and cantilever in θ_{6a}

Eq. 13.67 $\theta_{10}=-1.20\times 9.16\times 9800^3/(24\times 36283\times 2867.6\times 10^6)=-0.004143\,\text{rad.}$

Eq. 13.68 $\theta_{10a}=+1.20\times 10.30\times 10^6\times 9800/(3\times 36283\times 2867.6\times 10^6)=+0.000388\,\text{rad.}$

Cantilever deflection due to θ_{10} and free deflections due to visco-elastic self-weight and dead loads

Eq. 13.69 $\delta_{10}=(-0.004143+000388)\times 1500+[1.20\times 9.16\times 1500^4/(8\times 36283\times 2867.6\times 10^6)=-5.6+0.1=-5.5\,\text{mm}$

Final movement $\delta_{11}=\delta_8+\delta_9+\delta_{10}=0.2+3.3-5.5=$ **−2.0 mm** but less than $\delta_6=$ **+0.2 mm** < limit of $L'_{ef}/500=6600/500=13.2\,\text{mm}$.

13.4 CANTILEVER BEAMS

13.4.1 Use of prestressed concrete cantilever beams

Prestressed concrete (psc) cantilever beams are used less frequently than precast reinforced cantilever beams, and often as an 'add-on' cantilever to the design of the main beam. Figures 13.18 and 13.19 show examples of psc cantilevered rectangular or inverted-tee beams supporting solid units or hcu. Prestressed cantilever beams have the advantage that some of the top and bottom tendons may be debonded in the region of negative bending moments, enabling greater tensile stress in the top and greater compressive stress in the soffit unhindered by prestress, which might be in the order of $\sigma_t\approx-2\,\text{N/mm}^2$ and $\sigma_b\approx+15\,\text{N/mm}^2$ without debonding, or 50% of this if debonded. A minimum number of tendons are required in the bottom of the cantilever to control cracking and deflections. The top of the cantilever

Figure 13.18 Cantilevered precast beams and floor slabs at the General Post Office in Melbourne, Australia. (Courtesy Simon Hughes, Precast Concepts & Hollow Core Concrete, Australia.)

Figure 13.19 Cantilevered precast beams with half-joint connections give the same structural behaviour as a continuous beam supporting prestressed hollow core units.

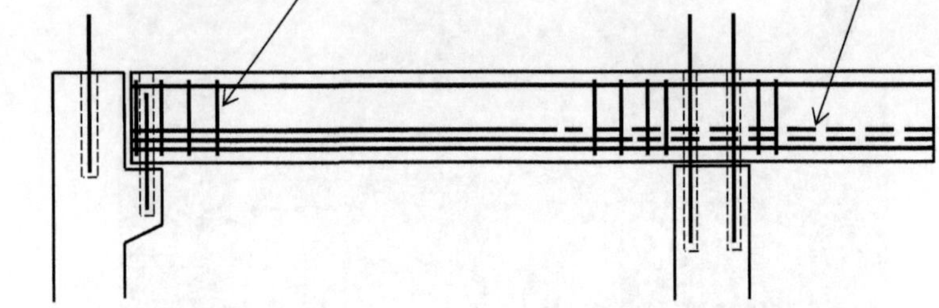

Figure 13.20 General arrangement for prestressed cantilever beams.

may contain pretensioned tendons in order to control camber in the main span and control maximum prestress σ_b and σ_t, but they may be debonded just short of the cantilever support. See Figure 13.20.

In this case, the top of the beam may be statically reinforced using up to 4 no. H16 to H25 giving a moment of resistance in the order of $M'_{Rd} \approx 100–400$ kNm, more than enough to cater for 2–3 m long cantilevers in 300–500 mm depth beams. Composite beams with additional rebars, such as 2 no. H16 stability tie bars, placed in the cast *insitu* infill above rectangular beams or to either side of the upstand in inverted-tee beams will increase the negative moment capacity. Steel mesh in a structural topping will increase moment capacity further, but by then the design of the cantilever will most probably be limited by long-term deflection.

The design of psc cantilever beams follows the procedures in Section 13.2 for

i. load cases I to III
ii. calculation of maximum service and ultimate bending moments
iii. limiting stresses in the top and bottom of the prestressed section
iv. long-term deflections and movement after finishes
v. beam stability and over-turning, and
vi. the procedures in Sections 7.2, 9.3 and 12.4 for serviceability moments, ultimate moments and shear and composite beams, respectively.

13.4.2 Effect of moment resisting connection at the cantilever support

End rotations of the main span at the cantilever support, e.g. long-term rotations θ_5, θ_6 and θ_{6a} in Eqs. 13.59–13.61 may be reduced owing to the flexural stiffness of a supporting column or wall, as shown in Figure 13.20. The effect may be small, as the rotations at installation are not affected, but it is worth looking at the analysis, particularly if deflections are large and the moment resisting connection is substantial.

Consider the end rotation θ_6 due to the gravity UDL in the main span, w, if the column-to-beam connection is fully rigid and the flexibility of the beam is I_{beam}/L, and the flexibility of the a column beneath the beam of height h is I_{column}/h. Then according to rigid frame formulae (ref. Kleinlogel, 1952) the end rotation of the beam reduces from $wL^3/24$ to $(wL^3/24 - 3wL^3/64\beta)$, all divided by $E_{cm}\ I_{beam}$ where $\beta = (I_{beam}\ h/I_{column}\ L) + 1$.

For example if the beam in Example 13.6 is rigidly attached to a 300×300 mm column (of equal strength of concrete) of height say 3.5 m, pin-ended at its base, and, as given in Example 13.6, $I_{beam} = 8449 \times 10^6$ mm^4, $L = 8500$ mm, $I_{column} = 675 \times 10^6$ mm^4, $L = 3500$ mm, then $\beta = 8449 \times 3500/675 \times 8500 + 1 = 4.15 + 1 = 5.15$. Cancelling through by $w\ L^3/E_{cm}\ I_{beam}$ the end rotation θ_6 reduces by a factor from 1/24 to $(1/24 - 3/64\beta) = 0.04167$ to 0.03256, i.e. a ratio of 0.78 or a reduction of 22%. If the bottom of the column is encastre, then $\beta = 3 \times 4.15 + 4 = 16.45$ and θ_6 reduces by a factor from 1/24 to $(1/24 - 3/16\beta) = 0.04167$–0.03027, i.e. a ratio of 0.73. Likewise θ_{6a}, due to the cantilever moment acting at the cantilever support at the top of the column, θ_{6a} reduces by a ratio 1/3 to $(\beta - 4)/3\ \beta = 0.333$–0.252, i.e. a ratio of 0.76, almost the same as for θ_6.

Of course, the opposing end rotations due to camber and cantilever moments will be reduced by similar amounts, but it shows that a rigid connection to a stiff column has an effect on θ_6 and hence the deflection at the end of the cantilever, especially as θ_6 is subject to long-term creep effects. As the beam-to-column connection will not be fully rigid, but semi-rigid as shown by the work on precast connections reported by Elliott (Elliott 2017), consider a typical semi-rigid factor of around 70% (i.e. 0.7 of the fully rigid end moment at the same rotation), the reduction ratios for θ_6 and θ_{6a} when supported by an encastre column will still be about 0.83. The implication of this is shown in Example 13.6.

13.4.3 Design of prestressed concrete cantilever beams

Th design involves checking the main span for the maximum service and bending moments M_s and M_{Ed} for case I loading (minimum cantilever load) and the maximum shear force in the shear plane at the cantilever support for case III loading. Given that a typical minimum service moment in the cantilever is 60–80 kNm, M_s will be about 30–40 kNm less than in a simply supported beam, and given that M_s per strand (near the bottom of a 600 mm depth beam) is in the order of 15 to 25 kNm per strand, it may be possible to use 2 or 4 fewer strands. Depending on the balance between dead and live loads M_s is usually critical over M_{Ed} for exposure class XC1, but well matched for exposure XC3.

The cantilever is designed using class II loading (in the same manner as for slabs with top tendons) for M_s', M_{Ed}' and V_{Ed}' at the cantilever support, checking Eqs. 13.25 and 13.26 for combined prestress (especially at the top fibre) and bending stress.

Eq. 13.25 $\quad f'_{b,max} = \sigma_b + M_s' / Z_{b,co} \leq 0.45 f_{cm}$

Eq. 13.26 $\quad f'_{t,max} = \sigma_t - M_s' / Z_{t,co} \geq -f_{ctm}$

If either of these is violated a number of strands in the central regions of the beam are debonded – preferably rather than deflecting, in order to decrease σ_b and/or increase σ_t. The latter may be critical if σ_t is in the order of –2 to –3 N/mm^2. After debonding the moments of resistance reduce to $M_{sR,d}$ and $M_{Rd,d}$ (suffix d for debond). A back-calculation is carried out to check the theoretical distance x_d from the simple support where $M_s = M_{s,d}$ in the main span, given as

$$F\ x_d - w_s x_d^2 / 2 = M_{sR,d} \tag{13.73}$$

The actual point for debonding is $x_d - l_{pt}$, where l_{pt} is the basic transmission length, typically 55 diameters for strand.

Similarly the distance $x_{Ed,d}$ for the ultimate moment where $M_{Ed}=M_{Ed,d}$ in the main span, given as

$$F_{Ed}x_{Ed,d} - w_{Ed}x_{Ed,d}^2 / 2 = M_{Rd,d} \quad (13.74)$$

The actual point for debonding is $x_{Ed,d}-l_{bpd}$, where l_{bpd} is the ultimate anchorage length, typically 100 diameters for strand.

The former service condition is usually the shortest and hence critical distance from the simple support. Finally, $V_{Rd,c,d}$ is calculated for the reduced number of tendons, and a check is made for $V_{Rd,cr,d}$ should M'_s exceed the cracking moment of resistance

$$M'_{cr,d} = (\sigma_b + f_{ctd}) Z_{b,co}$$

The ultimate moment of resistance of the cantilever M'_{Rd} is based on the inverted section analysis of the beam, where there are N_T' tendons in the tension zone, measured from the bottom of the beam. As the width of the soffit of a beam b' (particularly inverted-tee and I beams) may be much greater than at the top of the beam, the height to the neutral axis X' will be less than in a sagging mode and therefore the tendons will be stressed to a higher level, say around 1500 N/mm^2, which is a good starting point for the calculation of strains, as explained in Section 9.4.1.

Deflections are calculated using the same procedures as in Sections 13.3.2 and 13.3.3 and Figure 13.16. End rotations and camber in the main span are calculated using the total number of tendons, even though several tendons may be debonded near the cantilever support (the error in this assumption is tiny). The free deflections due to prestress in the cantilever δ_3 and δ_5 (Eqs. 13.57 and 13.62) are based on the reduced number of tendons after debonding.

Example 13.6

Carry out the design of the inverted-tee beam in Examples 7.10 and 9.5, as shown in Figure 6.10 (without the debonded strands) using the main span effective length $L=8.5$ m and cantilever $L'=2.0$ m. The self-weight of the beam is 7.688 kN/m, floor slab = 18.0 kN/m, other dead loads = 10.0 kN/m and imposed live load = 30.0 kN/m. Check the requirement for reduced strands in the main span, and for the combined stresses due to prestress and cantilever bending at the cantilever support. The bearing length at the cantilever is 300 mm. Calculate the long-term deflection and movement after finishes at the end of the cantilever. Also consider the implications of the cantilever support being a semi-rigid connection to the 300×300 mm column discussed in the text, where its effect on long-term end rotations is to reduce these by a factor of 0.83 (see Section 13.4.2.).

Solution

Maximum UDL $w_s=7.688+18.0+10.0+30.0=65.69$ kN/m, $w_{Ed}=89.56$ kN/m

Minimum UDL $w_{s,min}=7.688+18.0+10.0=35.69$ kN/m, $w_{Ed,min}=44.56$ kN/m

Maximum serviceability bending moment in the main span is based on load case I, where in the cantilever $M'_{s\,min}=35.69\times2.0^2/2=71.38$ kNm

Reaction at simple support

$F=65.69\times8.5/2-71.38/8.5=270.78$ kN

Shear span $a=270.78/65.69=4.122$ m from simple support

$M_s=270.78\times4.122-65.69\times4.122^2/2=$ **558.1 kNm** < 599.7 kNm from Example 7.10 using 26 strands.

The number of strands may be reduced by 4 no. from row 5 at y_s=250mm from the bottom, giving M_{sR}=562 kNm and M_{Rd}=860 kNm. Using 22 no. strands with an eccentricity z_{cp}=106.2mm gives σ_b=13.38 N/mm² in the soffit and σ_t=-2.07 N/mm² in the top at the cantilever support.

The maximum ultimate bending moment in the main span is based on load case I as follows:

$M'_{Ed\ min}=44.56\times2.0^2/2=89.12$ kNm
$F_{Ed}=89.56\times8.5/2-89.12/8.5=370.1$ kN
$a=370.1/89.56=4.133$ m.
$M_{Ed}=370.1\times4.133-89.56\times4.133^2/2=$ **764.8 kNm** <860 kNm using 22 no. strands (in fact only 20 no. strands are required for M_{Ed})

Check the maximum compressive stress in the bottom and top of the beam at the cantilever support from Eqs. 13.25 and 13.26

Using load case II, $M'_s=65.69\times2.0^2/2=131.38$ kNm

Eq. 13.25 $\quad f'_{b,max}=\sigma_b+M'_s/Z_{b,co}=13.38+131.38/31.671w=17.53\ \text{N/mm}^2$

$<0.45f_{cm}=23.85\ \text{N/mm}^2$ OK

Eq. 13.26 $\quad f'_{t,max}=\sigma_t-M'_s/Z_{t,co}=-2.07-131.38/25.356=-2.07-5.18$

$=-7.25\ \text{N/mm}^2<-f_{ctm}=-3.80\ \text{N/mm}^2$OK

Therefore the limiting stress σ_t should be$\geq-3.80--5.18=+1.38$ N/mm² by debonding. Due to the large number of permutations for the 22 no. strands a method of trial and error is used. The best procedure is to start at row 1 by debonding 4 no. strands, which increases σ_t to –0.04 N/mm², then debonding 4 no. strands from row 2 gives σ_t=+1.36 N/mm². The final number of strands is 14, i.e. 4, 0, 4, 2 in the bottom rows, 2 no. at the top of the boot and 2 no. at the top of the beam. Attempting to debond top strands instantly reduces σ_t to –1.45 N/mm² of course.

Using 14 no. strands as so $Z_{b,co}=30.98\times10^6$mm³, σ_b=+6.88 N/mm² and $M_{sR,d}$=330 kNm. The distance x_d from the simple support is given by

Eq. 13.73 $\quad 270.78x_d-65.69x_d^2/2=330$ kNm

Hence x_d=6.73m, or 1.77m from the cantilever support

The transmission length for 12.5mm diameter strands is l_{pt}=700mm, therefore debonding the sheaths starts at 6.73−0.700=6.03m from the simple support.

Using 14 no. strands as so $M_{Rd,d}$=583 kNm.

The distance $x_{Ed,d}$ is given by the quadratic in

Eq. 13.74 as $370.1x_d-89.56x_d^2/2=583$ kNm

Hence $x_{Ed,d}$=6.15m.

The ultimate anchorage length is l_{bpd}=1296mm, therefore the debonding sheaths commence at 6.15−1.296=4.85m from the simple support.

This is less critical than for the service condition and therefore the sheaths commence at 6.03m from the simple support and continue to the end of the cantilever.

The ultimate moment of resistance of the cantilever M'_{Rd} is based on the inverted section analysis of the beam, where there are N'_T=6 no. strands in the tension zone, i.e. 2 no. strands at 550, 350 and 200mm from the bottom of the beam. As the breadth of the beam is quite large, b'=600mm, it is likely that the height to the neutral axis X will be small and the strands will be stressed to a higher level, say around 1500 N/mm².

Referring to Section 9.4.1, the first estimate of X is found by considering $N_T'=6$ no. strands $A_{pT}'=558\,mm^2$ stressed to 1500 N/mm². $F_p'=558\times1500\times10^{-3}=837$ kN. Then $X'=837\times10^3/0.567\times45\times600\times0.8=68$ mm.

Prestrain at the cantilever support is found to be $\varepsilon_{po}'=1039/195000=0.005327$

Considering strands in top row 1, $\varepsilon_{p1}=0.005327+0.0035\times[(550/68)-1]=0.030136$

$f_{p1}'=1313+10186\times0.030136=1619$ N/mm² $>f_{p,max}=1516$ N/mm², therefore use $f_{p1}'=1516$ N/mm²

In row 2, $\varepsilon_{p2}'=0.005327+0.0035\times[(350/68)-1]=0.019842$

$f_{p2}'=1313+10186\times0.019842=1515$ N/mm²

and so on until $X'=67.9$ mm, $F_p'=831.1$ kN and $y_s'=230.3$ mm

$d'=600-230.3=369.7$ mm and $d_n'=0.4\times67.9=27.1$ mm

$z'=369.7-27.1=342.6$ mm

$M_{Rd}'=831.1\times10^3\times342.6\times10^{-6}=$ **284.8 kNm** >179.1 kNm

where $M_{Ed\ max}'=89.56\times2.0^2/2=179.1$ kNm

To complete the ultimate design the maximum shear force at the end of the main span at the cantilever support $M_{Ed\ max}'$ is given for load case III (maximum in main span and cantilever) as $M_{Ed\ max}'=179.1$ kNm. Then

$F_{Ed}=89.56\times8.5/2+179.1/8.5=401.7$ kN

V_{Ed} is calculated at $l_b/2+$boot depth 400 mm from the centre of support

$V_{Ed}=401.7-89.56\times(0.3/2+0.4)=352.4$ kN

At the intersection of the upstand and boot, with $\alpha_l=1.0$ because the critical shear plane is 2000 mm$>l_{pt2}$ from the end of the cantilever, then $V_{Rd,c,d}=508.1$ kN>352.4 kN.

Check the requirement for $V_{Rd,cr,d}$. The cracking moment of resistance is given as Eq. 13.75 $M_{cr,d}'=(6.88+1.77)\times30.98=268.1$ kNm$>M_s'=131.38$ kNm, therefore the section is not flexurally cracked.

Calculation of deflection and movement after finishes.

$P_{pm0}=2382.7$ kN. $P_{pmi}=2304.2$ kN. $P_{po}=1977.8$ kN. $P_{pm0}-P_{pmi}=78.5$ kN. $P_{pmi}-P_{po}=326.4$ kN. $\psi_l=0.71$. $\psi_{28}=1.20$. $E_{cm}(t)=32837$ N/mm². $E_{cm}=36283$ N/mm². $I_{c,co}=8449\times10^6$ mm⁴ (beam with 22 no. strands). $I_{c,co}=8326\times10^6$ mm⁴ (14 no. strands).

a. Deflection at transfer

Eq. 13.47 $\theta_1=+2382.7\times10^3\times106.1\times8500/(2\times32837\times8449\times10^6)=+0.003873$ radians (rad.)

Eq. 13.48 $\theta_2=-7.688\times8500^3/(24\times32837\times8449\times10^6)=-0.000710$ rad.

Eq. 13.49 $\theta_{2a}=+(7.688\times2000^2/2)\times8500/(3\times32837\times8449\times10^6)=+0.000157$ rad.

Eq. 13.51 $\delta_{1a}=+2382.7\times10^3\times106.1\times2000^2/(2\times32837\times8326\times10^6)=+1.8$ mm

Eq. 13.50 $\delta_1=[+0.003873\times2000]+1.8=+9.6$ mm

Eq. 13.53 $\delta_{2a}=+7.688\times2000^4/(8\times32837\times8326\times10^6)=+0.06$ mm

Eq. 13.52 $\delta_2=[(-0.000710+0.000157)\times2000]+0.06=-1.11+0.06=-1.0$ mm

Net deflection at transfer $=+9.6-1.0=+8.6$ mm

b. Deflection at installation

Eq. 13.54 $\theta_3=[1.71\times+0.003873]-78.5\times10^3\times106.1\times8500/(2\times36283\times8449\times10^6)=+0.006520$ rad.

Eq. 13.55 $\theta_4=1.71\times-0.000710-18.0\times8500^3/(24\times36283\times8449\times10^6)=-0.002717$ rad.

Eq. 13.56 $\theta_{4a}=1.71\times+0.000157+(18.0\times2000^2/2)\times8500/(3\times36283\times8449\times10^6)=+0.000602$ rad.

Eq. 13.57 δ_3=[+0.006520×2000]−[78.5×10³×106.1×2000²/(2×36283×8326×10⁶)]+1.71×1.8=+13.0−0.06+3.1=+16.1 mm

Eq. 13.58 δ_4 =[(−0.002717 +0.000602)×2000]+(18.0×2000⁴/(8×36283×8326×10⁶)+1.71×0.06=−4.2+0.1+0.1=−4.0 mm

Net (downward) deflection at installation=+16.1−4.0=+12.1 mm

c. Long-term deflections

To check the effect of a moment-resisting, semi-rigid column connection, it is necessary to separate the visco-elastic rotations between installation and long-term, as the moment restraint does not act until after installation. In the following, these rotations are marked as*

Eq. 13.59 θ_5=+0.006520+(1.20×2304.2×10³−326.4×10³)×106.1×8500/(2×36283×8449×10⁶)=+0.006520+0.003588*=+0.010108 rad.

Eq. 13.60 θ_6=−0.002717−[(1.20×25.688+(1+1.20)×10.0)]×8500³/(24×36283×8449×10⁶)=−0.002717−0.004409*=−0.007126 rad.

where self-weight of beam and slabs 7.688+18.0=25.688 kN/m and imposed dead w_2=10.0 kN/m

Eq. 13.61 θ_{6a}=+0.000602+[1.20×51.4+(1+1.20)×20.0+(1+2.00)×18.0]×10⁶×8500/(3×36283×8449×10⁶)=+0.000602 +0.001475*=+0.002077 rad.

where moment for the beam plus slabs is 51.4 kNm and dead loads is 20.0 kNm quasi-permanent moment due to live load is 0.3×30.0×2.0²/2=18.0 kNm (using ψ_2=0.3).

Eq. 13.62 δ_5=[+0.010108×2000]+[(1.20×2304.2×10³)−326.4×10³]×106.1×2000²/(2×36283×8326×10⁶)=20.2+1.7*=21.9 mm

Eq. 13.63 δ_6=[(−0.007126 +0.002077)×2000]+[1.20×25.688+(1+1.20)×10.0+(1+2.00)×9.0]×2000⁴/(8×36283×8326×10⁶)=−10.1+0.5*=−9.6 mm

Final deflection δ_7=21.9−9.6=**+12.3 mm**<limit of 24.1 mm

where the apparent effective length is found to be x'=7.48 m from the simple support. Then L'_{ef}=2×(2.0+(8.5−7.48))=6.04 m. The deflection limit is therefore 6040/250=24.1 mm.

Considering the effect of the moment resisting, semi-rigid column, where the visco-elastic component of the rotations marked* above are multiplied by 0.83, θ_5, θ_6 and θ_{6a} are reduced to +0.009498, −0.006376 and +0.001826 rads., respectively. Then

δ_5=+0.009498×2000+0.83×1.7=20.4 mm

δ_6=[(−0.006376 +0.001826)×2000]+0.83×0.5=−8.7 mm

Final deflection δ_7=20.4−8.7=**+11.7 mm**, a reduction of only 5%.

d. Long-term movement due to live load only

Eq. 13.64 θ_8=(1+1.20)×18.0×10⁶×8500/(3×36283×8449×10⁶)=+0.000366 rad.

Eq. 13.65 δ_8=[+0.000366×2000]+[(1+1.20)×9.0]×2000⁴/(8×36283×8326×10⁶)=0.7+0.1=**+0.8 mm**

e. Long-term rotation and movement due to changes in camber, visco-elastic deflections of dead loads and long-term movement due to live load

Eq. 13.66 δ_9=δ_5−δ_3=21.9−16.1=5.8 mm

Eq. 13.67 θ_{10}=−1.20×35.688×8500³/(24×36283×8449×10⁶)=−0.003575 rad.

Eq. 13.68 θ_{10a}=+1.20×71.38×10⁶×8500/(3×36283×8449×10⁶)=+0.000792 rad.

Eq. 13.69 δ_{10}=(−0.003575+0.000792)×2000+[1.20×35.688×2000⁴/(8×36283×8326×10⁶)=−5.6+0.3=−5.3 mm

Final movement δ_{11}=δ_8+δ_9+δ_{10}=0.8+5.8-5.3=**+1.3 mm**<limit of L'_{ef} / 500=6040/500=12.1 mm.

In comparison with floor slabs, it is seen in Examples 13.6 and 13.4 that the long-term cantilever deflection/span ratio is much smaller for the beam, i.e. 12.3/2000=1 in 163, than the slab, i.e. 15.5/1500=1 in 97. Of course, the span/depth ratios are different, 3.3 and 6.0 for the beam and slab, respectively, due to the greater beam loads, but the serviceability moment ratios are just below 1.0 for both beam and slab. However, if the cantilever span in the slab is increased to 2000 mm, the deflection/span is greater still, i.e. 25.1/2000 = 1 in 80. Furthermore, the ratio of long-term deflection to the limiting deflection is 0.51 for the beam but 0.84 for the slab. It may therefore be concluded that deflections in beams are generally less critical than their moments of resistance.

13.5 CONTINUOUS SPANS

13.5.1 Continuous spans in floor slabs

In the past 30 years, practical experience has been gathered in making the simply supported ends of psc solid units and hcu flexurally continuous at interior supports, both with and without structural toppings. Less attention has been given to psc beams in favour of reinforced beams, particularly where the continuity of negative bending moments over walls and columns is semi-rigid. See Elliott, Section 9.7.5 (Elliott, 2017). Figures 13.21 and 13.22 show examples of how continuity may be achieved in precast beams over column supports, using continuity top reinforcement in cast *insitu* joints or topping.

Continuity of floor slabs is common is seismic countries where the action is used more for structural integrity than for increased moment capacity. Continuous solid slabs, known as half-slab (Figure 12.2), are designed with a structural topping to create wet areas in residential buildings. The psc unit is 75–100 mm depth together with a 75–100 mm structural topping, depending on span. This construction is also common in short-span car parks,

Figure 13.21 Continuity of beams through deep slots in the columns in a hybrid prestressed – reinforced concrete frame.

Figure 13.22 Continuity of beams over single-storey columns in a reinforced beam frame with prestressed hollow core floor units.

where psc solid slabs, typically 3.6–4.8 m in length span between long-span (typically 16 m) transverse steel beams.

Whist double-tee floor units are not suitable for continuous construction, the so-called 'multi-rib' double-tee (or triple-tee) floor units have been specially developed with solid end details for continuous spans, as shown in Figure 2.21. The keyed sloping end face is designed to resist the compression from large negative bending moments at supports, working together with steel mesh and rebars in the topping. The negative restraint moment capacity M'_{sR} and M'_{Rd} must be considered at both the serviceability and ultimate limiting states, together with the ultimate shear capacity $V'_{Rd,c}$ in the presence of negative moments (using the same procedures as for shear at cantilever supports in Section 13.2.6).

Continuous hcu have been used in stadia design, for example Figure 13.23 shows a typical detail where site placed reinforcement of maximum diameter $6+h/25$ (h=depth of hcu in mm) is cast *insitu* (concrete f_{cki}=40 N/mm^2 minimum) into top opened cores, sometimes overlapping (but not physically touching) with factory placed prestressing tendons in the top of the hcu. The length of the rebars should be 2×bond length, or to the point of contraflexure under imposed live load only. At least three cores in a 1.2 m wide unit (four if the span is greater than 6 m) should be filled. Design procedures for restrained composite supports for hcu are published by FIB (FIB, 2000). The negative moment of resistance is relevant only to the imposed loads and is provided by an appropriate area of reinforcement calculated in the normal manner. Continuity leads to an increase in the span-to-depth ratio, typically from 30 to 35, with a small reduction (about 5%) in the required area of prestressing tendons. Due to the negative moment, shear capacity may be governed by the flexurally cracked capacity $V'_{Rd,cr}$.

Figure 13.23 Continuity of prestressed hollow core floor units over reinforced concrete beams, note the 1.5 m length of the slots in the top opened cores.

Although multi-spans, more than five continuous spans, may be used in short span units, as mentioned above for car parks, for longer span hcu of between 6 and 12 m span it is more common to limit the design to three continuous spans. In this case, the outer spans are more critical in the sagging mode, due to the simple support at the remote end, but the interior span is critical in the hogging mode and for the design of the area and curtailment of top steel over the supports.

The major benefit from continuous spans is summarised in the load vs. span graphs in Figure 13.24 for simply supported and continuous spans, both non-composite and composite. The data for continuous spans are based on the interior span of three continuous spans. The range of floor units chosen covers the practical range, i.e. 100 mm depth solid units with a 75 mm thickness structural topping, and 200 and 400 mm depth hcu with or without a 75 mm thickness structural topping. The greatest advantage in terms of span vs. imposed load is for the 200 mm depth unit at the greatest load (15 kN/m^2) where the fixed end moments are the greatest – the ratio of the continuous/simply supported span is from 1.17 to 1.23. For the 400 mm depth unit, the ratio is from 1.12 to 1.17. In the deeper units with larger cores, the self-weight of filled cores increases the mid-span moment and therefore reduces the advantage gained over simply supported units. For example for the 400 mm depth hcu spanning 14 m with 3 no. filled cores the moment due to the self-weight of the filled cores (2.44 kN/m^2) is 55 kNm out of a total M_s = 404 kNm.

Continuous composite slabs are more favourable than non-composite units because the latter require reinforced top opened cores, which for greater imposed loads may extend far into the main span, in some cases extending over the entire span, although the number of top opened cores may be reduced further from the supports. The number and diameter of rebars close to internal supports may also be large. Design and construction details are

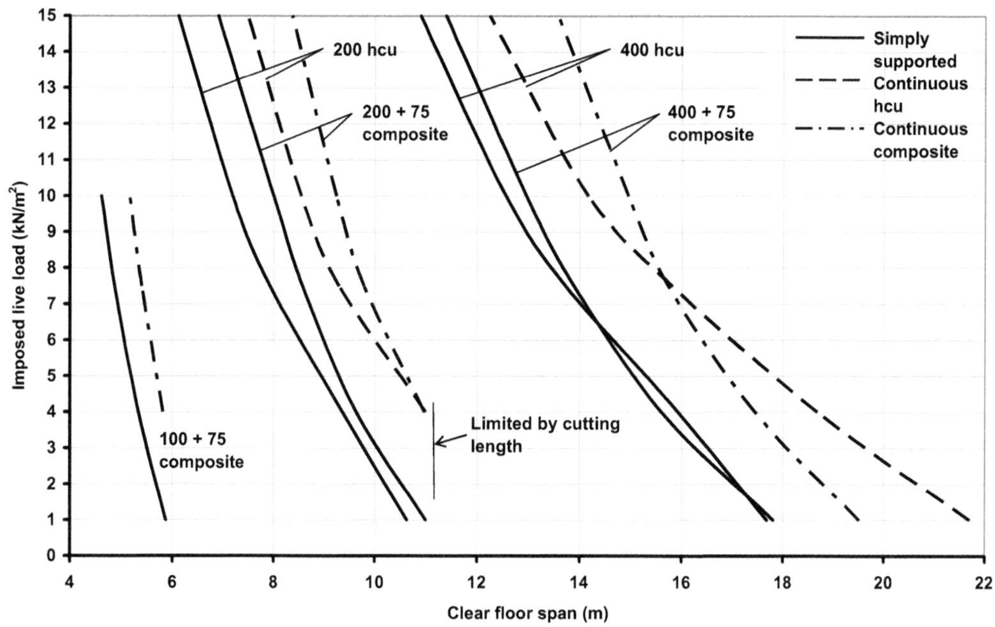

Figure 13.24 Load v span charts comparing continuous clear spans with simply supported clear spans for prestressed hollow core units and composite solid and hollow core slabs. Note that the clear span for the continuous spans is 300 mm less than the centres of bearing, but only 100 mm less for simply supported spans.

simplified for composite slabs as top opened cores are not necessary and the area of reinforcement in the topping is reduced thanks to the greater effective depth.

Referring to Figure 13.24 for the 200+75 mm depth composite slab, the ratio of the continuous/simply supported span is from 1.14 to 1.21, or 2 percentage points lower than the non-composite ratio because there are no fixed end moments due to the self-weight of the structural screed (but there are for finishes in the non-composite unit). Similarly, for the 400+75 mm depth composite slab, the ratio is from 1.10 to 1.19. For the solid 100+75 mm depth slab, the ratio is 1.09–1.14. Note that in Figure 13.24, the clear span for the continuous spans is 300 mm less than the centres of bearing, but only 100 mm less for simply supported spans. Therefore, 200 mm could be added to the continuous spans for a 'fair' comparison, adding 2–3 percentage points to the above ratios.

To cater for all loading conditions, Figure 3.4 shows five different loading conditions for five-span continuous design, and Figure 13.25 shows four conditions for three-span continuous design, with maximum and minimum imposed dead and live loads placed on the interior and exterior spans, resulting in critical bending moments and shear forces as follows:

Loading Case I. Maximum load in the interior span with minimum load (self-weight and imposed dead loads) on the exterior spans. This establishes the greatest sagging (positive) moment in the interior span. It also gives the maximum shear force at the simple support as the upward reaction from the negative moments is the least.

Loading Case II. Maximum load in the exterior spans with minimum load in the interior span. This establishes the critical distance into the interior span for negative top reinforcement.

Loading Case III. Maximum load in one exterior and the interior spans with minimum load in the other exterior span. If the exterior spans are of different length, Cases IIIa

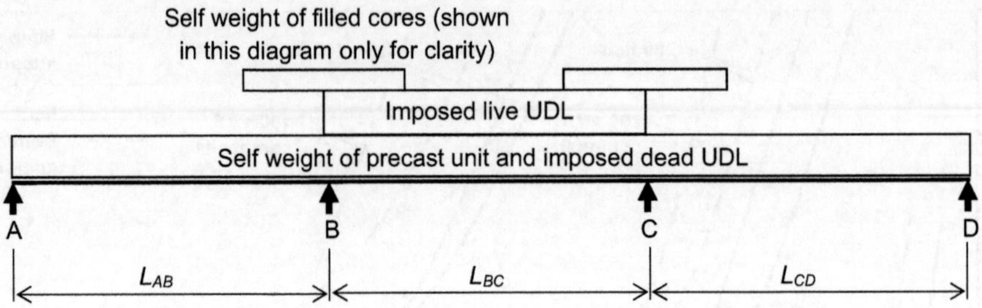

Case I. 'Min-max-min' with imposed live UDL on interior span only

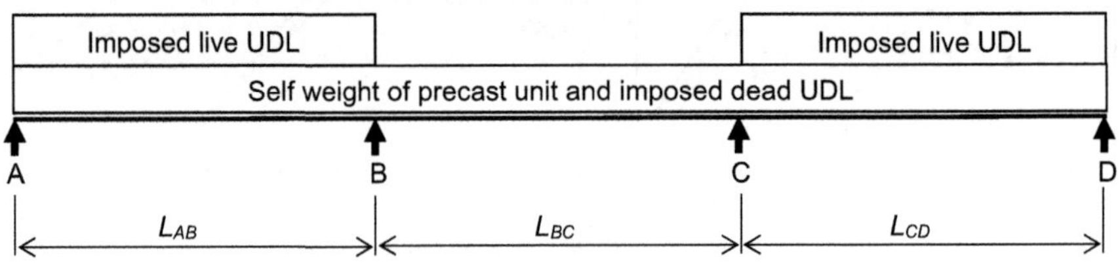

Case II. 'Max-min-max' with imposed live UDL on exterior spans

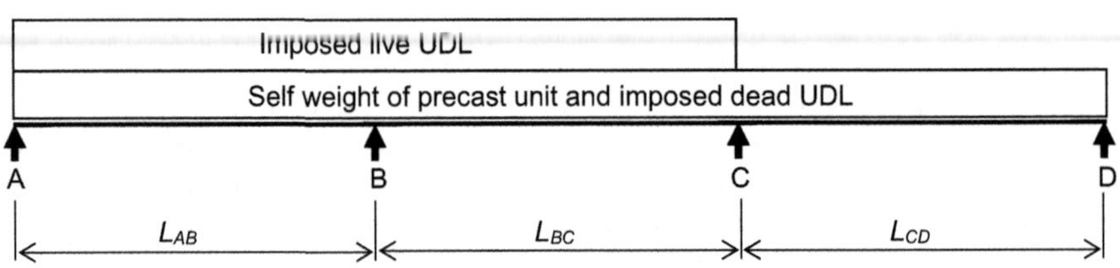

Case IIIa. 'Max-max-min' with imposed live UDL on two adjacent spans
Case IIIb similar but handed

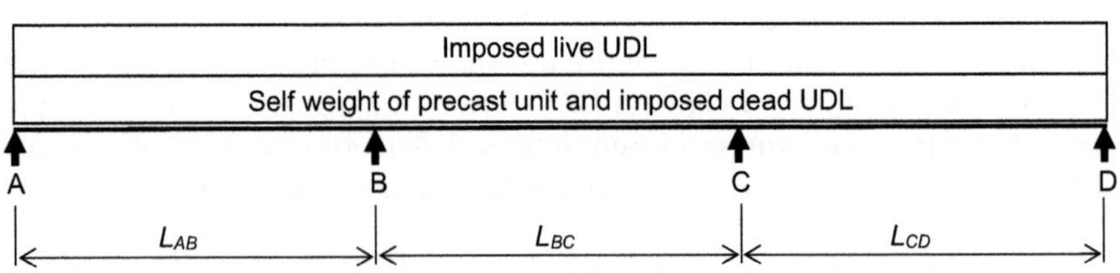

Case IV. 'Max-max-max' with imposed live UDL on all spans

Figure 13.25 Loading cases for interior and exterior spans for three-span continuous design.

and IIIb are for the maximum load in each exterior span in turn. This establishes the critical negative bending moment at the supports, and the greatest shear force in the interior span (due to the different fixed end moments).

Loading Case IV. Maximum load in all spans. This establishes the greatest support reaction.

It is usual to combine all load cases in bending moment diagrams, particularly when three effective spans are different, for both serviceability and ultimate limit states, as shown later in Figures 13.33 and 13.34 from Examples 13.7 and 13.8.

Section 3.3.1 introduced the load arrangements and partial safety factors for ultimate loads in continuous spans, commonly known as 'max-max-max', 'max-min-max', 'min-max-min', 'max-max-min', etc. for stage 2 imposed dead and live loads after the psc unit has been positioned in stage 1. For UDL loading for three continuous spans, these are summarised as

a. all spans carrying stage 2 design dead plus live loads γ_G $G_{k,}$+γ_Q Q_k, i.e. max{Exp. 6.10a; 6.10b}=max{1.35 $G_{k,j}$+ψ_0 1.5 Q_k; 1.25 $G_{k,j}$+1.5 Q_k}
b. alternate spans carrying γ_G G_k+γ_Q Q_k with other spans carrying the dead load only γ_G G_k with the same γ_G used for all spans, i.e. 1.35 G_k if Exp. 6.10a is maximum, otherwise 1.25 G_k.
c. two adjacent spans carrying γ_G G_k+γ_Q Q_k with other span carrying the dead load only γ_G G_k.

Fixed end moments due to the continuity of stage 2 bending moments are subjected to the same load factors.

13.5.2 Design of continuous spans in floor slabs

The precast floor units are simply supported in stage 1, known as the temporary stage. The moment capacity of the unit is checked for self-weight, a construction allowance typically 1.5 kN/m^2 (or any other construction plant, e.g. mobile elevated working platforms (MEWP) imposing point loads of up to 30 kN), plus the self-weight of the cast *insitu* structural topping (allowing for camber). This check is more critical than ordinary simply supported slabs because the requirements for the final prestress in a continuous design will be much less than in the simply supported case. For example comparing the worked Examples 7.1 and 13.7 (for the same 250 mm depth hcu, span and loading) the area of strands required in the simply supported span is 911 mm^2 compared with 539 mm^2 in the continuous spans.

For non-composite design rebars are site placed at the factory into top opened cores as shown in Figures 13.14, 13.23 and 13.26. The length of the opened cores is calculated from loading case III, shown in Figure 13.25, by adding the anchorage bond length l_{bd} plus the shift a_l if necessary due to shear forces (BS EN 1992-1-1, Fig. 9.2) to the theoretical cut-off point. For hcu with a small number of cores, e.g. 6 no.×140 mm width where it is desired to open and fill say 3 no. cores, it may be necessary to use 2 no. rebars per core to resist the maximum hogging moment M'_{Ed}. The 2 no. rebars may be curtailed to 1 no. at their theoretical curtailment point, and reduced to zero at a_l beyond the point where M'_{Ed}=0. In narrower cores, e.g. 11 no.×60 mm width, where only 1 no. rebar is used, the number of filled cores is reduced just once as explained above, or half of them curtailed to achieve a staggered (rather than sudden) reduction in capacity. Top cover to the rebars should satisfy the durability requirements, but as this is nearly always XC1 for the reason that finishes are

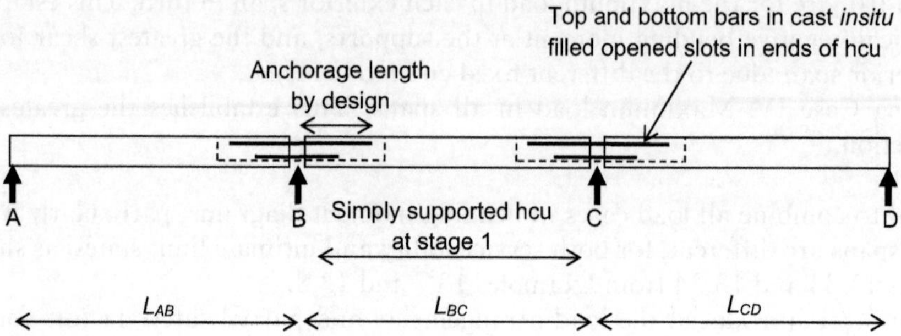

Figure 13.26 Reinforcing details for continuous non-composite units using additional top and bottom rebars in filled cores.

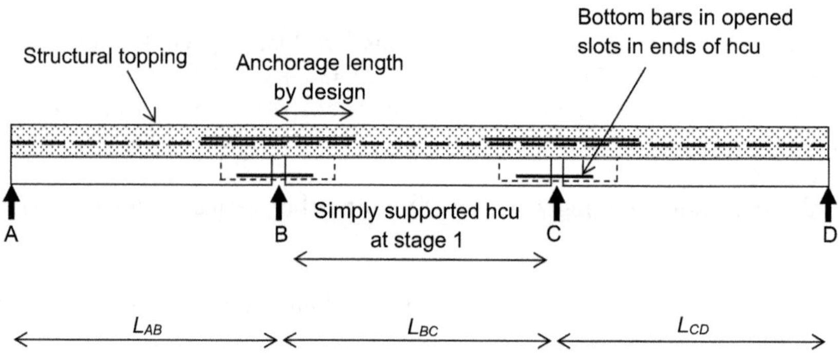

Figure 13.27 Reinforcing details for continuous composite slabs using steel mesh and additional rebars in the topping, additional bottom rebars in filled cores.

applied, the minimum cover of 20 mm is usually increased to allow for site tolerances, typically 25 mm. It is not recommended to place site rebars in the (often narrow) longitudinal gaps between adjacent units.

For composite design, rebars in the top opened cores are not required as stage 2 hogging moments are resisted by the steel mesh plus (if necessary) additional loose rebars placed on top of the mesh, as shown in Figure 13.27. The diameter of the additional bars should not, if possible, be greater than 12 mm in order to avoid complications with the top and bottom cover to the mesh, particularly if the depth of the topping is 75 mm. There is a construction difficulty here – if possible the mesh should be placed so that the bars are orientated with mesh bars parallel to the span. However, as the sheets of mesh may be lapped by upturning the sheets to nestle together, the bars will sit on top of the transverse mesh bars. For this reason, the top cover to the mesh should be about 35–40 mm to allow for H10–H12 bars with 25 mm cover. Interface loops are positioned as necessary (Figure 12.19) and designed as per Section 12.4. The requirements may be greater than for simply supported composite slabs due to the greater load-bearing capacity of continuous spans.

13.5.3 Design calculations for continuous floor slabs

13.5.3.1 Non-composite floor units

Stage 1. Let w_1=self-weight of precast floor unit, per unit width, w_{0c}=self-weight of wet *insitu* concrete in filled cores of length L_{0c} from the end of the span, and w_c=construction allowance, typically 1.5 kN/m^2.

The effective span L of the precast unit is taken to the centre of bearing=clear span+min{0.5 h; 0.5 l_b}. Referring to Figure 13.28a

$$M_{s1} = (w_1 + w_c)L^2/8 + w_{0c}L_{0c}^2/2 \le M_{sR} \tag{13.75}$$

After site rebars have been placed and the *infill* concrete in the top opened cores, in the longitudinal joints and between the ends of the precast units has hardened, the effective span for the continuous slab L_c is between the centres of supports=clear span+width of beam (or wall). For example, if supporting inverted-tee beams have a width of 600 mm, and l_b for the precast unit is 100 mm, L=clear span+100 mm and L_c=clear span+600 mm, the difference L_c-L=500 mm.

Stage 2 floor loads for the continuous spans are w_2=imposed dead loads, w_3=imposed live loads, per unit width. Referring to Figure 13.28b for the three spans AB, BC and CD of the composite slab, the final elastically distributed negative continuity moments in span BC are calculated as M_B and M_C. If M_B and M_C happen to be of equal negative magnitude then

$$M_{s2} = (w_2 + w_3)L_c^2/8 - M_B \tag{13.76}$$

When M_{s1} and M_{s2} are added together the construction allowance w_c is removed such that

$$M_s = w_1L^2/8 + w_{0c}L_{0c}^2/2 + M_{s2} \le M_{sR} \tag{13.77}$$

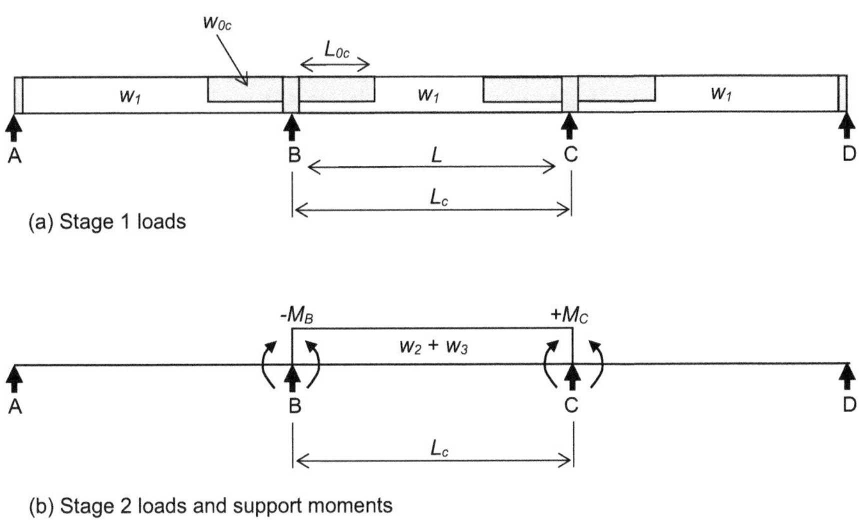

Figure 13.28 (a) Stage 1 loading due to self-weight of precast unit plus wet *insitu* concrete in filled cores and between ends of precast units. (b) Stage 2 loading due to imposed dead and live loads in interior span.

If M_B and M_C are unequal, the likely situation thanks to load cases IIIa or IIIb, $M_{s2,max}$ is not at mid-span, then the reaction at support B is

$$F_{B2}=(w_2+w_3)\ L_c/2-(M_B-M_C)/L_c \tag{13.78}$$

$$\text{Shear span } a_2=F_{B2}/(w_2+w_3) \tag{13.79}$$

$$M_{s2,max}=F_{B2}\ a_2-(w_2+w_3)a_2^2/2\ +M_B \tag{13.80}$$

Note that in some texts M_c is +ve according to global notation where clockwise moments are +ve.

For example, let $(w_2+w_3)=9.0$ kN/m, $L_c=10.0$ m, $M_B=-80$ kNm and $M_c=-50$ kNm. $F_{B2}=45.0-(-80--50)/10.0=48.0$ kN, $a=48.0/9.0=5.33$ m, $M_{s2,max}=48.0\times5.33-9.0\times5.33^2/2-80=48.0$ kNm. A quick check on the arithmetic is that at mid-span $M_{s2}=9.0\times10.0^2/8-65=47.5$ kNm.

Of course the final position and magnitude of $M_{s,max}$ also depends on w_1, w_{0c}, L and L_{0c}, a rather complex expression if the length of top opened cores is different at either end. A simple spreadsheet may be used to calculate M_s at say intervals of span/50, particularly if line loads and point loads are involved.

Ultimate bending moments M_{Ed1} and M_{Ed2} are calculated in the same manner using the appropriate load factors given in Section 13.5.1.

13.5.3.2 Calculation of fixed end moments and final negative moments in continuous spans

Fixed end bending moments (FEM) are first calculated according to the flexural stiffness of each span of the slabs, using the compound value for the second moment of area, i.e. 4 $E_{cm}\ I_{c,co}/L_c$ for continuous spans and 3 $E_{cm}\ I_{c,co}/L_c$ for a continuous span with a simply supported span with a pinned end. [The reader is reminded that the end rotation θ of any continuous element of stiffness EI and span L subjected to a bending moment M is given as θ=M/(4 EI/L). For a pinned-ended element is θ=M/(3 EI/L). Also if M is acting at the remote end of the element then θ=M/(2 EI/L).]

Referring to Figure 13.28 for the three spans, the maximum service moment in the interior span BC is calculated for load case I, i.e. dead load only on the exterior spans AB and CD, with the full dead and live loads in the interior span. Load case III is used to calculate the maximum negative moment at the interior supports, and to check whether the section is negatively flexurally cracked, which will nearly always be the case, for the purpose of calculating the shear capacity $V_{Rd,cr}$. Load case II is not required for service moments.

Ultimate loads are applied using the same load cases as service loads. Load cases II and III are also used for calculating the minimum interior span moments and maximum support moments for the purpose of extending top rebars into the top opened cores into the interior span.

Referring to Figure 13.29a, the initial FEMs (clockwise is +ve) at interior supports are

$$\text{At B}\quad \text{FEM}_{\text{BA}}=+(w_2+w_3)L_{c,AB}^2/8 \tag{13.81}$$

$$\text{At B}\quad \text{FEM}_{\text{BC}}=-(w_2+w_3)L_{c,BC}^2/12 \tag{13.82}$$

$$\text{At C}\quad \text{FEM}_{\text{CB}}=+(w_2+w_3)L_{c,BC}^2/12\ (\text{or }-\text{FEM}_{\text{BC}}) \tag{13.83}$$

$$\text{At C}\quad \text{FEM}_{\text{CD}}=-(w_2+w_3)L_{c,CD}^2/8 \tag{13.84}$$

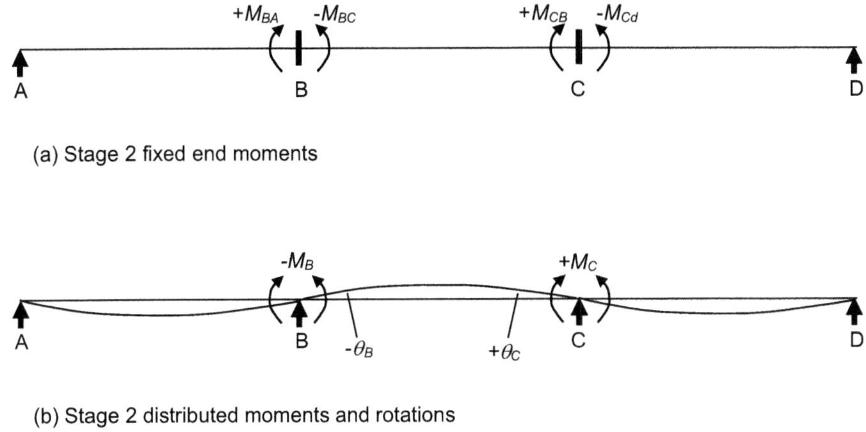

Figure 13.29 (a) Stage 2 fixed end moments (FEM). (b) Final distributed support moments and rotations.

The denominator of the FEM at the interior support B of the exterior span is 8 (not 12) because when the fixed end moment $-L_{c,AB}^2/12$ at the exterior support A is released to become a pinned end 50% carry-over moment $+L_{c,AB}^2/24$ is distributed to B. Then $\text{FEM}_{BA} = +L_{c,AB}^2/12 + L_{c,AB}^2/24 = +\ 3L_{c,AB}^2/24 = +L_{c,AB}^2/8$. Similarly for FEM_{CD}.

The 'out of balance' or restoring moments M_1 and M_2 at supports B and C, respectively, are

$$M_1 = -\ (\text{FEM}_{BA} + \text{FEM}_{BC}) \tag{13.85}$$

$$M_2 = -\ (\text{FEM}_{CB} + \text{FEM}_{CD}) \tag{13.86}$$

(Note that suffixes 1 and 2 are used for supports B and C in the following analysis.)

For example, using the same loads and spans as in the previous exemplar, $\text{FEM}_{BA} = 9.0 \times 10.0^2/8 = +112.5$ kNm, $\text{FEM}_{BC} = 9.0 \times 10.0^2/12 = -75.0$ kNm. Then $M_1 = -(+112.5 - 75.0) = -37.5$ kNm. $M_2 = -\ (+75.0 - 112.5) = +37.5$ kNm.

The stiffness terms for the rotation θ_1 and θ_2 of the supports at *B* and *C* due to out-of-balance moments M_1 and M_2 are as follows. Assuming E_{cm} and $I_{c,co}$ are the same for all spans, and letting shorthand $\text{EI} = E_{cm}\ I_{c,co}$, the rotational stiffness of support *B* due to M_1 acting on the pin-ended span *BA* is $k_{11} = M_1/\theta_1 = 3\ \text{EI}/L_{BA}$. Similarly, the rotational stiffness of support *B* due to M_1 acting on the continuous span *BC* is $k_{11} = M_1/\theta_1 = 4\ \text{EI}/L_{BC}$. Then

$$k_{11} = 3\ \text{EI}/L_{BA} + 4\ \text{EI}/L_{BC} \tag{13.87}$$

The rotational stiffness of support C due to M_1 acting on the continuous span BC is $k_{12} = M_1/\theta_2 = 2\ \text{EI}/L_{BC}$. Then

$$k_{12} = 2\ \text{EI}/L_{BC} \tag{13.88}$$

$$k_{21} = 2\ \text{EI}/L_{BC} \tag{13.89}$$

$$k_{22} = 3\ \text{EI}/L_{CD} + 4\ \text{EI}/L_{BC} \tag{13.90}$$

For example, following the previous worked examples for the 250 mm depth hcu, let $EI = E_{cm}\ I_{c,co} = 36283 \times 1296 \times 10^6 = 47 \times 10^{12}$ Nmm², and $L = L_c = 10.0$ m. $k_{11} = k_{22} = 3 \times 47 \times 10^{12}/10000 + 4 \times 47 \times 10^{12}/10000 = 32900 \times 10^6$ Nmm/rad. or 32900 kNm/rad. $k_{12} = k_{21} = 2 \times 47 \times 10^{12}/10000/10^6 = 9400$ kNm/rad. Note that $k_{22}/k_{21} = 3.5$.

Giving the stiffness matrix $$\begin{bmatrix} M_1 \\ M_2 \end{bmatrix} = \begin{bmatrix} k_{11} & k_{12} \\ k_{21} & k_{22} \end{bmatrix} \begin{bmatrix} \theta_1 \\ \theta_2 \end{bmatrix} \quad (13.91)$$

Yielding $$\theta_1 = \frac{M_2 - k_{22}\, M_1/k_{21}}{k_{12} - k_{22}\, k_{11}/k_{21}}$$

$$\theta_2 = \frac{M_1 - k_{11}\theta_1}{k_{21}} \quad (13.93)$$

Continuing the exemplar. $\theta_1 = +37.5 - 3.5 \times -37.5/(9400 - 3.5 \times 32900) = 168.75/-105750 = -0.001596$ rad. anti-clockwise. $\theta_2 = -37.5 - (32900 \times -0.001596)/9400 = 15.01/9400 = +0.001596$ rad. of course.

The final distributed moments are equal to the sum of the rotations times flexural stiffness at each end of the span, plus the original fixed end moment, which is still present. Referring to Figure 13.29b, for support B, or end 1, M_B is equal to rotation θ_1 at that end of span BC times the stiffness there, i.e. $M = 4EI\ \theta_1/L$, plus the rotation θ_2 of the remote end times the stiffness there, i.e. $M = 2EI\ \theta_2/L$, and finally plus the original FEM_{BC}, as

$$M_B = 4EI\ \theta_1/L_{BC} + 2EI\ \theta_2/L_{BC} + FEM_{BC} \text{ (–ve, anticlockwise)} \quad (13.94)$$

$$M_C = 4EI\ \theta_2/L_{BC} + 2EI\ \theta_1/L_{BC} + FEM_{BC} \text{ (+ve, clockwise)} \quad (13.95)$$

The ultimate support moments $M_{Ed,B}$ etc. are calculated using the same procedures using ultimate load factors according to BS EN 1990, Exp. 6.10a or 6.10b.

Referring to BS EN 1992-1-1, clause 5.3.2.2 (4) where a beam or slab is continuous over a support which provides no restraint to rotation, the moment at the support may be reduced due to a local restraint at the cusp of the support by an amount ΔM_{Ed} where

Exp. 5.9 $$\Delta M_{Ed} = F_{Ed,sup}\ t/8 \quad (13.96)$$

where $F_{Ed,sup}$ = ultimate design support reaction (from both sides of the support)
t = width of the support, e.g. thickness of wall, width of beam

Continuing the exemplar. $M_B = (4 \times 47 \times 10^{12} \times -0.001596/10000) + (2 \times 47 \times 10^{12} \times 0.001596/10000) - 75.0 = -30.0 + 15.0 - 75.0 = -90.0$ kNm, anti-clockwise causing an anti-clockwise rotation at B, leading to a sagging moment in the exterior span AB. By inspection $M_C = +90$ kNm. Note that the initial FEMs were 112.5 and 75.0 kNm, so the final negative moment is roughly their average, to be expected for equal spans.

To reader may note that in this example the k and θ terms and M_B and M_C moments all contain the common factor EI/L, which may be eliminated from the calculation. Thus $k_{11} = k_{22} = 3 + 4 = 7$ and $k_{12} = k_{21} = 2$. EI θ_1/L=+37.5-3.5×-37.5v(2−3.5×7)=168.75/−22.5=−7.5. $M_B = 4 \times -7.5 + 2 \times 7.5 - 75.0 = -30.0 + 15.0 - 75.0 = -90.0$ kNm.

End reactions due to distributed moments

$$F_B = (M_B - M_C)/L_{BC} \quad (13.97)$$

$$F_C = -V_B \quad (13.98)$$

These moments and shear forces are added to the free Stage 2 values. In the example, the sagging moment in the interior span is M_{s2}=112.5−90.0=+22.5 kNm, i.e. only 25% of the negative moment at the support.

The negative support moments are not redistributed in psc elements, owing to the lack of ductility when the tendons in the tension zone of a psc element are already stressed close to $f_{p,max}$, about 1516 N/mm². The UK NA to BS EN 1992-1-1, clause 5.1.3 (1)P (NA to BS EN 1992-1-1, 2004) states that for case IV loading (all spans fully loaded) the support moments should be reduced by 20% with a consequential increase in span moments. In the above example the moment redistribution is 0.2×90=18.0 kNm, increasing M_{s2}=22.5+18.0=+40.5 kNm. However, if the loading arrangement in this example is case I, and the dead loads acting on the exterior spans is say 4.5 kN/m (50% of the total) the support moments reduce to M_B=−67.5 kNm and M_{s2}=112.5−67.5=+45.0 kNm, almost the same result. If load case III is used (maximum in 2 adjacent spans) M_B=−97.5 kNm and M_C=+60.0 kNm, then M_{s2}=+34.3 kNm, and is not critical.

13.5.3.3 Distributed secondary moments due to relaxation and creep in prestressed elements

Known as 'secondary', or parasitic bending moments, and relevant only to psc elements, positive bending moments develop at the support due to creep deformations and long-term relaxation of tendons. The effect is manifest in the upward camber of long span, lightly loaded floor or roof slabs, exacerbated by high temperatures or differential temperatures on the top and bottom surfaces. The ends of the psc elements attempt to rotate negatively (in the same direction as camber, but opposing gravity loads), thus breaking the bond with the cast *insitu* infill over the supports, possibly within several months (or one hot summer) after construction. It is therefore necessary to provide rebars near the bottom of the precast units spanning across the infill joint. This is a feature common in psc bridge beam design.

The following expressions are for the growth effects of creep, which have a much greater effect than for long-term relaxation, by a factor of 3–4. Referring to Figure 13.30, secondary fixed end bending moments develop at supports B and C in the exterior spans as

$$\mathrm{FEM}_{BA}=-\mathrm{FEM}_{CD}=-1.5\times[\Psi\, P_{pmi}-(P_{pmi}-P_{po})]\, z_{cp} \tag{13.99}$$

and at supports B and C in the interior span as

$$\mathrm{FEM}_{BC}=-\mathrm{FEM}_{CB}=[\Psi\, P_{pmi}-(P_{pmi}-P_{po})]\, z_{cp} \tag{13.100}$$

where

$$\Psi=1-e^{-\psi 28} \tag{13.101}$$

is the visco-elastic creep growth coefficient after installation and hardening of cast *insitu* infill at the ends of the precast units, and ψ_{28} is the creep coefficient over this interval of time. As explained in Section 11.2.1 this is typically ψ_{28}=1.20, then $\Psi=1-e^{-1.2}=0.70$.

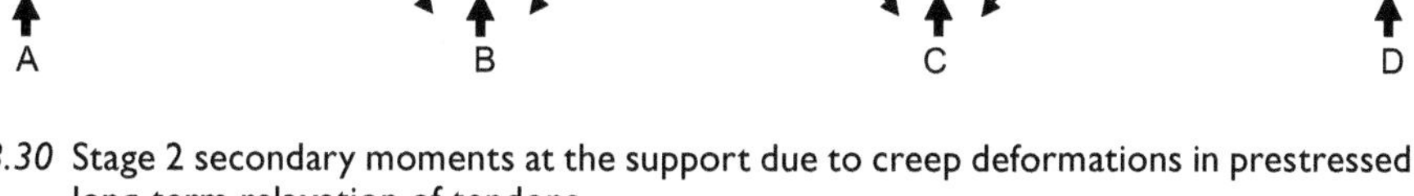

Figure 13.30 Stage 2 secondary moments at the support due to creep deformations in prestressed unit and long-term relaxation of tendons.

For the same loading case fixed end gravity moments are due to imposed dead loads only. These moments are distributed as above, resulting in moments and shear forces of opposite sign. The final moments are restrained using rebars placed about 20 mm above the bottom of top opened cores, typically 2 or 3 no. H10 to H16 bars per unit×about 600–800 mm embedment in each side of the support, or at the bottom of the longitudinal joint between solid planks units.

Continuing the exemplar: $P_{pmi}=1010.2$ kN and $P_{po}=860.2$ kN. $\text{FEM}_{BA}=-1.5\times(0.7\times1010.2-150.0)\times82.2\times10^{-3}+4.5\times10.0^2/8=-68.7+56.2=-12.5$ kNm and $\text{FEM}_{BC}=+45.8-37.5=+8.3$ kNm. After distribution the final support moment is $M_B=+54.9-45.0=+9.9$ kNm, and the ultimate $M_{EdB}=1.25\times9.9=12.4$ kNm.

13.6 DESIGN OF REINFORCEMENT IN OPENED CORES OR TOPPING

13.6.1 Design of reinforcement in top opened cores and/or with top pretensioned tendons

See Section 13.2.2 and Eqs. 13.10–13.24.

13.6.2 Design of reinforcement in topping

Mesh reinforcement is provided to cater for minimum area $A_{s,min}$

9.2.1.1 (1) $A_{s,min}\geq0.13\%\ b\ t$ but not less than $0.26\ f_{ctmT}\ b\ t/f_{yk}$

For example, let $t=75$ mm, $f_{ckT}=25$ N/mm², $f_{ctmT}=2.56$ N/mm², $f_{yk}=500$ N/mm²

$A_{s,min}=\max\{0.0013\times1000\times75;\ 0.26\times2.56\times1000\times75/500\}=100$ mm²/m, use A142 mesh.

Additional reinforcement in the form of loose rebars is provided to cater for the ultimate negative moment M'_{Ed}. Bars are typically placed at 200 mm spacing to align with the longitudinally orientated bars in the mesh

b'=width of precast unit at soffit
d'=depth of unit+topping – axis distance to bars

f_{ckT} for the *insitu* concrete topping

$$K' = M'_{Ed} / f_{ckT} b' d'^2 \quad (13.102)$$

$$z' = d' \min\{0.95; 0.5 + \sqrt{(0.25 - K'/1.134)}\} \quad (13.103)$$

$$A'_s = M'_{Ed} / 0.87 f_{yk} z' - \text{area of mesh}^* \quad (13.104)$$

*Area of mesh is calculated over the actual width of precast unit.

Anchorage length of the additional rebars, including the 'shift' distance d', is according to Section 13.2.2.

13.6.3 Design of positive moment reinforcement at bottom of cores

Ultimate positive moment=M''_{Ed}

b'' = width of precast unit at soffit
d'' = depth of hcu+topping – depth of bottom flange – 20 mm cover – bar radius.

f_{cki} for the *insitu* concrete infill in the cores

The procedure is the same as in Section 13.3.2, except that the distance to the curtailment point = distance to the point of zero moment plus the anchorage length plus the shift d'' due to shear. The factor $\alpha_2 = 1$ is to allow for the possibility that the infill concrete may not be homogeneous between the bars and the inside surface of the core.

13.6.4 Service stress due to prestress plus negative flexural stress

For non-composite units, see Section 13.2.3 and Eqs. 13.25 and 13.26.

For composite slabs, see Section 13.2.4.2 and Eqs. 13.32 and 13.33.

The flexural stresses $M_s'/Z_{b,co}$ and $M_s'/Z_{t,co}$ are calculated at the end of the transmission zone l_{pt} plus the gap between the end of the unit and centre of support, where the negative moment reduces to $M'_{s,lpt}$.

In both cases, the combined stress at the top of the unit is likely to be critical, particularly in long spans where the negative moment is much greater than the positive mid-span moment.

All section properties are based on the uncracked section modulus. If the tension at the top is exceeded, the shear capacity is based on the cracked section $V_{Rd,cr}$ or $V_{Rd,crc}$ according to Section 13.2.6, plus the shear capacity of filled cores $V_{Rd,t}$.

Example 13.7

Calculate M_{sR} at the supports and mid-span for the interior slab in a three-span continuous 250 mm depth prestressed hcu shown in cross section and elevation in Figure 13.31. The section properties, material data and prestress are as given in Example 7.1. Exposure class to the bottom and top surfaces is XC1. The effective span of the hcu is $L = 9.8$ m. After *insitu* concrete at the ends of the hcu has hardened the centre-centre span of the slabs is $L_c = 10.0$ m. As there are no tendons in the top of the hcu the negative moment of resistance due to continuity is to be provided by placing rebars of strength $f_{yk} = 500$ N/mm^2 into the top of 140 mm wide opened cores. The slots are filled with *insitu* concrete of the same strength as the precast unit $f_{cki} = 45$ N/mm^2 and known density 24.5 kN/m^3. The distance to the curtailment of these bars in the interior span (point E) is determined in Example 13.8 (ultimate design). Use self-weight of grouted unit = 3.63 kN/m^2, imposed dead loads = 3.30 kN/m^2, ceiling 0.5 kN/m^2 and imposed live load = 7.5 kN/m^2.

Using the four load combinations Case I to IV, given in Figure 13.25, check M_{sR} against the characteristic combination of the positive (sagging) design service moment M_s and the negative hogging M_s'. Plot graphs of bending moment distributions for M_{sR} and the critical M_s, taking $l_{pt} = 605$ mm (from Example 6.1).

Calculate the requirement for secondary positive reinforcement allowing for creep of prestress at the ends of the psc unit.

Given that the results in Table 13.2 are from further analysis of the same hcu used in Example 13.7 in both three-span and five-span continuous floor slabs, discuss the differences in mid-span and support moments for the two arrangements.

Solution

At supports

M_{sR} is calculated according to Example 7.1 except that the negative bending moments due to continuity will increase creep losses. As the critical load case for the positive M_s is case I, the negative support moment M_s' is required for imposed dead loads (excluding services), of $1.2 \times 3.30 = 3.96$ kN/m.

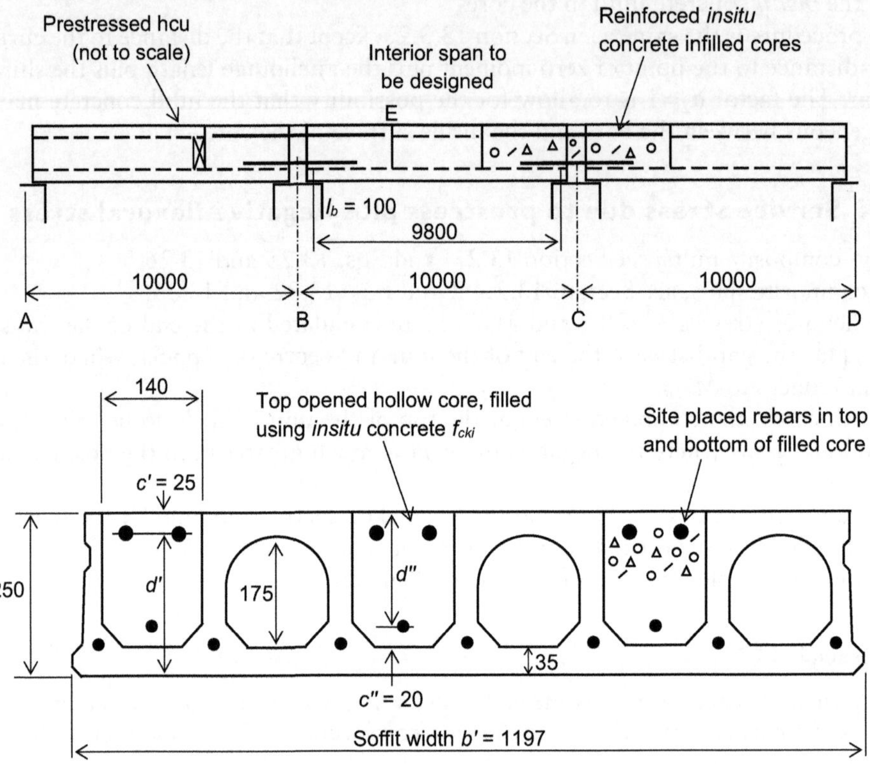

Figure 13.31 Elevation and cross section of the three-span continuous prestressed concrete hcu designed in Examples 13.7–13.9, showing position and extent of reinforced concrete filled cores. Distance to point E to be determined from Example 13.8.

Table 13.2 Maximum service bending moments at mid-span and at the supports of interior and exterior floor spans in three-span and five-span continuous slab arrangement

		Interior span[a] *(kNm)*	*Exterior span (kNm)*	*Load arrangement*
Mid-span	3-span	+142.4		Alternate min-max-min
	3-span		+189.9	Alternate max-min-max
	5-span	+161.5	+187.8	Alternate max-min-max-min-max
Support	3-span	−90.6	−90.6	Alternate min-max-min & max-min-max
	3-span	−150.6	−150.6	Two span max[b] over interior support
	5-span	−71.5	−95.4	Alternate max-min-max-min-max
	5-span	−136.1		Two span max[b] over second interior support
	5-span		−155.7	Two span max[b] over first interior support
Support FEM	$wL_c^2/12$	−113.0 −38.0	−113.0 −38.0	Max UDL Min UDL

The spans, self-weight and imposed dead and live loads are the same as in Example 13.7.

Maximum (max) and minimum (min) UDL, w = 13.56 and 4.56 kN/m, respectively. L_c = 10.0 m.

All mid-span moments include M_{sI} = +63.5 kNm for stage I self-weight and filled cores.

[a] This is the central interior span in five-span slabs

[b] 'Two span max' means two adjacent spans with maximum UDL and minimum UDL in next span(s).

Eq. 13.81 $FEM_{BA}=3.96\times10.0^2/8=+49.5$ kNm, $FEM_{BC}=-33.0$ kNm

Eq. 13.85 $M_1=-M_2=-(+49.5-33.0)=-16.5$ kNm

Eq. 13.87 $k_{11}=k_{22}=3+4=7$, and $k_{12}=k_{21}=2$

Eq. 13.92 $EI\,\theta_1/L=(+16.5-3.5\times-16.5)/(2-3.5\times7)=74.25/-22.5=-3.30$.

Eq. 13.94 $M_B=(4\times-3.3)+(2\times3.3)=-13.2+6.6-33.0=-39.6$ kNm.

Now, following the procedures in Example 7.1 where $M'_{s0}=M_B$
Due to M'_{s0}. $\sigma_{cp0}=-\,39.6/15.766=+2.51$ N/mm^2

Then $\sigma_{cp}=11.90+2.51=14.41$ N/mm^2

Loss of prestress due to creep $\Delta\sigma_{p,c}=139.49$ N/mm^2, shrinkage $\Delta\sigma_{p,s}=73.51$ N/mm^2 and relaxation $\Delta\sigma_{p,r}=25.80$ N/mm^2

$\sigma_{po}=1158.8-139.49-73.51-25.80=920.0$ N/mm^2 (compared with 1020.7 N/mm^2 simply supported).
$P_{po}=920.0\times911\times10^{-3}=838.1$ kN

Final prestress at mid-span. $\sigma_b=11.66$ N/mm^2 and $\sigma_t=-1.90$ N/mm^2

M_{sR} is critical at the bottom fibre, then
$M_{sR}=(11.66+3.80)\times10.701=$ **165.44 kNm**

At mid-span
M_{sR} is calculated according to Example 7.1 except that the negative bending moments due to continuity will also reduce the bending stress and increase creep losses at mid-span. The positive moment is based on a continuous span $L_c=10.0$ m, as $8.32\times10.0^2/8=104.0$ kNm. The net moment is $104.0-39.6=64.4$ kNm.
Due to M'_{s0} $\sigma_{cp0}=-64.4/15.766=-4.08$ N/mm^2
Then at mid-span $\sigma_{cp}=(1072607/1055654)\times11.90-4.08=8.01$ N/mm^2
Loss of prestress due to creep $\Delta\sigma_{p,c}=77.50$ N/mm^2, shrinkage $\Delta\sigma_{p,s}=73.51$ N/mm^2 and relaxation $\Delta\sigma_{p,r}=27.47$ N/mm^2
$\sigma_{po}=1177.4-77.50-73.51-27.47=999.0$ N/mm^2 (compared with 1020.7 N/mm^2 simply supported).
Final prestress at mid-span. $\sigma_b=12.66$ N/mm^2 and $\sigma_t=-2.06$ N/mm^2
M_{sR} is critical at the bottom fibre, then
$M_{sR}=(12.66+3.80)\times10.701=$ **176.18 kNm**

Calculation of positive service moments M_{s1} and M_{s2} in the interior span BC
It is necessary at this stage to decide on the number of filled cores. Based on the experience in Example 13.1 this is taken as 3, giving the same self-weight 1.245 kN/m^2.
It is shown later in Example 13.8 that the ultimate negative moment $M'_{Ed}=-214$ kNm. Taking the lever arm $z'=197$ mm from Example 13.1 $A'_s=214\times10^6/(197\times0.87\times500)\approx2500$ mm^2. Therefore use 6 no. H25 top bars (2946 mm^2), 2 no. in each core. The 6 no. bars may be curtailed to 3 no. at the appropriate curtailment point. Also from Example 13.8, the length of the top opened cores in the interior span is 3.82 m.
Therefore the mid-span service moment $M_{s1}=1.2\times3.63\times9.8^2/8+1.2\times1.245\times3.82^2/2=52.3+10.9=63.2$ kNm

Stage 2 calculations are carried out using Eqs. 13.76–13.95 for load cases I, II, III and IV, and are summarised in Table 13.3.

From load case I, $M_{s,max}=142.4$ kNm$<M_{sR}=176.18$ kNm

From load case III, the negative moment $M'_{s,max}=-150.6$ kNm

Combined prestress plus flexural bending stress is calculated at the end of the transmission length l_{pt} plus half the gap between the ends of the precast units. As the combined stresses in the top of the section will certainly exceed fctm, this check is carried out first, after which the flexurally cracked section properties, including the transformed area of top bars in opened cores, are calculated.

The negative bending moment $M'_{s,lpt}$ is at $l_{pt}=0.605$ m plus half the gap between the ends of the hcu=0.1 m, i.e. at 0.705 m from the centre of support B.

Total reaction at support B, $F_B=75.3$ (from Table 13.3)$+4.56\times9.8/2+1.49\times3.8=$ 102.3 kN. The total UDL=4.56+1.49+13.56=19.41 kN/m

Then $M'_{s,lpt}=102.3\times0.705-19.41\times0.705^2/2-150.6=+67.3-150.6=-83.3$ kNm

At the top fibre the combined tensile stress=prestress+flexural stress, where (from above) $\sigma_t=-1.90$ N/mm²

Eq. 13.26$f'_{t,max}=-1.90-83.3/10.060=-10.18$ N/mm²$\ll-3.80$ N/mm², therefore the section is flexurally cracked and the shear capacity is based on $V'_{Rd,cr}$.

To calculate the effective flexurally cracked section properties, the area of top bars at the support (see above) $A'_s=6\times491=2946$ mm². Taking the short-term transformed area $A'_s\ (m-1)$, where $m=E_s/E_{cm}=200000/36283=5.51$. Then $A'_s\ (m-1)=13290$ mm².

Height to centroid from bottom of transformed area of rebars in three filled cores $y'_{b,co}=124.3$ mm. $I'_{c,co}=1601\times10^6$ mm⁴. $Z'_{b,co}=1601/124.3=12.88\times10^6$ mm³ and $Z'_{t,co}=$ $1601/(250-124.3)=12.74\times10^6$ mm³.

Cracking moment of resistance $M'_{cr}=-f_{ctm}\ Z'_{t,co}=(-3.80-1.90)\times12.74=-24.2$ kNm$>$ $M'_{s,lpt}=-83.3$ kNm

In the cracked section the transformed area of top bars is $A'_s\ m=16235$ mm²

Height to centroid from bottom of flexurally cracked section=63.6 mm

Section modulus of cracked section at bottom fibre $Z'_{b,co}=465.0/63.6=7.306\times10^6$ mm³

The effective value for $Z'_{b,co,ef}=7.306+[(12.88-7.306)\ (24.2/83.3)^2]=7.77\times10^6$ mm³

The combined compressive stress=prestress at the end of the transmission length l_{pt}+flexural stress $M'_{s,lpt}\ /\ Z'_{b,co,ef}$

Eq. 13.25$f'_{b,max}=11.66+83.3/7.77=11.66+10.74=22.40$ N/mm²<23.85 N/mm²

If the long-term values are used, where the transformed area of top bars includes the creep factor $1+\psi_{28}$ of 2.20, the effective value for $Z'_{b,co,ef}$ is now 9.77×10^6 mm³, and is less critical. The compressive flexural stress at the end of the hcu is 150.6/9.77=15.41 N/mm². In fact, the limit could be taken as 0.567 $f_{cki}=25.5$ N/mm² as the prestress is not active.

Finally, the effective value for $Z'_{b,co,ef}$ in the 1200×250 mm solid section between the ends of the hcu is 8.06×10^6 mm³, giving a stress=150.6/8.06=18.68 N/mm²<25.5 N/mm².

Calculation of positive secondary bending moment due to creep and design of bottom bars

Use $P_{pmi}=1010.2$ kN, $P_{po}=838.1$ kN and $\psi_{28}=1.20$. $P_{pmi}-P_{po}=172.1$ kN.

Eq. 13.101 $\Psi=1-e^{-1.2}=0.70$

Eq. 13.98 $FEM_{BA}=-FEM_{CD}=-1.5\times(0.70\times1010.2-172.1)\times82.2\times10^{-3}+4.56\times10.0^2/8=$ $-66.0+57.0=-9.0$ kNm

Eq. 13.99 $FEM_{BC}=-FEM_{CB}=+44.0-38.0=+6.0$ kNm

After distribution the final service and ultimate support moments are

$$M_B=+52.7-45.6=+7.1\text{ kNm}$$

$$M_{EdB}=1.25^*\times7.1=+8.88\text{ kNm}$$

*1.25 is used because Ep. 6.10b is the critical ultimate load combination for all other load cases.

Effective depth to bottom bars (with 20 mm cover from the bottom of the cores) is $d'' = 250 - 35 - 20 -$ say $4 = 191$ mm, then

$K'' = 8.88 \times 10^6/45 \times 1200 \times 191^2 = 0.005$, then $z'' = 0.95\ d''$

$$A_s'' = 8.88 \times 10^6 / 0.95 \times 191 \times 0.87 \times 500 = 112 \text{ mm}^2$$

The area of 3 cores is 60980 mm², then $A_{s,min} = 0.13\% \times 60980 = 79$ mm²

Use 3 no. H8 bars (150 mm²)

The distance to zero moment x' from the support is based on ultimate dead load = 1.25×(4.36+4.56) = 11.15 kN/m, by solving the quadratic $55.75\ x' - 11.15\ x'^2/2 + 8.88 = 0$. Hence $x' = 156$ mm. This is so short the anchorage length should be considered. f_{bdi} is taken for 'poor' casting conditions around rebars in the bottom of cores

Eq. 13.17 $f_{bdi} = 2.25 \times 0.7\ f_{ctdi} = 2.79$ N/mm²

where $f_{ctdi} = 0.7 \times 0.3 \times 45^{2/3}/1.5 = 1.77$ N/mm²

Table 13.3 Summary of fixed end moments, rotational stiffness factors, rotations at supports, negative support moments and positive mid-span moments in interior span for serviceability limit state for the three-span continuous hcu in Example 13.7

		Load case			
		I	*II*	*III*	*IV*
		min-max-min	*max-min-max*	*max-max-min*	*max-max-max*
Stage I w_I	kN/m	4.36	4.36	4.36	4.36
Cores w_{0c}	kN/m	1.49	1.49	1.49	1.49
Length of cores	m	3.82	3.82	3.82	3.82
M_{sI}	kNm	63.2	63.2	63.2	63.2
$w_{2,AB}$	kN/m	4.56	13.56	13.56	13.56
$w_{2,BC}$	kN/m	13.56	4.56	13.56	13.56
$w_{2,CD}$	kN/m	4.56	13.56	4.56	13.56
FEM_{BA}	kNm	57.0	169.5	169.5	169.5
FEM_{BC}	kNm	−113.0	−38.0	−113.0	−113.0
FEM_{CB}	kNm	113.0	38.0	113.0	113.0
FEM_{CD}	kNm	−57.0	−169.5	−57.0	−169.5
θ_I at B	rad.	0.002381	−0.005592	−0.001340	−0.002403
θ_2 at C	rad.	−0.002381	0.005592	−0.001318	0.002403
M_B	kNm	−90.6	−90.6	**−150.6**	−135.6
M_C	kNm	90.6	90.6	75.6	135.6
Reaction F_B	kN	67.8	22.8	75.3	67.8
Shear span a	m	5.000	5.000	5.553	5.000
Total M_{s2}	kNm	78.9	−33.6	58.5	33.9
Total M_s	**kNm**	**142.4**	29.6	121.7	97.1

Rotational stiffness factors $k_{II} = k_{22} = 32923$ kNm/rad. $k_{I2} = k_{2I} = 9406$ kNm/rad.

Critical values in bold.

The breakdown of the maximum MB is −45.6 kNm (imposed dead) and −105 kNm (live load).

Eq. 13.15 $l_{b,rqd}$=0.25×0.87×500×8/2.79=311 mm

Eq. 13.14 α_2=1-0.15 (35+20−8)/8=0.11<0.7

Eq. 13.13 l_{bd}=0.7×311=217 mm

Curtailment length into the cores from the centre of support is $x'+d''$=217+191=408 mm.

Figure 13.32 shows the complete bending moment diagrams for the characteristic design moments M_s, M_s' and M_{sR}.

Discussion - on the service bending moments at mid-span and at the supports of interior and exterior floor spans in three-span and five-span continuous slab arrangement given in Table 13.2. All mid-span moments include M_{s1}=+63.5 kNm for the simply supported self-weight of the hcu and (wet *insitu* concrete) filled cores.

Starting with a single interior span with encastre supports for maximum UDL the support fixed-end moment (FEM)=−113.0 kNm and the mid-span moment $M_s = wL_c^2 / 24 + M_{s1}$=56.5+63.5=+120.0 kNm. The flexibility of continuity at supports in the three-span decreases the support moment to -90.6 kNm (a difference of 22.4 kNm) such that M_s increases to +120.0+22.4=+142.4 kNm (Figure 13.25, Case I). In the exterior span, where one end is simply supported with zero support moment M_s increases to approximately +142.4+0.5×90.6=+187.7 kNm, but because the maximum moment is not a mid-span then M_s=+189.9 kNm (Figure 13.25, Case II). This would be the critical design value to be checked against M_{sR}. The maximum support moment of −150.6 kNm is for the load case III in Figure 13.25. This is 1.33 times the FEM, a ratio that is simply a function of the maximum and minimum UDL's in the three spans. Note that the maximum possible ratio of final moment/FEM is 1.50 for a two-span slab with the maximum UDL in both spans, i.e. $-wL^2/12$ increases to $-wL^2/8$ after moment distribution.

In the five-span slab, moment distribution in the central interior span is now affected by the flexibility of the first and second interior supports, so one would expect a reduction in support moments. Indeed, this decreases the support moment to −71.5 kNm (a difference of 41.5 kNm) such that M_s increases to +120.0+41.5=+161.5 kNm. See Figure 3.4, case 2. By the same procedure as above (for three-span) M_s In the exterior span increases to +187.8 kNm, almost the same value as above (+189.9 kNm, as found in standard moment distribution texts). The maximum support moment of −155.7 kNm is for load case 4 in Figure 3.4. This is 1.38 times the FEM, a ratio that is again a function of the maximum and minimum UDL's in the five spans. As expected, the maximum interior support moment of −136.1 kNm is not critical. It is informative to note that if all five spans are subjected to the maximum UDL (Figure 3.4, case 1 plus stage 1) the maxima are −142.7 kNm and +166.9 kNm, both 10% less than the maximum moments in Table 13.2.

Example 13.8

Calculate M_{Rd} at mid-span for the three-span continuous 250 mm depth prestressed hcu shown in cross section and elevation in Figure 13.31. The section properties, material data, prestress, spans and loads are as given in Example 9.1 and 13.7. Check M_{Rd} against the ultimate positive (sagging) moment M_{Ed} and the negative hogging M_{Ed}'. Plot graphs of bending moment distributions for M_{Rd} and the critical M_{Ed}, taking l_{bpd}=1230 mm (from Example 9.1).

Solution

At mid-span

M_{Rd} is calculated according to Example 9.1 except that the prestrain is ε_{po}=999/195000=0.005123 (0.005234 in Example 9.1) and therefore M_{Rd}=**247.2 kNm** (247.3 kNm in Example 9.1).

The mid-span ultimate moment $M_{Ed1}=1.25\ M_{s1}=79.0$ kNm

Stage 2 calculations are carried out using the same procedures as in Example 13.7 and are summarised in Table 13.4.

From load case I, $M_{Ed,max}=194.5$ kNm$<M_{Rd}=247.2$ kNm

From load case III, the negative moment over the centre of the support $M'_{Ed,max}=-214.4$ kNm. But the moment at the cusp of the 300 mm width support is reduced by $\Delta M'_{Ed}$

The total reaction at support B. $F_{BC}=107.2$ kN (from Table 13.4)$+5.44\times10.0/2+1.87\times3.82=141.5$ kN plus $F_{BA}=19.19\times10.0/2+214.4/10.0=117.4$ kN. Total $F=259.0$ kN.

Eq. 13.96 $\Delta M'_{Ed}=259.0\times0.3/8=+9.7$ kNm

Net $M_{Ed}'=-214.4+9.7=-204.7$ kNm

$b'=1197$ mm and $d'=250-25-$say $12=213$ mm

$K'=204.7\times10^6/(45\times1197\times213^2)=0.0838$

$z'/d'=\min\{0.95;\ 0.5+\sqrt{(0.25-0.0838/1.134)}\}=0.92$

$z'=0.92\times213=196$ mm

$X'=(213-196)/0.4=42.5$ mm

The compression height is $0.8X'=34$ mm$<$bottom flange 35 mm, therefore the section is rectangular

$A'_s=204.7\times10^6/(196\times0.87\times500)=2400$ mm^2

Use 6 no. H25 bars (2946 mm^2)

Note that if 3 no. 32 mm diameter bars were chosen, $d'=209$ mm and $A'_s=2470$ mm^2, 3% over, but also 32 mm is a rather large diameter for this situation.

Negative M'_{Rd} is calculated as follows

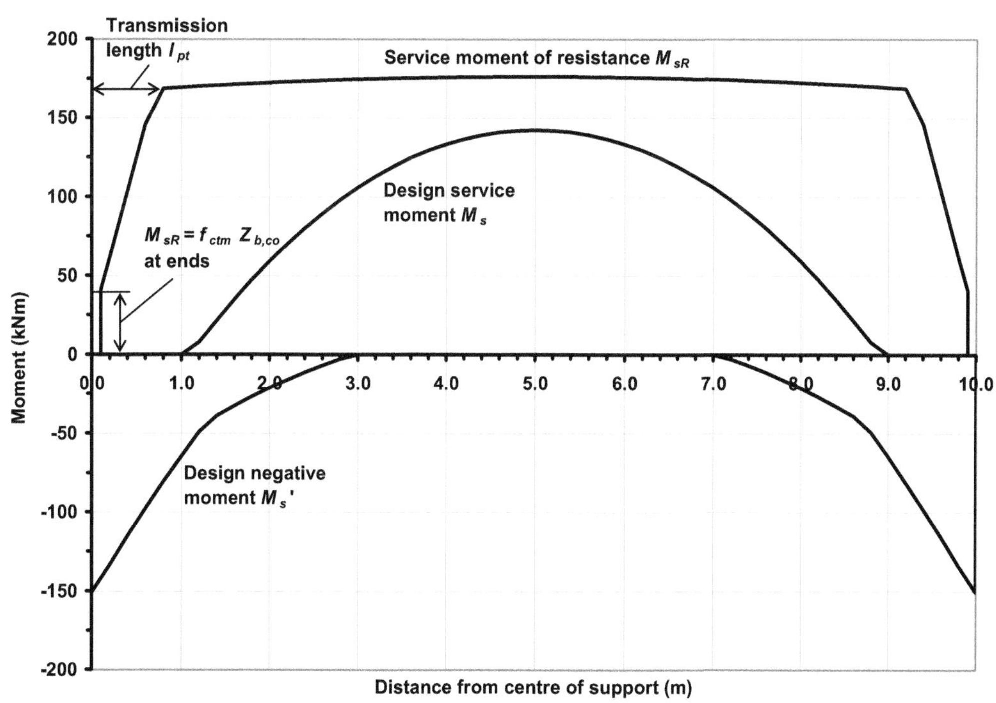

Figure 13.32 Service design bending moments and moments of resistance for the continuous hcu designed in Example 13.7.

$F_s' = 0.87 \times 500 \times 2946 \times 10^{-3} = 1281.5$ kN
$X' = 1281.5 \times 10^3 / (0.567 \times 45 \times 0.8 \times 1197) = 52$ mm
$0.8X' = 42$ mm $> h_{fb}$ 35 mm, but the reduction in width for the 3 open cores is only $3 \times 94 = 282$ mm, such that the change in the net depth to the centroid of compression will be negligible.
$z' = \min\{0.95 \times 213; 213 - 0.4 \times 52\} = 192$ mm
$M'_{Rd} = 1281.5 \times 192 \times 10^{-3} = 246.0$ kNm > 204.7 kNm

As the 6 no. bars are to be curtailed to 3 no., the anchorage length l_{bd}

σ_{sd} = actual design stress $= 0.87 \times 500 \times 2400/2946 = 354$ N/mm²
$f_{bd} = 2.25 \times 1.0 \times 1.0 \times (0.7 \times 3.80/1.5) = 4.00$ N/mm²
$\eta_1 = 1$ for good casting condition $h \leq 250$ mm
$\eta_2 = 1$ for bar $\phi \leq 32$ mm diameter
$l_{bd} = l_{b,rqd} = 0.25 \times 354 \times 25/4.00 = 554$ mm as $\alpha_1 = \alpha_2 = \alpha_3 = \alpha_4 = \alpha_5 = 1$
Curtailment length from centre of support $= l_{bd} + d' = 554 + 213 = 767$ mm

The number of top bars is reduced to 3 no. where $A_s' = 1473$ mm², $F_s' = 640.8$ kN, $x' = 26$ mm, $z' = 202$ mm and $M'_{Rd} = 129.5$ kNm.

This occurs at $x = 670$ mm from the centre of support.
As the bars are fully stressed $l_{bd} = 0.25 \times 0.87 \times 500 \times 25/4.00 = 679$ mm
Curtailment length from centre of support $= x + l_{bd} + d' = 670 + 679 + 213 = 1562$ mm.

The position of zero moment from the centre of support is $x = 3600$ mm, then curtailment length from centre of support $= x + d' = 3600 + 213 = 3813$ mm. The length of the top opened cores is rounded to $l_{core} = 3.82$ m.

Figure 13.33 shows the complete bending moment diagrams for the ultimate design moments M_{Ed}, M'_{Ed}, M_{Rd} and M'_{Rd} and the curtailment point E to the reinforced top opened cores and the curtailment point for reducing the number of top bars.

13.7 NEGATIVE MOMENTS AND TOP REINFORCEMENT FOR FIRE RESISTANCE ≥ R90

Referring to BS EN 1992-1-2, clause 5.7.3 (BS EN 1992-1-2, 2004), for continuous solid slabs, excluding cantilevers, the area and curtailment distances of the top reinforcement for fire resistance ≥ R90 (90 minutes) should extend for a distance $x \leq 0.3\ L_c$. Refer to the Code's clause 5.6.3 (3) and Fig. 5.6 (reproduced in Figure 13.34). There is no requirement for R60 resistance. Although the clause does not specifically refer to hcu, the filled cores and area of topping in composite slabs raise the proportion of solid material close to that of solid units, and therefore the following is recommended.

The area of top reinforcement required at distance x from the support is

Exp. 5.11 $$A_s(x)' = A_s' \text{ at the support} \times (1 - 2.5x / L_c) \quad (13.105)$$

At $x = 0.3\ L_c$, $A_s(x)' = 0.25\ A_s'$ at the support
L_c is the greater effective length of the main or adjacent span.

Subtract the area of mesh per 1.2 m wide unit in composite slabs.

In two-span continuous slabs where no restraint to bending at the end supports is provided either by design or inadvertently in construction, such as tie bars in the ends of hcu lapped to torsionally stiff edge beams, the minimum area of top reinforcement should be $A_s' \geq 0.005\ A_c$. For example in the 250 mm depth hcu in the worked examples $A_s' = 0.005 \times 168000 = 840$ mm².

Table 13.4 Summary of fixed end moments, rotational stiffness factors, rotations at supports, negative support moments and positive mid-span moments in interior span for ultimate limit state for the three-span continuous hcu in Example 13.8

		Load case			
		I	*II*	*III*	*IV*
		min-max-min	*max-min-max*	*max-max-min*	*max-max-max*
Stage 1 w_{Ed1}	kN/m	5.44	5.44	5.44	5.44
Cores w_{Ed0c}	kN/m	1.87	1.87	1.87	1.87
Length of cores	m	3.82	3.82	3.82	3.82
M_{Ed1}	kNm	78.98	79.0	79.0	79.0
$w_{Ed2,AB}$	kN/m	5.69	19.19	19.19	19.19
$w_{Ed2,BC}$	kN/m	19.19	5.69	19.19	19.19
$w_{Ed2,CD}$	kN/m	5.69	19.19	5.69	19.19
FEM_{BA}	kNm	71.2	239.9	239.9	239.9
FEM_{BC}	kNm	−160.0	−47.5	−160.0	−160.0
FEM_{CB}	kNm	160.0	47.5	160.0	160.0
FEM_{CD}	kNm	−71.2	−239.9	−71.2	−239.9
θ_1 at B	rad.	0.003775	−0.008185	−0.001806	−0.003401
θ_2 at C	rad.	−0.003775	0.008185	−0.002180	0.003401
$\mathbf{M_{Ed,B}}$	kNm	−124.4	−124.4	**−214.4**	−191.9
$M_{Ed,C}$	kNm	124.4	124.4	101.9	191.9
Reaction $F_{Ed,BC}$	kN	96.0	28.5	107.2	96.0
Shear span a_{Ed}	m	5.000	5.000	5.586	5.000
Total M_{Ed2}	kNm	115.5	−53.3	85.0	48.0
Total $\mathbf{M_{Ed}}$	**kNm**	**194.5**	25.7	164.0	127.0

Rotational stiffness factors $k_{11}=k_{22}=32923$ kNm/rad. $k_{12}=k_{21}=9406$ kNm/rad.

Critical ultimate load combination according to BS EN 1990 is Exp. 6.10b. Critical values in bold.

13.8 DEFLECTIONS IN CONTINUOUS SLABS DUE TO GRAVITY LOADS AND NEGATIVE MOMENTS

13.8.1 Deflection in main span in non-composite units

These are calculated in the same manner as for simply supported spans, as given in Section 11.2.1, at least in terms of the use of creep factors ψ_1, ψ_{28} and 0.8 ψ_∞, with the addition of negative (upward) deflections due to stage 2 fixed end moments. For the latter load case I is used, with maximum loads in the interior span and minimum imposed dead loads in the exterior spans. For multiple spans, e.g. five spans, the alternate loading pattern is used. The camber and deflection due to self-weight of the psc unit, and any wet *insitu* topping, is based on the effective span L. All other long-term deflections and movement after finishes are based on the centres of supports, i.e. continuity span L_c.

As expected from the magnitude of negative continuity moments, deflections in continuous spans are rarely critical, as the design of interior span is clearly limited by the magnitude of the support moments. Even where the exterior spans are relatively shorter than the interior, say ½ of the interior span, deflections are not critical. For example in the continuous slab design in Example 13.7 the upward immediate (no creep or quasi-permanent ψ_2 factors)

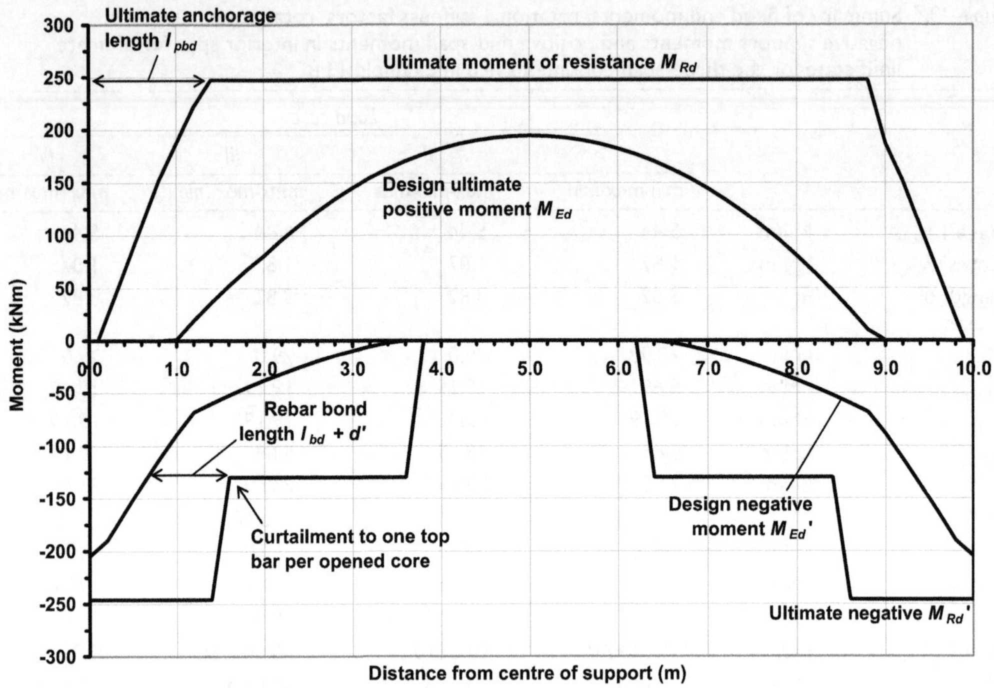

Figure 13.33 Ultimate design bending moments and moments of resistance for the continuous hcu designed in Example 13.8. Excludes the secondary positive moments due to creep of the prestressed unit.

deflection due to support moments of –90.6 kNm is $-ML^2/8\ EI = -90.6 \times 10^6 \times 10000^2/8 \times 36283 \times 1296 \times 10^6 = -24$ mm. This is about ⅔ of the downward deflection due to gravity loads $5wL^4/384\ EI = 5 \times 13.56 \times 10000^4/384 \times 36283 \times 1296 \times 10^6 = +37$ mm, so the net deflection is only +13 mm.

Another difference compared with simply supported spans is that the visco-elastic increase in upward camber after installation is reduced by the positive continuity restraint based on the creep growth function Ψ in Eqs. 13.99–13.101. The restraint is not an external moment, but a virtual restraint of the internal moment $[\Psi\ P_{pmi} - (P_{pmi} - P_{po})]\ z_{cp}$. The fixed end moments that develop due to this force, given by Eqs. 13.100 and 13.101 are distributed to give virtual positive support moments M_B and M_C that have the equivalent effect of reducing the increase in camber by $+M_B\ L_c^2/16\ E_{cm}\ I_{c.co}$ plus $+M_C\ L_c^2/16\ E_{cm}\ I_{c.co}$. For example in the continuous slab design in Example 13.7 $M_B = -M_C = +54.9$ kNm, and so the restraining deflection is $+54.9 \times 10^6 \times 10000^2/8 \times 36283 \times 1296 \times 10^6 = +14$ mm.

Long-term deflections after the *insitu* concrete in the top opened cores has hardened are based on the second moment of area $I_{c,co}$, which should include the section properties of the reinforced filled cores, at least to the end of the filled cores. For example in Example 13.7 with 3 no. filled cores the gross second moment of area $I_{c,co} = 1602 \times 10^6$ mm^4, an increase of 23% over the basic hcu section over a length of 0.38×span.

13.8.2 Calculations for camber and long-term deflection

Calculations for camber and self-weight at transfer and installation are the same as in Section 11.2.1, Eqs. 11.1–11.5.

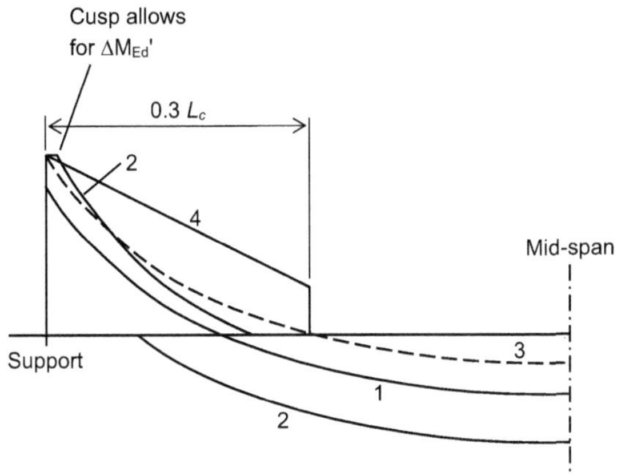

Key to bending moment diagrams

1. For actions in a fire situation at time $t = 0$
2. To be resisted by tensile reinforcement according to BS EN 1992-1-1, e.g. min-max-min load case I
3. In fire conditions
4. Resisting moments according to area of tensile reinforcement in Eq. 13.105 (from BS EN 1992-1-2, Exp. 5.11)

Figure 13.34 Envelope of resisting bending moments over supports for fire conditions, adapted from BS EN 1992-1-2, Fig. 5.6.

Calculations for the long-term camber are from Eqs. 11.6 and 11.7, minus the restraint by the creep growth function Ψ for the internal moment $[\Psi\ P_{pmi}-(P_{pmi}-P_{po})]\ z_{cp}$ in Eqs. 13.99–13.101. After distribution, the restraint moments are $+M_B$ and $+M_C$ and the equivalent downward camber for each of these moments is

$$\delta_{5a}=+M_B\ L_c^2/16\ E_{cm}\ I_{c,co} \tag{13.106a}$$

If the support moments are equal $\delta_{5a}=+M\ L_c^2/8\ E_{cm}\ I_{c,co}$ (13.106b)

Then δ_{5a} is added to $\delta_4+\delta_5$ in Section 11.2.1.

Long-term deflection due to self-weight, infill in joints and top opened cores, imposed dead and live loads are from Eqs. 11.8 to 11.10 minus the restraint from end moments $-M_B$ and $-M_C$ which have to be calculated individually for self-weight (including filled cores) $-M_{B1}$, dead $-M_{B2}$ and quasi-permanent live loads ψ_2-M_{B3}. All deflections are based on the centres of bearing continuity span L_c.

$$\delta_{6a}=\left[\psi_{28}-M_{B1}+\left(1+\psi_{28}\right)-M_{B2}+\left(1+0.8\ \psi_{\infty}\right)\ 5\ \psi_2-M_{B3}\right]L_c^2\ /\ 16E_{cm}I_{c,co} \tag{13.107a}$$

If the support moments are equal the denominator in Eq. 13.107a is 8 (13.107b)

The additional term for the visco-elastic deflection of filled cores, of self-weight w_{0c} over a length l_{core} from the supports is

$$\delta_{6b}=\psi_{28}\ w_{0c}l_{core}^2\left(3L_c^2-2l_{core}^2\right)/\ 48E_{cm}I_{c,co} \tag{13.108}$$

Then δ_{6a} and δ_{6b} are added to $\delta_6+\delta_7+\delta_8$ in Section 11.2.1.

Long-term movement due to visco-elastic self-weight and dead load deflections after finishes are applied is the same as in Section 11.2.1, minus the restraint from end moments $-M_B$ and $-M_C$ which have to be calculated for self-weight $-M_{B1}$, dead $-M_{B2}$ and quasi-permanent live loads $\psi_2 - M_{B3}$.

$$\delta_{13a} = \left[(1+\psi_{28})\ \psi_2 - M_{B3}\right] L_c^2 / 16 E_{cm} I_{c,co} \tag{13.109a}$$

If the support moments are equal the denominator in Eq. 13.109a is 8 (13.109b)

or $$\delta_{14a} = \delta_{13a} + \left[\psi_{28}\left(-M_{B1} + -M_{B2}\right)\right] L_c^2 / 16 E_{cm} I_{c,co} \tag{13.110a}$$

If the support moments are equal the denominator in Eq. 13.110a is 8 (13.110b)

Then δ_{13a} and δ_{14a} are added to δ_{13} and δ_{14} in Section 11.2.1.

Example 13.9

Calculate the long-term deflection and the deflection after applying finishes for the 250 mm depth prestressed hcu shown in cross section and elevation in Figure 13.31. The section properties, material data, prestress, spans and loads are from Example 13.7. The net deflections at transfer and at installation of the precast unit are from Example 11.1. The floor slab is for use in offices. Check the Code's limits if the floor finishes are brittle.

Solution

From Example 13.7, considering the prestress at the support the final pretensioning force is P_{po}=838.1 kN. Then $P_{pmi}-P_{po}$=172.1 kN

From Example 11.1, the net deflection at transfer=−12.9 mm, and net (upward) deflection at installation=−40.9+19.8=−21.1 mm.

Long-term deflections. After the *insitu* concrete between the ends of the prestressed hcu and in the top opened cores has hardened, the increase in camber is

Eq. 11.7 δ_5=−40.9−{[(1.20×1010.2×10³)−172.1×10³]×82.2×10000²/(8×36283×1296×10⁶)}=−40.9−22.7=−63.6 mm

Eq. 13.101 Ψ=1−e$^{-1.2}$=0.70

Eq. 13.98 FEM$_{BA}$=−FEM$_{CD}$=−1.5×(0.70×1010.2−172.1)×82.2×10^{-3}=1.5×−44.0=−66.0 kNm

Eq. 13.99 FEM$_{BC}$=−FEM$_{CB}$=+44.0 kNm

After distribution the final secondary moments at the support is M_B=−M_C=+52.7 kNm

Eq. 13.106b δ_{5a}=+52.7×10⁶×10000²/8×36283×1296×10⁶=+14.0 mm

Net camber δ_5=−63.6+14.0=−49.6 mm

Long-term deflection due to the restraint of self-weight, infill in joints and top opened cores, imposed dead and live loads.

Table 13.5 gives the support moments for M_{B1}=−43.6−9.7=−53.3 kNm, M_{B2}=−45.6 kNm and M_{B3}=−45.0 kNm

Eq. 13.107 δ_{6a}=[1.20×−53.3+(1+1.20)×−45.6+(1+2.00)×0.3×−45.0]×10⁶×10000²/8×36283×1296×10⁶=−54.4 mm

Eq. 11.5a, Eqs. 11.8 to 11.10 $\delta_4+\delta_6+\delta_7+\delta_8$=19.8+[(1.20×4.36+(1+1.20)×4.56+(1+2.00)×2.25]×[5×10000⁴/(384×36283×1296×10⁶)]=19.8+60.9=+80.7 mm

Visco-elastic deflection due to filled cores of self-weight w_{0c}=1.2×1.245=1.50 kN/m and length l_{core}=3.82 m

Eq. 13.108 δ_{6b}=1.20×1.50×3820² (3×10000²-2×3820²)/48×36283×1296×10⁶=+3.1 mm

Final deflection δ_{10}=−49.6+80.7+3.1−54.4=**−20.2 mm**<10000/250=40 mm

Long-term movement is the greater of δ_{13} or δ_{14}.

Eq. 11.12 $\delta_{13}=5\times(1+1.20)\times0.3\times9.0)\times10000^4/384\times36283\times1296\times10^6=+16.4$ mm
Eq. 13.109b $\delta_{13a}=[(1+1.20)\times0.3\times\text{-}45.0]\times10^6\times10000^2/8\times36283\times1296\times10^6=-7.9$ mm
Net $\delta_{13}=+16.4-7.9=+8.5$ mm
Eq. 11.13 $\delta_{14}=16.4+[5\times1.20\times(4.36+4.56)\times10000^4/384\times36283\times1296\times10^6]+(-49.6-40.9)=+16.4+29.6-8.7=+37.3$ mm plus 3.1 mm for the filled cores $=+40.4$ mm
Eq. 13.110b $\delta_{14a}=-7.9-[1.20\times(-53.3-45.6)]\times10^6\times10000^2/8\times36283\times1296\times10^6=-7.9-31.5=-39.4$ mm
Net $\delta_{13}=+40.4-39.4=+1.0$ mm $<\delta_{13}=+8.5$ mm $<<$ $10000/500=20$ mm.

Example 13.10

Calculate the crack width in the top of the hcu in Examples 13.7 and 13.8. Cover to top bars $=25$ mm. $E_s=200000$ N/mm^2. Use $\psi_1=0.5$ for office loading. $\psi_{28}=1.20$ for long-term creep factor after hardening of *insitu* infill.

Solution

Refer to Section 11.3 for general procedures, except the stress in the rebars is based on the frequent value $M\ (d-y_{b,cr})/I_{c,co}$ for the cracked section.

$\alpha_e=200000/36283=5.51$
Long-term $m=5.51\times(1+1.20)=12.127$

In the cracked section the transformed area of top bars is

$A_s'\ m=2946\times12.127=35716$ mm^2

Height to centroid from bottom of flexurally cracked section $y_{b,cr}=86.8$ mm
Second moment of area $I_{c,co}=782.9\times10^6$ mm^4
From Table 13.3, for load case III, maximum frequent negative moment at support $=-45.6\times0.5\times-105.0=-98.1$ kNm
Stress in top bars $\sigma_s=12.127\times98.1\times10^6\times(213-86.8)/782.9\times10^6=191.8$ N/mm^2

$A_{ct,eff}=1197\times(250-86.8)/3=65120$ mm^2
$\rho_{eff}=A_s/A_{ct,eff}=2946/65120=0.0452$
$(1+\alpha_e\ \rho_{eff})=1+5.51\times0.0452=1.249$

Eq. 11.34b $\varepsilon_{sm}-\varepsilon_{cm}=[191.8-(0.4\times3.80\times1.249/0.0452]/200000=749\times10^{-6}$
But at least $\varepsilon_{sm}-\varepsilon_{cm}=0.6\times191.8/200000=575\times10^{-6}$
Eq. 11.37 $s=3.4\times25+(0.8\times0.5\times0.425\times25/0.0452)=179$ mm
where $k_1=0.8$ for high tensile rebars
Eq. 11.33 $w_k=749\times10^{-6}\times179=$ **0.13 mm** <0.3 mm

Example 13.11

Calculate negative support moments M'_{Ed} and M'_{Rd} at the supports for the three-span continuous 250 mm depth prestressed hcu made composite with a 75 mm depth topping shown in cross section and elevation in Figure 13.35. The section properties, material data, prestress, spans and loads are as given in Example 13.8, except the screed is replaced by the topping. The precamber of the hcu at installation may be taken from Example 13.9 as -21 mm. The self-weight of the topping is therefore 2.14 kN/m^2. Use $f_{ckT}=25$ N/mm^2 for the *insitu* concrete topping.

Calculate the requirements for steel mesh and additional rebars in the topping, and for the secondary positive reinforcement allowing for creep of the psc unit.

Table 13.5 Summary of fixed end moments, rotational stiffness factors, rotations at supports, negative support moments in interior span for self-weight, and imposed dead and live loads for the three-span continuous hcu in Example 13.9

		Load case I			
		Self-weight pcu	*Filled cores*	*Imposed dead*	*Imposed live*
$w_{2,AB}$	kN/m	4.36	1.49	4.56	0
$w_{2,BC}$	kN/m	4.36	1.49	4.56	9.00
$w_{2,CD}$	kN/m	4.36	1.49	4.56	0
FEM_{BA}	kNm	54.5	12.2	57.0	0
FEM_{BC}	kNm	−36.3	−8.1	−38.0	−75.0
FEM_{CB}	kNm	36.3	8.1	38.0	75.0
FEM_{CD}	kNm	−54.5	−12.2	−57.0	0
θ_1 at B	rad.	−0.000773	−0.000173	−0.000808	0.003189
θ_2 at C	rad.	0.000773	0.000173	0.000808	−0.003189
M_B	kNm	−43.6	−9.7	−45.6	−45.0
M_C	kNm	43.6	9.7	45.6	45.0

Rotational stiffness factors $k_{11}=k_{22}=32923$ kNm/rad. $k_{12}=k_{21}=9406$ kNm/rad.

Solution

To determine the number and length of top opened cores, first calculate the positive secondary bending moment due to creep and design the bottom rebars. From Example 13.7, Eq. 13.98 $FEM_{BA}=-FEM_{CD}=-1.5\times(0.70\times1010.2-172.1)\times82.2\times10^{-3}=-66.0$ kNm

The imposed dead load=1.2×2.0=2.4 kN/m. Then the net $FEM_{BA}=-FEM_{CD}=-66.0+2.40\times10.0^2/8=-36.0$ kNm

Eq. 13.99 $FEM_{BC}=-FEM_{CB}=+44.0-20.0=+24.0$ kNm

After distribution the final service and ultimate support moments are $M_B=+28.7$ kNm and $M_{EdB}=1.25\times28.7=+35.9$ kNm

Effective depth to bottom rebars (with 20 mm cover from the bottom of the cores) is $d''=250+75-35-20-8=262$ mm, then

$$K''=35.9\times10^6/25\times1200\times262^2=0.0175\text{, then } z''=0.95\,d''$$
$$A_s''=35.9\times10^6/0.95\times262\times0.87\times500=331\text{ mm}^2$$

Use 2 no. H16 bars (402 mm²)

The self-weight of 2 top opened cores=0.83 kN/m², or ultimate self-weight=1.24 kN/m per unit width.

The total ultimate dead load=1.25×1.2×(3.63+2.14+2.00)=11.65 kN/m.

The distance to zero moment x' from the support is by solving the quadratic $58.25\,x'-11.65\,x^2/2+35.9=0$. Hence $x'=582$ mm. Curtailment length into cores $x'+d''=582+191=773$ mm, rounded to 800 mm.

Stage 2 calculations are carried out using the same procedures as in Example 13.8 and are summarised in Table 13.6.

Note that for the composite slab, from Example 12.1, $I_{c,c,co}=2819.9\times10^6$ mm⁴. Then

$$E_{cm}\,I_{c,c,co}=36283\times2819.9\times10^6=102.3\times10^{12}\text{ Nmm}^2$$
$$k_{11}=k_{22}=3\times102.3\times10^{12}/10000+4\times102.3\times10^{12}/10000=71600\text{ kNm/rad}$$
$$k_{12}=k_{21}=2\times102.3\times10^{12}/10000/10^6=20460\text{ kNm/rad.}$$

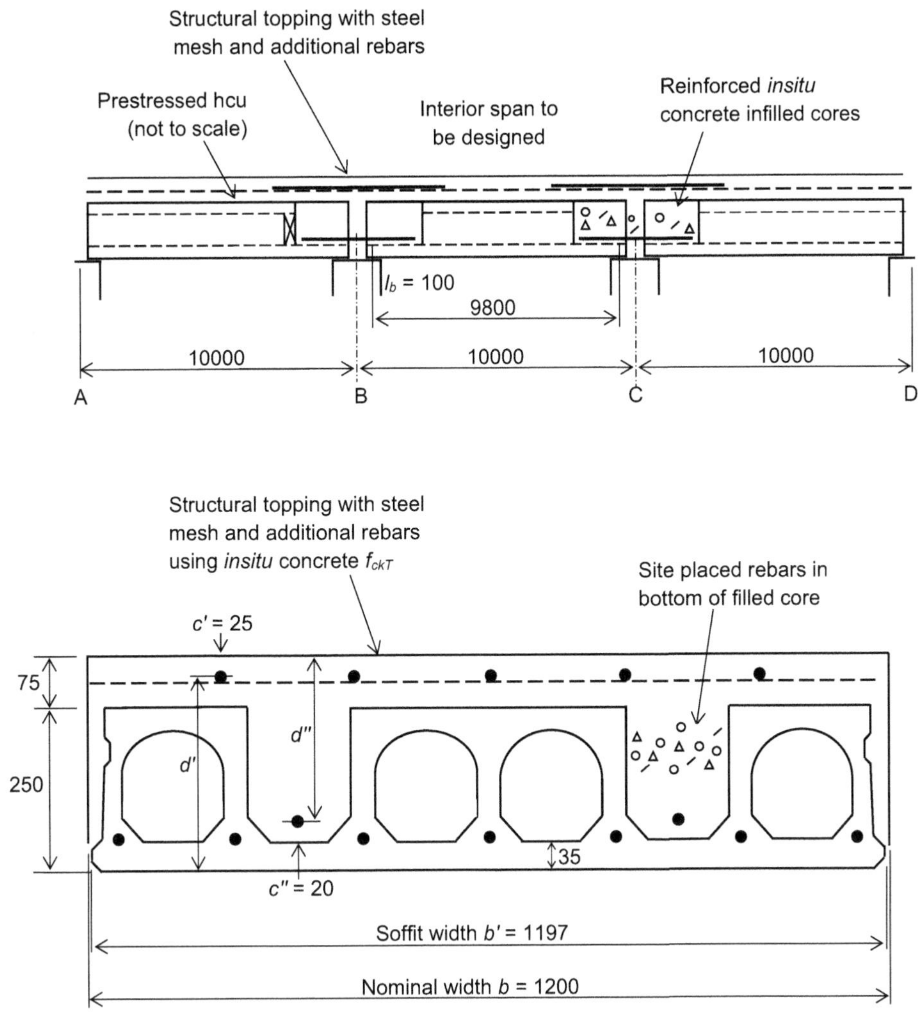

Figure 13.35 Elevation and cross section of the three-span continuous composite prestressed concrete hcu designed in Example 13.11, showing the position and extent of steel mesh and rebars in the topping and filled cores.

From load case III, the negative moment over the centre of the support $M'_{Ed,max}$=−187.5 kNm. But the moment at the cusp of the 300 mm width support is reduced by $\Delta M'_{Ed}$

The total ultimate reaction at support B. F_{BC}=93.7 kN (from Table 13.6)+8.65×10.0/2+1.24×0.8=137.9 kN plus F_{BA}=16.5×10.0/2+187.5/10.0+1.24×0.8*=102.2 kN. Total F=240.1 kN. *assume cores opened same in span AB.

Eq. 13.96 $\Delta M'_{Ed} = 240.1 \times 0.3/8 = +9.0$ kNm

Net $M'_{Ed} = -187.5 + 9.0 = -178.5$ kNm

$b' = 1197$ mm and $d' = 250 + 75 - 25 -$ say $10 = 290$ mm

$K' = 178.5 \times 10^6/(45 \times 1197 \times 290^2) = 0.0394$

$z'/d' = \min\{0.95;\ 0.5 + \sqrt{(0.25 - 0.0394/1.134)}\} = \min\{0.95;\ 0.964\} = 0.95$

$z' = 0.95 \times 290 = 275.5$ mm

$A'_s = 178.5 \times 10^6/(275.5 \times 0.87 \times 500) = 1490$ mm^2

Table 13.6 Summary of fixed end moments, rotational stiffness factors, rotations at supports, negative support moments and positive mid-span moments in interior span for ultimate limit state for the three-span continuous composite hcu in Example 13.11

		Load case			
		I	*II*	*III*	*IV*
		min-max-min	*max-min-max*	*max-max-min*	*max-max-max*
Stage I w_{EdI}	kN/m	8.65	8.65	8.65	8.65
Cores w_{Ed0c}	kN/m	1.24	1.24	1.24	1.24
Length of cores	m	0.80	0.80	0.80	0.80
M_{EdI}	kNm	108.5	108.5	108.5	108.5
$w_{Ed2,AB}$	kN/m	3.00	16.50	16.50	16.50
$w_{Ed2,BC}$	kN/m	16.50	3.00	16.50	16.50
$w_{Ed2,CD}$	kN/m	3.00	16.50	3.00	16.50
FEM_{BA}	kNm	37.5	206.2	206.2	206.2
FEM_{BC}	kNm	−137.5	−25.0	−137.5	−137.5
FEM_{CB}	kNm	137.5	25.0	137.5	137.5
FEM_{CD}	kNm	−37.5	−206.2	−37.5	−206.2
θ_1 at B	rad.	0.001956	−0.003544	−0.000611	−0.001344
θ_2 at C	rad.	−0.001956	0.003544	−0.001222	0.001344
$\mathbf{M_{Ed,B}}$	kNm	−97.5	−97.5	**−187.5**	−165.0
$M_{Ed,C}$	kNm	97.5	97.5	75.0	165.0
Reaction $F_{Ed,BC}$	kN	82.5	15.0	93.7	82.5
Shear span a_{Ed}	m	5.000	5.000	5.682	5.000
Total M_{Ed2}	kNm	108.7	−60.0	78.8	41.2
Total M_{Ed}	**kNm**	**217.3**	48.5	187.3	149.8

Rotational stiffness factors $k_{11}=k_{22}=71600$ kNm/rad. $k_{12}=k_{21}=20460$ kNm/rad.

Critical ultimate load combination according to BS EN 1990 is Exp. 6.10b. Critical values in bold.

For the steel mesh in the topping, $A'_{s,min}=\max\{0.13\%\times1000\times75;\ 0.26\times2.56\times1000\times75/500\}=100\,\text{mm}^2/\text{m}$. Use A142 mesh

Area of additional bars $=1490-1.2\times142=1320\,\text{mm}^2$ per unit width

Use 7 no. H16 bars (1407 plus 170 = 1577 mm²)

Negative M'_{Rd} is calculated as follows

$$F'_s=0.87\times500\times1577\times10^{-3}=686.0\text{ kN}$$
$$x'=686.0\times10^3/(0.567\times45\times0.8\times1197)=28\text{ mm}$$
$$z'=\min\{0.95\times290;\ 290-0.4\times28\}=275.5\text{ mm}$$
$$M'_{Rd}=686.0\times275.5\times10^{-3}=189.0\text{ kNm}>178.5\text{ kNm}$$

Curtailment of additional bars where $M'_{Ed}=M'_{Rd,mesh}$ for the mesh only is

$$F'_s=0.87\times500\times170\times10^{-3}=74.0\text{ kN}$$
$$z'=0.95\times290=275.5\text{ mm}$$
$$M'_{Rd}=74.0\times275.5\times10^{-3}=20.4\text{ kNm}$$

The distance from the support is based on load case II (max-min-max) where M'_{Ed} is small. For this load case w_{Ed}=11.65 kN/m, F_{Ed}=58.2 kN, and M_B=$-M_C$=−97.5 kNm. Using these data the distance to where M'_{Ed}=−20.4 kNm is x=1.570 m. The curtailment distance for the H16 bars from the centre of support is $x+d'$=1570+290=1860 mm.

The designer may wish to curtail 3 no. H16 bars to reduce the effect of a sudden change in stiffness at this point. The number of bars is reduced to 4 no. where A'_s = 804 mm^2, F'_s = 349.7 kN, X'=14 mm, z'=275.5 mm and M'_{Rd}=96.3 kNm. This occurs at x=700 mm from the centre of support, and as the bars are fully stressed l_{bd}=0.25×0.87×500×16/2.69=646 m m. The curtailment length from centre of support=$x+l_{bd}+d'$=700+646+290=1636 mm.

REFERENCES

BS EN 1990. 2002. Eurocode 0- Basis of structural design, +A1:2005, BSI, London, UK.

BS EN 1992-1-1. 2004. Eurocode 2: Design of concrete structures - Part 1-1: General rules and rules for buildings, +A1:2014, BSI, London, UK.

BS EN 1992-1-2. 2004. Eurocode 2: Design of concrete structures, Part 1-2: General rules - Structural fire design, +A1:2019, BSI, London, UK.

Elliott, K. 2017. Elliott, K. S. 2017. *Precast Concrete Structures*, 2nd ed., CRC, Taylor & Francis, London, UK, 694 p.

Kleinlogel, A. 1952. *Rigid Frame Formulas*, Frederick Ungar Publishing Company, New York.

FIB. 2000. *Bulletin 6, Special Design Considerations for Precast Prestressed Hollow Core Floors, Guide to Good Practice*, Fédération Internationale du Béton, Lausanne, Switzerland, 180 p.

NA to BS EN 1990. 2002. UK National Annex for Eurocode 0- Basis of structural design, +A1:2005, BSI, London, UK.

NA to BS EN 1991-1-1. 2002, Eurocode 1: Actions on Structures - Part 1-1: General Actions - Densities, self-weight, imposed loads for buildings, +A1:2019, BSI, London, UK.

NA to BS EN 1992-1-1. 2004. UK National Annex for Eurocode 2: Design of concrete structures - Part 1-1: General rules and rules for buildings, +A1:2014, BSI, London, UK.

Chapter 14

Precast prestressed concrete walls

14.1 APPLICATIONS OF PRESTRESSED CONCRETE WALLS

Precast prestressed concrete walls are used in a variety of applications and building types. For example: retaining walls, boundary walls, fire walls, blast walls and external envelope walls. The wall panels can be used with steel, *insitu* or precast framed structures, most commonly in industrial and agricultural buildings. Prestressed walls are often an economic solution due to the reduction in reinforcement quantities, reduction in concrete usage when hollow core walls (hcw*) are used and the efficient long line production methods available. The walls can be designed as horizontal spanning, vertical spanning and as cantilevers. Precast tanks can also be constructed above and/or below ground for liquid storage using prestressed wall panels – this is a common application for slurry storage in the agricultural industry. This chapter focuses on prestressed design for soil and grain retaining walls, boundary walls and non-loadbearing walls in buildings.

*the abbreviation hcw for hollow core walls is used to distinguish between these and hollow core units (hcu) used in floors and roofs.

14.2 RETAINING WALLS

Prestressed wall panels are widely used to retain soil, grain, aggregates and many other materials. When designing walls to retain materials, it is very important to note the effect that heaped material has on the pressure applied to the walls. The intention to store heaped material needs to be clarified and the level of overfill above the top of wall was agreed upon during the early design stage.

The wall panels require a medium/high axial prestress with a relatively small eccentricity to control sideways camber at installation in the absence of an opposing deflection due to self-weight. To avoid any tensile stress, the eccentricity z_{cp} should lie within the middle third of the thickness h of the section, i.e. $z_{cp} \leq h/6$.

14.2.1 Vertical spanning retaining wall

In the following Example 14.1 a solid prestressed wall was used to retain the backfill material at one level of an underground car park. The wall spanned vertically from an *insitu* capping beam to the ground floor level hollow core floor slab. Figure 14.1 shows the location of the walls within the structure. The corbel at the top of the wall was added with a

DOI: 10.1201/9781003319450-14

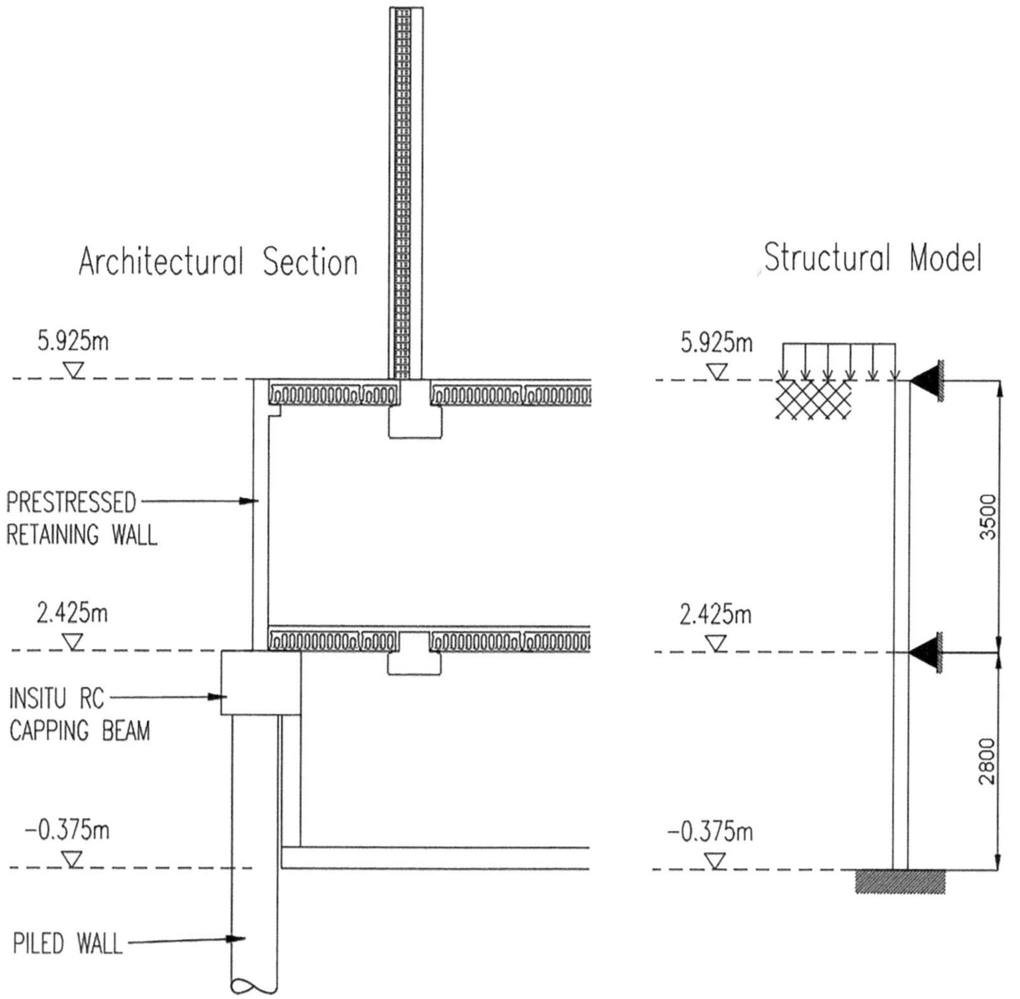

Figure 14.1 Vertical spanning prestressed retaining wall.

secondary pour onto the fill face. The corbel reinforcement was tied into the mould with the prestressed strand but also projected above the fill level of the mould. The concrete for the main body of the wall was poured, then a shutter was set onto the filled mould and the corbel was poured. In some locations, the hcu spanned onto this corbel for support. The eccentricity of this load and the moment generated were ignored in the design as it provides a beneficial effect by, very slightly, reducing the moment on the wall stem when under load from the retained material (Figures 14.2 and 14.3).

The wall was designed as simply supported and restrained at its top and bottom by the floor diaphragm which was constructed from a 250 mm depth hollow core units (hcu) with a 75 mm depth structural topping. The horizontal loading applied to the retaining walls was transferred through the floor diaphragm into the shear wall stability system for the main structure. The effective span was taken as 3.50 m, with no reduction taken for the bearing length when calculating the shear forces and moments.

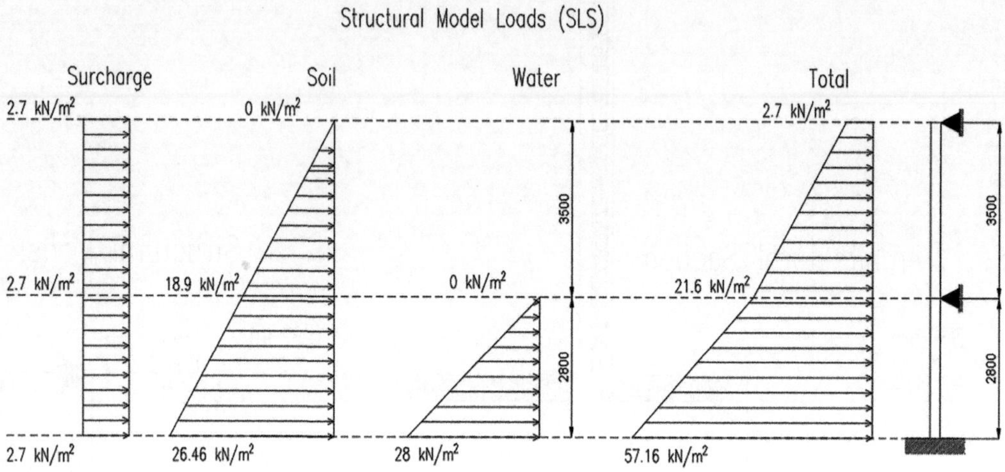

Figure 14.2 Structural model loading.

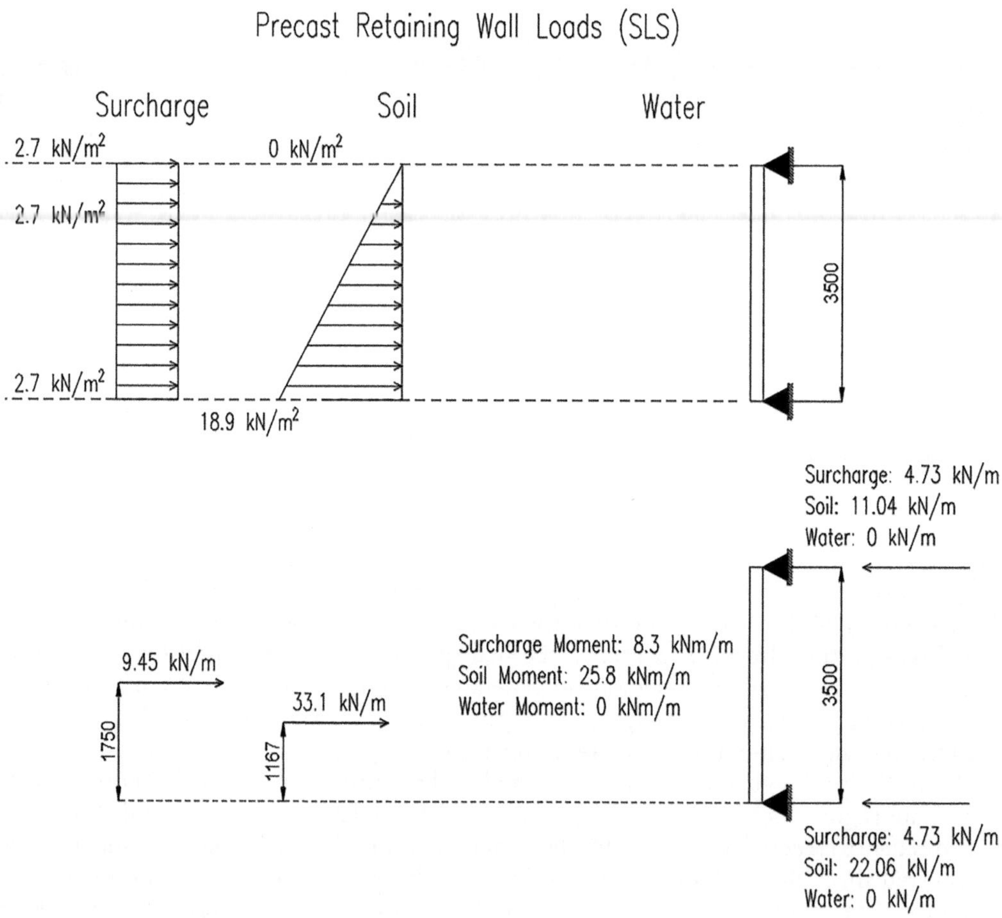

Figure 14.3 Loading on prestressed wall.

Saturated density of retained material $\gamma_{sr}=20$ kN/m^3
Taking a conservative approach, assume that the moist density is equal to the saturated density.
Variable surcharge load $s_l=10$ kN/m^2
Soil type: Medium-dense well-graded sand
Effective angle of shearing resistance $\phi'_{r,k}=35.1°$ is provided by client

Using Rankine's theory:

Active pressure coefficient $K_a = (1-\sin(\phi'_{r.k}))/(1+\sin(\phi'_{r.k}))=0.27$
Surcharge force $F_{sur}=s_l\,K_a \times 3.5$ m$=9.5$ kN/m
Surcharge force at top and bottom of wall $F_{sur,t\&b}=F_{sur}/2=4.7$ kN/m
Soil force $F_{soil}=\gamma_{sr}\,K_a \times (3.5\text{ m})^2/2=33.1$ kN/m
Soil force at bottom of wall $F_{soil,b}=F_{soil}\times 2.3$ m/3.5 m$=21.8$ kN/m
Soil force at top of wall $F_{soil,t}=F_{soil}\times 1.2$ m/3.5 m$=1$ 1.4 kN/m
Using partial safety factors for soil $\gamma_G=1.35$ and surcharge $\gamma_Q=1.5$
Ultimate linear shear force per m run
at base of wall $v_{Ed}=1.35\,F_{soil,b}+1.5\,F_{sur,t\&b}=36.9$ kN/m
at top of wall $v_{Ed}=1.35\,F_{soil,t}+1.5\,F_{sur,t\&b}=22.0$ kN/m
Linear moment per m run
due to surcharge $m_{sur}=F_{sur}\times 3.5$ m/2$=8.3$ kNm/m
due to soil $m_{soil}=F_{soil}\times 1.2$ m$\times 2.3$ m/3.5 m$=25.8$ kNm/m
Total service linear moment $m_s=M_{sur}+M_{soil}=34.1$ kNm/m
Ultimate linear moment $m_{Ed}=1.35\,M_{sur}+1.5\,M_{soil}=47.3$ kNm/m

Design of the prestressed wall element
The prestressed wall panels are manufactured 2400 mm wide

Maximum service moment $M_s=34.1$ kNm/m$\times 2.4$ m width$=81.8$ kNm
Maximum ultimate moment $M_{Ed}=47.3\times 2.4$ m$=113.5$ kNm
Ultimate shear at base of wall $V_{Ed}=36.9$ kN/m$\times 2.4$ m width$=88.5$ kN
Ultimate shear at top of wall $V_{Ed}=22.0\times 2.4=52.8$ kN
The exposure class for the wall panel in the car park is XC3.

Referring to the established procedures for the serviceability and ultimate moments of resistance, and the ultimate shear capacity given in Sections 6.2.2, 6.2.3, 7.2.4, 9.3.1, 10.2.2 and 10.2.3, the resistance provided by the prestressed wall panel is as follows.

Example 14.1

Calculate the serviceability and ultimate moments of resistance M_{sR} and M_{Rd} for the 200 mm thickness prestressed solid wall panel shown in elevation in Figure 14.1 and in cross-section in Figure 14.4. The initial prestressing force may be taken as 0.7 of characteristic strength of 'standard' 7-wire helical strand of $f_{pk}=1770$ N/mm^2. Manufacturer's data gives relaxation Class 2 detensioned at 20 hours after 50°C mean temperature curing. Relative humidity at transfer and in service$=70\%$. Geometric and material data given by the manufacturer are as follows:

Area$=476826$ mm^2; $I_c=1593.72\times 10^6$ mm^4; $y_b=100$ mm; $f_{ck}=40$ N/mm^2; $f_{ck}(t)=28$ N/mm^2; $f_{pk}=1770$ N/mm^2; $E_p=195$ kN/mm^2; $A_p=52$ mm^2 per 9.3 mm dia. strand; cover to strand$=30$ mm. Gravel aggregates. Cement CEM I grade 52.5R.

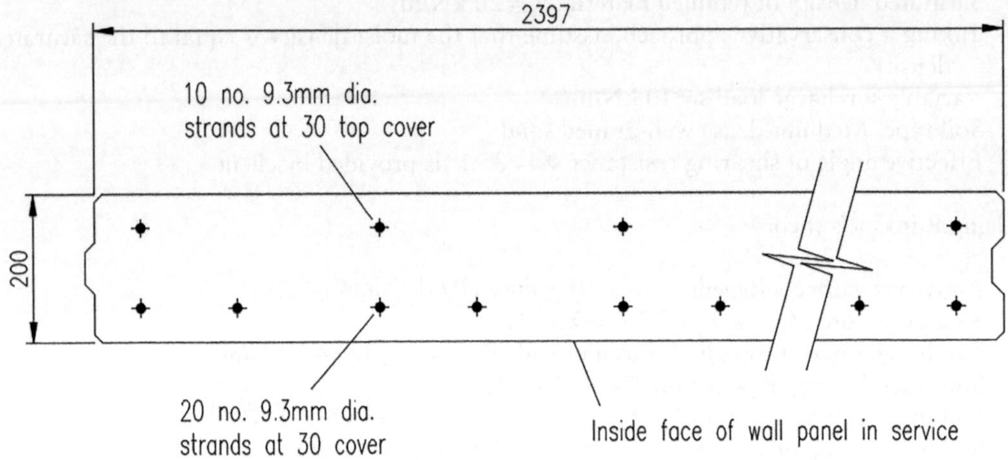

Figure 14.4 Cross-section of nominally 2400 mm width × 200 mm depth prestressed concrete wall panel to Examples 14.1 and 14.2.

Solution

Material properties

$f_{cm}=40+8=48$ N/mm^2; $f_{cm}(t)=28+8=36$ N/mm^2

$E_{cm}=22\times(48/10)^{0.3}=35220$ N/mm^2; $E_{cm}(t)=35220\times(36/48)^{0.3}=32308$ N/mm^2

$f_{ctm}=0.3\times40^{2/3}=3.51$ N/mm^2

$f_{ctm}(t)$ is not required as no tension is present (see below)

Section properties

Table 14.1 shows the strand pattern and applied prestress forces. The eccentricity $z_{cp}=21.8$ mm is within the kern of the unit, i.e. $z_{cp}<Z_T/A_c=33.3$ mm and therefore no tension exists at the outside surface (top, as cast) at transfer.

$Z=1593.72\times10^6/100=15.937\times10^6$ mm^3

$Z_z=1593.72\times10^6/21.8=73.162\times10^6$ mm^3

Section properties of compound section with the transformed area of tendons

$m-1=(195000/35220)-1=4.537$

$A_{c,co}=476826+4.537\times1560=483903$ mm^2; $y_{b,co}=99.7$ mm; $I_{c,co}=1623.9\times10^6$ mm^4; $Z_{b,co}=16.291\times10^6$ mm^3

Service moment of resistance M_{sR}

Prestress

Eq. 6.2 $\sigma_{pi}=0.7\times1770=1239.0$ N/mm2

$P_{pi}=1239.0\times1560=1932840$ N

Eq. 6.3 $\Delta\sigma_{pr}=1239.0\times0.66\times2.5\times e^{(9.1\times0.7)}\times(20/1000)^{(0.75\times(1-0.7))}\times10^{-5}=4.95$ N/mm^2

$P_{mo}=(1239.0-4.95)\times1560=1925116$ N

$\sigma_{cp}=(1925116/476826)+(1925116\times21.8/73.162\times10^6)=4.61$ N/mm^2

$\Delta\sigma_{el}=195000\times4.61/32308=27.83$ N/mm^2

Eq. 6.5 $\sigma_{pm0}=(1239.0-4.95-27.83)=1206.2$ N/mm$^2<0.75\times1770=1328$ N/mm^2 OK

$P_{pm0}=1206.2\times1560=1881705$ N

Check transfer stress at the inside face (i.e. the bottom face as cast)

Eq. 6.6a $\sigma_b(t)=(1881705/476826)+(1881705\times21.8/15.937\times10^6)=6.52$ N/mm$^2<0.6\times28=16.8$ OK

Table 14.1 Strand pattern with pretensioning forces and eccentricity for 200 mm thickness solid wall panel

	Row 1 (bottom as cast)	*Row 2 (top as cast)*	*Total*
Number in each row	20	10	30
Area in each row A_p (mm^2)	1040	520	1560
Axis height from bottom a (mm)	34.65	165.35	
Initial pretensioning force $P_{pi}=\sigma_{pi} A_p$ (kN)	1288.6	644.3	1932.8
P_{pi} a (kNmm)	44649	106532	151180

Common data. Diameter of strand=9.3 mm. Initial prestress $\sigma_{pi}=0.7\times 1770=1239$ N/mm^2. Cover to strands (bottom and top)=30 mm. Mean axis height $y_s=151180/1932.9=78.2$ mm. Eccentricity $z_{cp}=100-78.2=21.8$ mm.

Maturity of concrete during curing
Mean temperature during 20 hours curing=50°C
Eq. 7.13 $t_T=(20/24)\ e^{-[4000/(273+50))-13.65]}=2.96$ days
Factor for Cement Class R=1
Eq. 7.12 $t_{oT}=2.96\times[9/(2+2.96^{1.2})+1]^1=7.65$ days

Long time losses to life using $RH_s=70\%$ with two sides exposed

Eq. 7.25 $h_o=2\times$area/sides=actual thickness=200 mm
$\alpha_1=(35/48)^{0.7}=0.80$. $\alpha_2=(35/48)^{0.2}=0.94$. $\alpha_3=(35/48)^{0.5}=0.85$
Eq. 7.19 $\varphi_{RH}=[1+(1-(70/100)\times 0.80/(0.1\times 200^{1/3}))]\times 0.94=1.325$
Eq. 7.10 $\beta(f_{cm})=16.8/\sqrt{48}=2.425$
Eq. 7.11 $\beta(t_0)=1/(0.1+7.65^{0.2})=0.624$
Eq. 7.21 $\beta_H=((1.5\times(1+(0.012\times 70)^{18})\times 200)+(250\times 0.85))=526$ days
Eq. 7.20 $\beta_c(t_i,t_0)=[(20833-1)/(526+20833-1)]^{0.3}=0.993$
Eq. 7.18 $\varphi(t_i,t_0)=\varphi_{RH}\ \beta(f_{cm})\ \beta(t_0)\ \beta_c(t,t_0)=1.325\times 2.425\times 0.624\times 0.993=1.99$
Eq. 7.22a $\sigma_c=(1881705/476826)+(1881705\times 21.8/73.162\times 10^6)=4.51$ N/mm^2
Denominator in Exp. 5.46=1.054
Loss of prestress due to creep $\Delta\sigma_{p,c}=195000\times 1.99\times 4.51/(35220\times 1.054)=$ 47.1 N/mm^2

Concrete shrinkage

Eq. 7.28 $k_n=0.85-0.001\ (200-200)=0.85$
Eq. 7.26 $\beta_{RH}=1.55\times(1-(70/100)^3)=1.018$
Eq. 7.29 $\beta_{ds}(t,t_s)=(20833-1)/[(20833-1)+0.04\times 200^{1.5}]=0.995$
For cement grade 42.5R, class R, $\alpha_{ds1}=6$ and $\alpha_{ds2}=-0.11$
Eq. 7.31 $\varepsilon_{cs}=\beta_{ds}(t,t_s)\ k_n\ \varepsilon_{cd,o}=0.995\times 0.85\times 0.85\times(220+110\times 6)\times e^{-0.11\times 48/10}\times 1.018\times 10^{-6}=0.000380$
From above $\varphi(t,t_0)=1.99$
Eq. 7.32 $\Delta\sigma_{p,s}=0.000380\times 195000/1.054=70.29$ N/mm^2

Strand relaxation

Eq. 7.33 $\mu=1206.2/1770=0.681$
Eq. 7.34a $\sigma_{pr}=1206.2\times 0.66\times 2.5\times e^{(0.681\times 9.1)}\times[(500000/1000)^{(0.75\times(1-0.681))}]\times 10^{-5}=43.34$ N/mm^2
Eq. 7.34b $\Delta\sigma_{p,r}=0.8\times 43.34/1.054=32.90$ N/mm^2

Final prestress in the strands

Eq. 7.35 $\sigma_{po}=1206.2-47.1-70.3-32.9=1055.9$ N/mm^2
$P_{po}=1055.9\times 1560=1647204$ N

Table 14.2 Moments of resistance and shear capacity for 200 mm thickness solid wall panel

	Design values	*Resistances*	*Utilisation ratio*
Service M_s and moment of resistance M_{sR} (kNm)	81.8	93.0	0.88
Ultimate M_{Ed} and moment of resistance M_{Rd} (kNm)	113.5	235.2	0.48
Maximum ultimate V_{Ed} and uncracked shear capacity $V_{Rd,c}$ (kN)	88.5	777.6	0.11
Service M_s and cracking moment of resistance M_c (kNm)	81.8	119.7	

Final prestress and moment of resistance at the inside face (bottom as cast)

Eq. 7.1 $\sigma_b=(1647204/476826)+(1647204\times 21.8/15.937\times 10^6)=5.71$ N/mm$^2<0.45\times 45=20.25$ N/mm^2
Top stress is not critical as it is compression
Eq. 7.37 $M_{sR}=(5.71+0)\times 16.291=$ **93.0 kNm** $>M_s=81.8$ kNm.

Ultimate moment of resistance M_{Rd}
The f_p v ε_p relationship for strand with $f_{pk}=1770$ N/mm^2 is shown in Figure 9.13.

$f_{cd}=0.85\, f_{ck}/1.5=0.567\times 40=22.67$ N/mm^2
$A_{pT}=1040$ mm^2
$y_{sT}=30+9.3/2=34.65$ mm
Eq. 9.19 $d=200-34.65=165.35$ mm
$\varepsilon_{po}=1055.9/195000=0.005415$
Eq. 9.17 $\varepsilon_p=0.005415+0.0035\ ((165.35/X)-1)$ Eq. 1
Eq. 9.13 $F_p=f_p\times 1040$
Eq. 9.14 $F_c=22.67\times 0.8\times 2400\ X=43520\ X$
Then for equilibrium: $X/f_p=1040/43520=0.0239$ Eq. 2
Eq. 9.18 $f_{LOP}=0.9\times 1770/1.15=1385$ N/mm^2
$f_{pd}-f_{LOP}=1770/1.15-1385=154$ N/mm^2
Eq. 9.18 $\varepsilon_{LOP}=1385/195000=0.007104$
Eq. 9.20 If $\varepsilon_p>0.007104$, then $f_p=1385+[154\times(\varepsilon_p-0.007104)/(0.0222-0.007104)]$ Eq. 3
Combining Eqs. 1–3 gives: $657.0\ X^2-20918\ X-92636=0$
Then $X=35.8$ mm, $d_n=0.4$ X $35.8=14.3$ mm and $z=165.35-14.3=151.0$ mm
From Eq. 1. $\varepsilon_p=0.018090$
From Eq. 3. $f_p=1497$ N/mm$^2<$limit at $\varepsilon_s=0.02=1516$ N/mm^2
Eq. 9.28 $M_{Rd}=1497\times 1040\times 151.0\times 10^{-6}=$ **235.2 kNm** $>M_{Ed}=113.5$ kNm
Table 14.2 shows the design moments, the moments of resistance and the ultilisation ratio (i.e. design/resistance). As expected, for solid units under exposure conditions of XC3, the wall panel is clearly critical at the serviceability limit state.

Anchorage bond length to full $M_{Rd}=651+0.19\times 9.3\times(1497-1055.9)/1.96=1048$ mm, where $l_{pt2}=651$ mm (see Example 14.2), $\alpha_2=0.19$, σ_{po} at the support$=1055.9$ N/mm^2 and $f_{bpd}=1.2\times 0.7\times 3.51/1.5=1.96$ N/mm^2.

Example 14.2

Calculate the uncracked shear capacity $V_{Rd,c}$ at the face of the support for the 200 mm thickness solid wall panel shown in cross-section in Figure 14.4 using the section and material properties in Example 14.1. Nominal bearing length at the composite slab $l_b=250+75=325$ mm. The width of panel at the centroidal axis is 2377 mm (ignoring the infill between panels). Check the cracking moment of resistance to determine whether the flexurally crack shear capacity $V_{Rd,cr}$ is required.

Solution

$f_{ctm}(t) = 3.51 \times 36/48 = 2.63$ N/mm²
$f_{ctd}(t) = 0.7 \times 2.63/1.5 = 1.23$ N/mm²

Eq. 10.6 $f_{bpt} = 3.2 \times 1.23 = 3.93$ N/mm²

$\sigma_{pm0} = 1206.2$ N/mm² (from Example 14.1)

Eq. 10.4 $l_{pt2} = 1.2 \times 0.19 \times 1206.2 \times 9.3/3.93 = 651$ mm

$P_{po} = 1647.2$ kN (from Example 14.1)

Prestress at NA (including $\gamma_{p,fav} = 0.9$)

Eq. 10.3 $\sigma_{cp} = 0.9 \times 1647.2 \times 10^3/476826 = 3.11$ N/mm²

$f_{ctd} = 0.7 \times 3.51/1.5 = 1.64$ N/mm²
$l_x = l_b + y_b = 325 + 100 = 425$ mm

Eq. 10.2 $\alpha_l = 425/651 = 0.653$

$b_w = 2377$ mm

Eq. 10.1b $V_{Rd,c} = (⅔ \times 2377 \times 200) \times [\sqrt{(1.64^2 + 0.653 \times 3.11 \times 1.64)}] \times 10^{-3} =$ **776.6 kN>>** $V_{Ed} = 88.5$ kN.

Table 14.2 shows the design shear forces, shear capacity and the ultilisation ratio (i.e. design/capacity), clearly not critical. The cracking moment of resistance at the inside face $M_c = 16.291 \times (5.71 + 1.64) = 119.7$ kNm $> M_s = 81.8$ kNm, therefore the section is not flexurally cracked and $V_{Rd,cr}$ is not required.

Calculations for the ultimate bearing capacity of wall panel at the edges of the composite hollow core slab, where $f_{ckT} = 30$ N/mm², are found to be in the order of 1800 kN, and therefore not critical.

Similarly, the long-term sideways camber of the wall panel is found to be approximately 3.2 mm, which is completely negated by the soil and surcharge pressures, and therefore not critical.

14.2.2 Horizontal spanning retaining wall

Figures 14.5 and 14.6 show a typical steel-framed grain store with precast prestressed wall panels. The panels span horizontally between the main steel columns. The panel at the base of the wall is subjected to the highest pressure from the retained material and in some cases this panel is cast with projecting bars along its bottom edge to tie into the *insitu* floor slab. The panels are stacked on a foundation to the design height required.

Figure 14.7 shows a section and elevation of a grain store constructed using 250 mm thick prestressed hcw, the units were manufactured with a 9-core profile.

Table 14.3 shows the strand pattern and applied prestress forces. The eccentricity $z_{cp} = 28.0$ mm is within the kern of the unit, i.e. $z_{cp} < Z_T/A_c = 51$ mm and therefore no tension exists at the outside surface (top, as cast) at transfer.

Two cores were removed from the slip-forming machine to create a solid zone in the panel, this solid zone was used to post-fix the restraint clips to the wall panels using concrete screws. A typical restraint clip detail can be seen in Figure 14.8 and they are also visible in service on Figure 14.6. The clips are cranked to suit the thickness of the column flange and fixed to the wall panel using a post-fixed anchor such as a concrete screw or alternatively a cast-in fixing such as a socket or channel.

The clips are designed to restrain the wall panels against wind loading during construction and when the grain store is empty. During installation, the horizontal joint between the wall panels is bedded with structural mortar to make the building weather-tight and also to provide a shear key. The vertical joint between panel ends is typically finished with a mastic seal.

The following calculation determines the pressure applied to the retaining walls in a grain store that is designed for level fill material, i.e. no heaped material is permitted.

Figure 14.5 Internal view of grain store.

Figure 14.6 External view of grain store, restraint clips visible at the ends of the prestressed concrete wall panels.

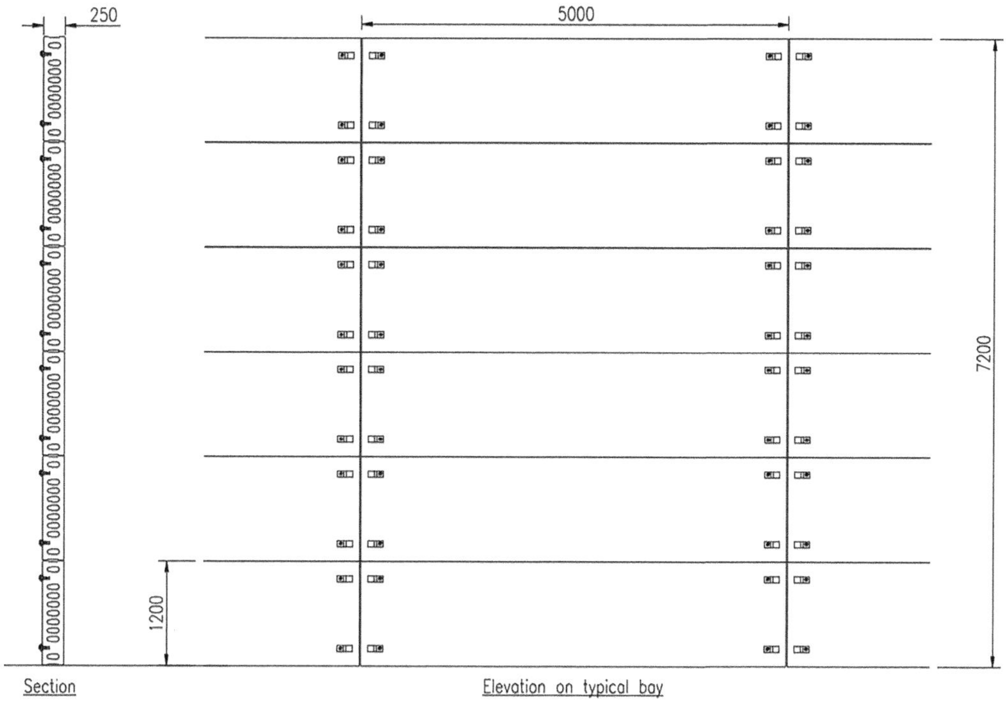

Figure 14.7 Example grain store geometry.

Table 14.3 Strand pattern with pretensioning forces and eccentricity for 250 mm thickness hcw panel with 9 no. cores

	Row 1 (bottom as cast)	*Row 2 (top as cast)*	*Total*
Number in each row	12	6	18
Area in each row A_p (mm^2)	624	312	936
Axis height from bottom a (mm)	34.65	215.35	
Initial pretensioning force $P_{pi}=\sigma_{pi} A_p$ (kN)	773.1	386.6	1159.7
P_{pi} a (kNmm)	26789	83248	110037

Common data as Table 14.1. $y_s = 110037/1159.7 = 94.9$ mm. $z_{cp} = 122.9 - 94.9 = 28.0$ mm.

Retained material

Grain general (dry[b]) as per BS 5502, Part 22 Table 7 (BS 5502-22, 2003)
[b] reference to grain as dry corresponds to the moisture content of 14% (m/m) or less
Density of material $\gamma = 8$ kN/m^3
Internal angle of shearing resistance $\phi = 25°$
Using Rankines theory, active pressure coefficient $k_a = (1 - \sin\phi)/(1 + \sin\phi) = 0.406$

Retaining wall data

Design height of retaining wall $h = 7200$ mm
Effective span of wall panels $l_e = 4900$ mm
Base panel height $b = 1200$ mm

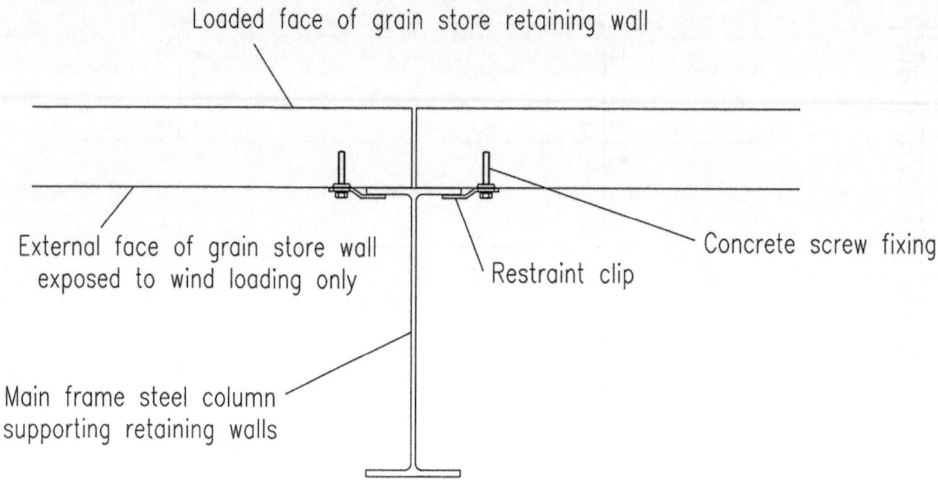

Figure 14.8 Typical restraint clip detail.

Load analysis
Referring to Figure 14.9

Maximum horizontal pressure $p_m = k_a \gamma h = 23.4$ kN/m^2
Pressure at top of base panel $p_1 = p_m - [(p_m/h) \times b] = 19.5$ kN/m^2
Average pressure $p_a = (p_m + p_1)/2 = 21.45$ kN/m^2

Design output

Service moment $M_s = p_a b\, l_e^2 / 8 = 77.2$ kNm (strictly for a trapezoidal load distribution $M_s = 77.3$ kNm, but may be ignored)
Ultimate moment $M_{Ed} = 1.35\ M_s = 104.2$ kNm
Ultimate shear $V_{Ed} = 1.35\ p_a\ b\ l_e/2 = 85.1$ kN (ditto for trapezoidal load 87.7 kN)

The capacity of the prestressed panel is calculated in the same way as in Examples 14.1 and 14.2 and compared against the design shear and moments from the analysis.

Table 14.4 shows the design moments and shear forces, the moments of resistance, shear capacity and the ultilisation ratio (i.e. design/resistance). As expected, the hcw panel is critical at the serviceability limit state. The cracking moment of resistance at the inside face $M_c = 99.1$ kNm $> M_s = 77.2$ kNm, therefore the section is not flexurally cracked and $V_{Rd,cr}$ is not required.

Transverse load distribution is possible by transferring load through the shear keys between adjacent walls. The capacity of the shear key is checked in accordance with BS EN 1168:2005+A3:2011 Clause 4.3.3.2.3. The following calculation checks the 250 mm hcw in the example

Cylinder strength, precast unit $f_{ck} = 40$ N/mm^2
Cylinder strength, joint infill $f_{ckj} = 40$ N/mm^2
Inside surface (as cast top) flange thickness $h_t = 40$ mm
Outside surface (as cast bottom) flange thickness $h_b = 35$ mm
Sum of flanges $h_f = h_t + h_b = 75$ mm
Net width of joint between 250 mm thickness units $h_j = 230$ mm

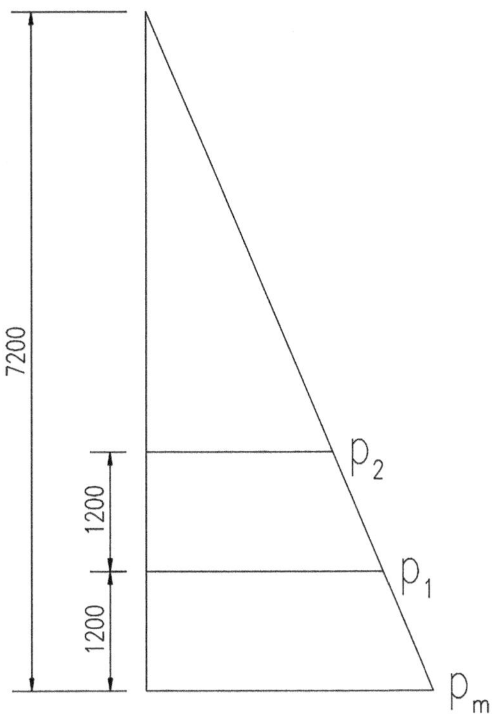

Figure 14.9 Pressure diagram for hcu wall panels.

Table 14.4 Moments of resistance and shear capacity for 250 mm thickness hcw panel with 9 no. cores

	Design values	*Resistances*	*Utilisation ratio*
Service M_s and moment of resistance M_{sR} (kNm)	77.2	81.0	0.95
Ultimate M_{Ed} and moment of resistance M_{Rd} (kNm)	104.2	186.5	0.56
Maximum ultimate V_{Ed} and uncracked shear capacity $V_{Rd,c}$ (kN)	85.1	294.7	0.29
Service M_s and cracking moment of resistance M_c (kNm)	77.2	99.1	

Precast unit

Mean tensile strength $f_{ctm}=0.3f_{ck}^{2/3}=3.51$ N/mm^2
5% fractile value $f_{ctk005}=0.7\ f_{ctm}=2.46$ N/mm^2
Partial safety factor $\gamma_c=1.5$
Design tensile strength $f_{ctd}=f_{ctk005}/\gamma_c=1.64$ N/mm^2
Shear capacity of flange $v_{Rdj1}=0.25\ f_{ctd}\ h_f=30.7$ kN/m

Joint

f_{ctmj} and f_{ctdj} as f_{ctm} and f_{ctd} above
Shear capacity of joint $v_{Rdj2}=0.15\ f_{ctdj}\ h_j=56.5$ kN/m
Critical shear capacity $v_{Rdj}=\min(v_{Rdj1}, v_{Rdj2})=30.7$ kN/m

Using the guidance given in Annex C of BS EN 1168, the maximum pressure and loading applied to the bottom panel can be distributed to the panels above that are supporting less load. For example, the average pressures are compared across the bottom two panels and the peak loading on the bottom panel is treated as a line load.

Load analysis – second panel from bottom

Design height $h_2=6000\,\text{mm}$
Horizontal pressure at bottom of panel $p_1=k_a\,\gamma\,h_2=19.5\ \text{kN/m}^2$
Horizontal pressure at top of panel $p_2=p_1-[(p_1/h_2)\,b]=15.6\ \text{kN/m}^2$
Average horizontal pressure $p_{a2}=(p_1+p_2)/2=17.5\ \text{kN/m}^2$
Equivalent line load, bottom panel $w_{eq}=p_a\,b=25.7\ \text{kN/m}$
Equivalent line load, second panel $w_{eq2}=p_{a2}\,b=21.0\ \text{kN/m}$
Peak line load to distribute $w_d=w_{eq}-w_{eq2}=4.7\ \text{kN/m}$

This value is less than the calculated value of $v_{Rdj}=30.7\ \text{kN/m}$, by inspection no issues distributing the line load through the shear key.

From Figure C.1 of Annex C BS EN 1168, the percentage of the load supported on the directly loaded panel=45%. This load is multiplied by 1.25 in accordance with Clause C.4 d)

Service load on bottom panel $w_{s,ls}=w_{eq2}+(0.45\,w_d)=23.1\ \text{kN/m}$
Ultimate load on bottom panel, $w_{Ed,ls}=w_{eq2}+(1.25\times0.45\,w_d)\times1.35=32.0\ \text{kN/m}$

Design output with transverse distribution

Service moment; $M_s=w_{s,ls}l_e^2/8=69.5\ \text{kNm}$
Ultimate moment; $M_{Ed}=w_{Ed,ls}l_e^2/8=95.9\ \text{kNm}$
Ultimate shear force; $V_{Ed}=w_{Ed,ls}\,l_e/2=78.3\ \text{kN}$

Design output with no distribution

Service moment $M_{sR}=77.2\ \text{kNm}$
Ultimate moment $M_{Ed}=104.2\ \text{kNm}$
Ultimate shear force $V_{Rd,c}=85.1\ \text{kN}$

With transverse distribution, the service load case can be reduced by approximately 10% and the ultimate load cases reduced by approximately 8%. This distribution of loading may allow a reduction in the thickness of the wall and/or a reduction in the number of prestressed strands required in the design.

14.3 BOUNDARY AND SECURITY WALLS

Precast prestressed wall panels used in conjunction with an *insitu* base are an efficient and common solution for the boundary walls of prisons and other secure facilities. The walls are manufactured with holes through the section to allow fixing of the site reinforcement for the *insitu* base. Figure 14.10 shows a wall and base section, the wall is 175 mm thick and 5.2 m in height above ground level, the base is 2.5 m wide and 600 mm deep. The geometry shown is a fairly typical arrangement for this type of construction although the base geometry will be governed by the wind loading and any ground-bearing pressure limitations specific to the site. The calculation of the bearing pressures and the design of the *insitu* base is outside the scope of this book.

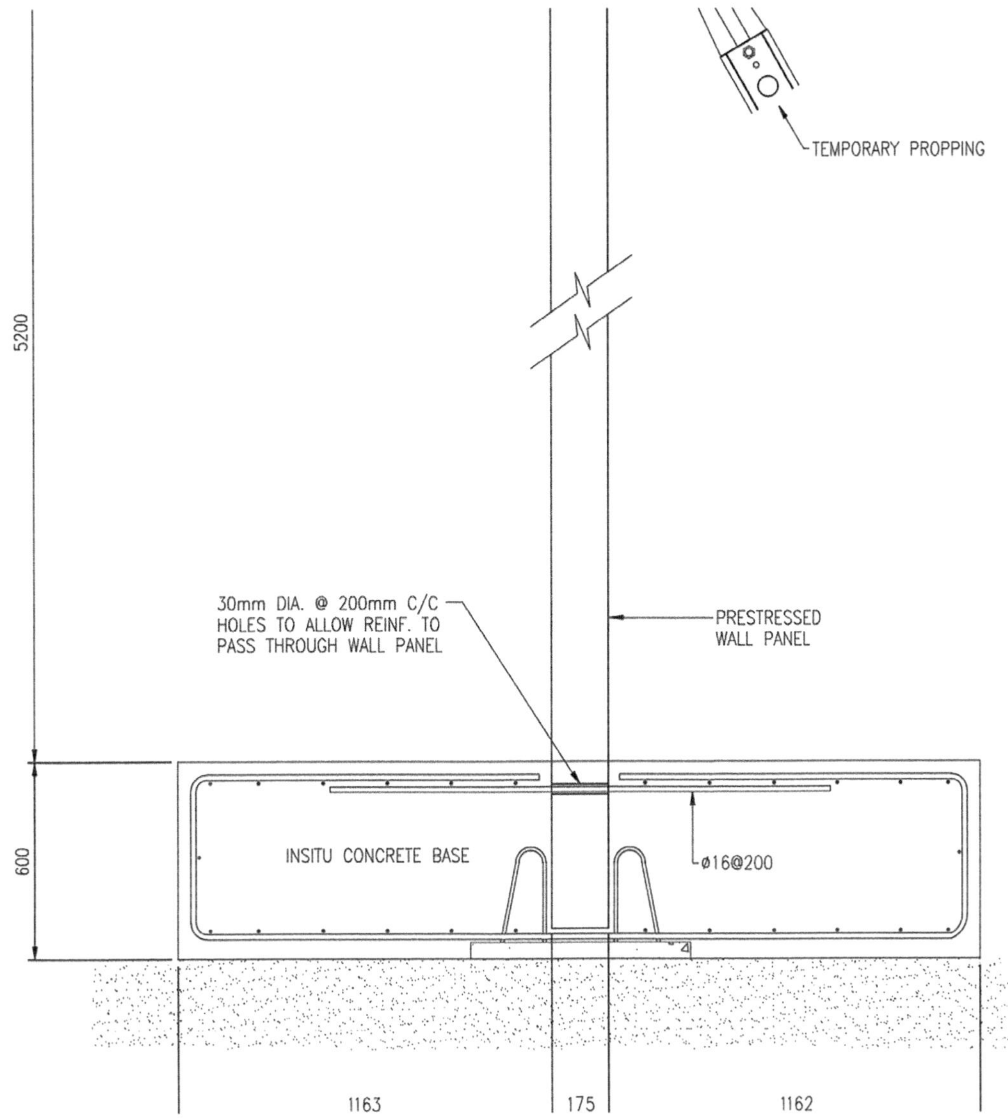

Figure 14.10 Section through boundary wall and base.

Wind load design is in accordance with BS EN 1991, Part 1-4 (BS EN 1991-1-4, 2005) and NA to BS EN 1991, Part 1-4 (NA to BS EN 1991-1-4, 2005).

The following calculation for the wind load considers the worst case at a free edge using Zone A and B coefficients from Table 7.9 of BS EN 1991-1-4. During the design process, the wind load is calculated at the different zones along the length of the wall, with the loading being reduced as you move away from a free edge. The detailed design process can allow a reduction in the strand pattern in the walls and a reduction in the width of the *insitu* base. Normally the wall thickness and base depth are kept consistent. Note that the wall and base need to be checked for the temporary wind load case during construction, the erection sequence may result in sections of wall being subjected to free edge conditions. The temporary wind loading can be reduced with the seasonal factor and the probability factor being adjusted in accordance with the time of year the erection is taking place and the return period respectively. Both of these values are 1.0 in the calculation for the permanent wind load case. The directional factor can be used to reduce the wind loading in the permanent and the temporary case if there is certainty regarding the orientation of the walls, a conservative approach is to use 1.0.

Height of wall panel $h_p = 5200\,\text{mm}$
Width of wall panel $b_p = 1800\,\text{mm}$
Reference height $z = 5200\,\text{mm}$
Figure NA.1. Fundamental basic wind velocity $v_{bmap} = 25.9\,\text{m/s}$
Site altitude $A_{lt} = 50\,\text{m}$
NA.2.5. Altitude factor $c_{alt} = (1 + A_{lt}/1000\,\text{m}) = 1.1$
NA.2.4. Wind velocity with altitude correction $v_{b0} = \text{v}_{bmap}\ \text{c}_{alt} = 27.2\,\text{m/s}$

The season factor relates to temporary wind loading but is shown for reference in this calculation.

Table NA.2. Season factor $c_{season} = 1.0$
A conservative approach is taken regarding the directional factor
Table NA.1. Directional factor $\text{c}_{dir} = 1.0$, assume worst case

The probability of exceedance relates to temporary wind loading but is shown for reference in this calculation. The value used for p below represents a return period of 50 years which will result in a probability factor of 1.0.

Referring to BS EN 1991, Part 1-6 (BS EN 1991-1-6, 2005)
Table 3.1. Probability of exceedance $p = 0.02$
Probability factor $\text{c}_{prob} = [(1 - K\ \ln(-\ln(1 - p)))/(1 - K\ \ln(-\ln(0.98)))]^n = 1.0$
Basic wind velocity $v_b = c_{prob}\ c_{dir}\ c_{season}\ v_{b0} = 27.2\,\text{m/s}$
NA.2.18. Air density $\rho = 1.226\,\text{kg/m}^3$
Basic velocity pressure $q_b = 0.5\ \rho v_b^2 = 0.453\ \text{kN/m}^2$
Distance to shoreline $l_{sho} = 2\,\text{km}$
Figure NA.7. Exposure factor for Country Terrain $c_e\ (z) = 2.22$
Figure NA.2. When orography is not significant $c_o = 1.0$
Peak velocity pressure $q_p(\text{z}) = c_e(\text{z})\ q_b = 1.01\ \text{kN/m}^2$
From Table 7.9
Net pressure coefficient Zone A $c_{pa} = 2.3$
Net pressure coefficient Zone B $c_{pb} = 1.4$
Zone A length $A = 0.3\ h_p = 1560\,\text{mm}$
Zone B length $B = b_p - A = 240\,\text{mm}$
Wind load $W_w = (c_{pa}\ A + c_{pb}\ B)\ q_p(\text{z})\ h_p = 20.5\ \text{kN}$

Design output – wall stem

Service moment $M_s = W_w\ h_p/2 = 53.4\ \text{kNm}$
Ultimate moment $M_{Ed} = 1.5\ M_s = 80.1\ \text{kNm}$
Ultimate shear force $V_{Ed} = 1.5\ W_w = 30.8\ \text{kN}$

The capacity of the precast prestressed panel is calculated in the same way as the previous examples and compared against the design shear and moments from the analysis.

Table 14.5 Strand pattern with pretensioning forces and eccentricity for 1880mm width × 175mm thickness solid wall panel

	Row 1 (bottom as cast)	*Row 2 (top as cast)*	*Total*
Number in each row	18	18	36
Area in each row A_p (mm^2)	936	936	1872
Axis height from bottom a (mm)	34.65	140.35	
Initial pretensioning force $P_{pi}=\sigma_{pi} A_p$ (kN)	1159.7	1159.6	2319.3
P_{pi} a (kNmm)	40184	162755	202939

Common data as Table 14.1. y_s=202939/2319.3=87.5mm. z_{cp}=87.5−87.5=0mm (by definition for symmetrical prestress).

Table 14.6 Moments of resistance and shear capacity for 1880mm width × 175mm thickness solid wall panel

	Design values	*Resistances*	*Utilisation ratio*
Service M_s and moment of resistance M_{sR} (kNm)	53.4	57.3	0.93
Ultimate M_{Ed} and moment of resistance M_{Rd} (kNm)	80.1	168.0	0.48
Maximum ultimate V_{Ed} and uncracked shear capacity $V_{Rd,c}$ (kN)	30.8	603.5	0.05
Service M_s and cracking moment of resistance M_c (kNm)	53.4	72.7	

Table 14.5 shows the strand pattern and applied prestress forces. The eccentricity z_{cp}=0mm and therefore no tension exists at the outside surface (top, as cast) at transfer.

Table 14.6 shows the design moments and shear forces, the moments of resistance, shear capacity and the ultilisation ratio (i.e. design/resistance). As expected, the hcw panel is critical at the serviceability limit state. The cracking moment of resistance at the inside face M_c=72.7 kNm>M_s=53.4 kNm, therefore the section is not flexurally cracked and $V_{Rd,cr}$ is not required.

Figure 14.11 shows the reinforcement cage for the *insitu* base installed and ready for the prestressed wall panels to be erected. Figure 14.12 shows the wall panels in position, propped off kentledge blocks and ready for the *insitu* base to be poured.

14.4 NON-LOAD-BEARING WALLS IN BUILDINGS

Prestressed wall panels provide a robust solution for internal partitions and external envelopes. They are typically manufactured in thicknesses from 100mm to 250mm with an interlocking tongue and groove side profile.

14.4.1 Fire walls

Prestressed wall panels can be used as fire walls to provide compartmentation in buildings with fire resistance of up to 240 minutes achievable for non-load-bearing and load-bearing walls. Tables 5.3 and 5.4 in BS EN 1992, Part 1-2 (BS EN 1992-1-2, 2004) provide details on the criteria that satisfy the EI and REI requirements. Figure 14.13 shows a fire wall in a whiskey distillery constructed from prestressed wall panels. In this building, the wall panels were installed inside the flanges of the steel columns. The fire resistance of the panels was EI 120. The horizontal joints were bedded with a structural mortar and the geometry of the

Figure 14.11 Reinforcement cage for wall.

Figure 14.12 Prestressed wall panels propped off kentledge blocks.

Figure 14.13 Whiskey distillery fire wall.

Table 14.7 Fire resistance of non-load-bearing prestressed walls

Wall thickness (mm)	*Fire resistance (minutes)*
60	EI 30
80	EI 60
100	EI 90
120	EI 120
150	EI 180
175	EI 240

joint was detailed and constructed in accordance with the guidance given in Clause 4.6 of BS EN 1992-1-2. Fire resistance mastic was used to seal the vertical interface between the steel column and the precast wall. Table 14.7 gives the fire resistance EI provided by wall thickness.

Figure 14.14 shows a paper mill that was constructed using prestressed hcw attached to a steel frame. The walls were typically stacked from foundation level on top of each other and restrained back to the steel structure using two connection types. The standard detail used for the majority of walls is shown in Figure 14.15. This consisted of a folded steel plate and an L-bolt that was located into an RSA welded to the steel column. The folded plate restrained the wall panels and also acted as a shutter for the thixotropic grout that was used to fill the vertical joint between the ends of the wall panels. The grout infill was a requirement in the fire strategy for the bottom 7 m of the building, above the 7 m level the vertical joint was infilled with mineral wool. Figure 14.16 shows the parapet level restraint details. The wall panels were manufactured with preformed holes to allow

Figure 14.14 Prestressed hollow core units envelope walls.

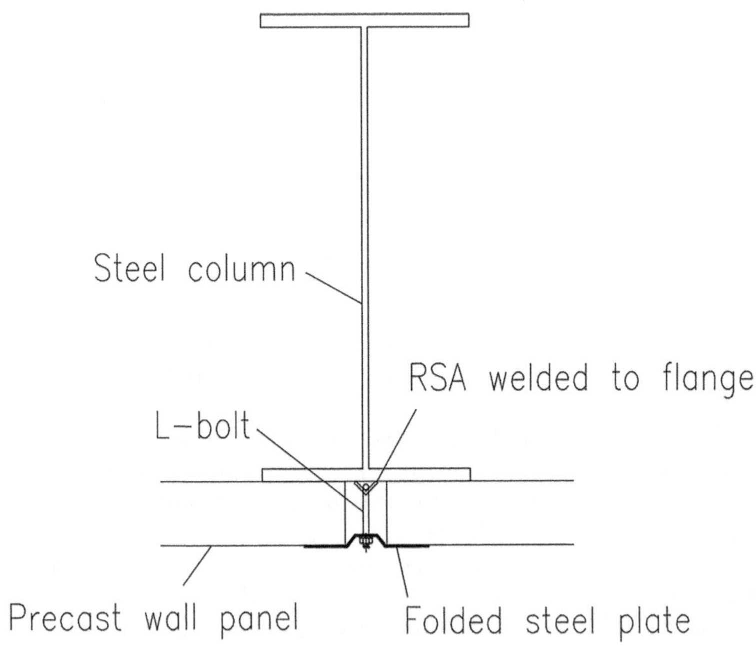

Figure 14.15 Typical restraint connection.

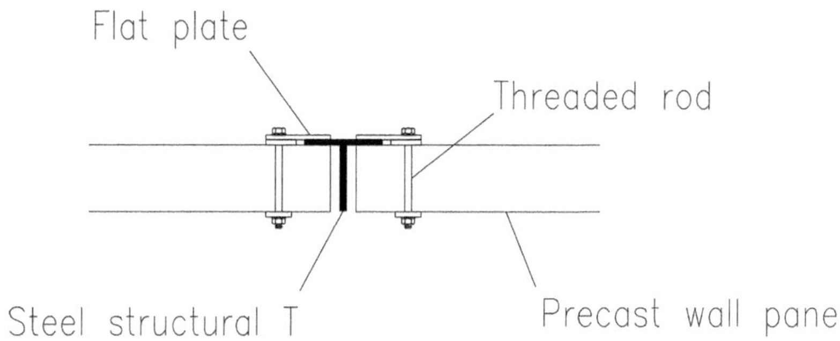

Figure 14.16 Parapet level restraint connection.

Figure 14.17 Parapet level wall panel being installed.

a threaded rod to pass through and make the connection to the supporting member with a flat steel plate.

The walls were lifted using bespoke steel brackets with a lifting pin running through preformed holes in the wall, the design of the brackets meant that the walls could be delivered to site flat on the trailer, pitched to the vertical position using the brackets and lifted into position in one operation. Figure 14.17 shows a wall being lifted into position on the building.

REFERENCES

BS 5502-22. 2003. Buildings and structures for agriculture - Code of practice for design, construction and loading, +A1 2013, BSI, London, UK.

BS EN 1168. 2005. Precast Concrete Products - Prestressed concrete hollow core units, +A3:2011, BSI, London, UK.

BS EN 1991-1-4. 2005. Eurocode 1: Actions on structures - General actions - wind actions, +A1:2010, BSI, London, UK.

BS EN 1991-1-6. 2005. Eurocode 1. Actions on structures - General actions. Actions during execution, BSI, London, UK.

BS EN 1992-1-2. 2004. Eurocode 2: Design of concrete structures, Part 1-2: General rules - Structural fire design, +A1:2019, BSI, London, UK.

NA to BS EN 1991-1-4. 2005. UK National Annex to Eurocode 1: Actions on structures - Part 1-4, +A1:2010, BSI, London, UK.

Chapter 15

Off-site benefits and temporary works

15.1 BENEFITS OF PRECAST CONSTRUCTION

15.1.1 Off-site benefits

The terms off-site manufacture and modern methods of construction are often used to describe the process and systems employed to design, manufacture and erect precast elements and structures on site. The construction industry uses precast concrete in its many and varied forms in a wide range of applications and building types. Developers, building contractors, engineers, architects and the end users can all benefit from the advantages provided by precast and prestressed concrete elements and the structures created using this form of construction. These benefits range from improved Health & Safety, factory environment quality control, construction programme reduction and compliance with the demands of statutory regulations such as fire resistance and energy performance.

A full understanding of the benefits of precast concrete requires a holistic approach to be taken by designers and specifiers – this will allow the true value and impact of using precast to be assessed.

A number of the benefits of precast structures can be seen in Figure 15.1. This shows an inner city site, no scaffolding is required, edge protection is visible on the top of the walls, windows and aluminium cladding have been installed off site and the building footprint is extremely close to the site boundary maximising the available area.

15.1.2 Health & safety

Traditional build construction is labour intensive with a high number of site-based operatives and a considerable amount of equipment and materials required on site. The use of precast in buildings reduces the number of operatives on site with only a small fitting crew required to erect the precast elements; the operatives, equipment and materials that would be used on the construction site are essentially used at the factory production site. A relatively small amount of concrete, grout and reinforcement materials are required on site to connect the precast elements. The site operatives that are based on site are there for less time when building with precast therefore reducing risk.

Any residual Health & Safety risks associated with casting concrete are moved from the site environment to a factory environment, which allows for any risks to be managed in a much more controlled and effective manner.

A precast structure can be constructed without the need for scaffolding and this minimises the number of operatives working at height and also minimises the time that operatives spend working at height. Precast elements such as walls, floors and stairs can have

DOI: 10.1201/9781003319450-15

Figure 15.1 Student accommodation precast structure in Leeds.

edge protection systems mounted to them at ground level prior to being erected therefore providing a very safe system of work.

The installation of windows and doors in precast cladding and sandwich panels allows the installation of these items to be carried out at ground level, removing the need for this trade to work at height.

The casting of mechanical and electrical (M&E) services into precast elements removes the need for tracking of walls on site, which reduces manual work with power tools and the production of dust and noise (tracking is the term used to describe the process of cutting tracks into walls to receive post installed conduits typically for electrical services).

Temporary works required on precast structures are usually very simple to design and easy to erect and control on site. Walls and columns for example are propped using proprietary push-pull props (props with both compression and tension capacity) designed using simple methods. Precast floors, on the rare occasion that they need to be propped, typically use 'acrow' type touch props at midspan.

15.1.3 Quality control

Precast concrete manufacturers may acquire, among others, certification to ISO9001 Quality Management System, ISO14001 Environmental Management System, ISO45001 Occupational Health & Safety Management and BES6001 Responsible Sourcing of Construction Products. These certificates are issued by the auditing body once the prescribed standards are met. Each certificate is valid for a certain duration after which the precast company must go through the recertification process.

Figure 15.2 Recycled water filtration system from cement slurry by Jungdo Company Ltd., Korea.

The concrete used for the precast products must be sourced from an accredited batching plant, where third-party accreditation of the batching plant can be provided by British Standard Institute (BSI) or Quality Scheme for Ready Mixed Concrete (QSRMC).

When precast products have a Eurocode product standard they can be CE marked by certified auditors to demonstrate compliance with the design and manufacturing standards set out for that product. For example, BS EN 1168 is the product standard for Hollow core slabs that takes precedence over other related codes; this standard contains design, tolerance and other specific requirements that are not provided in the guidance in other Eurocodes such as BS EN 1992 and BS EN 13369.

All reinforcement used in the precast products is sourced from companies that are certified by the Certification Authority for Reinforcing Steels (CARES). CARES is an international provider of assured certification for reinforcement bars and prestressing strand.

15.1.4 Sustainability

Precast concrete is manufactured from aggregates that are available in almost every country in the world and precast manufacturing facilities are often situated in close proximity to the market they supply into. The use of locally available aggregates reduces transportation distances and subsequently reduces energy usage. There is also the option to use recycled aggregates in some products. A tonne of precast product uses 218 kg of secondary materials and by-products and produces only 6 kg of waste that goes to landfill (The Concrete Centre, 2010).

Water recycling is a well-established process at precast manufacturing sites, with 85% of water being reused at production facilities. Figures 15.2 and 15.3 show a solid-liquid

Figure 15.3 Recycled slurry separation unit by Jungdo Company Ltd., Korea.

separation unit using a filter press to produce re-useable mixing water from cement slurry. As the filter pressure increases solids form within the chambers forming a cake, which is removed allowing the mixing water to be released.

Durability and long design life of precast concrete are assured due to the combination of the enhanced tolerances for cover achieved in a quality-controlled factory environment and a range of workable concrete mixes of grade C40/50 and above. With typical reinforcement design and details, for XC1 internal environments a 120 years design life can be easily achieved for precast elements and for XC3 external environments a 60 years design life can be easily achieved. Precast concrete insulated sandwich panels illustrate how precast can provide an extremely durable solution for both the inside and the outside walls of the building.

When used in an XC1 internal environment, precast prestressed hollow core units (hcu) can provide a design life of 120 years. These units are versatile with excellent load span capacities, making them suitable for use in residential, commercial and industrial building structures. The hollow core section profile reduces material usage and helps to reduce loading into the supporting superstructure and substructure, this makes the floor a key component in creating a sustainable and structurally efficient building. Hcu have a relatively low self-weight and can achieve a very high span/depth ratio. Long spans allow flexibility in room layouts and a reduction in the supporting structure for such as columns and beams. The large open plan space created gives excellent future proofing should the room layouts or the use of the building change.

The thermal mass of concrete can play a role in the heating and cooling of a building. In simple terms, the concrete absorbs heat during the day to cool the building and releases stored heat at night to warm the building. This natural process reduces the need for energy-consuming heating and cooling systems. The very high standard of aesthetic finish that can be produced with precast concrete means it is particularly suited to being exposed to view, this exposure maximises the effectiveness of the concrete to contribute thermally and help to control the internal temperature of the building.

Precast concrete structures can provide an extremely air-tight construction when adequate sealing is provided at junctions with other elements such as windows and doors. Low air permeability test results can help as part of the energy performance assessment to demonstrate compliance with the UK Building Regulations, Part L (The Building Regulations, 2010a) or the design specification requirements.

Precast elements can have proprietary bee, bird and bat boxes cast into them to provide an integrated wildlife habitat.

15.1.5 Fire resistance

Concrete as a material is classified as A1 according to EN 13501-1 (BS EN 13501-1, 2018). This means it is completely non-combustible. It does not burn and does contribute to fire. Precast concrete elements provide an excellent means of satisfying the fire resistance requirements of UK Building Regulations, Part B (The Building Regulations, 2010b). A typical fire strategy in a multi-storey building will require a compartmentation of 120 minutes to prevent fire spread. A structure built with a hcu floor and precast walls can provide fire resistance REI120 without the need for any testing or additional fireproofing materials. Criteria R E and I are all covered by the guidance for design given in BS EN 1992, Part 1-2 (BS EN 1992-1-2, 2004) and the UK National Annex (NA to BS EN 1992-1-2, 2004). See Chapter 5, Section 5.5 for further details.

15.1.6 Acoustic performance

Precast concrete dividing floors and walls provide an effective means of preventing the passage of airbourne sound to adjacent rooms. External façades constructed of precast cladding or sandwich panels provide effective resistance to the passage of external airbourne sound to the internal areas of the building. Robust Details (Robust Details Ltd, 2021), provide guidance on effective forms of construction using precast concrete elements with other building materials and forms of construction to achieve the necessary acoustic performance required in the Building Regulations without the need for acoustic testing.

15.1.7 BIM (Building Information Modelling)

Precast concrete elements and structures are very well suited to take advantage of the benefits provided by a collaborative BIM approach in the design process. The precast is modelled using the latest 3D software, typically the model is built up from the architectural intent for the building using the project architect's model and drawings. The precast model can be combined with other models to check for clashes and as a means of design development in a project; typically the 3D models from the architect, the *insitu* substructure designer, the Mechanical Electrical Plumbing (MEP) subcontractor and the steel fabricator are used together in an Industry Foundation Class (IFC) model, the IFC contains the combined 3D model information from each of the design team members.

Once the model is complete the General Arrangements (GA) and all other drawings are generated from it for approval by the client and design team. The precast model is used to extract the information required to create the production drawings, ensuring that the geometry and cast-in items, etc. exactly match the source model. Quantities for procurement can also be extracted from the model to populate material order sheets. The model can also be used as a project management tool to produce the delivery & erection sequence for site and subsequently create the production planning programme to meet the required site date for each individual piece of precast.

It is possible to cast embedded services such as electrical conduits, lightning protection, ducting and plumbing pipework into precast elements such as walls, hcu, columns and beams. The details for all cast-in assemblies required by other trades are added to the model and scheduled on the production drawings.

Windows and doors are not added to the model but a reference and dimensions for the opening will be shown on the elevations so it is clear what window or door is to be fitted. The geometrical details required at the cill or threshold, jambs and header are modelled to ensure the interface with the window is correct. The details and dimensions required for restraints, fixings, EPDM (Ethylene Propylene Diene Monomer is a synthetic rubber used as a waterproofing membrane around window and door frames), etc. are typically provided as 2D information from the window supplier and this sets out the geometry in the 3D precast model.

The precast wall panel shown in Figure 15.4 is an extract from the 3D model. All the components required to manufacture the panel are included in the model. This panel contains the reinforcement, the M&E backboxes (metal or plastic boxes that house switches, etc.) and conduits, the lifting anchors, grout sleeves and couplers for on site connections and the temporary works propping sockets. The volume of concrete required to manufacture the panel can also be extracted from the model, this will remove the volume from voids created by grout sleeves.

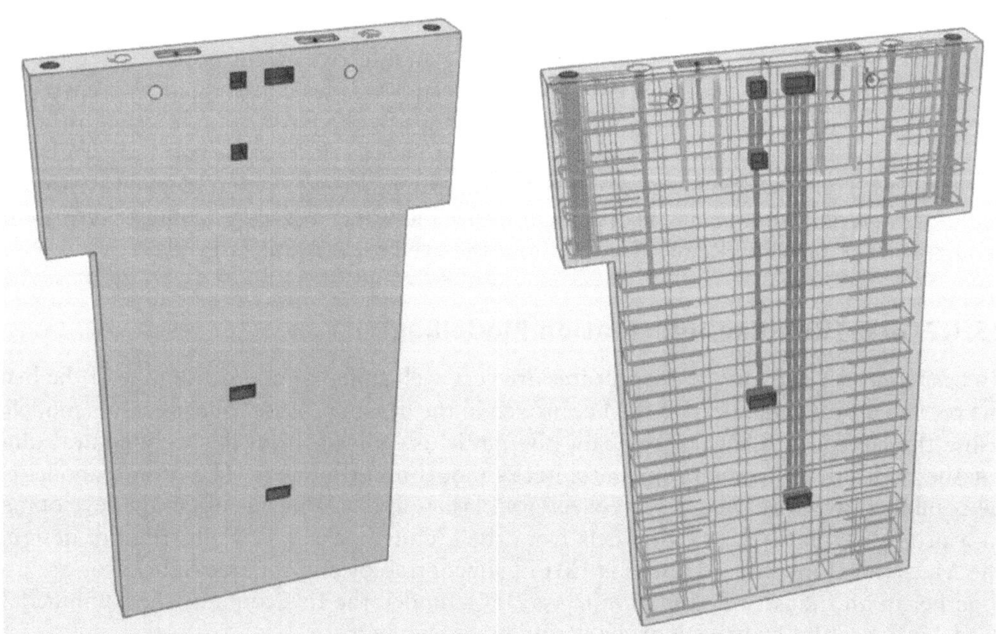

Figure 15.4 Precast wall panel from BIM model.

15.1.8 Programme duration and site logistics

Precast concrete is often associated with speed of construction and reduced on site time and costs. There are always a number of points and options to consider in order to ensure that these benefits are realised, in particular when more than one form of construction is being used to build the structure. Taking a carpark as an example – the foundations are *insitu* concrete, lifts and stair cores are precast, the frame is steel and the floors are hcu. Typically, design workshops are held by the main contractor and sub-contractors to ensure all interfaces are reviewed and construction details are suitable to build the structure in the sequence and durations in the contract programme. Adequate planning is critical at all stages of a project to ensure a successful outcome with respect to programme. Prior to starting on site, there needs to be an understanding of every task, each interface, handover between sub-contractors, access to craneage, site logistics and any other factors that will affect the build programme. The precast build speed gives an advantage but the overall fit out and follow on trades also need to be thought through to really make the most of the programme savings that are possible.

The choice of the particular type of precast elements used on a scheme will be governed by the specific requirements, for example precast and prestressed slabs have a high speed of erection and can cover large areas each day. Site durations can be significantly reduced compared to all other floor types and forms of construction. Temporary propping is rarely required for prestressed slabs so access to the floor and follow-on site operations can proceed without delay or obstruction. Once the longitudinal joints and ends of the slabs have been concreted and cured adequately, typically 72 hours is allowed but this can be reduced depending on concrete mix used, the floor can be handed over to the main contractor and used as a working platform for follow on trades with the use of MEWPS, scissors lifts, etc. possible dependent on their self-weight and the slab capacity.

The soffit of hcu is formed on steel beds so provides a high quality, smooth, durable surface that is suitable for direct application of a painted finish or it can left untreated and exposed to view when used for example as a carpark floor. This reduces the cost and time required to apply finishes to the soffit and also has the benefit of maximising the use of the concretes thermal mass potential.

Brownfield sites, in particular city and town centre sites with little space for materials, are very well suited to precast construction, and in some cases, it is easily the most viable option. The precast is brought to site on timed deliveries and lifted from the lorry directly into its final position on site, minimising the need for storing material on site.

On sites where the building line is at or very close to the site boundary, the use of precast construction can completely remove the need for external scaffolding.

Integration with other trades and pre-fabrication systems helps to maximise the reduction in overall site duration. For example, bathroom pods and flat pack kitchen systems can be programmed for delivery to site and craned into position as each level of precast structure is erected to help reduce the overall build programme.

Precast is compatible with other forms of construction and can be readily integrated with steel frames, *insitu* concrete structures, masonry and timber frame buildings. Stair cores and lift shafts are regularly used with the aforementioned types of construction to provide a rapid build solution that can be used as part or all of the stability system for the building. The precast cores will also provide immediate access and egress to the building during the construction phase. Quite often precast provides an economic and effective means of satisfying the requirement for firefighting cores or lift shafts when used in conjunction with lightweight forms of construction.

15.2 ERECTION OF PRECAST ELEMENTS AND STRUCTURES

15.2.1 Temporary works

Most people think of temporary works as the propping that is often seen on building sites where concrete structures are being erected. In fact, this on site propping is generally the final stage in the temporary works design process for precast elements and structures.

Temporary work design must be considered carefully for all stages of the process: propping and bracing of mouldage, lifting anchors for demoulding, lifting and spreader beams, lifting anchors for handling/turning in the yard, stillages for storage of the precast in the yard, toast racks and A-frames for transportation of the precast to site on lorries, lifting anchors for site erection, temporary edge protection, abseiling anchor points, on site propping during erection and any other requirements for the project that are of a temporary nature.

The following sections will discuss the design of lifting anchors and the design and control of onsite propping. These two forms of temporary works are high risk and must be adequately assessed. The design and the design-checking processes must be documented and carried out by suitably qualified and experienced people. A failure of a lifting anchor or an onsite prop could result in serious injury or a fatality.

15.2.2 Lifting anchors

There is a wide range of proprietary lifting anchors for precast concrete available on the market. The variety of lifter types can be bewildering but the principles of their design and use can be put into two basic categories

1. Lifting anchors in unreinforced concrete
2. Lifting anchors in concrete with specific reinforcement details for the stressed applied during lifting operations

It is critically important that engineers who are specifying lifting anchors understand the principles of how the anchor works and the limitations and suitability of each type of anchor for the specific lifting operation. The manufacturers and suppliers of proprietary lifting anchors provide technical data for the safe use of their products but be aware that the use of this information requires the engineer responsible for the temporary works design to be fully aware of all the details surrounding the lifting operation.

15.2.3 Load applied to each anchor

The way in which a unit is lifted will influence the load that is applied to the lifting anchors, each lifting load case must be considered for the precast element to ensure all lifting operations are safe to perform.

Some lifting arrangements are statically indeterminate, and in these situations, it must be assumed that the lifters are not equally loaded. Figure 15.5 shows a slab being lifted with 4 no. lifting anchors cast into it. These anchors need to be designed to consider that only 2 no. anchors are supporting the load; this is common practice on stair flights and landings unless a compensating chain rig is used. Lifting beams with pulleys and compensating chain rigs are two common methods of providing a statically determinate lifting system.

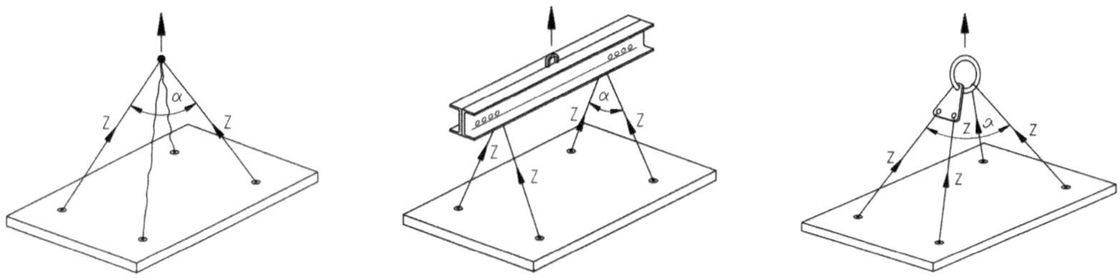

Figure 15.5 from left to right: statically indeterminate lifting arrangement, lifting beam, compensating chain rig. (Courtesy of Construction Fixing Systems, UK.)

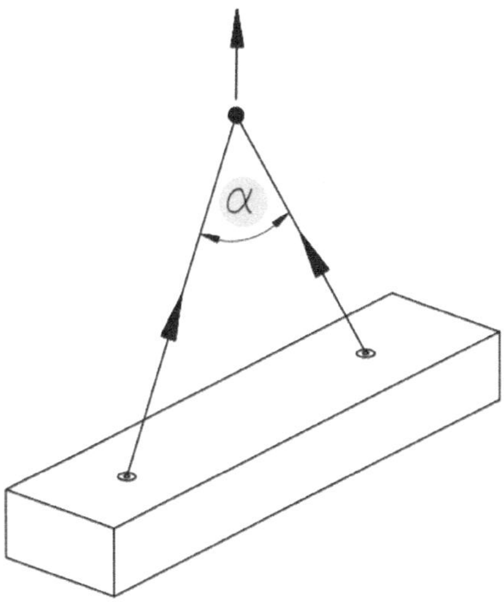

Spread Angle α	Spread Coefficient z
0°	1
15°	1.01
30°	1.04
45°	1.08
60°	1.16
70°	1.26
90°	1.41

Figure 15.6 Chain angle coefficients. (Courtesy of Construction Fixing Systems, UK.)

15.2.4 Chain angle

The angle of the chains is governed by the length of the chain to the crane hook and the position of the lifting anchors on the precast element. The appointed person writes the lift plan and determines the length of the chains required to satisfy the chain angle used in the design of the lifting anchors. It is important for the designer of the anchors to communicate the requirements for the chain angle; the lift plan will be based on this value, and it is good practice for the angle to be noted on the GA drawings used for site erection. Figure 15.6 gives coefficients to be used with the calculated chain angles. The chain angle affects the tension force applied to the lifting equipment; as the angle increases away from vertical, the tension force increases proportionally.

15.2.5 Dynamic factors

When lifting precast elements, a dynamic action is applied to the lifting anchors based on the type of crane or machine used. The magnitude of this effect is determined by the length and type of cable or chain used and the hoisting speed. Typical values are provided in Table 15.1. If there are any doubts regarding the factor to be used, the supplier of the lifting anchors should be contacted for further guidance.

Table 15.1 Dynamic factors

Lifting equipment	*Dynamic factor*
Tower, overhead and portal cranes	1.2[a]
Mobile crane	1.3
Lifting and moving on flat terrain	2.0–2.5
Lifting and moving on rough terrain	3–4

[a] In precast factories and if specific provisions are made at the building site, lower values may be appropriate.

Table 15.2 Formwork adhesion coefficients

Formwork type	*Adhesion coefficient,* q_{adh} *(kN/m²)*
Oiled steel formwork	1
Varnished timber formwork	2
Rough formwork	3

15.2.6 Adhesion to formwork

Additional forces can be applied to the lifting anchors when demoulding due to adhesion to the formwork. The type of formwork used and the release oil will affect the value of the force that needs to be considered. Typical values are provided in Table 15.2 for flat surfaces, profiled surfaces can have greater adhesion forces and the demoulding process may need to take this into account. Occasionally, additional lifters specifically for breaking the adhesion to the formwork are cast into the element.

15.2.7 Lifting methodology

The simplest way to lift precast elements is by using two equally loaded lifting anchors, the placement of the lifters symmetrically about the centre of gravity (CofG) ensures equal loading. All 3D software packages can provide the position of the CofG for any type and shape of precast element. Note that 2D software, spreadsheets or hand calcs can provide the CofG for wall panels like the example in Figure 15.7. If it is not possible to position the lifting anchors such that the load is distributed equally between them, the alternative is to calculate the worst case loading on the anchors and select the lifter type and capacity based on this value. The same lifting anchor is used throughout the precast element. Different anchor capacities are to be avoided in the same element as this creates the risk of the wrong anchor being placed during manufacture at the location with the worst case loading.

It is not always possible to use only two lifting anchors; when it is necessary to lift with three or more anchors the lifting equipment needs to be arranged to ensure the loading applied to each anchor is within its capacity. The use of compensating chain rigs, pulleys, lifting beams and block and tackle equipment are some of the methods to control the loading onto each lifting anchor. This is quite a complicated topic and outside the scope of this book, but the main point is that designers need to have an awareness regarding the practicalities of using multiple lifting anchors and seek out guidance from experienced personnel when this situation arises.

Once all of the relevant points have been considered the selection of the lifter is actually quite simple, but it is essential that the lifting of the precast element is considered carefully at each stage, i.e. demoulding, handling in the yard, transportation, and erection on site.

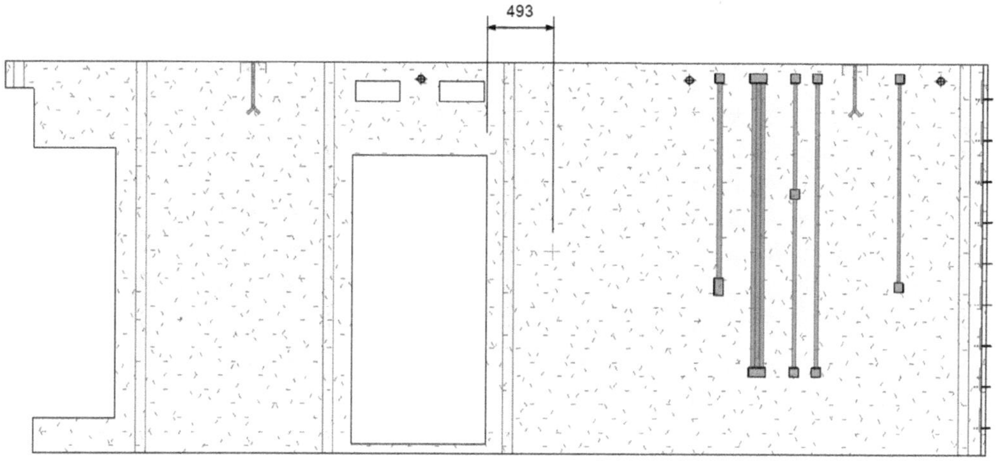

Figure 15.7 Centre of gravity position on wall panel W-A64-06.

These different stages may all use the same lifting anchors but any one of them could apply the maximum stress to the lifters, the concrete or the reinforcement.

The wall panel shown in Figure 15.7 and all precast elements require a number of points to be considered when designing the lifting anchors and the elements reinforcement. An example of a checklist is given below.

1. How is the precast element manufactured and demoulded?
2. What is the concrete strength at demoulding?
3. What is the adhesion coefficient at demoulding?
4. If the element is pitched from horizontal to vertical what reinforcement quantity is required for pitching, this may occur at low concrete strengths
5. Does the element need to be turned or rotated in the factory as part of the finishing process?
6. How will the element be transported to site?
7. How will the element be erected on site?
8. What is the chain angle coefficient at each stage?
9. What is the dynamic factor for the craneage being used at each stage?

   Specific points relating to the wall panel shown in Figure 15.7 are

10. The CofG needs to be accurately determined due to the irregular geometry and the door opening in the panel
11. There are a considerable number of M&E backboxes on the right-hand side that will limit where lifting anchor can be positioned
12. The door header has two penetrations through it to allow for the routing of M&E ducting, the reinforcement design and details in the header will need to account for these holes and deal with the handling stresses during lifting.

15.2.8 Lifting anchor design

The wall panel shown in Figure 15.7 is manufactured on a tilting table with oiled steel formwork. The demoulding cube strength is 15 N/mm^2, mould adhesion is 1 kN/m^2, the panel is not pitched and does not require turning in the yard, panel is sent to site upright on an

A-frame, panel is erected on site directly from the A-frame. Chain angle α in the factory is maximum 90°, dynamic factor in the factory is 1.1. Chain angle α on site is maximum 60°, dynamic factor on site is 1.3.

Try lifting anchor CFS-FE-75-320 (erection and tilting anchor)

See the manufacturer's data in Figures 15.8 and 15.9 and Table 15.3. Use concrete cube strength of minimum 15 N/mm^2.

Wall panel outline dimensions

Thickness 200 mm
Height 2885 mm
Length 7368 mm

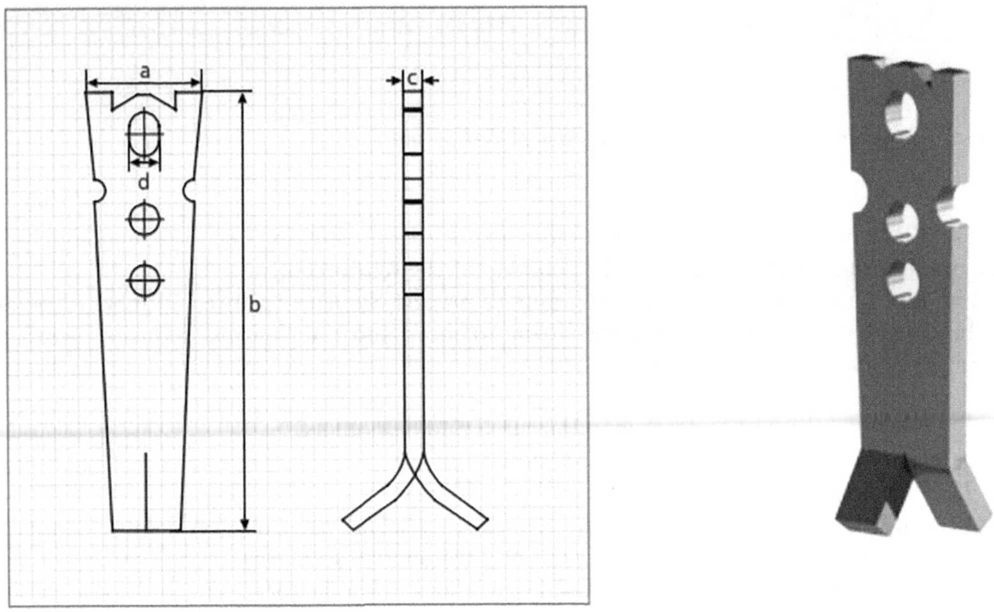

Figure 15.8 Quick-lift erection and tilting anchor. (Courtesy of Construction Fixing Systems, UK.)

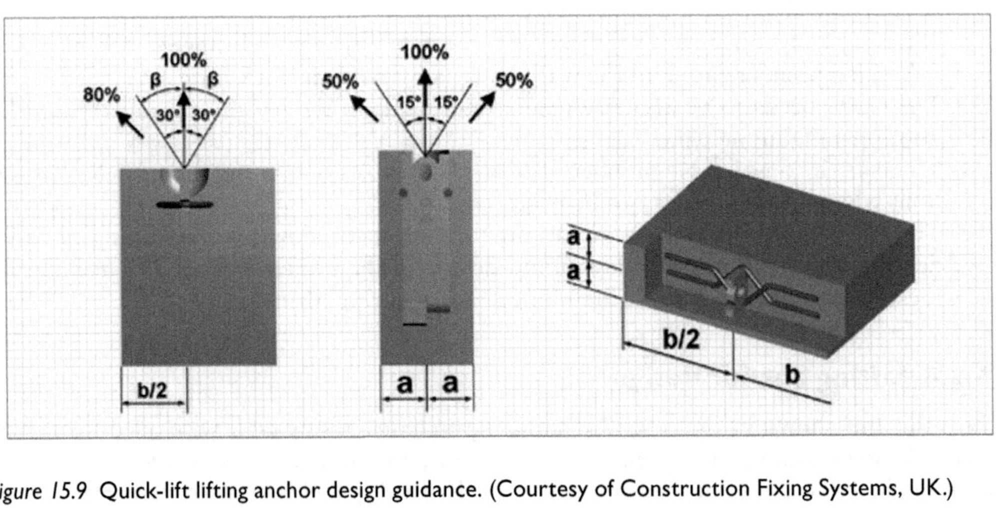

Figure 15.9 Quick-lift lifting anchor design guidance. (Courtesy of Construction Fixing Systems, UK.)

Table 15.3 FE lifting anchor capacity

		Minimum thickness of wall		*Load capacities*		
	Minimum spacing	*With tail reinf.*[a]	*Without tail reinf.*	*Axial or angled load ≤ 30°*	*Angled load 45° > β ≤ 30°*	*Tilting load*
	b	*2a*	*2a*			
Part No.	*mm*	*mm*	*mm*	*kN*	*kN*	*kN*
CFS-FE-14-200	700	100	100	14.0	11.2	7.0
CFS-FE-25-150	550	120	120	25.0	20.0	12.5
CFS-FE-25-230	800	120	120	25.0	20.0	12.5
CFS-FE-40-270	950	150	150	40.0	32.0	20.0
CFS-FE-50-290	1000	160	180	50.0	40.0	25.0
CFS-FE-75-320	1200	175	200	75.0	60.0	37.5
CFS-FE-100-390	1500	200	250	100.0	80.0	50.0

Minimum edge distance = *b*/2. Minimum concrete cube strength 15 N/mm².

[a] Tail reinforcement is a U-bar hooked through the lifting anchor that extends into the body of the concrete element.

The first operation is to break the adhesion to the mould surface, this can be carried out while the panel is in the as cast horizontal position on the mould prior to tilting or once the mould has been tilted and the panel is almost vertical. In this example, the panel is manufactured on a tilting table, and therefore the majority of the self-weight will be supported on the mould – a simple conservative approach is to use 25% of the self-weight in the calculation. In the worst-case situation when the panel is horizontal the self-weight load applied to the lifters must be calculated, for example in some cases the panel may have deep headers over door openings that will lead to a greater percentage of load on the lifting anchors.

1. Lifting operation in factory, break adhesion to mould (ignores openings in panel)
   Adhesion to mould $f_{adh} = 2.885\text{ m} \times 7.368\text{ m} \times 1\text{ kN/m}^2 = 21.3\text{ kN}$
   25% of unit self weight $G_{25} = 83.5 \times 0.25 = 20.9\text{ kN}$
   Chain angle factor; $f_c = 1.41$
   Factored load; $G_{fs} = (G_{25} + f_{adh})\, f_c = 59.5\text{ kN}$
   Number of lifting anchors; $n = 2$
   Load on each lifting anchor; $F_{la} = G_{fs}/n = 29.8\text{ kN}$
   Use worst-case tilting capacity of anchor
   Lifting anchor capacity; $F_{cap} = 37.5\text{ kN}$; Lifting anchor capacity OK

2. Lifting operation in factory, vertical from tilting table
   Unit self weight; $G_{unit} = 83.5\text{ kN}$
   Dynamic factor; $f_d = 1.1$
   Chain angle factor; $f_c = 1.41$
   Factored load; $G_{fs} = G_{unit}\, f_d\, f_c = 129.5\text{ kN}$
   Number of lifting anchors; $n = 2$
   Load on each lifting anchor; $F_{la} = G_{fs}/n = 64.8\text{ kN}$
   Lifting anchor capacity; $F_{cap} = 75\text{ kN}$; Lifting anchor capacity OK

3. Lifting operation on site, vertical from A-frame
   Unit self weight; $G_{unit} = 83.5\text{kN}$
   Dynamic factor; $f_d = 1.3$
   Chain angle factor; $f_c = 1.16$
   Factored load; $G_{fs} = G_{unit}\, f_d\, f_c = 125.9$ kN
   Number of lifting anchors; $n = 2$
   Load on each lifting anchor; $F_{la} = G_{fs}/n = 63.0$ kN
   Lifting anchor capacity; $F_{cap} = 75$ kN; Lifting anchor capacity OK

15.2.9 Design and control of onsite temporary works

When designing temporary works, Health & Safety is always the first and most important consideration. The control over temporary works on site is an integral part of the design process, and thought needs to be given to how best to communicate with the site team to mitigate any residual risk. For precast temporary works, using wall panels as an example, the use of props and other restraint fixings occurs as each wall is installed by the crane. The type of prop, the number of props, their position and connections are all given on the temporary works drawings.

The people involved in each stage of the temporary work process must be qualified and have adequate experience to perform the role assigned to them. It is the responsibility of the Designated Individual within the company to appoint the following personnel in writing.

a. Temporary Works Designer (TWD) – e.g. Company engineer
b. Temporary Works Design Checker (TWDC) – e.g. Company engineer or independent engineer
c. Temporary Works Co-ordinator (TWC) – e.g. Project Manager
d. Temporary Works Supervisor (TWS) – e.g. Site Manager

The following is a general example of a procedure for control of temporary works.

15.2.10 Temporary works procedure

The purpose of this procedure is to ensure design, installation, maintenance and dismantling of temporary site propping is carried out in a safe and controlled manner. Primarily, this is achieved by moving from site-based control to a permit to release system controlled by the TWD. Figure 15.10 shows a typical permit to unload, with the required information provided on it to safely control the removal of propping. This ensures that site pressures do not affect the control of temporary works and props are released using a safe and engineered process.

Only the TWD may specify and authorise the removal of propping.

The structure is to be erected in accordance with delivery and erection sequence.

It is the responsibility of Site Manager/TWS to implement the use of prop permits and follow the delivery and erection sequence.

All temporary works designated individuals are appointed by the designated individual.

Selection is based on criteria as required in BS 5975 (BS 5975, 2019).

It is a requirement that all individuals must have attended and passed, Safety in the Management & Coordination of Temporary Works Institute of Occupational Safety and Health (IOSH)

15.2.10.1 Design

- All propping is designed and specified by the TWD.
- All propping must be designed, have a prop procedure, maintenance, inspection and release procedure, all in accordance with the latest revision of Delivery & Erection Program.
- Where required, TWD will specify minimum ground bearing pressures or bearing plates.

<table>
<tr><td>Site - Bellevue Brae Rd</td><td>Contract - 22-0971</td><td>PERMIT NO. - 017</td></tr>
<tr><td colspan="3"><u>Permit to UNLOAD</u></td></tr>
<tr><td colspan="3"><u>(Temp Works must not be unloaded unless Both Stage 1 and stage 2 have been signed off)</u></td></tr>
<tr><td colspan="3">Stage 1: - All Safety measures in accordance with Method Statements, Risk Assessments, and Task Sheets have been executed and checked, e.g. Hole Protection, Leading Edge, etc.</td></tr>
<tr><td colspan="3">Stage 2: - All necessary checks have been carried out and the Temporary Works described may be unloaded in accordance with the Method Statements, Risk Assessments and Task Sheets</td></tr>
<tr><td colspan="2">TWS Signature J. Maunsell</td><td>Date 10/06/2021</td></tr>
<tr><td colspan="3">Prop Locations</td></tr>
<tr><td>Drawing Nos.</td><td>Floor/Level</td><td>Grid Lines</td></tr>
<tr><td>22-0971-CCP-BA-04-DR-Y-1517</td><td>Level 17 walls</td><td>BA-09 to BA-14</td></tr>
<tr><td></td><td></td><td></td></tr>
<tr><td colspan="3"><u>All Drawings to be marked up with props that are to be removed</u></td></tr>
<tr><td colspan="3">Grout Curing</td></tr>
<tr><td>Drawing Nos.</td><td>Fill - Time/Date</td><td>Cure - Time/Date</td></tr>
<tr><td>22-0971-CCP-BA-04-DR-Y-1017</td><td>03/06/2022</td><td>10/06/2022</td></tr>
<tr><td></td><td></td><td></td></tr>
<tr><td colspan="3"><u>All grout sheets for these walls to be submitted</u></td></tr>
<tr><td colspan="3">Temperatures</td></tr>
</table>

Figure 15.10a Typical permit to unload

Day time - Coldest	**Night time - Coldest**
19 degrees	12 degrees
Notes/Instructions	
Level 17 perimeter concrete stitching has been completed on 20/05/2022 on gridlines BA-09 to BA-14. All vertical connections on walls on Level 17 have been filled with Parex GP grout. All push-pull props between GL BA-09 & GL BA-14 on Level 17 walls to be removed. (highlighted on drawing attached)	
Have Marked up Drawings been included? YES	
Have filled in Grout Sheets been included? YES	
TWS Signature J. Maunsell	**Date** 10/06/2021
TWD Signature Mark Magill	**Date** 11/06/2021

Figure 15.10b Typical permit to unload

- The design team will produce a propping drawing for each location where propping in the temporary condition is required.
- Propping drawings will include component mark numbers, prop reference, length and fixings, including length and size.
- TWD will issue Brief Design, methodology and release procedure. This will form part of the project Method Statement.
- Temporary propping may only be released on written authorisation of TWD.
- Temporary Works Design is checked by a suitably qualified checker TWDC.
- Procedure controlled by permit to UNLOAD, permits must be signed by TWS with verifying documentation provided and countersigned by TWD before releasing any props.

15.2.10.2 Procurement

- No deviation from TWD requirements is permitted without written authorisation. This includes any change in manufacturer or supplier of any product.

15.2.10.3 Construction

- TWD is responsible for TW Design and provides The TWC responsible for implementation.
- TWD is to provide all information to Project Manager/TWC and Site Manager/TWS.

- Site Manager/TWS to inform all operatives on site and ensure all aspects of site propping procedure including signage are implemented and contained within the Safety Method Statement.
- All site personnel will receive and sign for Temporary Works Method Statement covering temporary propping requirements.
- Temporary Propping Procedure includes Project specific Method Statement and forms part of site induction.
- TWS will ensure that props and prop fixings conform to prop specifications, drawings and procedure.
- Props will be installed per unit as noted on prop drawing. Crane will not be released until props are secured and safe.
- Warning notices will be placed on each prop stating

15.2.10.4 Danger structural propping do not remove

- TWD will determine and specify the inspection/maintenance regime. TWS will inspect and records will be kept on file in site office.
- Request for prop release by TWS must include the props to be removed, marked up on a Temporary Works Drawing or clearly identified props on photographs all submitted with the "PERMIT TO UNLOAD".
- Following review of information received that confirms all criteria have been met the TWD will sign the "PERMIT TO UNLOAD" to authorise which props may be released.
- Using this information, the TWS will either physically identify props that can be released or give a copy of the drawing and the "PERMIT TO UNLOAD" to his designated representative to release the props
- Props may be removed.
- Propping must be inspected and adjusted where required by TWD propping inspections.
- TWD must specify time periods and adjustments within the Design Brief as described in the Method Statement.
- Compliance of site propping will be in accordance with Temporary Works Inspection Checklist
- No deviation is permitted from this procedure.
- In addition to Temporary Props there may be steel angles and plates used as Temporary Support to structure and adjacent walls – these are to follow the same process install/secure/tag before crane release and follow Temporary Prop Procedure

15.2.10.5 Unplanned propping requirements

As is the case with even the most well-designed construction activities circumstances can arise when temporary propping may be required that has not formed part of the pre-project design and planning. In this situation, all the conditions noted in the above procedure apply.

Propping must be specified by TWD and not from site. It is the responsibility of the TWS to get the relevant information to the TWD for an engineering solution and then to input requirements on site.

Any elements that cannot be installed must be placed back on the ground until a propping drawing or sketch is received from the TWD. Elements must not be left in position on temporary props or suspended from the crane while awaiting the engineer's solution.

15.2.11 Temporary works design in accordance with BS 5975

Design of propping for precast elements is carried out using the guidance given in BS 5975. Loading onto props and their connections comes from a number of sources. These are typically wind pressure, an allowance for verticality deviation and the placement of other elements onto the propped element. The latter two cases are covered by guidance in Clauses 19.2.4 and 17.4.3.4 of BS 5975, respectively.

A load case for the working condition is given in Clause 17.5.1.9. It is unrealistic for the loads from Clause 17.4.3.4 to be applied at the same time as the peak wind loading, which is based on the maximum for the return period. A minimum working wind load of 0.2 kN/m^2 is recommended for use when for example installing precast slabs onto propped precast walls.

The following is an example of a design statement and calculation for simple push-pull propping. The statement prior to the calculation provides information on the design check category and key points for the site team to incorporate into the Method Statement.

15.2.12 Design Statement for push-pull props used on stair core wall panels

1. Design Check Category (reference BS5975, Table 2)
   Category 1.
   Independence of checker – in house by company engineer.
2. Design Engineer Site Visit Requirements.
   No specific requirements.
3. Erection Hold Points.
   At each level, all vertical and horizontal joints between walls must be grouted before the next level of walls is erected.
   Landing slabs and walls cannot support props until the grouted connections have reached a minimum cube strength of 20 N/mm^2, typically at 1 day for 'Parex GP' grout or similar approved and 2 hours for 'Parex Speedsure S' grout or similar approved.
   No props or temporary supports may be released without written permission from TWD in accordance with the temporary works procedure.
4. Checking of props to ensure they are tight.
   No checks are required once vertical wall joints are grouted and horizontal slab joints are concreted. A daily check is required until grouting and concrete works are completed at each level.

15.2.13 Design of push-pull propping

15.2.13.1 Wind load

Refer to BS EN 1991, Part 1-4 (BS EN 1991-1-4, 2005) and the UK National Annex NA to BS EN 1991-1-4 (NA to BS EN 1991-1-4, 2005), for wind actions on structures,

Wall panel height $h_p = 2650$ mm
Site orography not significant $c_o = 1$
Ignore nearby buildings or obstructions
Distance from ground to base of wall $h_b = 3000$ mm
Reference height above ground for wind pressure $z = 5.65$ m
For free standing walls, reference height = wall height (Clause 7.4.1 (2) of BS EN 1991-1-4)

Terrain type taken as 'Country'
Distance to shoreline L_{shore}=100.0 km
Fundamental basic wind velocity (taken from the wind map for the UK in Figure NA.1 of NA to BS EN 1991-1-4) $v_{b,map}$ = 21.7 m/s
Site reference height to sea level A = 120.0 m
Altitude factor $c_{alt} = 1 + 0.001 \times 120 = 1.12$
Fundamental value of the basic wind velocity $v_{b,0} = 1.12 \times 21.7 = 24.3$ m/s
Directional factor $c_{dir} = 1$
Season factor $c_{season} = 1$
Distance inside town = 5.0 km
Return period T_{py} = 2 years
Probability factor $c_{prob} = 0.776$
Basic Wind Velocity $v_b = 0.776 \times 24.3 = 18.9$ m/s
Reference wind pressure $q_b = 0.613 v_b^2 \times 10^{-3} = 0.218$ kN/m²
Exposure Factor $c_e(z) = 2.0$
Peak Velocity Pressure $q_p = 2.0 \times 0.22 = 0.436$ kN/m²
Force coefficient – Table 17 BS5975;, $c_f = 2.0$;
Wind pressure $f_w = c_f\, q_p = 0.87$ kN/m²
Wall panel height; h_p = 2650 mm;
Length of wall panel; l_p = 8672 mm;
Number of props; n_p = 2 (equally loaded)

Care must be taken to ensure props are positioned on the wall panel so they are equally loaded. To ensure equal loading the spacing of the props should be checked based on the tributary area they support. This principle may also be applied when using more than two props. If it is not possible to avoid one of the props supporting a greater load than the others, this worst case is used in the design and all props and connections are sized for this load.

15.2.13.2 Prop connection to slab

Figure 15.11 shows the wall and prop geometry showing the connection position of the prop head on the wall and the prop base on the floor slab. These two critical dimensions are provided on the temporary works drawings so the site team install the props at the correct angle. Continuing the example above:

Horizontal distance from wall; a_h = 2000 mm;
Height to prop position on panel; b = 2530 mm;
Angle of prop to horizontal; $\alpha_p = \tan^{-1}(b / a_h) = 52°$;
Prop angle < 60° OK

15.2.13.3 Prop design

Moment applied to wall; $M_{aw} = f_w\,(h_p\, l_p)\, h_p/2 = 26.6$ kNm
Length of prop; $c = \sqrt{(b^2 + a_h^2)} = 3225$ mm
Load on prop due to wind; $W_p = [((M_{aw}/b)/n_p)/a_h] \times c = 8.5$ kN

Referring to Figure 15.12, Number 3 push-pull prop compression capacity at 3225 mm extension is at least 20 kN. Note that Figure 15.12 shows a typical push-pull capacity graph, although each prop manufacturer provides capacity data for their products.

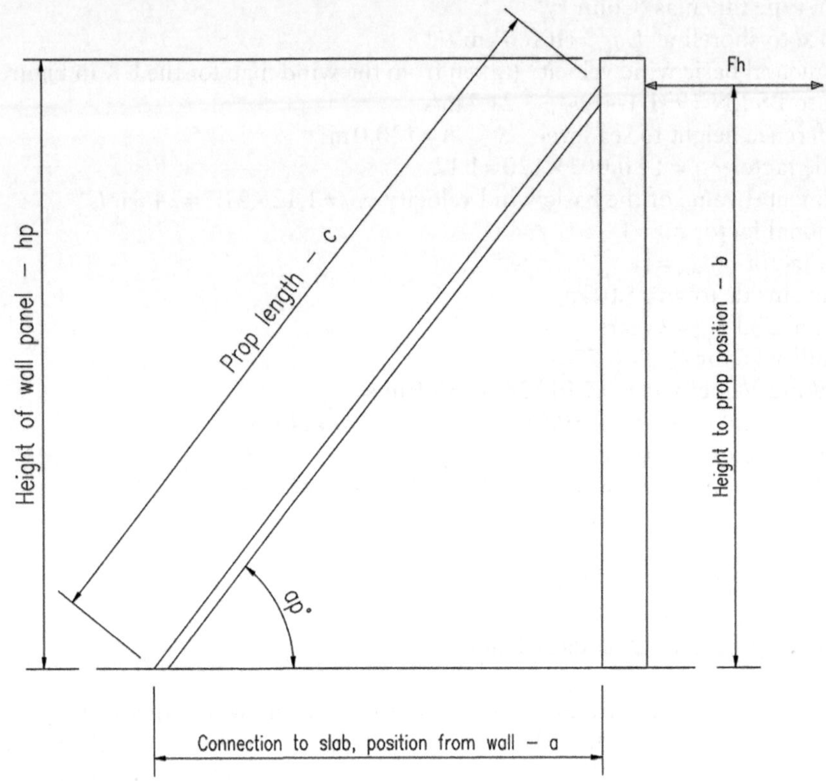

Figure 15.11 Wall and prop geometry

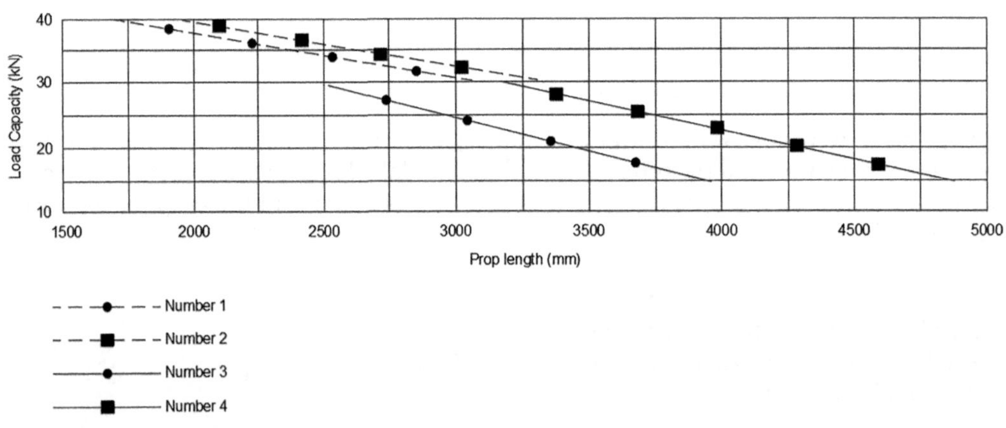

Figure 15.12 Push-pull prop capacity graph

Note that this example prop and connection design does not take into account the additional moment from the self-weight of the panel due to verticality deviation according to BS5975, Clause 19.2.4.

When designing the temporary works for precast sandwich panels, it is very important to consider the additional moment applied to the structural leaf due to the eccentricity of the suspended outer façade leaf. This can be a significant additional moment due to the projecting geometry of some architectural facades.

15.2.13.4 Design of fixing cast into wall panel for prop head connection

Consider a M16 solid rod socket – tension and shear capacity safe working loads for the cast-in fixings are based on a 120 mm edge distance, minimum grade of concrete is C35/45. A reinforcement U-bar is present around the socket.

15.2.13.5 Wind direction putting prop in tension

Design for ultimate limit state with $\gamma_w = 1.5$;
Tension load on prop connection; $F_b = \gamma_w (M_{aw}/b)/n_p = 7.9$ kN
Tension capacity M16 solid rod socket; $T_c = 28.6$ kN; from Table 15.4
Pull out capacity OK
Shear load on prop connection; $F_v = \tan(\alpha_p)\ F_b = 10.0$ kN
Shear capacity M16 solid rod socket; $V_c = 21.9$ kN; from Table 15.4
Shear capacity OK
Note that a U-bar is required to achieve this shear capacity, Figure 15.13 shows the necessary reinforcement detail referred to as a shear U-bar in Table 15.4
Component check; $\rho_{cc} = (F_b/T_c) + (F_v/V_c) = 0.73 < 1.2$ fixing OK

Table 15.4 Capacity of solid rod sockets

Part No.	H_{ef} (mm)	Length of cross pin (mm)	Edge distance m (m) $c_1 = c_2$	$c_3 = c_4$	Element thickness (mm)	Shear U-bar ϕ (mm)	L_b (mm)	Concrete strength grade C35/45 N_{Rd} (kN)	V_{Rd} (kN) Shear U-bar No	Yes
CFS-LSRB(S)-10-50	38	50	80	55	80	6	100	16.1	6.1	12.3
CFS-LSRB(S)-12-50	34	75	90	55	80	6	100	17.0	6.1	12.3
CFS-LSRB(S)-12-75	59	75	125	90	100	8	130	23.1	12.8	21.9
CFS-LSRB(S)-16-75	56	75	120	80	100	8	130	28.6	11.3	21.9
CFS-LSRB(S)-20-75	51	90	120	75	100	8	130	28.0	10.5	21.9
CFS-LSRB(S)-24-100	70	100	160	100	130	12	200	40.7	16.8	49.2

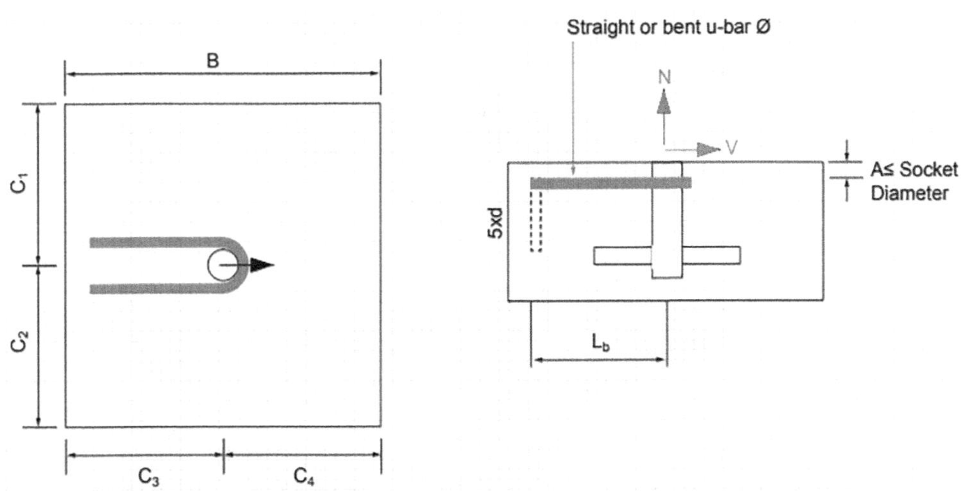

Figure 15.13 Solid rod socket design guidance. (Courtesy of Construction Fixing Systems, UK.)

A similar check is carried out for the connection at the base of the prop. This may be a different type of cast-in or post-fixed detail dependent on the type of floor or member that is supporting the prop base.

A typical fully precast apartment structure is shown in Figure 15.14 where push-pull props supporting the crosswalls and sandwich panels can be seen prior to the installation of the next level of hcu. Bathroom pods have been lifted into place and positioned at approximately their final location. The propping drawings will include the bathroom pods drawn in their location so that the props can be positioned and designed to avoid clashes. Figure 15.14 illustrates one of the major benefits of precast concrete construction – minimal and very simple propping is required to erect a precast structure.

15.2.14 Build sequence and temporary works

The design of temporary works to ensure an efficient build methodology on site is essential to maximise programme savings. The following example demonstrates a way to construct the lift shafts and stair core area of an apartment scheme in a safe and practical manner.

The lift shaft in this project is manufactured as separate walls that are bolted together in the precast yard and the propping is installed prior to dispatch to site. The F-shape formed is stable when erected without the need for any additional site fixed props or other temporary works. The props shown are used to brace the walls during lifting to prevent any movement that would cause a fitting or alignment issue with the F-shape assemblies above and/or below. Figure 15.15 shows the F shape assembly as it looks in the BIM model. The brackets visible on the headers of the door walls are provided to support the temporary steel

Figure 15.14 Precast apartment structure during construction.

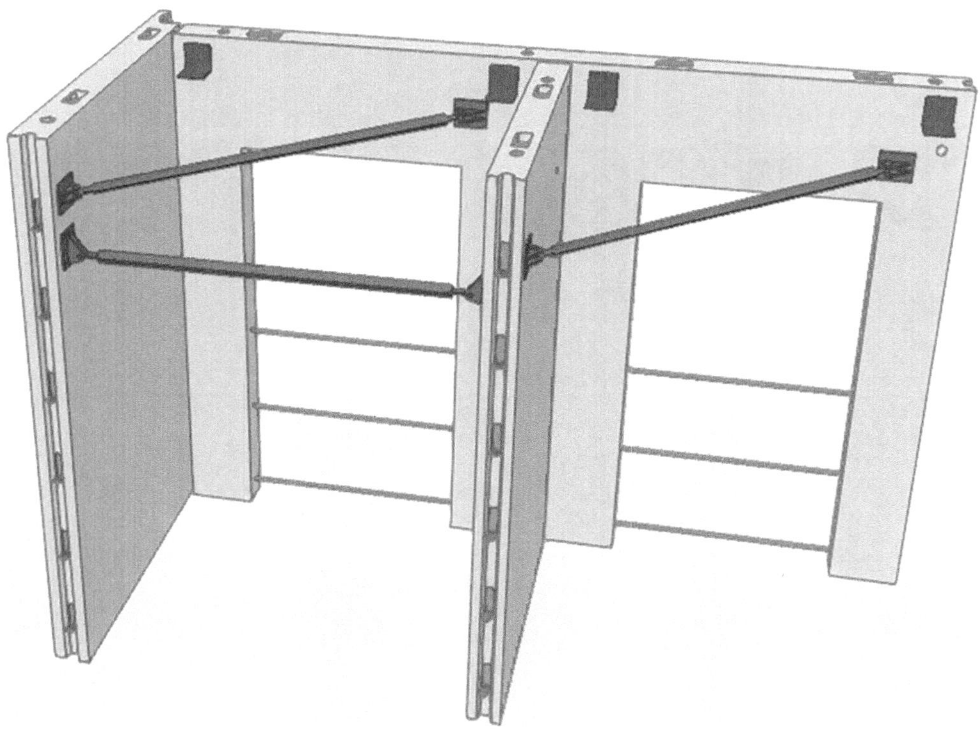

Figure 15.15 F-shape assembly from BIM model.

mesh floor that is placed within lift shafts to allow finishing works to be carried out on the internal vertical and horizontal joints between the precast elements.

The bars cast into the walls that cross the door opening can serve two functions – edge protection barriers for the lift shaft doors are fixed to these bars and, if required as part of the design, the bars will brace the vertical pier sections on each side of the door during demoulding and handling.

The erection sequence allows for 18 no. pieces to be fitted without the need to prop off any hcu. All propping and restraint plates are designed and detailed to be either wall to wall or wall to solid precast slabs. This ensures the crane is kept working and installation does not have to stop to allow *insitu* concrete works associated with the hcu to cure.

Using 'Speedsure' grout, or a similar approved rapid curing grout, in the connections between wall-to-wall and wall-to-landing slabs it is possible to build this area in 2 days.

The temporary platforms can be seen as grills at the top of the lift shafts in Figure 15.17. Similar temporary platforms (not shown on this snapshot) were used to cover each of the large openings visible in the floors. The hcu are not shown in Figure 15.16, they are hidden from view, when installed they form the floor in front of the lifts and in the small room at the top right corner.

Referring to Figure 15.17, the erection sequence for the walls is as follows

1. F-shape lift shaft wall assembly, walls C3, C7 and C8
2. Wall C4 – 1 no. restraint plate to wall C7, 1 no. restraint plate to wall C8 1 no. prop wall to slab L1
3. Wall C5 – 1 no. restraint plate to wall C4, 1 no. prop wall to slab L1

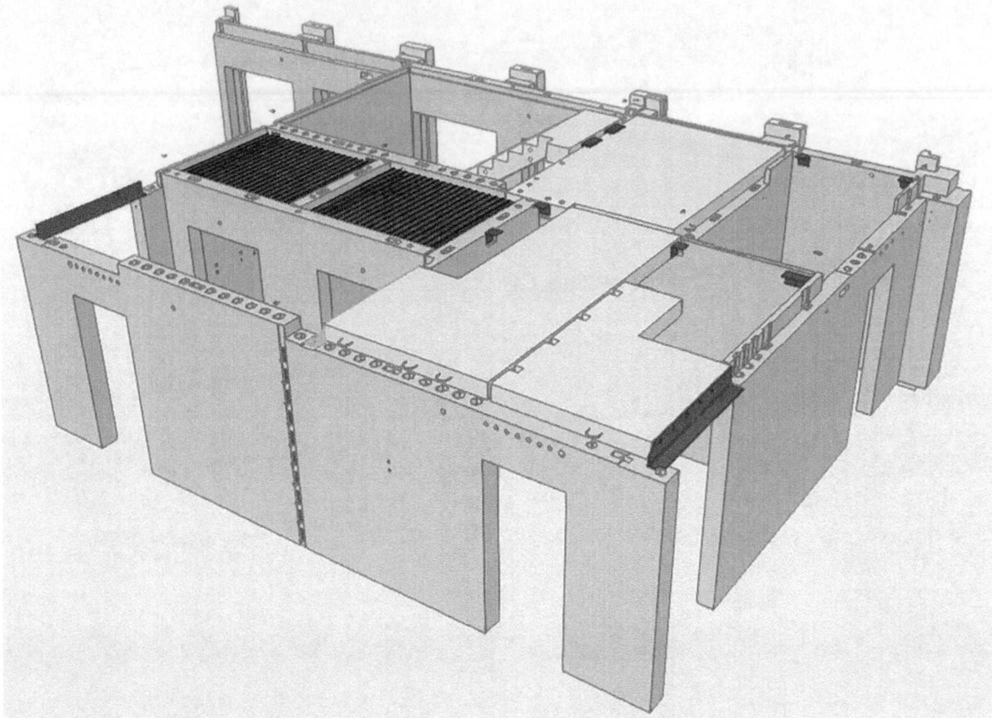

Figure 15.16 3D view of core and lift shaft area.

4. Wall C2 – 1 no. restraint plate to wall C4, 1 no. prop wall to slab RC1, 1 no. prop wall to wall C5
5. Wall C6 – 1 no. prop wall to slab RC1, 1 no. prop wall to wall C3
6. Wall C1 – 1 no. restraint plate to wall C3, 1 no. restraint plate to wall C4
7. Sandwich panel N3 – 1 no. restraint plate to wall C2, 1 no. restraint plate to wall C5, 1 no. prop wall to slab L1
8. Sandwich panel N2 – 1 no. restraint plate to panel N3, 1 no. restraint plate to wall C1, 1 no. prop wall to wall C1
9. Install next level of slab L1
10. Install next level of slab RC1
11. Install next level of slab RC2
12. Install stair flight SF1
13. Install stair flight SF2
14. Install corridor beam CB1

The 4 no. hcu can now be installed in front of the lift shafts and the small room in the top right corner of the snapshot to complete this section of the building. Figure 15.17 shows the plan view of the area with element references used for the erection sequence and the location of the prop and restraint plates.

The restraint plates used for temporary support are typically manufactured from 10 to 12 mm thick steel plate and the top of the walls are recessed where they are located. These recesses prevent the plates and the head of the bolts used to fix them from clashing with the next level of walls and slabs when they are erected.

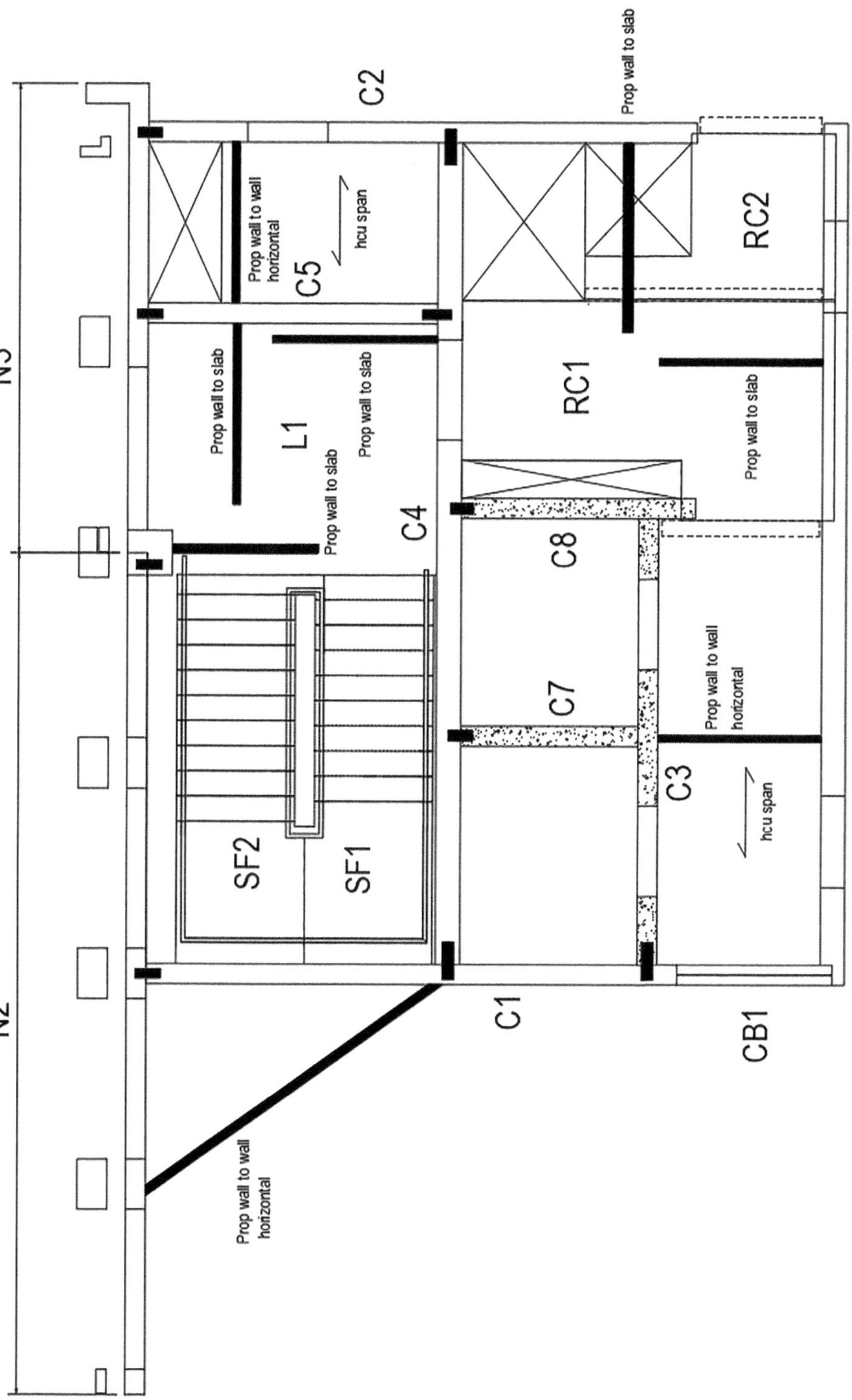

Figure 15.17 Erection sequence plan.

15.3 HOLLOW CORE UNITS (HCU) INSTALLATION

The fitting of hcu can be carried out using lorry-mounted off-loaders like the one shown in Figure 1.20. This is a suitable and cost-effective method for slabs at ground and first floors on sites that allow close access to the building. The slab in the photo is a ribbed slab but hcu is fitted in the same manner. Figure 1.20 also shows a clamp being used to lift the slabs, clamps are usually restricted to low-level fitting such as the building in the photo. The most common method of fitting hcu is by tower crane or mobile crane. Figure 15.18 shows hcu being installed onto a steel frame. The slabs are being fitted using lifting anchors cast in during production, see Figure 4.19. The slab also has safety chains wrapped around it at each end to act as a secondary means of supporting the slab in the unlikely event of an issue with the lifting pins. The slab is landed onto its bearing as shown in Figure 15.19, the safety chains are removed and the slab is moved into its final position butted up to the adjacent slab. Slings and chains can be wrapped around the hcu to lift them but neither of these are preferred methods, the chain causes damage to the sides and soffit of the slabs and both methods require the slab to be moved into its final position using a crowbar to manually slide it on its bearing into its final position bitted up to the adjacent slab. Another method is the use of lifting hooks placed into the ends of the hollow cores, this is no longer a common method in the UK and Ireland and it is rarely if ever used on sites now.

The installation works shown in Figure 15.18 uses netting fixed to the underside of the steel frame to prevent falls from height, Figure 15.19 shows installers wearing harnesses and connected to fall arrest lanyards to prevent falls from height. The edge protection is provided by a proprietary barrier system and scaffolding.

When installing long span hcu, the sequence of installation and/or temporary works may need to be considered due to the torsion applied to supporting beams and their

Figure 15.18 Mobile crane fitting hcu onto steel frame.

Figure 15.19 Installation of hcu with lifting anchors.

connections. For edge beams, temporary propping can be installed to prevent excessive torsion forces causing a problem, the props are released once the hcu have their ends concreted and are connected to the beams. For internal beams the sequence of the installation may be arranged to load each side of the supporting beams one hcu at a time to prevent excessive torsion, this is not always practical on site so it is more common to provide temporary bracing or propping to internal beams so that each bay of hcu can be fully installed.

Where a structural topping is being poured to create a composite hcu, temporary propping of the hcu is sometimes required to maximise the load-carrying capacity of the available structural slab depth. Typically this will be a single line of props at midspan. It is common practice to concrete the longitudinal joints and slab ends of the hcu prior to casting the structural topping.

Figure 15.20 shows a hcu layout drawing for an apartment block. The main supporting structure is constructed from precast crosswalls, with steel beams supporting Slab Marks 5, 6, 13, 14 and 15. The production detail drawing for Slab Mark 1 is shown in Figure 15.21, the details for open cores, solid ends, holes, notches, lifting pins and strand patterns are all given on these drawings. The drawings are used by the production team to create the specific details required for each hcu. The notches in the ends of the hcu allow vertical tie bar reinforcement to be installed on site. Figure 15.22 shows the bearing details in Section 4-4 and 6-6 on the hcu layout drawing, the correct bearing length at installation is very important to ensure the hcu are safely supported in the temporary state. If the noted bearing cannot be achieved the amount of bearing that can be achieved is measured and the slab is returned to ground level, the engineer responsible for the design is consulted to determine if the amount of bearing that can be achieved is satisfactory. If the bearing is not adequate, temporary

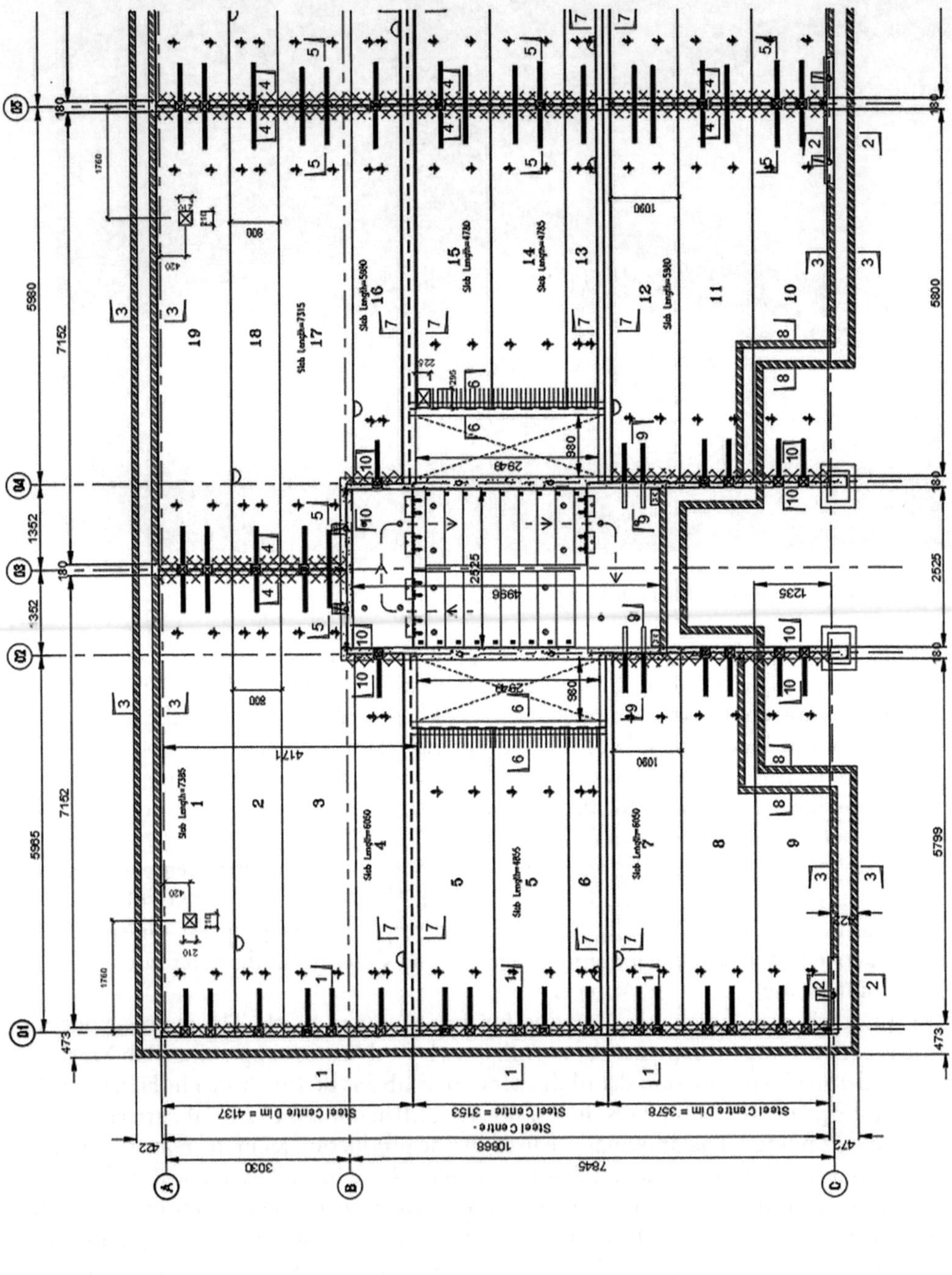

Figure 15.20 Plan drawing of hcu layout for an apartment building.

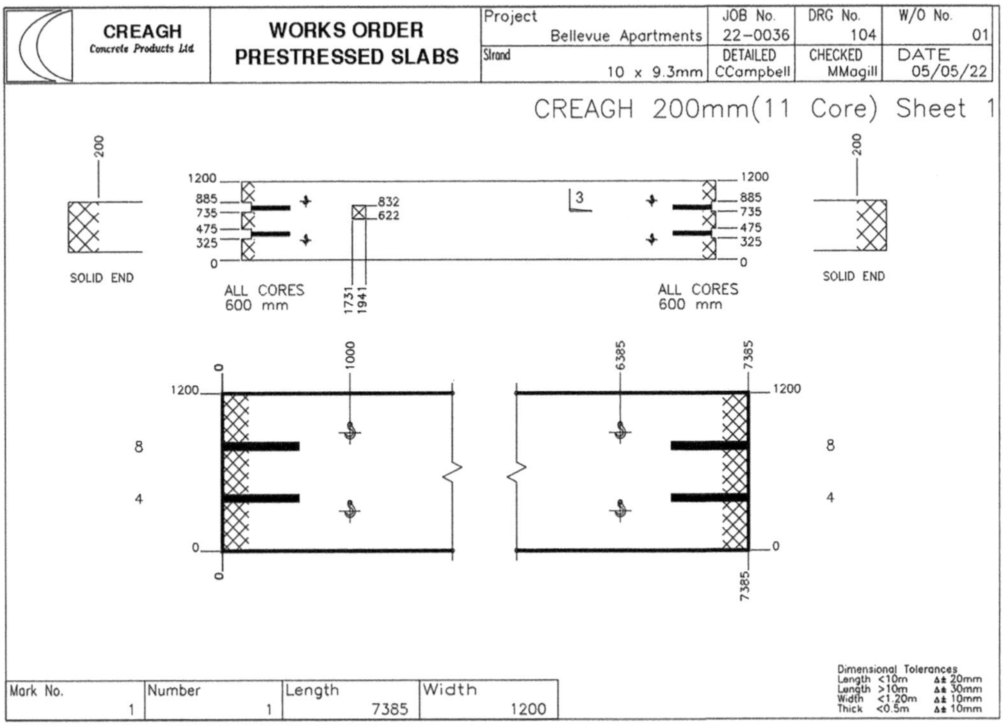

Figure 15.21 Production detail drawing for hcu, slab mark 1 on Figure 15.22.

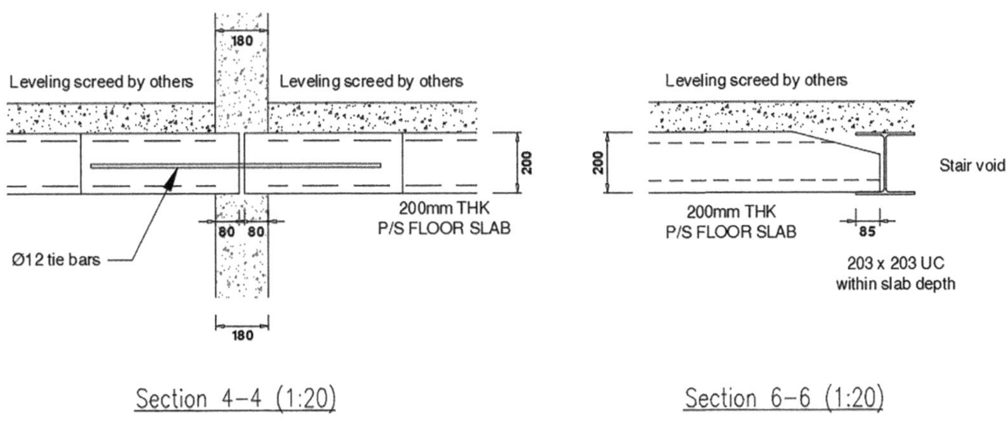

Figure 15.22 Bearing details for hcu onto precast wall and steel beam.

works can be designed to support the hcu and the permanent works design is checked to ensure that once the *insitu* concrete infill to longitudinal joints and the ties at the slab ends are in place the bearing is adequate.

The maximum number of planned lifts per day for hcu is usually 40. This would generally be the target for slabs of 150–250 mm deep with spans of approximately 6–10 m. For example, 40 no. 150 mm deep, 1.2 m wide slabs at 6.5 m long would cover 312 m^2 per day; 40 no. 250 mm deep, 1.2 m wide slabs at 10 m long would cover 480 m^2. These values will vary depending on the site and obviously longer slabs will cover larger areas with the same number of lifts. However, the logistics of deliveries via lorries to site also has to be factored in, for example a 450 mm deep, 1.2 m wide slab at 13.5 m long weighs approximately 8.6 tonnes, this means there will only be 3 slabs on a standard load, for 150 mm deep slabs at 6.5 m, 13 slabs could be delivered on a standard load. The number of loads that can be delivered to site will vary depending on location, access and other logistic factors but the average would be seven loads per day. The height of the building has an effect on installation but as the lifting speeds for hcu on tower cranes can comfortably be 45 m/min or more the additional time for fitting onto medium rise buildings is not that significant, the fitting of hcu down through fully erected steel frames tends to be a much slower process than lifting up onto tall structures.

REFERENCES

BS 5975. 2019. Code of practice for temporary works procedures and the permissible stress design of falsework, BSI, London, UK.

BS EN 1991-1-4. 2005. Eurocode 1: Actions on structures - General actions - wind actions, BSI, London, UK.

BS EN 1992-1-2. 2004. Eurocode 2: Design of concrete structures, Part 1-2: General rules - Structural fire design, +A1:2019, BSI, London, UK.

BS EN 13501-1. 2018. Fire classification of construction products and building elements - Classification using test data from reaction to fire tests, BSI, London, UK.

NA to BS EN 1991-1-4. 2005. UK National Annex to Eurocode 1: Actions on structures - Part 1-4, BSI, London, UK.

NA to BS EN 1992-1-2. 2004. UK National Annex to Eurocode 2: Design of concrete structures, Part 1-2: General rules - Structural fire design, BSI, London, UK.

Robust Details Ltd. 2001. *Part E Robust details. 4th edition. Unit 14*, Shenley Pavilions, Milton Keynes, UK.

The Building Regulations. 2010a. *Approved Document Part L - Conservation of Fuel and Power, Volume 1, 2021 edition*, HM Government, London, UK.

The Building Regulations. 2010b. *Approved Document Part B - Fire Safety, Volumes 1 and 2, 2019 edition*, HM Government, London, UK.

The Concrete Centre. 2010. *Concrete Credentials, Sustainability*, MPA - The Concrete Centre, London, UK.

Chapter 16

Case studies

16.1 INTRODUCTION TO CASE STUDIES

The market for precast concrete structures has been expanding, developing and changing over the last 30 years in the UK and Ireland. Precast concrete has an increasing share of the market in virtually every type of building being constructed. Precast elements are widely used in various functions for structures that are built for a huge range of uses: residential, office, retail, hospitals, schools, car parks, stadia, agricultural and industrial to name a few. These buildings can also consist of various forms of construction such as steel frame, masonry, timber frame and *insitu* frame, but they all have an area or part of the structure that benefits from the inclusion of precast concrete. The use of precast can be a benefit to Health & Safety, sustainability, programme, cost and many other possible advantages.

The size of the market for fully precast structures has grown steadily in the last 10–15 years so that a noticeable trend has started to develop within the construction industry. More multi-storey buildings in the residential, hotel and student accommodation sectors are being planned and delivered as fully precast structures each year. The use of precast in the custodial sector has been commonplace for many years. The possibly unsurpassed suitability of precast design, manufacture and erection methods to the repetitive modular form of prisons makes for a very natural fit.

A range of structural types can be built using precast concrete. The most common types are beam and column frames such as offices and car parks, crosswall frames such as apartments, boxes such as lift shafts, and wall frames such as stair cores. Sandwich panels and cladding are also commonly used with steel and *insitu* frames as well as precast frames. Tables 1.1–1.3 in Chapter 1 give data related to the type of precast structures recently erected in the UK and Ireland. The case studies will expand on this data.

16.2 PRECAST CAR PARK STRUCTURES

16.2.1 Benefits of precast car park construction

Precast prestressed concrete is particularly well suited for car park construction, with benefits such as a long design life of 50 years or greater, and inherent fire resistance with periods up to 2 hours being achievable with standard product details. Architectural concrete can fulfill a structural role as well as provide a durable and aesthetic external façade. Long-span prestressed hcu give clear open areas ideal for car park layouts and traffic circulation. The snapshot from the 3D model in Figure 16.1 shows the columns, beams and stability cores with precast prestressed hollow core units (hcu) hidden from view. This illustrates the large open plan arrangement possible when using precast and prestressed concrete.

DOI: 10.1201/9781003319450-16

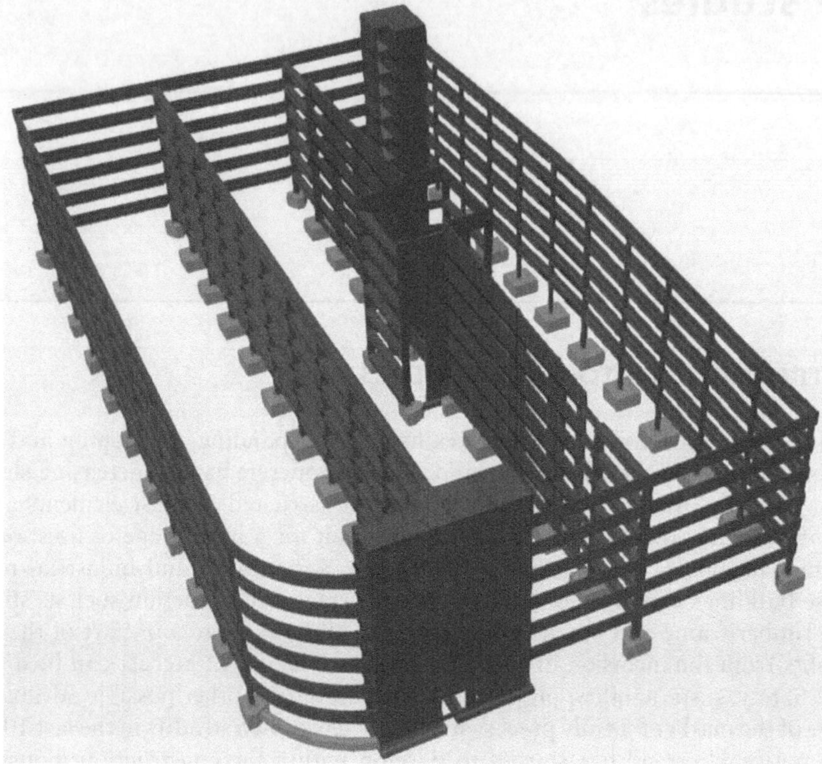

Figure 16.1 3 model extract of precast car park frame structure at Blackpool Hospital, UK.

16.2.2 Design statement

The following design statement describes the typical details for the precast concrete-framed car park shown in Figure 16.1. The particular feature of this building is the curved spandrel beams at one of the corners. The erection programme for the precast structure was 18 weeks, equating to approximately 1230 sq.m per week using a team of 8 specialist fixers and 14 labourers, working with 2 cranes.

The design statement

- General – the structure consists of six storeys. The building footprint comprises eleven bays of 7.2 m length and three perpendicular bays of 15.7 m each. The building footprint is approximately 79.2 m×47.1 m, there is a slight reduction in net area due to one corner having a curved perimeter. There is a single access ramp positioned close to the centre of the building, a movement joint alongside the ramp subdivides the car park into two independent structures. There are two stair cores in the building, one at each end of the independent structures. The total number of car parking spaces is 1120.
- Columns – these are manufactured as three-storey precast concrete elements with corbels on all faces for internal columns and corbels on three faces of perimeter columns. Columns are designed to be braced and tied to the floor slabs at each level. Ground-level columns were cast with a steel base plate; the holding down bolt connection and the base plate were designed to remove the need for any propping on site. Figure 16.2 shows the three-storey columns with two of the internal beams erected.

Figure 16.2 Multi-storey columns and internal beams for the car park shown in Figure 16.1.

- Walls – manufactured as single-storey precast elements and designed as reinforced concrete shear walls within the braced structure. Precast walls form the two stair cores, lift shafts and ramp structure.
- Main deck beams – typically inverted T-beams that span simply supported between columns. They are designed as a wholly precast reinforced section to minimise the need for temporary propping during construction.
- Spandrel beams (loadbearing) – manufactured with a continuous projecting nib to support the floor slabs and designed as simply supported spanning between columns or walls. To avoid the development of torsional stresses in the beams, the ends of the hcu slabs are anchored to the spandrels via couplers concreted into the ends of open cores in the slabs. The spandrels have to be temporarily propped prior to installing the hcu and the props remain in place until the slab end anchorage is completed and adequately cured. The top section of the spandrel also acts as a vehicle impact barrier around the car park perimeter.
- Spandrel beams (non-loadbearing) – these span parallel to the hcu units, simply supported spanning between columns or walls. They do not require temporary propping as there are minimal torsional effects. They are tied to the structural topping using projecting reinforcement break out bars. The top section of the spandrel also acts as a vehicle impact barrier around the car park perimeter. The non-loadbearing spandrels are up to 15.7m long. Figure 16.3 shows the installation of a white concrete spandrel beam, designed to be lifted with slings using a spreader beam.
- Deck slabs – typically 400mm precast prestressed concrete hcu spanning between main deck beams and/or perimeter spandrel beams. The units are tied to both beams at their ends. The hcu are designed to act as composite slabs with a 100mm structural topping. Maximum clear span is 15.17m.

Figure 16.3 Lifting of a 15.7 m long white concrete spandrel beam for the car park shown in Figure 16.1.

- Stairs & landings – flights span simply supported between landings and are tied at their ends to the landings. The landings span perpendicular to the flights and are supported by rolled steel angles (RSA) bolted to the precast walls. Landings are tied to the supporting walls.
- Lateral stability – the structure is designed to withstand the wind forces and equivalent horizontal forces due to geometric imperfections given in BS EN 1992, Part 1, clause 5.2 (BS EN 1992-1-1, 2004). All lateral loads are resisted by vertical cantilever shear walls forming the stair cores, the lift shafts and the access ramp structure. The horizontal forces are transferred to the shear walls via the diaphragm action of the car park deck slabs. The car park is designed for stability as two separate structures on each side of the movement joint. Consideration is given to stability for both temporary and permanent conditions.
- Robustness – disproportionate collapse is mitigated by the use of horizontal and vertical ties. Where a tied solution is not practical or possible the precast units are designed as key elements as defined in BS EN 1991, Part 1-7 (BS EN 1991-1-7, 2006).

16.2.3 Design of curved spandrel

The design of the curved spandrel beam and its connections to the hcu was one of the specific engineering challenges on this project. The beam had an outside radius of 11.5 m and spanned 12 m between supports. The torsion applied at the connections between the spandrel ends and its supports created a structural problem that could not be resolved by the typical methods of connecting precast elements. The solution adopted to resolve this

Figure 16.4 Back propping of curved spandrels for the car park shown in Figure 16.1.

problem was the back propping of all spandrels to ground level. The car park structure was built to full height, with the spandrel and floor slabs bearing onto props as shown in Figure 16.4.

To remove the excessive torsion from the spandrel, it was necessary to make a rigid connection into the hcu and the structural topping. Reinforcement bars were anchored into the hollow cores close to the bottom of the spandrel and into the structural topping to form a moment connection between the spandrels and the composite slabs. Figure 16.5 shows the connection details between beam and slab.

The back propping solution was chosen as follows:

1. It resolved the problem created due to torsion
2. It was simple and quick to erect on site
3. It allowed the erection of the precast to proceed without delay
4. It satisfied the practicalities of pouring the structural topping

Note for point 4, the preference is to pour structural toppings from the top level down in car park construction for the practical reason that this prevents any concrete spillages or grout leakage from ruining a finished surface below it. Figure 16.5 shows the curved spandrels under construction viewed from outside the building with temporary propping in place. Figure 16.6 is a view from inside the building, the *insitu* toppings have been poured, the curved spandrels are fully connected to the floor slab at each level, the back propping has been removed (note the remaining props that can be seen are non-structural and are there to support temporary edge protection).

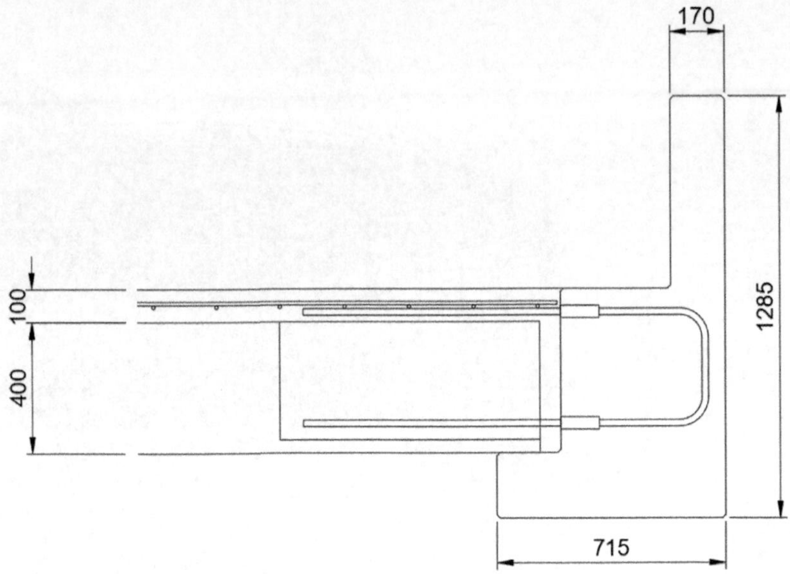

Figure 16.5 Curved spandrel connection to precast composite floor slab for the car park shown in Figure 16.1.

Figure 16.6 View from inside showing curved spandrels for the car park shown in Figure 16.1.

16.3 DESIGN OF PRECAST STRUCTURES USING FEM SOFTWARE

The use of finite element method (FEM) software to analyse a precast structure's behaviour can produce a very efficient and economic design. The stiffness of all of the elements can be utilized and lateral loading can be distributed to minimize concentration of compressive axial loads and tension or uplift forces. Hand calculation or simple analysis methods can sometimes be overly conservative, which can result in problematic concentrated loads and in particular excessive tension or uplift in stability walls. The high compressive forces can cause a requirement for increased wall thicknesses and high tension forces require impractical numbers of starter bars.

There are numerous FEM software packages available on the market. Users can model virtually any type of element or structure with different material properties and shapes. This software is a very powerful tool for engineers, but it needs to be used carefully. The output from an analysis model can be altered significantly by how it has been set up or built, for example: a change in support definition, assigning different boundary conditions to the members at a multiple joint interface, modifying the stiffness of a wall with a large opening and numerous other inputs can all be edited by the user. This amount of freedom needs to be controlled by experience, a thorough understanding of expected results and a results verification process. The following is a brief overview of a number of the key points when using FEM software to design precast structures.

16.3.1 FEM model set up

Support conditions for the precast structure must be properly defined. For example, if a precast structure is modelled as being supported on a rigid support instead of a flexible support, the load distribution will change. This is important when working with *insitu* concrete transfer structures and in particular with cantilever transfer structures. In such cases, the degree of flexibility of the *insitu* structure needs to be determined and provided by the *insitu* concrete designer; alternatively, the *insitu* structure can be modelled by the precast designer within their modelling software. The importance of this design co-ordination should not be underestimated as it adds accuracy to how the *insitu* transfer structure will deflect and redistribute the forces in the analysis model in a more realistic way. Figure 16.7 shows the analysis model for a precast structure erected on top of an *insitu* cantilever transfer structure. The supports can be seen at the base of the model; in this example, the *insitu* structure was not modelled with the same software as the precast model. The stiffness of the supports was provided by the *insitu* designer from their model, and these were used in the precast model.

Modelling of a precast structure can be subdivided into two general types – the first is framed structures with columns and beams and the second is shear wall type structures. Framed structures are typically modelled with member elements while shear wall type structures would be modelled using panel members, different software packages may use different terminology.

Floors are usually modelled as a rigid or semi-rigid diaphragm. The design of the diaphragm is generally undertaken separately and justified using hand calculations or simple methods of analysis. The internal forces transferred through the floor diaphragm into columns, beams and walls are extracted from the model. The interface between each element is checked to ensure the design shear does not exceed the shear capacity at any of the connections. It is important to note that hcu are typically modelled as a one way spanning slab, and it is critical to properly define the correct span and any release or boundary conditions.

Figure 16.7 Analysis model for precast structure.

Horizontal wall-to-wall joints are almost always modelled as pinned about the minor axis. Precast walls are typically of a width that permits only a single central row of connections across the horizontal joint thus limiting the fixity that can be achieved. Columns are typically modelled as pinned at the column-to-column connections. Multi-storey columns are pinned where they connect to foundation and/or other columns.

Vertical wall-to-wall joints can be modelled as rigid if the walls are interconnected and there is sufficient vertical shear capacity in the joint. In cases where the joint does not have sufficient vertical shear capacity it needs to be modelled as released. The design of the joints between walls and other elements is critical, the behaviour of the structure will be affected by these details, and there is the potential to change the load distribution and subsequently the internal forces in the structural members.

Figure 16.8 shows the results from a precast shear wall stability system modelled with three of the walls interconnected as a C-shape and Figure 16.9 shows the same layout with all of the walls released vertically. The load distribution changes considerably based on the releases used, the peak reactions are reduced by approx 20% by modelling the three walls as a connected C-shape. This will affect the axial load on the walls and the connections to the foundation and/or transfer structure.

16.3.2 Horizontal joints

The axial compression forces from the model are used to check the capacity and design the horizontal joint interfaces between *insitu*, precast, steel, etc. The axial load outputs are used to check the capacity of the element and design the reinforcement content for the wall

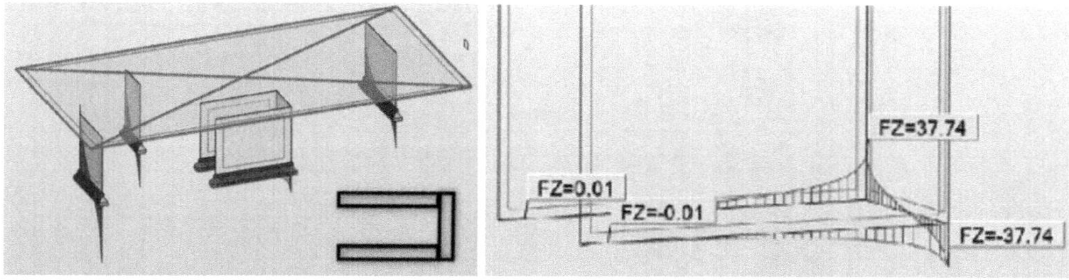

Figure 16.8 Connected wall structure.

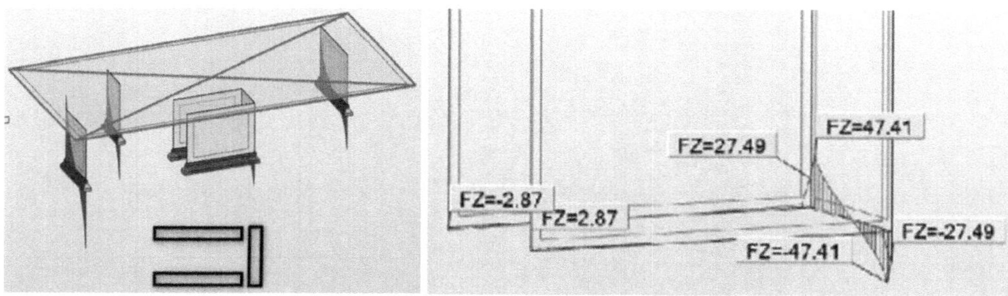

Figure 16.9 Released wall structure.

or column in question. Any tension force or uplift is typically dealt with by connecting the walls to the foundations using starter bars in grout sleeves or proprietary wall shoes. These connections can also provide the necessary vertical ties for robustness.

16.3.3 Vertical joints

These are the most critical joints when using FEM software to analyse precast structures, the performance of the vertical joint has the greatest influence on the load distribution and therefore requires considerable focus and care when modelling.

Once the forces in the vertical joints have been extracted from the analysis model, each joint is checked to ensure the connections can be designed to withstand the applied forces. Proprietary connection systems such as loop boxes are available from a number of suppliers. The manufacturers of these systems provide technical data stating the capacity available for specific geometry details, concrete strengths and other relevant conditions. If proprietary systems do not provide the capacity required, enhancement of the shear capacity at the vertical interfaces between walls is possible with a bolted connection. If a situation arises where the forces extracted from the analysis model exceed the capacity at some of the joint connections, an alternative panelisation of the walls needs to be modelled so the vertical joints that are a problem are moved to a position where the forces are reduced. Alternatively, if it is practical, multiple elements can be cast as one single element; common examples of this are lift cores cast as box shapes and also walls cast as L and T-shapes. A further option exists, although not preferred due to the additional site works and aesthetic implications; the precast walls are designed to have a section of *insitu* concrete cast between them. The walls are manufactured with the required projecting reinforcement or cast in couplers and the infill section between them is shuttered and concreted on site to provide a monolithic connection between walls.

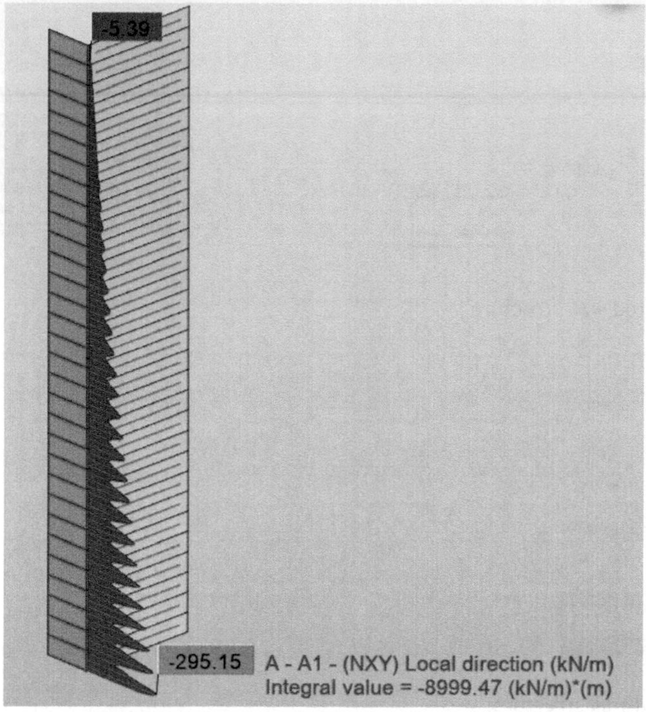

Figure 16.10 Extract from analysis model.

The methods above will alter the distribution of the loads and can provide a solution when vertical shear capacity in joints is causing a problem or a limitation in the design of a precast structure.

Figure 16.10 shows an extract from the analysis model of a precast apartment building. The elements selected are part of the stability system for the structure, and the displayed forces are the shear forces at the vertical interface between two wall panels. The building is 32 storeys in height and there are large shear forces developed in the joint that require high capacity connections at the lower levels. Moving up the building the forces reduce to the point where their magnitude allows the use of proprietary loop box connections. Figure 16.11 shows the design principle used for loop boxes.

Table 16.1 gives typical capacities for loop box connections, using size 140 loop boxes at 250 mm vertical spacing with grade C40/50 wall/grout strength would give a vertical shear capacity of 44.2 kN×(1000/250 mm) = 176.8 kN/m. Any vertical joints that have V_{Ed} in excess of the loop box capacity will need to be connected using bolted connections or one of the other methods described above.

An example of a bolted connection detail is shown in Figure 16.12. This method uses a coupler, grout sleeve, threaded rod, plate washer and nut. The vertical joint, grout sleeve and recess pocket are filled with a high strength, non-shrink, pourable grout.

16.3.4 Multiple layer elements

When dealing with multiple layer elements, such as sandwich panels where typically only the inner leaf is load bearing, it is important to consider how best to account for the effect on gravity loading when setting up the analysis model. There are several methods to efficiently model these walls. One method is to model the geometry of the inner leaf of the sandwich

Table 16.1 Vertical shear capacity of loop boxes

	Vertical shear resistance V_{Rd} (kN/pair of boxes) for concrete grades					
Wall/grout strength	*C25/30*	*C30/37*	*C35/45*	*C40/50*	*C45/55*	*C50/60*
Size = 60–120 mm	14.5	16.9	19.4	22.1	24.6	27.1
Size = 140 mm	31.5	37.0	42.5	44.2	45.1	46.0

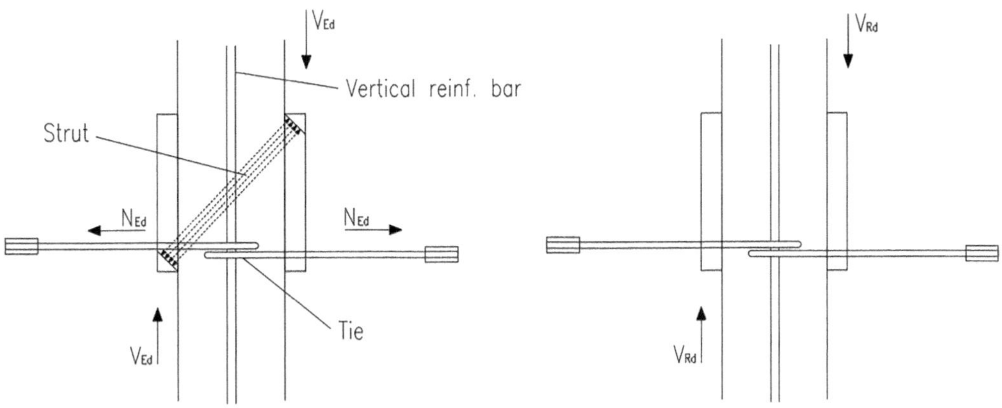

Figure 16.11 Strut and tie model for vertical shear.

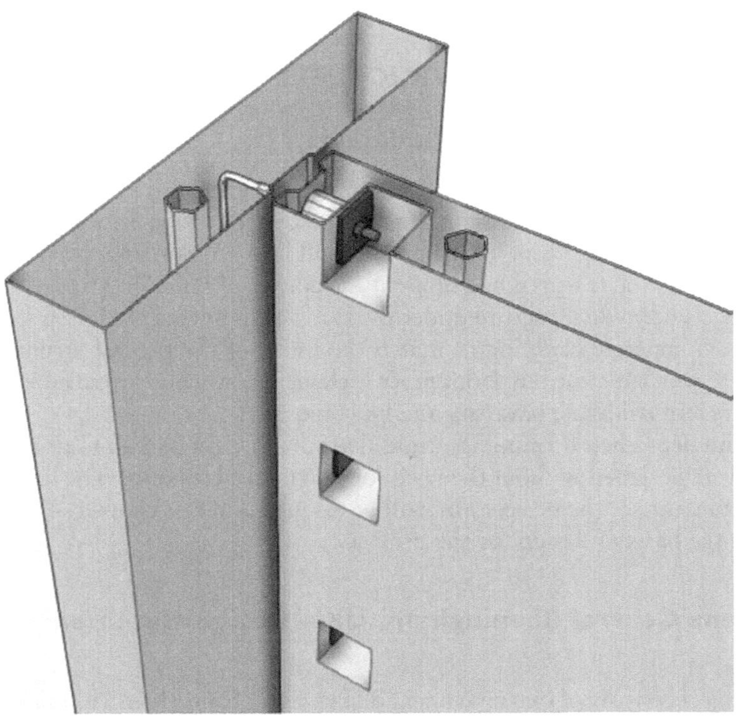

Figure 16.12 Bolted connection between wall panels.

panel as it is and modify its material properties, manually inputting a denser material that accounts for the self-weight of the non-structural leaf and the self-weight of any windows, doors or cladding within the panel. This works very well for sandwich panels that have a repetitive profile but care must be taken to identify any changes to the profile geometry and thickness of the non-structural leaf as this can significantly alter the total self-weight of the panel. Another method, of course, is to simply use the commands and functions within the software package and apply the self-weight loading from the non-structural outer leaf, windows, doors, etc. to the structural load-bearing leaf. The decision on what method to use for setting up the model generally comes down to the speed or level of accuracy required when modelling.

16.3.5 Model verification

As part of the design process, it is crucial to verify the model outputs. The checks below are a minimum and each project should be reviewed for any unusual details, geometry or applied loading. Specific checks need to be identified on a project-by-project basis.

General checks in the model must be undertaken to ensure no obvious anomalies have occurred such as very large deflections or unexpected high concentration of forces.

The first check is typically to ensure all gravity loads have been applied to the structure by checking overall reactions in a hand check against software results.

The lateral distribution of forces must also be checked. In simple terms, the software will calculate the position of the applied loads and the centre of rigidity of the structure in order to distribute the loads. As previously stated, the modelling of walls being connected or released plays a key role in this load distribution. It is imperative to carry out hand checks to ensure that any connected wall arrangements are justified as behaving monolithically and also that the construction details will allow walls that are assumed to be released to behave as independent panels.

16.4 PRECAST RESIDENTIAL STRUCTURES

16.4.1 Multi-storey apartment buildings

The residential multi-storey project at Arena Central in Birmingham, UK, was a fully precast crosswall structure erected on top of an *insitu* concrete transfer structure. Figure 16.13 shows the general arrangement plan for the building, comprising two blocks A and B. Figure 16.14 shows the featured balconies at the corner of Block B.

The two levels of *insitu* structure under the transfer slab were clad with single skin architectural precast concrete cladding to match the finish of the precast structure above. The superstructure was built entirely from precast elements, namely: crosswalls, sandwich panels, hcu, stairs and landings, parapets and balconies.

The 250 mm depth hcu forming the rigid floor diaphragm had a maximum clear span of 9.95 m and were designed without the need for a structural topping. The balconies and their design were a feature of the project and will be discussed in this chapter. The following is an extract from the basis of design for the project.

16.4.2 Arena Central Birmingham, UK – Design statement

1.0 General
1.1 The project consists of two apartment blocks rising from an *insitu* concrete transfer slab
1.2 Block A is the tallest at 62.32 m (22 storeys), Block B is 48.04 m (17 storeys)
1.3 Footprint of Block A is 38.2 m×16.85 m, footprint of Block B is 41.7 m×16.85 m

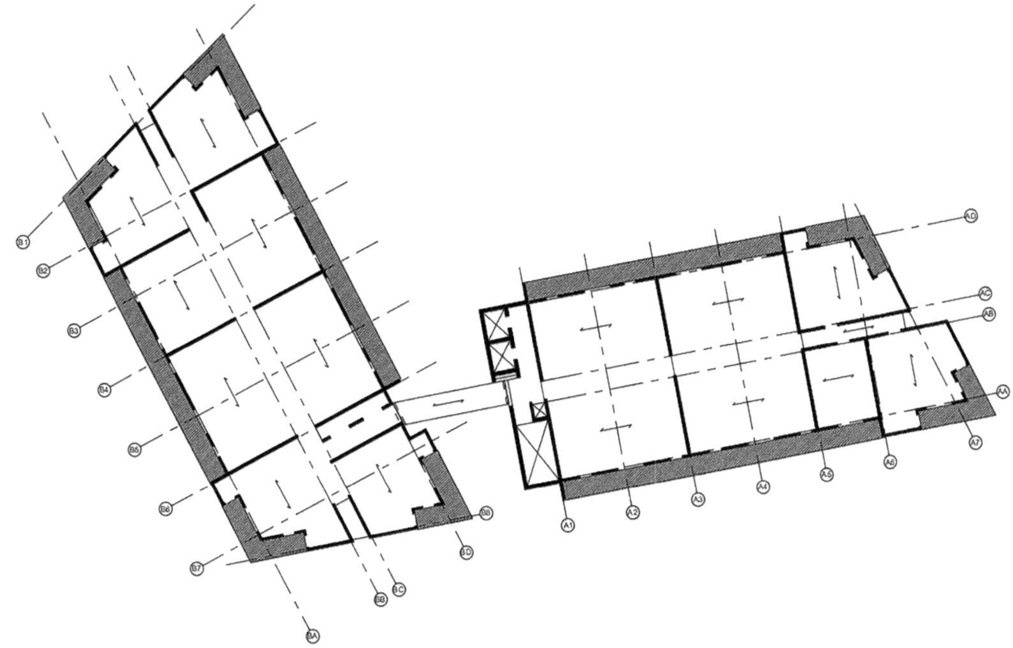

Figure 16.13 Block A (right hand side) and Block B floor plans for Arena Central Birmingham, UK.

Figure 16.14 Block B in foreground, corner balconies supported on outer leaf, at Arena Central Birmingham, UK.

1.4 Each block is designed as an independent structure
1.5 A movement joint is provided on the Block A side of the link bridge connecting the blocks
1.6 All construction is new build

2.0 Standards
2.1 BS EN 1990 (EC0) Basis of structural design
2.2 BS EN 1991 (EC1) Actions on structures
2.3 BS EN 1992 (EC2) Design of concrete structures
2.4 BS EN 1993 (EC3) Design of steel structures
2.5 National Annexes to relevant Eurocodes
2.6 PD6687 Background paper to the National Annexes to BS EN 1992-1 and BS EN 1992-3
2.7 BS EN 13369 Common rules for precast concrete products
2.8 BS EN 14992 Precast concrete products – Wall elements
2.9 BS EN 1168 Precast concrete products – Hollow core slabs
2.10 BS EN 14843 Precast concrete products – Stairs
2.11 BS EN 12354 Building acoustics
2.12 BS 8297 Design and installation of non-loadbearing precast concrete cladding
2.13 BS 8110 Structural use of concrete
2.14 BS 8500 Concrete – Complementary British Standard to BS EN 206
2.15 BS 5975 Code of practice for temporary works procedures and the permissible stress design of falsework
2.16 Non-contradictory published codes of practice and standards

3.0 Acoustic performance – calculated values BS EN 13369, clause 4.3.5 and BS EN 12354-1, Annex B and C.
3.1 Weighted sound reduction index (R_w) of 160 mm thickness precast concrete solid wall. R_w=54 dB
3.2 R_w of 200 mm solid wall. R_w=58 dB
3.3 R_w of 250 mm solid wall. R_w=61 dB
3.4 R_w of 250 mm hcu. R_w=52 dB

4.0 Thermal performance
4.1 Sandwich wall panels and insulated cladding panels to provide a U-value of 0.19 W/m^{2}K
4.2 Any reduction in cavity and insulation thickness at door thresholds and window returns to be included within the door and window critical junction calculation.
5.0 Structural Element Philosophies
5.1 Shear walls are designed as vertical cantilevers
5.2 All walls are designed as pinned and braced perpendicular to the minor axis
5.3 All floor slabs are designed as simply supported
5.4 Stair flights span simply supported between landings/supports
5.5 Landings span simply supported between supports perpendicular to stair flights
6.0 Lateral Stability
6.1 The prestressed hcu are designed to act as a rigid diaphragm at each level
6.2 The floor diaphragm is tied to and transfers horizontal loading to the shear walls at each level
6.3 All lateral loads are resisted by shear walls
6.4 Stair cores, lift shafts, crosswalls and the load-bearing leaf of external sandwich panels are used as shear walls to provide stability

7.0 Robustness

7.1 The building is considered as Class 3 according to BS EN 1991, Part 1-7

7.2 Disproportionate collapse mitigated by provision of vertical and horizontal ties and key element design

8.0 Loading

8.1 Per manent, Apartment Levels
 8.1.1 Raised access floor = 0.6 kN/m^2
 8.1.2 75 mm depth screed = 1.85 kN/m^2
 8.1.3 Ceiling and services = 0.6 kN/m^2

8.2 Permanent, Roof Level
 8.2.1 Ballast = 1.6 kN/m^2
 8.2.2 Single ply membrane = 0.2 kN/m^2
 8.2.3 Insulation = 0.1 kN/m^2
 8.2.4 75 mm thickness screed = 1.85 kN/m^2
 8.2.5 Ceiling and services = 0.6 kN/m^2

8.3 Permanent, Balconies
 8.3.1 SDL = 0.75 kN/m^2 (as per Design Brief)

8.4 Variable, Levels 2 - 17/22
 8.4.1 Apartments (including partitions) = 2.5 kN/m^2
 8.4.2 Balconies = 2.5 kN/m^2 (as per Development Brief)
 8.4.3 Corridors (including partitions) = 4.0 kN/m^2

8.5 Variable, Roof 18/23
 8.5.1 Access = 1.5 kN/m^2
 8.5.2 Snow = 0.7 kN/m^2

8.6 Wind (in accordance with BS EN 1991-1-4)
 8.6.1 Basic wind speed = 21.7 m/s
 8.6.2 Site altitude = 120 m
 8.6.3 Nearest distance to the sea >100 km
 8.6.4 Peak velocity pressure = 1.28 kN/m^2

8.7 Second order effects and geometric imperfections. Equivalent horizontal forces are applied to the structure due to calculated second order effects and geometric imperfections

8.8 Balcony wall and roof parapet. Horizontal load of 0.74 kN/m applied at 1100 mm height

9.0 Serviceability

9.1 Long term deflection of stairs, landings and floor slabs limited to span/250

9.2 Minimum natural frequency of floor slabs 4 Hz

9.3 Internal exposure class for all elements is XC1

9.4 External exposure class for all elements is XC3

9.5 Crack width limited to 0.3 mm

9.6 Durability in accordance with BS 8500

9.7 Design life for all precast elements is 60 years

9.8 Fire resistance REI 120 for wall panels (wall exposed on one side only, compartmentation as per the consultant's Fire Strategy Report

9.9 Fire resistance REI 120 for floor slabs

9.10 Fire resistance R 60 for stair flights and landings

9.11 Wind acceleration limited to threshold of perceptibility equal to 1.5% g = 0.14715 m/s^2

9.12 Building sway limited to height/750 for the entire precast structure and height/300 over any single storey

10.0 **Material Characteristics**
10.1 Typical precast concrete strength for load-bearing walls C50/60, density 2450 kg/m^3
10.2 Typical precast concrete strength for hcu is grade C40/50, density 2400 kg/m^3
10.3 Typical precast concrete strength for solid slabs, stairs and landings is grade C32/40, density 2450 kg/m^3
10.4 Typical *insitu* concrete strength is grade C32/40, density 2400 kg/m^3
10.5 Typical grout strength is grade C40/50, density 2100 kg/m^3
10.6 Typical structural mortar strength is grade C40/50, density 2095 kg/m^3
10.7 Typical reinforcement strength is 500 N/mm^2 (B500B Type 2 deformed)
10.8 Tie reinforcement strength is 500 N/mm^2 (B500C Type 2 deformed)
10.9 Prestressing strand strength is 1860 N/mm^2

11.0 **Surface Finish**
11.1 Internal precast wall elements that are exposed to view have a surface finish that is suitable for direct decoration, reference Type C description given in Clause 6.2.7.3 of BS 8110, Part 1 (BS 8110-1:1997)
11.2 External precast cladding and the external leaf of each type of sandwich panel is to have an approved benchmark sample with colour, consistency and finish agreed by the client
11.3 Surface finish of precast slab soffits covered by a suspended ceiling is Type A, reference description given in Clause 6.2.7.3 of BS 8110, Part 1.
11.4 Surface finish of precast stair and landing soffits is Type C, reference description given in Clause 6.2.7.3 of BS 8110, Part 1.

12.0 **Manufacturing Tolerance**
12.1 BS EN 13369 Common rules for precast concrete products
12.2 BS EN 14992 Precast concrete products – Wall elements
12.3 BS EN 1168 Precast concrete products – Hollow core slabs
12.4 BS EN 14843 Precast concrete products – Stairs
12.5 BS 8297 Design and installation of non-loadbearing precast concrete cladding
12.6 National Structural Concrete Specification for Building Construction 4th Edition
13.0 **Erection Tolerance**
13.1 BS EN 13670 Execution of concrete structures
13.2 National Structural Concrete Specification for Building Construction, 4th Edition
13.3 BS 5606 Accuracy in building

16.4.3 Balcony design

The balconies were an important architectural feature and an engineering design challenge on the project. The corner balconies and the straight balconies used two different methods for support and restraint back to the main structure.

16.4.4 Corner balconies

The corner balconies were supported on the outer leaf of the sandwich panels; this required both the inner and outer leaf of the sandwich panels at these locations to be designed as load bearing. To support the loading and satisfy the fire performance requirements, the outer leaf was increased in thickness to 160mm. Restraint was provided by using a bolted connection between the outer load-bearing leaf and the inner load-bearing leaf as shown in Figure 16.15, the inner leaf is tied back into the hcu and floor diaphragm through the horizontal tie reinforcement details.

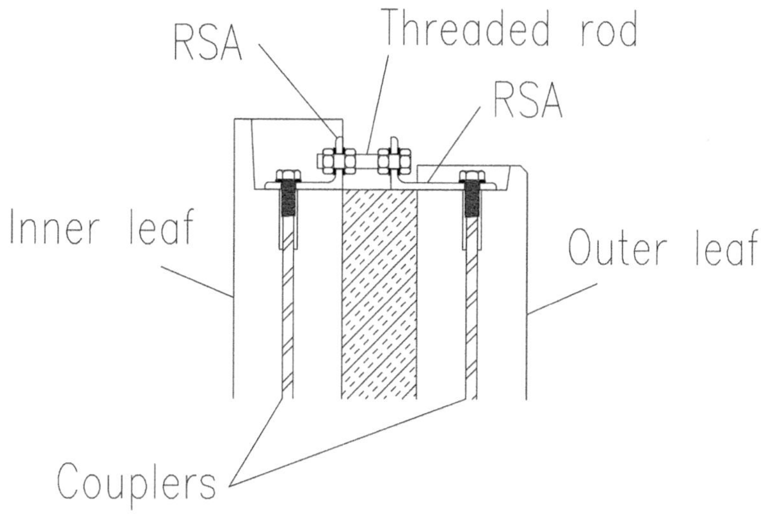

Figure 16.15 Restraint connection between inner and outer leaf at Arena Central Birmingham, UK.

A steel beam cast within the balcony to cantilever in one direction at the corner, and a reinforced concrete beam was formed within the beam to cantilever in the other direction. Although the balcony slab was cast as one unit with the integral beams, it makes it easy to understand how the design works by thinking of them as separate beams providing the support for balcony slabs spanning as shown in Figure 16.16. The inside edge of the balcony is supported on the outer leaf, shown hatched in Figure 16.16, the circles on the walls indicate the position of vertical ties on the inner and outer leaves and also bolted connections between the balcony and the outer leaf. The back-spans for the cantilever beams within the balcony slab run along the top of the hatched outer leaf.

The upstand walls visible in Figure 16.17 were cast onto the balcony slabs in a second stage of manufacture prior to being sent to site. The glass balustrades were installed on site at ground level prior to lifting the balconies into position.

16.4.5 Straight balconies

The straight balconies were supported on the inner leaf of the sandwich panels and restrained back to the hcu and floor diaphragm using steel plates. The balconies were manufactured with steel beams cast into them at approximately 1000 mm centres along their length. The beams then projected from the balconies across the width of the cavity on the sandwich panels. Each steel beam was fabricated with a welded end plate designed to make the connection to the sandwich panel. The balconies were installed onto the building by landing the end plates onto RSA's cast into the sandwich panels. The RSA supported the vertical load. The endplates were manufactured with oversize holes that fitted over threaded rods projecting from the sandwich panel. A plate washer was used to complete the bolted connection between the endplate and the sandwich panel. The threaded rods were cast into the sandwich panel, and this anchorage detail was designed and reinforced to deal with the tension forces generated by the cantilever balcony. Figure 16.18 illustrates the details of the connection method. The end plate was isolated from the RSA by a structural thermal break to ensure the Psi value for this junction satisfied the thermal performance requirements. The Psi value is a measure of heat lost through a junction such as the connection of an external balcony to an internal structural wall.

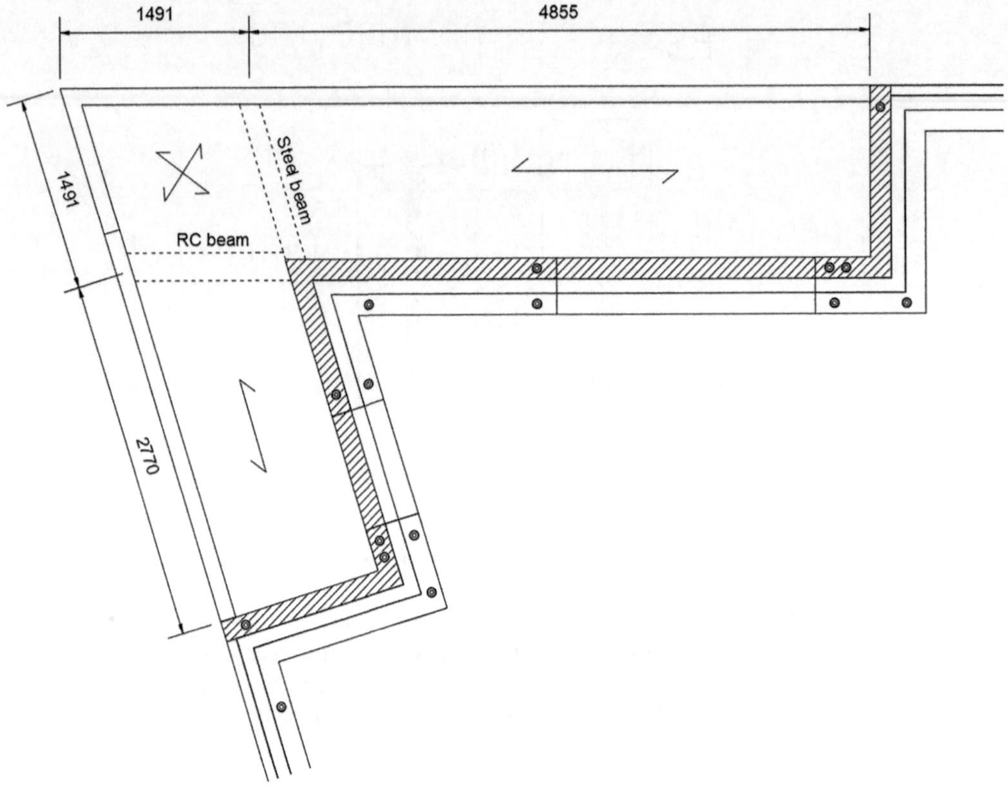

Figure 16.16 Largest corner balcony on the project – Block B at Arena Central Birmingham, UK.

Figure 16.17 Corner balcony being installed at Arena Central Birmingham, UK.

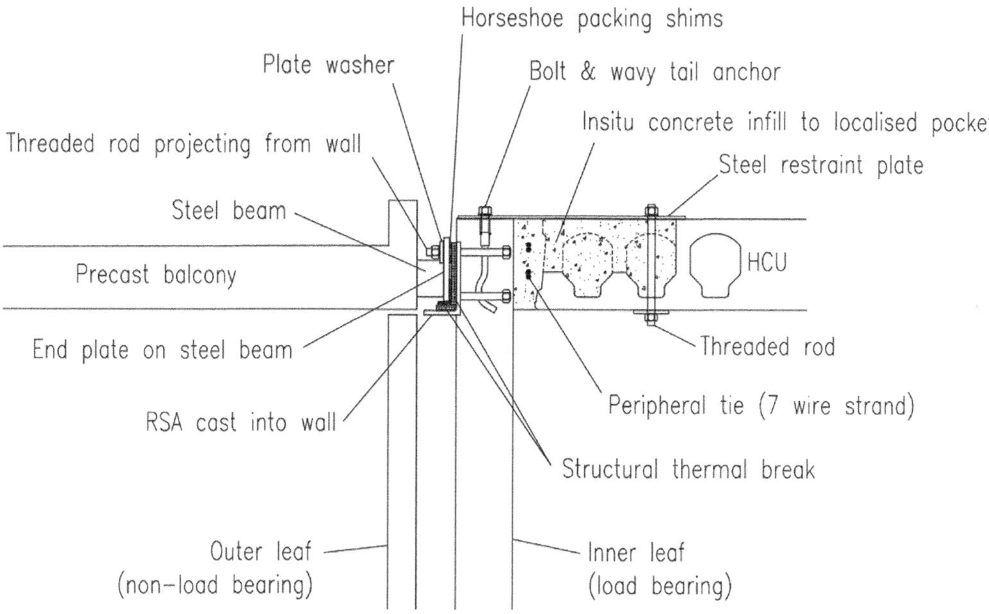

Figure 16.18 Precast cantilever balcony connection.

The top of the sandwich panel was tied to the hcu using steel plates and bolted connections. A cast-in wavy tail anchor was used to make the bolted connection to the wall. A threaded rod with nuts and a plate washer was used to form the connection to the slab. During the production process, the hcu had side pockets pre-formed along their lengths to align with the connection points in the wall. The peripheral tie strand was positioned as shown in Figure 16.18. All plates and bolted connections were installed, and the side pockets and the joint between the wall and slab were infilled with *insitu* concrete. Figure 16.19 shows the straight balconies and corner balconies on one elevation of Block A with a number of the corner balconies on Block B visible on the left of the photograph.

16.5 PRECAST EDUCATIONAL BUILDINGS

16.5.1 Manchester Metropolitan University, Birley Fields, UK

The project consisted of an almost fully precast structure with *insitu* concrete toppings to the precast prestressed floor units. The only structural members not manufactured from concrete were two steel box girder transfer beams at second floor level and five steel trusses supporting the sports hall roof.

The architectural intent was for the concrete columns, beams and slab soffits to be exposed to view throughout the building; this required an extremely high quality of finish. The exposed concrete surfaces also ensured the benefits of the thermal mass of the structure could be utilized. The precast plate flooring units had heating and cooling pipework cast into them during manufacture. The exposed concrete soffits were a necessity to maximise the efficiency of this method for internal temperature control. Figure 16.20 shows the

Figure 16.19 Block A straight balconies and corner balconies at Arena Central Birmingham, UK.

Figure 16.20 Structure nearing completion at Manchester Metropolitan University, UK.

structure nearing completion. The entrance area to the right-hand side of the building where the crane is positioned was the final phase of the build – his area featured the Spanish Steps and a 17 m high circular column. The Spanish Steps were a communal seating and congregation space in the entrance foyer that resembles large terrace units.

16.5.2 Build statistics

The precast structure consisted of hcu, plate flooring, beams, columns, walls, Spanish Steps terrace units, stairs and landings The total weight of the precast elements supplied to the project was 26,170 tonnes. The heaviest single piece weighed 28.7 tonnes. There were 5936 tonnes of *insitu* concrete poured to complete the construction of the floor slabs. Quantities for each of the 2240 precast product type are as follows: 327 no. columns, 546 no. beams, 325 no. wall panels, 55 no. stair flights, 64 no. 'landing' slabs, 36 no. terrace units, 753 no. plate flooring units, and 134 no. hollow core units giving a total floor area of 16,333 m^2. The erection phase was competed in a 20-week site programme. This included the erection of the steel box girders and trusses.

16.5.3 Precast connections

The stability of the structure was provided by the four stair cores. These were positioned symmetrically on the rectangular footprint of the building close to each corner. The precast walls forming the cores were designed as vertical cantilever shear walls. The walls were panelised to span vertically from floor to floor, being pinned and restrained at each floor level.

16.5.4 Connections at horizontal joints

The precast walls were connected to the *insitu* foundation with starter bars and spirol tube grout sleeves cast vertically into the walls. Spirol tube grout sleeves are helical ribbed metal formers used to provide an idented key inside the precast unit suitable for anchoring reinforcement bars. Figure 16.21 shows a wall on the back of a trailer with

Figure 16.21 Grout sleeves in wall at Manchester Metropolitan University, UK.

Figure 16.22 Starter bars at foundation level at Manchester Metropolitan University, UK.

Figure 16.23 Walls erected and propped off kentledge blocks at Manchester Metropolitan University, UK.

the grout sleeves visible at its base. Figure 16.22 shows the starter bars projecting from the *insitu* foundation level, the walls are lowered into position over the top of the starter bars which align with the grout sleeves. The connection is completed by filling the grout sleeves with a high flow non-shrink grout. This connection is designed to resist the tension forces generated from the lateral loading applied to the shear walls through the floor diaphragm. Figure 16.23 shows the walls installed and propped off precast kentledge blocks.

At first floor level and above, the connection at the horizontal joint was formed using a proprietary wall shoe. Figure 16.24 shows the alternative principles that can be used

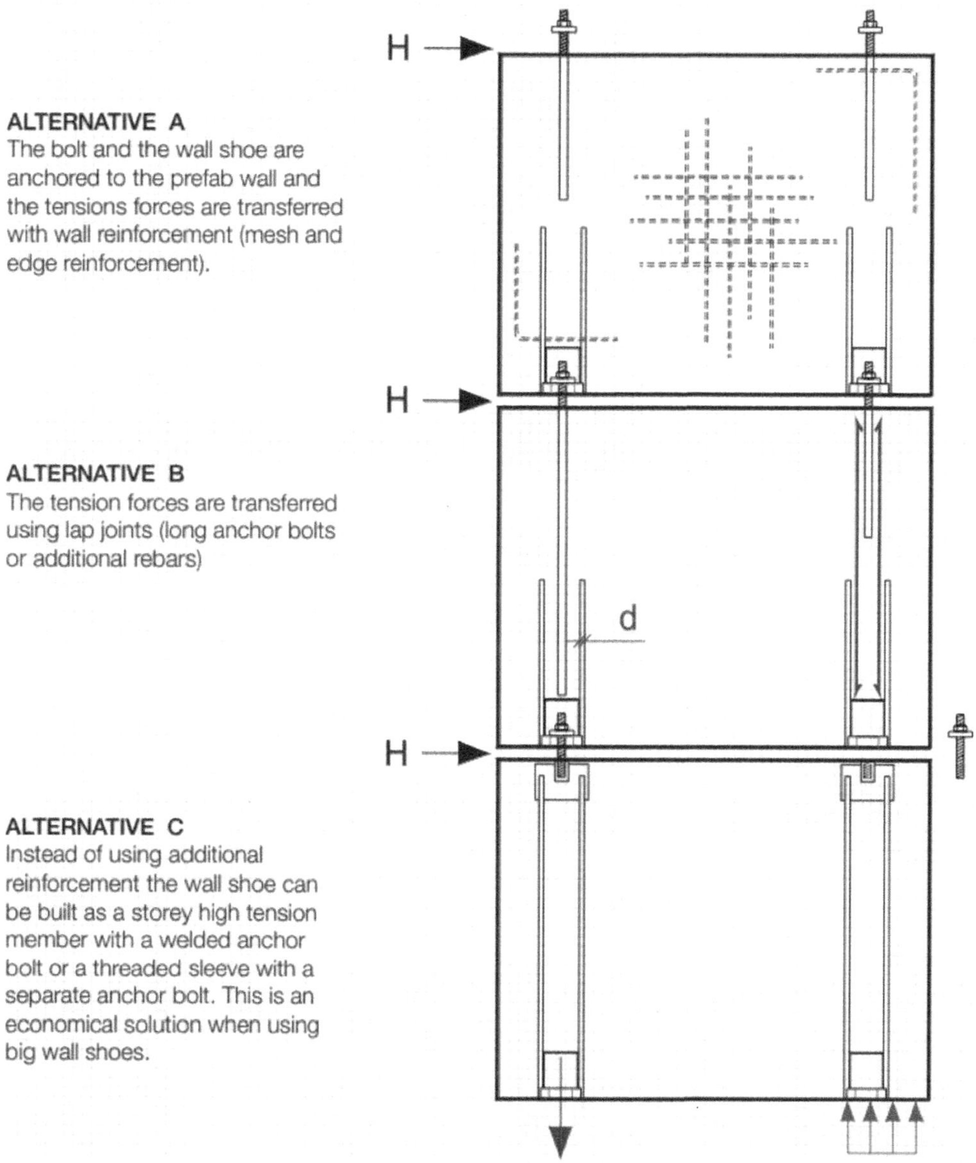

Figure 16.24 Wall shoe connection principles. (Courtesy of Construction Fixing Systems, England.)

Figure 16.25 Wall shoe threaded rod connection at top of wall at Manchester Metropolitan University, UK.

Figure 16.26 Wall shoe boxed out connection at bottom of wall at Manchester Metropolitan University, UK.

when designing this connection. Figures 16.25 and 16.26 show the projecting threaded rod and box out details, respectively. The threaded rod is cast in or screwed into a female thread in the top of the wall below the horizontal joint. A boxed-out steel anchorage assembly is cast into the bottom of the wall above the horizontal joint. On site a plate washer and nut are used to mechanically fix the walls together. The boxed out section is then infilled with grout to compete the connection. The manufacturers of these wall shoe systems provide technical data on tension capacity, reinforcement details, minimum wall dimensions, etc.

16.5.5 Connections at vertical joints

The vertical connection between panels is provided by loop boxes. The loop boxes transfer load across the vertical joints to allow the wall panels to be designed as a monolithic shear wall rather than individual sections. The shear forces are checked as discussed in Section 16.3 to ensure the loop boxes have adequate capacity. Figure 16.27 shows loop boxes projecting from the edge of the wall. The wall that is installed adjacent to this wall will have corresponding load boxes cast in along its height.

On site, a continuous vertical reinforcement bar, as shown in Figure 16.28, is fixed between the loop boxes and the joints are filled with a pourable non-shrink grout. Figure 16.29 shows the completed stair core with the individual precast walls connected together using loop boxes in the vertical joints and wall shoes in the horizontal joints at the levels above foundation.

Figure 16.27 Loop boxes on vertical joint at Manchester Metropolitan University, UK.

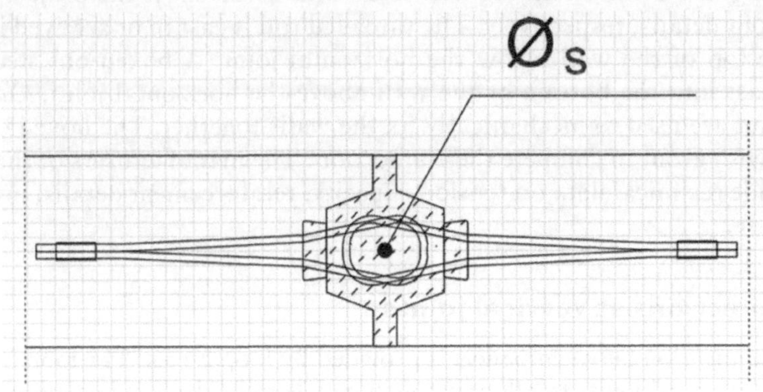

Figure 16.28 Loop box joint with vertical reinforcement bar. (Courtesy of Construction Fixing Systems, England.)

Figure 16.29 Wall panels connected with loop boxes and wall shoes at Manchester Metropolitan University, UK.

REFERENCES

BS EN 1991-1-7. 2006. Actions on structures, General actions - Accidental actions, +A1:2014, BSI, London, UK.

BS EN 1992-1-1. 2004. Eurocode 2: Design of concrete structures - Part 1-1: General rules and rules for buildings, +A1:2014, BSI, London, UK.

BS 8110-1. 1997. Structural Use of Concrete - Part 1: Code of practice for design and construction, BSI, London, UK.

Index

For Product Safety Concerns and Information please contact our EU representative GPSR@taylorandfrancis.com Taylor & Francis Verlag GmbH, Kaufingerstraße 24, 80331 München, Germany

Batch number: 10519288

Printed by Printforce, the Netherlands